COAL HANDBOOK

ENERGY, POWER, AND ENVIRONMENT

A Series of Reference Books and Textbooks

Editor

PHILIP N. POWERS
Professor Emeritus of Nuclear Engineering
Purdue University
West Lafayette, Indiana

Consulting Editor
Energy Management and Conservation
PROFESSOR WILBUR MEIER, JR.
Head, School of Industrial Engineering
Purdue University
West Lafayette, Indiana

1. Heavy Oil Gasification, *edited by Arnold H. Pelofsky*
2. Energy Systems: An Analysis for Engineers and Policy Makers, *edited by James E. Bailey*
3. Nuclear Power, *by James J. Duderstadt*
4. Industrial Energy Conservation, *by Melvin H. Chiogioji*
5. Biomass: Applications, Technology, and Production, *by Nicholas P. Cheremisinoff, Paul N. Cheremisinoff, and Fred Ellerbusch*
6. Solar Energy Technology Handbook: Part A, Engineering Fundamentals; Part B, Applications, Systems Design, and Economics, *edited by William C. Dickenson and Paul N. Cheremisinoff*
7. Solar Voltaic Cells, *by W. D. Johnston, Jr.*
8. Energy Management: Theory and Practice, *by Harold W. Henry, Frederick W. Symonds, Robert A. Bohm, John H. Gibbons, John R. Moore, and William T. Snyder*
9. Utilization of Reject Heat, *edited by Mitchell Olszewski*
10. Waste Energy Utilization Technology, *by Yen-Hsiung Kiang*
11. Coal Handbook, *edited by Robert A. Meyers*
12. Energy Systems in the United States, *by Asad T. Amr, Jack Golden, Robert P. Ouellette, and Paul N. Cheremisinoff*

Additional Volumes in Preparation

COAL HANDBOOK

Edited by

ROBERT A. MEYERS

TRW Energy Systems Group
Redondo Beach, California

MARCEL DEKKER, INC. New York and Basel

LIBRARY OF CONGRESS CATALOGING IN PUBLICATION DATA

Main entry under title:

Coal handbook.

(Energy, power, and environment ; 11)
Includes bibliographies and index.
Contents: Coal structure / Robert A. Meyers – Coal sampling and analysis / Gregory Gould and Jan Visman – Underground mining / Wilhelm J. Euler – [etc.]
1. Coal. I. Meyers, R. A. (Robert Allen), [date]. II. Series.
TP325.C5174 553.2'4 81-7824
ISBN 0-8247-1270-6 AACR2

Marcel Dekker, Inc.
270 Madison Avenue, New York, New York 10016

Current printing (last digit):
10 9 8 7 6 5 4 3 2

PRINTED IN THE UNITED STATES OF AMERICA

Preface

The *Coal Handbook* is, in a sense, a gift from the coal scientific and engineering community to scientists, engineers, economists, business people, and planners who have need of a single comprehensive and authoritative source book on coal science and technology.

The eleven chapters of the handbook present coal science and technology, starting with basic definitions of the structure of coal (Chapter 1) and methods of sampling and analysis (Chapter 2), through the technology and equipment of underground and surface coal mining (Chapters 3 and 4) and size reduction (Chapter 5). Chapters on coal cleaning (Chapter 6), utilization (Chapter 7), combustion (Chapter 8), and the conversion of coal to coke (Chapter 9), along with chapters on gaseous fuels (Chapter 10) and liquid fuels (Chapter 12), complete the presentation. An extensive glossary of technical terms is available at the back of the handbook.

This book is intended to be particularly useful to scientists and engineers who are experts in one or two coal technology disciplines but require ready access to a wider range of coal technology. Scientists and technologists just entering the coal area cannot only familiarize themselves with the complete technological field but can obtain an in-depth introduction to specific areas. The government manager, legislator, or planner will be in a better position to evaluate and select among planning and legislative alternatives after reading this handbook, and media personnel, charged with the responsibility of interpreting energy technology for their readers, will now have a much-needed compendium of this type. For these readers, the glossary can be consulted for definition of the specialized vocabulary of the coal community, while the detailed index will allow ready entry into specific contents needed for assessment.

Of course, no single volume can hope to be all things to all people. Each author was asked to present the information in his field essential to a basic understanding of the technology. The chapter authors were not only given this difficult responsibility, but also required to present the information within severe limitations on chapter length. Readers are left to their own judgment as to the success of their endeavors.

Robert A. Meyers

Acknowledgments

Thanks are in order to the many participating firms and research institutes that provided facilities and encouragement to our chapter authors:

Battelle Columbus Laboratories
Ford, Bacon and Davis Engineers
Fuel Engineering Company of New York
Inland Steel Company
Institute of Gas Technology
Kennedy Van Saun Corporation
North American Mining Consultants, Inc.
The Ralph M. Parsons Company
TRW Systems and Energy
Vortek Systems, Ltd.

Special thanks are given to TRW Systems and Energy and especially Mr. B. Dubrow, Manager of TRW's Chemistry and Chemical Engineering Laboratory. The individual acknowledgment sections of each chapter cite a large number of additional firms and institutes which have supported various reviews and contributions.

My thanks go also to Verna Melough, my secretary, who edited and prepared portions of the manuscript and especially to my wife, Ilene, who graciously shared our weekends and evenings with the preparation of this book and provided constant and loving encouragement.

R. A. M.

Contributors

*Ronald F. Ayers** Battelle, Columbus Laboratories, Columbus, Ohio

Richard E. Barrett Battelle, Columbus Laboratories, Columbus, Ohio

*Ronald R. Bevan*** Kennedy Van Saun Corporation, Danville, Pennsylvania

William W. Bodle Institute of Gas Technology, Chicago, Illinois

William DuBroff Inland Steel Company, East Chicago, Indiana

Wilhelm J. Euler Occidental Research Corporation, Lexington, Kentucky

Gerard C. Gambs Ford, Bacon and Davis, Inc., New York, New York

Robert D. Giammar Battelle, Columbus Laboratories, Columbus, Ohio

Gregory Gould Fuel Engineering Company of New York, Thornwood, New York

Herbert R. Hazard Battelle, Columbus Laboratories, Columbus, Ohio

Michael O. Holowaty Inland Steel Company, East Chicago, Indiana

Jack Huebler Institute of Gas Technology, Chicago, Illinois

---

*Mr. Ayers is currently affiliated with the Allis-Chalmers Corporation, Milwaukee, Wisconsin.

**Mr. Bevan is currently affiliated with the Youghiogheny and Ohio Coal Company, St. Clairsville, Ohio.

George L. Landsly Inland Steel Company, East Chicago, Indiana

Arthur Levy Battelle, Columbus Laboratories, Columbus, Ohio

Robert A. Meyers TRW Systems and Energy, Redondo Beach, California

James B. O'Hara The Ralph M. Parsons Company, Pasadena, California

Richard G. Phelps Inland Steel Company, East Chicago, Indiana

Jan Visman Vortek Systems Ltd., Edmonton, Alberta, Canada

Myles A. Walsh Ford, Bacon and Davis, Inc., New York, New York

D. Glanville Williams North American Mining Consultants, Inc., New York, New York

Contents

COAL HANDBOOK

1 Coal Structure

ROBERT A. MEYERS / TRW Systems and Energy, Redondo Beach, California

I. INTRODUCTION

Literally, thousands of studies have appeared in the scientific literature describing various facets of the structure of coal. In these articles, coal is described as rocklike and worthy of petrological examination, a solid gel, a storehouse of fossilized plant material and mineral matter, a complex mixture of organic chemicals and organic polymers, and so on.

In this chapter, a serious attempt is made to cut through the vast amount of information available on one or another aspect of coal structure and get down to the minimum information necessary for a comprehensive understanding of the structure of coal. Simplifications are necessarily made but with a minimum loss of scientific precision.

The chapter opens with a discussion of the origin of coal, classification of coal by rank, and the relationship of coal rank to the relative amounts of the major elemental components. Following this, the structure of coal is presented, first on a visual basis with illustrative photographs and then on an atomic and molecular basis.

II. CLASSIFICATION OF COAL BY RANK

Coal deposits were formed from plant materials by the action of heat acting through long periods of time, supplemented by the effect of pressure. The first stage of coalification involves the formation of peat, then lignite, followed by subbituminous coal, bituminous coal, and finally anthracite. Plants and trees contain about 50% water and have a *dry-basis* heat content of about 8000 to 9000 Btu/lb [1]. The chemical composition of the plant kingdom consists mainly of varying ratios of the multiring highly oxygenated organic compounds cellulose and lignin. The coalification process can be regarded as a progressive elimination of oxygen in the form of carbon dioxide and water, such that the concentration of carbon in the remaining material increases. Thus, the 30 to 50% dry mineral-matter-free oxygen content found in plant material progressively reduces to 20 to 25% for lignite, 10 to 20% for subbituminous coal, 5 to 10% for bituminous, and to as low as 1% for anthracite. There is a concurrent increase in mineral-matter-free heat content from 4000 to 4500 Btu/lb for plant materials to near 15,000 Btu/lb for some bituminous and anthracite coals.

Classification of coal by rank as presently utilized in the United States is shown in Table 1, together with ultimate analysis carbon/hydrogen/oxygen ratios for the various ranks. The American Society for Testing and Materials (ATSM) ranking system classifies coal according to fixed carbon or volatile matter for medium-volatile bituminous through anthracite coals, on the basis of both fixed carbon and heat content for high-volatile A bituminous coals and on the basis of heat content for lignite through high-volatile B bituminous

TABLE 1 Classification of Coals by Rank

	ASTM classification method		Ultimate analysis ratio[b]		
Class	Limits of fixed carbon[a] (% w/w)	Limits of heat content[a] (Btu/lb)	Carbon	Hydro-gen	Oxygen
I. Anthracite	86-98	-	90-97	3-5	1-3
II. Bituminous			85-90	4-5	5-10
low-volatile	68-86	-			
medium-volatile	69-78	-			
high-volatile A	<69	≤14,000			
high-volatile B		13-14,000			
high-volatile C		11,500-13,000			
III. Subbituminous		8,300-11,500	75-85	5	10-20
IV. Lignitic		6,300-8,300	70-75	4-5	20-25
(plants and trees)	25-30	4,000-5,000 [1]	50-60	5-6	35-40

[a]Mineral-matter-free (mmf) basis. To convert to MJ/kg, multiply by 0.00232.

[b]Dry mineral-matter-free (dmmf) basis, normalized for exclusion of sulfur and nitrogen.

coal. The British National Coal Board has established a ranking system based on coal volatile matter and coking properties and there is an international system of coal classification based on volatile matter, heat content, and coal caking properties [2].

III. PHYSICAL AND CHEMICAL STRUCTURE

To the naked eye, coal appears to be a compact stratified material. It is chemically and physically complex, having a physical microstructure that is discernibly derived from plants, and a chemical structure containing a wide variety of polymeric organic compounds and crystalline minerals.

The structure of coal is presented visually in the following section, beginning with photographs of features visible to the naked eye and apparent to the touch, and progressing in ever-smaller dimensions through fossilized plant structures discernible in two-dimensional perspective by optical microscopy (10^{-4} to 10^{-6} m) and in three dimensions by the electron microscope (10^{-6} to 10^{-7} m or 10,000 to 1000 Å), culminating in a model representation of the molecular structure of the coal organic matrix (atomic and molecular features a few angstroms in size). The chemical structure of coal will be built up starting with the elemental composition, arrangement of the elements as molecules, and molecules as a polymeric and crystalline continuum.

A. Visual Properties

Coal petrography is based on the macroscopic and microscopic appearance of coal as well as secondary differentiations made on the basis of coal hardness and density. It is these latter properties that allow mechanical separation of visibly differentiated coal components.

1. *Macroscopic Properties*

Blocks of bituminous coal are shown in Fig. 1 as seen by the naked eye. Certain structural features are at once evident. Brilliant shiny bands can be seen as a single 3-cm-thick layer (labeled V) or as thin 2 to 5-mm sheets alternating with bands of dull appearance (labeled C—the combined effect is a bright look). A less regular area of gray (labeled D) is also seen as well as a charcoal-like region (labeled F) in the top portion of Fig. 1a. Coal petrography begins here, in the consideration and identification of the composition of these and other structural features. These regions are termed "lithotypes." The brilliant shiny band is vitrain, the gray area is durain, the bright region is clarain (alternating bands of vitrain and durain), and the charcoal-like area is fusain.

This block of coal, and coal in general, can be crushed and the fragments separated by lithotype. The vitrain fragments are very shiny, clean to the touch, and if equal force has been applied, noticeably smaller in size than either durain or clarain. The durain fragments are dull gray in color, and the largest in particle size, as durain is the hardest lithotype. The fusain material is soft and charcoal-like and crushes to a fine-grained powder that soils the hands if picked up. The term "durain" is often used for dull coal and the term "fusain" for fossil charcoal.

FIG. 1 Photographs of bituminous coal; scale bar, 5×10^{-2} m: (a) fusain (F), vitrain (V), and clarain (C) lithotypes; (b) clarain (C) and durain (D) lithotypes. [Courtesy of R. J. Helfinstine and R. D. Harvey, Illinois Geological Survey, Urbana, Ill.]

TABLE 2 Coal Petrographic Classification [3]

Megascopic		Microscopic
Rock type	Appearance	Dominant maceral groups
Simple		
vitrain	Brilliant black luster, very friable into small cubes, clean to the touch	Vitrinite
fusain	Charcoal-like, very friable into soiling black powder, difficult to wet	Inertinite
Complex		
clairain	Bright with alternating bands of vitrain (brilliant) and durain (dull), fairly friable	Vitrinite and exinite
durain	Dull gray color, very hard, difficult to crush	Inertinite and exinite

2. *Microscopic Properties*

When the lithotypes are polished, etched, or cut into thin sections and viewed in an optical microscope at magnifications of roughly 50 to 1000×, a number of microscopic petrographic units can be seen. These are related to the plant origin of coal and are termed "macerals." There are so many types of macerals that three major maceral groups have been designated: vitrinite, exinite, and inertinite. The dominant maceral-group composition of each macroscopic lithotype discussed above is shown in Table 2.

Relatively high magnification photomicrographs of a thin section of bituminous coal are shown in Fig. 2. Figure 2a, taken in transmitted light, shows the three maceral groups. The black irregular shaped lenses (mainly upper left) are inertinite. The finely laminated and globular material (center) is vitrinite. The globule is about 15×10^{-6} m thick. The white lenticular particles (largest are at the bottom) are exinite. The small elliptical white spots ($1\text{-}3 \times 10^{-6}$ m in diameter) are holes in the thin section where particles of pyrite and quartz occurred but were plucked out of the coal during preparation of the specimen. Figure 2b was taken in reflected light and is of the same specimen section. The detailed structure of the vitrinite (the gray matrix of the photograph) is not resolved in reflected light, but the structure of the inertinite and exinite particles are most evident by this method of examination. Inertinite appears white (mainly upper left) and exinite is dark gray (largest segment at the bottom) under reflected light.

Individual macerals have been found to have significant density differences. For coals of the same rank, the density of macerals decreases in the order: fusinite, micrinite, vitrinite, exinite [4]. Maceral densities for specific ranges of carbon content are shown in Table 3. Thus, macerals can be concentrated, according to specific gravity, into components enriched in specific macerals by various float-sink separation techniques.

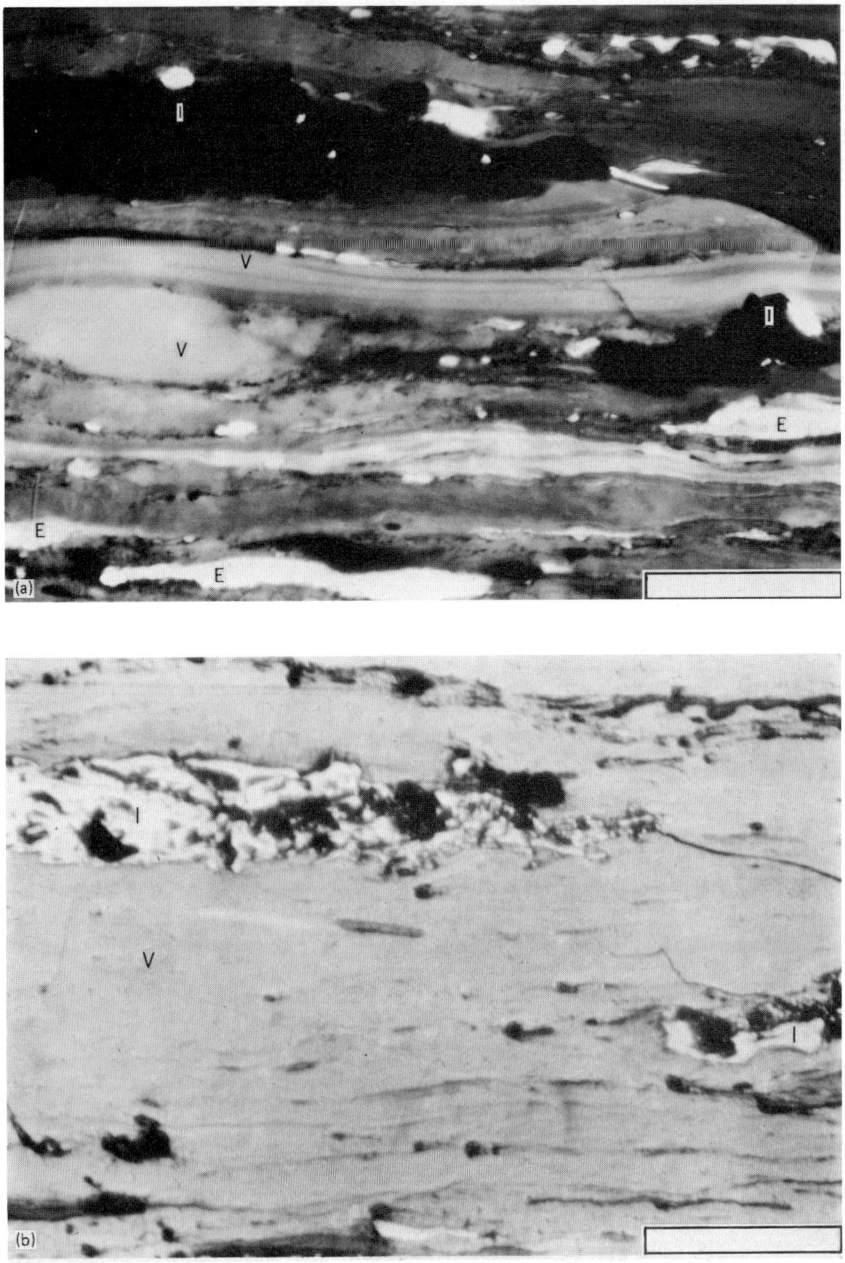

FIG. 2 Photomicrographs of a polished thin section of Davis Member bituminous coal from southern Illinois; scale bar, 40×10^{-6} m: (a) taken with transmitted light (I, inertinite; V, vitrinite; E, exinite); (b) taken with reflected light. [Courtesy of R. J. Helfinstine and R. D. Harvey, Illinois Geological Survey, Urbana, Ill.]

TABLE 3 Some Properties of Macerals

	d (calc)	Carbon content (%)	Reference
Vitrinite	1.3	84-89	5
Fusinite (inertinite)	1.5	92-95	6
Exinite	1.2-1.3	86-89	5
Micrinite (inertinite)	1.4	87-90	5

Scanning electron microscopy (SEM) offers a way to view three dimensional microscopic details that are thousands or even hundreds of angstroms in size (10^{-7} to 10^{-8} m). Many important coal microstructural features, such as cellular components and cell-contained mineral microcrystals, can be viewed in this way. These units are only two to three orders of magnitude larger than the organic molecules making up the coal molecular structure.

A SEM view of the top of a fresh fracture of a high-volatile C bituminous coal sample is shown in Fig. 3a at 30× magnification. The coalified (carbonized) plant fibers run from left to right.

A side view of the stacked fibers looking into the fiber openings is shown in Fig. 3b. The scale bar measures 20×10^{-6} or 200,000 Å. It can be seen that the cell walls are on the order of 20,000 Å thick, while pore openings of 500,000 Å in radius are ubiquitous and openings as small as 50,000 Å in radius can be seen clearly. These pores are relatively large compared to pore radii identified in adsorption studies, which vary from 75,000 to less than 20 Å [7]. The cell structure is seen to be compressed and distorted in the region shown in Fig. 4a with little pore structure remaining. A fresh fracture surface of coalified plant debris is shown in Fig. 4b. Some possible pore openings can be seen in the middle right section of the figure with radii of 1/10 to 1/100 of the scale bar or 1000 to 10,000 Å.

Pores of radii 1000 to 75,000 Angstroms as shown in Figs. 3b and 4b total roughly one-third to two-thirds of the pore volume of bituminous coals, while the total porosity volume generally centers around 10% of the coal volume [7,8].

The intimate association of minerals in coal with the organic coal matrix is shown in Fig. 5a, where deposits of micrometer-size pyrite crystals can be seen within the coalified plant fiber. A closer look at some pyrite mineral matter is shown in Fig. 5b, where individual pyrite crystals and a spherical crystal assembly (known as a framboid) are seen.

B. Chemical Structure

The major elemental components of coal are carbon, hydrogen, and oxygen; the minor elements include sulfur, silicon, nitrogen, aluminum, iron, calcium, magnesium, potassium, sodium, and titanium. Essentially all of the remaining elements occurring in the earth's crust have either been identified in trace amounts or are likely to be present in coal. Carbon is by far the most abundant

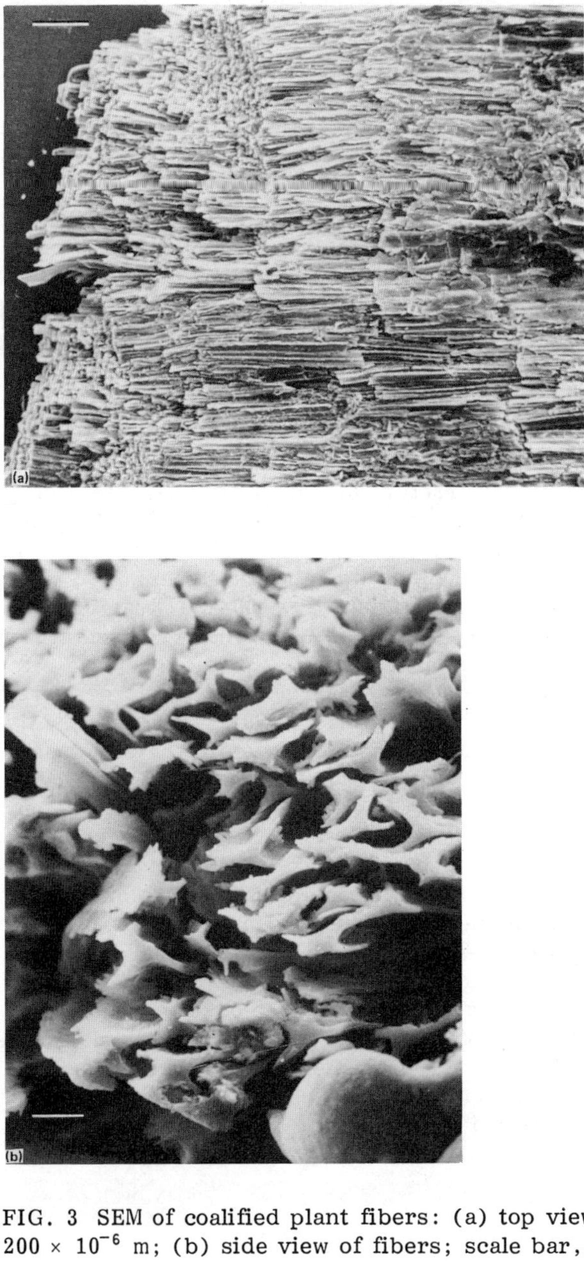

FIG. 3 SEM of coalified plant fibers: (a) top view of fibers; scale bar, 200×10^{-6} m; (b) side view of fibers; scale bar, 20×10^{-6}. [Reprinted from R. T. Greer, Coal Desulfurization, *Adv. Chem. Ser.* 64:3 (1977), by courtesy of the publisher.]

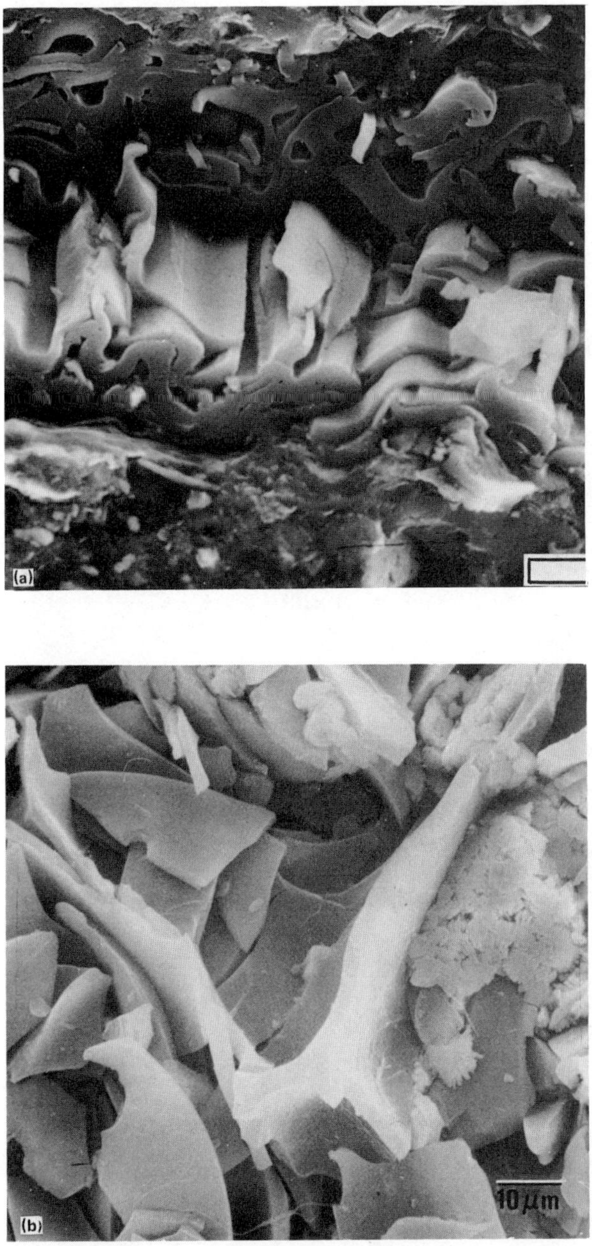

FIG. 4 SEM of compressed coalified plant fibers and debris: (a) top and side of compressed fibers; scale bar, 15×10^{-6} m; (b) plant debris; scale bar, 10×10^{-6} m. [Reprinted from R. T. Greer, Coal Desulfurization, *Adv. Chem. Ser.* 64:3 (1977), courtesy of the publisher.]

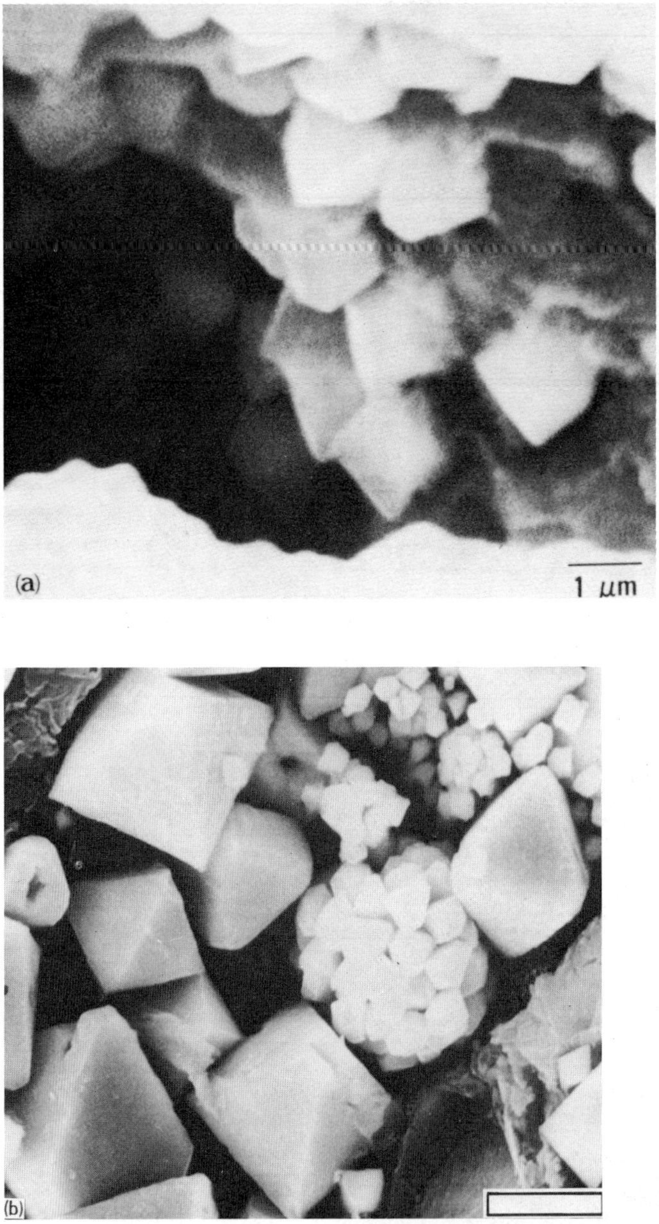

FIG. 5 SEM of pyrite crystals: (a) crystals attached to upper surface of fiber opening; scale bar, 1×10^{-6} m; (b) individual crystals and crystal assembly; scale bar, 5×10^{-6} m. [Reprinted from R. T. Greer, Coal Desulfurization, *Adv. Chem. Ser.* 64:3 (1977), by courtesy of the publisher.]

element present and in its molecular form, combined with small amounts of hydrogen and oxygen, provides the structural integrity and most of the heat content and reducing power of coal.

The elements are combined to form chemical compounds which are either inorganic or organic in nature and thus form the two coal subdivisions: the mineral matter (crystalline inorganic compounds) and the organic coal matrix. The elemental composition of the mineral matter includes all the elements listed, while the organic matrix consists mainly of polymers of carbon, hydrogen, and oxygen with small amounts of sulfur and nitrogen.

The intimate association of the mineral components with the fossilized plant cell remains (comprised of organic carbon) was shown in the preceding section. The mineral matter content of coal varies from near zero for pure vitrinite maceral, to 10% w/w for face samples or clean coal, to 40 to 50% for some run-of-mine coal.

The organic structure of coal and the composition of minerals present in coal are described in the following sections.

1. Organic Structure

As shown in Table 1, the dry mineral-matter-free element weight ratios found in coal vary from about 70 to 97 parts for carbon, 3 to 5 parts for hydrogen, and 1 to 25 for oxygen (when normalized for the absence of sulfur and nitrogen). The sulfur and nitrogen contents of the organic phase range from 1 to 2 parts.

When calculated on an atomic ratio basis, it can be seen that hydrogen is an important component of the coal structure (Table 4). The atomic hydrogen/carbon ratio decreases with increasing rank. Similarly, the oxygen/carbon ratio decreases with increasing rank. The amounts of nitrogen and sulfur atoms are always small and thus although often critical in determining coal use, these elements are not important in the basic structure of coal.

The atoms are bonded together to form various organic ring structures, aliphatic chains, and functional groups. Thus far, almost every organic compound that has been sought by researchers has been found to be a component of coal. Unfortunately, the complexity of coal causes a great deal of "noise" in the use of spectroscopic methods which provide nondestructive coal analysis (e.g., infrared, nuclear magnetic resonance, ultraviolet and fluorescence spectroscopy). Many analytical studies have depended on break-

TABLE 4 Atomic Ratios by Rank

	Atomic ratio				
Coal rank	C	H	O	N	S
Anthracite	100	50	2-5	1-2	1/2-1
Bituminous	100	70	4-8	1-2	1/2-1
Subbituminous	100	80	10-20	1-2	1/2-1
Lignite	100	80	20-25	1-2	1/2-1

ing down the coal matrix to release various organic compounds which are then analyzed by spectroscopic, chromatographic, or wet chemical means. However, the degradation products are difficult to relate absolutely to specific coal structural precursors. Thus, the molecular structure of coal is not known with any degree of certainty, but reasonable guesses can be made.

The following organic molecules (Fig. 6) are thought to be among the important components in the organic coal structure: benzene, cyclohexane, cyclohexanone, methylene, napthalene, phenol, phenyl ether, polymethylene, phenanthrene, and dihydroaromatics (9); carbazole, benzthiophene, and pyridine units are probably minor components. Functional groups such as hydroxy, carboxy, amino, and thiol have been identified.

These organic structures are bonded together to form cross-linked polymers and there may be some relatively unpolymerized organic molecules trapped in the matrix. The molecular weight of the polymeric chains, between cross-links, appears to range from about 500 to 1500, with the molecular weight of the chains between cross-links decreasing as rank increases [10]. Thus, anthracites can be thought of as being more tightly cross-linked than subbituminous coals; and so on. The organic structure of coal would appear to have a form similar to that of a cross-linked "thermoset" polymer, and the mineral matter can be thought of as analogous to the "filler" which is often added to plastics to improve properties.

A model of a postulated bituminous coal polymer chain segment [11] (molecular weight about 800) is shown in Fig. 7 (prepared with Dreiding molecular models). The dark points, where the tubes connect, represent atomic nucleii (carbon, oxygen, sulfur, or nitrogen), the tubes correspond to the bonds between atoms, and the ends of tubes signify centers of hydrogen atoms (except the oxygen atom in ring h). The length of the polymeric chain is on the order of 15 to 20 Å. The figure shows the molecular model in a conformation illustrative of the type of molecular shape that can be assumed for this portion of the coal structure.

Rotational movement within the polymeric segment is quite restricted because of both cross-links and the degree of crystallinity present in coal. This crystallinity has been measured by numerous researchers using x-ray scattering techniques [12-14]. As coal rank increases, the number of fused aromatic rings increases and the degree of crystallinity approaches the structure of graphite, which involves parallel stacking of condensed aromatic rings (lamellae). For bituminous coal, as shown in the molecular model, x-ray studies indicate that the most frequent crystalline packing involves two to three layers lying in parallel planes separated by distances of about 3.6 Å. A molecular model showing such stacking is shown in Fig. 8 for the model polymeric segment of Fig. 7. The additional aromatic rings shown in the figure are attached to polymeric segments similar in nature to the one shown. If these polymeric units extended out in planes very roughly parallel to the molecular segment shown, the voids encompassed by the crystalline region, on the one hand, and molecular segments and cross-links, on the other hand, could account for micropores which have been measured in coal with radii of 2 to 6 Å [8]. As both cross-links and crystalline zones become more frequent, as in anthracite, microporosity becomes the dominant pore volume feature [8].

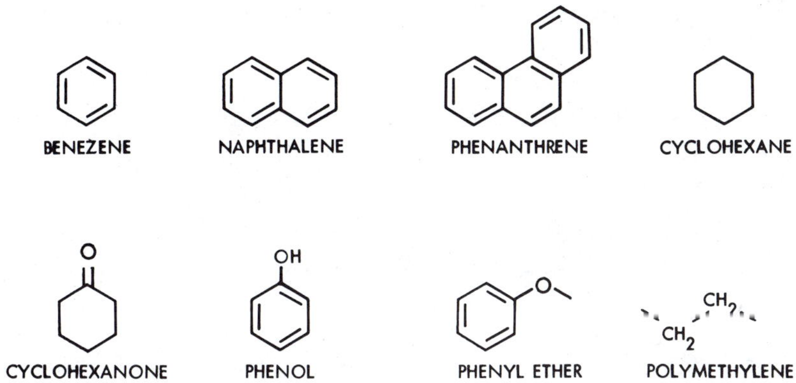

FIG. 6 Some organic moities thought to be important in the structure of coal.

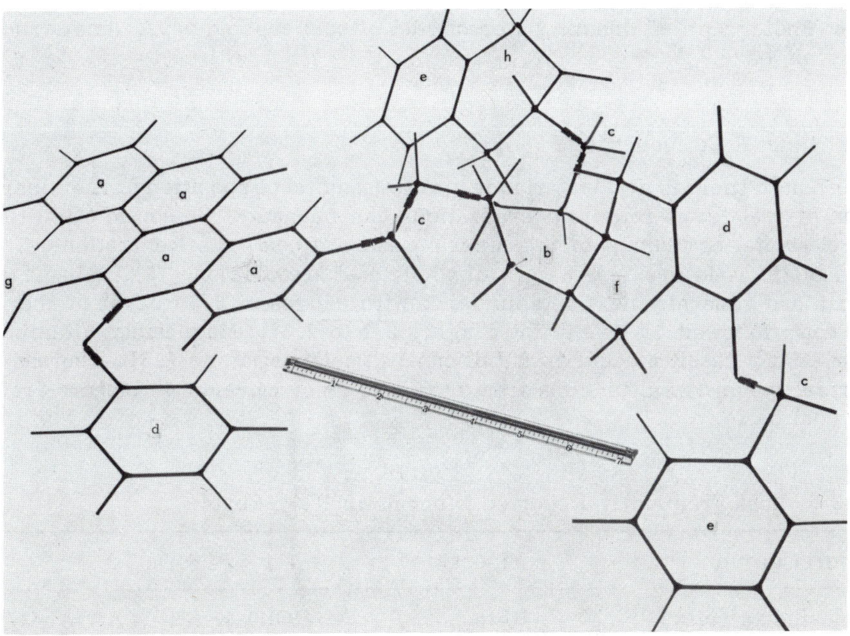

FIG. 7 Photograph of Dreiding stereomodel of coal organic structure: (a) condensed aromatic rings; (b) cyclohexane; (c) methylene or polymethylene; (d) phenylether; (e) benzene; (f) cyclohexane; (g) thiol; (h) cyclohexanone; scale bar 7 Å.

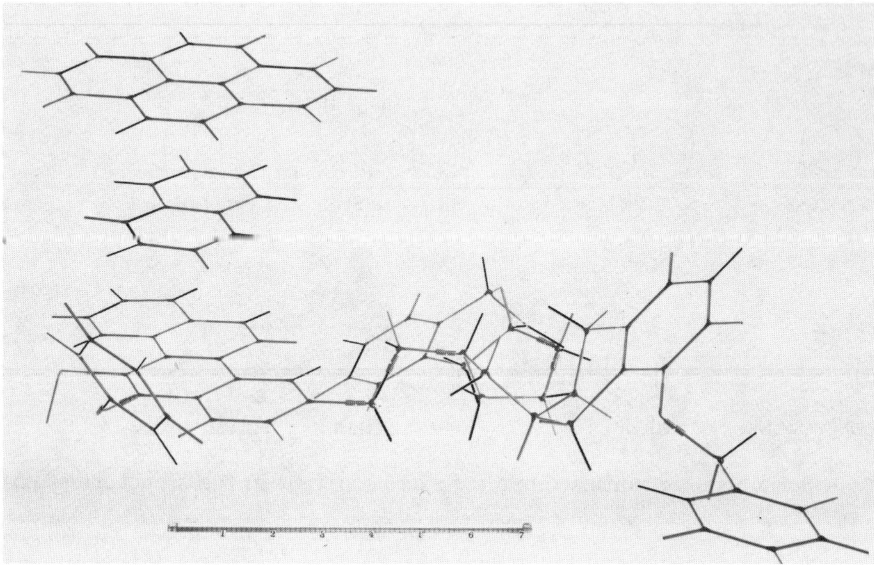

FIG. 8 Photograph of Dreiding stereomodel of coal organic crystalline structure; scale bar, 7 Å.

2. *Mineral Components*

Although there is a wide variance in the elemental composition of the mineral portion of coal, some rough generalizations can be made. Silicon is often the major elemental component of the mineral matter of coal. Concentrations of 1 to 4% of the moisture-free weight of whole coal are common [15]. Aluminum and iron are generally next in abundance with ranges of 0.3 to 2.5% of the whole coal, followed by calcium at roughly 0.1 to 2.5%. Magnesium, titanium, socium, and potassium occur in significantly smaller amounts [15], whereas inorganic (mineral) sulfur runs from near 0 to 2% or more on a moisture-free whole coal basis.

TABLE 5 Most Frequently Reported Minerals in U.S. Coals

Mineral group	Mineral	Formula
Aluminosilicate (clay)	Illite	$(OH)_4K_2(Si_6 \cdot Al_2)Al_4O_{20}$
	Kaolinite	$(OH)_8Si_4Al_4O_{10}$
Sulfide	Pyrite	FeS_2
Carbonates	Dolomite	$CaCO_3 \cdot MgCO_3$
	Ankerite	$2CaCO_3 \cdot MgCO_3 \cdot FeCO_3$
	Calcite	$CaCO_3$
Silica	Quartz	SiO_2

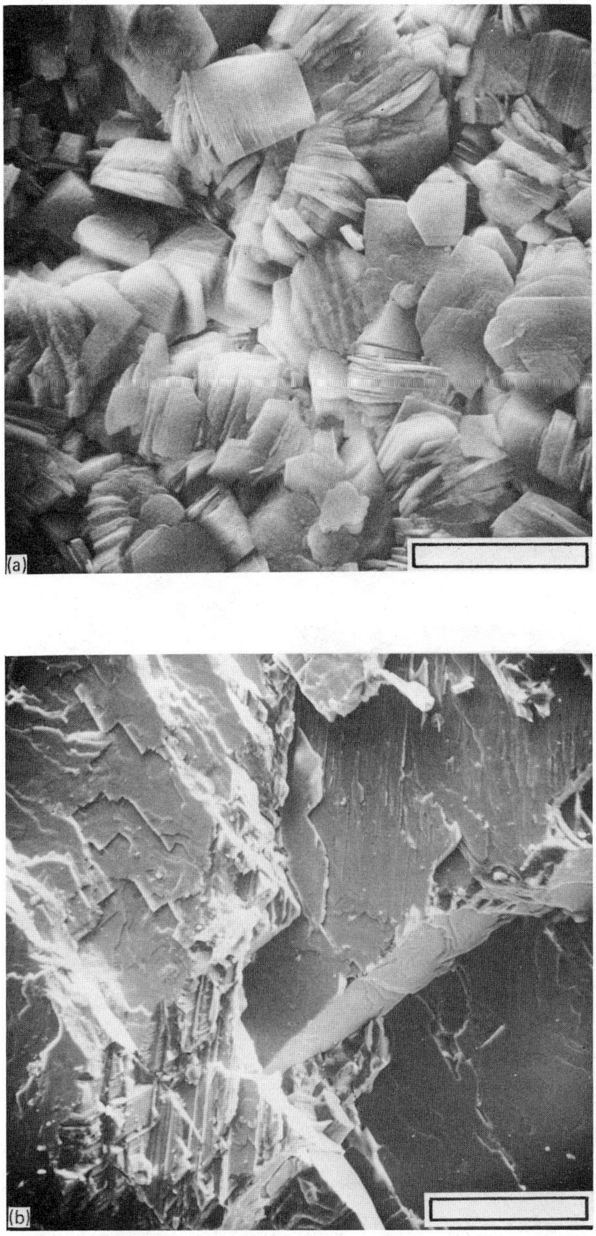

FIG. 9 SEM of minerals from coal: (a) kaolinite "books" from a cleat in the Herrin (No. 6) Coal Member; scale bar, 10×10^{-6} m; (b) calcite from a cleat in the Herrin (No. 6) Coal Member; scale bar, 100×10^{-6} m. [Reprinted from H. J. Gluskoter, *Energy Sources*, 3(2):125 (1977), by courtesy of the publisher.]

FIG. 10 SEM of sulfide minerals from coal: (a) framboids of pyrite in a low-temperature ash; scale bar, 5×10^{-6} m; (b) sphalerite (ZnS) along a cleat in the Herrin (No. 6) Coal Member; scale bar, 50×10^{-6} m. [Reprinted from H. J. Gluskoter, *Energy Sources*, 3(2):125 (1977), by courtesy of the publisher.]

These elements are combined with each other and with carbon and oxygen to form a wide range of minerals. The mineral matter and the organic coal matrix are so intimately mixed that it is difficult to accomplish physical separation for quantitative determination of the mineral composition of coal. Further, the mineral system is complicated by the large number of both stoichiometric and nonstoichiometric compounds. Nevertheless, it is known that several dozen minerals are present in coals and that only a few occur in major amounts.

The clay minerals are regarded as being the most common inorganic components of coal and of nearby strata [16]. Illite (a shale mineral) and kaolinite (a kaolin mineral) are the major minerals in this group. The clay minerals all contain water bound within their crystal lattices. For example, kaolinite contains 14% water and illite 4.5% bound water [16]. The major mineral groups, together with a listing of the most common minerals and their formulas, are shown in Table 5, in rough order of amount present U.S. coals. Electron photomicrographs of kaolinite and calcite from Illinois coal samples are shown in Fig. 9; both minerals were located at cleats (vertical fractures) in the Herrin (No. 6) Coal Member. The sulfide minerals pyrite and sphalerite are shown in Fig. 10. Mineral matter is highly variable in coal, and often a normally trace mineral is found in larger quantities in certain geographical regions. Sphalerite (zinc sulfide) is an example. Normally present only in trace amounts, coals from northwestern Illinois contain up to 1% of this mineral [17].

ACKNOWLEDGMENTS

I wish to thank the scientists, from many leading institutions and companies, who supported the preparation of this chapter through providing photographs or photomicrographs, or by consultation or review. These include R. J. Helfinstine and R. D. Harvey of the Illinois Geological Survey; B. Cecil and R. W. Stanton of the U.S. Geological Survey, now of the Exxon Development Company; and Professor R. T. Greer of Iowa State University.

REFERENCES

1. K. Howlett and A. Gamache, Forest and Mill Residues as Potential Sources of Biomass, Vol. 6, Mitre Technical Report No. 7347, Energy Research and Development Administration, Contract No. E (49-18)-2081, Washington, D.C., May 1977.
2. H. H. Lowry, ed., *The Chemistry of Coal Utilization*, Vol. 1. John Wiley & Sons, Inc., New York, 1945, pp. 27-29, 58.
3. M. C. Stopes, *Fuel*, 14:4 (1935).
4. H. H. Lowry, ed., *The Chemistry of Coal Utilization, Supplementary Volume*. John Wiley & Sons, Inc., New York, 1963, p. 110.
5. C. Kroger and J. Bakenecker, *Brennst.-Chem.*, 38:82 (1957).
6. F. J. Huntjens, H. N. M. Dormans, and D. W. Van Krevelen, *Brennst.-Chem.*, 39:S67-70 (1958).
7. R. G. Jenkins and S. C. Mitchell, *Fuel*, 57:394 (1978).

8. H. Gan, S. P. Nandi, and P. L. Walker, Jr., *Fuel*, 51:272 (1972).
9. N. C. Deno, *Fuel* 59(10):694 (1980).
10. J. W. Larsen and J. Kovac, in *Organic Chemistry of Coal*, J. W. Larsen, ed. ACS Symposium Series No. 71. American Chemical Society, Washington, D.C., 1978, pp. 36-49.
11. W. Wiser, cited in *Organic Chemistry of Coal*, J. W. Larsen, ed. ACS Symposium Series No. 71. American Chemical Society, Washington, D.C., 1978, p. 29.
12. R. E. Franklin, *Proc. R. Soc. Lond.*, 209:196 (1951).
13. P. B. Hirsch, *Proc. R. Soc. Lond. Ser. A*, 226:143 (1954).
14. J. T. Kwan and T. F. Yen, *Am. Chem. Soc. Div. Fuel Chem. Prepr.*, 21(7):67 (1976).
15. R. S. Mitchell and H. J. Gluskoter, *Fuel*, 55:90 (1976).
16. H. J. Gluskoter, in *Trace Elements in Fuel. Advances in Chemistry Series 141*, American Chemical Society, Washington, D.C., 1975, pp. 1-22.
17. H. J. Gluskoter, *Energy Sources* 3(2):125 (1977).

2 Coal Sampling and Analysis

GREGORY GOULD / Fuel Engineering Company of New York, Thornwood, New York

JAN VISMAN / Vortek Systems Ltd., Edmonton, Alberta, Canada

I. BULK MATERIALS GENERAL SAMPLING THEORY

A. Basic Sampling Concepts

Sampling theory is an effort to develop a rational understanding of that which constitutes good sampling practice. Conceptually, the goodness of a sampling method is the degree of perfection with which the identical composition and attributes of the entire population are obtained in the sample.

The sampling of lumpy bulk heterogeneous materials such as ores and minerals has always presented problems somewhat apart from those associated with sampling of discrete units, such as manufactured items. The principles of statistics and probability as used with sampling of the latter category simply did not yield a verifiable explanation accounting for known facts and phenomena, such as segregation. It is no less confusing, then, that the precision of sampling procedures in either case is judged by examining the frequency distribution that is generated when the procedure is applied again and again to the same population. It is stated in the American Society for Testing and Materials (ASTM) coal sampling method, for example, that the general-purpose method is intended to provide a precision such that, if the same consignment were to be sampled over and over, the results in 95 of 100 cases would fall within ±1/10 of the average ash.

When some property or attribute of a large volume of material is to be measured, a small portion believed to be a representative sample of the entire volume is collected for testing. The sample value will generally differ from the true, unknown value of the full lot or consignment. This difference, called the "sampling error," has a frequency distribution with a mean value and a variance. "Variance" is a statistical term defined as the mean square of errors. Its square root is the more broadly known statistic called the "standard deviation," or standard error, of sampling.

It is necessary to estimate the magnitude of the sampling error before results can be reported with any degree of assurance or precision. The sampling error depends on both the characteristics of the material being sampled and the manner of sampling. The influence of these two factors is expressed in terms of variance components that contribute to the overall sampling variance. Sampling error can thus be expressed as a function of the sampling variance or sampling standard deviation, each of which, in turn, is directly related to the material and the specifics of sample collection.

B. Sampling Characteristics—Material Constants

Materials may vary over wide ranges, and the circumstances under which the samples are collected can also vary widely, but the causes of variation in the sample value are limited. Two factors are inherent in the nature of lumpy bulk heterogeneous materials: random variation and segregation. These are statistical properties in the nature of material constants which will be denoted A and B, respectively, and are just as much properties of the material as are its heating value, bulk density, or chemical composition. These material constants, A and B, are variance components that can be determined by test as described in Appendix A-1 of ASTM Standard D2234, Collection of a Gross Sample of Coal [1], to which frequent reference will be made. A and B can also be estimated in advance for a coal for which the average composition (ash content and particle size range) and the variability of ash content from shipment to shipment (or car to car) are known. The nomogram in Fig. 1, together with the procedures outlined there, can be used for this purpose.

Every sampling operation consists essentially of either extracting one sample from a given quantity of material or of extracting from different parts of the lot a series of small portions or "increments" that are combined into one "gross sample" without prior analysis. The latter method is known as "sampling by increments" and will be the focus of our attention in the following discussion. The former method can be regarded as a special case of incremental sampling in which the number of increments is 1.

C. General Equation for the Bulk Materials Sampling Variance

It is known empirically that the precision of sampling is a function of the size of increments collected and the number of increments included in a gross sample, improving as both are increased, subject only to the constraint that increment size not be small enough to cause selective rejection of the largest particles present. Recognition of this was evidenced in the specification of minimum number and weight of increments in the ASTM standard method for collection of coal samples adopted some 25 to 30 years ago. The manner in which sampling is performed as it relates to the precision of the sample thus depends upon the number N of increments collected from all parts of the lot and the size w of the increments. N and w are operating variables that can, within certain limits, be regulated by the sampler.

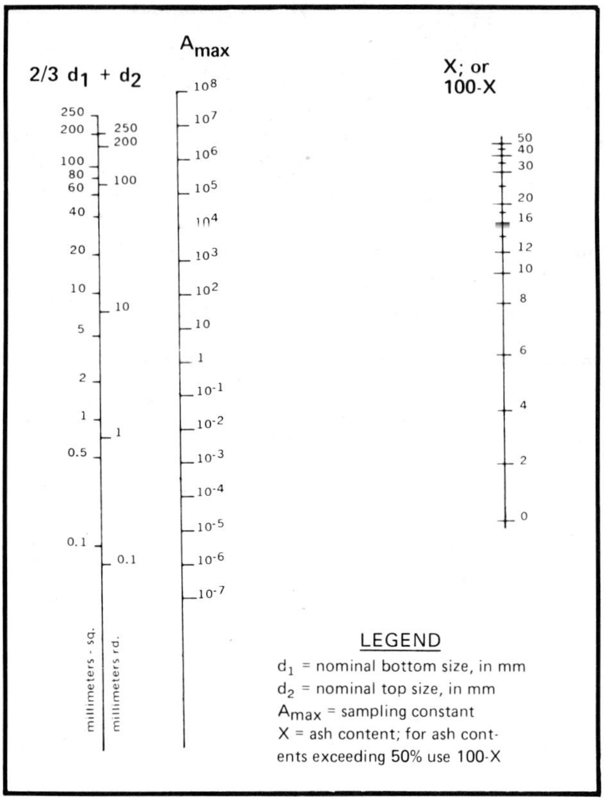

Procedure for Use of Nomogram

1. Estimate A_{max}. The value of d_1 should be the size of the screen opening through which passes 15% or less of the sample. For crushed run of mine and resultants this can often be taken as zero, without seriously underestimating the value of A. The value of d_2 can usually be taken as the nominal top size of the coal. It should be the size of the screen opening upon which a total of less than 5% of the sample would be retained. Draw a line from the $d_1 + d_2$ scale on the left to the X scale on the right. X = nominal percent ash of the coal to be sampled. Read A_{max} from intersection of line with A_{max} scale.

2. Compute the estimated value of B from the variance of the ash of (at least 25) individual lots of not less than single carloads from the following two equations:

$$s^2 = [\text{Sum } p^2 - (\text{Sum } p)^2/n]/(n-1) \quad (1)$$

where s^2 = variance of individual ash content (p)
p = ash % of individual lots
n = number of lots

$$B = ms^2 \quad (2)$$

where B = sampling constant B
m = number of increments constituting the sample from which the individual ash content (p) was found
s^2 = as in equation (1)

3. Find the variance of sampling from:

$$S^2 = [A/w + B]/N \quad (3)$$

where S^2 = sampling variance of gross sample to be collected
A = sampling constant A_{max} found from nomogram*
B = sampling constant B found from equation (2)
w = size (weight, in grams) of increments that will be collected to form the gross sample*
N = number of increments that will be collected for gross sample

Note*: If w is measured in lb, divide A_{max} by 454.

FIG. 1 Procedure for estimating sampling variance from estimates of sampling constants A and B.

Visman [2], in his general sampling theory, has shown that the overall variance of sampling can be expressed in terms of the four variables A, B, N, and w according to the equation

$$S^2 = \frac{A/w + B}{N} \tag{1}$$

This equation operates independently of the shape of the parent distribution of the variate. It applies generally for first-order estimates of the upper limit of the sampling variance. The degree of segregation z is described by the equation $z^2 = B/Am$, where m is the weighted-average number of particles per unit of measurement (lb, kg). It should also be noted that gross sample size is defined by Nw.

D. Propagation of Random Variance and Segregation Variance

One theory of sampling materials that are nonrandomly distributed is known as "stratified sampling" or sometimes as "representative sampling." The procedure involved consists of collecting separately a sample from each stratum of the lot, for each of which its own mean differs from the mean of all the strata combined. With this theory, the precision of sampling is expressed in terms of the sum of the "within-strata" and the "between-strata" variances, taking into account the average size (weight) of the strata.

Incremental sampling has been perceived as a form of stratified sampling in which the strata are imaginary because there are no physical boundaries associated with them and they are simply identified with the portions from which the individual increments are collected. The "within-strata" and "between-strata" variances in this case become a function of the size and number of increments. In this context they are sometimes confused with the random variance and segregation variance sampling constants, A and B, of which they give only a biased estimate (depending on the size and number of increments used).

The meaning of random sampling error goes back to the classical experiment wherein a number of black and white balls are mixed in a vase and a sample consisting of one or more balls is withdrawn at random. The random error occurs when the hand collecting the sample selects by chance a white instead of a black ball, or vice versa. The sampling variance is the random variance if and only if the black and white balls are perfectly mixed. Any segregation causes the sampling variance to increase. The advantages of this approach are that (1) it applies to all states of segregation, including stratification; and (2) no detailed information is needed about the state of segregation such as is required with stratified sampling.

E. Experimental Evidence from Model Populations

To demonstrate the concepts discussed here, a model population will be used, generated by the Visman sampling board, which will be described presently. There are certain limitations associated with the use of a finite model population because the theoretical variance of a sample of given size drawn from an infinite population is always larger than if a sample of the same size is drawn from a finite population having otherwise identical characteristics. In actual practice all populations are finite, but this does not invalidate the theoretical estimate of the variance, provided stipulation is made that it is an estimate of the maximum value that this variance may attain for an infinite population. The same problem is encountered when samples are drawn from a segregated population. Increments that straddle the boundaries between strata contribute less to the sampling variance estimate than do those that are drawn wholly from individual strata. Consequently, the latter variance estimate is always larger than the former.

The sampling board just referred to simply consists of a 10-in. × 10-in. wire screen with 10 openings to the linear inch. This, together with a supply of lead pellets, can be used to create populations to be sampled. The screen provides a total of 10,000 openings, and this represents the total number of individuals in our model population. Upon it we distribute lead pellets in whatever patterns we may desire, creating a population of pellet and nonpellet individuals corresponding to the black and white balls we spoke of earlier. If we distribute 2500 pellets on the screen, we have a binomial population that is exactly 25% pellet (black balls, if you will) and 75% nonpellet (white balls), with a mean of 0.25 and a variance $S^2 = p(1 - p) = 0.1875$.

The samples collected from this population are not removed from the screen; instead, they are counted. This constitutes sampling with replacement, a deviation from actual practice that is unimportant as long as the sample collected is small enough relative to the total lot or population so that its presence or absence would not have any consequential effect on the mean of the total population. A sample is taken by placing a frame over the screen with a window in it that is located at desired sampling stations and the number of pellets within the boundaries of the window are counted. The size of the window can be varied as desired, and thus both the size of increments and the number of increments collected per gross sample can be varied at will. Samples can be collected either systematically at fixed stations marked off on the screen or they can be collected randomly at stations for which the coordinates are obtained from a table of random numbers.

The procedure that has been followed to investigate the effects of variations in the model population distribution has been to collect several gross samples from a given population of known conditions in regard to randomness and segregation. The same number of increments have been used for each gross sample, but the size of increments has been varied between samples. With increments larger than single-unit size, it is possible for both pellet and nonpellet members of the population to be included in an increment. When increments larger than unit size have been collected, the counts have been divided by the total number of units (increment-weight equivalency) corresponding to the increment size. It may be helpful to compare this with the laboratory analysis of a sample increment containing a multiple number of particles, the result yielding an average analysis of all the particles present.

Theoretically, the pellets can be distributed in any conceivable manner to generate populations representing conditions ranging from ideal random mixtures to complete segregation. Such a laborious process, carried out by Visman in his original experiments, can now be easily simulated by computer, generating anything from an ideal random distribution by means of its random

FIG. 2 Completely segregated population distribution.

number function, to a completely segregated distribution, without the mechanical complications associated with physically implementing generation of the populations and performing the counts for the sampling.

Figures 2 and 3 represent a perfectly segregated population and an ideal random population, consisting of 2500 pellet and 7500 nonpellet members, that have been generated precisely this way. Located in the lower left-hand quadrant of Fig. 2, for example, are 2500 asterisks representing the 2500 lead pellets. The other three quadrants are vacant, representing the remaining 7500 nonpellet members of the model population. This corresponds to a binomial population designated by p = 0.25. Increments collected from this mixture will be either 100% pellets or 100% nonpellets, except for those increments that straddle the boundary between the areas. This is of little consequence as long as the increments are small compared with the "patch" of 2500 pellets. In Fig. 3 the 2500 asterisks are distributed in an ideal random distribution pattern throughout the entire population of 10,000. However, this is still a binomial population with a mean of 0.25 and a variance of 0.1875.

FIG. 3 Ideal elemental unit random population distribution.

Replicate sampling of these two population distribution extremes reveals important basic properties of the segregation variance and the random variance and their effect on the overall variance of sampling. With the completely segregated population distribution, the overall variance of sampling has no random variance component because there is no randomness in this model population distribution. Equation (1) reduces to

$$S^2 = \frac{B}{N} \tag{2}$$

Similarly with the ideal random population distribution, the overall variance of sampling has no segregation variance component, because there is no segregation in this model population distribution. Equation (1) reduces to

$$S^2 = \frac{A}{Nw} \tag{3}$$

The unique capability of the computer to generate such pure model population distributions clearly simplifies investigating the relationships of segregation variance and random variance to the overall variance of sampling by thus isolating one from the other. Once this is accomplished, the manner in which they function together in the sampling of real populations that are neither completely segregated nor ideally randomly distributed is more easily understood.

As was pointed out earlier, it is known empirically that the precision of sampling improves as both the size of the increments collected and the number of increments collected are increased. We have also noted that the precision of sampling can be expressed in terms of the variance or standard deviation. Discovery of the true relationship between the variance of sampling and increment size is therefore of fundamental importance to an understanding of sampling precision.

In the computer-simulated repeat of the original Visman experiments, these populations were sampled with random spacing and with systematic spacing of increments throughout the populations using increment sizes of 1 pellet, 9 pellets, and 81 pellets, according to the program in Table 1.

Linear regression with replication and correlation analysis on the double logarithmic transform of the results of these samplings, together with analysis of variance of the linear regression, confirms that there is a simple inverse relationship between increment size and the random variance conponent of the sampling variance and that the segregation variance component of the sampling variance is independent of increment size. Tables 2 to 5 summarize the results of this statistical treatment of the data.

The independence of the segregation variance from increment size is reflected in the zero slope of the regression line associated with the completely segregated distribution. The locus of points representing this relationship, plotted on double log coordinates, is a horizontal line at the binomial variance of the model population. The variance of the unit increment size is equal to

TABLE 1 Sampling Regime for Sampling Experiments

Random increment spacing			Systematic increment spacing		
Increment size	Number of gross samples	Number of increments per gross sample	Increment size	Number of gross samples	Number of increments per gross sample
Completely segregated population distribution					
1	10	25	1	10	25
9	10	25	9	10	25
81	10	25	81	10	25
Ideal random population distribution					
1	10	25	1	10	25
9	10	25	9	10	25
81	10	25	81	10	25

TABLE 2 Sampling Experiment Results

Completely segregated distribution:
Random spacing of increments

Results of linear regression and correlation

Slope of the regression line	0.013
Correlation coefficient r^2	0.0097
Confidence range of the slope	-0.037 to +0.0623

At the 95% confidence level, this slope is not significantly different from 0.000.

Anova of least-squares linear regression with replication

Source	Sum of squares	df	Mean square	F
Regression line	0.0155	1	0.0155	
Variation unaccounted for by regression line	0.0252	1	0.0252	0.43
Residual error	1.5647	27	0.0580	
Total	1.6054	29		

Note: The variation unaccounted for by the regression line is not significant. The table value of F 05/1/27 = 4.21

TABLE 3 Sampling Experiment Results

Completely segregated distribution:
Systematic spacing of increments

Results of linear regression and correlation

Slope of the regression line	-0.0349
Correlation coefficient r^2	0.4538
Confidence range of the slope	-0.050 to -0.020

At the 95% confidence level, this slope is significantly different from 0.000.

Anova of least-squares linear regression with replication

Source	Sum of squares	df	Mean square	F
Regression line	0.1179	1	0.1179	
Variation unaccounted for by regression line	0.0063	1	0.0063	1.26
Residual error	0.1355	27	0.0050	
Total	0.2597	29		

Note: The variation unaccounted for by the regression line is not significant. The table value of F 05/1/27 = 4.21.

TABLE 4 Sampling Experiment Results

Elemental unit ideal random distribution:
Random spacing of increments

Results of linear regression and correlation

Slope of the regression line	-1.015
Correlation coefficient r^2	0.9837
Confidence range of the slope	-1.066 to -0.965

At the 95% confidence level, this slope is not significantly different from -1.000.

Anova of least-squares linear regression with replication

Source	Sum of squares	df	Mean square	F
Regression line	99.5312	1	99.5312	
Variation unaccounted for by regression line	0.0.0246	1	0.0246	0.41
Residual error	1.6200	27	0.0600	
Total	101.1758	29		

Note: The variation unaccounted for by the regression line is not significant. The table value of F 05/1/27 = 4.21.

TABLE 5 Sampling Experiment Results

Elemental unit ideal random distribution:
Systematic spacing of increments

Results of linear regression and correlation

Slope of the regression line	-1.011
Correlation coefficient r^2	0.9706
Confidence range of the slope	-1.079 to -0.943

At the 95% confidence level, this slope is not significantly different from -1.000.

Anova of least-squares linear regression with replication

Source	Sum of squares	df	Mean square	F
Regression line	98.7526	1	98.7526	
Variation unaccounted for by regression line	0.2125	1	0.2126	2.07
Residual error	2.7761	27	0.1028	
Total	101.7413	29		

Note: The variation unaccounted for by the regression line is not significant. The table value of F 05/1/27 = 4.21.

the binomial variance of the model population, given by the equation $S^2 = p(1 - p) = 0.1875$.

The negative unity slope of the regression line that is associated with the ideal random distribution reflects a relationship such that for a given percentage increase in increment size, the sampling variance decreases by a similar percentage. The locus of points representing the size/random variance relationship, plotted on double log coordinates, is a straight line sloping downward from the variance of the elemental unit increment size at a 45° angle.

It is noted that systematic spacing of increments yields comparable results to random spacing of increments with both population extremes, which is a matter of substantial practical importance.

This is a somewhat different analysis of the data than Visman used, but it directs attention to the simplified techniques that can be used for making acceptable estimates of the random variance and the segregation variance sampling constants, A and B. For those interested in pursuing the mathematical statistical derivation of these relationships, the work of W. G. Cochrane [3], A. J. Duncan [4,5], and J. Visman [5,6], listed in the references at the end of this chapter, are recommended.

F. Practical Ramifications of the General Equation

The practical ramifications of these relationships are the manner in which they apply to real bulk materials sampling where the population is neither ideally randomly distributed nor completely segregated. Then both components of the sampling variance are present, and the sampling variance of individual increments is the sum of the two components A/w and B. In a partially segregated population, the segregation variance will always be smaller than the parent elemental unit random variance; and the locus of the points representing the sampling variance relationship, which is the sum of these two curves, becomes a concave curve, sloping downward from the elemental unit random variance, approaching the segregation variance asymptotically with increasing increment size, when plotted on double log coordinates (see Fig. 4). Thus, the segregation variance represents the limiting value that the sampling variance of individual increments may be made to approach by increasing increment size (i.e., by reducing the numerical value of the A/w term).

Once the improvement in sampling precision associated with a given increase in increment size reaches a point of little practical consequence, further meaningful improvement in sampling precision can be achieved only by collection of multiincrement gross samples (i.e., increasing N), reducing the sampling variance according to Eq. (1), $S^2 = (A/w + B)/N$.

The optimum point where the most appropriate choice lies between increasing the size of increments and increasing the number of increments is not critical but generally will be somewhere in the region where $A/w < B$. Obviously, in practical sampling situations, minimum increment size is limited to the point where selective rejection of top size would occur. Most machines for mechanical sampling of coal are limited to collection of minimum-sized increments where the A/w term is much smaller than the B term, and the only real choice for adjusting the sampling variance to satisfy sampling precision requirements is in varying the number of increments.

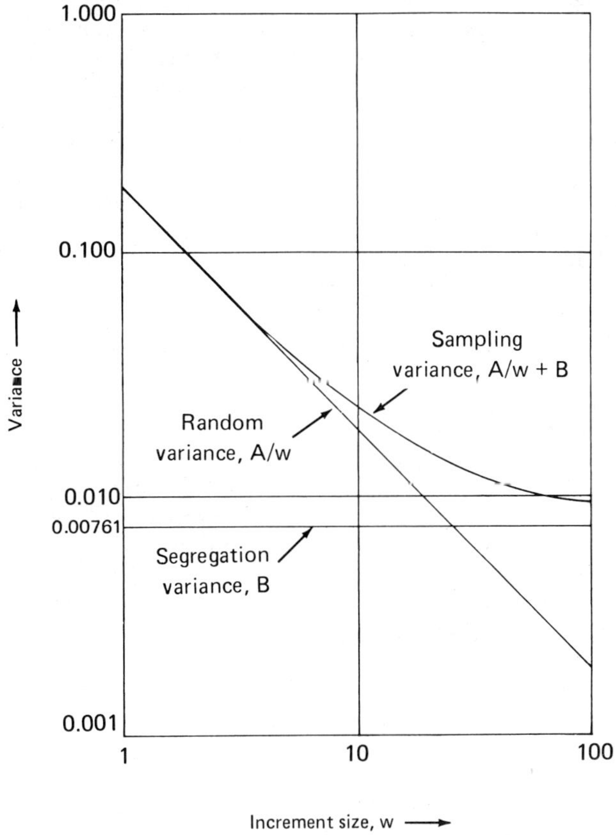

FIG. 4 Relationship of random variance and segregation variance to overall variance of sampling.

A point of some importance in the design of sampling regimes is that random spacing is difficult to achieve with either mechanical sampling or manual sampling. However, systematic spacing of increments, as noted above, produces comparable results to random spacing of increments. It follows that systematic spacing of increments and random spacing of increments are interchangeable modes of sampling provided that appropriate safeguards are applied to the systematic spacing of increments to preclude the collection of increments from coinciding with systematically occurring highs and lows in the material.

G. Determination of Sampling Constants A and B by Test

The method for determining sampling constants A and B described in Appendix A-1 of ASTM D2234 represents a substantial investment in time and effort and is therefore unlikely to be economically attractive except where large tonnages of coal are involved. One source of uncertainty encountered in attempting to apply this method is the question of what constitutes an elemental unit of the material to be sampled. It is well to recall that this method yields an estimate of the upper limit of the sampling variance and thus yields results that are generally conservatively high (i.e., safe) with regard to the minimum size and number of increments required for a given sampling precision, given that later the character of the coal does not deteriorate in a manner reflecting greater variability. Precise identification of elemental unit size (weight) is therefore not essential. It is suggested that it be taken as minimal 10 times the weight of the weighted-average particle size. For a 0×2 in. coal, for example, it might be taken as 1 lb or 0.5 kg.

Conceptually, the ASTM procedure is simple enough, and involves collection of two sets of 30 increments from a stopped belt conveyor. One increment of each set is a very small increment to furnish data for the random variance component. The other increment of each set is a large increment, to furnish data for the segregation variance component. The procedure is to use a two-section belt divider constructed of light-gauge steel plate assembled to form two compartments when forced into the coal on the belt. The bottom edges of the plates are contoured to fit the belt curvature. The smaller of the two compartments should partition off a length of belt corresponding to three times the nominal top size of the coal, and the other section should partition off a length of belt corresponding to about 20 times the nominal top size of the coal. The coal collected from the smaller of the two partitions is not the small increment for the purposes of the test. It is removed from this coal after the coal has been collected. To facilitate removal of the coal from the belt from between the belt divider plates, the bracing holding the divider together should be clear of the coal on the belt, and the best type of tool to use is a hoe. Also, the job can generally be made easier if a temporary chute is installed at the sampling location to discharge the coal directly into bags or other receiving container. The separate containers for the large and small increments should be carefully marked even though the difference in quantity would seem to be obvious, and care taken to see that identities are not lost during later processing for analysis.

Detailed step-by-step procedures are given in the ASTM method, including illustrative example with calculations that can be easily followed. Too much emphasis cannot be placed, however, on proper organization and physical arrangements for carrying out the collection of the large and small increments, including additional temporary lighting if needed, foolproof procedures for starting and stopping the belt to ensure the safety of the sampling personnel, and means for carrying out the increments to facilities for processing them.

II. PRECISION AND ACCURACY

A. Definitions

In commerce, unit prices invariably reflect the relationship between some desirable attribute or combination of attributes and quantity. These parameters are expressed in terms of numerical values obtained by measurement. The accuracy of these measurements is thus of paramount importance. They need to be accurate-enough approximations of true values that the residual errors are not a matter of economic consequence. Broadly, on a lay level, numerical information tends to be taken as discrete units, fixed absolute values, not as approximations, and the concept of residual errors of measurement tends to be ignored. In this climate, precision and accuracy generally fail to receive appropriate expression in the terms of contract clauses or in the settlement of disputes. The reality of the imperfection that is always present in all measurements is thus obscured if not lost altogether.

Nevertheless, the usefulness of measurements depends entirely on their precision and accuracy, and definition of these terms is important. ASTM Committee E11 defines precision as "the degree of agreement among individual test results obtained under prescribed similar conditions. . . ." This is a conceptual definition that provides no limitations on how precision should be expressed. Precision is commonly expressed inversely by the imprecision of results in terms of their standard deviation or their variance.

Precision, by definition, does not include any systematic error or bias, but accuracy as defined by ASTM does. ASTM defines accuracy of a test as "the degree of agreement of individual test results with an accepted reference value. . . ." Defined in this manner, accuracy includes both the random error of precision and any systematic error as well. This definition, again, is a conceptual one that does not restrict how accuracy should be expressed. Accuracy, similar to precision, is often expressed inversely by the inaccuracy of results in terms of the standard deviation or variance. The effect of systematic error on this standard deviation is to inflate it. In the measurement of coal quality for commercial purposes, accuracy expressed in this manner is generally of less interest than is systematic error itself. When systematic error is reduced to a magnitude that is not of practical importance, the specification of precision and accuracy can become meaningful parameters for defining truly representative sampling and for interpretation of test results.

B. Precision in Terms of Variance, Repeatability, and Reproducibility

Most of the ASTM standard methods on sampling and testing coal and coke, at present, are silent with regard to systematic error and bias. With the exception of the precision of general-purpose sampling, ASTM Committee D05 specifies precision and accuracy of analysis in coal and coke testing standard methods in terms of a repeatability interval and a reproducibility interval without specification of any statistical confidence level. Repeatability is defined as the difference between replicate determinations performed in duplicate within the same laboratory, by the same analyst, using the same equipment and the same prescribed methods. Replicate determinations are determinations performed separately and independently of each other. Reproducibility is defined as the difference between results obtained by different laboratories performing their analyses by the same prescribed methods on splits of the same sample.

The specification of repeatability and reproducibility intervals, without specification of a statistical confidence level, weakens the precision and accuracy specifications to the extent that this leaves open to question the magnitude of the underlying variance. If, for example, the repeatability interval is never to be exceeded, the variance would have to be zero. From a practical standpoint, this is impossible. Furthermore, the variances (standard deviations) are of direct importance in regard to the details of performing sampling and testing operations because the overall variances can be partitioned into components associated with identifiable sources of variation. This permits assessment of the relative importance of specific details with regard to precision and accuracy. With regard to reproducibility, for example, there is a component of random variance that affects the degree of agreement between laboratories but does not affect the degree of agreement within laboratories (repeatability). Virtually nothing is known about this component of variance except that it exists, and the standard methods do not address this factor directly. However, recognition of it is evidenced in the standard methods by specification of reproducibility intervals that are universally larger than would be accounted for by the variances associated exclusively with the specified repeatability intervals.

Under the general sampling theory, discussed in Sec. I.C, sampling variance was introduced and explained. In the overall accuracy of results, the sampling variance is but one component, but it is the largest single component. This is a matter of major importance that is frequently missed by the uninitiated. The full statement of sampling precision for general-purpose sampling in ASTM D2234, paragraph 7.1.1, reads as follows: "The general-purpose sampling procedure is intended for a precision such that if gross samples are taken repeatedly from a lot or consignment and one ash determination is made on the analysis sample from each gross sample, 95 out of 100 of these determinations will fall within plus or minus one tenth of the average of all the determinations. Under some conditions, as detailed in 6.2 and 6.2.2, Condition C and D and Type II, this precision may not be obtained." This statement merits critical examination, in terms of expected performance, to put matters in proper perspective.

The language "intended for a precision" clearly removes any guarantee that such precision will be achieved, and the closing sentence emphasizes conditions where it may not be obtained. But assuming that this precision were obtained, then for a 10% ash coal the statistical probabilities would be 1 chance in 20 that the results of a second sampling conducted immediately by the same people using the same methods would differ from the results of the first sampling by as much as 1.4% ash. At present, when multiseam raw steam coals ranging as high as 30% ash are not unknown, such precision could result in a corresponding difference as large as 4.2% ash. Nevertheless, in the ordinary commerce of coal, seldom is such performance anticipated, nor is there any regard for the sampling characteristics of the coal, as discussed in Sec. I.C.

Keller [7] and Aresco and Orning [8] have shown that for coal sampled in accordance with ASTM general-purpose sampling procedure and prepared and analyzed in accordance with ASTM specifications, the components of the overall variance of the final results are in general on the order of 80% attributable to sampling, 15% attributable to sample preparation, and less than 5% attributable to laboratory analysis. Toward fulfillment of precision requirements, a similar allocation of attention, facilities, and costs would be appropriate. Unfortunately, seldom is this the case. Emphasis is most frequently placed on the laboratory, its manpower, its facilities, and its methods, not on sampling.

TABLE 6 Precision of ASTM Standard Methods for Coal

Parameter	ASTM standard	Conditions	Within-laboratory repeat-ability interval	Equivalent standard deviation 95% proba-bility level	Between-reproduc-ibility interval	Equivalent standard deviation 95% proba-bility level
Moisture (%)	3173	<5%	0.2	0.072	0.3	0.108
		>5%	0.3	0.108	0.5	0.180
Total moisture (%)	3302	—	0.30	0.108	0.5	0.180
Ash (%)	3174	No carbonates	0.2	0.072	0.3	0.108
		Carbonates	0.5	0.180	1.0	0.361
		12%	0.5	0.180	1.0	0.361
Volatile (%)	3175	Anthracite	0.3	0.108	0.6	0.216
		Bituminous	0.7	0.253	1.0	0.361
		Lignite and peat	1.0	0.361	2.0	0.722
Total sulfur (%)	3177	<2%	0.05	0.018	0.10	0.036
		>2%	0.10	0.036	0.20	0.072
		Coke	0.03	0.011	0.05	0.018
Btu	2015	Dry basis	50	18	100	36
FSI	720	No spec.	—	—	—	—
Ash softening temp. (°F)	1857		50	18	100	36
GI (points)	409		2	0.72	3	1.08
Carbon (%)	3178		0.3	0.108	—	—
Hydrogen (%)	3178		0.07	0.025	—	—
Nitrogen (%)	3179		0.05	0.018	—	—
Forms of sulfur (%)	2492	Sulfate	0.02	0.007	0.04	0.104
		Pyritic <2%	0.10	0.036	0.40	0.144
		Pyritic >2%	0.10	0.36	0.40	0.144
Chlorine (%)	2361		0.03	0.011	0.06	0.022

TABLE 7 Precision of ASTM Standard Methods for Coal Ash

Parameter	ASTM standard	Conditions		Within laboratory repeatability interval	Equivalent standard deviation 95% probability level	Between-labs reproducibility level	Equivalent standard deviation 95% probability level
Major and Minor Elements (ranges in %)	3682	SiO_2	10–60	2%	0.722%	4%	1.443%
		Al_2O_3	5–30	1%	0.361%	3%	1.082%
		Fe_2O_3	5–30	1%	0.361%	2%	0.722%
		CaO	≥2	1.0%	0.361%	1%	0.361%
		CaO	<2	0.25%	0.090%	0.5%	0.180%
		MgO	0.3–10	0.20%	0.072%	—	—
		MgO	≥2	—	—	0.6%	0.216%
		MgO	<2	—	—	0.2%	0.072%
		Na_2O	≥5	0.5%	0.180%	0.7%	0.253%
		Na_2O	<5	0.1%	0.036%	0.2%	0.072%
		K_2O	0.3–3	0.1%	0.036%	0.2%	0.072%
		MnO_2	<0.2	0.02%	0.007%	0.07%	0.025%
		P_2O_5	0.3–3	0.2%	0.072%	0.4%	0.144%
		TiO_2	0.5–2	0.2%	0.072%	0.4%	0.144%
Trace Elements (ranges in ppm)	3683	Be	0.5–5	200 ppb	72 ppb	500 ppb	180 ppb
		Cr	5–50	3 ppm	1.1 ppm	5 ppm	1.8 ppm
		Cu	5–50	2 ppm	720 ppb	3 ppm	1.1 ppm
		Mn	10–300	3 ppm	1.1 ppm	—	—
		Mn	<50	—	—	6 ppm	2.2 ppm
		Mn	50–500	—	—	20 ppm	7.2 ppm
		Ni	<5	1 ppm	360 ppb	3 ppm	1.1 ppm
		Ni	5–30	3 ppm	1.1 ppm	—	—
		Ni	10–100	—	—	9 ppm	3.2 ppm
		Pb	10–100	2 ppm	720 ppb	9 ppm	3.2 ppm
		V	10–100	5 ppm	1.8 ppm	—	—
		V	10–50	—	—	9 ppm	3.2 ppm
		V	50–100	—	—	20 ppm	7.2 ppm
		Zn	<50	3 ppm	1.1 ppm	4 ppm	1.4 ppm
		Zn	50–100	8 ppm	2.9 ppm	17 ppm	6.1 ppm

Accuracy is commonly expected in magnitudes corresponding to the precision of ASTM laboratory analytical procedures alone, which negates the large contributions of sampling and sample preparation to the imprecision of the final result. This lack of perspective is probably accountable for more disputes in the commerce of coal than is any other single factor. The simple truth that the best laboratory work can do more than produce an accurate analysis of the material delivered to the laboratory is so fundamental that to mention it at all seems the ultimate in trivia. Yet it needs to be emphasized and reemphasized because it so frequently escapes consideration.

Tables 6 and 7 list the current repeatability and reproducibility intervals of the ASTM standards for analysis of coal and coal ash most commonly performed in the regular commerce of coal, together with the equivalent standard deviations derived on the basis of the 95% probability level.

III. BIAS AND TESTING FOR BIAS

A. Perceptions of Bias

As of 1979 there was no American National Standard or ASTM standard method for testing for bias. Testing for sampling bias has proved to be a highly controversial subject which has slowed development of a standard method. ASTM Committee D5 is working on the development of a standard, and some of the considerations upon which they are deliberating are covered in this section. The International Standards Organization and the British Standards Institute have both published methods which are based on some but not all of the concepts discussed below.

Bias error generally does not receive the same degree of theoretical analysis that is afforded random error. It is defined as a systematic error and as an offset from a true, theoretical, or accepted reference value. It occurs in a manner that eludes the laws of chance and is therefore perceived as being a nonstatistical problem. A classic example would be an analytical balance out of tare during a series of weighings, causing each weight to be incorrectly registered 0.1 g higher than the actual weight. This would shift the mean of the series of weighings by 0.1 g but would not affect the random error. As a matter of fact, the calculation of variance can often be simplified by coding the data, that is, by adding or subtracting a constant to or from each data item.

However, the problem is not entirely a nonstatistical one. The foregoing example is misleading if it suggests that there is direct information, such as the discovery that a 100-mg weight had been inadvertently left on the balance pan, thus removing all doubt of the source and the magnitude of the systematic error. That is an entirely different matter. Where direct specific information is not available, which is more commonly the case, the systematic error itself would have to be determined from measurements (estimates), and random error would thus be a factor.

Unlike random errors, which randomly occur in both positive and negative directions around the mean and thus tend to cancel each other, systematic error occurs consistently in one direction or the other and does not exhibit a tendency toward self-cancellation. Stated another way, the average of random errors, taking sign into account, tends to converge on a value of zero, whereas the average of systematic errors, taking sign into account, tend to converge on some finite nonzero value.

Returning to the series of weighings above, if it were known only that the accepted reference value was 1 g, 1 g could be subtracted from each weight that had been recorded and the mean of the differences could be computed. Suppose, for the sake of argument, that 20 weighings were made, and that the mean of the differences was 0.100 g and the variance was 0.04. In this case, as the following section will clarify, we could be certain (95% certain only) that a systematic error existed and equally certain that it did not exceed 0.171 g. This is quite a different thing from a fixed known systematic error of 0.100 g. Random error and statistical probability, do indeed play a vital role in the measurement of bias.

B. Statistical Significance Versus Practical Significance

The issue of testing for bias has been further confounded by a failure to distinguish between matters of statistical significance and matters of practical importance. This is a distinction of fundamental importance, as will be explained. It has led to ambiguity in terminology and the need for more precise definitions. Accordingly, the term "bias," as used herein, is defined as a systematic error that is of practical importance. Practical importance is further defined as a status of economic, legal, social, or scientific importance.

The measurement of systematic error is carried out by taking the differences of replicate results and applying statistical tests for significance to the mean of the differences, designated by $\bar{d}$. Accordingly, reference to the mean throughout this section will always denote the mean of the differences, unless specifically stated otherwise. From a statistical standpoint, to detect a systematic error it is necessary to reduce the precision limits of the mean to a value less than some multiple of the standard deviation of the differences, designated by s_d. This multiple is a function of the level of risk. When the mean is equal to or greater than the precision limit, it is statistically significant. Thus, the precision limit defines the boundary that is the threshold of statistical significance. We shall call the precision limit the "threshold of statistical significance," and will use the abbreviation TSS to designate it. The mathematical expression for the TSS is

$$\mathrm{TSS} = \pm \frac{ts_d}{\sqrt{n}} \tag{4}$$

where TSS = threshold of statistical significance
t = symmetrical two-tailed t-distribution t value
s_d = standard deviation of individual differences
n = number of differenced pairs

It can be seen from this equation that TSS can be varied at will by the number of differences that are used and the probability level chosen for t. It is pertinent, then, that any magnitude of $\bar{d}$ other than zero can be shown to be statistically significant at any level of probability simply by choosing enough observations (differences).

When $\bar{d}$ is statistically significant, the presence of systematic error is confirmed. Simple detection of systematic error, however, is insufficient. When the presence of systematic error is confirmed, it is important to know the upper limit that it is unlikely to exceed. Conversely, if it is not confirmed, it is important to know the upper limit of systematic error that could go undetected. In applying the statistical test of significance to the hypothesis that the mean is zero, $H_0:\bar{d} = 0$, there are risks of concluding that there is significant systematic error when in fact there is not, and of concluding that there is no systematic error when in fact there is. These mistakes are called type I and type II errors. They are discussed in detail later. The upper limit of the minimum detectable systematic error is defined only when both types of error are taken into account. This will be called the "threshold of detectable systematic error," abbreviated TDS, to distinguish it from the precision limit, TSS, which defines the boundary of statistical significance. Without going into the derivation, for the moment, the mathematical expression for the TDS is

$$\text{TDS} = \frac{(t_\alpha + t_\beta)s_d}{\sqrt{n}} \quad (5)$$

where TDS = threshold of detectable systematic error
t_α = symmetrical, two-tailed, t-distribution t value
t_β = unsymmetrical, one-tailed, t-distribution t value
s_d = standard deviation of individual differences
n = number of differenced pairs

For the untared balance example, TSS, by Eq. (4) would be $2.093(0.04/20)^{\frac{1}{2}} = 0.094$ at the 95% probability level. Since the mean of the differences $\bar{d} = 0.10$ is greater than TSS, it is certain (95% certain) that a systematic error exists. Taking the 95% probability level for both type I and type II errors, TDS from Eq. (5) would be $(2.093 + 1.729)(0.04/20)^{\frac{1}{2}} = 0.171$. Thus, it is equally certain that the systematic error does not exceed 0.171.

Systematic error, as specified in the foregoing definition, must be of a magnitude that is of practical importance in order to be classified as bias. It follows that bias is confirmed when TDS is equal to or greater than the boundary which is the lower limit of practical importance. The term "threshold of materiality," designated TM, will be used for this boundary. Accordingly, the mathematical representation of bias is given by

$$b = \text{TDS} = >\text{TM} \quad (6)$$

where b = bias
TDS = threshold of detectable systematic error
TM = threshold of materiality

The widespread failure to recognize the distinction between matters of statistical significance and matters of practical importance has lead to an equally widespread practice of interpreting the results of tests for bias sterilely on the basis of statistical significance alone, without proper experimental design of the test for detection of systematic error of a magnitude that is of practical importance (bias as now defined) or for taking into account the risks of type I and type II errors. These are matters of fundamental importance and constitute errors of omission that render confusing, or worse misleading, and sometimes excessively expensive, the findings of most tests for bias in coal sampling systems as currently executed.

The classification of differences by the probability of their occurrence according to the laws of chance error is a technological advance of great importance because it removes the matter from the realm of subjective judgment and makes it scientifically and mathematically objective. But the laws of chance error provide no scientific or mathematically objective criteria for judging the practical or economic impact of systematic errors on such matters as acceptable levels of emission or the profitability of commercial transactions. Such criteria relate to considerations entirely apart from and independent of the laws of chance error. The criterion of profitability, for example, is a straightforward, hardheaded business judgment which is the responsibility of business management, not of the statistician or engineer responsible for the experimental design and execution of the test for bias.

C. Type I and Type II Errors

Turning now to the derivation of the formula for the TDS (the threshold of detectable systematic error), one does not need to understand statistical mathematics to gain a conceptual grasp of this boundary and its relationship to type I and type II errors, which have their origins in statistical theory. However, administrative authority, responsible for maintaining operations within legal limits or for the equitable and profitable conduct of commerce does need a conceptual understanding of it in order to grasp the intrinsic limitations and implications of test results.

The concepts involved are developed in the following paragraphs with the diagrams of Fig. 5, the data listed in Table 8, and a minimum use of statistical mathematics. The data of Table 8 represent the ash content of 10 sets of paired increments from a bias test. The observations in each pair are related by a "natural tie," here characterized by a positive correlation for which the correlation coefficient is 0.9798. This correlation is created by collecting the increments from the same point on the coal load-out belt: one collected by the regular mechanical sampling system, and the other collected by the stopped-belt method. They are listed together with the corresponding within-pairs differences and differences squared. It is the population of within-pairs differences upon which attention will be focused. This parallels the earlier example of bias in connection with weighings on an analytical laboratory balance. Statistical equations derivations, and computations associated with the values listed at the bottom of Table 8 are included in the appendix at the end of the chapter.

Figure 5 depicts the frequency of occurrence of means of differences at various locations of interest. The height of the curve above any given point on the x axis represents the relative frequency of occurrence of that value. The shape of the curves reflects the tendency of results of replicate samplings to cluster around a central value, occurring more frequently near the central value than for values farther removed. The curve of Fig. 5a represents the frequency distribution for the means of differences of 10 pairs

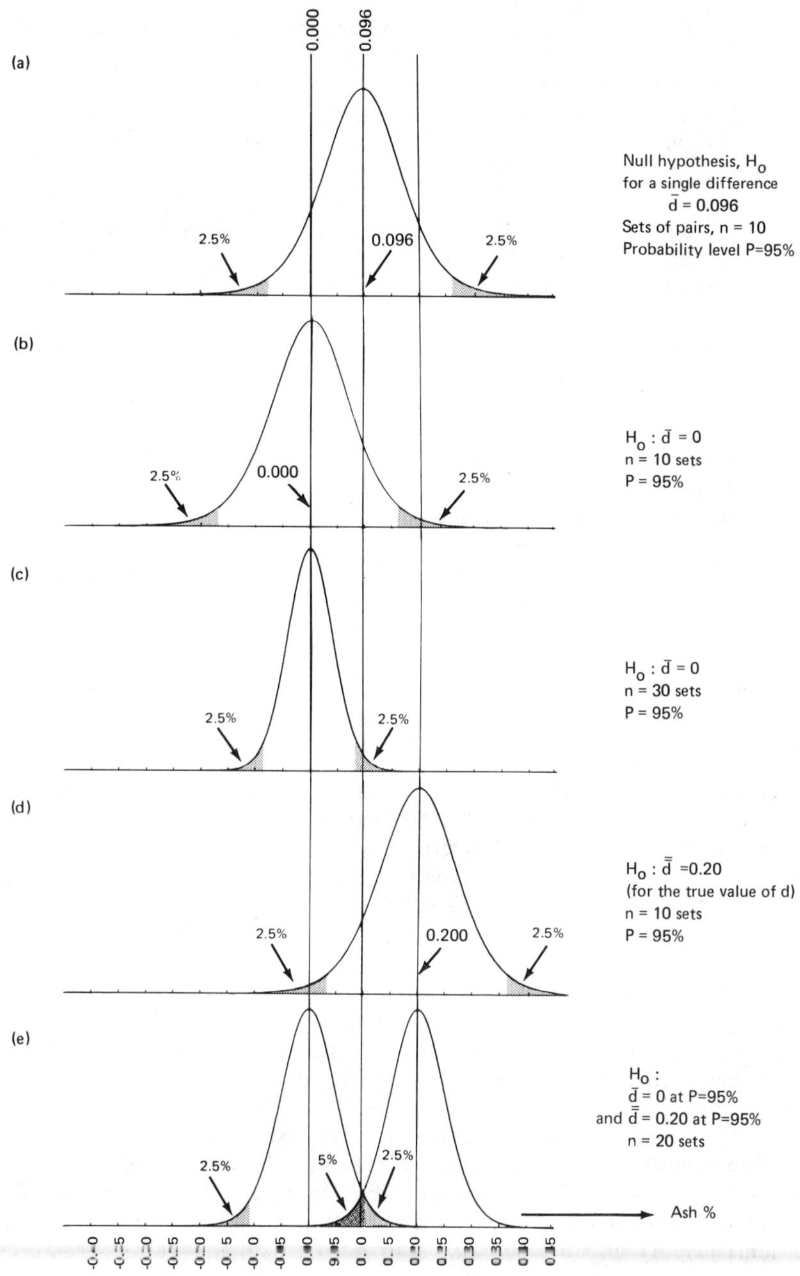

FIG. 5 t-Distribution curves of the mean of differences within pairs, d.

TABLE 8 Bias Test Ash Data

Sample number	Sampling system, X_1	Stopped belt, X_2	Difference, $d = X, -X_2$	$(\text{Difference})^2$, $(X, - X_2)^2$
1	10.41	10.47	-0.06	0.0036
2	11.90	11.53	0.37	0.1369
3	11.58	11.26	0.32	0.1024
4	12.27	12.24	0.03	0.0009
5	11.27	11.55	-0.28	0.0784
6	12.35	12.40	-0.05	0.0025
7	13.33	13.37	-0.04	0.0016
8	11.66	11.34	0.32	0.1024
9	13.90	13.54	0.36	0.1296
10	10.37	10.38	-0.01	0.0001
Sum			P = +0.96	Q = 0.5584

Number of sample sets, n	10
Mean of the differences, $\bar{d}$	+0.096
Standard deviation of individual differences, s_d	0.2276
Table value of t_α 05/9	2.262
Table value of t_β 05/9	1.833
Threshold of statistical significance, TSS	0.163
Threshold of detectable systematic error, TDS	0.295
Standard error of sampling and analysis, $s_{\bar{x}}$	0.0509
Correlation coefficient between X, and X_2, r	0.9798

of observations. Note that this curve is centered on the mean of the differences of the sample, 0.096. Recall that the exact value of the mean of the differences of the population is unknown. The sample mean is only an estimate of the population mean. In this respect it should be noted that this curve represents the t distribution, as opposed to the normal or Gaussian distribution, because it is based on a finite sample, not on the entire population.

The area under each curve in Fig. 5 represents probability or cumulative chance of occurrence of values between ordinates. The unshaded area is 95% of the total area under the curve and the ordinates of the unshaded area represent the limits within which one may be 95% certain that the true population mean is located. The shaded areas in the tails are 5% of the total area, 2.5% in each tail. The process of performing a statistical test of significance proceeds in two steps: (1) setting up a hypothesis, for example that the population mean does not differ from zero; and (2) ascertaining the probability of a sample drawn from the population having a mean differing from the population mean by as much as the observed $\bar{d}$, exclusively as a result of chance errors. If, at the above 5% level for which one is willing to take the risk of being wrong, it is found that chance errors would account for the difference, the "null" hypothesis is accepted. Thus, the unshaded area is called the "acceptance region" and the shaded area is called the "rejection region."

In Fig. 5a, the mean of differences that fall in the unshaded area are accepted as not being significantly different from the estimate of the population mean, 0.096. In other words, one will accept chance errors as being accountable for values that differ from 0.096 but which fall in the acceptance region. By the same token, chance errors alone would not account for values that fall in the rejection region. Such values, accordingly, would be judged as statistically significant (i.e., different from $\bar{d} = 0.096$). It is common engineering and scientific practice to use the 95% confidence level, but the choice is optional and may be set at any level corresponding to the risks one is willing to take. It should be noted, however, that a 99% confidence level (i.e., a 1% risk of being wrong) would broaden the range of the acceptance region. Thus, it would require a larger difference before the null hypothesis would be rejected. Note that zero easily falls within the 95% acceptance region under the curve of Fig. 5a.

There is a relatively subtle difference between the diagrams of Fig. 5a and b. The curve of Fig. 5b has been located so that it is centered on a value of zero instead of 0.096. All else is exactly the same as before. Because the curve is symmetrical, centering it on zero merely has the effect of shifting the area of interest from one side of the center line 0.096 to the other, but the area under the curve is exactly the same because it also represents the frequency distribution for the means of differences of 10 pairs of observations. The shift to the left represents setting up a null hypothesis that the population mean $\bar{\bar{d}}$ is zero. The appropriate notation is $H_0{:}d = 0$. This represents a change in the point of reference. For Fig. 5a, one would say that zero is not significantly different from 0.096. For Fig. 5b one would say that 0.096 was not significantly different from zero. In either case, the probability that the mean of the differences, $\bar{d}$, is caused by chance errors, is the same.

The change in shape between the curves of Fig. 5b and c is obvious. The curve of Fig. 5c depicts the frequency of occurrence of means of differences for 30 pairs of observations, instead of 10 pairs of observations. The number of sets of observations is exclusively responsible for the change in shape of the curve. Once again the shaded areas under the curve represent 5% of the total area. Note how the change in shape of this curve changes the significance status of a mean of differences as large as 0.096. It now falls in the rejection region. According to the risk we are willing to take, we would now reject the hypothesis that the population mean was zero, because chance errors alone would not account for a difference as large as 0.096. One may intuitively see that as a larger number of observations (differences) are included in the sample mean, the less often would large differences be likely to occur. The change in shape of the curve covering a narrower range reflects this change in probability of occurrence of a given magnitude. Thus, all one needs to get significance for any magnitude at any level of probability is enough measurements.

In conventional practice, the need for enough measurements is often ignored. The results of the test are interpreted on the basis of an arbitrary number of observations without regard for the correct number of observations to reconcile statistical significance with material importance. This is the flaw in the conventional wisdom of judging the presence or absence of bias on the basis of statistical significance alone. One of the difficulties associated with conventional practice occurs when sulfur is selected as one of the variable constituents for which the bias test will be conducted. Sulfur results frequently reflect statistical significance when ash and Btu do not because sulfur tends to exhibit a smaller coefficient of variation. This apparent anomaly can be resolved by an experimental design which adjusts the number of observations according to that required to detect a magnitude of systematic error for which the TDS is not less than a specified TM.

Practical methods and equipment for sampling and measurement are not absolutely perfect. The point is not that we must detect and remove all bias. Our aim is to reduce it to acceptable levels where, in each case, it does not exceed a predesignated magnitude. Then the test for bias can be designed to confirm the presence of bias when the probability of a bias of that magnitude exists. By the nature of the problem, it cannot be proved that there is no bias. It can only be shown that it is unlikely that there is a bias greater than a given magnitude. This is the TDS boundary. It is therefore important to know how unlikely it is that there is a bias greater than this. It will be shown how the odds of such an event can be estimated and controlled.

In Fig. 5d, the now-familiar frequency density curve of Fig. 5a has been centered at still another location. This time it has been centered at 0.20% ash, denoting an hypothesis $H_0:\bar{d} = 0.20\%$ ash, in other words, that the population mean is 0.20. The shaded areas still represent 5% of the total area, and it will be seen that 0.096% ash easily falls within the acceptance region. Accordingly, this means that at this level of risk, the mean of differences of 0.096% ash is not significantly different from 0.20% ash. But it was found earlier from the curve of Fig. 5b that 0.096 was not significantly different from zero. This brings us to the type I and type II errors mentioned earlier. These terms are used to specify false rejection and false acceptance of the null hypothesis, respectively. Initially, it was assumed that the population mean $\bar{d}$ was zero. Restricting consideration to that proposition alone ignores the possibility that the population mean might have been a value other than zero. This is a possibility that cannot be ignored because the cumulative impact of bias over time can be very large.

Suppose, for the sake of argument, that the population mean was 0.20% ash but the hypothesis was accepted that it was zero. Compare the curves of Fig. 5b and d. This would constitute an error of false acceptance and is called a type II error. Suppose, on the other hand, that the population mean was zero but the hypothesis that it was zero was rejected. This would constitute an error of false rejection called a type I error. The test for bias can be structured to limit the risks of both type I and type II errors at preselected levels of probability. A method for doing this is presented below.

Turning, once again, to Fig. 5, the curves of Fig. 5e represent two distributions, one centered on a population mean of the differences equal to zero and the other centered on a population mean of the differences equal to 0.20% ash, which we shall now identify as the threshold of materiality TM. Recall that in making measurements, it is not known with certainty where the true value is located. What one needs to be sure of is that if the null hypothesis is accepted, the risk of an undetected bias as large as TM = 0.20% ash (a type II error) is limited to 5% or whatever level of risk is chosen for type II error. This corresponds with the left tail in the curve centered on 0.20 in Fig. 5e. In addition, one needs to be sure that if the null hypothesis is rejected, the risk of there being no bias of any practical consequence (a type I error) is limited to 5%, or whatever level of risk is chosen for Type I error. This corresponds with both tails of the curve centered on zero in Fig. 5e. Conceptually, what the method does is adjust the shape of the two curves so that the rejection regions of the two curves (the left-hand tail of the curve centered on 0.20 in Fig. 5e representing 5% and the right-hand tail of the curve centered on zero in Fig. 5e representing 2.5%) touch each other, back to back. Mathematically, this is done by solving two simultaneous equations in which the common parameter is the number of observations.

D. Bias Test Method Limiting Type I and Type II Errors

The definition of bias as a systematic error resulting in an offset from a true, theoretical, or accepted reference value constrains methods of testing for sampling bias to statistical procedures for comparison of natural pairs, that is, pairs of observations that have a natural tie as evidenced by the fact that they are positively or negatively correlated. Several different situations arise in regard to bias in coal sampling systems. In respect to these, bias may be classified as relative bias or as absolute bias, depending on the nature of the reference value upon which the bias measurement is based. For example, where there is a legitimate question as to whether the results obtained by two different sampling systems are substantially the same, the matter is one of relative bias. More generally, the question is whether the results obtained by a given sampling system are biased with respect to an intrinsically bias-free reference value, and the matter is one of absolute bias.

By the nature of the problem, matters of relative bias are likely to involve comparisons of gross sample results. Matters of absolute bias, on the other hand, where the measurement of bias is based on comparison with intrinsically bias-free reference values, for reasons of cost if no other, are normally based on increment-by-increment comparisons. An important feature of this is that the sampling variance of individual increments is nominally N times greater than the sampling variance for gross samples (where N is the number of increments collected per gross sample). As illustrated earlier, the variance of the measurements is a critical factor governing the boundaries within which the true value may be expected to be located with preassigned levels of risk. The importance of this is not a matter that can be ignored with impunity. It has a direct bearing on the validity of test result interpretation.

In the testing of coal sampling systems for absolute bias, two distinctly different situations arise. In one of these the logical reference value is the population mean of the feedstream itself, and its component save and reject streams are accessible in their entirety. In the other situation, the reference value is restricted to an estimate of the population mean obtained by a parallel but separate and intrinsically bias-free sampling method. The method to be described appears to lend itself with equal facility to all relative and absolute bias situations, the reference value or the means of obtaining the reference value simply being different.

In the testing of individual components or subsystems of a sampling system downstream of the primary sample cutter, both the save and reject portions of the entire sample stream are usually accessible and a positive bias in one is accompanied by a directly related negative bias in the other. One has the option, therefore, of sampling both the save and the reject streams by intrinsically bias-free methods or collecting the entire streams in paired segments for processing and laboratory analysis. In either case the relationship between the opposing effects is proportional to the ratio of division (cutter ratio) of the sample cutters involved. If the cutter ratio C is taken as the fraction of the feedstream that is the save portion, the estimate of the bias b associated with the component or subsystem may be expressed as follows:

$$b = \frac{\Sigma(X_s - X_r)}{n}(1 - C) \tag{7}$$

where b = observed bias
X_s = within-set save-portion value
X_r = within-set reject-portion value
C = cutter ratio
n = number of sets

In the testing of a complete sampling system, the test for bias is structured around pairs (sets) of increments. In each pair X_s represents the member from the save stream of the sampling system and X_r represents the reference member collected by an intrinsically bias-free method (usually a stopped-belt increment). Ideally, each pair should reflect sampling errors exclusive of variations in coal quality. For this reason they must be collected as much as possible from the same point in the coal stream. This establishes a strong natural tie and minimizes differences that would be associated with variations in the coal from one point to another within a lot. Bias associated with a complete system may be expressed by a similar equation without the cutter ratio term since, in this case, bias in the sampling system member of each set is not associated with a related opposite effect in the reference member of each set. Thus, the equation reduces to

$$b = \frac{\Sigma(X_s - X_r)}{n} \tag{8}$$

Where two sampling systems are being compared on the basis of gross sample results, once again it is imperative that the gross samples within each pair be collected from and represent the same lot or consignment. Under these conditions, Eq. (5) applies equally appropriately to relative bias, except that the subscripts of X_s and X_r might be changed to X_a and X_b.

It is the dependence of the test for bias on measurements subject to random error that precludes establishing that there is no bias. It can only be shown that there is not likely to be a bias of more than a certain magnitude. For optimum cost effectiveness this magnitude is limited to the maximum bias that can be tolerated. This is the value that we have identified in Eq. (6) as the threshold of materiality TM. The ultimate use to which the results of a test for bias will be put merits consideration at this point because the TM should be chosen before the test begins, regardless of whether the test is being conducted to resolve a buyer/seller dispute or for internal control and policy objectives.

The test for bias includes the following essential steps:

1. Pretest inspection
2. Choice of test specifications
3. Establishment of detailed procedures for conduct of the test
4. Preliminary test increment (sample) collection, processing, and analysis
5. Determination of number of observations required
6. Final test increment (sample) collection, processing, and analysis
7. Statistical analysis and interpretation of results

Rarely can the pretest inspection be omitted. Occasionally, a test for bias is conducted for the express purpose of determining "as-is" performance, and the pretest inspection then assumes secondary importance. Even then, a pretest inspection provides documentation of what constitutes "as is" conditions and should still be carried out. In all other situations, no test should be carried out until all visible malfunctions, maladjustments, and delayed maintenance have been remedied.

The choice of test specifications includes (1) choosing the variable constituents for which the test will be conducted, (2) assignment of the TM for each constituent, and (3) selection of the levels of risk for type I and type II errors. The choice of variable constituents or attributes of coal for which the test will be carried out can include any variable for which a given incremental change is identifiable as being of practical importance, economic or otherwise. In this respect size is generally unsuitable, because breakage between sampling points, during increment collection, and during processing tends to produce unpredictable and unreliable results.

Each variable constituent or attribute for which the test will be carried out requires assignment of a TM for that variable. Moisture is such an elusive substance that a physically separate test for this constituent is often the best approach. But as a practical matter, each constituent or attribute represents a test that can often be viewed on a stand-alone basis. Exclusive of moisture, all constituents should be evaluated on dry basis. Most constituents are affected by errors in size distribution associated with size selectiveness. Screen tests to obtain size distribution information, particularly in the tails of the size distribution, can sometimes prove helpful in troubleshooting a system found to be performing unsatisfactorily. Notwithstanding, size consist itself is not generally suitable as a test variable.

The choice of levels of risk has a profound impact on the cost of conducting the test for bias. In general, levels of risk for type I and type II errors of less than 5% are associated with a sufficiently large increase in the number of observations required to render the test for bias economically unattractive. By the same token, levels higher than this are likely to reflect commercial or operating risks that are unattractive. The 95% probability level (i.e., the 5% level of risk for both type I and type II errors) is generally a good trade-off between these considerations.

The establishment of detailed procedures for conduct of the test requires careful planning the duties of each member of the staff, and exactly how they will be performed. It includes the provision of all necessary facilities and equipment for collection of increments, processing, packaging, and transporting them from collection sites to processing points and conveyance to the laboratory. Special safety precautions are often necessary and must be carefully thought out to make them both foolproof and in full compliance with all safety regulations. Extra lighting may be needed as well as temporary protection from weather. Means for communicating with persons remotely located from each other is usually necessary.

Seldom are less than 10 data pairs adequate for a bias test, and the preliminary test increment collection event can usually be planned around the collection of between 10 and 20 sets. The activities of several people always have to be coordinated and the initial stages of this event provides a means for training personnel in the details of their duties. There should be no hesitancy about discarding increment sets about which there is any question. In the initial planning stages realistic time frames should be estimated for processing and laboratory analysis. These samples often represent a substantial extra load on the laboratory facilities that will be used, and shoddy work under the pressure to get out results cannot be tolerated.

When all the analytical results are available, tabulate the data by pairs similar to Table 8, compute the differences, and calculate the standard deviation of the individual differences, taking sign into account. The total number of observations that will be required for each variable constituent can now be estimated using Table 9. Compute the g factor for each variable. This is done by dividing the TM of each variable constituent by its corresponding standard deviation of an individual difference. Enter the column in Table 9 corresponding to the risk levels chosen in step (2) and select the number of observations from the first row for which the table value of the g factor is equal to or less than the g value calculated. Plan and execute the final increment collection event around the number of observations for that variable requiring the maximum number of observations. It is wise to collect two or three extra increment sets as security against having to discard any—or the possibility of adverse changes in the magnitude of the standard deviation of the individual differences as the test proceeds. Before applying standard statistical tests of significance, recheck the number of pairs of observations required using Table 9 as before and discard, by random selection, data pairs that are in excess of the number of observations found to be needed (even though the integrity of the results of these pairs is unimpaired).

The foregoing procedure provides an experimental design for the test that ties statistical significance to magnitudes of practical importance at preselected levels of risk for type I and type II errors. Standard statistical procedures for evaluation of the results can now be applied because the experimental design under which the test for bias was conducted assures that a finding of statistical significance confirms a bias of practical consequence. Either the t test or analysis of variance and the F test can be used as $t^2 = F$ in the special case of paired observations. The derivation of this relationship is presented in the appendix to this chapter. It should be noted that in order to establish nonsignificance, the observed mean of the differences, $\bar{d}$, has to be substantially smaller than the TM. This should not come as a surprise, although it frequently does.

Mere certification of sampling systems as unbiased, without qualification, is insufficient because it can only be shown that there is not likely to be a bias of more than a certain magnitude (TDS). The language in which the findings are reported should therefore reflect this state of affairs to avoid misinterpretation of the findings at administrative nontechnical levels. Certi-

TABLE 9 g Factor for Limiting Risks of Type I and Type II Errors

Number of paired observations	Risk levels: Type I 10% Type II 1%	Type I 5% Type II 1%	Type I 5% Type II 5%
7	1.922	2.113	1.659
8	1.730	1.896	1.506
9	1.585	1.734	1.389
10	1.472	1.607	1.295
11	1.380	1.505	1.218
12	1.303	1.420	1.154
13	1.238	1.348	1.099
14	1.182	1.200	1.051
15	1.132	1.231	1.009
16	1.089	1.183	0.971
17	1.050	1.141	0.938
18	1.015	1.102	0.907
19	0.983	1.067	0.880
20	0.954	1.036	0.855
21	0.928	1.007	0.832
22	0.904	0.980	0.810
23	0.881	0.955	0.790
24	0.860	0.933	0.772
25	0.841	0.911	0.755
26	0.822	0.891	0.739
27	0.805	0.873	0.724
28	0.789	0.855	0.710
29	0.774	0.838	0.696
30	0.754	0.823	0.684

fication should also be accompanied by a statement of (1) the mean levels of each variable constituent that prevailed during conduct of the test, (2) the nominal sizing of the coal, and (3) some indication of the preparation (washing) to which the coal has been subjected, since these influence the sampling constants, A and B, of the coal (see Sec. I.C) and may affect the magnitude of bias observed. Finally, the certification should include a statement of the flow rates that prevailed in the sampling system and note whether or not intermittent flows may have affected the results.

E. Practical Ramifications of Bias and Type I and Type II Bias Errors

Type II bias errors tend to be rather insidious, because a real bias goes undetected and the economic impact remains hidden in regular books of account. Furthermore, type II bias errors may disadvantage either party in a buyer/seller relationship. If the type II error is associated with an undetected bias that overstates quality, the buyer is penalized to the extent that he pays more than the commodity is worth. If the type II error is associated with an undetected bias that understates quality, the seller is penalized to the extent that he sells his product for less than it is worth. Type II bias errors are therefore not exclusively a buyer concern nor exclusively a seller concern.

Type I bias errors can be equally as damaging as type II bias errors if payment of reparations or adjustment is made; similarly, they are not exclusively a buyer concern nor exclusively a seller concern. If the type I bias error is associated with an accepted bias that overstates quality, payment of reparations or adjustments by the seller would have the same cumulative disadvantage to the seller as a type II bias error, differing only in the clearly visible nature of such adjustments. Conversely, if the type I bias error is associated with an accepted bias that understates quality, payment of reparations or adjustments by the buyer would impose the same kind of cumulative disadvantage on the buyer as a type II bias error, again differing only in the clearly visible nature of such adjustments.

There exists a common misconception of bias as a constant value equal to the value observed in a test for bias in each separate event of an ongoing series of events, such as a series of shipments of coal. This is not necessarily the case. Moisture differences alone may be enough to alter bias from one shipment to another. Furthermore, the measurement of bias, as illustrated earlier with the analytical laboratory balance example, is an estimate of a true value the magnitude of which can be ascertained only approximately, lying with preselected certainty between given boundaries. These aspects of the matter create serious pitfalls for arrangements that apply fixed adjustments to each shipment of a series of shipments to compensate for bias. Sooner or later a situation arises where the combined effects of the fixed adjustment applied to a given shipment and of random error cannot be reconciled with historical relationships, thus undermining the credibility of the arrangement. In general, it is far better practice when bias is found to seek a remedy for the cause rather than to treat the symptom by exacting or paying adjustments.

IV. SAMPLING AND SAMPLE PREPARATION

A. Types of Sampling Classified by Purpose

Classification of the purposes for which the information obtained will be used offers a useful way in which to organize the procedures of sampling, because this is the single most important factor governing selection of equipment and procedures. ASTM has been consolidating sampling specifications for many years, and most of the ASTM sampling procedures are now incorporated in the Standard Methods for Collection of a Gross Sample of Coal, Designation D2234. The only other stand-alone sampling standard is the Standard Method of Sampling and Fineness Test of Pulverized Coal, Designation D197.

There are specific sampling precautions and details incorporated in a number of other ASTM standards, such as the Standard Method of Sieve Analysis of Coal, Designation D410, but these are not themselves sampling standards. This does, however, emphasize the point that the purposes for which the sampling is conducted are an overriding consideration regarding the details of equipment and procedures that are appropriate to a given objective. Committee D5 is working on additional methods that may become stand-alone standards or be incorporated into existing ones, extending their scope. Procedures are currently under consideration for manual sampling, sampling from coal piles, and for sampling coal in situ in the mine, for example.

1. *Exploration*

One important objective for which there are no dedicated or specific standard methods relates to exploration of reserves in the ground. This is done for the most part by core drilling, although along outcrops, coal is sometimes sampled by channel sampling techniques. Sampling in connection with exploration falls largely in the province of the geologist and mining engineer. It is their special expertise that is required to plan the location, spacing, depth, and size of core drills required to assess the quality and quantity of coal in the deposit being explored.

2. *Production*

Preparation plant performance testing and routine quality control in mining operations and preparation plants require sampling coal both in situ and at various stages of processing following removal from the bed. Other than channel sampling for sampling coal in situ, and the sampling of coal slurries, the sampling techniques for quality control purposes and preparation plant performance are covered in the ASTM sampling methods of D2234. The testing of preparation plant performance, however, may require complex sampling programs for the sampling of many coal streams with widely different sampling properties involving collection of sample increments for which the timing has to be tightly coordinated. Such sampling almost always depends upon manual sampling with a variety of sampling implements, often in locations with difficult if not inadequate access. The complexity of such activities defies standardization.

3. *Commerce*

The standard methods of ASTM D2234 apply most directly to the commerce of coal in which the purpose of the sampling is valuation for determination of price or conformance with purchase and acceptance specifications. Commerce is carried on in consignments or lots that are clearly definable as distinct units and can be sampled as such, but the size of such lots varies widely, from single truckloads to shiploads of 30,000 to 50,000 tons. Nevertheless, the ASTM general-purpose procedures stipulate the minimum weight and minimum number of increments required, together with precautions, increment types, conditions of increment collection, and increment spacing.

4. *Utilization*

Optimization of coal combustion in power plants and processes is a function of the many variable constituents of coal. Sampling is conducted to determine efficiency, heat inputs, and special operating needs. The effect of fineness on the combustion of pulverized coal is dramatic, and the special problems associated with collection of an unbiased sample of pulverized coal resulted in adoption of ASTM D197 to cover this particular situation. So-called "operating" samples are often collected from the feedstreams to power plant boilers on a shift or daily basks for the calculation of heat balances and operating efficiency. Another objective of operating samples is to document compliance with air pollution emission regulations based on fuel composition. ASTM D2234 is applicable to all sampling related to utilization except for pulverized coal.

B. Manual Sampling

The ASTM sampling standard states: "Coal is one of the most difficult of materials to sample, varying in composition from noncombustible particles to those which can be burned completely, with all gradations between." Its composition can vary substantially from the top to the bottom of the seam, from side to side, and from one end to the other, within an unmined bed. This variability is often bewilderingly increased by mining, preparation, and handling. Natural processes of segregation account for systematic variations between side to side, end to end, and top to bottom of individual cars, between cars, from one side of a conveyor belt to the other, and between one location and another in coal piles. The challenge in sampling coal, therefore, is to collect a relatively small portion of a shipment or consignment that accurately enough represents the quality. This requires that the sample increments be collected in such a manner that no piece, regardless of size, shape, density, or position or velocity relative to the sampling implement, is selectively collected or rejected.

No matter how manual sampling is done, it is a dirty, physically demanding job that also requires the proficiency, integrity, and reliability expected of the laboratory analyst. Sampling conditions are seldom ideal and may have to be done in cramped or awkward positions. It is usually done out in the open, where conditions seem always too hot, too cold, too wet, too dry, or too windy—almost never just right. The range of circumstances under which samples must be collected, together with the intractible naturally difficult sampling characteristics of coal, set manual sampling apart as one beset by difficulties and uncertainty.

D2234 states: "The wide variety of coal handling facilities preclude the publication of detailed procedures for every sampling situation." There are two considerations involved with the principle of ideal sampling (that every particle in the entire mass to be sampled have an equal opportunity to be included in the sample): (1) the dimensions of the sampling device, and (2) proper use of the sampling device.

The opening of the sampling device must be $2\frac{1}{2}$ to 3 times the top size of the coal to meet ASTM D2234 requirements. Except for this, ASTM provides no guidance. The International Standards Organization has established design criteria for several types of hand tools (see Fig. 6) that can be used for manual sampling. The main considerations are that the width be not less than specified and the device be able to hold the minimum increment weight specified without overflowing.

In harmony with the foregoing principle of ideal sampling, the most favorable situation occurs when coal passes a sampling point in a stream, on a conveyor belt, for example, or when it falls from a chute. Systematic spacing of increments collected from a stopped belt is universally accepted as the

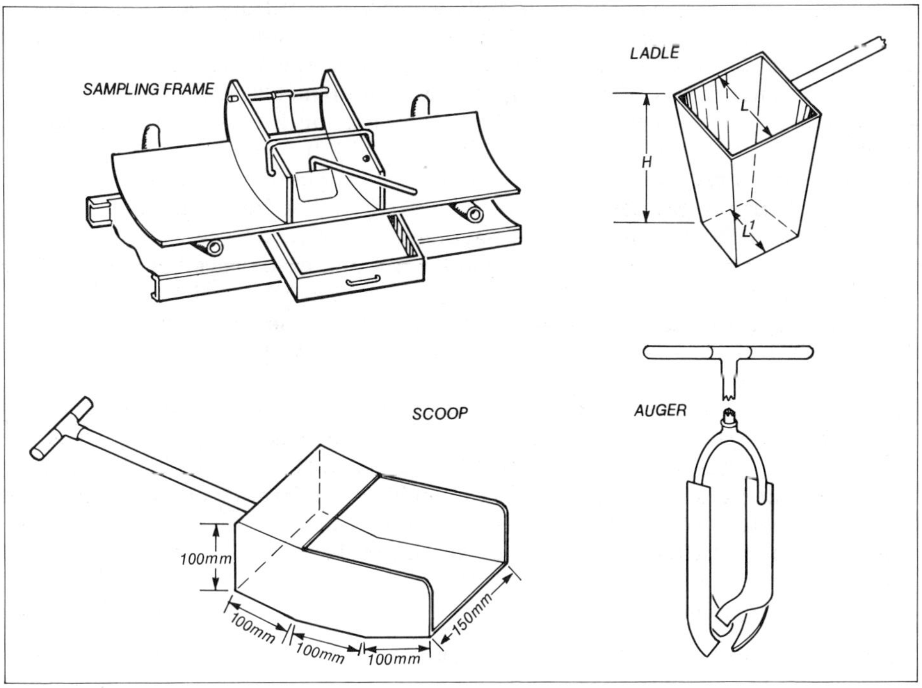

FIG. 6 ISO manual sampling tools. [Reprinted from *Power*, p. 65, Apr. 1978.]

reference method of sampling that is intrinsically bias-free. Stationary sampling, that is, sampling coal at rest in piles, or in transit in trucks, railcars, barges, and ships, suffers decreased reliability to an indeterminate degree. "Stream sampling" and "flow sampling" are terms usually reserved for the collection of sample increments from a free-falling stream of coal as opposed to the collection of increments from a stopped belt. This mode of sampling is recognized as second best only to stopped-belt sampling provided that increment collection is achieved by cutting the full stream. The collection of increments from the sides of a moving belt is sometimes loosely called "flow sampling" or "stream sampling," and this terminology therefore should not be accepted as assurance that increments were collected from a free-falling stream. Such procedures would rank as less reliable than collection of increments from a free-falling stream. Increment collection is classified by ASTM according to type, spacing, and condition, with the stipulation that the highest possible classification that is feasible be used.

1. Sampling Coal in Motion

For manual sampling, the best place to obtain a flow sample is from a free-falling stream discharging from a chute or a belt conveyor head pulley. One advantage of sampling coal in motion is that the sampler cannot clearly view the texture and appearance of the coal, so he is not influenced subjectively in the performance of his duties. Sampling from the sides of moving conveyors involves potential danger to the sampler as well as the possibility of bias caused by segregation. Conveyors moving faster than about 300 ft/min present a serious hazard. Coal larger than $1\frac{1}{2}$ in. top size, or coal heaped to a depth of more than about 8 in. on a conveyor belt, exhibits a tendency to segregate. Coal that passes from one belt to another at an angle invariably becomes more segregated, with a greater predominance of coarse particles on the far side and a greater predominance of fine particles on the near side.

When the falling stream is more than 1 ft thick or is more than about 2 ft wide, the forces involved tend to be greater than can be resisted with handheld equipment and simple mechanical assists should be considered. Streams of coal with a top size larger than 3 in. also generally fit in this category. A pivoted scoop with the needed mechanical advantage is useful. Slide gates, a dropout section at the bottom of a scraper conveyor, or a flop gate in a vertical chute are other possibilities. In the event that such alternatives are not feasible, partial stream cuts are permissible, although less reliable. Partial stream cuts should be made systematically at different points in the stream to proportionately represent all parts of the stream. This requirement becomes extremely difficult if flow rates are irregular.

2. *Sampling Coal at Rest*

There are numerous situations where coal must be sampled at rest despite the sacrifice in reliability associated with it. A major problem with sampling coal at rest is that an inevitable and unknown degree of segregation will prevail, and there is no possible way to penetrate all parts of the entire mass in such a way that every particle has an equal opportunity to be included in the sample.

The commonest situation where coal must be sampled at rest arises where the coal has to be sampled in railcars. The alternatives for sampling from hopper cars are, top, bottom, and a combination of the two. Preferences for one over another cannot be factually supported. As coal is loaded into hopper cars, it suffers segregation related to the loading process that is not necessarily obvious, and the degree to which it will affect results of samples is unpredictable. Sometimes the segregation is clearly evident, such as when cars are loaded from a stream that enters the car from the side, causing the large pieces to be shot to the far side while fines dribble down the near side. When cars are flood-loaded in unit train facilities, segregation is likely to occur more subtly, and differences in texture and appearance may not be clearly visible.

Partly due to segregation but due also to other causes, the moisture content of coal in hopper cars is not uniform throughout the mass, and it is constantly changing. From the moment the coal enters the car, changes start taking place. When there is a significant amount of surface moisture present, some of it begins migrating downward immediately, and soon moisture in the coal at the bottom of the car can be substantially higher than at the top. The difference in moisture at different levels may become more pronounced as time passes, owing to the effects of both evaporation and precipitation. Heavy rainfall can cause the moisture content to be much higher at the top and sides than at the bottom. It makes a difference of no small consequence if freezing conditions prevail.

In "car-top sampling," as the term implies, only the coal near the top surface has an opportunity to be included in the sample, which violates the basic principle of ideal sampling. Curiously enough, sample collection that violates this principle does not always produce biased or inaccurate samples. But it does increase the uncertainties regarding the accuracy of the results. This calls for an added element of caution in judgments based on such samples. Spacing of increments should be uniformly and systematically located in positions over the entire consignment, considering a string of railcars as a continuous ribbon. A necessary precaution to such spacing of increments is that they not inadvertently be collected predominantly from any given location relative to the dimensions, in plan view, of the rail cars, such as all from one side, the center line, or the ends. If cars vary substantially in size, the number of increments per car should be varied proportionately or approximately so. Care must be taken not to tip the sampling scoop, allowing any of the coal collected to roll off it, as this can introduce bias.

The alternative of sampling the coal as it is discharged from the bottom of hopper cars, called "bottom sampling," is seen by some as an improvement over car-top sampling, primarily because the coal is in motion. The benefits of the coal being in motion in this situation are illusory. It is physically impossible to penetrate the full depth of the coal cascading out of the car. Attempts to penetrate the stream invariably result in overflowing of the sampling scoop. Increment collection is limited to the exposed surface at the sides of the car. As pointed out previously, in this location moisture is often higher than for the entire contents of the car. Flow rates are highly variable. Disproportionate amounts of coarse coal are often collected because the coarse particles segregate and roll down the exposed surface. Sometimes the sampler waits until the coal flow stops momentarily, then collects the increment. This effectively reduces the procedure to sampling coal at rest from the side of a conical pile, which is beset with a maximum of segregation.

Sampling coal at rest suffers the handicap that the sampler is under the subjective influence of the texture and appearance of the coal. For this reason meticulous adherence to predesignated systematic spacing of increments is essential. Nowhere is this more important than sampling in shipholds and barges. Here the working surface is usually very irregular, and localized segregation associated with these irregularities will cause bias if increments tend to be collected mostly from the valleys or mostly from the peaks. In shiphold sampling it is conventional practice to collect increments from different levels in separate sampling sorties, and predesignated spacing of these should also be observed.

Sampling storage piles presents similar problems. In conical piles the natural process of segregation results in a central core in which fines predominate and a gradation of sizes down the sides from predominantly fines at the top to predominantly coarse around the periphery at the base. Judgment of what spacing of increments over the conical surface will accurately represent the overall quality of the entire mass is always problematical. A regime of increment spacing over the entire surface holds more promise than any sampling regime that limits increment collection to a midpoint down the side or just at the top and the bottom.

Storage piles vary in size from less than 100 tons to 1 million tons or more, and these very large piles are usually stocked out by bulldozers or scrapers, which serve to compact the pile as it is built up. This is an important aspect of proper stockpiling to prevent spontaneous combustion. These large piles are often characterized by wide variations in quality at different levels in elevation and different locations in plan view caused by stocking out coal from different sources, different seams, and different mines. If at all possible, coal piles should be moved to be sampled. How the pile is moved will then govern the details of how it is sampled. Arrangements should be planned to optimize sampling as much as possible with collection of increments from a falling stream. Where it is infeasible to move a pile, there is no choice but to sample it in situ. The sampling of large piles in place can be done by core drilling or augering equipment. Another approach is to expose coal at various depths and locations with a bulldozer so that it can be sampled manually. Sampling in place is often done in connection with taking a physical inventory by aerial survey or ground survey. Drilling then has the advantage that it can provide information on pile depth, bulk density, and coal quality if the necessary procedures are integrated into the sampling plan.

C. Mechanical Sampling

Large coal users and the coal industry alike are strangely uncritical of the performance of machine sampling, which enjoys a largely unearned, almost mystical reputation for trouble-free perfection. Nevertheless, many machine systems simply automate the deficiencies so often attributed to manual sampling. Although it is generally recognized that coal is difficult to handle as well as difficult to sample, this imposes harsh demands on mechanical sampling systems, which must do both. Mechanical sampling systems can make no special claim, per se, as to reliability or accuracy. Yet the psychological advantage of machine sampling is undeniable, but it frequently causes the suspension of critical judgment. This unfortunate circumstance simply serves to perpetuate the many serious deficiencies that exist.

The presumption that machine sampling effectively removes the human element from the sampling process is deceptive. Machine maintenance and operation require human attention, and the human element simply has an effect at a different level. The ever-increasing size of shipments, together with larger and larger flow rates, necessitates ever-more-frequent recourse to automated mechanical sampling systems. But malfunctions, maladjustments, and mechanical failures in such systems also multiply the cumulative effect of error.

The psychological advantage of machine sampling has stimulated interest in machines for sampling under conditions where manual sampling has been the only economically feasible sampling means. The superiority of such machines for such service has not been established.

A wide variety of mechanical devices are now in use. They include flow-through cutters, bucket cutters, reciprocating hoppers, augers, slotted belts, fixed-position pipes, and rotating spoons. Documenting the performance of these devices relative to their capability for producing unbiased samples is a very difficult, challenging proposition. The ASTM sampling standard stipulates that mechanical sampling systems be tested for bias, but it does not specify how this is to be done. This is not an oversight. Development of a standard procedure has defied all efforts to date. Meantime, the vacuum this creates has led to many sincere but incomplete, faulty, and/or inadequate test for bias. This subject is discussed in detail in Sec. III.

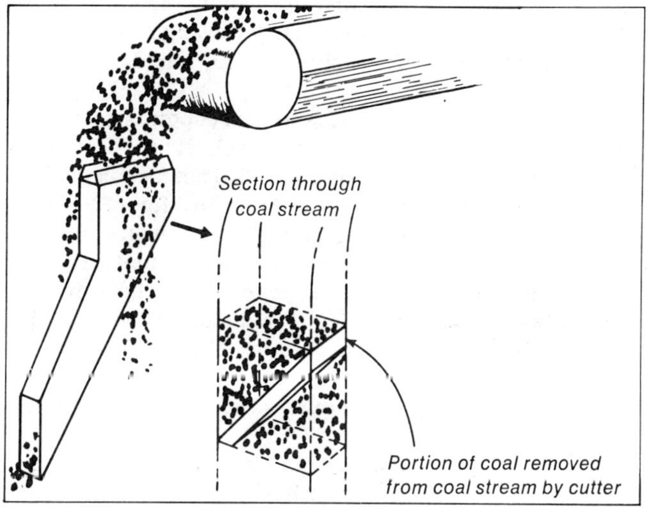

FIG. 7 Cross-stream primary cutter. [Reprinted from *Power*, p. 40, July 1978.]

1. *Generic System Design*

Mechanical sampling systems typically collect the primary increments and perform at least part of the sample preparation by crushing and dividing it down to the 4-mesh or 8-mesh stage of reduction specified in ASTM Standard Methods of Preparing Coal Samples for Analysis, D2013.

2. *Conventional Cross-Stream Primary Cutter System*

Conventional design of most mechanical sampling systems for large tonnages use some form of cross-stream primary cutter to divert the primary increments from the main stream of coal (see Fig. 7). A major advantage of these systems is that they sample coal from a moving stream, and most of them satisfy the principle that every particle in the entire mass has an equal opportunity to be included in the primary increments. There are two important disadvantages common to this type of system: (1) increments are inordinately large compared to the minimum weight specified by the ASTM standards, and (2) the quantity of coal collected on each pass of the sample cutter at the last stage of sample division tends to be too small. The first inflates the cost of the system; the second adversely affects sampling precision and sampling bias.

The inordinately large primary increment problem stems from fundamental engineering limitations of the design concept. In part, this arises from ASTM specifications that restrict cutter speeds to not more than 18 in./sec. The intent of this restriction is "to prevent segregation and rejection due to disturbance of the coal stream." However, for this purpose restrictions on cutter speed would be more appropriately referenced to the component of impact velocity perpendicular to the direction of coal flow in the plane in which the cutter is moving. This is an area in which some change in the

ASTM specifications may be anticipated. Structural limitations are also a factor influencing the choice of cutter speeds, especially in large systems where acceleration and deceleration forces of massive moving parts have to be considered.

In more common terms a cutter speed of 18 in./sec corresponds to about 1 mi/hr. In this perspective it is easily recognized as a relatively slow speed. Specification of minimum effective cutter widths of $2\frac{1}{2}$ to 3 times maximum particle size translates into typical cutter widths of 8 to 12 in. for a 2-in.-top-size coal commonly used in large power plants and electric utility stations. The combination of these parameters accounts for primary increments ranging from a few hundred pounds to half a ton or more, depending on the rate of flow, belt width, and coal top size. Considering that minimum increment size specified by ASTM D2234 for a 2-in.-top-size coal is 6 lb, this is clearly excessive, although unavoidable. The physical size of towers, conveyors, drives, crushers, bins, and other system components must be correspondingly large and costly.

The large primary increments associated with this design approach require high ratios of sample division to reduce sample quantity to practical amounts for final preparation. This usually results in the use of both secondary and tertiary cutters and either one or two stages of crushing. There are many variations in design of secondary cutters, many of which are simply scaled-down versions of the cross-stream primary cutters. They operate at slower speeds, generally not above 12 in./sec. Parallel slot riffles have not been used extensively in the sample division section of mechanical sampling systems because of their susceptibility to plugging and excessive headroom requirements, particularly in multistage division.

Secondary cutters often traverse the coal stream from front to back as opposed to cross-stream. A mechanically inverted variation of this is the moving hopper/fixed position cutter. Another variation is the slotted belt shown in Fig. 8. An advantage of the slotted belt is that it requires minimum headroom.

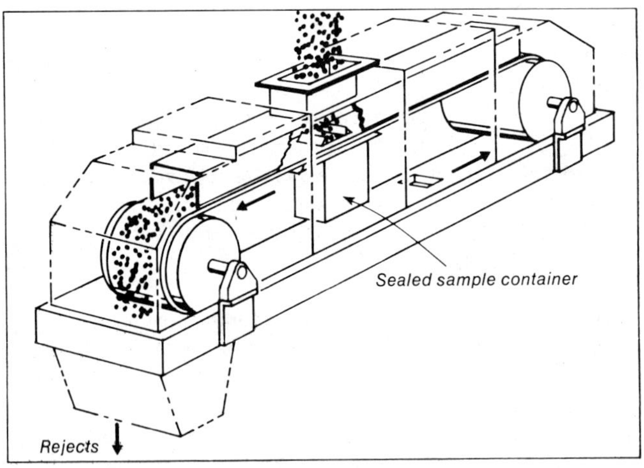

FIG. 8 Slotted-belt secondary cutter. [Reprinted from *Power*, p. 42, July 1978.]

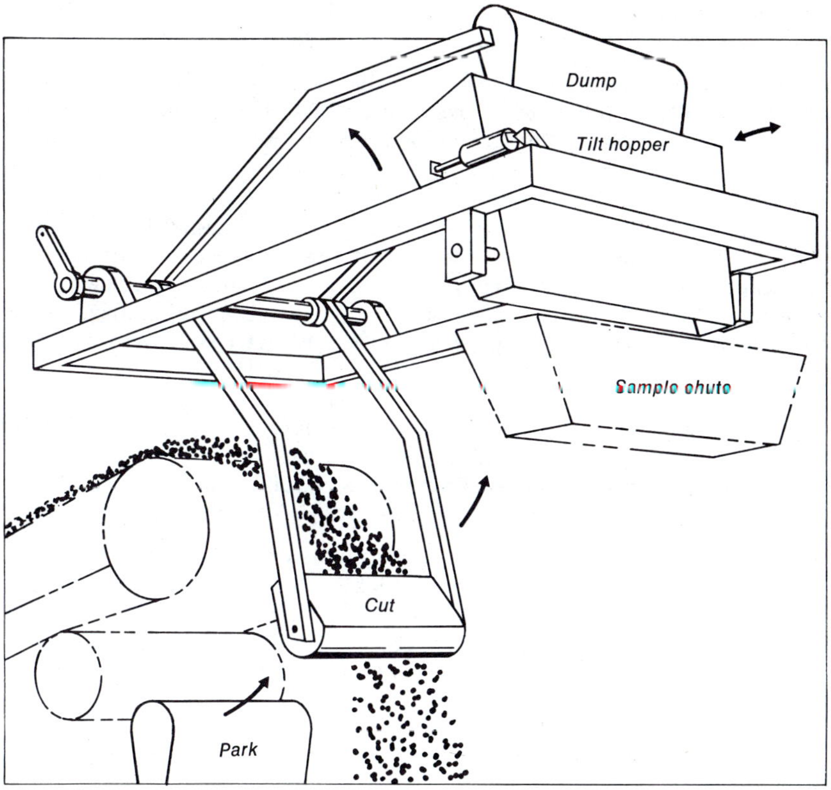

FIG. 9 Swing-arm primary cutter. [Reprinted from *Power*, p. 40, July 1978.]

The high division ratios required in these systems has led to designs where, in some cases, the final cuts are as small as a few grams of 8-mesh material. Debate continues over what minimum weight of cut should be specified. The problem is most acute at the tertiary-cutter stage, where bias in mechanical sampling systems is frequently experienced. Some experts think that the lower limit at this stage should not be less than about 200 g. Quantities less than this increasingly tend to plug downstream chutes and pipes for lack of sufficient mass to overcome plugging forces.

3. Alternatives to the Cross-Stream Primary Cutter System

a. Swing-Arm Cutter One of the more innovative primary cutter designs, which reduces the component of impact velocity perpendicular to the direction of coal flow in the plane in which the cutter is moving, is the swing-arm or arcial cutter (Fig. 9). This design achieves this goal by passing through the stream at an obtuse angle instead of at a right angle. Available evidence appears to confirm the impact velocity concept and that it can be moved at substantially higher speeds than a cross-stream cutter without causing bias associated with disturbance to the coal flow. The swing-arm sampler shown in Fig. 9 is capable of collecting primary increments that weigh only about

230 lb from a 2500-ton/hr stream. The comparable increment weight for a conventional cross-stream cutter would be about 900 lb.

b. (Visman) Pipe Sampler This is an off-the-belt sampler requiring minimal headroom which collects increments from within the stream cross section by means of one or two pipes. The sampling pipes are mounted at an obtuse angle on a horizontal axle positioned at right angles to the direction in which the coal flows.

Increments are collected through an orifice that is located in the bottom wall of the sampling pipe. This orifice is protected from collecting sample until after the pipe has passed through the vertical position as it swings against the direction in which the coal moves. The orifice is again protected from collecting sample before the pipe leaves the coal stream, by a protruding edge which causes the coal particles momentarily to dam it off.

Increments thus collected leave the pipe the next time it reaches the vertical position and is stopped. The increment slides through the pipe into one (or two) containers. The installation is run with a timing device to regulate the number of increments according to the required sampling precision. It lends itself to regular bias testing and ready adjustment.

c. Rotary Car Dumper Sampler This system, also called the tube sampler, consists of several large-diameter tubes, each with one or two openings, which are attached to a rotary car dumper. The openings are located in the path of the coal as it is discharged from the railcar. It typically collects 100 to 300 lb from a railcar. Thus, if increments were collected from every car of a 10,000-ton train, the total sample collected would weigh about 20 tons. This is about half that which would be collected from a 20,000-ton lot by a cross-stream cutter sampling from a stream flowing at a nominal rate of 2500 tons/hr.

There is compromise, however, associated with this reduction in sample size. There is evidence that the tubes collect coal predominantly from the top and far side of the car in a fixed relationship to the full contents of each car. This precludes every particle from having an equal chance of being included in the sample, making the system susceptible to bias related to some segregation patterns, such as side loading of railcars. The tubes overflow during increment collection, which also precludes every particle from having an equal chance of being included in the sample and is another potential source of bias. The physical dimensions of the machine result in spacing of increments in clusters as opposed to uniform systematic spacing of increments. This is likely to inflate the variance of sampling and potentially may contribute to bias. The tubes are susceptible to plugging with wet and frozen coal and oversize lumps. This requires constant visual observation and manual clearing when plugging occurs. This is much more of a problem with the tube sampler than with conventional cross-stream cutter systems, where the coal usually passes through grizzlies, primary crushers, or breakers of some kind before reaching the cutter. The overall effect of these potential sources of bias with this machine depends on (1) how cars are loaded, (2) the top size of the coal, (3) the range of variation in coal quality with particle size, and (4) the condition of the coal with regard to moisture and freezing. When these conditions are favorable, the tube sampler has been found to perform satisfactorily.

d. Spoon Sampler This device might be considered a distant cousin of the tube sampler. It consists of one or more pipes, arranged like the spokes of a wheel. Openings located at the tips collect the sample as the device is rotated through coal on a moving belt. This machine can be designed to collect very small primary increments, smaller, in fact, than ASTM specifications for minimum weight of increments. Below minimum increment weight, selective rejection of top size tends to become a problem, and this is potentially a source of bias with such equipment. The spoon pipes also overflow during increment collection and are subject to the same plugging problems as the tube sampler.

e. Augers Many solids, such as metals, soils, and bales of textile fibers, are sampled by drilling and core-sampling methods, and efforts to adapt similar techniques to the sampling of coal at rest is not illogical. The auger drill appears to offer the most promise as a sampling device for penetrating a stationary mass of coal and withdrawing material from its interior. The reliability of this device for obtaining unbiased samples is unproven. At least two specialized auger-sampling machines, designed for sampling from trucks, are commercially available in the United States. One of these uses a 10-in. auger and is intended for sampling uncrushed run-of-mine coal. A truck-mounted portable version is also being used for sampling from railcars, but no data on performance have been published. This machine incorporates a stage of sample division without particle size reduction. ASTM procedure permits reduction of very large primary increments by resampling provided that at least six increments are collected from each primary increment, and these increments conform with minimum-increment-weight specifications for the top size involved. It would not appear that there is any practical means for fulfilling this requirement with this equipment. There are also no performance data with respect to its capability to withdraw coal from the lower portions of the sample hole in equal proportions to the coal withdrawn from the upper portions of the sample hold. Nevertheless, these devices enjoy the psychological advantage of machine sampling, and their use appears to be spreading.

4. *A Serious Problem Common to All Mechanical Sampling Systems*

The maintenance of uniform flow rates to the crushers and cutters of mechanical sampling systems is a serious problem common to all designs. The reason this is serious is that it affects moisture losses, uniformity of crusher product size and composition, and uniformity of sample division cuts. The best systems rely on a bin-and-feed step before and after crushers and before cutters. But even with this sophistication, it is difficult to provide the flexibility needed for large variations in lot size or dumping rates. Without such refinements, cutters invariably cycle through their traverses, busily collecting nonexistent increments, when coal is not flowing. Or they collect increments of widely varying weight, corresponding to variations in coal flow rates, thereby giving disproportionate representation to some parts of the lot relative to others.

Variable flows also affect the performance of most crushers that selectively crush the more-friable material first. This selective crushing causes segregation by composition downstream of the crusher, with the ash content shifting as the feed rate tails off. This can be a source of bias, particularly if such variations get in step with cutter traverses. Similarly, moisture losses can be adversely exaggerated. Crushers are usually responsible for more moisture losses in sampling systems than is any other component. The hammer mill, which is most commonly used for the sample crusher, is surprisingly effective as a fan and is capable of pumping large volumes of air. Thus, as coal flow tails off, the amount of air sweeping across each unit weight of coal increases, and moisture losses increase correspondingly.

The bin-and-feed configuration before and after crushers acts as an air seal preventing the flow of air through the rest of the system. To further alleviate this problem, a pressure balancing pipe is often connected across the crusher between the entrance throat and the receiving bin following the crusher. By equalizing the air pressure differential across the mill, air flows are minimized and the recirculating air tends to seek an equilibrium saturation level, inhibiting moisture losses. Since air flow is proportional to the square of the tip speed, reduction in mill revolutions per minute (rpm) can also help minimize the effects of air circulation. Generally, reduction of mill speeds below 1200 rpm is not practical because of the adverse effect on mill capacity and particle size reduction. Reduction in rpm requires smaller screen holes or closer bar spacing to obtain the same particle size reduction. Hangups, plugging, blinding, and failure to produce specified particle size are perennial problems in sampling systems, and specification of oversize mills is good design practice.

D. Sample Preparation

Sample preparation is defined in ASTM D2013, Preparing Coal Samples for Analysis, as the "process that may include air drying, crushing, division, and mixing of a gross sample for the purpose of obtaining an unbiased analysis sample." Thus, sample preparation is not just simply a matter of dividing down a gross sample. Accomplishing the task in a manner that results in an unbiased analysis sample is far more difficult than is usually perceived. Moisture losses are a perpetual problem.

Virtually all sample preparation is accompanied by some moisture loss unless the gross sample increments are weighed as they are collected and all are air-dried and reweighed before crushing and division. Paragraph 8.4.2 of ASTM D2013 specifies this procedure as the referee method. Outside design and technical circles, this is a little recognized problem associated with the mechanical sampling of coal wherein a part of the sample preparation is an automated part of the system. This is a matter over which buyer and seller are generally very unforgiving, even though they themselves may be responsible for imposing conditions that preclude the practical or economic feasibility of weighing, air drying, and reweighing each increment before crushing and division. Time and cost factors almost universally are considered to outweigh the benefit, and accommodation of the attendant problems is not readily attainable.

Sample preparation, when properly executed, is the source of the second largest component of the overall variance of sampling: sample preparation and analysis. Although it is not specified as a requirement, under the precautions of Section 6 of ASTM D2013 it is stated that "it is necessary that the variance of sample division and analysis . . . be not more than 20 percent of the total variance of sampling, division, and analysis." Since the total variance is usually neither known nor feasible to obtain, it is imperative for actual practice to adhere rigidly to the procedures and precautions of the standard method. This includes not prolonging the air-drying procedure past the time necessary to bring the sample moisture into equilibrium with the air in the room in which further reduction and division are to be made. It also includes performing the operations as quickly as possible, protecting samples from moisture losses while awaiting preparation, and not dividing samples after crushing to quantities less than specified for that particle size. Where division is performed manually, specified riffles must be used. With mechanical division, sample cuts must be equally distributed throughout the entire gross sample, and at each stage of division a minimum of 60 cuts is required. The nonreferee method provides for dividing out a 500-g moisture sample at the 4-mesh stage of reduction for two-stage total moisture determination in conformance with ASTM D3302.

In the final stage of sample preparation the sample is pulverized to 100% passing a No. 60 sieve. Sample pulverizers are characteristically driven by relatively small motors and tend to slow down under load. This is often a source of undetected difficulties because particle size is a function of mill speed. Samples should be routinely checked on a No. 60 sieve to assure that particle size reduction is being properly attained. Power auger feed to sample pulverizers is highly desirable as a means of achieving consistent pulverization to -60 mesh. Pulverizer screens are also notoriously fragile and should be inspected very frequently for ruptures. This imposes no operational handicap, because the pulverizer must be thoroughly purged and cleaned out between samples to avoid possible contamination of following samples anyway.

There are two ways in which the total moisture can be obtained. Either an analysis sample can be prepared from the moisture sample, or the regular analysis sample can be used for this purpose, provided that the moisture analysis is performed on the analysis sample within a few hours after the air dry sample is prepared. When separate analysis samples are used for moisture and for the other determinations that are made, moisture determinations must be made on both. The moisture obtained on the regular analysis sample is then used only for the purpose of calculating results to the dry basis. The moisture determined on the analysis sample prepared from the separate moisture sample is combined with the air-dry loss to determine the total moisture, and the total moisture is then used for calculation of as-received basis results.

V. LABORATORY ANALYSIS

A. Standard Methods

Some of the principal variables in terms of which coal quality is expressed are measured by empirical tests that do not directly measure the variables themselves, but rather the behavior of coal under fixed conditions. Standardization of these fixed conditions is essential for obtaining comparable results between determinations performed within a laboratory and between determinations performed by different laboratories. The source of most of these standard methods used in the commerce of coal is the American Society for Testing and Materials, and the committee that has jurisdiction over these standards is Committee D05 on Coal and Coke. There follows a brief summary catalog of these tests, together with a few special tests not currently promulgated by ASTM.

B. Proximate Analysis

ASTM D3172 covers the methods of analysis associated with the proximate analysis of coal and coke. This is a combination of three determinations and the calculation of a fourth. The proximate analysis is defined as "an assay of the moisture, ash, volatile matter, and fixed carbon as determined by prescribed methods." Moisture, volatile matter, and ash are all determined by subjecting the coal to prescribed temperature levels for prescribed time intervals. The losses of weight are by stipulation called moisture and volatile matter. The residue remaining after ignition at the final temperature is called ash. Fixed carbon is the difference of these three values summed and subtracted from 100. In low-volatile materials such as coke and anthracite coal, the fixed carbon value equates approximately to the elemental carbon content of the sample. The ASTM procedures covering these analyses are D3173, D3174, and D3175.

C. Equilibrium Moisture

ASTM d1412 covers another moisture test, the results of which are taken to be an estimate of the bed moisture present in the coal in the seam before mining. It is the moisture of coal at 96 to 97% relative humidity and 30°C. This is defined as the equilibrium moisture.

D. Single-Stage Total Moisture

ASTM D2961 is a test method for Total Moisture in Coal Reduced to No. 8 Top Sieve Size. Total moisture includes both the surface moisture and the residual moisture remaining in the sample after determining the air-dry loss. In the single-stage method the coal is not taken through that intermediate stage. This is classified as a limited-purpose method, recognizing that certain coals will give results that reflect varying levels of oxidation. The procedure is to dry the coal for 1½ hr at 107°C. High-rank Appalachian coals tend to be less affected by this than lower-rank Western coals. The method specifies minimum sample weight of 500 g. Therefore, throughout it involves a scale-up of equipment to much larger sizes than those normally used in the laboratory.

E. Two-Stage Total Moisture

ASTM D3302 is the standard method for Total Moisture in Coal. It covers the measurement of total moisture in coal as it exists at the site, at the time, and under the conditions that it is sampled. It is applicable to all coals, regardless of rank, as mined, processed, shipped, or utilized in normal commerce, except slurries, sludges, or pulverized products under 0.5-mm size. In this procedure the gross sample is first air-dried to constant weight (defined as a weight loss of not more than 0.1% per hour). The sample is then prepared to a 60-mesh pulp and the residual moisture determined. The total moisture is calculated in accordance with the following formula:

$$M = \frac{R(100 - A)}{100 + A} \tag{9}$$

where M = total moisture, %
A = air-dry loss, %
R = residual moisture, %

F. Residual Moisture

ASTM D3173 covers the method for Moisture in the Analysis Sample of Coal and Coke. The analysis sample is the laboratory pulp, on which laboratory analyses are performed, that has been pulverized to pass a 250-μm (No. 6) sieve. This moisture is determined on portions of about 1 g and is used for calculating other analytical results to the dry basis. It is also used in conjunction with the air-dry loss, when it is determined, to calculate results to the as-received basis.

G. Ash

ASTM D3174 covers the test method for Ash in the Analysis Sample of Coal and Coke. This method is used for determining the ash content of coal for the proximate analysis and the ultimate analysis. It is not applicable to the preparation of ash for determination of the ash constituents. Determinations are made on approximately 1-g portions which are brought to a temperature of 700 to 750°C. Use of the dried coal from the moisture determination is permitted. The method stipulates ignition to constant weight (defined as ±0.001 g). This is a clumsy procedure for the production laboratory, where time is of the essence. Commercial laboratories usually determine the necessary time after the final ignition temperature is reached to achieve this result and do not cool, weigh, reheat, cool, and reweigh each sample to determine whether this point has been reached.

H. Volatile Matter

ASTM D3175 covers the method for Volatile Matter in the Analysis Sample of Coal and Coke. This method determines the weight percent of gaseous products, exclusive of moisture, which are released upon heating the sample under specified conditions of time and temperature, in the absence of air. The procedure is the least precise of the determinations associated with the proximate analysis. It is used for the classification of coal by rank, for estimating coke yield, and for an indication of burning characteristics.

I. Total Sulfur

ASTM D3177 covers the test methods for Total Sulfur in the Analysis Sample of Coal and Coke. There are three alternative procedures, known as the Eschka method, which is generally considered to be the reference method; the bomb washing method; and the high-temperature combustion method. Eschka mixture is a mixture of calcined magnesium oxide and anhydrous sodium carbonate. The sample is ignited and intimately mixed with Eschka mixture, forming soluble sulfates from which insolubles are filtered. The sulfates are then precipitated with barium chloride. The barium sulfate precipitate is filtered, ignited, weighed, and back-calculated to the elemental sulfur equivalent.

The bomb washing method follows the same final steps after the sample has been ignited in an oxygen bomb. The repeatability interval of this method is specified the same as that for the Eschka method, but laboratories frequently experience difficulty in achieving this. The nuances of the method require special care in the rate at which the gases are released from the bomb and in polishing the bomb during the washing process. The high-temperature combustion method consists of burning the sample in a resistance-heated tube furnace in a pure oxygen atmosphere and collecting the sulfur oxides in hydrogen peroxide. The final step is titration with sulfuric acid. This method is not as precise as the other two. The repeatability interval of this method is double that for the Eschka and bomb washing methods.

J. Btu Heating Value

Either ASTM D3286 or ASTM D2015 can be used for determination of the heating value. D3286 covers the test method for Gross Calorific Value of Solid Fuel by the Isothermal-Jacket Bomb Calorimeter and D2015 covers the test method for Gross Calorific Value of Solid Fuel by the Adiabatic Bomb Calorimeter. The repeatability intervals of both methods are the same, but the human factor weighs heavily in the isothermal procedure, and laboratories generally experience difficulty in matching the precision of the adiabatic calorimeter procedure. Most commercial laboratories use the automatic adiabatic calorimeter with programmer and digital thermometer. It requires less attention from the laboratory technician, who is freed for other activities and the human element in the determination is minimized. This determination yields the higher heating value because the heat of vaporization of the moisture and water of combustion is liberated by condensation in the bomb. American engineering practice is to make a correction in heat-balance calculations to account for the fact that in normal combustion in a furnace the moisture and water of combustion are not condensed but are discharged from the boiler at an elevated temperature.

The heating value of the pure coal substance is a function of rank. ASTM D388 covers the standard classification of coals by rank according to their degree of coalification in ascending order from lignite to anthracite. This classification takes into account fixed carbon and calorific value on a mineral-matter-free basis and agglomerating character determined by examination of the residue from the volatile determination. In high volatile, medium-volatile, and low-volatile bituminous coals, the moisture and ash-free Btu (MAF Btu) commonly reported by commercial laboratories tends to increase with decreasing volatile from approximately 14,700 Btu/lb to approximately 15,600 Btu/lb. In the lower-rank coals the MAF Btu decreases with decreasing rank, approaching roughly that of wood in the range 6000 to 8000 Btu/lb for lignites.

K. Ultimate Analysis

ASTM D3176 covers the standard method for Ultimate Analysis of Coal and Coke. The ultimate analysis includes the determination of elemental carbon, hydrogen, sulfur, and nitrogen, together with the ash in the material as a whole. Oxygen is calculated by difference. The method references separate ASTM standard methods for each of these determinations as follows: D3173 for moisture, D3174 for ash, D3177 for sulfur, D3178 for elemental carbon and hydrogen, and D3179 for nitrogen. The determination of moisture, ash, and sulfur are discussed above. Determination of elemental carbon and hydrogen is performed by combustion in a closed system, collecting the products of combustion in an absorption train. The percentages by weight of carbon and hydrogen are back-calculated from the gain in weight of the absorption bulbs. This method gives the total percentages of carbon and hydrogen in the coal, including any carbon in the carbonates and any hydrogen in free moisture and in water of hydration of silicates. Reporting of the hydrogen and oxygen in the free water (moisture) associated with the sample may optionally be included or excluded from the hydrogen and oxygen values. The test method recommends making a notation of the option used as the basis for reporting results. Hydrogen in the water of hydration of silicates is generally ignored. The standard ASME heat balance separately accounts for heat losses associated with moisture and with the hydrogen in the coal. It is preferable, therefore, that the hydrogen and oxygen associated with the free moisture not be included in the hydrogen and oxygen values and that moisture be reported as a separate item.

L. Nitrogen

ASTM D3179 covers the test for Nitrogen in the Analysis Sample of Coal and Coke. The test provides for determination of the nitrogen by the Kjeldahl-Gunning method and an alternative method. The repeatability interval for both procedures is the same. Both procedures convert the nitrogen to ammonia and differ in the final titration. The Kjeldahl method distills the ammonia into a sulfuric acid solution and back-titrates the excess acid with sodium hydroxide solution. The alternative method titrates the ammonia directly with sulfuric acid. The nitrogen oxides associated with combustion processes are believed to originate primarily with the nitrogen in the fuel as opposed to the nitrogen in the air. The determination of the nitrogen content of fuels may therefore become a matter of increasing importance.

M. Softening Temperature of Ash

ASTM D1857 covers the test method for Fusibility of Coal and Coke Ash. In this test, the temperatures are observed at which triangular cones prepared from the ash begin to deform and then pass through specified stages of fusion when heated at a specified rate. The test procedure provides for performance of the test in a controlled atmosphere. It is first performed in a mildly reducing atmosphere and is then repeated with a second set of cones in an oxidizing atmosphere. The critical temperature most commonly referenced in the evaluation of the properties of coal is the softening temperature. This is defined as the temperature at which the cone has fused down to a spherical lump in which the height is equal to the width at the base. The significance of performing the test in reducing and oxidizing atmospheres is that most oxides of metals exhibit higher fusion temperatures in their highest state of oxidation. In practical combustion there invariably are zones where the supply of oxygen is depleted or carbon dioxide is reversibly reduced to carbon monoxide in the presence of excess hot carbon. This can produce a reducing atmosphere in a hot zone where ash particles in an incipient state of fusion begin to melt. Slagging, fouling, and clinkering difficulties have been found to correlate with the fusibility of the coal ash.

N. Free Swelling Index

ASTM D720 covers the test method for the Free Swelling Index of Coal. The test consists of carbonizing the coal in a specified silica crucible with a cover under specified conditions of temperature. The coke button remaining at the end of the test is compared with standard profiles classified by numbers from 1 to 9 in $\frac{1}{2}$-unit increments. A value of zero is assigned to a residue that is noncoherent. The highest button is assigned a value of 9. The test is an indication of the caking characteristics of coal burned in a fuel bed. It is also an indication of the state of weathering of coal mined near outcrop lines. Weathered coal generally exhibits a loss in heating value due to partial oxidation. A button several free swelling index (FSI) numbers lower than is normally exhibited by such a coal is a confirming indication of weathering.

O. Hardgrove Grindability Index

ASTM D409 covers the test method for the Grindability of Coal by the Hardgrove Machine Method. This is a test of the relative ease of pulverization of coals in comparison with coals chosen as standards. Standard coal samples for calibration of the test procedure are supplied by ASTM. The test consists of grinding a specially prepared sample in a pilot-scale ball mill of specified dimensions and determining the change in size consist by sieving. The results are converted into the equivalent Hardgrove grindability index. High grindability index (GI) numbers indicate easy-to-grind coals. There is a rough relationship between volatile content and grindability in the low-, medium-, and high-volatile bituminous rank coals. Among these the low-volatile coals exhibit the highest GI values, often over 100. The high-volatile coals range in GI from the mid-50's down to the high 30's. There is also a correlation of sorts between friability and grindability. Soft, easily fractured coals generally exhibit relatively high GI values. ASTM has two methods for measuring friability, the Drop Shatter Test, D440, and the Tumbler Test, D441, which should be used where more than a rough indication of friability is needed.

P. Forms of Sulfur

ASTM D2492 covers the test method for Forms of Sulfur in Coal. The test determines the three commonly recognized forms of sulfur: sulfate sulfur, pyritic sulfur, and organic sulfur. The sulfate sulfur is determined by extraction in dilute hydrochloric acid in which pyritic and organic sulfur forms are insoluble. The pyritic sulfur is determined by extraction in dilute nitric acid and analyzing for iron present in the pyrite. Organic sulfur is determined by difference, subtracting the sum of the sulfate and pyritic sulfurs from the total sulfur.

Pyrite, which is much heavier than coal, frequently occurs in discrete particles which can be liberated by crushing. As a result, much of the pyritic sulfur in coal can be removed by gravimetric separation in a medium that is dense enough for the coal to float and the pyrite to sink. Very broadly, about half the pyritic sulfur in coal can be removed in this manner, although coals vary widely from this benchmark. As a first approximation, the test for forms of sulfur provides a basis for judging the potentials for a coal to be cleaned to environmentally acceptable compliance specifications.

Q. Chlorine

ASTM D2361 covers the test method for Chlorine in Coal. Chlorine is believed to be a factor in fireside fouling problems. Although the validity of the values is uncertain, generally accepted fouling classification of coal, according to total chlorine content, is as follows:

Total percent chlorine in coal	*Fouling type*
<0.2	Low
0.2-0.3	Medium
0.3-0.5	High
>0.5	Severe

The test is performed by burning the coal mixed with Eschka mixture in an oxygen bomb followed by potentiometric titration with silver nitrate solution or by a modified Volhard colorometric titration. The repeatability interval for either procedure is the same.

R. Mercury

ASTM D3684 covers the test method for Total Mercury in Coal by Oxygen Bomb/Atomic Absorption Method. Mercury was identified as a potentially dangerous environmental contaminant, largely by reason of the process of concentration in the food chain. It was later discovered that the form in which mercury exists alters its toxicity. Nevertheless, mercury remains a sensitive matter and the test for total mercury is required primarily to evaluate the environmental influence that may be associated with the use of coal. The test method consists of burning the sample in an oxygen bomb with dilute nitric acid and determination of the mercury by flameless cold vapor atomic absorption.

S. Elemental Analysis of Coal Ash

ASTM D2795 and ASTM D3682 cover the test methods for Analysis of Coal and Coke Ash. The procedures of D2795 cover determination of nine major elements commonly determined: silicon dioxide, aluminum oxide, ferric oxide, titanium dioxide, phosphorus pentoxide, calcium oxide, magnesium oxide, sodium oxide, and potassium oxide. A combination of spectrophotometric, chelatometric, and flame photometry is used. The procedures of D3682 cover determination of the same elements plus manganese dioxide by atomic absorption.

The chemical composition of coal ash is recognized as an important factor in fouling and slagging problems and in the viscosity of coal ash in wet bottom and cyclone furnaces. Appalachian coals with high-iron ash, above 20% ferric oxide, typically exhibit ash softening temperatures under 2200°F. With the onset of intensive development of Western U.S. coal reserves, utilization of these coals has been plagued with slagging and fouling problems. This has been traced to a pattern of higher-alkali oxides than iron oxide in the coal ash. As oxides, most ash elements have high melting points, but they tend to form complex compounds with relatively low melting points. High-calcium low-iron ash coals of the West exhibit a pronounced tendency to produce low-melting-temperature slags when sodium exceeds about 4%.

T. Trace Elements in Coal Ash

ASTM D3683 covers the test methods for Trace Elements in Coal and Coke Ash by Atomic Absorption. The procedures cover determination of beryllium, chromium, copper, manganese, nickel, lead, vanadium, and zinc. The release of certain trace elements to the environment associated with the combustion of coal and disposal of its ash, which contains a wide range of trace elements, has become a matter of concern. The determination of these elements is expected to become an increasingly important aspect of coal analysis.

U. Plasticity and Fluidity

ASTM D1812 and D2639 cover methods for measuring the plastic properties of coals. D1812 covers the method for Plastic Properties of Coal by the Gieseler Plastometer and D2639 covers the method for Plastic Properties of Coal by the Constant-Torque Gieseler Plastometer. Plastic behavior is of practical importance for semiquantitative evaluation of metallurgical coal and coal blends used in the production of coke for the steel industry. When bituminous coals are heated in the absence of air through a temperature range spanning 300 to 550°C, volatile materials are released and the solid coal particles soften, becoming a more-or-less plastic mass that swells and eventually resolidifies. The plastometer stirs a sample of coal during the heating process, and the temperatures at which the rate of movement of the stirrer conforms with certain criteria are recorded. It is a mechanical test under rigidly controlled conditions.

V. Washability

There is no ASTM standard covering the conduct of washability testing. In simplest terms, the laboratory simulates the gravimetric separation of light and heavy fractions that occurs in full-scale gravimetric separating processes. A complete washability study consists of a series of float/sink tests, carried out on predesignated size fractions into which the coal is divided by screening. Solutions or suspensions of different specific gravities are used for the separations. This is performed repetitively with higher and higher specific gravity solutions, separating the size fraction into gravity fractions. The coal is generally of lower density than the mineral matter (ash), and these gravity fractions thus generally exhibit higher and higher ash content. Each size fraction and each gravity fraction is dried, weighed, and analyzed. The analysis is usually made for ash and sulfur content but is frequently extended to include heating value and other variables.

The resulting data base represents a substantial amount of detail, which is then reduced by computation of cumulative ash, sulfur, or other variable constituents, for each gravity of each size fraction or composited size fractions. From these data four curves are generated: (1) a cumulative float curve, (2) a cumulative sink curve, (3) a coal characteristic curve, and (4) a specific gravity curve. The coordinates of the first three curves are the cumulative percent yield (float) reporting to cleaned coal, usually placed at the left on the y axis; the cumulative percent reject (sink), usually placed at the right on the y axis; and the cumulative percent by weight of the variable constituent, which is placed on the x axis. The coordinates of the specific gravity curve are the cumulative percent yield and specific gravity. The coal preparation engineer uses these curves to predict performance of the full-scale plant, taking into account the efficiency of the facilities that will be used.

Increasing emphasis on the environmental impact associated with the use of coal has brought the washability study into wide use as producers explore the practical and economic feasibility of sophisticated coal washeries to meet compliance steam coal specifications. In the past, washability studies were performed mainly on metallurgical-grade coals.

W. Froth Flotation

With decreasing particle size, the practical feasibility of gravity separation diminishes because the efficiency of separation drops off. Particularly in low-volatile coals, a substantial proportion of the coal may be lost in these fines if other means are not applied for separating them from the slimes (clays) and recovering them. Froth flotation is one technique that is used for this purpose. There is no ASTM standard method for froth flotation testing. Laboratory methods are limited to scaled-down froth flotation cells which are operated in batch mode. This differs from commercial processes, which are continuous processes. In principle, froth flotation consists of bubbling air through a suspension of fine coal and water to which various chemical agents have been added to improve the processes. The separation occurs by reason of a preferential physical attachment of air bubbles to the coal. The coal particles float to the top and are removed. The procedure is sometimes repeated, reprocessing the float fraction several times to simulate multistage full-scale froth flotation in rougher, cleaner, and recleaner circuits.

APPENDIX: STATISTICAL COMPUTATIONS, EQUATIONS, AND DERIVATIONS

There is a unique statistical relationship between the members within pairs when duplicate sets of observations are related by a natural tie (e.g., observations belonging to the same pair derive from a common source, whereas each pair derives from a significantly different source). The nature of this relationship is such that the statistical parameters t and F equate according to the formula $t^2 = F$. The mathematical relationships of the two-way analysis of variance (ANOVA) are as shown in Table 10. The derivation of these relationships follows.

When the general equation for the variance between columns and the error variance of a two-way ANOVA are applied to R pairs or duplicates, the variance of differences within pairs (errors caused by sampling and analysis, for example) is measured by CMS = EMS + RV_c, on the premise that the "true" variance between columns $V_c = 0$ (i.e., observations between columns, within one row, are from a common source). Hence, EMS = $e/(C - 1)(R - 1) = e/(R - 1)$.

It can be shown that for n such sets of pairs,

$$\frac{n_{min}}{t^2} = \frac{2s^2}{(TSS)^2}$$

where the standard error of sampling $s = [(Q - P^2/n)/2n(n - 1)]^{\frac{1}{2}}$ and the boundary of statistical significance TSS = $ts\sqrt{2/n}$. (Note that $s_d{}^2 = 2s^2$.)

Derivation of F for n Sets of Pairs

For n sets of paired observations X_1 and X_2, the sum of squares between columns, using the notations of Table 10, is

$$b = \frac{[(\Sigma X_1)^2 + (\Sigma X_2)^2]}{n} - \frac{[\Sigma(X_1 + X_2)]^2}{2n}$$

$$= \frac{(\Sigma X_1)^2 + (\Sigma X_2)^2 - 2\Sigma X_1 X_2}{2n}$$

$$= \frac{[\Sigma(X_1 - X_2)]^2}{2n} \qquad \text{with df} = 1$$

The total sum of squares

$$a = \Sigma X_1^2 + \Sigma X_2^2 - \frac{[\Sigma(X_1 + X_2)]^2}{2n} \qquad \text{with df} = 2n - 1$$

The sum of squares between rows

TABLE 10 Two-Way Analysis of Variance

Source of variation between	Sum of squares, SS	df	Mean square, MS	Expected MS
Columns, C	$b=(A+B+\cdots)/R-M/CR$	$C-1$	$b/(C-1)=CMS$	$CMS=EMS+RV_c$
Rows, R	$c=(D+E+\cdots)/C-M/CR$	$R-1$	$c/(R-1)=RMS$	$RMS=EMS+CV_r$
Error	$e=a-(b+c)$	$(C-1)(R-1)$	$e/(C-1)(R-1)$	$EMS=EMS$
Total	$a=(\Sigma X)-M/CR$	$CR-1$	—	—

A,B, ··· = sum of columns 1,2,...
D,E, ··· = sum of rows 1,2,...
M = sum of all observations
C,R = number of columns, rows
M/CR = correction term
X = individual observations

True variance estimates

True variance, V_c = (CMS - EMS/R)
True variance, V_r = (RMS - EMS/C)

$$c = \frac{\Sigma(X_1 + X_2)^2}{2} - \frac{[\Sigma(X_1 + X_2)]^2}{2n} \qquad \text{with df} = n - 1$$

The error sum of squares

$$e = a - (b + c)$$

$$= \Sigma X_1^2 + \Sigma X_2^2 - \frac{\Sigma(X_1 + X_2)^2}{2} - \frac{[\Sigma(X_1 - X_2)]^2}{2n}$$

$$= \frac{\Sigma(X_1 - X_2)^2 - [\Sigma(X_1 - X_2)]^2/n}{2} \qquad \text{with df} = n - 1$$

$$= \frac{Q - P^2/n}{2} \qquad \text{when using the notation of Table 8}$$

Thus, the error variance, which for paired observations is related by a natural tie, represents the variance of sampling and analysis:

$$s^2 = \frac{Q - P^2/n}{2(n - 1)}$$

It follows that $F - CMS/EMS = (n - 1)P^2/(nQ - P^2)$.

Derivation of t for n Sets of Pairs

(For symbols, refer to Table 8.) We have for $\bar{d}$, the mean of the differences $X_1 - X_2$,

$$\bar{d} = ts\frac{\sqrt{2}}{n} = \frac{ts_d}{\sqrt{n}}$$

$$t^2 = \frac{(\bar{d})^2 n}{2s^2} \quad \text{where } \bar{d} = \frac{P}{n}$$

$$s^2 = \frac{(Q - P^2/n)}{2(n - 1)}$$

$$t^2 = \frac{P^2/n}{2(Q - P^2/n)/2(n - 1)}$$

$$t^2 = \frac{(n - 1)P^2}{nQ - P^2} = F$$

The official F table demonstrates, in the first column where the F values are listed for df = 1 of the larger variance, that $F = t^2$ for t values of the same df, in the two-tailed t-distribution table.

REFERENCES

1. *1978 Annual Book of ASTM Standards*, Part 26. American Society for Testing and Materials, Philadelphia, 1978.
2. J. Visman, A General Sampling Theory, *Mater. Res. Stand.* 9(11):9 (1969).
3. W. G. Cochran, *Sampling Techniques*, 2nd ed. John Wiley & Sons, Inc., New York, p. 234.
4. A. J. Duncan, Comments on "A General Sampling Theory," *Mater. Res. Stand.* 11(1):25 (1971).
5. J. Visman, A. J. Duncan, and M. Lerner, Further Discussion: "A General Theory of Sampling," *Mater. Res. Stand.* 11(8):32 (1971).
6. J. Visman, Discussion 3: "A General Theory of Sampling," *Mater. Res. Stand.* 7(3):345 (1972).
7. G. E. Keller, Determination of Quantities Needed in Coal Sample Preparation and Analysis, Society of Mining Engineers of AIME, Preprint No. 65F13, 1965.
8. S. J. Aresco and A. A. Orning, A Study of the Precision of Coal Sampling, Sample Preparation and Analysis, Society of Mining Engineers of AIME, Preprint No. 65f38, 1965.

3 Underground Mining

WILHELM J. EULER / Occidental Research Corporation, Lexington, Kentucky

I. INTRODUCTION

Coal is a mineral of sedimentary origin. The decomposed plants from which it originates were originally deposited in horizontal layers of fairly consistent thickness. Subsequent erosion and tectonic movements created the varying thickness of seams and overburden and the more-or-less disturbed and faulted formation of coal-bearing strata found today. The uniformity of the deposits, their size, and the early and massive consumption of coal promoted the development of specialized equipment and methods to recover this resource.

Approximately 70% of demonstrated U.S. minable reserves are located at a depth which, with today's technology, can be mined only by underground methods. Therefore, the application of underground excavation methods and the technological advancement of these methods will be of utmost importance if coal is to become the major energy source that many experts predict it will once the oil supply starts to diminish.

Historically, only two centers, North America and Europe, have developed a coal mining and equipment technology. In most other areas large-scale coal mining utilizing modern methods and equipment started only recently, in some cases only after the start of the energy crisis. These countries have adopted existing mining techniques and equipment with minor changes.

North American underground mining technology are presented in the first portion of this chapter and European methods are discussed in the second portion.

II. COAL MINING IN NORTH AMERICA

Most of the North American mining technology base was originated in the United States. Here, as opposed to many other countries, the mining industry has always been privately owned.

The number of large close-to-the-surface coal deposits lead to very selective mining, usually only one or two of the best seams. Seams closest to the surface of acceptable thickness and quality in favorable bedrock conditions were mined first. Areas where subsidence damage to the surface could cause additional costs to the operator are either avoided or laid out and mined in such a way that no subsidence occurs. This type of layout and mining usually means that a large percentage of coal remains in the ground, so that recovery rates are only in the range 25 to 50%.

A. Mine Safety and Legal Requirements

Until recently, the mining industry has been very loosely regulated. Attempts by government to regulate the industry have primarily been responses to underground disasters and health problems. Therefore, they deal almost exclusively with the health and safety aspects of coal mining. The only addition to this are environmental regulations enacted during recent years which deal with such problems as acid mine water discharge, the effects of subsidence, and the interruption of aquifers.

Historically, the introduction of mine safety legislation was sparked by large mine disasters which focused public attention on this area. Even today, the results of investigations into accidents and health problems often initiate new safety regulations.

At present, these regulations are rather extensive, dealing with a broad spectrum of health- and safety-related items. Their main purpose is the reduction of accidents and improvement of working conditions.

The regulations deal with the basic areas of:

1. Mine fires
2. Mine gases
3. Rock falls (roof and rib control)
4. Operational and functional safety of equipment
5. Mine rescue
6. Mine planning
7. Dust and noise

Each of these areas will be discussed in the sections to follow.

1. Mine Fires

Mine fires can have a number of causes: spontaneous combustion, conveyor belts working in coal fines, electrical sparks or electrical malfunction, smoking, mine gas explosions, welding, or cutting.

As a result, pertinent regulations deal in detail with underground ventilation requirements, and demand that all openings be regularly patrolled, that conveyors be maintained free of coal fines, and that all electrical equipment be tested and certified for use in gaseous mines (equipment working in the immediate vicinity of a mining operation must be explosion-proof). Equipment that is not certified as explosion-proof must remain a specified distance away from any mining operation and has to be positioned in a fresh air stream. Smoking and smoking material are prohibited in coal mines, and welding or cutting can take place only after gas tests with certified testing devices have proven that the air is free of gas, and only if fire-extinguishing equipment has been placed in the immediate vicinity.

2. Mine Gases

The most common gas is methane. It is a by-product of the carbonization process of coal and its concentration depends primarily on the degree of mineralization. This mineralization or carbonization process is advanced to the highest level in anthracite coal, which means that mines producing this grade of coal have the lowest degree of methane liberation.

Methane is lighter than air, and once mixed and diluted, will not concentrate again. Because of the large amounts liberated from both solid and mined coal, the law deals very explicitly with it. The maximum allowable methane concentration is set at 1%, well below the explosive range of about 5 to 15%. Supervisory as well as mining equipment operating personnel are required to check the air with certified devices at regular intervals. Additionally, mining machines are equipped with monitors which automatically disconnect the power supply if the methane level becomes too high.

The explosion of a methane mixture is usually a localized process, which, however, can create an air/coal dust mixture that is of far greater danger

FIG. 1 Mine rescue team member equipped with self-rescue breathing apparatus. [Courtesy of the National Mine Service Company.]

than a small methane explosion. The methane explosion cannot only prepare the proper mixture of air and coal dust but can also create the necessary high-ignition temperature. A subsequent coal dust explosion can easily affect the entire mine, because coal dust can be found in sufficient quantities in most underground openings. Coal dust explosions are therefore the kind of mine disaster that usually produces the most devastating results.

Safety regulations are aimed at reducing this danger by mixing coal dust with rock dust to convert it into uncombustible material. Certain percentages of rock dust are required in any coal dust accumulation underground.

Other gases, for example carbon monoxide, are produced as a by-product of explosions or mine fires. Only when the danger of a mine fire is suspected or detected is the mine air monitored for its carbon monoxide content. The occurrence of this gas is relatively rare. However, because of its deadliness in concentration of only a few hundredths of a percent, each worker is required to carry a self-rescuer (Fig. 1) which chemically binds the gas. This provides personnel with breathing protection for about 1 hr. Other gases, such as hydrogen sulfide or carbon dioxide, occur only under special conditions and are monitored as circumstances require.

3. *Rock or Roof Falls*

Rock or roof falls are one of the most frequent causes of injury or death. Here, too, the law deals in great detail with this major problem. An approval procedure for mine plans is used to assure that safety requirements are met. This plan has to comply with general roof-support requirements as well as special local conditions. Frequent visits by federal and state mine inspectors assure compliance with the mine plan.

Additional measures against roof falls are personnel training, modified equipment design (canopy over operator seats, remote controls, etc.), and operating procedures in regard to unsupported or dangerous roofs.

4. *Operational and Functional Safety of Equipment*

Other areas that regulatory agencies are involved are in the operational safety of equipment. For example, specifically designed cutoff switches or levers are attached and mining equipment is fitted with the above-mentioned methane monitors with automatic power cutoff. Sometimes other, more long-term effects are considered in the law, as for example airborne dust or noise levels, which have to be kept below certain maximum levels.

5. *Mine Rescue*

Each mine is required to have a rescue team that is trained and equipped to work in the sometimes dangerous underground environment in the event of a mine disaster (Fig. 1).

6. *Mine Planning*

Mine planning control for safety feature design provides a very extensive means to improve safety. Escapeways, sufficient air supplies, safety pillars, roof support, and equipment selection are important features of mine planning.

7. *Dust and Noise*

Dust and noise created by the mining equipment usually affect the human body noticeably only after years of exposure. However, the effects of black lung or silicosis can be very severe and physically disabling.

Regulations limiting the allowable maximum level of airborne dust to 2 mg/m^3 are aimed at this common problem. Portable samplers attached to the clothing or (in the near future) equipment-mounted units provide the necessary inspection data.

B. Selection of the Mine Site

A number of considerations go into the general assessment of the economic viability of a mining site. These include coal quality, geologic conditions, location, transportion, shipping, availability of facilities, and availability of labor.

The size of a property in tons recoverable and the quality of the coal can be expressed as a dollar value. This value has to be compared with the required investments for development and operation of the mine (surface facilities, coal treatment, shafts, slopes, shipping facilities) and the shipping costs to determine the economic viability.

A second factor that can have a major impact on the value of a mining property are the geologic conditions. A varying thickness can require different sets of equipment, faults could necessitate additional shafts or slopes, and water or strata that are too hard or too soft can make mining too expensive or impossible.

The market can be so far away that shipping costs, particularly in small quantities, become too high, especially if no rail or barge shipping facility is nearby. If a coal cleaning plant from another mine can be utilized, the operation of a smaller mine might be made economical.

The availability of personnel is often a determining factor, since most mine sites, particularly those in the Western United States, are located far away from the populated centers. Costs for relocation and housing can amount to substantial sums.

The foregoing considerations, plus the data obtained from an exploration program, are combined with such factors as equipment cost and expected production to supply the information necessary for a detailed economic analysis. This analysis can then be used as a basis to obtain the necessary capital.

C. Exploration

The first step in evaluating a proposed mining venture is to carry out an exploration program. This is done to establish the size and quality of reserves as well as to provide information on probable mining conditions. A core-drilling program is designed with core holes or checkpoints positioned according to a pattern that allows a projection of the continuance of coal seam. The number of core holes can sometimes be reduced if the coal outcrops. The spacing of the checkpoints depends on the general continuance of the strata in the given area. If portions of the deposit are suspected to be eroded or highly faulted, the check points have to be spaced considerably closer than in a well-defined undisturbed formation.

Certain general guidelines are commonly used to establish the density of these checkpoints. The most widely used has been established by the U.S. Geological Survey:

Measured resources	*Indicated resources*	*Demonstrated resources*	*Inferred resources*
At checkpoints $\frac{1}{2}$ mil apart	At checkpoints $1\frac{1}{2}$ mil apart	Measured plus indicated resources	Checkpoints 2 mil apart

Drilling is usually extended to a distance several feet below the seam. A core started several feet above the seam and ending several feet below allows judgment on the roof and floor condition during mining. The price for the holes can vary by several hundred percent, depending on the difficulty encountered in moving the drilling equipment, on the site, water supply,

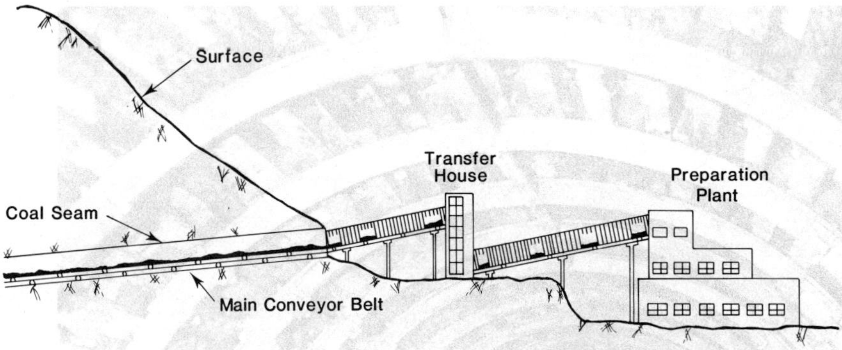

FIG. 2 Drift mine.

the hardness or fracturing of the material to be drilled, voids drilled into, the size and length of the holes and core, and the gas or water deposits encountered.

D. Mine Design

Once a resource is determined to be minable, the means to gain access to the deposit are selected. Three different design methods are available: the drift mine, the slope mine, and the shaft mine.

1. *The Drift Mine*

If a seam outcrops at the surface and is reasonably horizontal, the mine is commonly developed directly from the surface (Fig. 2). The first step consists of an entry system starting at the outcrop point. Only a small amount of work is necessary to prepare the actual mine entrance.

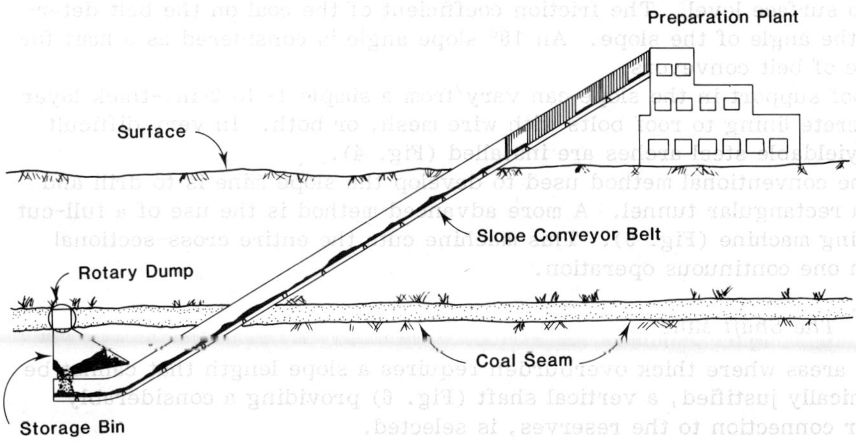

FIG. 3 Slope mine.

FIG. 4 Steel arches installed in mine. [Courtesy of the Kaiser Steel Corporation.]

2. *The Slope Mine*

The slope mine access to a coal deposit is by means of an inclined tunnel (Fig. 3). In most cases a conveyor belt is installed to elevate the coal from mine to surface level. The friction coefficient of the coal on the belt determines the angle of the slope. An 18° slope angle is considered as a limit for the use of belt conveyors.

Roof support in the slope can vary from a simple 1- to 2-in.-thick layer of concrete lining to roof bolts with wire mesh, or both. In very difficult cases yieldable steel arches are installed (Fig. 4).

The conventional method used to develop the slope mine is to drill and blast a rectangular tunnel. A more advanced method is the use of a full-cut tunneling machine (Fig. 5). This machine cuts the entire cross-sectional area in one continuous operation.

3. *The Shaft Mine*

In areas where thick overburden requires a slope length that cannot be economically justified, a vertical shaft (Fig. 6) providing a considerably shorter connection to the reserves, is selected.

FIG. 5 Tunneling machine with disk cutters and conveying systems. [Courtesy of Jarva, Inc.]

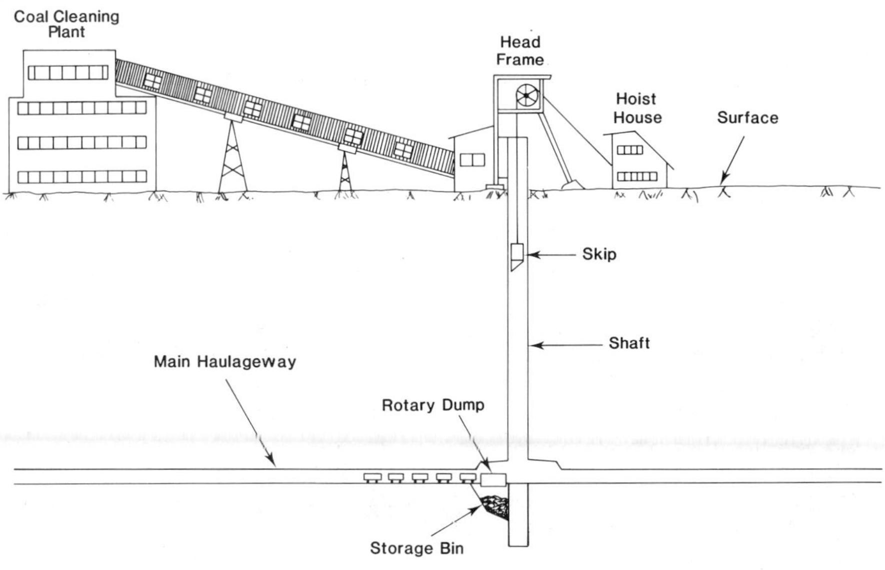

FIG. 6 Vertical shaft mine.

Modern shafts are usually developed with a circular cross section and lined with concrete. Two basic methods are used to sink shafts. The conventional method is to drill and blast and to remove the muck with orange-peel loaders and cylindrically shaped muck buckets. The second, very recently developed method, utilizes large-diameter shaft-boring equipment.

Round ventilation shafts of about 5 to 10 ft diameter excavated by mechanical boring equipment have been used for many years. New developments allow shafts of 20 ft and more in diameter to be developed with this boring equipment. This means of shaft sinking is considerably faster, therefore reducing the lead time required for mine development.

Basically, a shaft borer is similar to a large tunnel-boring machine. A cutting head equipped with roller bits continuously excavates the entire shaft cross section at once (Fig. 7). The removed fine muck is pumped as a slurry to the surface or dropped through a pilot hole for underground removal. In many cases sprayed-on concrete is used as the initial lining.

In cases where difficult ground conditions exist, such as friable, easily caving material, or sand connected with a water inflow, a special freezing method is used. Holes are drilled into the ground at a distance of 7 to 10 ft from the shaft. Steel casing is installed in the upper section of these holes or all the way to the water-bearing strata. A refrigerated brine is continuously pumped through these holes, providing a circle of frozen cylinders through which the shaft-sinking process proceeds.

In other cases, a grouting method is used. Holes are drilled into the shaft walls and a cement/water mixture is injected under high pressure to stop the water flow and to stabilize the rock formation.

Under such difficult conditions steel shaft lining is sometimes used. Cylindrical flanged sections 1 to $2\frac{1}{4}$ in thick, sometimes called Tuebbing sections, are bolted and sometimes welded, to provide a watertight seal. The installation of this metal lining is done after the shaft has been developed with an initial asphalt and concrete lining. The asphalt provides a flexible barrier between the concrete and the rock wall in case ground movements occur as a result of the mining activity. Once the shaft is finished, the steel lining is installed from the bottom up. Concrete vibrated into place is used to fill all remaining openings between the steel and the initial concrete lining. Several different versions of this steel lining are available today, some of which are made with concrete-filled cores.

4. *Combination*

There are only a few cases where a mine is developed as strictly a shaft, slope, or drift mine. Because coal mines in the United States usually confine their extraction operation to one or two seams only, mining activities spread very rapidly into a large area. This leads to the necessity of providing additional openings for ventilation. Bore holes 4 to 8 ft in diameter are a fast and economical way to provide additional intake or exhaust connections to the surface. Often, these shafts are equipped with an emergency hoisting facility for use in a disaster.

To reduce unproductive underground travel time, regular shafts with hoisting facilities for workers and materials are sometimes added to an extended drift or slope mine.

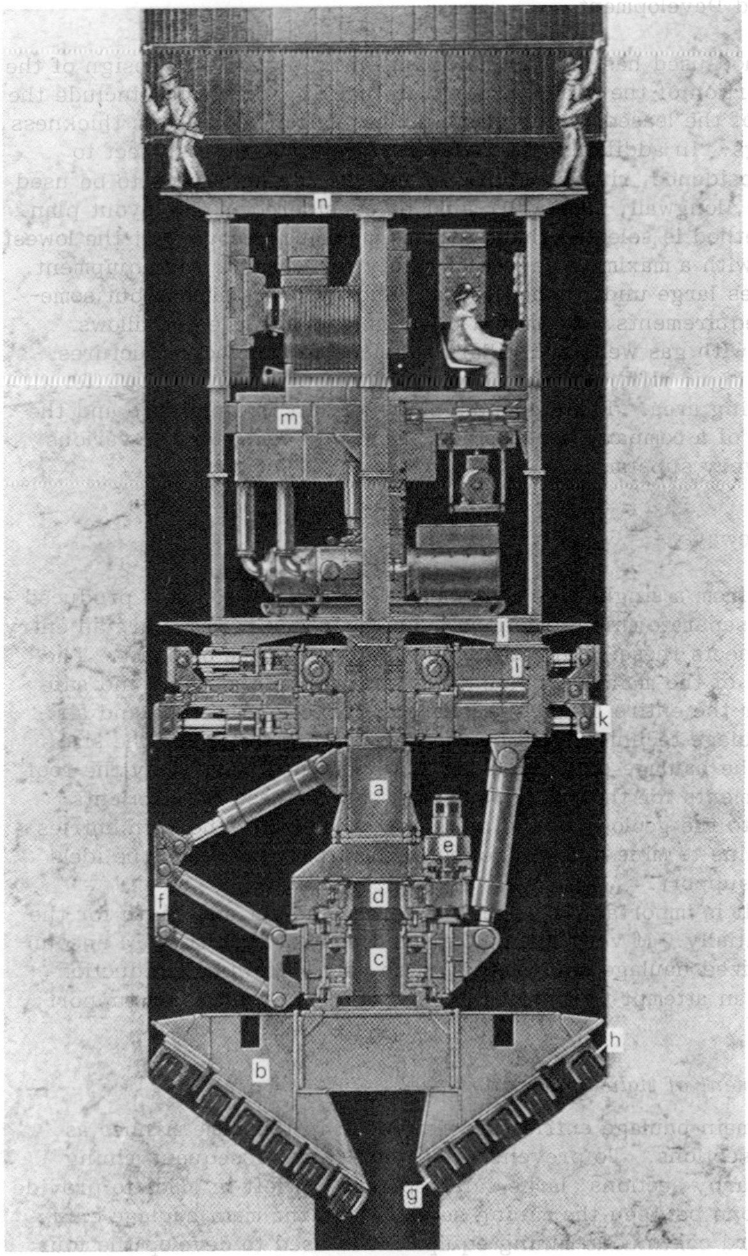

FIG. 7 Schematic of a shaft-sinking operation with a shaft borer: installation of steel rings and concrete lagging. (a) center Kelly; (b) cutter head; (c) cutter head support; (d) transmission; (e) hydraulic motors; (f) guide shield; (g and h) disk cutter bracket; (i) anchoring device; (k) anchoring jack; (l, m, and n) work platforms. [Reprinted courtesy of Verlag Glückhauf GmbH, Essen, West Germany.]

E. Underground Development

The access method used has little or no effect on the layout and design of the underground portion of the mine. Criteria influencing the design include the shape and size of the leased mining area, faults, variations in seam thickness, and outcroppings. In addition, gas wells, surface structures subject to damage from subsidence, rivers, aquifers, and the mining method to be used (room and pillar, longwall, shortwall) must be considered in the layout plan.

A mining method is selected that promises maximum recovery at the lowest possible price, with a maximum degree of safety for workers and equipment. Longwall requires large undisturbed areas. Shortwall has similar but somewhat reduced requirements. Room and pillar is more flexible and allows mining in areas with gas wells and pillars required for surface structures. To optimize a design, different mining methods may be selected for different parts of the mining area. This depends, however, on the total size and the capital reserves of a company because equipment and costs for the various methods differ very substantially.

F. Main Haulageways

The production from a single mine is usually the combined tonnage produced by a number of separate mining sections. The main haulageways are an entry system that connects these mining sections with the surface or shaft. The key requirement of the main haulage entries is to provide a stable and safe passageway over the entire life of the mine for workers, material, and air. The selected haulage technique will be the determining factor for the size and number of the haulage entries. Size, however, is governed by the roof support requirements for the particular seam. Since these requirements vary according to the geology of the coal seam, the size of the main entries will vary from mine to mine and will reflect a compromise between the ideal and the cost of support.

Therefore, it is important to select the correct height and width for the entry system initially. If very difficult ground control problems are encountered, smaller-sized haulage equipment may be used, allowing a reduction in entry size in an attempt to minimize ground control problems or support requirements.

1. *Development of Main Haulage Entries*

Commonly, main haulage entries are developed in the same manner as room and pillar sections. To prevent any impact from subsequent mining operations in nearby sections, large safety pillars are left in place to provide a stabilization zone between the mining sections and the main haulage entry system. Standard continuous mining equipment is used to develop the four to eight entry systems. If the seam thickness is sufficient, production similar to a mining section can be achieved, allowing for an early cash flow.

2. *Roof Support*

Proper roof support in these entries is vital. The single-seam operation utilized in most cases requires very long entry systems, because the operation spreads very rapidly into a large area. Roof-support requirements must

FIG. 8 Steel arches with roof bolts. [Courtesy of the Kaiser Steel Corporation.]

be considered very carefully to be cost-effective and to obtain a maximum effect with a minimum of effort. As a result it is often necessary to vary the type of roof support along the entry. It is not economical to support the entire entry length for the worst conditions expected. On the other hand, an undersupported entry will almost always result in excessively high operating and maintenance costs.

Roof bolts (Fig. 8) are the most widely used support in coal mines. For difficult areas, support sets consisting of steel arches, wooden props, and wooden or steel beams and yieldable all-steel sets or concrete are used.

Mined coal seams can vary in thickness from $2\frac{1}{2}$ to 20 or 30 ft in underground mines. When mining narrow seams, it is often necessary to remove roof or floor material to achieve enough entry height for the haulage equipment. This, together with roof support costs, causes many companies to limit the number of haulage entries. Only one or two entries of the initial system are equipped and maintained as haulage entries. The rest serve only as walkways and air intake.

3. *Rail Haulage*

Coal cars (Fig. 9) commonly range in capacity from 8 to 25 tons. To reduce the number of trips, coal cars of the largest possible size are commonly selected. Smaller cars are usually employed in thin-seam operations.

The selection of large cars, in most cases, allows the use of only one track per entry. As a result, at least two entries are required for an un-

FIG. 9 Coal cars used in underground mining. [Courtesy of Difco, Inc.]

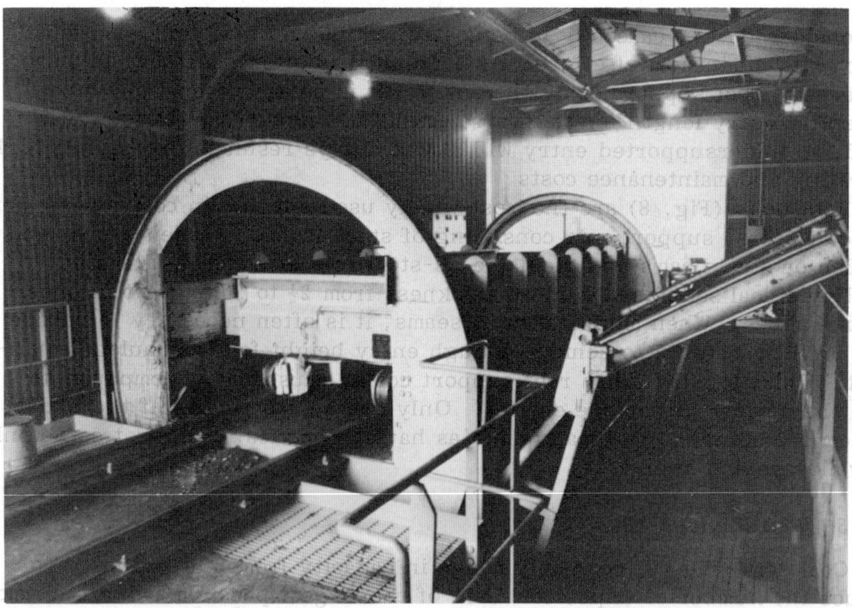

FIG. 10 Rotary car dump. [Courtesy of Difco, Inc.]

restricted flow of full and empty cars. A strict sequencing, restricting the capacity of the haulage system, is required if only one entry is utilized. Usually, spur tracks are provided in this case to maintain as much capacity as possible. Radio communication through a central dispatcher backed by a telephone system are generally used to direct a mine haulage operation.

In most cases the trains are propelled by direct-current (dc) trolley locomotives because of the high torque of the series-wound motor. Diesel locomotives are seldom used, in spite of improved safety aspects (no sparks between trolley pole and wire, no danger of electrocution because of accidental contact with the trolley wire, no stray electricity that could set off electric blasting caps) because of union resistance.

The majority of the approximately 2400 underground coal mines in the United States are either slope mines or drift mines. The drift mines haul the coal directly to the outside, where cars are unloaded by either utilizing bottom dump cars or a rotary dump (Fig. 10). In the majority of cases slope mines unload their cars at the bottom of the slope and convey the coal by belt conveyor to the surface. Only in a few cases is rope haulage (pulling cars with a hoist and a steel rope) practiced since the capacity of this system is obviously quite limited.

4. *Belt Haulage*

Belt conveyor systems (Fig. 11) allow the elimination of many of the problems encounted with track haulage systems. Belt widths in main haulage systems range from 36 to 52 in, depending on production. Because of their size, entries can be kept small and shaped for maximum self-supporting ability. This reduces support and maintenance requirements considerably. These systems can negotiate grades of up to 18°, whereas track haulage is limited to between 3 and 5% slope (influenced by the length of the grade, wheel bearings used, type of steel wheels, grease in bearings, condition of tracks and wheels, water on rails, radius of curves, and proper gauge of tracks). Another key advantage is the possibility of supervising several belts and belt junction points by means of one person utilizing television cameras. This reduces substantially the total number of people required to operate a belt system. Labor costs are therefore lower with belt systems in general.

Belt conveyor systems transport bulky material poorly. Thus, an additional system is needed to haul supplies and workers into the mine. A track or monorail system or trackless haulage equipment is used for this purpose. The requirements for this additional system in terms of capital investment and number and size of units is minimal. The rail size in this supply system can usually be reduced from 60 to 120 lb/hd for mainline systems to 40 to 60 lb, as commonly used on secondary tracks. No coal cars are required: only supply cars, personnel cars, and lightweight locomotives are needed.

5. *Future Trends*

The latest trends in main haulage entry development are marked by the use of tunneling machines. Slopes, as well as the first main entries, have been developed with this type of equipment (Fig. 5).

The entry cross section is circular and therefore provides excellent features for maximum self-supporting capability. This means that costs for

FIG. 11 Conveyor belt in an underground coal mine. [Courtesy of the Dowty Corporation.]

support material as well as subsequent maintenance requirements will be reduced substantially. Reduction in the time required for mine development to a fraction of that required for a standard continuous-miner operation may be realized utilizing tunneling machines.

However, there are problems connected with the use of this type of equipment. The first is the setup time. The system requires not only a considerable amount of time for the installation of the tunnel boring machine itself, but also for the conveyor, car haulage and switching, ventilation, and roof-support system. The experimental development of such a tunneled entry in West Virginia by the U.S. Bureau of Mines (USBM) has been accomplished with an 18-ft-diameter cross section. All the operational support functions are positioned in this entry. The machine is incapable of making sharp turns, to that cross-cuts cannot be developed initially.

The high advance rate of boring/tunneling machines requires special measures to control methane liberation. A vacuum pipe system connected to horizontal boreholes in the coal seam are used in the West Virginia experiment, plus a ventilation system supplying 50,000 ft^3/min of air. This is more than double the requirement of many standard entry systems.

Considering the setup time and the fact that the system requires special permission to be given by the regulating agency in such areas as electrical layout, ventilation, and escapeways, only a very long entry system on the order of several thousand feet will justify utilization. However, if the system could be developed and simplified to the point where it could be used in the development of section entries, a more general application would be possible.

Automation of rail haulage systems is beginning to get attention. Research and development work sponsored by various governmental agencies is now being done. This would allow the operation of trains through automated and remote controls, removing workers from this dangerous activity.

Another very promising system, not only for mainline haulage but for coal haulage from the face to the surface is the slurry pipeline. Slurry pipeline systems for use in the underground coal mines are in the experimental stage in the United States. Their utilization cannot be considered as an isolated function for mainline haulage purposes only. The capital required for the purchase of crushers, injection feeders, and water-removal equipment has to be considered. Therefore, consideration should be given to crushing and injecting the coal at the coal face, where it is mined. In this way, all coal-handling facilities could be simplified, reducing labor requirements to essentially a maintenance force. The basic system is well known and extensively used on large surface installations. Most of the equipment and technology has been explored to a considerable extent. European and Soviet coal mines have been using this method for quite some time.

G. Room and Pillar Mining

1. *Development*

About 95% of the U.S. coal produced by underground mines is mined by the room and pillar method. A room and pillar mining section is initiated by the development of a number of parallel entries. The entry height is usually determined by the thickness of the coal seam. Fourteen feet is considered the most that can be mined with modern continuous-mining equipment. If the seam is more than 14 ft thick, coal is left on the roof or floor, depending on the various bedrock conditions or the strength of the coal.

The entry width is determined by either the size of the equipment and/or the mining conditions (which might force a narrower entry for better roof and/or floor control). It varies between 10 and 50 ft, with the most common size about 20 to 25 ft. These parallel entries are usually connected to each other by cross-cuts spaced at about 40 to 100 ft center to center (Fig. 12). The cross-sectional configurations of these cross-cuts are similar to those of the entries.

If ground control problems force the entries to be kept narrow, angled "fish-bone" cross-cuts are developed to ease the operation and maneuvering of large mining equipment (Fig. 12).

The general layout of a room and pillar mining section is influenced by a number of factors. The size of the pillars created by the system of entries and cross-cuts depends on the thickness of the overburden, the allowable subsidence at the surface, aquifers in the upper strata, and the composition and strength of the immediate roof and floor material. As the thickness of overburden increases, the size of these pillars is also increased. Rock mechanics engineering is applied to determine the size of these pillars. If the compressive strength of the rock material next to the seam is low relative

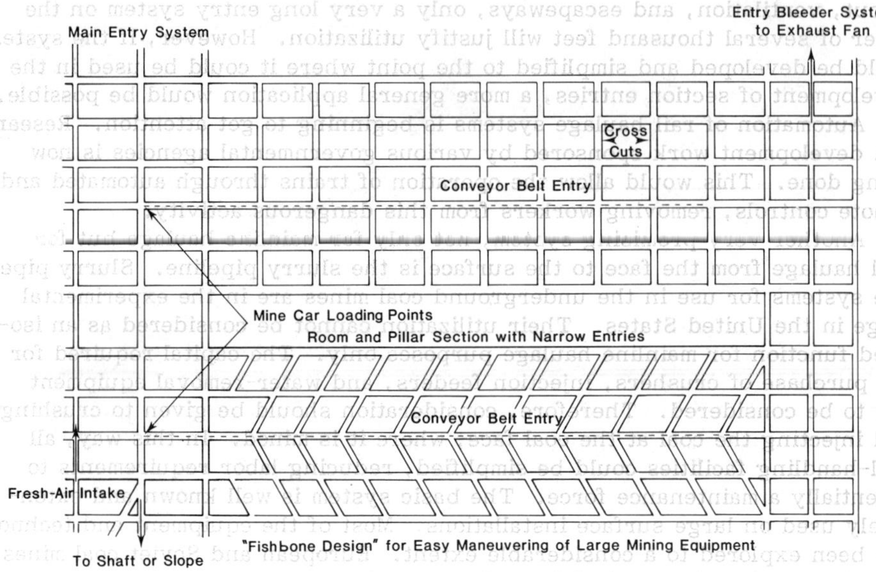

FIG. 12 Two fully developed room and pillar sections.

to the weight of the overburden, subsidence factors are incorporated into the design. Eventual closure of even very narrow openings cannot be avoided under these conditions.

This means that ultimately the thickness of the overburden sets a limit to the applicability of the room and pillar mining method. At greater depths, only longwall will be an economically feasible method (see the next section). The applicability limits for room and pillar mining are even narrower if the bedrock material is soft. If there are no restrictions on subsidence, the final step of the room and pillar method, removal of the pillars, can be undertaken.

The mining equipment most often used for room and pillar sections consists of a continuous miner that cuts and loads the coal, shuttle cars (usually two), or special extensible or flexible conveyors that haul the coal to the section belt conveyor, one or two roof bolters, and a belt feeder next to the belt tailpiece. A belt conveyor (panel or section belt) is used to haul the coal to a loading station for loading into mine cars or onto a mainline belt conveyor.

The crew usually consists of one foreman, one continuous miner operator and helper, two shuttle car operators, one roof bolter operator and helper, one mechanic, one general laborer, one belt conveyor patrol man, and one loading station operator. Thus, 11 workers are usually required to operate a section for each production shift. Maintenance personnel are usually centrally dispatched according to the required work. Production ranges between 250 and 1200 tons per shift. Average production ranges from 350 to 600 tons per shift.

On a room and pillar section with a conventional production cycle, the equipment consists of a cutting machine (Fig. 13), a coal drill, a loader, shuttle cars, and roof bolters. The coal is undercut near the floor to ease

FIG. 13 Cutting machine for conventional room and pillar sections. [Courtesy of the Goodman Equipment Corporation.]

and simplify the blasting operation. The remaining upper portion of the seam is drilled and blasted. The coal is then loaded onto the shuttle car by the mobile loader.

The labor requirements for this operation are higher than those of a continuous-mining section, because of the divided loading-cutting, drilling, and shooting cycle. The operation is less continuous because of this additional cycle. The productivity per worker-shift is lower than that of a continuous-miner section. The method is applicable only in areas with good roof conditions because of the stresses caused by the blasting operation.

Basically, mine ventilation is provided by a main exhaust fan on the surface. The pressure differential created is used to channel fresh air in the desired quantities through the underground openings. The law provides the basic requirements, with a minimum of 9000 ft^3/min at the last open cross-cut and 3000 ft^3/min at the face of a room and pillar section. This is only the minimum and might have to be increased if mine gas or dust concentrations exceed the allowed maximum level.

The coal is hauled away by trailing cable-equipped shuttle cars of between 5 and 10 tons capacity. These cars, usually two per continuous miner, are maneuvered under the conveyor boom of the mining or loading machine and are maintained in this position until the loading limit is reached. The coal is then carried to a belt feeder. The belt feeder, usually equipped with a crusher, allows the shuttle car to discharge quickly without overloading the section belt conveyor.

Attempts are being made to replace shuttle cars with continuous-haulage systems which allow a more efficient operation of the mining machine. However, so far the equipment has not penetrated the market.

During the development phase, several parallel entries and the connecting cross-cuts (Fig. 12) are developed simultaneously. The continuous miner (Fig. 13) advances 20 ft by sumping the cutting head near the roof into the coal face (Fig. 14). The cutting head is then lowered hydraulically, cutting the remaining coal down to the floor. This operating cycle is repeated until the operator positioned on the machine reaches the edge of the supported roof. According to mining law, he cannot proceed beyond this point. Since the distance from the cutting head to the seat of the operator measures an average of around 20 ft, this is the maximum advance during a single mining

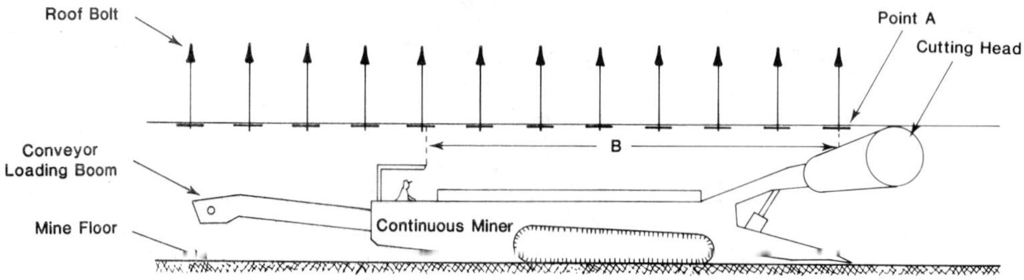

FIG. 14 Advance of a continuous miner. As a maximum the continuous miner can advance distance B (about 20 ft). At the point A the operator has to move to a different entry because he reached the end of the supported roof section.

cycle. Difficult roof conditions might force the operator to a shorter cut of only 5 or 10 ft. If the entry is wider than the cutting head (which is commonly the case), the miner has to shift to the other side of the entry to cut the remainder using the same procedure. After the mining cycle has ended, the miner will move to a parallel entry or cross-cut to perform the same operation.

A roof bolter enters the mined-out area after the continuous miner finishes its cycle. Temporary roof support must be positioned before the unit can begin to install roof bolts in the newly excavated area. This temporary support can consist of wooden or hydraulic props or of hydraulic support units as attached to some of the newly developed roof bolters. Once the new area is secured, the installation of the mechanical or resin bolts can begin. A pattern approved for the section is used to position the bolts (e.g., 4 by 4 ft). In difficult cases, additional support of wooden props, together with wooden or steel beams, are installed.

About 30% of U.S. coal is mined with the conventional method of mining. A slot several inches wide, 8 to 12 ft deep, is cut near the floor over the entire width of the entry. The remaining upper coal is then drilled and shot. A loading machine similar to a continuous miner without a cutting head loads the coal onto a shuttle car. This means that the complete operating cycle is expanded, converting the coal removal and loading process into two separate operations.

The higher skill requirements, the necessity for good roof conditions, the dangers and problems involved with explosives, and the often lower productivity make this method in many cases inferior to continuous mining.

2. *Mining of Pillars*

When a section is fully developed, the final process of pillar mining begins. This consists of slicing the pillars to a point just before the roof caves in. Coal fenders, stumps, cribs, posts, and roof bolts are used to provide support for this operation.

It is important that the roof caves, in the mined-out and abandoned area, to avoid excessive abutment pressures from large roof sections overhanging the operating area. Such pressure can result in excessive convergence and roof-fall problems, or it can lead to sudden breakage of a large portion of the roof, resulting in roof caving in the adjacent mining area. If the roof does not break at the appropriate intervals, caving might have to be induced through removal of the remaining supports or through weakening of the roof by drilling and shooting.

Clearly, the proper caving of the roof is a vital necessity for a successful execution of this mining process.

3. *Ventilation*

Continuous liberation of methane, particularly from the freshly mined coal, plus the airborne dust created by the cutting and conveying action, require the face area to be constantly ventilated by several thousand cubic feet of air per minute. The law requires a minimum volume of air, which in many cases is insufficient because of excessive methane liberation. Sufficient air has to be supplied to maintain the methane content below 1%.

Air supply is achieved by partitioning off a 3- to 5-ft wide section of the entry. The partition is made up from burlap or plastic sheets fastened to the roof and floor. This partition has to be maintained at a 10-ft distance from the face. The pressure differential created by the main ventilation fan on the surface forces the air to the face and back through the return channel created by the partition, which is usually called a line curtain. This scheme assures that gas as well as dust is moved away from the personnel in the face area.

In cases where the mining machine occupies the entire cross section (boring miner), the line curtain is replaced by electrically powered fans and tubing.

The ventilating air is channeled into the mining section through the intake entries on one side and returned through parallel entires of the other side. The law requires that belt entries and entries equipped with trolley wire are ventilated with a separate flow of fresh air.

The main airstream is guided through the intake entries channeled along the pillars being mined, with a portion of it forced through the caved area into the bleeder system. This is done to avoid any gas accumulation in the gob. The remainder of the main airstream is then channeled into the return air entry system.

4. *Equipment*

The continuous miner is an electrically powered crawler-mounted unit. It can be equipped with cutting drums of varying width of between 8 to 16 ft. Some makes are equipped with an ascillating cutting head, which expands and contracts during rotation. The cutting head or cutting drum is attached to the main machine body, which is equipped with a coal-loading device usually consisting of gathering arms resembling the claws of a lobster in shape and movement. A continuous miner is shown in Fig. 15, with cutting head raised and gathering arms and chain conveyor loading through the middle of the unit. Pillars can be seen to the side of the unit. The chain conveyor transfers the coal, via an articulated boom, onto the shuttle car.

A second type used in specific cases of suitable seam height is the borer miner. In contrast to the continuous miner, the cutting head of the borer

FIG. 15 Continuous miner. [Courtesy of the National Mine Service Company.]

FIG. 16 Borer-type continuous miner. [Courtesy of the Joy Manufacturing Company.]

miner is equipped with two intersecting cutting heads which rotate around a horizontal axis positioned parallel to the line of advancement. The roof and floor section not cut by the boring heads is removed with a cutting chain (Fig. 16). The advantages claimed for this unit are curved ribs, which leave only a small amount of roof unsupported. Under a variance, some mines are allowed to advance a considerable distance (100 ft or the length of a pillar if entries are developed narrow, around 14 ft) without the need to move the equipment to another entry for the installation of roof support. Idle moving time is therefore reduced and the advance rate is improved by a substantial margin. The key problem is limited height adjustment, which reduces application.

Shuttle cars or conveyors haul the coal from the mining machine to the section belt. Shuttle cars are electrically powered rubber-tired vehicles. A fully loaded shuttle car is shown in Fig. 17. The power is supplied via a trailing cable. A chain conveyor at the floor of the unit allows an even and more effective distribution of the load as well as a quick discharge. Capacities range from 3 to 10 tons.

The shuttle car commonly discharges into the belt feeder. The belt feeder (Fig. 18) is a crawler-mounted surge bin positioned at the end of the section belt. It receives discharged coal from the shuttle car and loads it at an appropriate rate onto the belt conveyor. It is equipped with a double outboard

FIG. 17 Loaded shuttle car. [Courtesy of the Joy Manufacturing Company.]

FIG. 18 Feeder breaker for continuous-mining sections. [Courtesy of the Stamler Corporation.]

chain conveyor and in most cases with a crusher for the reduction of oversized coal.

Continuous-haulage systems utilize a variety of conveyor types (e.g., outboard chain conveyors, extendable or telescopic belt conveyors, and flexible conveyors capable of overcoming relatively sharp turns). Considerable moving time is required for these systems in the cyclic operation of a room and pillar section. These units will find acceptance only in applications where the mining machine can advance a considerable distance.

Roof bolters with one or two drill heads are the principal means of installing roof support. They are rubber-tire-mounted vehicles equipped with a drilling device and a dust collector (Fig. 19). A hole of about $1\frac{1}{2}$ to $1\frac{1}{4}$ in. is drilled into the roof. Different-length drill steels are used to achieve maximum length of the hole with the fixed stroke of the drilling head. The different-length drill steels are interchanged until a desired depth is achieved. The roof bolt is then inserted and torqued to the desired value. Roof bolts are installed in approved patterns. It is common to install bolts at a 4-ft spacing. Dust supression at the roof drill is accomplished by either absorbing the dust by vacuum through a hole in the center of the drill steel or by water fed through the drill steel.

Two types of roof bolts are used: the friction bolt and the resin bolt. The resin bolt was introduced only a few years ago and has found very rapid acceptance in the mining industry. It embeds a steel rod over its entire length in a very quick setting resin. It provides resistance against lateral as well as horizontal movement. It is usually superior in controlling roof movement compared to the friction bolt. The friction bolt, considerably cheaper than the resin bolt, has only a short mechanically expandable section at its upper end, which is tightened by turning the bolt. The function of roof bolts is mainly to clamp the different layers of roof material together to stiffen them to a stronger package of self-supporting roof. The clamping action is also working against horizontal movement by increasing friction between the layers.

5. Future Trends

One of the key problems with room and pillar mining is the alternating operating cycle. The requirement to remove the continuous miner after it has cut a maximum of 20 ft (Fig. 14) severely decreases productivity and increases maintenance requirements. An automated extraction system was developed under a USBM contract. The continuous miner was equipped with a chocklike support, built-in roof bolters, and integral ventilation channels. This allows installation of the support over the operating unit or immediately behind it, provides for easier automation of the unit, and better utilizes advanced continuous-conveying technology.

Walking or drawler-mounted hydraulic supports are under development for the final mining of the pillars. They would replace the wooden or hydraulic posts presently utilized or replace coal fenders and thereby improve the recovery rate.

FIG. 19 Roof drill. [Courtesy of the FMC Corporation.]

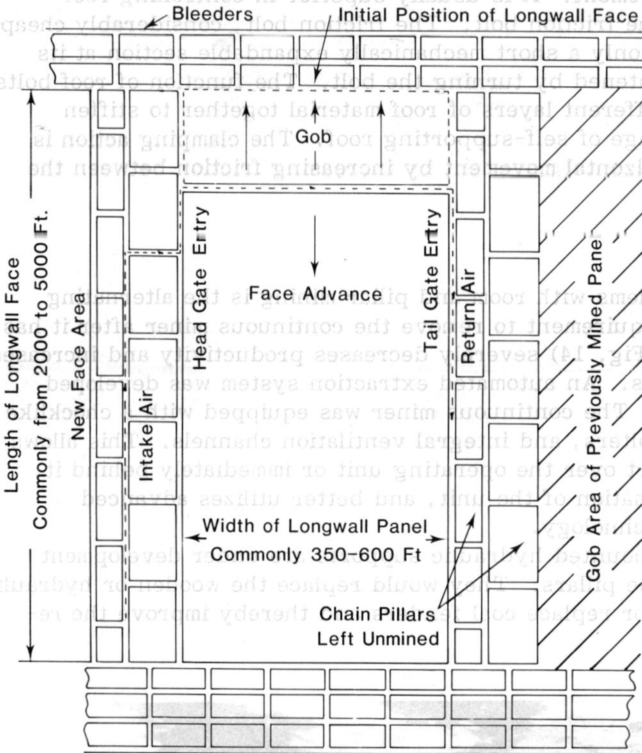

3 - Entry System Utilizing the Yield Pillar Concept Flow of Ventilating Air with Partial Flow Through the Gob

FIG. 20 Longwall mining section.

H. Longwall Mining

Longwall mining is viewed by a number of experts as a means of reversing the present trend of declining productivity. It allows better ground control and requires less development work per ton of coal mined. A production range of about 500 to a maximum of almost 4000 tons per shift points out the significantly higher capacity of this method.

1. *Development*

A normal longwall section is created by the development of two entry systems of two to four entries each, 350 to 600 ft apart. These entries are developed in the common room and pillar fashion (Fig. 20) to a length of between 2000 and 5000 ft. At the end they are connected by an entry system (bleeder system) developed perpendicular to the two initial systems.

2. Mining of the Panel

The coal-getting equipment removes coal in bidirectional slices along the entire coal face. The longwall crew moves the flexible conveyor into the newly excavated area by pushing it with hydraulic cylinders which are attached to the base of the supports and the face conveyor. The conveyor is advanced in a continuous motion from one end of the face to the other. Supports are advanced one by one following the advancement of the conveyor. This is done by first lowering the support hydraulically. The same cylinders that pushed the conveyor now pull the supports forward. In this new position the units are raised again and set up against the roof. As the supports move, the roof behind them is allowed to cave.

The entire seam section between the two entry systems is so removed, leaving only the chain pillars behind. Recovery rates of between 60 and 80% are achieved in such longwall operations. Efforts are made by government and industry to improve the system and its recovery rate through refined entry systems and development methods.

3. Ventilation

In most mines longwalls are operated retreating. The ventilation system for this type of operation is roughly similar to that of room and pillar section. An airstream of about 20,000 to 50,000 ft^3/min is channeled through one entry system (usually the head entries) into the section. The entire flow is directed along the face, with a portion of it forced through the gob. The remainder then returns through the tail entry system. The course that the main airstream follows has a U shape. The only difference compared with a room and pillar section is that the longwall face is a single straight front which does not require separate air flows channeled into individual entries by line curtain or fans and tubing.

4. Rock Mechanics

Longwall, as a deep mining method, must deal heavily with rock mechanical and ground control technology. An area mined with longwall techniques creates a system of forces and problems significantly different from those developed by other methods of mining. The forward and side abutment loads of a longwall section impose stress onto the entries that requires extensive measures for ground control (Fig. 21). So far many of the steps taken have proven inadequate, resulting in considerable entry support problems, which in turn affect production. Considering the size of a longwall panel, the number of entries is less than for a room and pillar section. This reduces entry maintenance and the amount of support material needed. As a result, however, each of the entries is of greater importance and has to be maintained as a means for the passage of air, workers, and materials and to provide the two separate escapeways required by law. The thickness of overburden has only a limited impact on the ground control problems experienced at the longwall face. The entries, however, are in a different situation, because of abutment loads and the required service life. Ground forces have a considerable time span to effect entries, which is not the case with the continuously advancing face.

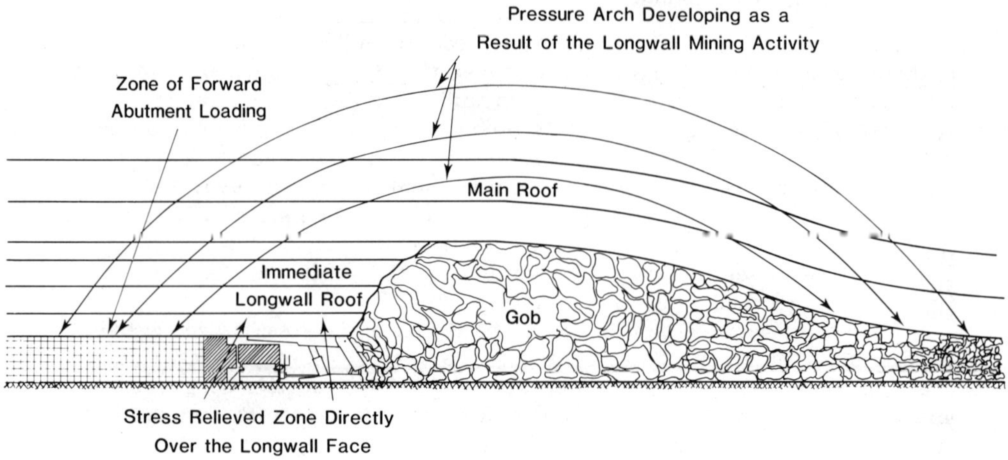

FIG. 21 Cross section through an active longwall panel.

Experience shows that ground control problems in entries are relatively small during the mining of the first panel. The mining of the second panel, however, allows the combination of the forward abutment load of the operating panel with the side abutment load of the previously mined panel. These two forces meet over the entry system next to the mined-out panel, often creating severe ground control problems. The opposite entry system usually experiences very few serious ground control problems.

The traditional means utilized for ground control in mining sections are roof bolts, cribs, and wooden posts. Roof bolts are an excellent means to control a stable roof. In a longwall entry with rapidly progressing convergence, however, additional means are necessary.

The commonly used cribs provide little or no support at the beginning and substantial resistance after a certain amount of convergence has taken place. This results in serious initial deterioration of the roof and floor strata. It would be uneconomical if not impossible to install a support that completely prevents convergence. The task of a support is therefore to maintain a sufficiently large opening for the duration of the mining process. This is best done by a support that provides resistance sufficient to slow convergence to the desired degree. Hydraulic support provides a quick and continuous supporting force at an adjustable rate, allowing a controlled yielding. However, this is expensive, requires a high degree of maintenance, and is therefore used only in the immediate vicinity of the advancing longwall. Yielding steel friction sets are a more economical solution. These sets can usually be reused, have a high capacity, and are therefore more advantageous than wooden props.

5. *Equipment*

Longwall equipment consists of three major components: the hydraulically powered roof supports, the chain conveyor, and the coal-cutting machine.

a. Face Roof Supports Three basic roof support designs are available today. These are the chock, the frame, and shield-type supports. There are four key requirements of longwall roof support: (1) sufficient support capacity, (2) prevention of roof and gob material from entering the face area, (3) support close to the face, and (4) sufficient cross-sectional area for ventilation, travel, and equipment. These requirements are best met by the recently introduced shield-type supports (Fig. 22). This design also reduces maintenance requirements because of its excellent shielding effect, protecting the attached hydraulic gear and the operating personnel. The shield support improves general roof conditions because it is a laterally stable support, designed to create a forward thrust during yielding. The small amount of forward movement during yielding helps to maintain the integrity of the roof strata.

b. Face Conveyors Face conveyor designs are distinguished primarily by the position and number of conveying chains. The conveying chain in most cases consists of a round link chain and can be arranged as a double outboard, a double outboard plus a center strand, a double center, and a single center chain (Fig. 23). The chains are made of high-strength steel at a range of around 75,000 to 80,000 psi.

In the past, the strength of these chains was often the determining factor for the maximum possible face length. This was caused by the outboard chain design, requiring the chain to be guided inside the side section of the conveyor pan. Any increase in chain size resulted in a correspondingly higher

FIG. 22 Shield support. [Courtesy of the Dowty Corporation.]

FIG. 23 Center chain design face conveyor. [Courtesy of the Dowty Corporation.]

FIG. 24 Complete set of longwall face equipment. [Courtesy of the National Mine Service Company.]

conveyor. Since height is very critical for the application of certain coal-getting equipment, chain sizes were not increased substantially until the introduction of center chain conveyors. This type of conveyor, with its high-strength chain and heavy-duty pan sections, increased reliability of the equipment and allowed the face length to be increased up to 1200 ft or more.

The connection between the individual pan sections and the design of the conveyor accessories were additional steps necessary to improve reliability. The design of accessories attached to a face conveyor depends largely on the type of coal-getting equipment utilized. The drum-type shearer and the plow or planer each require specially designed attachments.

The typical attachments for a shearer-equipped longwall face are a spill plate, a shearer guide rail, or guide tube and a wedge-shaped ramp plate to to scoop up spilled coal on the face side of the conveyor. A complete set of longwall face equipment with attachments is shown in Fig. 24. On some of the newer designs the shearer rides on these accessories because wear on the pan line side sections and breakage of pan connections created considerable downtime problems. The spill plate accommodates the power supply cable and the water hose, which are pulled along the face by the shearer. Shearers are propelled along a round link chain which is anchored to the two conveyor drives. This chain is being gradually replaced by a rack-and-pinion type device which is attached to the conveyor pans.

The plow operates in a manner that alters the requirements toward conveyor accessory design. Some of the more recent designs place all the guides required for the plow on the face side of the conveyor, whereas older versions require a gob-side guidance system. The heavy round link chain used to pull the plow is channeled through compartments along either the face or gob side of the conveyor, depending on the plow design.

Conveyors of less than 200 ft are usually equipped with only one drive at the discharge end. Longer faces require a drive on each end of the conveyor. A drive consists of a heavy drive frame which accommodates the drive sprocket for the conveyor chain. It is powered by a squirrel-cage motor of between 100 and 300 hp output. The motor shaft is connected to a hydraulic coupling, which in turn transfers the power through a gear-type speed reducer to the drive sprocket. Conveying capacities range to a maximum of about 1200 tons/hr, the average being about 500 to 700 tons/hr.

The stage loader (Fig. 25) is a conveyor usually similar in design to the face conveyor. It is located in the head gate entry in a position perpendicular to the face conveyor and serves as a connecting link to the belt conveyor. This stage loader has two functions:

1. To elevate the coal to belt conveyor level
2. To provide a flexible link between the advancing face conveyor and the stationary belt conveyor.

This flexibility is achieved by either allowing the face conveyor to advance along the stage loader or by telescoping the stage loader over the belt tailpiece. The stage loader is advanced by hydraulic jacks attached to the entry gate supports or by a crawler-mounted belt tailpiece.

FIG. 25 Stage loader. [Courtesy of the Joy Manufacturing Company.]

FIG. 26 Longwall shearer cutting coal. [Courtesy of Mining Progress Inc.]

c. Coal-Getting Equipment on Longwall Faces The two different types of equipment used for this purpose in U.S. coal mines are shearers and plows.

The shearer is a self-propelled, rather complex piece of equipment that is equipped with skids or rollers riding on the face conveyor or its accessories (Fig. 26). It is engaged with its drive sprocket to a high-strength chain stretched out and tensioned along the entire face. The new rack-and-pinion type of arrangement, which is gradually replacing the chain devices, eliminates the dangers created by the swinging or breaking chain. Cutting drums are equipped with cutting bits which are reinforced against wear with carbide tips. Spiral vanes help to move the loosened coal toward the conveyor. Built-in water sprays reduce airborne dust created by the cutting action.

The drums are usually attached such that their position relative to the shearer body can be raised or lowered. In some cases the entire shearer body is raised or lowered. This adjustment allows the mining operation to follow the varying course of the seam.

Plows (Fig. 27) are used in low- to medium-size seams (2 to 5 ft thick). The unit consists of steel construction equipped with carbon-tipped bits. This passive steel unit is engaged to a guiding system on the face conveyor. An endless round link chain (breaking strength ranging between 60 to 85 metric tons) powered by synchronized electric drives on each end of the face conveyor pulls the plow body at speeds between 120 and 420 ft/min along the face.

FIG. 27 Longwall plow undergoing height adjustment. [Courtesy of Mining Progress Inc.]

For the cutting process the plow has to be forced against the coal face. This is done by hydraulic cylinder attached to the gob side of the face conveyor and to the base of the face supports, or by a separate hydraulic prop. Forces of between 1 and 3 tons are applied per cylinder.

A plow drive is attached to each drive frame of the face conveyor. Only 30 to 60% of the drive power supplied to the plow is used for the cutting and loading of coal; the remainder is lost in friction. This means that the power loss is considerably higher than that of a shearer, which uses 75 to 85% of its power for the removal of the coal. As a result, rather large drives are required at the face ends. The resulting reactive forces on the drives are counteracted by large and complex hydraulic anchoring devices. The anchoring devices usually consist of a specially designed roof support unit which anchors itself between the roof and floor. This allows advancement of the drives under tension.

To adapt the plow to a different seam height, extensions are bolted (Fig. 27) to the top of the main body, or manually adjustable extensions are utilized.

d. Operational Features Both plow or shearer units are designed to work bidirectionally, transversing over the entire length of the face. The shearer takes a cut between 2 and 3½ ft deep while the plow slices 1 to 6 in. off the coal face. In lower seams (up to about 5 ft) the plow is equal or superior in production capabilities, whereas in thicker seams, the high-horsepower shearer can outproduce a plow. Therefore, the plow is used mostly in low-seam operations.

I. Shortwall

Shortwall is a mining technique similar to longwall, but with the panel width reduced by approximately 50%. Commonly, shortwall panels are only 150 to 250 ft wide.

1. Equipment

Standard continuous miners and shuttle cars are utilized instead of shearers, plows, or chain conveyors. Modified hydraulic longwall supports are used as face supports.

2. Development

A system of three to four entries is developed on each side of the panel. These entry systems are connected at the bleeder system. The system is mined by retreating. This positions the face supports at the far end of the panel.

3. Mining of the Panel

Coal is removed by a continuous miner making 8- to 10-ft-wide cuts along the face. Usually, two shuttle cars are employed to transfer the coal from the miner to the panel belt. Because only one shuttle car can pass at the face, the second one has to wait its turn outside the face area. This causes delays for the continuous miner. When the mining machine reaches the tail entry, it trams in the backward position to the head gate to start a new cut.

The ventilation scheme is the same as for longwalls, with the air moving from the head gate entry system along the face to the tail entry system and a portion forced through the gob area into the bleeder system.

4. *Pros and Cons of Shortwall Mining*

Since the supports require the largest portion of the equipment investment, a major cut in capital outlay is achieved by the reduction of the face length. The investment requirements are further reduced because already existing continuous-mining equipment can often be utilized.

However, the operational procedure and the equipment configurations have limited this method to use on deposits with favorable bedrock conditions. Additionally, the amount of development work per ton of mined coal has to be increased because of the smaller panel size.

The supports (Fig. 28) are designed with an extra-long extended canopy to allow the large continuous miner to operate under supported roof. The extended canopy supports the newly exposed area only partially and with a very low capacity. Therefore, a stable roof with high self supporting ability is required for the application of this method. Further, the miner can cut in only one direction, toward the tailgate entry. After the cut is completed, the miner cannot turn around, because of space restrictions. The miner must tram backward to the headgate idle. In this mode of operation, workers avoid the airborne dust from the mining machine; however, productivity is substantially reduced. The cyclic operation caused by the shuttle cars reduces the production capacity even further. As a result, average production is similar to that of standard room and pillar sections. The recovery rate, compared to room and pillar sections, is improved only marginally. The considerable investment for face supports has led to very limited application of this method.

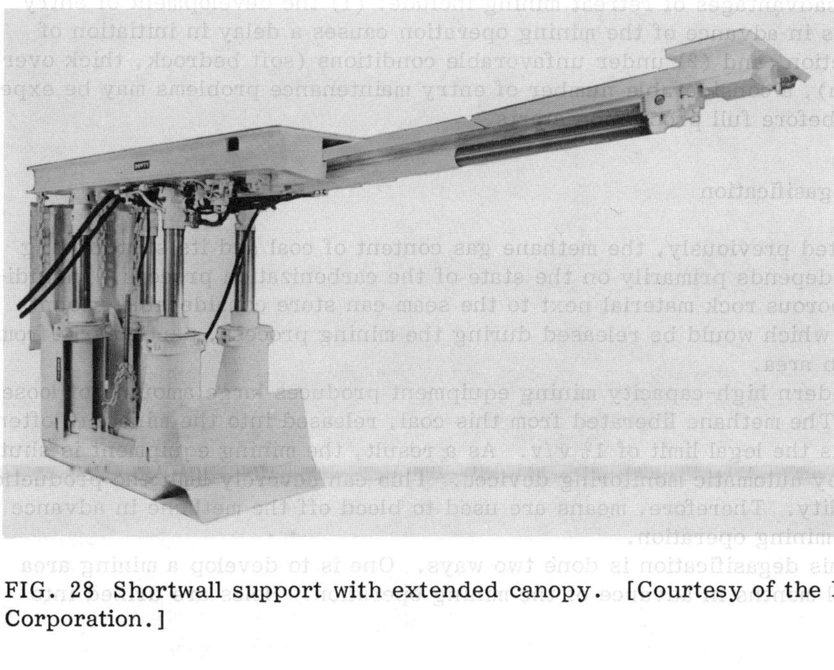

FIG. 28 Shortwall support with extended canopy. [Courtesy of the Dowty Corporation.]

J. Advance and Retreat Mining

The advancing method is applied when development work is behind schedule or where excessive convergence is experienced before the mining operation has begun.

Advance mining offers the following advantages: (1) quick startup of full production, and (2) shorter entry service life, because the entries do not have to be ready before the mining process begins. The following disadvantages are encountered with this method: (1) high cost for packwalls; (2) abutment pressures have a full impact on the entries, possibly leading to excessive entry support and maintenance problems; and (3) methane liberation may be higher, because no bleeding has taken place during the entry development process and the danger of spontaneous combustion from air leakage through the packwalls into the sealed gob area is greater.

Retreat mining is carried out by first developing a mining area through a system of entries (Fig. 20). After completion of development, the final extraction of coal starts at the farthest point out. The roof material is allowed to cave into the excavated area right behind the ongoing mining operation. As the mining operation retreats, it leaves behind an increasing area of caved material. Safety pillars are left at the finishing point to prevent any abutment loads from the mined-out area damaging the main haulage entries.

Retreat mining is the dominant method for U.S. longwall operations. This procedure has the following advantages: (1) it reduces entry maintenance, because the entries are abandoned shortly after they come under the influence of the forward abutment load; (2) it allows proper bleeding of the gob area by forcing air through the bleeder system; (3) it reduces the danger of spontaneous combustion; and (4) it assures that entry development will not hinder the face operation and allows positive control of the airstream around the solid coal block.

Disadvantages of retreat mining include: (1) the development of entry systems in advance of the mining operation causes a delay in initiation of production, and (2) under unfavorable conditions (soft bedrock, thick overburden), a considerable number of entry maintenance problems may be experienced before full production starts.

K. Degasification

As stated previously, the methane gas content of coal and its surrounding strata depends primarily on the state of the carbonization process. In addition, porous rock material next to the seam can store considerable amounts of gas which would be released during the mining process, particularly from the gob area.

Modern high-capacity mining equipment produces large amounts of loose coal. The methane liberated from this coal, released into the mine air, often exceeds the legal limit of 1% v/v. As a result, the mining equipment is shut down by automatic monitoring devices. This can severely limit the production capability. Therefore, means are used to bleed off the methane in advance of the mining operation.

This degasification is done two ways. One is to develop a mining area several months in advance of the mining operation. Holes are drilled into the panel and connected to a pipeline and a vacuum pump. A sufficient reduction of the methane level has often been achieved with this method.

The second system operates similarly except that holes are drilled from the surface. This allows startup of the degasification prior to any mining activity underground. However, drilling costs are considerably higher because a hole must be drilled through the entire overburden.

In many cases, methane gas is discharged into the exhaust air of the mine ventilation system, or directly into the atmosphere if the hole was drilled from the surface. Since the advent of energy crises and rising natural gas prices, it is often financially attractive to pump the gas directly into the public distribution system. This not only pays for the degasification efforts, but may result in additional revenue.

L. Subsidence

With today's environmentally conscious society, subsidence becomes a very important issue. Areas mined with the room and pillar method frequently experience subsidence decades after the operation has ended.

Longwall mining is a more favorable method in regard to subsidence. Ground movements initiated by longwall mining are completed in a relatively short period. The settling process is quite consistent, even over large areas, allowing surface structures to move as a whole, limiting damage. Only the border areas and areas above the chain pillars might experience serious problem.

M. Spontaneous Combustion and Roof Bounces

The process that leads to spontaneous combustion is not fully understood and requires more research. However, a few facts are known. High-volatile coal is more susceptible to spontaneous combustion; anthracite therefore presents the lowest probability. It has been found that certain temperature and oxygen-supply conditions are required to initiate this process. As a result, measures are directed toward a reduction of temperature through ventilation of gob areas (this also reduces the methane content). However, as mining progresses to greater depth, this becomes more difficult, because rock temperatures increase and a greater overburden weight causes a tighter sealing of the mined-out areas.

Roof bounces are an effect in which rock or coal under pressure expands with explosive speed into an excavated area. When the loading on very strong layers of rock exceeds the capacity of the rock formation, it causes a sudden break. This results in a quick and considerable loading of formations next to the excavated area. The sudden squeezing of these formations results in an explosionlike removal of material. The process is more likely to occur at greater depths, where the high overburden weight provides the necessary loading.

The European coal industry, mining at greater depth and faced with this problem more frequently, has developed a limited method to detect and combat this problem. In this detection method, holes are drilled into the coal face with hand-held augers. If the amount of coal removed from a hole exceeds its approximate volume, a tension area is found. If such an area is identified, additional holes are drilled to relieve the tension.

N. Future Trends

Presently available equipment types are being refined in terms of reliability, output, safety, flexibility, and applicability. Revisions in the mining cycle to achieve better equipment utilization can be foreseen. Rising labor costs will provide additional incentive for the ongoing efforts of developing automated and remotely controlled equipment.

Some changes can be expected in equipment operation and application. Hydraulic mining, utilizing high-pressure water jets, might replace the cutting bit at some time in the future. Hydraulic conveying and remotely controlled continuous miners with advanced roof-support systems will be able to advance for large distances, probably doubling or tripling production. This is the result of elimination of the intermittent cutting cycle caused by the change-over and shuttle car operation.

Pillar mining might be improved through the utilization of hydraulic self-advancing roof supports which would reduce the skill requirements of operators and improve production and recovery rates, because no coal pillars would be needed for roof support.

Longwall has made some progress toward automation. Roof support and shearer guidance systems are under development which would be particularly helpful in low seams, where workers are rather slow.

Other areas for future development are thick-seam and steep-seam mining.

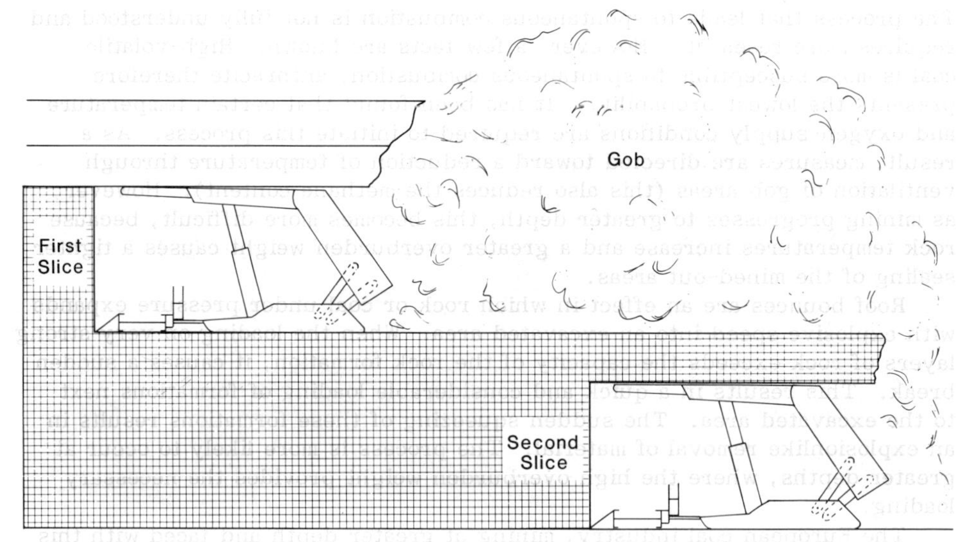

FIG. 29 Multy slice thick-seam mining. European experience shows that an average of $1\frac{1}{2}$ years delay should be allowed between slices for the consolidation of the upper slice gob material.

FIG. 30 Longwall equipment for mining thick seams.

Thick-seam mining operations in the United States have a very low recovery rate. Foreign methods to mine thick seams with longwall equipment have been investigated by U.S. companies. Two basic systems are considered: multy slice mining and sublevel caving.

In multy slice mining, a coal seam is removed by mining in several layers, starting at the top. Once the first layer is mined, the caved roof is allowed to settle and consolidate for a period of several months to several years before the next lower slice is removed (Fig. 29). The caving method mines a slice at the bottom of the seam. The longwall supports are equipped with

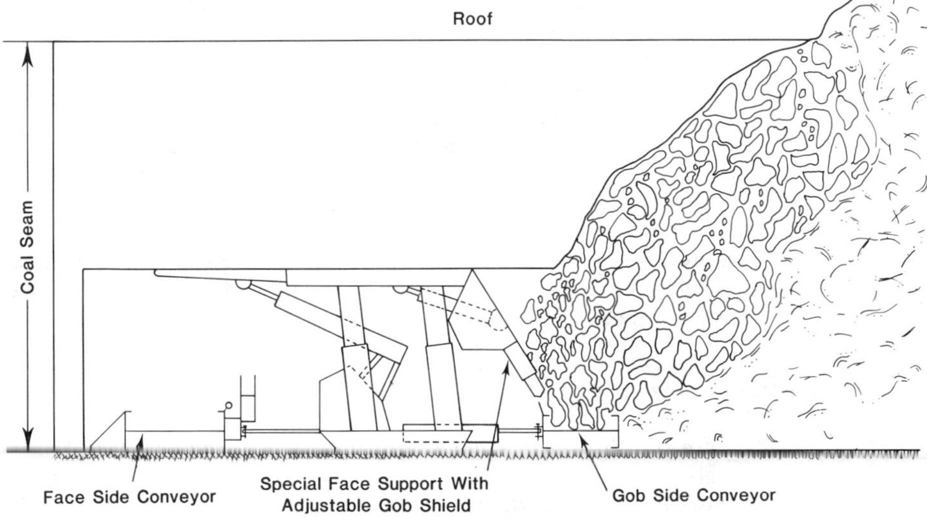

FIG. 31 Thick-seam mining with the sublevel caving method.

adjustable shields at the gob side. An opening in or under the shield allows the coal from the gob to flow into a second conveyor installed in the rear end of the face supports (Figs. 30 and 31).

Roughly defined steep seams range in inclination from 35 to 90°. Mining of these seams with longwall equipment is practiced rather extensively in Europe and Russia but not in the United States. The efficiency of current systems has caused some U.S. companies to consider its application here. The coal-getting equipment in this application can consist of specially designed shearers or plows, or a system of water nozzles removing the coal with high-pressure water jets.

III. MINING IN EUROPE

Most present-day mining in Europe occurs under 2000 to 4000 ft of overburden, as more easily mined coal deposits have been depleted. At this depth most mines are developed as shaft mines (Fig. 32). All personnel, material, and coal have to be hoisted through these shafts. Considering the two factors of hoisting capacity and required length of shaft, a considerable investment is necessary to reach the coal-bearing strata. The required development and maintenance of an extensive net of underground entries or tunnels requires huge investments. Openings at this depth have to be equipped with costly supports, and periodic reworking and repair is necessary.

Mines not only extend horizontally but also vertically through the development of new levels. The life of the mines is thus extended considerably, and surface installations can be amortized over a longer period.

The more limited reserves have forced companies into mining less favorable deposits, and European governments require that all possible deposits be mined to conserve the nation's energy resources. These factors and the large percentage of inclined seams and faults make mining very difficult and costly. The population density and the heavy surface buildup cause additional expense in the form of payments for subsidence damage to surface structures. Therefore, backfilling is frequently practiced to reduce subsidence. The close spacing of faults often severely limits the size of a mining section, forcing frequent moves and excessive development work.

The thickness of the overburden results in very high ground pressure. This would require extremely large pillars if the room and pillar method was applied. Additionally, support is required for any opening, adding prohibitive costs to a multiple-entry room and pillar operation.

As a result, single-entry longwall operations requiring the minimum number of entries and allowing maximum recovery of resouces is the mining method almost exclusively practiced.

A. Mine Design

Shaft mines dominate the European coal mining industry. Shafts 20 to 30 ft in diameter, with circular cross section, lined with masonry, concrete, or steel, are the dominant means of gaining access to the coal-bearing strata. They are often extended beyond the last mining level to provide for future expansion. As in the United States, shafts are developed by drilling, blast-

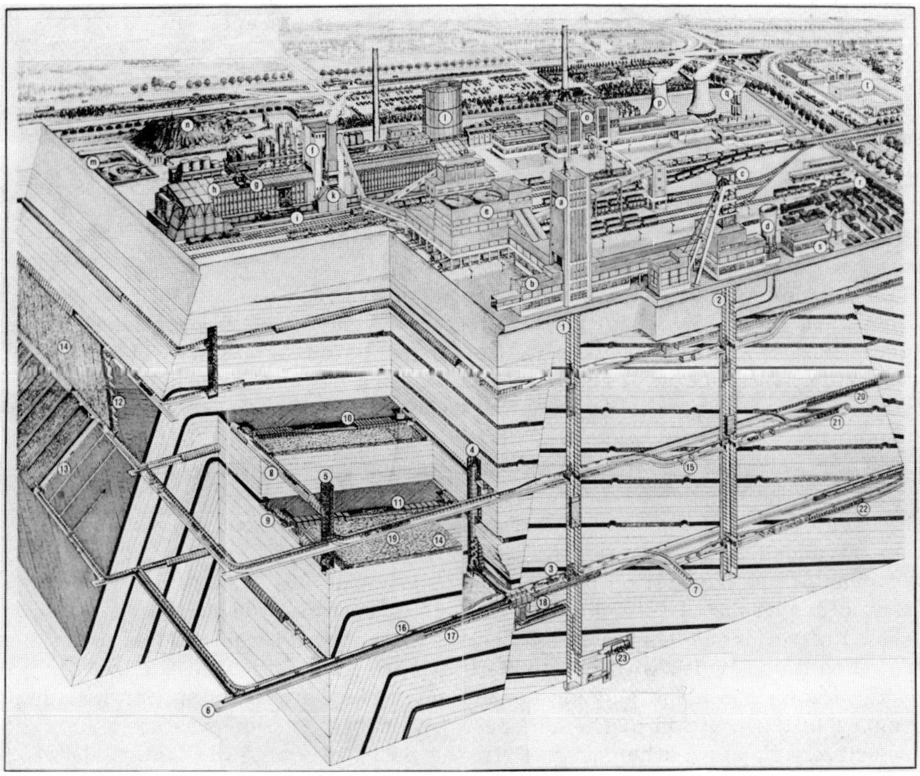

FIG. 32 German multilevel, multiseam shaft-type coal mine. Underground facilities: (1) main shaft with skip hoisting; (2) exhaust ventilation shaft with multistage cage; (3) third-level station; (4) blind shaft with cylindrical storage bin; (5) blind shaft with car-hoisting facilities; (6) main entry; (7) main entry; (8) section or panel entry; (9) road heading machine; (10) longwall section with plow; (11) longwall section with shearer; (12) longwall section in a steep seam mined manually with air picks; (13) longwall section in steeply pitching seam with plow; (14) mined-out gob area; (15) ventilation lock; (16) belt conveyor as main haulage system; (17) mine car haulage; (18) storage bin and skip-loading facilities; (19) supply haulage with a monorail; (20) supply haulage with mine cars; (21) monorail system as personnel carrier; (22) worker-trip cars; (23) pump station.

Surface facilities: (a) hoisting tower with overhead hoist; (b) shaft building; (c) head frame; (d) main exhaust fan and diffuser; (e) coal preparation plant with loading facilities; (f) coking coal silo; (g) container vehicle for filling of coke ovens; (h) coke oven battery; (i) coke watering car; (k) coke quenching tower; (l) gas tank; (m) water-treatment plant; (n) refuse pile; (o) power plant; (p) cooling tower; (q) water tower; (r) supply storage area; (s) sawmill; (t) training and teaching center. [Reprinted courtesy of Verlag Glückauf GmbH, Essen, West Germany, copyright 1976.]

ing, and excavating or by large-diameter shaft-boring equipment. Shaft boring is more frequently used, particularly on the smaller and shorter subshafts (Winze), which connect the different levels but do not extend to the surface.

Haulage in the shafts is usually accomplished by hoisting of the filled mine cars on multistage cages or by skips. Pumping of coal slurry is also done in special cases.

D. Underground Development (General Considerations)

The complex system of forces and the resulting rock mechanical problems developed by mining activities at different levels result in significant differences between European and U.S. underground development. The rock mechanical interaction of the extraction operations at the various levels require that all deposits be mined as completely as possible. Pillars left after mining create zones of extreme and often unmanageable ground control problems, as well as a high probability of roof bounce.

Since the number of entries is kept to a minimum because of cost, no bleeder systems are provided. If retreat mining is practiced, only two entries are advanced into a new mining area.

Panels are laid out as large as possible. The large-panel layout is used as another means of reducing the number of entries. Mined-out panels are sealed off to prevent spontaneous combustion through the removal of oxygen.

The main levels, with extensive entry systems, are used for coal, supply, and personnel haulage and for ventilation. They are often spaced with little regard to the position of the coal seams, because the deposits are reached selectively through other means. In the past, 165- or 330-ft intervals were selected while increasing ground pressures and development and maintenance costs forced an increase of this distance to 660 to 990 ft. The temperatures at greater depth increase substantially, requiring large volumes of air for cooling. As a result, entry cross sections at these levels have to be increased.

1. *Main Level Entry Development*

Most of the main entries are equipped with very durable and expensive support, usually consisting of heavy steel arches, masonary, or preformed concrete blocks with layers of wood between to provide a limited amount of flexibility and yielding.

Three different methods of development or excavation are used. The conventional method is to drill, blast, shoot, and load. Drill jumbos, throw loaders, or mucking machines and support erection equipment are utilized in these operations.

The second method is to employ road heading machines (Figs. 33 and 34) for excavation and support installation. This crawler-mounted machine is equipped with an articulated boom that excavates the material with its cutting head by removing small sections at a time.

The third method utilizes a full-cut tunneling machine which excavates the entire face with rotating roller-type bits. This type of equipment requires

FIG. 33 Road heading machine. [Courtesy of Paurat.]

substantial and costly preparations and is utilized only for large development projects.

Both means of mechanical excavation, especially the full-cut tunneling machine, allow very rapid advance. This less shock-intensive method of rock removal also reduces the problem of excess material, breaking out beyond the desired cross section, as experienced in blasting operations.

2. *Sublevel Development*

Since the large spacing between the main levels allows only a limited number of seams to be reached, additional means must be provided. Subshafts (Winze) connecting the main levels in coal-rich areas, sublevels developed from these shafts, and rock slopes are used to selectively open up additional mine sections.

FIG. 34 Voest-Alpine road heading machine. [Courtesy of Voest-Alpine.]

C. Underground Haulage

Belt conveyors, monorail systems, ski lifts, and mine cars are used to haul coal, material, and workers. Coal is hauled to the subshaft, where it is discharged into a bin positioned inside the subshaft. These bins, consisting of a metal cylinder with spiral inner sections, usually end in a high-capacity loading station at the main level. Here the coal is loaded into mine cars of between 2 and 5 metric tons capacity.

The mine cars are small enough to allow several tracks in each entry. This permits unhindered flow of empty and full cars. Complex control systems are often used to monitor and direct the position and movement of trains.

In many cases, coal cars are hoisted to the surface and discharged by rotary dumps. However, underground belt haulage has captured part of the market. This has the advantage of low space requirements, which is important at greater depth, and high ground pressures. The belt systems discharge into large bins at the shaft, where the coal is hoisted by skips in an often fully automated operation.

D. Rock Mechanics

The ground control problems experienced as a result of the overburden thickness and multiseam operation influence not only the size and shape of entries and entry supports but also the mining sequence, the operating method, the panel design, and the extent to which a seam is mined. The mining sequence is usually from top to bottom. No pillars are left between panels. Computer models and extensive underground measurements are often used to predict the impact and interaction of the various extraction operations.

E. Mining Methods

1. *Longwall Mining*

Considering the cost of development, the ground control problems, the limited size of the reserves, and the problems and costs caused by subsidence, the longwall mining method is crucial to economical and safe extraction. The longwall method minimizes the number of access entries, allows complete extraction under severe conditions, results in an even and predictable subsidence, and is the safest method available.

2. *Longwall Entry Development*

The longwall entries are developed with high self-supporting ability. Rock has to be removed, in most cases, to achieve the desired width/height ratio. Many of these panel entries are still developed by drilling, shooting, and loading. However, smaller mechanical excavating equipment, capable of cutting rock as well as coal, is being rapidly introduced.

The smaller and more flexible road heading machines are utilized in a number of operations. They can excavate coal and rock selectively, require a reasonable amount of preparation and assembly time, and can follow the often-sharp turns in panel entries. Contrary to U.S. operations, entry development is usually not a profitable process. The rock material that has to be removed and the heavy support material required make it a rather costly part of the mining operation.

3. *Backfilling*

Many coal deposits are located under highly populated areas. Extensive surface buildup, large structures, rivers, and canals force companies to use every means at their disposal to reduce subsidence.

Pneumatic stowing or hydraulic backfilling are used to achieve this reduction. Refuse material from the cleaning plant is hauled back into the mine and usually deposited in the mined-out area directly behind the longwall face supports. Hydraulic backfilling is done by utilizing gravity to wash the material into the gob area. Pneumatic stowing is more generally applicable. A properly operating pneumatic backfilling operation can reduce subsidence up to about 50%.

4. *Ventilation*

Ventilation systems in a multilevel, multiseam operation are very complex and have to be closely monitored to detect air currents short-cutting through fractured strata sections.

The system consists of one or several exhaust shafts with large fans. The intake air usually enters through hoisting shafts to the deepest point in the mine. From here it is channeled upward through all mining sections to the upper main level. Here, in many cases, the currents are united again and directly exhausted to the main fan. The air increases in temperature as it moves through the mine, providing some aid to the fan through natural uplift.

The high rock temperatures in deep mines can substantially increase air temperatures. Some European countries have adopted a maximum-temperature regulation. If the maximum ambient temperature in a section exceeds the set limit, the working time has to be reduced. The equipment industry has developed large air-conditioning units to combat this problem.

F. Underground Efficiency

The price of coal in Europe is in many cases two to five times that of the world market or U.S. price. A number of factors contribute to this. On the face operation of a retreat mining section, between 10 and 15 people are employed on U.S. longwalls. A European operation requires about the same amount of people for the basic mining operation. However, the single-entry method requires a packwall, which in turn requires a crew of between three and six workers. Entry maintenance often requires several additional crews. Depending on the severity of the convergence problem, a dozen workers might have to be added.

If stowing is practiced, another crew of up to five or six is necessary. These additional operations result in considerable supply requirements, which in turn means more haulage equipment and more people.

In spite of the fact that the basic face production is easily comparable or in many cases better than U.S. figures, overall production rates are rather low, 2.5 to 5 metric tons per worker-shift. U.S. underground production per shift is around 8 metric tons.

G. Health and Safety

Health and safety regulations are similar to U.S. regulations. Often, however, regulations are more detailed and require tighter control of the underground environment with regard to dust, gas monitoring, and equipment safety.

Multiseam operations frequently lead to a fracturing of large strata sections. This, in turn, allows air to enter regions of unmined coal. Spontaneous combustion then produces the highly poisonous carbon monoxide gas, endangering underground crews. To detect this danger, a system of monitoring devices is positioned in the exhaust airways of mine sections. To combat spontaneous combustion, sections are sealed, flooded with water, or flushed with nitrogen. Underground workers are equipped with carbon monoxide filters. Mine rescue teams with oxygen breathing apparatus are sometimes employed to complete sealing work.

The most common gas danger, methane, is controlled through increased air volumes. Prebleeding of sections through boreholes is practiced with increasing frequency.

The danger of a coal dust explosion from methane ignition is the most-feared disaster. Several methods are employed to reduce this danger. The first and basic method is to mix limestone dust with coal dust to convert it into a noncombustible mixture. The second method is to distribute a highly hygroscopic salt, which absorbs moisture from the air and converts coal dust into mud, which eventually hardens into a semisolid substance. Mining sections are individually protected by dust or water barriers. A certain amount of limestone dust or water in nonflammable plastic containers is placed on supports located in the upper portion of an entry cross section. Should a methane explosion occur, spreading will be prevented because the barrier supports are constructed for collapse under the force of the explosion. The falling rock dust or water will then cool the explosion flame.

Dust as a danger to the human respiratory system is handled in several ways. The first means is to apply spray water at the point of origin. A second relatively frequent method is water infusion at coal faces. A third method, developed in the recent past, is redesign of mining machines to produce a coarser material and therefore reduce the creation or production of dust. Additionally, respirators are worn on a voluntary basis.

In the past, shaft-sinking and entry development operations have had the highest concentration of dust known to cause silicosis. Today, these areas are under control through the use of proper ventilation schemes, wet or vacuum drilling equipment, air filtering systems, and plenty of spray water. This is all possible because the areas are relatively small and air volumes are of manageable proportions.

Coal faces with large amounts of air and dust and fast-moving mining equipment are far more difficult to control. Since modern technology has not yet supplied the means to sufficiently control or reduce airborne dust on production faces, monitoring of all underground employees is practiced. Regular checkups are used to find signs of silicosis and to monitor its progress.

H. Trends

Future improvements in European mining are aimed toward reducing the labor-intensive nature of the mining operation. Another major effort is being made in the area of increasing equipment reliability.

Several means are used to reach the desired goals. To reduce the labor intensiveness of coal mining, a change from many small-production sections to a few high-production sections is being made. This reduces the number of people but requires higher investments for more sophisticated and higher-capacity equipment.

The second major means to reduce labor requirements is mechanization and automation. Progress is being made toward automation of the face operation, for example advancement of the support and control of the cutting horizon of the coal-getting equipment. However, many experts agree that presently available technology does not provide the means to fully replace human beings in the difficult and rapidly changing underground environment of a face operation.

4 Surface Mining of Coal

GERARD C. GAMBS and MYLES A. WALSH / Ford, Bacon and Davis, Inc., New York, New York

I. INTRODUCTION

A. Summary

Surface mining excavates the earth and rock occurring above the coal seam and places this overburden to one side as spoil. The coal thus exposed is loaded into trucks and hauled away. The spoil surface is then graded to a desirable configuration or to a contour suiting local environmental and land-use requirements, and planted with useful grasses, shrubs, young trees, or other suitable vegetation. The mined lands are reclaimed and restored, as part of the overall mining plan, to productive uses such as residential and commercial building sites, pasture, farmland, forest, or recreation. Much land thus reclaimed and vegetated after surface mining becomes more useful to humankind than prior to mining (Figs. 1 and 2).

A brief tabulation of some of the estimated additional costs that the coal industry faces as a result of recent legislation and regulations show that the coal is rapidly being priced out of the marketplace. Examples of these new additional costs are shown in Table 1.

B. Importance of Surface Coal Mining in the United States

About 60% of the coal mined in the United States is currently produced by surface mining from seams found relatively close to the earth's surface. Advances in surface-mining technology and the development of larger-capacity

FIG. 1 Example of reclamation: an Illinois wheatfield that was once mined land. [Courtesy of National Coal Association.]

FIG. 2 Example of reclamation: a Kentucky mountainside mined and reclaimed for use as pasture. [Courtesy of Ford, Bacon & Davis, Inc.]

TABLE 1 Estimated Additional Cost That the Coal Industry Faces Due to Recent Legislation and Regulations

	Additional cost per ton of coal (1978 dollars)
1. Surface Mine Regulation	\$ 6.52-\$17.17
2. Hazardous Materials Act	\$ 5.00-\$10.00
3. Stack-gas scrubbing to remove 90% of SO_2 with high-sulfur coals	
Capital	\$15.00-\$18.00
Operating	15.00- 30.00
Subtotal stack-gas scrubbing	\$30.00-\$48.00
Total additional cost for items 1-3	\$41.52-\$75.17

equipment have made possible the production of increasing amounts of the nation's coal reserve by surface mining. Approximately 141 billion tons of coal, or about one-third of the demonstrated coal reserve base of the United States, are potentially minable by surface mining [1].

Productivity of workers at surface coal mines in the United States is about three times greater than that of workers at underground mines, making surface mines substantially less labor-intensive. Surface mining also has the advantages of making better use of the available coal resource; in actual practice surface mining recovers about 85% of the coal area being mined, in contrast to about 60% for underground mining.

Surface mines are normally developed in less time and at less capital cost than are underground mines. The lead time necessary for establishing medium- to large-size surface coal mines is approximately 3 to 6 years. However, in Wyoming and a few other states it takes 8 to 10 years to acquire permits from state and federal agencies needed to develop a surface coal mine. Equivalent underground mines require periods of 5 to 10 years and sometimes more to bring them to full production. Similarly, the capital cost of surface mine development is less: about \$20 to \$40 per annual ton of salable coal production capacity for surface coal mines compared with about double these sums for medium- to large-size underground mines.

The development of any surface mine should be preceded by a technical and economic evaluation of all the contributory factors, such as mine location; environmental impact of surface mining on the neighborhood; probability of acceptance of surface mining by the local community and concerned citizen groups; geology, hydrology, and terrain; overburden type; recoverable coal reserve potential; coal selling price and market; sulfur content; overall quality and long-term market potential for this quality; optimum mine production rate; local infrastructure; public highway network; railroads; waterways; operating expenses of nearby mines and experience with unions; local skills; and the availability of equipment and financing for mine development [2].

II. SURFACE-MINING METHODS

A. Mine Planning and Design [3]

Successful mine planning requires an abundance of information about the coal property. Trenching and core drilling are necessary for data about the coal seams, overburden, and geology. Preliminary exploration drilling is followed by drilling of additional holes to delineate the coal seams and overburden to obtain sufficient information about the coal property for determining its optimum production rate. Drill data are plotted on topographic maps so that the coal seams relate to their overburden and terrain. Computer storage of data at an early stage of exploration and development is desirable, as it facilitates selection of the optimum mine design by computer trial runs of various design possibilities and their alternatives.

1. *Sampling and Testing*

Loan agencies financing mines, buyers of coal, and others interested in mining projects frequently wish to examine drill cores. It is therefore prudent and good technical practice to split all cores in half longitudinally, sending only one-half to the laboratory for analysis, preserving the remainder to show visitors this physical evidence of the coal encountered by drilling. Since surface mining requires handling of the overburden, drill cores of the overburden are also used as samples for chemical analysis and for mechanical testing for predicting the stability of the walls of the pit during and after mining. Sometimes certain strata of the overburden, such as the shale parting between coal seams and the strata underneath, produce acid when disturbed by mining. This environmental hazard, if detected early by laboratory analysis of drill cores, can be avoided by proper mine design. Test pits provide bulk samples of coal, overburden, and shale or fire clay located directly underneath the coal seams. A few surface coal mines are able to recover clays economically for brickmaking. One major manufacturer of bricks in the United States produces high-quality clay from surface coal mines for making refactory bricks for furnaces. At some coal-mining pits it is possible to recover sand, gravel, and building stone from the overburden. Generally, however, most coal people overlook or neglect the by-product possibilities of surface coal mining when considering mine design.

2. *Mapping*

Satellite imagery and aerial photography of coal properties are useful at the earliest stages of exploration and preliminary mine planning. Aerial photographs taken in color at the start of exploration and mine development provide an excellent legal record of the actual environmental conditions, vegetation, drainage, and the general overall appearance of the surface prior to mining.

Maps are essential for mine planning, for showing property boundaries, and for mine-operating purposes. Small mines frequently use government topographic maps enlarged photographically to the scale of 1 in. equals 400 ft. These are particularly useful for application for mining permits. Larger operations utilize aerial photogrammetry for mapping. Property boundaries, sites of drill holes, trenches, test pits, access roads, and all human-made and natural surface features are photographed using color, infrared, or black-and-white aerial photography. Photogrammetric data thus obtained, together with electronic or other surveying, are used for stereoplotting of maps to the scale needed for mine planning and operating purposes. Computer drafting applied to the photogrammetric data and drill logs produces necessary plotted profiles and cross sections, stratigraphic sections, and other engineering refinements. The digitizing of photogrammetric data, together with drill log and test pit information stored on punchboards or tapes, facilitates immensely the designing of the optimum mine plan for a coal property. Various mine designs and mining sequences and their alternatives are tested by trial runs on the computer and evaluated for technical and economic feasibility.

3. *Product Engineering*

Consumers of coal generally prefer and are willing to pay a premium for a product that is consistently of good quality. Surface mining produces raw coal usually variable in quality because of waste-rock contamination in the pit and objectionable sulfur and other minerals in the coal bed and associated strata. This raw coal quality problem should be recognized at the earliest possible stage of product engineering, as possibilities for solution are generally limited to the following:

1. Negotiating coal sales contracts that pay well for raw coal of relatively consistent quality (not a likely possibility).
2. Mining plan, pit design, and equipment that produce raw coal of relatively consistent quality meeting the specifications of coal sales contracts paying premiums and penalties based upon quality (possible for some good-quality coal seams mined by skillful operators, as practiced by many family-owned small mines in Appalachia).
3. Mine-mouth coal produced primarily for a utility designed specifically to burn raw coal from the mine (mine and utility planned together, as is sometimes done for large mines associated with utilities).
4. Coal preparation plant facilities designed to clean raw coal from surface coal mines. The clean coal produced is usually of consistent quality, but its quantity of marketable product is about 20 to 40% less than the raw coal received from the mines. Mining costs per ton of clean coal are therefore increased accordingly. Furthermore, the capital costs of a modern coal preparation plant and facilities designed to treat 1,000,000 tons of raw coal per year are $6 to $20 million, to which must be added the cost of operating the plant and the cost of disposing of the large quantities of waste products of coal preparation.

Large surface mines usually include in their design facilities for a preparation plant that beneficiates the coal, thereby assuring long-term coal sales contracts (Fig. 3). These contracts often assist negotiations for financing the mine and the coal preparation plant.

Most small mines with limited capital operate without coal preparation facilities other than a small tipple that crushes and screens the coal to produce standard sizes of coal lumps and screenings. Some small mines are able to ship raw coal to another plant for treatment.

FIG. 3 Modern coal preparation plant in West Virginia. In such plants, coal is crushed, washed, sized, and dried before being shipped to consumers. [Courtesy of the Consolidation Coal Co.]

Site selection of the tipple, washer, or the more sophisticated coal beneficiation plant is governed largely by the following factors:

1. Recent regulations in the United States require that the plant be self-contained: that no liquids or wastes flow from the plant area to the surroundings. Air quality regulations are also severe.
2. Plant property controlled by the mining company, with ample space and terrain suitable for stowing plant wastes for the life of the mine.
3. Accessibility to market at reasonable cost by railroad, barge, highway, or belt conveyor, or any combination of these transportation systems.
4. Central location in relation to the coal properties so that the transportation of raw coal to the plant is minimal and mostly downhill, if feasible.

4. *Machinery Use*

Surface-mining machinery and equipment range from bulldozers, front-end bucket loaders, scrapers, and trucks frequently used on highway construction and similar earth-moving projects, to gigantic power shovels, draglines, and bucket wheel excavators. The biggest machines are mostly used in the relatively flat terrain of the Midwest (Midwest Basin) and West (Mountain States) of the United States, where coal seams are nearly horizontal and may extend over large areas at relatively shallow depth. Under these favorable natural conditions for surface mining, large-scale mines often utilize mammoth machines that operate in the same area for decades. The big shovels and draglines excavate the overburden to expose the coal seam so that it can be loaded into trucks by smaller shovels or front-end loaders. These large-

FIG. 4 A surface-mining shovel, as tall as a 20-story building, takes a 140-yd^3 bite of earth and rock—about 210 tons—in a coal mine in Illinois. [Courtesy of the Bucyrus Erie Co.]

scale surface mines provide and maintain their own private highways for hauling coal from the pits in trucks of up to 250-ton capacity and larger, too big for use on public roads.

The gigantic stripping shovels excavate overburden from within the pit. They use the top of coal seams as exposed by digging as a floor or roadway, over which they move ponderously on crawler treads for short distances as excavations progress (Fig. 4). The shovel bucket digs overburden along the highwall or solid side of the pit, pivots with loaded bucket elevated and extended, dumps its load of spoil on the other side of the trench where coal has already been extracted or is nonexistent, and reverses pivot direction to return the bucket to dig again. This digging cycle requires about 1 min, during which the largest shovel manufactured excavates 180 yd^3 of overburden. This rate of overburden removal exposes an average of about 5 to 11 tons of coal per minute, depending upon the machine availability and the stripping ratio of the coal property.

Draglines excavate from a leveled surface above the pit perched almost precariously at the crest of the highwall. The machine is stable, however, because its base is a huge tub provided with a "walking" mechanism moving along the pit edge. The dragline's boom, up to 360 ft in length, suspends, by means of wire ropes, a three-sided, open-ended bucket which is dropped into the overburden area being excavated and pulled by dragging toward the machine. With the filled bucket suspended from its boom, the dragline rotates on its axis toward the pit until the boom reaches the other side, where the bucket discharges by tipping. By swinging in the reverse direction, the empty bucket is returned to the solid side of the pit to dig again (Fig. 5).

The complete digging cycle for large draglines is slightly less than that for large shovels, about 50 sec. Draglines, however, can have larger-capacity buckets; the largest has a bucket capacity of 220 yd^3. The productivity (availability) of large draglines averages about 80% of scheduled hours. At most surface coal mines, the large shovels and draglines are scheduled to operate year-round about 92% of scheduled hours.

Shovels and draglines are usually preceded along the solid wall of the pit by vertical or horizontal drills, which bore holes for explosives to break, heave, and loosen the overburden for easier digging.

Bucket wheel excavators are sometimes used where the overburden consists of large volumes of uniformly soft, granular material and the solid wall of the pit is a long mining face. A large rotating wheel equipped with buckets and mounted on a long boom digs the solid wall of the pit, dislodging at U.S. mines 30 to 60 yd^3/min, discharging spoil on built-in conveyor belts that transport across the pit to spoil piles about 450 ft distant. The bucket wheel excavator, like a huge shovel, operates from within the pit, and all parts, including the machinery and superstructure, are mounted on a platform that can rotate 360°. In the United States there are currently only about 11 large-size bucket wheel excavators operating at surface coal mines [4].

The overburden depth of mines in which bucket wheel excavators can operate is restricted to about 100 ft. Excavators are sometimes used, as in Illinois, to work in tandem with stripping shovels in the same pit. The bucket wheel excavator removes the top soft-earth portion of overburden in highwall mines and the lower, hard rock, is excavated by stipping shovels (Fig. 6).

Monster-size bucket wheel excavators popular in Europe are considerably larger, with a capacity up to three times more than those used in the United

FIG. 5 A surface-mining walking dragline swings an 85-yd^3 bucket on the end of a 275-ft boom in western Kentucky. [Courtesy of the Marion Power Shovel Co.]

FIG. 6 A bucket wheel excavator, 12 stories tall and almost 450 ft long, removes soft overburden in Illinois, depositing this on top of coarser spoil across the pit, so that the mined lands can be regraded with bulldozers and easily replanted. [Courtesy of the General Dynamics Corp.]

States. This is principally because of the favorable soft-earth overburden conditions of European lignite and brown coal mines [5].

Surface coal mining in the mountainous Appalachian area of the eastern United States generally utilizes much smaller equipment than that described above. Steep slopes and rugged terrain necessitate the use of modest-size mobile machines for overburden removal, such as bulldozers, front-end bucket loaders, scrapers, trucks, and relatively small shovels and draglines. The capital cost of mine development using these equipment types is relatively low; this is conducive to small-scale surface mining such as that popular in eastern Kentucky, West Virginia, and Pennsylvania.

In times of fuel crisis, small-scale surface coal mines are a special asset to the United States because they can be developed in the Appalachian area within 3 to 6 months provided that mining permits are readily available and the sales price of coal is attractive. Historically, the Appalachian area has been quick to increase production when coal prices rise substantially.

During normal times, it requires 6 months to 2 years for development of small surface coal mines utilizing bulldozers and equivalent small-scale equipment. These mines generally use trucking contractors for coal haulage in

25-ton-capacity trucks furnished by the contractor, who utilizes public highways for hauling coal to market.

Equipment for surface mining is directly related to the type of mining method used for exposing the coal seam.

B. Types of Surface Coal Mining [6-8]

The types of surface mining currently in effect are principally area mining, which generally uses equipment of medium size to the largest types available, and contour mining, which generally uses smaller, more-mobile equipment. Variations of area mining include mountaintop removal, multiple-seam, and pitching-seam mining.

Auger mining is often associated with contour mining. Each of these surface mining methods involves the removal of the overburden strata overlying the coal seam for the purpose of exposing and extracting the coal. The basic concept of surface mining is to expose the coal seam by removal of its overburden as long as this is economically feasible.

Open-pit mining of coal is a form of area mining that has adopted some features of large quarries and open-pit metal mines. The open-pit method is becoming increasingly important in surface coal mines in the western United States and is herein treated as a separate type of surface coal mining.

1. Area Mining

a. Flatlands and Gently Rolling Terrain In the Midwest (Midwest Basin) and West (Mountain States) of the United States many of the coal seams and their overburden strata lie nearly horizontal beneath flat to gently rolling topography. These seams, where not too deep, can often be recovered economically by area mining.

Area mining usually begins at or near the coal outcrop where overburden is thin, and continues toward thicker overburden areas, often to the digging limit of the excavating machines. These machines for overburden removal are usually large stripping shovels or draglines, which dig long, parallel trenches or pits across the coal property. Overburden excavated from the solid highwall of a new trench is cast into the previous trench from which coal has already been extracted. By this process, area mining progresses from trench to trench, much as a farmer plows from furrow to furrow [7].

Coal exposed during these excavations is recovered by relatively small equipment, usually a conventional loading shovel or front-end bucket loader, which heaps coal into trucks for hauling to a nearby coal preparation plant, or to a unit train-loading silo or barge-loading facility for shipping to market.

Area mining of the foregoing type is well developed in the Midwest (Indiana, Illinois, and western Kentucky) at surface mines producing 1,000,000 to 3,000,000 tons of coal annually—more at some mines.

b. Mountaintop Removal Area mining is also practiced on a smaller scale on mountaintops of Appalachia, where many of the ridges, knobs, and tops of mountains or foothills are underlain by coal seams. These seams, sometimes up to seven in number, are exposed by systemically removing their overburden and parting between seams.

FIG. 7 Excavating coal seams at mountaintop level at this West Virginia surface coal mine creates large areas of level ground suitable for crops, housing, or community development—a rarity on the steep slopes of Appalachia. [Courtesy of the National Coal Association and the Cannelton Coal Co.]

(1) Conventional Mountaintop Removal. Area mining of mountaintops usually begins with excavation of overburden along the coal outcrop, hauling the spoil in dump trucks or scrapers to the head of a hollow or to a ridge having no coal. Since mountaintops in Appalachia are often long ridges, the first cut along the outcrop will be parallel to the long axis of the ridge. A series of cuts parallel to the first excavation are made in sequence, placing the spoil mostly on the solid bench of the previous cut and the excess as head-of-hollow or ridge fill. Frequently, but depending upon local geology and favorable topography, parallel cuts are continued into the mountainside until the entire mountaintop ridge is removed. Reclamation of the mined area usually flattens the peak of the mountain, making its surface more accessible for pasture, building sites, or other use (Fig. 7).

(2) Cross-Ridge Mountaintop Removal. Cross-ridge mountaintop removal mines in a direction perpendicular to the long axis of the ridge of the mountaintop. By excavating a series of cuts or trenches of overburden that repeatedly cross the ridge, the peak of the mountain is leveled and reclamation develops a flat area attractive for dwellings and many other uses [9].

Area mining by mountaintop removal is popular in eastern Kentucky, where about 75% of all surface-mined coal is produced from mountaintops. Most mountaintop removal mines are of medium size, producing 300,000 to 1,000,000 tons of coal annually; a few mines, however, exceed 1,000,000 tons.

FIG. 8 Example of contour mining: exposing coal at its outcrop on a mountainside in Appalachia. In the background, a front-end bucket loader excavates the exposed coal seam into waiting trucks. The spoil in the foreground will be pushed back into the pit, graded, and planted to restore the mountainside.

2. *Contour Mining*

Contour mining is used principally at the coal outcrop on hilly to steep mountainsides of Appalachia. The often-flat-lying continuous coal beds occurring in mountainous Appalachia have led naturally to mining technology that follows the coal outcrop along the mountainside, taking a series of cuts as deep as economics will allow. Mining operations along a single seam may employ bulldozers, front-end bucket loaders, scrapers, and sometimes small power shovels and draglines. Overburden removal starts at the coal outcrop and follows this continuously along the mountainside. The excavation produces a long, narrow bench with highwall extending along the side of the mountain, appearing as a contour line for horizontal coal seams, from which contour mining derives its name [7].

Excess spoil from excavations of overburden is sometimes stowed permanently at the head of a hollow on the mountainside, or temporarily set aside on the mined bench for use later as backfill against the highwall. Environmental regulations for mountain slopes greater than 20° require reclamation that obscures the highwall created during mining.

The coal exposed by overburden removal is loaded and hauled away in dump trucks as the excavation of bench and highwall progresses. Successive cuts are made into the mountainside, increasing the highwall height and the volume of spoil handled until an economic limit is reached or until the bulldozers, front-end loaders, trucks, and small shovels used can no longer excavate efficiently because of physical limitations or reclamation requirements (Fig. 8).

The state of the art of contour mining has vastly improved in recent years, so that after reclamation, minimal scars of mining remain on the hillside. This is due mostly to improved systems of overburden removal and spoil placement to bury highwalls progressively as mining advances along the hillside, such as overburden removal by block cut and by haulback.

During the past decade contour-mining technology has progressively improved from the conventional method, evolving roughly in the following order:

1. Conventional contour mining
2. Head-of-hollow fill, storage of excess spoil, reducing spoil placement on the outslope, and avoiding the possibility of spoil slides
3. Block cut contour mining with various modifications
4. Haulback—controlled placement of spoil

These methods, however, increase contour mining costs 20 to 30% over the former conventional contour-mining method now outlawed on steep slopes after many years use because it perches spoil sometimes precariously on mountainsides and leaves exposed highwalls. Recent federal regulations prohibit many former practices of conventional contour mining; however, improved methods of contour mining have evolved.

Contour mining is generally the business of the small surface-mine operator, usually a family-owned business of 5 to 10 workers. Production is relatively small in scale, often producing less than 100,000 tons of coal annually per mine. Productivity per worker-day is high, however, averaging about 40 to 50% more than the average for all surface mines of United States. This is because of operating skills, high wages or profit sharing, and cooperation among workers of family-run operations. Contour mines are more numerous than any other surface mine type in the United States; however, as a group, the produce the smallest total amount of coal.

3. *Auger Mining*

Auger mining is sometimes done before backfilling the final highwall of contour mining. The coal auger extracts coal by boring underneath the final highwall, operating on the same principle as an auger used to bore a hole in wood.

The augering process is mostly a salvage operation, recovering additional coal tonnage when overburden removal of the highwall reaches its economic limit. About 25 to 35% of the coal in the seam extending under the final highwall is recovered for the penetration distance of the coal auger, which averages about 150 ft under the highwall.

Since only three to six workers are required to operate an auger, productivity per worker-day is high, and auger coal production costs are correspondingly low. The success of auger mining, however, depends largely upon the skill of the mine operator and his synchronization of augering with other surface-mining activities at the highwall.

FIG. 9 Example of open-pit coal mining: thick-seam coal at this mine in the West is loaded by shovel into a truck. In the background, a bulldozer and scraper remove overburden to expose more coal for mining.

Auger mining supplies about 15,000,000 tons of coal per year in the United States, only about 4% of the surface-mined coal production. Coal reserves in the United States recoverable by augering by the present state of the art are about 5 billion tons. The future of augering is clouded, however, because recent surface-mining regulations do not allow the final highwall to be exposed for long periods, leaving little time for augering in the pit before reclamation activities bury the coal seam with spoil [8].

4. Open-Pit Coal Mining

Some Western states of the United States have coal seams up to 100 ft and more in thickness, the mining of which is done much like a rock quarry [10].

The relatively thin overburden is removed and hauled away in trucks or scrapers, exposing a large area of coal for open-pit operations. It is the great thickness of the coal seam or seams that makes open-pit coal mining feasible (Fig. 9).

At the thick coal seams of the Powder River Basin of Wyoming, some individual open-pit mines produce up to 24,000,000 tons of coal annually. The favorable topography and geology of this area make it feasible to create a large open pit. Power shovels strip the overburden and also load the coal, serving a dual function. Trucks are used to haul overburden and coal. This shovel-truck system of mining develops an open pit without hazardous highwalls, having a series of 35- to 40-ft-high benches. The overburden at each bench is loaded by shovel into trucks and dumped as fill according to a carefully designed reclamation plan, so that toxic spoil becomes buried at the bottom of mined-out areas. The use of trucks to deliver spoil to specific areas eliminates rehandling of materials or leveling of spoil banks. Exposing a large expanse of coal in the open pit gives flexibility to the mining process, making it possible to excavate, stockpile, and blend coal to produce uniform quality. Also, the large open pit makes conditions less hazardous for workers and equipment [11].

III. PRODUCTION STATISTICS AND FORECAST—SURFACE MINING OF COAL

A. Increase in Surface Mining, 1920-1979

The surface mining of coal in the United States has greatly increased in magnitude from the early 1900s; annual surface coal production was 9 million tons in 1920, 1.6% of the total coal production of 569 million tons in that year. This may be compared to the present production, with surface mining accounting for 469 million tons, over 60% of the total 770 million tons of coal mined in 1979, based on preliminary production data for 1979.

B. Reasons for Increased Coal Production by Surface Mining

The trend toward surface mining of coal built up slowly during the period 1930-1969 but has been accelerating rapidly since then. This was brought about by three major factors:

1. The rapid decline in productivity in underground mines since 1969
2. The rapid increase in surface mining in the Western states
3. The lower cost of mining coal by surface mining

The productivity decline in underground mines caused the peak tons per worker-day, nearly 16 in 1969, to drop to 8.3 tons per workers-day in 1978, a decline of nearly 50%. Actual tons of underground coal mined during this period (1969-1978) declined from 347 million to 243 million. At the same time, surface mining of coal increased rapidly and more than doubled, from 197 million tons in 1969 to 422 million tons in 1978. Total coal production during this 9-year period increased by only 19%, from 560 million in 1969 to 665 million tons in 1978, but increased by 38% to 770 million tons in 1979.

Even more interesting is that this increase in surface mining has been occurring principally in the Western states, and the current projection by the U.S. Department of Energy (DOE) is that this tend will continue.

C. Carter Energy Plan and Surface-Mined Coal

The Carter National Energy Plan of April 29, 1977, projected that of a total production of 1265 million tons of coal to be produced in 1985, 638 million would be from the Western and Midwestern states and 627 million tons would come from the Appalachian states. In a revised DOE estimate of June 1978, the estimated tonnage of surface mined coal by 1985 is 541 to 686 million tons, 55 to 58% of the total of 991 to 1188 billion tons.

The June 1978 report, "Federal Coal Leasing and 1985 and 1990 Regional Coal Production Forecasts," by the Leasing Policy Development Office of the Department of Energy projects coal production by type of mining and by states. The DOE projections of U.S. coal production in 1985 and 1990 show the following trend:

1. Total coal production is to increase from 688 million tons in 1977 to a range of 991 to 1188 million tons by 1985 and a range of 1114 to 1856 million tons by 1990.
2. Eastern surface-mined coal is to decrease from a level of 264 million tons in 1977 to about 196 million tons in 1985 and to about 146 million tons in 1990.
3. Western surface-mined coal is to increase from a level of 153 million tons in 1977 to a range of 347 to 487 million tons by 1985 and to a range of 419 to 1008 million tons by 1990.
4. Total surface-mined coal in the United States is to increase from a level of 417 million tons in 1977 to a range of 541 to 686 million tons in 1985 and to a range of 560 to 1154 million tons in 1990. The medium case in the 2 years is 634 in 1985 and 865 million tons in 1990.
5. Total U.S. deep-mined coal is to increase from a level of 271 million tons in 1977 to a range of 449 to 503 million by 1985 and to a range of 554 to 702 million tons by 1990.

It should be noted that the total U.S. underground coal production level has remained in the range 271 to 366 million tons per year during the 25-year period from 1953 to 1977. In 1977, it was 271 million tons, having been in a steady decline since 1967, when it was at a level of 349 million tons.

The DOE medium projections for surface mining of coal through 1990 by regions are shown in Table 2.

The magnitude of the switch to surface-mined coal is seen in Table 3.

D. Department of Energy Forecast for Surface-Mined Coal

The DOE submitted a report to the Interior Department on "Federal Coal Leasing and 1985 and 1990 Regional Coal Production Forecasts" in June 1978. The National Coal Association summarized the report as follows:

Case 1. Low Western coal production, resulting in low national coal consumption, and a shift of coal production to the East

Case 2. Medium Western coal production, resulting in an intermediate forecast

Case 3. High Western coal production, resulting in high national coal consumption and a major regional production shift to the West

The DOE's national and regional coal production forecasts for each of these three cases for 1985 and 1990 are shown in Table 4 for all coal produced in the United States by surface and underground mines.

TABLE 2 DOE Projected Surface Mining of Coal—Medium Case (Millions of Tons)

Region or state	Actual 1977	Projected 1985	Projected 1990
East of Mississippi	264	196	146
West of Mississippi			
Eastern Northern Great Plains	12	22	23
Powder River Basin			
Montana	29	86	205
Wyoming (total)	45	215	320
Colorado	8	17	18
Utah	0	2	2
New Mexico	11	21	56
Central West	14	6	4
Gulf (Texas)	17	58	80
Arizona	11	7	8
Northwest	6	4	3
Subtotal West	153	438	719
Total U.S.	417	634	865

TABLE 3 Surface Mining of Coal in the United States (Millions of Tons)

Year	Total	West	East
1970	264	34	230
1977	417	153	264
1985			
Low	541	347	194
Medium	634	438	196
High	686	487	199
1990			
Low	560	419	141
Medium	865	719	146
High	1154	1008	146

TABLE 4 DOE Coal Production Forecast, June 1978 (Millions of Tons)

Region	1977	1985 Low	1985 Medium	1985 High	1990 Low	1990 Medium	1990 High
Appalachia	390	427	440	455	396	445	479
Midwest	133	183	204	213	264	312	327
Total East	523	610	644	668	660	757	806
Northern Great Plains	86	244	328	374	292	552	800
Rest of West	79	137	145	146	162	212	250
Total West	165	381	473	520	454	764	1050
National total	688	991	1117	1188	1114	1521	1856

The DOE report analyzed Western coal production in detail to determine the significance of their projections for federally owned coal. The report concluded that:

1. The majority of Western coal production is to meet demands of coal consumption in states west of the Mississippi River.
2. Taking into account the identified 1985 and 1990 coal production commitments of the mining industry in effect today, and comparing these with projected production requirements in the years under review, there will be a shortfall of production in the Western states in 1985 of from 31 to 149 million tons.
3. These production shortfalls do not necessarily indicate that a new coal-leasing program is needed. The shortfalls could be met by expansion of existing approved mining operations, development of existing nonproducing leases, expanded development of nonfederal coal, additional leasing, or any combination of these policies.
4. The extent of any new required leasing in many regions will depend on enforcement of diligent development regulations on existing federal leases.

The DOE report also includes a detailed description of the methodology and assumptions used in preparing the forecasts.

Coal ownership patterns are broken down by region, with particular attention devoted to the Wyoming-Montana Powder River Basin. Among the assumptions that drive the various conclusions are the assumption that future production will take place in mining areas of least cost, regardless of ownership; the assumption that shortfalls will be met wherever possible with nonfederal coal; and the assumption that future production will be limited to areas surrounding existing leases.

An attempt is made to assess practical limiting factors, including economies of scale for various mine sizes and the economic penalties associated with the development of nonfederal coal.

All areas are not discussed in equal detail. More detailed analysis is promised. On a regional basis the study concludes with respect to federal Western coal:

Powder River Basin—"Federal coal is significant in all three . . . scenarios, ranging from a low of 50% to 100% of the production to meet the projected shortfalls."

Colorado—Relatively large amounts of reserves exist in contiguous nonfederal blocks, and therefore "additional data is needed to analyze the degree to which federal coal is required."

Eastern Northern Great Plains—High degree of interspersing between federal and nonfederal reserves, mostly lignite, make simultaneous development of both inevitable.

Wyoming (non-Powder River Basin)—Checkerboarding ownership of mineral leases means that major surface mines are "unlikely" to be developable without including federal reserves.

Northwestern New Mexico, Utah—Development will require the use of federal coal because of ownership patterns.

The overall conclusion of the study is that development of Western coal seems necessarily to include federal reserves. These reserves would, of course, have to be under lease at the time of development. As a result, either expanded development of existing leases, or new leases, or (more likely) both will in fact be required to meet the Western production requirements outlined in the study.

As an example of the coal production in one of the Western states, Wyoming's coal production of 44 million tons in 1977 was at one time projected to reach 250 million tons by 1985. That figure has been scaled down to 150 million tons, with most of it being surface-mined coal.

In our opinion, it will be impossible to expand the U.S. coal industry to a level of 1 to 1.3 billion tons by 1985. This fact is slowly being recognized by our leaders in Washington and we will soon start to see lower estimates of our coal production being forecast for 1985. For example, figures of 1000 million tons, including 100 million tons for export by 1985, are now being circulated. But unless an authority begins to understand what even this lower level of production means in terms of the job to be done, even this lower forecast will not be attained. Let us give one example of the magnitude of the job to be done. If the present (1979) coal production level of 770 million tons is to be increased to 1000 million tons by 1985, we would have to increase mine capacity by 230 million tons plus the mine capacity that will be depleted at the rate of about 3% per year for Eastern coal capacity or 15 million tons/year, which is about 90 million tons of capacity in 6 years. The total additional new capacity is, therefore, 320 million tons. If we assume that 300 million tons will be Western coal, this will require 60 new mines of 5 million tons/year capacity in the Western states. The balance of 20 million tons/year could be obtained by developing 5 new 2 million tons/year underground mines and 5 new 2 million tons/year surface mines in the East. This schedule, which calls for constructing 70 new large mines in the next 6 years, to 1985, needs to be compared with the number of new large coal mines that have started producing coal since 1972.

According to the 1978 Keystone Coal Industry Manual, only 44 U.S. coal mines produced 2 million tons/year or more in 1977. Of these, only 6 mines produced more than 5 million tons/year. Only 13 of these mines started production in 1972 or later, and only 4 of the mines produced over 5 million tons/year each in 1977. If we exerted a superhuman effort and removed all

TABLE 5 Annual U.S. Coal Production (Millions of Tons)

Year	3% Annual increase	2% Annual increase	Actual
1975	635	635	648
1976	654	648	679
1977	674	661	691
1978	694	674	658
1979	715	687	770[a]
1980	736	701	
1981	758	715	
1982	781	729	
1983	804	744	
1984	829	759	
1985	853	774	

[a]Preliminary.

TABLE 6 NCA Forecast of U.S. Coal Consumption and Production (Millions of Short Tons)

	1977[a]	1978[b]	% Change 1978/1977	1979[b]	% Change 1979/1978
	Consumption				
Electric utilities	474.8	481	+1.3	510	+6.0
Coking coal	77.4	71	-8.3	75	+5.6
General industry and retail	67.4	68	+0.9	72	+5.9
Total domestic	619.6	620	-1.0	657	+6.0
Canada	17.2	15	-12.8	16	+6.7
Overseas	36.5	25	-31.5	31	+24.0
Total consumption	673.3	660	-2.0	704	+6.7
	Production				
East	522.5	464	-11.2	507	+9.3
West	166.1	182	+9.6	206	+13.2
Total	688.6	646	-6.2	713	+10.4
	Imports				
Coke	1.8	~5	—	~6	—
Steam coal	1.8	~3	—	~4	—

[a]U.S. Department of Energy.

[b]NCA Economics Committee estimates.

Source: Coal Outlook, Dec. 25, 1978.

the roadblocks and obstacles to developing all the new coal mines that we need, we would probably still fall short of this revised and lower forcast by 1000 million tons/year by 1985. Since we see no hope that anyone in Washington understands the problem, we believe that it will be impossible to attain a level of coal production of 1000 million tons by 1985.

If we are able to achieve a 2 to 3% net increase per year from 1979 through 1985, we would reach a production of 774 to 853 million tons by 1985. It is interesting to note that the average increase in coal production during the period 1960-1978 was 2.6% per year. The actual increase for 1976 was 2.6% over the 1975 production, and the 1978 production was lower than 2975 because of the miners' strike in 1978. This is shown in Table 5.

We therefore believe that because of the wishful thinking in Washington and the lack of understanding of the magnitude of the problem, we will probably be producing only 800 to 850 million tons of coal per year by 1985. The only thing that will change this is a sincere recognition by our leaders that they—Congress and the Administration—are the reason why the coal industry cannot produce as much coal as the country needs in order to survive; and having recognized this face, that they—Congress and the Administration—have to pass the necessary legislation that will allow the coal industry to do the job it is capable of.

In December 1978, the economic panel of the National Coal Association issued its forecast for 1979 coal production in the United States. The forecast showed that a record coal production of 713 million tons is expected for 1979. Preliminary figures for 1979 show production at 770 million tons, exceeding the forecast by 57 million tons, or 8%. The production of Western coal, which will be mostly surface-mined coal, will be responsible for the increased production level.

Western coal production in 1979 is expected to increase 13% over the 1978 levels and 24% over the 1977 level.

On the other hand, Eastern production will grow only 9% over the 1978 level and will not reach the 1977 level.

The NCA forecast for 1979 compared with actual 1977 and estimated 1978 production is shown in Table 6.

IV. IMPACT OF SURFACE COAL MINING ON THE ENVIRONMENT

Surface coal mining disturbs at least 153,000 acres each year; however, the industry actually reclaims more land each year than it disturbs. A total of about 1.5 million acres, 0.1% of the area of the United States, has been disturbed, of which about one-third has been reclaimed. Abandoned land mined prior to state reclamation laws will be reclaimed under the Federal Surface Mining Control and Reclamation Act of 1977, funds for which are assured through an assessment on surface-mined coal of 35 cents/ton, 15 cents/ton for underground-mined coal, and 10 cents/ton for lignite.

The area of the United States disturbed by coal mining represents about one-fifth of that potentially minable by surface mining using current technology, containing coal reserves estimated at 27 billion tons, enough to supply the nation for 45 years at the present rate of coal consumption [12].

Approximately 190,000 acres in the West would be disturbed by the year 2000 to satisfy current projections of U.S. coal production, exclusive of proposed coal gasification projects. Even assuming optimistic projections of 41 gasification plants and total coal production in the West of about 1 billion tons of coal annually by the year 2000 (about six times current production for this region), the environmental impact of surface coal mining would be relatively slight. Only about 0.3% of the land would be affected, and 0.2% of the cropland and 0.2% of the grazing animals would be temporarily displaced [13,14].

The surface coal mining industry actually uses a relatively very small area of the United States compared to that used by other activities. For example, some single counties in the West have areas up to 12 million acres. The federal government has set aside about 50 million acres for the military. One-third of the United States—an area approximately as large as everything east of the Mississippi—is made up of public land.

The United States Surface Mining Control and Reclamation Act of 1977 should henceforth substantially reduce the impact of surface mining on the land. This act, for example, requires containment of any silt eroded from mined lands; it also requires all mines to preserve and restore topsoil on land disturbed by mining. The stringency of these requirements contrasts sharply with practices of U.S. agriculture, which uses the most modern agricultural technology, but which loses about 4 to 5 billion tons of topsoil annually. In this nation an estimated 200 million acres of land used for crops have been either totally ruined by soil erosion or have been so severely eroded that the land is only marginally suitable for agriculture. This may be compared with 1.5 million acres disturbed by surface coal mining in the United States to date or less than 0.5% of the acreage ruined by agriculture, taking into account acreage reclaimed by the coal-mining industry [15-17].

Surface mining is currently being done responsibly without damaging the land and water permanently. Most mine operators today utilize the best available technology for restoring mined lands to a condition as good, if not better, than that prior to mining. In addition, some prominent mining companies have been systematically rehabilitating mined lands since World War II. After vegetation, the lands are often used by subsidiaries for commercial agriculture. As an example, AMAX Coal Company, through its subsidiary, Meadowlark Farms, Inc., on land that was farmland before and after mining, produced 1.8 million bushels of grain and raised 1.1 million pounds of live cattle and hogs during 1979 on company-controlled land (Fig. 10) [18].

A. Environmental Impact Statements

A National Environment Policy Act (Nepa) was signed into law (PL 91-910) on January 1, 1970. It obliges each federal agency to prepare an Environmental Impact Statement prior to each major "action" recommendation or report on legislation that may alter significantly the human environment. A broad range of public and private activities are affected by this law, including coal mining.

An Environmental Impact Statement for coal mine development can be very involved, costly, and time consuming. As an example, applications to the government by four private mining companies to develop certain coal resources in the eastern Powder River Coal Basin of Wyoming resulted in a massive five-volume draft environmental statement prepared by an interagency team composed of representatives of seven principal departments and agencies of the

FIG. 10 Example of reclamation: a herd of Charolais cattle in Indiana grazes on reclaimed surface-mined land. [Courtesy of the National Coal Association.]

federal government. Although the total area to be surface-mined by industry was less than 20,000 acres, the government study area contained approximately 5,000,000 acres, involving formal requests for comment on the environmental impact of the proposed mine development from 23 agencies of the federal government, 17 offices of the State of Wyoming, 8 boards of county commissions, and 22 private organizations of concerned citizens' groups and others [10].

This Powder River mine development was delayed abnormally (about 3 years), largely over environmental issues and opposition to mining from preservationist groups. Other major coal-mining projects have not been so fortunate; an example is the proposed Kaiparowits mine-mouth coal power complex for southern Utah, which was finally abandoned by its sponsors after 13 years of endeavor largely because of opposition from concerned citizens' groups and governmental requirements regarding environmental matters, which made the project both too time consuming and economically questionable.

B. Surface Coal Mine Operation Permits, Licenses, and Approvals—Western United States [11]

Surface-mine development requires a whole series of mining permits and approvals from government agencies before mine construction can be undertaken. For example, mine development in the western United States requires the following operation permits, licenses, and approvals:

	Federal	Probable federal agency*
1.	Environmental impact statement (EIS)	EPA, BLM, and others
2.	Federal mining permit	USGS
3.	Railroad right-of-way	BLM
4.	Railroad certificate of public convenience and necessity	ICC
5.	Power-line right-of-way over public lands	BLM
6.	Telephone right-of-way	BLM
7.	Pipeline relocations	USFS
8.	Oil and gas well plugging	USGS
9.	Water discharge permit	EPA
10.	Air pollution permit	EPA
11.	Solid waste disposal permit	EPA
12.	Road construction	BLM
13.	Special land use permit/land exchange-plant sale	USFS
14.	MESA identification	USFS

	State and local	Probable state or local agency
1.	State mining permit	Department of Environmental Quality
2.	State mining license	Department of Environmental Quality
3.	Power-line right-of-way over state lands	Commission of Public Lands
4.	Telephone right-of-way over state lands	Commission of Public Lands
5.	Oil and gas well plugging	Oil and Gas Commission
6.	Water rights—dam, wells, ponds, lakes	State Engineer
7.	Water discharge permit	Department of Environmental Quality
8.	Air quality permit to construct and operate	Department of Environmental Quality
9.	Road construction permits	State and county
10.	Building permits	State and county

*EPA, Environmental Protection Agency; BLM, Bureau of Land Management; USGS, U.S. Geological Survey; ICC, Interstate Commerce Commission; USFS, U.S. Forest Service.

C. Surface Mining Permits—Eastern United States

Surface-mine development in the eastern United States requires extensive permits and approvals of government agencies, requiring the mining industry to submit much information about mining projects before development may begin. For example, the "application for surface mining" in West Virginia has 24 entries, covering such items as identification of the applicant, names and addresses of every officer and partner of the mining organization, reputed owners of all the surface area within 500 ft of the lands to be disturbed by mining, operator's legal right to conduct operations on the land, fees payable to the state, reclamation bond (minimum $10,000), and reclamation tax ($60 per acre). West Virginia supplies a checklist to assist each applicant, requesting "yes-no" answers to 74 questions. Another 46 questions are asked on another checklist for "SMA number," and the applicant is requested to submit a "mining and reclamation plan," covering details of the proposed mining sequence, construction of haulageways, blasting, topsoil handling, placement of overburden, hydrological balance, and mining plans. Information in detail, including maps, is requested about preplans to grade and reclaim mined lands.

D. Environmental Impact of Surface Coal Mining Operations

1. *Exploration*

Much of the damage described below can be eliminated partly or entirely by proper planning to avoid disturbing the environment. Reconnaissance explorations using rugged off-road vehicles create trails and primitive roads that can change wildlife habitat, harm vegetation, and increase erosion. The ruts of access roads for reconnaissance drilling harm some plants and make rough land surfaces that often induce erosion. Drill holes sometimes penetrate several separated water-bearing strata, causing degradation of water resources through comingling of waters of different quality. Exploration along coal outcrops often involves trenching, which can considerably disturb the surroundings. In the past, drilling, trenching, and test pitting could be done without an exploration permit, and disturbed areas were often not substantially reclaimed, particularly if explorations were unsuccessful. Today's modern mine operator is careful to minimize environmental disturbance during exploration activities. Any areas disturbed are reclaimed promptly. When making arrangements with landowners to explore their lands, the proprietors are assured that any environmental disturbance caused by explorations will be temporary only [19].

2. *Mine Development*

The construction phase of mine development can cause considerable environmental disturbance. An influx of construction workers may severely overtax the local community and its facilities and create serious socioeconomic problems in the neighborhood. Site preparation, construction of the tipple or coal preparation plant and coal storage silos, erection of large earth-moving equipment, construction of haulage roads, grading, and ditching can cause environmental changes lasting for the life of the mine. Wildlife patterns are likely to be altered temporarily. Some of the infrastructure constructed during mine development, such as main roads, electric power lines, and railroads, may serve for the duration of mining and longer. Changes made to the environment during mine development, therefore, can be lasting and must be planned carefully as an integral part of the overall mine development plan. This plan will include rehabilitation and restoration of the mined lands and surroundings so that the area is not degraded permanently. Mined lands can become more productive after mining than before mining (Fig. 11).

3. *Mining*

Stripping of the overburden from the coal seam is devastating to the environment at the surface area of the digging site. Vegetation is removed; microflora and microorganisms are destroyed; soil, subsoil, and underlying strata are ruptured and displaced; wildlife must scatter or die; air quality is temporarily degraded; land use changes dramatically; and the surface is exposed to oxidation and mineralogical alteration, weathering, and general topographic changes. This havoc to the environment covers only a limited area and is usually temporary, because the lands are systemically restored.

The environmental impact of surface mining is relatively minor when compared with the violence and ravages of nature, such as hurricanes, forest fires set off by lightning, flash floods or gully busters of the normally arid

FIG. 11 Example of reclamation: alfalfa produced on a land surface mined in the early 1950s in Harrison County, West Virginia. [Courtesy of the National Coal Association.]

regions of the West, or volcanos such as Mount St. Helens which belch hot lava, ashes, and sulfurous and other hazardous fumes on a tremendous scale—all damage that nature eventually repairs.

The principal areas of impact on the environment by surface mining are as follows:

a. Hydrology All new coal-mining operations are operated in such a way as to completely prevent acid mine water discharge. Some older prelaw mining operations may still be discharging gradually decreasing amounts of acid mine water. There has been improvement in the elimination of acid mine water drainage and discharge during the past 15 years. Stream quality can deteriorate from acid mine drainage. When augering, precautions must be taken not to penetrate and drain former underground mine workings. Mine waters can also contain objectionable trace elements, dissolved solids, precipitates, colloids, and sediments. Removal of overburden may also change the groundwater level and deteriorate the quality of groundwater, as follows:

1. Drain usable water.
2. Change the direction of aquifer flow, lower the water table, drain local wells of the neighborhood.
3. Create channelways that allow contaminated waters to mingle with aquifers of good quality.
4. Disrupt alluvium aquifers.

Surface mining can, however, benefit the local hydrology. Spoil piles often act as sponges, absorbing large quantities of water for future plant growth and use, retarding and diminishing runoff, making good aquifers for supplying flow to nearby streams on a steady basis.

b. Topsoil and Microorganisms in Soil Current federal regulations require removal of the topsoil and replacing it on the backfilled area, or segregation in a separate stockpile for later redistribution. Removal changes the natural soil characteristics, pulverizes, breaks down the soil structure, compacts, and generally disrupts normal organic input. Microorganisms are retarded and destroyed by the changes in the soil and organic matter. Erosion is encouraged by the removal and handling of topsoil.

c. Vegetation All vegetation is removed prior to excavation; construction of plant sites, roads, and railroads; and at topsoil storage sites. The mechanical process of destruction and removal of vegetative cover temporarily degrades the air quality and increases the dust in the area. Mine operators periodically wet the haul roads and take other measures to suppress dust.

d. Wildlife

(1) Short-term impact. The scalping of vegetation and removal of overburden displaces and sometimes destroys wildlife in the limited area disturbed by surface mining. Wildlife habitat is temporarily destroyed and the seasonal wildlife cycle is interrupted in the year excavated. Birds, many mammals, game animals, and other mobile wildlife will leave safely. Sedentary animals, especially those that burrow or hibernate, are likely to be destroyed by mining activities. Many types of small mammals, reptiles, and invertebrates will lose their homes and lives. Drainage of streams, ponds, and swamps prior to

mining destroys aquatic life, such as fish, amphibians, and aquatic invertebrates. On a short-term basis the displacement or destruction of wildlife can be severe for the relatively small area of excavations. However, as the land surface is reclaimed and the wildlife habitat is restored, equivalent animals will repopulate the area mined.

Some forms of vegetation upon which wildlife depend for food and habitat may take a long time to grow. In some cases, where spoil from mining fills the natural drainage, such as head-of-hollow fill, the wildlife habitat after mining will not be the same as that prior to mining because of changes in the topography and drainage. If the habitat will not support certain forms of wildlife, some animals will not return but will remain in the nearby undisturbed surroundings.

Most mine operators plant vegetation and restore the terrain so that the land becomes a better habitat for both wildlife and humankind than it was prior to mining. Wild geese and ducks are now common at mined lands in the midwestern United States. Deer are sometimes so abundant at and near mined lands of Appalachia that it is difficult to keep the mines operating during hunting season because so many mining personnel take leave to go hunting.

Many mined lands have, by design, become forests of valuable tree species in which animals and plants thrive. Other lands have been restored to become pastures for domestic animals at the periphery of which wildlife abounds. Mined lands are being reclaimed for the benefit of both wildlife and people.

(2) Long-term impact. Mining can and does cause some long-lasting wildlife habitat impairment or changes of habitat. A few wildlife species may be unable to adjust to these changes, and they do not return to the restored lands, but live elsewhere in the neighborhood. Generally, however, given enough years after restoration of the mined lands, the long-term impact of mining is favorable for wildlife. Long-term impacts of surface mining on wildlife can be minimized by careful consideration of wildlife presence when making plans for mining. For prompt return of wildlife, the mine design should compensate for possible disruptions to wildlife during mining, such as:

1. Habitat impairment and the degree to which individual animals or species will tolerate environmental change.
2. Change in local migratory patterns and breeding grounds.
3. Potential displacement of animals and the impact of this displacement on animals living on adjacent lands. Will displacement produce overcrowding, strife among the animal population, overgrazing, and degradation of the remaining habitat?
4. Loss of feeding grounds. Proper land management requires rehabilitation of mined lands to recreate feeding grounds so that they are attractive for the return of displaced species.
5. Potential sheet erosion and silt and sediment pollution. Mining plans must provide silt structures to avoid lasting steam pollution and destruction of aquatic life, burying of spawning beds, reduction of light transmission because of murky waters, alteration of temperature gradients, and general destruction of aquatic habitat by changes in stream profiles through sedimentation. The gradual weathering of spoil can extend the term of possible silt formation in mined areas, and mining plans should include long-range plans to offset any potential silt increase as the spoil weathers.

6. Toxic minerals and substances exposed during removal of overburden, such as acids, materials of high alkalinity, and dilute concentrations of heavy metals. These materials can adversely affect wildlife by creating a hostile environment and, in some cases, destruction of species. Even small quantities of toxic materials will suppress wildlife productivity and growth rate. Toxic waste pollution can continue for long periods. For this reason, mine design should allow for possible harmful substances caused by the weathering of spoil piles. Mine design must provide for the burial of all toxic materials.
7. Highwalls up to 200 ft high that are hazardous to human beings and to wildlife and domestic animals. In many instances, mine design reduces or buries the final highwall, so there is no hazard.
8. Spontaneous ignition due to rapid oxidation of exposed coal seams and stockpiles. Fires are also caused by carelessness, lightning, and forest or prairie fires. Provision for these hazards in mine design virtually eliminates them.

e. Archaeology Artifacts of historic sites will be destroyed unless they have been systematically investigated before being disturbed by mining. The mining plan should include systematic archaeological studies of the area to be mined. Such studies by the mining industry often benefit the community and create an appreciation for historic values.

f. Aesthetic Rehabilitation of mined lands can and is being done so that the land is aesthetically more pleasing after mining than before. This is particularly evident now in the Midwest, where mined lands are used extensively for country clubs and recreation areas. Some tracts have been donated or sold by mining companies to states for use as state parks or state forests. Small lakes or ponds sometimes form as a result of mining, enhancing the beauty of the restored lands and increasing their value in excess of their original purchase price before mining.

The removal of overburden, however, is disruptive to the landscape and aesthetically repugnant for a temporary period until the lands are restored. It is beneficial to the community and to the mining company to include in the mining plan consideration of aesthetic values.

V. EFFECT OF FEDERAL LEASING AMENDMENT ACT ON SURFACE COAL MINING

The effect of the Federal Coal Leasing Amendment Act on surface coal mining is to slow down and inhibit the surface mining of coal in the Western states. Data supplied by National Coal Association in July 1978 relative to the Interior Department's New Coal Leasing Program showed that some of the regions still contemplated production from federal leases which currently are without mining plans. In addition, the projected Western states production plans for 1990 require that over 10 billion tons of federal coal reserves will need to enter production between 1985 and 1990.

The Department of the Interior (DOI) issued a draft paper on July 18, 1978, which described the Department's proposed coal-leasing program based on preferences identified by Secretary Andrus on June 30, 1978.

A. Background and General Policy

The basic objectives of the Department of Interior's Coal Management System are to carry out President Carter's Environmental and Energy messages and to meet the policies outlined in the Federal Coal Leasing Amendments Act, the Surface Mining Control and Reclamation Act, and the Federal Land Policy and Management Act. The basic goals of the program are to:

1. Assure that sufficient quantities of federal coal are produced to help meet the objectives of the National Energy Plan.
2. Establish land-use planning and effective enforcement of environmental laws to assure that federal coal is produced in an environmentally acceptable manner and in a way that is responsive to local communities and private land owners affected by federal coal development.
3. Assure that federal coal is produced in an economically efficient manner, with a fair economic return to the United States for all coal that is produced.
4. Emphasize consultation and cooperation with state governments in planning the development and leasing of federal coal.

Many of the elements of a leasing and management system have been established by recent statutes. These laws have strengthened the Interior Department's ability to deal with the potential problems involved in the development of federal coal resources.

Under these laws, no mining that disturbs the surface can take place unless it meets stringent standards for protection of water and air quality, for restoration of land contours and revegetation, and for a variety of other important measures.

To protect the financial interests of the United States and to ensure that the government receives a fair return for its property, the laws require that all leases must be issued after competitive bidding and upon payment of fair market value. If surface mining is involved, the lessee must pay a royalty of not less than 12.5% of the value of the coal. The laws foreclose speculation in federal coal by mandating that new leases are forfeited unless production commences in 10 years from the date of lease issuance. Other elements, such as acreage limitations and prelease issuance review by the Department of Justice of antitrust considerations, will work to maintain competition in the coal industry.

The proposed program incorporates the objectives and requirements of all these statutes. These elements will operate within a system that determines when additional leasing is necessary to achieve national coal production goals. The program includes four major elements: (1) a system for evaluating federal coal demand and determining the amount of additional production that will have to be stimulated by federal coal leasing; (2) a planning system and environmental protection standards to decide, in consultation with state and local governments, industry, and the public, which federal coal reserves should be made available for production; (3) procedures for conducting sales and issuing leases; and (4) postlease issuance enforcement of terms and conditions.

B. Review of Federal Leasing Programs by National Coal Association

The coal industry, through the National Coal Association, has reviewed the proposed Department of the Interior's program. The Department has expressed the strong desire to succeed in implementing a program that will in fact effectively resume the orderly leasing of federal coal, and at the same time survive any future attacks by environmental or other public interest groups, such as ranchers and landowner associations in the West. The National Coal Association review of the proposed program discloses, in their words, ignorance or failure on the federal government draftsmen's part to understand or take into account the way in which coal is mined or marketed, and the practical problems of coal development.

The issues that were decided upon by the Secretary of the Interior, the options that he selected, and the coal industry's responses are as follows, as reported by Carl E. Bagge, President of the National Coal Association, to members by memo of July 27, 1978, p. 3-9.

a. *Issue I—Need for Leasing/Leasing Systems Choice Option Selected.* Lease to meet Department of Interior calculated leasing level, which is to be derived from Department of Energy production goals plus input from state, local, industry, and interest groups.

Comments by National Coal Association on the above Issue I choice were as follows:

The statements of the Secretary's decisional options fails adequately to reflect at least two extremely important questions:

First, the underlying decision documents discuss in only general and optimistic terms the very difficult question of how the federal government will decide in which coal region leasing will occur. In fact, interregional choices will be extremely difficult to resolve at the state level. In the past, such issues have produced virtual confrontations between the Department of the Interior and various governors. The assumption that the Department of the Interior will, in fact, be able to reach timely accommodation with the states thus appears unwarranted. This may be particularly true if the states learn, as the National Coal Association has, that among the options rejected by the Secretary of the Interior was the option to allow states to determine levels of leasing through veto of federal leasing sales. The significance of the option selected versus the options rejected is also great with respect to the on-going development of the DOI-DOE relationship. The option to set leasing levels to meet or exceed DOE goals was rejected. The actual effect of the Secretary's choice to set levels to meet DOE goals as adjusted by DOI analysis and other input can, therefore, only be inferred to reflect an intent to create a mechanism to justify leasing at levels lower than *DOE* goals.

b. *Issue II—Tract Selection and Planning Problems.* This issue was deferred pending completion of issue papers. Its resolution is vitally important to potential lessees, and is complicated and controversial within the Department of Interior.

Comments by the National Coal Association on this Issue II were as follows:

This issue and the options to be presented under it are really the heart of the whole program. Questions addressed to Department of the Interior personnel indicate possibly serious failure on their part to appreciate the mechanical difficulties involved in juggling their own new standards and criteria, and applying them to new areas under consideration for leasing. Of perhaps greater importance, they seem even less clear about the manner in which this process will be applied to those areas previously subject to examination and nomination under the EMARS program. It seems that the most prime potential leasing areas, i.e., those previously subject to nomination, may well be subject to the greatest uncertainties in working out the new program. In addition, the Department of Interior has now created within Bureau of Land Management (BLM) a new leasing policy office based in Washington, D.C. The relationship between the state directors of the BLM and this office is tenuous at best. Lack of cooperation at the state level may well occur on a state-by-state basis, and as a result even well-intentional BLM efforts directed from Washington may not in fact be able to receive timely implementation in the field. The National Coal Association review goes on to say: "Finally, the decision documents of the Department of the Interior discuss in detail under 'Planning Responsibilities' the intention by the Department to screen all available federal coal lands in a successive process that will eliminate from availability for leasing in sequence, (i) areas proscribed by statute, (ii) areas 'otherwise unsuitable' (e.g., 'alluvial valley floors, critical wildlife areas, visual rentention zones'), and (iii) lands 'unlikely to be reclaimed.' The application of this screening is theoretically to be accomplished with public input. However, Department of Interior representatives concede that in those areas where the Department's on-going Management Framework Plan (MFP) program is well advanced, there will be little opportunity for further public input by industry as this newly devised screening process takes place."

c. *Issue III—Sale and Pre-Sale Procedures*

1. *Bidding systems—tract questions.* The option selected by the Department of Interior was to retain discretion to use either single or intertract bidding system.

Comments by National Coal Association were as follows:

The problem appears to be one of Interior's own making. EMARS created a system driven by indications of interest from industry or interest groups for or against nominations. Under such a system, the question of how many tracts to offer for lease would in effect decide itself. The difficulty of reaching a decision as to whether to offer a single tract or several tracts in a single sale arises from an internal consistency in the basic, policy-level assumption that the Department will in fact decide both where and how much coal is necessary for leasing. Any decision as to where coal should be leased necessarily implies a priority of geographical locations. Meeting such geographical priorities could well prove impossible, however, under an intertract leasing system since only a limited number of leases and associated LMU's could reasonably be expected to be concentrated in a given area.

2. *Maximum economic recovery.* The option selected was to calculate MER (Maximum Economic Recovery) and require mining on the basis of

all seams which "Collectively are profitable with consideration for social and environmental costs."

Comments from National Coal Association were as follows:

This is one of the most controversial decisions made on the leasing program—the person drafting the system displayed considerable ignorance of how coal is marketed by their selection of this option. They assume, for instance, that a collective price can be established for all coal seams at the time of lease sale.

C. Effect of 1971 Moratorium on Leasing of New Federal Coal Reserves

The 1971 moratorium on leasing of new federal coal reserves will have a disastrous effect on new coal production from such leases. For example, it is estimated that the moratorium will be kept on until at least 1980, thereby assuring that no coal will be produced by 1985 from any leases that could be made available in the early 1980s.

D. New Federal Leasing Program

On December 16, 1978, the New York *Times* noted that Interior Secretary Andrus had described a new program the day before which could reopen federal lands in the West to competitive coal leasing by mid-1980.

The proposed Federal Coal Management Program calls for a spring 1979 comment period and a midsummer 1979 decision by Secretary Andrus on how much coal from federal lands the nation will need to meet its energy requirements in the next 5 to 10 years.

These decisions would be followed by a regional identification of federal lands in which coal mining would not endanger the environment or the local economy. In cooperation with state and local governments, standards for mining operations would be established and decisions made on which of the federal areas should be leased to private mining companies.

After these steps have been completed, the way will be cleared for competitive bidding for the leases, which it is hoped could be scheduled to start by September 1980.

At the present time, because of a series of lawsuits by environmentalists as well as federal government indecision, there has been a moratorium on new mining permits and the issuing of new federal coal leases.

In the past year, there have been only 10 short-term lease sales made on an emergency basis to companies mining adjacent to federal lands, where a halt in mining operations would have meant widespread unemployment in the area.

The Interior Department intends to increase production on federal coal lands already leased and would require this to be done under the same environmental protection standards that will be set for the new leases. Secretary Andrus and the Coal Leasing Act of 1976 required companies holding old leases to start mining by 1986 at the latest. This same "diligence" requirement would be applied to new leases.

VI. EPA SULFUR OXIDE REGULATIONS EFFECT ON WESTERN COAL

If the U.S. Environmental Protection Agency (EPA) sticks to its proposal for full scrubbing regardless of the sulfur content of coal at new power plants, it is forecast that Western coal production will be lower than if the regulations were to allow for the use of low-sulfur-compliance coal without scrubbing.

A. Studies by NERA and ICF

Several studies by National Economic Research Associates, Inc. (NERA) and International Coal Federation (ICF) show this switch in future coal production away from the Western states (Table 7).

B. Estimated Cost of Stack Gas Scrubbing

The estimated cost of stack gas scrubbing is as follows:

Carrying costs	\$15 to \$18 per ton of coal
Operating costs	\$15 to \$30 per ton of coal
Total costs	\$30 to \$48 per ton of coal

VII. ECONOMIC EFFECTS OF GOVERNMENT REGULATIONS ON THE SURFACE COAL MINING INDUSTRY

The federal Surface Mining Control and Reclamation Act of 1977 is expected to increase the cost of surface-mined coal by substantial amounts. For example, Peabody Coal Company, the largest coal company in the United States as well as the largest surface-mining coal company, has prepared cost estimates as of November 1978 which show the effect of the federal draft permanent regulations and the additional costs to comply with these new regulations compared with whatever state rules existed before passage of the federal Surface Mine and Control Act of 1977.

A. Peabody Coal Company Estimates of Additional Costs

The Peabody Coal Company cost estimates show that the new federal strip-mine regulations will cost at least \$2.43 to \$5.58 per ton of coal exclusive of the cost of permitting, which is still under study. These cost estimates are based on actual mines being operated by Peabody in three states: Missouri, Illinois, and Kentucky. The details of the additional costs of the new federal surface-mine regulations are shown in Table 8.

These cost estimates were given to the Council on Wage and Price Stability by Peabody. A different White House agency, the Regulatory Analysis Review Group, has already sent the Interior Secretary a letter critical of the costs, which the Interior Department ignored in its draft regulatory analysis of the regulations.

TABLE 7 Comparison of ICF and NERA Regional Coal Projections for 1990 (Millions of Tons)

	Existing regulations		0.2 Floor		0.5 Floor	
	NERA	ICF	NERA	ICF	NERA	ICF
Appalachia	531	464	503	449	532	449
Midwest	313	274	385	318	326	290
West	666	785	604	735	643	782
total	1510	1524	1492	1502	1501	1521

Source: Coal Outlook, Sept. 18, 1978.

TABLE 8 Peabody Coal Company's Estimated Costs of Federal Surface Mine Regulations

	Additional Costs ($/ton)			
Description	Mine A, Missouri	Mine B, Illinois	Mine C, Kentucky	Weighted average
Permitting	—	—	—	—
Blasting	$0.177	$0.272	$0.279	$0.223
Prime farmland	4.080	0.160	0.549	2.351
Topsoil handling	0.122	0.821	0.716	0.422
Mine closures	0.078	0.370	0.162	0.138
Permanent and temporary stream diversions	0.006	0.045	0.066	0.034
Siltation structures	0.559	0.187	0.143	0.212
Revegetation	0.080	0.053	0.213	0.104
Cover for acid and toxic materials	0.379	0.064	—	0.452
Impoundment and embankments	0.022	0.051	0.052	0.036
Hydrologic monitoring	0.013	0.039	0.097	0.039
Fugitive dust	0.066	0.152	0.154	0.107
total	$5.582	$3.214	$2.431	$4.118

Source: Coal Outlook, Nov. 6, 1978; and Dec. 4, 1978.

Peabody's figures compare the cost of meeting the draft permanent regulations with whatever state rules existed before passage of the federal surface mine act.

The costs listed in Table 8 are exclusive of the costs of mine permitting, which can be substantial, especially when projects are delayed awaiting permit approvals. Some operators have estimated that filing for a surface mine permit in late 1979 and early 1980 will be 10 times more expensive than it is now. Two years and at least $100,000 will be needed to prepare a permit, according to preliminary estimates that mining operators have given the Mining and Reclamation Council of America (MARC).

Coal industry representatives, in testifying before an Office of Surface Mining (OSM) hearing on this Agency's draft Environmental Impact Statement (EIS), stated that an EIS for the permanent surface mining program should have been completed early enough to be used in formulating the regulations, rather than as justification for the rules after they were proposed.

The witness testified that a major function of an EIS is to discuss and analyze alternatives to the proposed rules—a job OSM neglected to complete. Industry is particularly concerned about the decision to use uniform national standards for application to all operations. A state-by-state or a regional approach are both reasonable alternatives that should have been analyzed and discussed in the draft EIS but which were not.

The industry spokesman also noted that the draft EIS fails to analyze fully all major adverse effects of the proposed federal action, and particularly glosses over negative socioeconomic effects, such as those stemming from anticipated closures of numerous small operations in Appalachia and the Midwest.

Additionally, the draft EIS and OSM proposed regulatory analyses almost totally ignore the effect that the proposed permanent regulations would have on underground mining.

B. Joint National Coal Association and American Mining Congress Surface Mining Committee Report on OSM Regulations

As additional information became available on the rules proposed for the permanent surface mining program, new estimates were made of the costs to the coal industry.

A joint National Coal Association-American Mining Congress Surface Mining Committee report showed that the total cost of surface mining regulations could range between $6.52 and $17.17 per ton of salable coal produced.

The NCA-AMC Committee did not attempt to show the overall cost for average mines; instead, it listed the highest and lowest costs found on a limited set of regulations for specific surface mines studied in two regions of the country—the Midwest and Appalachia—and for deep mines generally. These were reported in the December 4, 1978, issue of *Coal Outlook* [6], a coal industry publication.

The NCA-AMC Committee made a report on the regulatory analysis done by the Office of Surface Mining. It flailed OSM for ignoring costs that it decided had only "local impact," that OSM felt were mandated by the law itself, or that resulted from regulations developed too late for the regulatory analysis done by the OSM to consider.

TABLE 9 NCA-AMC Costs for Specific Mines for Rules Not Analyzed by OSM

	Cost range ($/ton)	
	Low	High
Midwest		
Segregation of A horizon or stockpiling of surface soil	$0.10	$0.29
Prime farmland	0.24	4.08
Cover for toxic materials	0.38	1.06
Mine closures	0.02	0.40
Blasting	0.17	0.28
Total these five rules	$0.91	$6.11
Appalachia		
Segregation of B horizon or stockpiling subsoil	$2.20	$5.50
Backfilling and grading	1.48	2.32
Valley fill drainage	0.10	1.51
Valley fill controlled placement where gravity placement now allowed	2.43	5.56
Valley-filled compaction in 18-in lifts where controlled placement already required	(0.60)[a]	(1.83)[a]
Roads	0.25	1.83
Coal-processing wastes	0.06	0.45
Total these six rules	$6.52	$17.17
	(4.69)[a]	(13.44)[a]
Deep mines, area irrelevant		
Subsidence	$0.3657	$0.8136
Topsoil	0.0665	0.6050
Roads	0.1438	1.2570
Total these three rules:	$0.5760	$2.6756

[a]West Virginia already bans gravity placement of spoil in valley fills and requires controlled placement in 4-ft lifts. The extra costs of the valley fills in West Virginia would stem mainly from a change in 18-in. lifts, a shift that is discretionary with the regulatory authority.

Source: Coal Outlook, Dec. 4, 1978.

OSM's decision to concentrate its analysis on the costs added to new mines wrongly ignored the higher costs and disruption existing mines would suffer, the NCA-AMC panel said.

According to the report in *Coal Outlook*, existing mines with short remaining lives cannot spread out the expense of new investment required by the regulations. This will force marginal mines out of business and put healthy mines into a financially precarious position wherein market slumps or labor problems could end the enterprise.

A joint committee, in their report of November 1978, wants OSM to:

1. Suspend the comment period on the proposed permanent regulations
2. Publish a revised draft of its regulatory analysis that covers the entire set of proposed rules
3. Publish revised draft permanent regulations and reopen the comment period for 60 days

The NCA-AMC costs for specific mines for rules not analyzed by OSM are shown in Table 9.

C. Regulatory Analysis Review Group Report on the OSM's Permanent Program

The White House Council on Wage and Price Stability Regulatory Analysis Review Group (RARG) prepared its report on the proposed permanent regulations of the Office of Surface Mining and gave an economic analysis of the proposed regulations.

As reported in the December 4, 1978, issue of *Coal Outlook* [6], the final RARG report concluded that the overall cost of the upcoming permanent program could reach $3 billion per year.

The RARG report estimates that 15 to 20% of Western coal reserves could be barred from recovery by rules protecting alluvial valley flows.

The cost of just surveying a potential Western site for alluvial valleys will cost $500,000 to $1 million and take 11 to 26 months according to the report.

The estimated cost of $3 billion was based on per ton costs developed by the American Consulting Engineers Council, which were then multiplied by national coal production figures. The engineers' report estimated an $8 to $12 per ton cost in Appalachia and $3 to $5 per ton elsewhere.

D. Electric Utility Forecasts of Additional Costs

The customers for surface-mined coal have also voiced their views on what effect the surface-mining regulations will have on the cost of coal they purchase. One of them is Public Service Co. of Indiana, which estimated that OSM surface-mining regulations will cost them $25 million more annually in coal cost and will cause a monthly increase of $1.25 for an average suburban customer.

This utility company, which receives 2.4 million tons annually for its Cayuga Generating Station from the Universal Mine of Peabody Coal Company, based its estimate on Peabody's figures that coal would increase $3 per ton, from $13 to $16 per ton or a 23% increase.

Peabody estimated that surface reclamation would cost the following for the period April 15-December 31, 1978:

1.	Contract cost of 11 dozers, one grader, and three trucks on two-shift operations	$3,693,000
2.	Additional labor of 30 workers plus fringe benefits	745,000
3.	Additional nonunion labor (two workers)	54,000

4.	Additional administrative and general costs	139,000
5.	Estimated sedimentation costs	24,000
6.	Additional equipment for well casings, meteorological stations, and surface-water monitoring stations	82,000
	Total for 8½ months	$4,737,000

The estimated tonnage of coal during this period is 1,622,000, which gives an average cost of $2.92 per ton or an estimated figure of $3.00 per ton additional costs.

Public Service Indiana burned approximately 9.4 million tons of coal in 1977, of which 8.1 million tons were from surface mines in Indiana and Kentucky. The total cost for fuel in 1977 was $163 million, about one-third of its operating costs. An additional annual increase of $25 million for the OSM surface mining regulations will increase this utility's costs by over 15%.

Another electric utility company, Detroit Edison Co., found that the new federal surface-mining regulations will increase the cost of their coal by an average of $2.38 per ton. The range in additional costs is from $0.65 per ton in Ohio to $5.00 per ton in eastern Kentucky.

The permanent program regulations were issued in January 1979 but will not become effective until states or the Office of Surface Mining implement them in late 1979 or early 1980.

The additional cost of energy caused by federal regulation of the coal-mining industry is eventually charged by the utility to each consumer of electricity. Therefore, any regulation that is inefficient, unwarranted, or excessive is costly and uneconomical to society at large and can be a major contributor to inflation.

E. Department of Energy's Review of the Surface Mining Control and Reclamation Act

The Department of Energy also leveled a highly critical attack on the Interior Department's coal surface mining and reclamation regulations, charging that they should be simplified and/or made more cost-effective, and identified five major problem areas, one of which could cost the coal industry as much as $3 per ton to comply.

As reported in the December 4, 1978, issue of *Inside D.O.E.*, the DOE policy chief, Al Alm, and deputy environment head, James Liverman, in a November 22, 1978, letter to Interior, wrote that DOE's detailed review of Interior's regulations, announced on September 18, had found five major areas where DOE foresaw problems:

1. *Fugitive dust or air pollution control.* DOE said that these controls duplicate the requirements under the Clean Air Act amendments.
2. *Disposal of excess spoil, or soil not needed.* DOE said that it would cost the coal industry up to $3 per ton of coal because of overly strict requirements for constructing fills or disposal areas.
3. *Identification of alluvial valleys in the West.* DOE said that this should be simplified in order to save up to 26 months and $750,000 for each permit involving an alluvial valley.

4. *Permit requirements.* DOE estimated that these would cost $19,000 per permit for small eastern mines because it requires surface- and groundwater monitoring even if not needed.
5. *Sedimentation control.* DOE estimated that the cost could be reduced from $116 to $103 million if OSM would require monitoring of effluent runoffs only when a mine is in operation.

F. Environmental Protection Agency's Regulations Under the Resource Conservation and Recovery Act

The Environmental Protection Agency on December 15, 1978, proposed guidelines and regulations and proposal on identification and listing of hazardous wastes. Fly ash may be on this list; if it is, it will be very expensive to dispose of fly ash and other coal ash in the future.

The Department of Energy started in December 1978 to prepare a nationwide assessment of the cost impact of the Resource Conservation and Recovery Act (RCRA) regulations on conventional and advanced coal-use technologies. This study was designed to gauge whether RCRA will inhibit future coal use.

According to the December 25, 1978 issue of *Inside D.O.E.* [13], the purpose of the undertaking by DOE is to give the Environmental Protection Agency, which proposed the regulations December 15, 1978 (published in the *Federal Register* of December 18, 1978), a full understanding of the implications of the standards on coal use and to provide DOE with an estimate of the effect that it will have on DOE programs.

DOE and many coal industry officials fear that the new standards—particularly the toxic substance leaching procedure proposed by EPA—will make all coal residues (incouding fly ash, scrubber sludge, and wastes from fluidized-bed combustion, gasification, and liquefaction) hazardous substances that would require costly disposal procedures. According to the *Inside D.O.E.* article, a preliminary estimate made in November 1978 by DOE indicated that RCRA would add from 12 to 38 cents per million Btu to the cost of new coal technologies and 42 cents per million Btu to the cost of conventional coal combustion. This last figure is equal to over $10.00 per ton of coal just for the proposed RCRA regulations.

To supplement the assessment, the RCRA task force was collecting geological maps showing wetlands and floodplains in an attempt to determine what areas will be off-limits to hazardous-waste disposal sites. Since the disposal sites cannot be located in such areas, the cost of transporting wastes such as fly ash to acceptable locations may be quite high, according to the *Inside D.O.E.* article.

Additional information on the hazardous materials aspect of fly ash and other residues was contained in the December 25, 1978, issue of *Coal Outlook*. This publication indicated that an even-greater problem could be associated with the disposal of coal ash and indicated that the new RCRA regulations would put a premium on coal that is low in radioactivity and trace-metal content. The article also suggested that the new rules would increase the value of low-ash coal.

The proposed regulations under RCRA will require that bottom ash, fly ash, and scrubber sludge would have to be tested at each power plant producing any of this waste.

Wastes found to be poisonous due to trace metals would face expensive disposal costs.

The high-volume wastes from power plants were put in a special category. The monitoring and upcoming performance standards for the special category would apply on a case-by-case basis for waste tested and found to be dangerous.

Even utility wastes that are not judged dangerous would still have to conform to new solid waste rules that prevent placement in floodplains and environmentally sensitive areas.

The general solid waste rules would make low-ash coal even more attractive than it is now, because utilities would have less waste to transport and store.

According to the *Coal Outlook* article, some utilities fear that the cost of the RCRA regulations could be as much as double the cost of EPA's air and water pollution rules combined

It is also speculated that up to 15% of coal reserves could fall into RCRA's radioactive category. Exact figures are not available in this category because EPA has not issued regulations defining what level constitutes radioactivity and because data on coal radioactivity have not been collected on a wide scale.

In addition to the extra handling costs for this special radioactive waste, utilities may find it difficult to locate disposal sites for anything labeled as radioactive waste.

Some estimates of the cost of disposing of coal ash under the RCRA regulations are as high as $100 per ton, which would add $10/ton to the cost of coal if it had 10% ash.

The EPA estimated that it would take up to 5 years to issue permits to all disposal sites after the regulations become effective in mid-1980.

The significance to surface mining of the RCRA regulations is as follows:

1. Uranium occurs in certain parts of some lignite deposits in the United States, particularly in Montana, Wyoming, North and South Dakota, Idaho, and New Mexico.
2. Some thin lignite deposits in the Dakotas and Montana are of commercial grade and have been mined for their uranium content.
3. It is reported by Nininger that some Pennsylvanian-age coals and other coals in the eastern United States contain up to 1/5 lb of uranium per ton in thin seams near the base of the coal bed and in the under lying clay.
4. Any restriction or added cost to the disposal of ash will have a depressing effect on coal production, and surface mining of coal will be restrained as a result.

G. Analysis by ICF, Inc., of Effect of Proposed Surface Mining Regulations on Coal Production

According to the January 8, 1979, issue of *Coal Outlook* [6],

> . . . the consulting firm ICF, Inc. predicted that the proposed surface mining regulations could result in a repeat of the coal buying panic of 1974. The regulations which ICF believes may cause the shortage of coal would require immediate compliance with the proposed permanent programs for all existing, non-conforming structures.

The consultant, ICF, Inc. made the warning based on the regulatory analysis done by the Office of Surface Mining. OSM said immediate compliance for existing, non-conforming structures would cost 102 million tons of production as mines in violation shut down.

Assuming the 102 million ton figure is accurate, ICF said producers could not expand production from other mines to fully compensate for the estimated loss.

H. Suit by Environmental Groups Against White House Council of Economic Advisors Charging Intervention

As a sequel to the surface-mining furor, it was reported in the New York *Times* on January 14, 1979, that

> . . . lawyers for three environmental groups have filed a lawsuit in Washington, D.C. challenging the right of President Carter's economic advisors to intervene in the regulation-writing procedure of the Interior Department in regard to rules to curb the environmental damage of surface mining.
>
> The three groups, the Natural Resources Defense Council, The National Wildlife Federation and The Council of the Southern Mountains, petitioned the Federal District Court on January 12, 1979 for an injunction that would bar officials of the President's Council of Economic Advisors from further contacts with the Interior Department on the surface mining regulations.
>
> If the injunction is granted, it would also forbid the department's surface mining regulators from giving any consideration to criticisms that the council had already made, after the formal close of a 60-day period for public comment on the proposed rules. Council officials reportedly told the rule's writers in closed meetings that many of the environmental structures might be inflationary, unnecessary, or both.
>
> The intervention of the council had precipitated a six month delay in enforcement of the new environmental curbs, infuriating environmentalists. With that delay, legislatures of coal producing states, which are to enforce the rules, would no longer have enough time to draft and pass surface-mining control laws based on the rules before February 3, 1979, when they were to take effect. Congress mandated a six-month delay should that deadline not be met.
>
> This suit is believed to be the first to challenge the authority of the President's closest staff in making behind-the-scenes efforts to influence government regulations.
>
> In complex and potentially costly legislation, such as the Federal Surface Mining and Reclamation Act of 1977, Congress commonly adopts broad objectives and tells the administrating federal agency to draft the detailed regulations, which then become the law. The stringency of the resulting regulations can mean hundreds of millions of dollars in added industrial operating costs.
>
> The rule-making process is controlled by the Administrative Practices Act. It specifies that all sides be given a chance to be heard, then calls for the regulation writers to be isolated from further contacts with

the agency involved. The lawsuit contends the law calls for the regulation writers to be isolated from the White House, too.

If the process is changed by White House intervention after the public hearing period has ended, the suit charges that the procedure would be 'tainted.' This is what the suit contends was the case in several meetings by council members with the rules' writers after the November 1, 1978 cutoff date.

The suit also charged the Council on Economic Advisors had sought and received comments from the coal industry in preparing its intervention and had thus become its 'advocate.'

Officials of the council denied that the Council had been 'enlisted' by the coal industry. They said that Charles L. Schultze, the Chairman of the Council, was carrying out President Carter's directive to monitor government regulations for inflationary effects. A spokesman for the council would not comment on the lawsuit which named Mr. Schultze and Lyle E. Gramley and William Nordhaus, Council members, as defendants.

Similar issues had been raised, and could end up in court, regarding alleged White House interventions in the preparation of regulations to limit the exposure of textile workers to cotton dust, to require special equipment on mass transit vehicles for the handicapped and to curb air polluting emissions.*

VIII. FUTURE TRENDS IN SURFACE COAL MINING

Future trends in the United States surface-coal-mining industry depend largely upon:

1. A positive national energy policy that assures the surface-coal-mining industry of a fair share of the energy market
2. Fair and equitable administration of the federal Surface Mining Control and Reclamation Act of 1977 and similar legislation affecting surface mining
3. Availability of large sums of investment capital needed to finance new mines
4. Development and installation of the mines with improved materials-handling systems for removal of overburden and excavation of coal

For the purpose of forecasting future trends of surface coal mining, it is assumed that the federal government will avoid industrial choas and take positive measures to balance national energy demand with domestic supply by encouraging coal production through energy policy and correcting confusion, uncertainty, and inequities that currently exist in the U.S. energy industry.

A sound and firm federal energy policy, fairly and justly administered, should establish the favorable investment climate needed for attracting the investment capital required for constructing increased surface-mining capacity at the rate of approximately 8% per year. This rate of increase of surface-mining capacity will satisfy the most optimistic coal production projections.

TABLE 10 Productivity of U.S. Surface Mines

Year	Tons per worker-day
1940	15.63
1945	15.46
1950	15.66
1955	21.12
1960	22.93
1965	31.98
1970	35.96
1971	35.69
1972	36.33
1973	36.67
1974	33.16
1975	26.69
1976	26.40
1977	26.90
1978	25.00

Source: National Coal Association, *Coal Facts 1978-1979.* Washington, D.C., p. 84.

Given favorable government actions that will encourage the surface-coal-mining industry to expand, the trend of the industry will be to vigorously develop and install mines that operate more efficiently than at present, with larger-capacity equipment, producing greater productivity per manday. The decline in the average productivity for surface mines appears to have leveled out and may have been reversed, as shown in Table 10.

Productivity increase is already apparent at mines of thick-seam coal in the West. In Appalachia the trend will be toward larger mines with systems of mining having greater capacity than at present, which remove mountaintops to deeper depths to expose an increased number of layers of coal beds. This trend toward multiple-seam mining by mountaintop removal is already in progress and is further encouraged by demand for increasing amounts of low-sulfur coal from Appalachia. Improved coal beneficiation plants will make surface mining of multiple seams more efficient, separating rock waste from the coal mechanically in the plant, rather than by the practice common to small mines of high-grading by selective mining of individual seams as exposed in the pit by overburden removal.

A. Compliance with the Federal Surface Mining Control and Reclamation Act of 1977

Many small mines will either increase in size or cease operating. This is because small mines will not be able to afford the staff and expense of satisfying the 10 to 16 federal, state, county, and local government agencies which, with much duplication of effort, regulate and monitor activities of surface coal mining. However, small mines may reappear during times of energy crisis when the sales price of coal is unusually high.

Generally, the trend in surface coal mining will be toward improving practices to comply with performance standards of the Federal Surface Mining Control and Reclamation Act of 1977, such as the following [15]:

1. Conduct operations to maximize the recovery of coal.
2. Reclaim mined lands contemporaneously with mining or as soon as possible thereafter.
3. Restore the land so that it will be capable of supporting its uses prior to mining.
4. Where natural slopes are greater than 20°, prevent placement of spoil and other materials on the downslope, bury the highwall, and restore the surface to approximately its original contour.
5. Assure that the quality and quantity of water will not diminish for uses adjacent to the mining area.
6. Prevent contamination of waters by burying or otherwise treating all acid-forming or toxic materials and combustibles.
7. Control erosion and pollution of the air and water by stabilizing the area; stabilize all waste piles.
8. Restore topsoil to the area affected by mining.
9. Establish a permanent vegetative cover on the area affected.
10. Minimize, where possible, adverse impacts of mining on wildlife, fish, and associated environmental values.

Requirements of regulatory agencies, together with increased demands for coal, will cause changes in surface-coal-mining technology that improve the continuity of operations. To illustrate, the Federal Surface Mining Control and Reclamation Act of 1977 requires reclamation contemporaneously with mining, or as soon as possible thereafter. Similarly, high interest rates and cash flow problems at many mines encourage loading and shipping coal to market as soon as possible after overburden removal. This combination of government forces and economic pressures cause the mine operator to consider improvements in mining methods that will maximize the flow of coal and spoil from the pit and at the same time accomplish reclamation concurrently with mining. As a result, government and industry are endeavoring to develop technology that will maximize the continuity of flow of materials to and from the pit through surface mining research and development, as follows [15]:

1. Excavating machinery that digs continuously, such as rotating wheels equipped with buckets that excavate constantly, depositing a steady stream of spoil by means of conveyor belts directly to the area to be reclaimed. This concept is in sharp contrast to draglines, shovels, scrapers, and dozers, which dig intermittently.
2. Continuous surface miners that utilize cutter drums, penumatic choppers or shearing devices to excavate and load coal continuously while moving across its surface.
3. Crosspit overburden conveyors that work in conjunction with draglines and facilitate the flow of spoil across the pit to a specific reclamation area.

4. Belt conveyor systems that transport coal via tunnels, through spoil or up ramps cut into the highwall; also conveyors for coal that go over highwalls and spoil banks.
5. Overburden belt conveyor systems that replace trucks in the block cut and haulback methods of mining.
6. Contour-mining concepts on a massive scale, which include timbering, woodchip and mulch production, topsoil preservation, acid mine water treatment, systems of spoil conveyors and augering that penetrates up to 1000 ft under the final highwall. Mining and reclamation would be continuous and complete, resulting in no need ever to return to the area to recover more coal.
7. Computer-logging systems for draglines that evaluate and improve dragline operator performance and increase machine productivity.
8. Multiseam mining by the haulback method. By mining several seams simultaneously, the number of working places for loading coal increases and the flow of coal from the pit can be more continuous.
9. Crawler-mounted hydraulic excavators with front-shovel or back-hoe arrangements for digging unblasted overburden, boulders, or semi-consolidated overburden.
10. Self-loading, four-wheel-drive, twin-engine, elevating scrapers for selectively excavating and segregating overburden, such as topsoil, subsoil, and other strata, including coal.
11. Monster-size bulldozers, weighing 90 tons and more; front-end loaders with buckets of capacity of 24 yd^3 and larger.
12. Shovel-truck overburden-stripping methods that reclaim to suitable contours without rehandling spoil.
13. Aluminum booms on draglines that increase productivity 7 to 15%. By using aluminum it is possible to operate booms of 360 ft and longer.
14. Horizontal drilling of blast holes at contour and area mines.
15. Automation drilling of blast holes.
16. Custom-programmed blast patterns that improve breakage of rock and produce minimum amount of shock, ground vibration, and noise.
17. Use of chemicals to eliminate haulroad dust.
18. Haulage trucks with heaped capacity exceeding 250 tons.
19. Improved species of shrubs, trees, legumes, and grasses for reclamation on different spoils.
20. Two-way radio and television communications systems and mobile radios for maintaining equipment operating efficiency, contact with workers, and safety.
21. Foam tires for haulage trucks, which eliminate tire failures.
22. New mining shovel configurations that increase digging forces.
23. Use of aerial photography to aid mine inspections and the detection of illegal mining activities.
24. Use of wheel tractor-scrapers instead of draglines in some areas for overburden removal and handling of topsoil. This trend is due to the relatively large capital cost of a dragline, which can exceed $25 million, and to the mobility of scrapers.
25. Inoculation of mine waters with strains of bacteria that inhibit or decrease acid formation.

B. Development of Improved Technology Through Research and Development

Considerable research and development is already being done by the U.S. Bureau of Mines to improve surface-mining productivity and efficiency; some of these projects are listed above. This trend toward research and development should continue by both government and by industry.

The coal industry and equipment manufacturers are reluctant to finance the cost and face the risk of developing new equipment designs for improving surface coal mining and reclamation. This trend of hesitancy by industry will continue until the United States establishes a meaningful energy policy, including the long-term role of coal.

ACKNOWLEDGMENTS

Liberal use has been made of coal-related statistics, tables, and other data prepared by the U.S. Bureau of Mines, U.S. Environmental Protection Agency, National Coal Association, Bituminous Coal Research, Inc., and the Society of Mining Engineers of the American Institute of Mining, Metallurgical and Petroleum Engineers, Inc. The use and appreciation of these valuable sources of information is acknowledged. The authors are also indebted to numerous colleagues in the surface coal mining industry, who contributed information used herein.

Ford, Bacon & Davis, Inc., engineers-consultants since 1894—before the Panama Canal—the digging of which inspired surface coal mine development in the United States—very kindly made available its research library and wealth of experience.

REFERENCES

1. National Coal Association, *Coal Facts 1978-1979.* Washington, D.C.
2. E. R. Phelps, in *Elements of Practical Coal Mining,* S. M. Cassidy, ed., Society of Mining Engineers of the American Institute of Mining, Metallurgical and Petroleum Engineers, Inc., New York, 1973, p. 377.
3. W. H. Weimer and W. A. Weimer, in *SME Mining Engineering Handbook,* A. B. Cummins, ed., Vol. 2. Society of Mining Engineers of the American Institute of Mining, Metallurgical and Petroleum Engineers, Inc., New York, 1973, pp. 17-122.
4. Ford, Bacon & Davis, Inc., Surface Coal Mining Machinery and Equipment, prepared for U.S. Bureau of Mines under Contract No. SO 241050, Bureau of Mines, Open File Report 107-76, Washington, D.C., Dec. 1975, p. 42.
5. *The Northern Miner*, World's biggest bucket wheel goes into operation, Toronto, Nov. 9, 1978, p. 12.
6. Ford, Bacon & Davis, Inc., and the Energy Bureau, *The Outlook for Coal,* R. W. Nash, ed. Production 27. New York, 1976.
7. E. C. Grim and R. D. Hill, Environmental Protection in Surface Mining of Coal, National Environmental Research Center, Office of Research

and Development, U. S. Environmental Protection Agency, Cincinnati, Ohio, EPA-670/2-74-09, U.S. Government Printing Office, Washington, D.C., Oct. 1974, p. 28.

8. Ford, Bacon & Davis, Inc., Technology of Auger Mining, prepared for Bureau of Mines under Contract No. SO241050, Washington, D.C., Oct. 1974, p. 47.
9. N. P. Chironis, Cross-ridge Mining of Mountaintop, a Better Technique for Appalachia, *Coal Age*, p. 74 (Dec. 1976).
10. D. P. Baker, Draft, Environmental Impact Statement Development of Coal Resources in the Eastern Powder River Coal Basin of Wyoming, 5 vols., prepared by Interagency Team, Dept. of the Interior, BLM, USGS, Dept. of Agriculture, USFA, ICC, Cheyenne, Wyo., 1974, pp. i-I, III-20.
11. H. W. Evans and J. Tate, Jr., Black Thunder Mine Planning and Environment, Preprint 75-F 128, Society of Mining Engineers of the American Institute of Mining, Metallurgical and Petroleum Engineers, Inc., Salt Lake City, Utah, 1975, pp. 12, 36.
12. U.S. Senate Interior and Insular Affairs Committee, Hearing Before the Subcommittee on Minerals, Materials and Fuels, Pursuant to S. Res. 45, A National Fuels and Energy Policy Study, 92nd Congress, First Session on S. 2777 and S. 3000, Serial No. 92-13, Part 3, U.S. Government Printing Office, Washington, D.C., 1972, p. 1020, 70-8670.
13. D. Pimental, E. C. Terhune, et al., *Science*, 194: 149 (1976).
14. D. Pimental, W. Dritschilo, J. Krummel, and J. Kutzman, *Science*, 190: 755 (1975).
15. N. P. Chironis, *Coal Age—Operating Handbook of Coal Surface Mining and Reclamation*. McGraw-Hill Book Company, New York, 1978, pp. 33, 418.
16. U.S. Department of the Interior, Office of Surface Mining Reclamation and Enforcement, Permanent Regulatory Program Implementing Section 501 (b) of the Surface Mining Control and Reclamation Act of 1977, Draft Environmental Statement, Washington, D.C., Sept. 1978.
17. B. H. Wills and W. M. Henebry, *Hydrocarbon Process*. 55:87 (Oct. 1976).
18. AMAX, Inc., 1979 Annual Report, 1980, p. 20.
19. National Academy of Science/National Academy of Engineering Study Committee, *Rehabilitation Potential of Western Coal Lands*. Bellinger Publishing Co., Cambridge, Mass., 1974, pp. 2, 4.

5 Size Reduction

RONALD R. BEVAN* / Kennedy Van Saun Corporation, Danville, Pennsylvania

---

*Present affiliation: Youghiogheny and Ohio Coal Company, St. Clairsville, Ohio.

I. INTRODUCTION

Size reduction of coal plays a major role in enabling run-of-mine coal to be utilized to the fullest possible extent for power generation, production of coke, and production of synthetic fuels. Run-of-mine (ROM) coal is the "as-received" coal from the mining process, whether the mining process be stripping, augering, continuous mining, short- or longwall mining, or any other method currently practiced. Because the types of mining processes are varied, and size reduction actually begins at the face in the mining operation, it is quite understandable that the characteristics of the products from these various processes differ widely. The type of mining process directly affects the top size and size consist of the mine product. In addition, the amount of out-of-seam dilution in the mine product is directly related to the mining process employed.

The term "run-of-mine coal" is used to designate the product from a mining process prior to any type of treatment. Once the ROM coal has been treated by the removal of tramp iron, screening, or crushing, the material is referred to as raw coal.

II. SIZE REDUCTION PRINCIPLES

In order to properly understand size reduction processes, one must first understand the basic principles and concepts of size reduction.

A. Size

The size of a particle can be defined in terms of a surface opening through which the particle will barely pass or will not pass at all. In sizing a material, the individual particles are presented to the surface openings numerous times.

One encounters problems in defining the size of a particle because a three-dimensional particle is presented to a two-dimensional opening. In the case of a narrow, elongated particle, the particle will be sized according to the orientation of the particle with respect to the surface opening.

ASTM Standard D431-44 defines the top size or upper limit of a quantity of material as the smallest sieve (smallest opening) upon which is retained a total of less than 5% of the sample [1]. In industry, the term "nominal top size" is used extensively when describing the output spectrum of a size reduction device. The nominal top size of a quantity of material is the smallest sieve (smallest opening) upon which is retained a total of 5 to 20% of the sample.

B. Reduction Ratio

Reduction ratio is the ratio between the feed top size and the product top size, or the ratio between the feed and product sizes at which a certain percentage of the material passes (i.e., 80% passing). For size reduction units that produce a product top size larger than 1 mm, size reduction ratios are normally of the order of 8:1. For units whose product top size is smaller than 1 mm, size reduction ratios can range to 200:1 and higher. If one tries to exceed the practical size reduction ratio of a unit, a decrease in the operational efficiency of the unit will result.

C. Grindability

Knowing the relative ease with which coal can be comminuted (reduced in top size) by mechanical action is an aid in properly designing a size reduction circuit. The term used to refer to this relative ease is the grindability index of the coal.

The most common grindability index used in conjunction with coal size reduction is the Hardgrove grindability index (HGI). This index is determined by grinding 50 g of 16 × 30 mesh dried coal in a standardized ball and race mill for 60 revolutions at an upper-grinding-ring speed of 20 ± 1 rpm. Then the sample is removed and sieved at 200 mesh to determine W, the amount of material in grams passing through the 200-mesh sieve [2]. The index was calculated from the formula

$$HGI = 13.6 + 6.93W \qquad (1)$$

From Eq. (1), one can deduce that as the resistance to grinding increases, the HGI will decrease.

In recent years, use of the formula above has been eliminated by means of a calibration chart. ASTM Standard D409 (1977) defines the procedures for obtaining the calibration chart for a given grindability machine.

D. Specific Energy

Specific energy is the energy per unit of throughput required to reduce a feed material to the desired product size. The specific energy for size reduction, SE, is proportional to the grindability index of the coal, GI, times a function of the reduction ratio, RR, as follows:

$$SE = GI \times f(RR) \qquad (2)$$

From the specific energy expression, it can be deduced that the total required energy increases with increasing throughput for a given feed material and reduction ratio. For a fixed throughput and material, the total energy required varies with a changing reduction ratio. This relationship between energy and reduction ratio is illustrated in Figs. 1 and 2. These figures show that this relationship is curvilinear. Flavel says that this relationship is exponential for crushers [3]. Figure 1 shows that this relationship is dependent upon the particular size reduction device. Although two different size reduction devices might have similar energy requirements in a given range of reduction ratios, outside this range the energy requirement for one device might approach a maximum reduction ratio limit asymptotically, whereas the energy requirements for another device will increase with increasing reduction ratio but at a much slower rate. Figure 2 shows that the mode of operation of a size reduction device affects the relationship between energy and reduction ratio. In practice, observations have been made that proper feeding to a crusher can usually achieve throughputs in excess of design, particularly when the feed rate is controlled by the power (energy) draw.

Figure 3 illustrates that the relationship between specific energy and grindability is dependent upon the type of device used. In addition, there

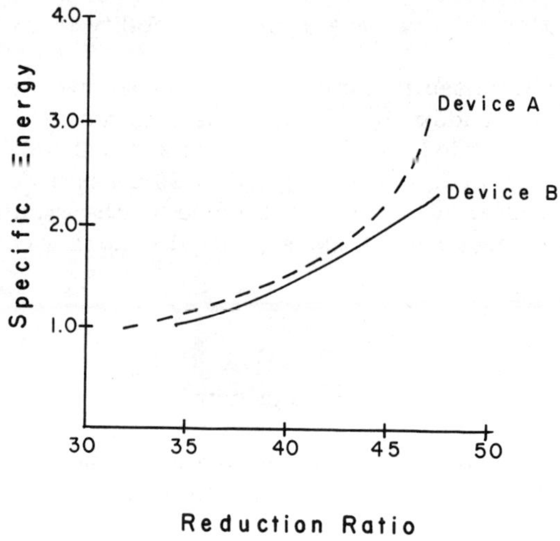

FIG. 1 Specific energy as a function of reduction ratio for two size reduction devices.

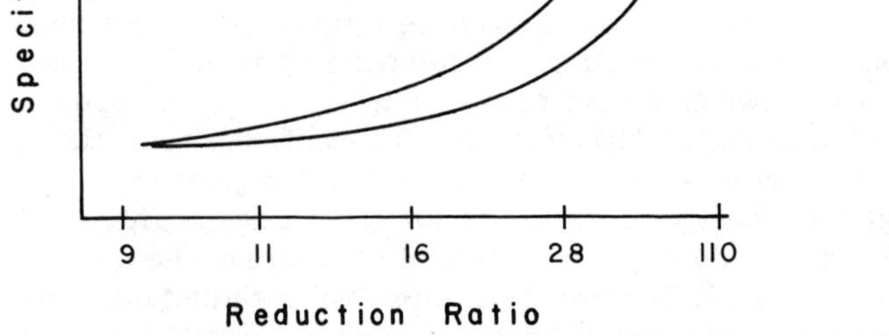

FIG. 2 Specific energy as a function of reduction ratio, open circuit versus closed circuit.

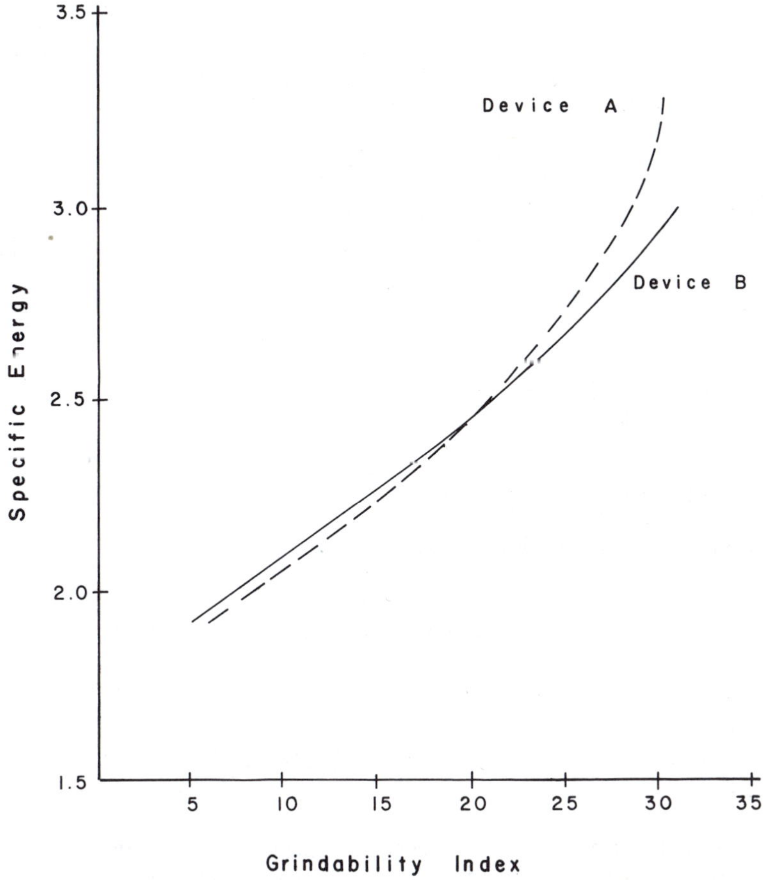

FIG. 3 Specific energy as a function of the coal's grindability index for two size reduction devices.

ranges in which one device is better suited than another, and vice versa. For a given application, pilot tests must be performed in order to acquire the parameters that satisfy the specific energy equation.

The moisture content of the coal feedstock to a size reduction unit is a factor that must be considered when considering the energy requirements of a size reduction unit (Fig. 4). Excessive surface moisture causes less efficient removal of fines from the pulverizing zone because of agglomeration of the fines.

The specific energy requirements, expressed as horsepower per ton per hour of product, will increase as the desired reduction ratio increases with decreasing desired product top size. Normally, a given type of size reduction unit will have a range of applicable specific energy requirements. Table 1 shows types of size reduction units. The figures listed were derived from various equipment manufacturers' product bulletins. These are ballpark

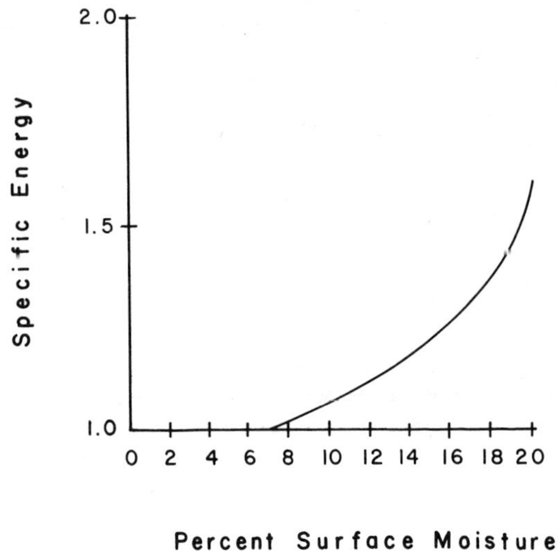

FIG. 4 Specific energy as a function of coal's surface moisture.

figures for various reduction ratios and coals of various grindability indexes. When impact and tumbling-type devices are used to produce top sizes that approach 1 μm in size, the specific energy requirements greatly exceed those listed.

E. Size Distribution

In designing size reduction circuits, the process engineer is concerned with the size distribution of both the feed and the product of a size reduction unit. More reliable selections could be made in designing these circuits if the feed

TABLE 1 Size Reduction Specific Energy Requirements

Size reduction device	Specific energy (hp/ton/hr)
Rotary breaker	0.1-0.3
Rotary breaker hammer mill	0.2-0.7
Roll crusher	0.1-0.6
Hammer mill	0.2-8.0
Impactor	0.5-20
Tumbling mill	1.0-20
Roller-type mill	10-15

and product of a size reduction unit would conform to a distribution in which the weight fractions related to a particular size. This would allow one to describe the size information in a functional form rather than as a set of numbers.

Fortunately, the size distribution of broken coal particles generally follows an exponential law similar to a normal or Gaussian distribution. Both the Rosin-Rammler and Schuhman distributions tend to represent the size distributions of broken coal.

1. *Rosin-Rammler Distribution*

In the Rosin-Rammler distribution,

$$R(x) = 100e^{(x/\bar{x})^n} \qquad (3)$$

where $R(x)$ = residue or the percentage of oversize particles of minimum particle size x
$\bar{x}$ = absolute size consist
n = size distribution constant
$\bar{x}$ = measure of the relative coarseness or fineness of the material
n = measure of the degree of dispersion of the particles [4]

When $\bar{x}$ is equal to x, the following is true.

$$R(\bar{x}) = 100e^{-1}$$

$$= \frac{100}{e}$$

$$= 36.79\%$$

In practice, $\bar{x}$ is the abscissa dimension where the plotted curve crosses the line $R(\bar{x}) = 36.79\%$.

The Rosin-Rammler distribution holds if a straight line results when the data are plotted as the log of the log 1/R versus the log x. When a straight line is obtained, the slope of the line is n. As n becomes smaller, there is a larger spread in the particle sizes. By knowing n and $\bar{x}$, a size distribution is completely specified. Rosin-Rammler graph paper is available with the necessary scaling of the ordinate and abscissa, which aids in the plotting of the size information.

2. *Schuhman Distribution*

In the Schuhman distribution,

$$P(x) = 100\left(\frac{x}{k}\right)^b \qquad (4)$$

where $P(x)$ = percentage of material finer than particle size x
b = size distribution parameter
k = size modulus, the theoretical value of x when $P(x)$ is equal to 100% [4]

If a straight line results when the data are plotted as the log of P(x) versus the log of x, the Schuhman distribution holds, and b is the slope of the straight line. For very small particle sizes, both the Rosin-Rammler and Schuhman equations yield the same results. From the Rosin-Rammler equation, the percentage passing size x is

$$P(x) = 100 - R(x)$$
$$= 100 - \left(100e^{-(x/\bar{x})^n}\right)$$
$$= 100\left(1 - e^{-(x/\bar{x})^n}\right)$$

For very small values of x,

$$e^x \simeq 1 + x$$

Therefore,

$$P(x) \simeq 100\left[1 - \left(1 - \frac{x}{\bar{x}}\right)^n\right]$$
$$\simeq 100\left(\frac{x}{\bar{x}}\right)^n$$

which is equivalent to

$$P(x) = 100\left(\frac{x}{k}\right)^b$$

3. *Translation*

An important concept when dealing with size distribution of feed and product samples from size reduction units is that of translation. Translation results when material is reduced in top size and the degree of dispersion of the particles as measured by n or b is the same for the entire product size distribution. When translation is demonstrated by a particular size reduction unit, one can project the size distribution of the product from the unit by specifying the desired top size and by knowing the degree of dispersion for the particular unit. In instances where information is not available on the product degree of dispersion for a size reduction unit, some manufacturers of coal size reduction equipment assume that the degree of dispersion of the feed material will hold for the product from the size reduction unit.

F. Size Reduction Circuits

Essentially all ROM coal must undergo some type of beneficiation prior to final utilization. No matter whether the run of mine coal undergoes a relatively simple sizing and crushing process or whether the coal quality is upgraded in an elaborate processing facility, size reduction of the coal is invariably involved. The final utilization of the coal must receive ample consideration in properly choosing the type and number of stages of size reduction to be employed.

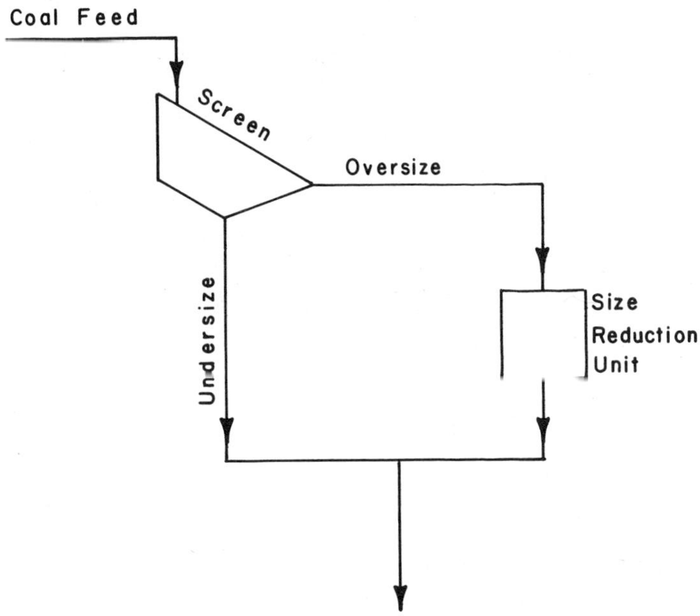

FIG. 5 Screening of feed material prior to size reduction.

In certain instances, coking coal and coal utilized for power generation is of sufficient quality whereby size reduction of the material is the only processing required. Size reduction of the ROM material will help ensure proper storage, handling, and marketing requirements for the coal producer, customer, and haulage agency. In instances where coal quality must be upgraded, ROM size reduction is used to maintain the proper size of feed to the processing unit, and to enhance the liberation of refuse from coal prior to cleaning.

According to the comminution principles, size reduction is performed most efficiently in a series of stages. There is no single machine that efficiently reduces large sizes to small sizes by repeated breakage. In addition, a particular type of machine is most efficient in acting on a certain size range of feed.

In instances where the production of an extreme amount of fines must be kept to a minimum, the material in the feed to the size reduction unit finer than the desired nominal top size of the product can be removed by screening prior to size reduction (Fig. 5). Another means to help minimize the production of fines is to close-circuit the size reduction system (Fig. 6). In a closed-circuit system, the internal or external circulating load should be high, and the classifying device (screen or other type of classifier) should be efficient, so that all the undersized material is removed from the circulating load and not returned to the size reduction unit. Besides reducing the production of fines, a closed-circuit system can help increase the output and reduce the specific grinding energy from a size reduction unit.

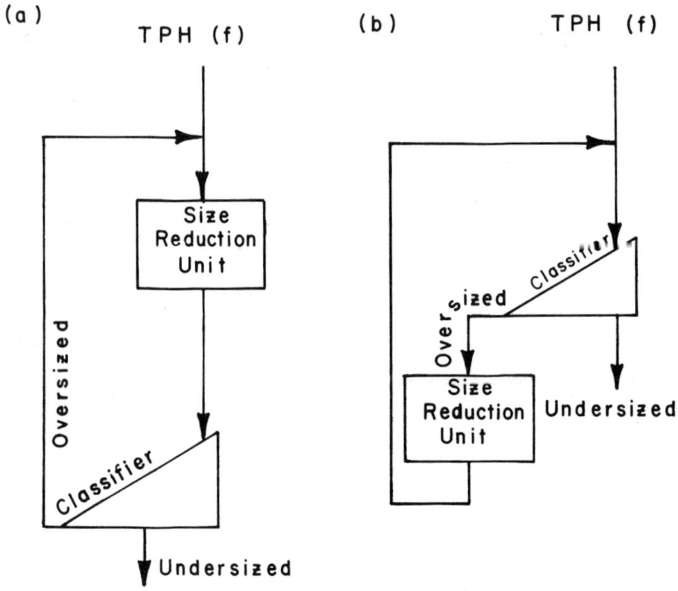

FIG. 6 Closed-circuit size reduction circuit: (a) classifying after size reduction. (b) classifying prior to size reduction.

For Fig. 6a, in which classification occurs after size reduction, the tons per hour to the classifier or size reduction device, TPH, can be calculated as follows:

$$TPH = \frac{TPH(f)}{(Eff)P(s)} \tag{5}$$

where TPH(f) = feed tonnage
Eff = classifier efficiency
P(s) = cumulative fraction of the size reduction device product less than the classifier setting

For Fig. 6b, in which classification occurs prior to size reduction, the tonnage to the classifier is equal to the tonnage of the size reduction device, TPH(r), plus the feed tonnage to the circuit, TPH(f). The tonnage to the size reduction unit, TPH(r), can be calculated as follows:

$$TPH(r) = \frac{(1 - Eff)G(s)}{(Eff)P(s)} TPH(f) \tag{6}$$

where Eff = classifier efficiency
G(s) = cumulative fraction of feed less than the classifier setting
P(s) = cumulative fraction of the size reduction device product less than the classifier setting

G. Applications

Initial ROM size reduction entails the reduction of the top size of the coal to a product with a nominal top size as fine as 1½ to 1 in. Normally, the nominal top size is in the range 3 to 8 in.

In general, four types of equipment are used for the size reduction of ROM coal: rotary breakers, roll crushers, hammer mills (ring mills), and impactors. Normally, ROM crushing produces the desired product in a single stage. However, there are cases where the size reduction of ROM coal entails multiple-stage size reduction units.

Quite often, size reduction is one of the last unit operations performed in the benefication of coal. Size reduction can help to ensure proper handling characteristics and maintain proper marketing specifications. Normally, the nominal top size of the product will be within the range 6 to ½ in., although reduction to a nominal top size of 1 mm can be achieved. Usually roll crushers, impactors, hammer mills, and ring mills are used in this application. The factors that influence the final product size from the benefication process include storage and handling requirements associated with the coal processing facility, transportation requirements and limitations, plus market requirements, which includes storage and handling limitations.

During the benefication of coal, the process engineer will probably be confronted with the problem of treating a middlings material. This material is not of sufficient quality to be included with the high-quality product, yet it contains potentially recoverable coal. If this material is simply recirculated through the cleaning circuit, little or no quality upgrading can be achieved. Liberation can be achieved if the nominal top size of the material is reduced, which permits recovery of the high-quality coal in the cleaning unit. Hammer mills, ring mills, impactors, and roll crushers in either open or closed circuits can be utilized to reduce the nominal top size of the middlings. Normally, reducing the nominal top size to the range 3/4 to 1/4 in. is practiced.

Infrequently, size reduction must be performed on the refuse or gangue material from the benefication process. Reducing the nominal top size of the refuse results in improving problems associated with handling and abrasion characteristics of the refuse material in the process installation. In addition, the increased surface area that results from the size reduction of this material aids in stabilizing refuse disposal areas. This application can be handled by the use of hammer mills, ring mills, and impactors.

Prior to final utilization of either raw or clean coal, some type of size reduction is usually required. The number of stages in the final size reduction process depends upon final specific utilization of the coal as well as the condition (nominal top size and size consist) of the coal to be utilized. For example, the size of the coal feed to a coking operation is coarse compared with the size of the coal feed to an electric-power-generating unit. Furthermore, the number of stages of size reduction involved in preparing a coal feed for a specific utilization will increase as the nominal top size and degree of dispersion of the size consist of the material increases.

Coal utilized for power generation and most conversion processes must undergo size reduction to produce a product with a nominal top size of less than 1 mm prior to final utilization. Equipment capable of producing such a product include rod and ball mills, roller-type mills, hammer mills, and impact-type mills.

III. SIZE REDUCTION DEVICES

As mentioned previously, size reduction actually begins with the mining process. The three major factors that influence the size consist of the mine product are the coal itself, the particular mining process, and the handling system utilized to convey the coal from the mine face or pit to the processing facility.

Since coal preparation engineers have, in general, little control over factors influencing the size consist of the mine product, their primary function in designing a size reduction system is to choose a device which will yield a product that meets the desired product specifications. Fortunately, the preparation engineer has numerous types of size reduction devices from which to choose. However, to make the best possible selection, the preparation engineer must have a solid understanding of the various devices.

A. Rotary Breaker

The rotary breaker is an autogenous size reduction device in which the feed material acts as a crushing or grinding media to effect size reduction. The rotary breaker causes the breakage of material along natural cleavages by lifting the material to a given height and then dropping the material against a hard surface. The rotary breaker is used in reducing the top size of run of mine or raw coal. The raw coal from the rotary breaker can be utilized directly, undergo further size reduction, or the quality of the product can be upgraded in a coal-washing facility.

The rotary breaker is a cylindrical, slow-speed device (Fig. 7) that rotates at a speed of 12 to 18 rpm [5]. The speed at which the unit rotates is expressed as a function of the critical speed, the lowest speed that will cause a small particle on the liner to centrifuge. Since the critical speed for a rotary breaker decreases as the diameter of the unit increases, the operating speed of the rotary breaker must also decrease as the diameter of the unit increases. The liners of a rotary breaker are actually screen plates that function as a peripheral discharge for properly sized material. To raise the coal for breakage, lifters are employed. Plows are used to enhance the flow of material from the feed to the discharge end.

The rotary breaker is able to maintain a uniform product top size while producing a minimum amount of fines because the undersize material is removed at the rate it is produced through the screen plate openings. In most installations, a grizzly screen is installed ahead of the rotary breaker so that the undersize material in the feed is removed before it can enter the breaker.

Rotary breakers can remove large rocks and tramp material when the coal is considerably softer than the tramp material by taking advantage of perferential breakage. Other advantages of the rotary breaker include minimum maintenance requirements and its ability to handle fluctuations in feed moisture, feed tonnage, and feed size consist.

The rotary breaker can reduce ROM or raw coal to a top size as small as $1\frac{1}{2}$ to 1 in., although the usual product top size is in the range 8 to 3 in. Usually, the maximum top size in the feed to a rotary breaker should not exceed 3 to 4 ft.

FIG. 7 Rotary breaker: (1) grizzly screen; (2) screen plate; (3) lifters; (4) rock discharge.

The size of a rotary breaker is based upon the diameter and length of the unit. The diameter of a rotary breaker ranges from 6 to 14 ft, with the length in the range 8 to 28 ft [6]—in other words, a length/diameter ratio between 1.33 and 2.00.

Manufacturers' power requirements for a rotary breaker range from a low of approximately 15 hp to a high of approximately 250 hp, with corresponding capacities in the range 60 to 1500 tons/hr. The capacity and power requirements for a particular unit are dependent upon the coal grindability, top size of the feed, and desired product top size. The power consumption of a rotary breaker can be approximated by the formula

$$P = \frac{d(75 + \pi L)}{44} \tag{7}$$

where P = horsepower
d, L = diameter and length of breaker screen, respectively, ft [7]

Several manufacturers have slightly modified the rotary breaker to produce the rotary breaker manner mill (Fig. 8). The difference between the two devices is the addition of a rotor at the discharge end of a conventional

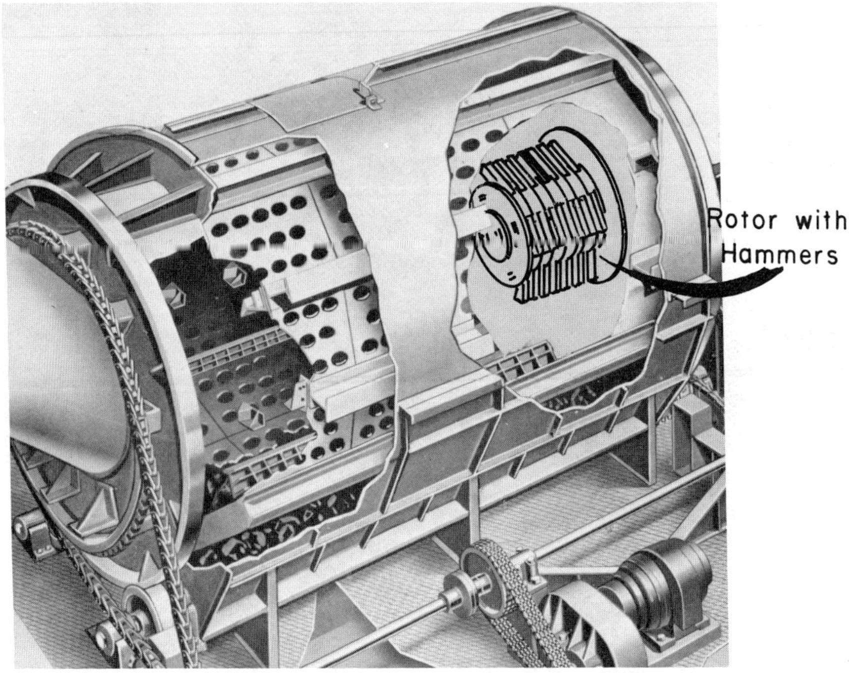

FIG. 8 Rotary breaker hammer mill.

breaker, which results in the material remaining within the breaker until it is of sufficient size to pass through the openings in the screen plates. The addition of the rotor helps facilitate breakage of harder fragments that would normally be discharged as tramp material, plus an increase in the capacity of a given conventional rotary breaker by as much as 20% [8]. The power increase required by the addition of the rotor is in the range 30 to 50 hp. This type of unit can be advantageous where available space is limited.

B. Roll Crusher

Roll crushers are size reduction devices that shear and/or compress the material that is to be reduced in size. The splitting and/or compressing action occurs between two hardened surfaces. The surfaces used can take the form of a rotating roll and a stationary anvil, or two equal-speed rolls rotating in opposite directions. The surfaces of the rolls can be smooth, corrugated, or toothed. The particular design of the teeth and corrugations is critical in keeping the production of fines and dust to a minimum [6]. In units where the size reduction occurs between two rolls, the teech on the rolls are designed to mesh. Roll crushers, particularly double roll crushers, unfortunately have a tendency to produce slabby-type particles.

The speed and diameter of the rolls, as well as the desired capacity, feed size consist, and desired product size consist, are critical in the proper sizing of these units. The clearance between the crushing surfaces can be approximated once the feed and product specifications are known. Then the roll speed must be set so that it is fast enough to prevent stalling or loss of capacity, yet not so fast that the feed material is thrown away from the crushing rolls. Depending on the particular unit and application, roll speeds can range from a low of approximately 60 rpm to a high of 250 rpm [8]. For most applications, the peripheral speed of the rolls is inversely proportional to the size of the feed material.

The "angle of nip," frequently used in conjunction with double roll crushers, is the angle formed by a spherical particle setting between two equal-diameter crusher rolls separated by a given distance. The angle of nip is a function of the coefficient of friction between the particle and crusher rolls. For a given size particle, the roll diameter and spacing must be chosen so that the angle of nip is large enough to prevent the particle from being thrown away from the crushing rolls. Hence, there is a constraining relationship among the top size of the feed, the diameter of the rolls, the gap setting, and the surface of the rolls. This is the reason three- and four-roll crushers are designed to conduct two-stage size reduction when large reduction ratios are desired.

Normally, the clearance between the crushing surfaces of roll crushers can be adjusted. Some units have been designed with the ability to adjust this clearance between the crushing surfaces while the units are operating.

Information presented in the report by Kestner et al. [9], plus information presented in literature supplied by McNally Pittsburg and that presented by Narasimhan and Sastri [10], indicate that the concept of translation holds for roll-type crushers. The information showed that the degree of dispersion of the particles, b, for the Schuhman distribution, remained relatively constant as the product top size decreased for a particular unit.

A single-roll crusher (Fig. 9) consists of one movable roll plus a curved anvil with renewable wear plates. The anvil is normally hinged at the top to permit the passage of tramp material without damage to the unit. Most units are also equipped with shear pins to prevent an overload of the drive unit. For a given reduction ratio, single-roll crushers are capable of reducing

FIG. 9 Single-roll crusher.

ROM material to a product with a top size in the range 8 to 3/4 in. in a single pass, depending upon the top size of the feed coal. Normally, 36-in. cubes in the feed are the maximum-size material that a single-roll crusher can accept.

Single-roll crushers can vary in size from 18 × 18 in. to 36 × 100 in. (roll diameter by roll length). This type of unit can handle capacities as large as 1500 tons/hr with maximum power requirements of 200 hp.

Double-roll crushers (Fig. 10) consist of two rolls that rotate in opposite directions. Normally, one roll is fixed while the other roll is movable against spring pressure. This permits the passage of tramp material without damage to the unit. The drive units are normally equipped with shear pins for overload protection.

Double-roll crushers rely more on shearing action than on compressing action, which results in the production of minimal amounts of fines [6]. Double-roll crushers can reduce ROM material (maximum top size of 36 in.) to a product with a top size in the range of 14 to 3/4 in. The roll size of these units can vary from 12 × 18 in. to 60 × 60 in. (roll diameter by roll length). Capacities can range between 10 to 2000 tons/hr with corresponding power requirements of 5 to 150 hp.

Roll crushers are also manufactured in three- and four-roll, two-stage units. The upper stage performs coarse crushing and the lower stage is utilized for fine crushing [6]. In the triple-roll unit, the upper stage consists of a single-roll crusher and the lower section is a double-roll unit. In the four-roll unit, both the upper and lower sections are double-roll units.

Normally, these two-stage units produce a product with a top size in the range of 4 to $\frac{1}{2}$ in. in a single pass. Generally, the rolls for these units can vary in size from 12 × 24 in. to 18 × 72 in. (roll diameter by roll length). The upper and lower rolls for these units have identical diameters and lengths. This type of unit can handle capacities up to 600 tons/hr with power requirements in the range 20 to 200 hp.

C. Hammer Mill

The hammer mill (Fig. 11) is a size reduction device in which the feed material is initially struck (impacted) by rotating hammers and then thrown (impacted) against grid breaker) plates. Then the material that is not fine enough to

FIG. 10 Double-roll crusher.

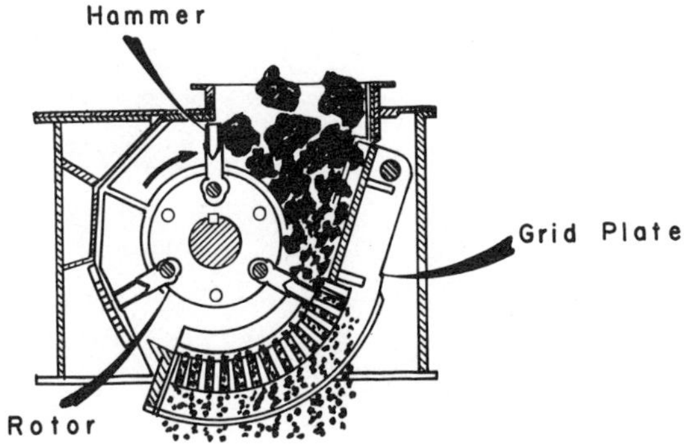

FIG. 11 Hammer mill.

pass through the grid plates at the bottom of the unit is further reduced in size as a result of the hammers grinding the material against the grid plates until the material is of sufficient size to pass through the openings. Normally, the size reduction due to impact occurs above the center line of the rotor, whereas the size reduction due to attrition (grinding) occurs below the center line of the rotor.

Impact occurs when a force is applied suddenly and quickly to a particle or object. As the speed of the crushing surface or particle to be crushed is increased (i.e., increase the applied force), potential for size reduction that can occur is also increased. It is postulated that an impact-type force causes selective failure primarily along the natural breakage cleavages of the material and results in a product whose particles are fairly rounded [6].

Results from tests conducted by Popelnik, Shkodkin, Donde, Ivanov, and Akulov claim that in coal size reduction the breaker plates are not involved in the size reduction of the coal. Rather, all size reduction results from impact with the hammers and grinding between the hammers and the grid plates [11].

The grinding action or attrition that occurs in the hammer mill results from the rubbing action of feed material between two hardened surfaces (i.e., the hammers and the grid or screen plates). This attrition will produce slivers, and a comparative high percentage of fines, as compared to other types of size reduction devices [6].

One way of reducing the fines production from a hammer mill is to remove the grid plates and then close-circuit the hammer mill with a screen or classifier. Close-circuiting, as shown in Fig. 6, means sizing the feed to a hammer mill or the product from a hammer mill on a screen or in a classifier, and returning the oversized fraction to the hammer mill for further size reduction.

The ring-type hammer mill (Fig. 12) is a modification of the hammer mill that was developed to help minimize fines production. In the ring-type hammer mill, the hammers have been replaced by rings (alternately toothed and plain) which are free to revolve. The rings cause size reduction in the attrition zone by a rolling compression rather than the grinding action that occurs with hammers [6].

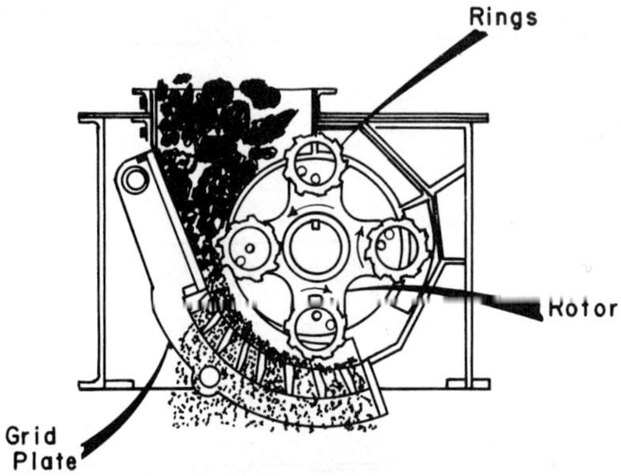

FIG. 12 Ring-type hammer mill.

Another method employed to reduce the production of fines in a hammer mill involves the use of an air-swept crushing circuit. By adjusting the ventilation rate and associated air classifier, the operator can determine how long the material remains in the milling zone and how many times the coarse fractions circulate around the circuit [12].

The hammer mill is a fairly versatile unit and can be used in the size reduction of ROM coal, washed or prepared coal, middlings, and refuse material. Hammer mills can be designed with side pockets for the capture and removal of tramp iron, which could severely damage a unit. Tramp iron can be removed from the feed prior to entry into the size reduction unit by positioning an electro or permanent magnet in close proximity to the feedstream.

Factors that affect the performance of a hammer mill include rotor speed, rotor diameter, feed coal characteristics, type, configuration, and condition of hammers, hammer clearance, feed rate, and grid openings.

The feed characteristics, particularly the moisture content of the feed, affects the performance. An increase in the moisture content in the feed to a hammer mill tends to reduce the capacity and increase the fineness of the product [13]. According to T. G. Callcott, the fines production in a hammer mill increases as the feed tonnage to a particular unit is increased [13]. Also, the capacity tends to decrease as the size of the raw coal increases, for a given reduction ratio [14].

The rotor speed has a significant effect on hammer mill performance. For a given size unit, the capacity is directly proportional to the rotor speed [11]. The fineness of the product is determined to a large extent by the intensity of the hammer blow. This, in turn, is directly proportional to the tip velocity of the hammer. If the rotor speed is too low, stalling occurs. However, if the rotor speed is too high, excessive fines are produced [15]. In addition, the capacity of the given unit can be increased by increasing the rotor diameter [12].

The clearance between the hammers and grid plates contributes to the product characteristics. As this clearance is reduced, there is an increase in the associated crushing in the attrition zone, particularly in the coarse end of the product size distribution [15]. Also, the capacity of a given unit will decrease as this clearance increases as a result of wear [14]. The grid openings normally have a large effect on the size of the product only when the size of the openings exceeds the hammer clearance [15].

To help reduce maintenance from wear, most units are equipped with reversible hammers. The effect of hammer wear on hammer mill performance has a greater effect as the rotor speed increases [15]. To help reduce maintenance and downtime due to excessive wear, the reversible hammer mill was developed. In this unit, each side or half is symmetrical with the other (Fig. 13). This permits rotation of the rotor and hammers in both the clockwise and counterclockwise directions. Periodic reversal of direction of rotation helps to minimize maintenance.

Changing the number of hammers used, within reasonable limits, has some effect on product characteristics. However, this effect becomes less appreciable at high rotor speeds [15]. Normally, operating a hammer mill with a full set of hammers will yield a slightly more uniform product, although more power is required [16]. It has also been noted that the weight of the hammers used does not significantly affect the reduction ratio and throughput of a given unit. However, lighter hammers help reduce the fines production and power consumption [11].

In a study conducted by Rakov, the hammers from a hammer mill were replaced with fish plates. This arrangement resulted in the reduction of the amount of the coarse fraction in the product without significantly increasing the production of fines. In addition, as a result of this modification, the capacity increased and the power consumption was reduced [17].

Tests have also been conducted by Yampolskii and Khesin with a vibrating hammer mill to help reduce certain operating problems. These tests indicated that vibrations helped reduce regrinding of the crushed product, helped eliminate problems associated with excessive feed moisture, and helped increase the capacity of the hammer mill [18].

Tests by Rakov [17] and by Dolganvo et al. [12] indicate that b, the degree of particle dispersion for the Schuhman distribution, shows translation

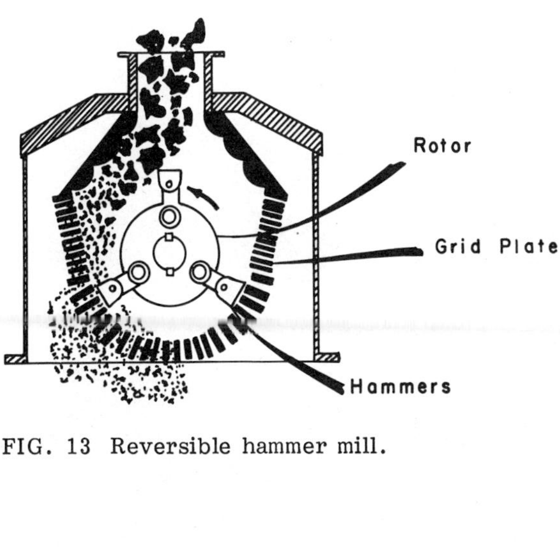

FIG. 13 Reversible hammer mill.

between the feed and product of a particular unit. In addition, information contained in Jeffrey Sales literature shows translation to be valid as the product top size decreases and as the rotor rpm increases.

Hammer mills are capable of receiving ROM or raw coal with a maximum top size as large as 60 in. Normally, though, most commercial units receive material with a maximum top size in the range 18 to 24 in. The maximum allowable top size to a hammer mill will decrease as the size of the rotor decreases.

The rotor size for a hammer mill is designated by the diameter and width. The size of a rotor will normally fall within the range 12 in. in diameter by 12 in. in width to 72 in. in diameter by 140 in. in width. However, some mills are available which fall on either side of this range.

Industrial units require drives that fall in the range 20 to 2500 hp. Lab units normally require drives in the range ½ to 5 hp. Most industrial units use rotor speeds in the range 400 to 1800 rpm. The slower rotor speeds are used with larger-sized, larger-capacity units. Normally, units whose rotor speed falls within the above-mentioned range will produce a product with a top size in the range 4 mesh to 2 in. To produce a product with a top size in the range 20 to 200 mesh, units with rotor speeds in the range 4000 to 8000 rpm are employed.

Capacities of hammer mills can be as low as 1/8 ton/hr to 2 tons/hr for the lab-type units or range from 5 to 2500 tons/hr for industrial units.

D. Impactors

Impactors are size reduction devices that strike (impact) or impart energy to the feed material, which is then thrown against a hard surface or against other particles in the feedstream. The required forces are applied suddenly and quickly. As the speed of the crushing surface or the speed of the particles to be crushed is increased, the potential reduction that can occur also increases. The major difference between an impactor and a hammer mill is the elimination of the grid bars employed in a hammer mill.

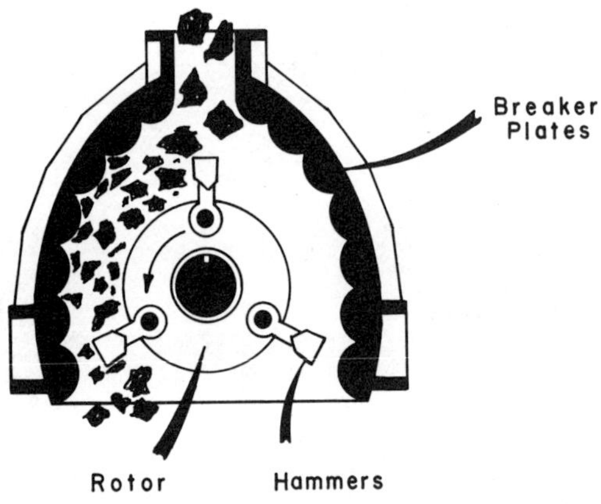

FIG. 14 Reversible-rotor-type impact mill.

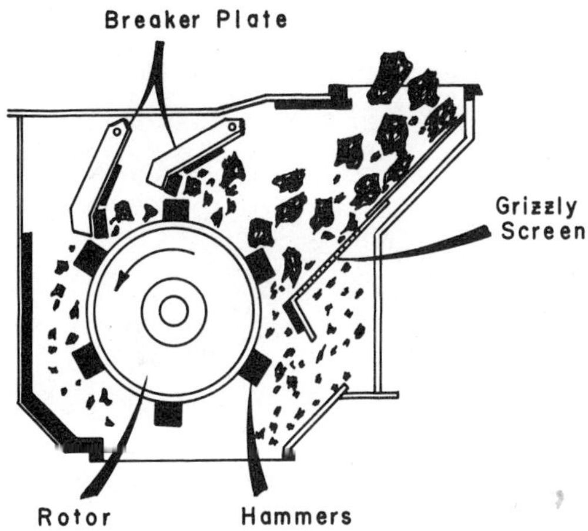

FIG. 15 Single-rotor impact mill.

Impact-type size reduction normally produces cubicle-shaped particles, a well-graded product, and breakage of material along natural cleavages (selective breakage). An impact mill applies its shattering power in direct proportion to the particle size [6].

Since grid bars are not used in an impact mill, the product from the impact mill will usually be coarser than the product from a hammer mill, and it is usually difficult to accurately control the top size of the product from an impact mill. Frequently, an impact mill will be put into closed-circuit with either a screen or an air classifier, as shown in Fig. 6. In such an arrangement, the product from the impact mill is classified in the appropriate unit with the oversized fraction being returned to the impact mill for further size reduction. When an air classifier is used, the operator uses the ventilation rate and classifier adjustments to control the output from the circuit [12].

In impact-type mills, deterioration of quality and quantity of the product occurs as a result of an increase in the moisture content of the feed [18]. If an air classifier is used to maintain proper product characteristics (maintain proper top size and help minimize fines production), ventilation air can be preheated and used to dry the feed material as well as convey it.

The rotor-type impact mill (Figs. 14 to 16) uses rotors or rotors equipped with beaters to effect size reduction. In these devices, the feed material drops into the path of the rotor, is hit and shattered by the rotor, and is then driven against the impact surface, where it is further reduced in size. The material rebounds into the path of the rotor and the cycle repeats itself until the product is discharged from the bottom of the impactor.

Rotor-type impactors can be manufactured with rotors that can be rotated in either the clockwise or counterclockwise direction. Periodic reversal of rotation helps to reduce maintenance requirements.

Factors that influence the performance of a rotary-type impact mill include rotor speed, rotor diameter, feed coal characteristics, and feed rate. These factors follow the same general relationships that were discussed in the section pertaining to hammer mills.

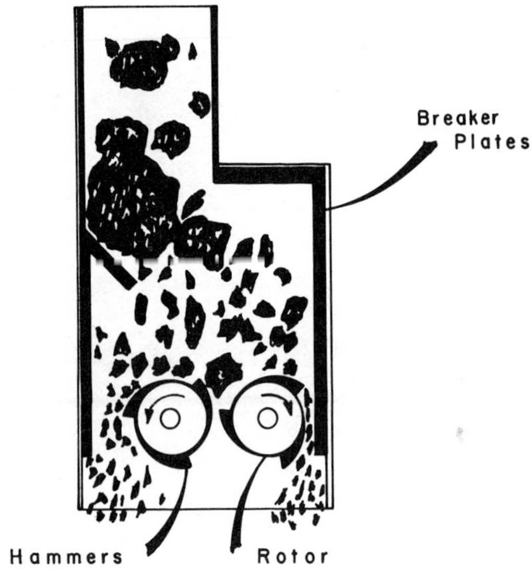

FIG. 16 Double-rotor impact mill.

The study conducted by Leonard and Clendenin [19] shows that translation holds between the feed and product of a given unit. Also, the study by Kestner et al. [9], plus the literature from International SMB Crushing Equipment, indicate that translation is valid as the rotor speed of a given unit is increased.

For the rotor-type impact mill, maximum feed size will range from 1/8 in. top size for lab units to 60-in. cubes for commercial units. The top size of the product can range from as large as 6 in. for industrial units to a top size of less than 400 mesh for lab units.

For industrial units, the rotor size normally falls within the range 18 in. in diameter by 18 in. in width to 72 in. in diameter by 120 in. in width. The rotor speed will normally fall within the range 200 to 1500 rpm. For industrial use, the capacities of such impactors range from 2 tons/hr to 1900 tons/hr with power requirements in the range 10 to 2100 hp.

Cage mills (Fig. 17) are internally fed size reduction devices which consist of either one, two, four, or six concentric cages with the adjacent cages spinning in the opposite direction at the same rotational speed. Usually, two motors are required, one for each direction of rotation. However, for single-cage mills, only one motor is required. Each cage consists of rows of rods supported on each end by a ring. These rods are normally 4 to 10 in. in diameter. The feed is introduced into the center of the revolving inner cage. Centrifugal forces throw the material outward through the cage toward the replaceable breaker plates. The feed material strikes the cages, other feed particles, and the breaker plates. After many rebounds, the product discharges through the bottom of the mill. The product can then be classified and oversized material returned to the cage mill. Wear of components in these units can be improved by periodically reversing the direction of rotation.

The principal means for controlling the amount of size reduction is to control the speed of the cages. However, the product size can be affected by the spacing of the rods in the cages if the spacing between the rods exceeds the cage clearance with respect to the breaker plates [15].

Sales literature by Steadman Foundry contains information that substantiates the concept of translation. As the number of rows in the mill increases and the spacing between the rods decreases, the top size of the product is reduced, but the degree of dispersion of the particles measured as b for the Schuhman distribution remain relatively constant.

A flared cage mill is simply a cage mill with a flared housing. It was designated to help eliminate clogging and plugging when handling wet, sticky material. A stud or peg mill operates essentially the same as the cage mill. Basically, the only difference is that these mills use studs or pegs to strike the material, as opposed to the rods in the cage mill.

Various sizes of cage mills can handle feed sizes from as small as $\frac{1}{2}$ in. to as large as 18 in. The product top size normally falls within the range of $1\frac{1}{4}$ in. to 150 μm. Cage mills used for lab work handle capacities of less than 100 lb/hr, whereas commercial units handle capacities from 1 to 3600 tons/hr. Normal power requirements fall within the range 5 to 1500 hp, with corresponding cage speeds of 1600 to 600 rpm. Commercial cage mills range in diameter from 18 to 95 in. The indicated diameter is the outside diameter of the largest cage in the particular unit.

A couple of equipment manufacturers produce a size reduction unit that performs multiple-stage size reduction. A hammer mill section is employed first. The product from this section is further reduced in size by impact and attrition. The process of impact and attrition can occur in either one or two sections. The product is then classified and the oversized material returned to the impact and attrition sections for further size reduction. In many instances the classifying air is preheated so that higher-moisture feed material can be handled successfully. Higher feed moisture will cause a reduction in capacity of the units.

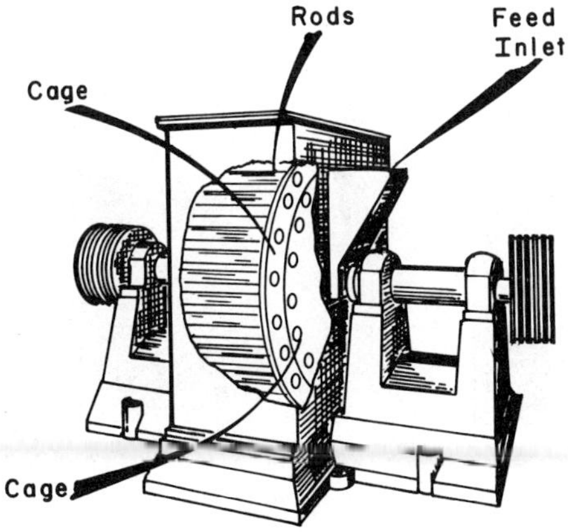

FIG. 17 Cage mill.

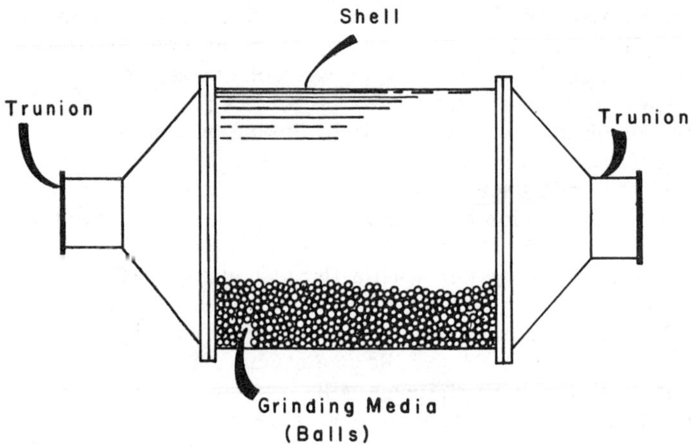

FIG. 18 Ball mill.

These types of devices can handle a feed top size of approximately 2 in. and produce a product that contains approximately 70% -200 mesh material. Capacities of these units lie between 1500 and 5400 lb/hr with power requirements in the range 20 to 400 hp.

E. Tumbling Mills

Tumbling-type mills (Figs. 18 and 19) are horizontal, cylindrical size reduction devices that tumble the material and the grinding media to effect size reduction. The grinding media can consist of forged steel, cast alloy, or ceramic balls, or hot-rolled high-carbon or alloy steel rods.

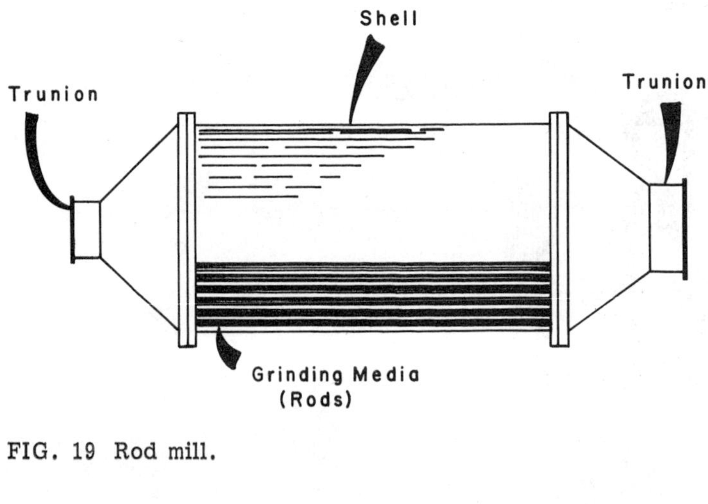

FIG. 19 Rod mill.

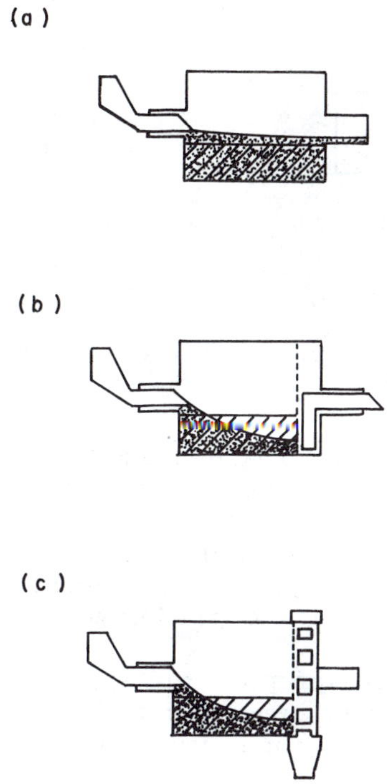

FIG. 20 Ball and rod mill discharge arrangements: (a) trunnion overflow arrangement; (b) grate discharge arrangement; (c) peripheral discharge arrangement.

As the mill rotates at a relatively slow speed, normally in the range 20 to 30 rpm [20], the material and grinding media are lifted approximately two-thirds of the way up the periphery of the shell and then cascade continually toward the center of the mill. In addition to size reduction on the larger particles from impact as the material and media are dropped, size reduction of the smaller particles can result from attrition and compression.

Of the various types of tumbling mills used for size reduction, the ball and rod mills are employed for the size reduction of coal. Both devices are essentially large lined drums that are supported by hollow trunnions at each end. The feed material can be introduced through one or both of the trunnions. The units are usually driven by a gear train located at one end of the unit. The inside of the drum and trunnion is lined with replaceable wear-resistant cast iron or rubber. Normally, the drum liner is fitted with lifters that enhance the lifting action of the feed material and grinding media.

Several arrangements can be used with tumbling-type mills to remove the product from the units. Units can be manufactured with overflow, grate, or peripheral discharge arrangements (Fig. 20). For an overflow discharge (Fig. 20a), the mill level is maintained slightly above the bottom of the trunnions so that the ground material can readily pass through the trunnions at the discharge end.

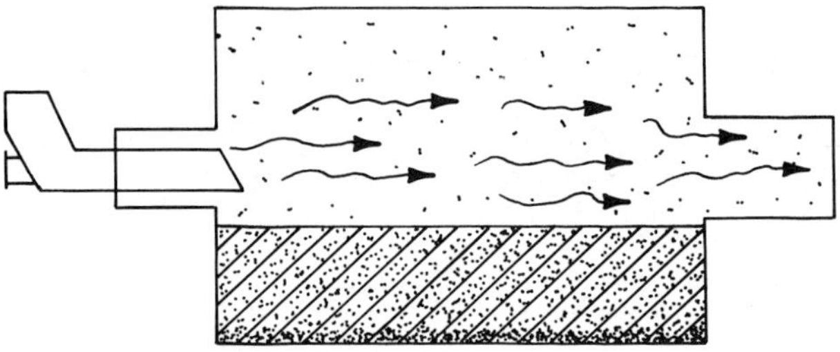

FIG. 21 Air-swept ball mill.

For the grate arrangement (Fig. 20b), discharge grates are fitted between the cylindrical mill body and the discharge chamber. The ground material first passes through the discharge grates and then passes through the trunnion at the discharge end of the unit.

In a peripheral discharge arrangement (Fig. 20c), the ground material leaves the unit through openings in the mill shell. The discharge openings are normally fitted with a perforated partition to keep the grinding media within the unit.

For most coal applications, tumbling mills are equipped with a special discharge. This arrangement permits the use of air sweeping in the units (Fig. 21). Most tumbling mill installations are air-swept systems in which the air, after sweeping the mill, will enter a classifier and the oversized fraction is returned for additional size reduction, while the undersized material proceeds to the final utilization process. Sweeping air is often preheated so that feed material with high surface moisture contents can be handled satisfactorily. Excessive surface moisture is detrimental because it results in agglomeration of the coal and coating of the grinding media and liners. The coating reduces the capacity and if severe enough can actually cause the capacity to approach zero. The agglomeration causes proper-size particles to be returned to the mill during classification, leading to lower capacity and overgrinding. One drawback to dry size reduction is that surface forces come into action to cause cushioning and media coating, which result in less efficient use of energy.

As the sweeping air rate through a mill is increased, the fineness of the product decreases and the output increases. The use of an air classifier in the milling circuit helps to reduce the effect of variations in sweeping air rates [21]. Increased sweeping air rates do not exhibit the above-stated results if the sweeping air does not contain sufficient heat to dry the coal to a low-enough moisture level [21]. In an air-swept system, progressive drying and size reduction occurs from the feed end to discharge end until the particles are of sufficient size to become entrained in the sweeping air [21]. In an air-swept closed-circuit system, the dry oversized material that is returned to the mill for further size reduction helps to reduce the average moisture content of the feed.

The use of a tumbling mill in a closed-circuit system minimizes the amount of super fines and helps to maintain tighter control of the top size. In addition, a properly operated closed-circuit system requires less grinding horsepower per ton of product than does an open-circuit system when producing material with a common product size particle, such as 80% passing 200 mesh [20].

As mentioned earlier, these units normally rotate at speeds from 20 to 30 rpm. The upper speed of these units is limited by the critical speed, which is the lowest rpm that will cause a small particle on the shell liner to centrifuge. This critical speed is a function of the mill diameter. As the mill diameter increases, the critical speed decreases. Usually, the speed of a ball or rod mill will not exceed 70 to 75% of the critical speed for the particular unit [22]. There is an optimum speed at which the energy consumption is at a minimum. Normally, this optimum speed is less for a rod mill than for a ball mill.

Essentially, all ball and rod mills are custom-designed to the particular application. The process engineer must consider the feed characteristics (particularly the top size, hardness, and moisture of the material), desired product consist, speed of the unit, size of the unit, and grinding media charge. The process engineer is faced with various trade-offs in selecting the proper unit. By increasing the size of the grinding media, a given unit is capable of reducing a larger or harder particle to a particular size. Conversely, as the size of the grinding media is decreased, the number of balls available for grinding increases, which results in an increased grinding capacity. The problem of handling larger or harder material with relatively small grinding media can be alleviated by increasing the mill diameter or increasing the mill speed [22].

As the length of a rod or ball mill is increased, the retention time also increases if feed to the mill is held constant; hence, the fineness of the product increases. If the retention time is held constant, the capacity increases as the length increases, assuming that mass throughput can be maintained. Longer mills can be designed to handle the mass throughput with two or three compartments (Fig. 22). These mills can contain larger grinding media in the first compartment, with each successive compartment containing smaller grinding media. This permits more efficient operation when producing a very fine product. Such mills are not common in coal grinding, except when the first compartment of a two-compartment arrangement is used as a tumbling dryer.

Rod mills preferentially grind the largest particles, which results in a product with a relatively narrow size distribution. These units are capable of handling feed material as large as 2 in. and produce a product that can range in size from 100% -4 mesh to 100% -20 mesh [20]. In sizing rod mills, the rule of thumb employed is that the length/diameter ratio of the unit should be between 1.3 and 2.0 [20]. The interaction between the size of the rods and the mass throughput determines the ratio. The maximum length of good-

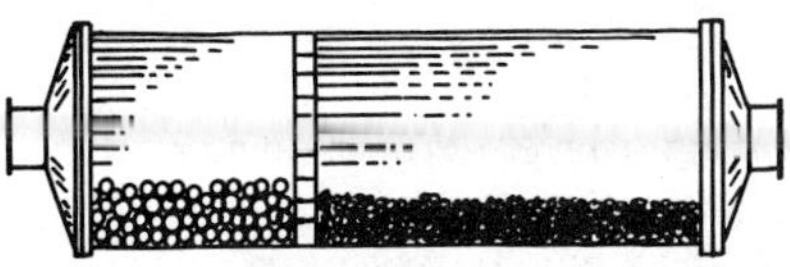

FIG. 22 Two-compartment ball mill.

quality rod available is 20 ft. The length of rod/diameter of the rod mill ratio necessary to avoid entangling the rods during operation is 1.25 to 1.4. Normally, the rod charge occupies 35 to 45% of the inner-shell volume [23]. New rods can range in size from $1\frac{1}{2}$ to 6 in. in diameter [20] and are normally 6 to 9 in. shorter than the internal distance between the head liners [22]. Rexnord has manufactured rod mills that require 2500-hp drives to handle capacities in excess of 300 tons/hr of coal to produce a 14-mesh top size product. Rod mills can be equipped with an overflow discharge, end peripherial discharge, or center peripherial discharge.

Ball mills are capable of producing material that is finer than the product from a rod mill. Ball mills normally handle a coal feedstock with a top size as large as 2 in. and produce a product with a top size that can range from 35 mesh to 10 μm [22]. Ball mills for coal size reduction are designed with the length/diameter ratio between 1.0 and 1.4. Ball diameter ranges in size from 3/4 to 4 in. [20], but coal mills normally use a maximum ball top size of 2 in.

For initial charging, various size balls are used. However, at periodic intervals, the mill is recharged, while operating, with the largest size ball that the system was designed to handle [6]. If properly maintained, a ball mill will produce a product of a given fineness, regardless of the length of mill service. A ball mill offers several other advantages. First, these units can handle tramp material without a loss of performance, and ball mills have a large reserve capacity which permits output responses independent of feed conditions.

Design calculations have been carried out for ball mills that would handle capacities as large as 200 to 300 tons/hr of product, with power requirements as high as 2500 to 6000 hp, depending upon the reduction ratio.

F. Roller Mill

For this discussion, the term "roller mill" will be used to include medium-speed mills such as roller mills, ring ball mills, and bowl mills (Fig. 23). In these types of mills, size reduction results from the material being ground (compressed) or rubbed between two hardened surfaces, with one surface rolling over the other. Normally, the grinding surfaces are rollers and a bowl (Fig. 23a), rollers, and a ring (race) (Fig. 23b), or balls and a ring (race) (Fig. 23c). Either springs or hydraulic cylinders are used to maintain proper pressure on the grinding elements. Centrifugal force feeds the material to the grinding area.

In a ball-ring type of mill, the ball diameter is approximately 20 to 35% of the ring diameter. For a roll-ring type of mill, the roll diameter is approximately 25 to 45% of the ring diameter [24].

As the material is reduced in size, the ground material is entrained in a stream of heated air, which drys and conveys the material to an internal air classifier. The oversized material is returned to the size reduction section for further treatment, while the undersized material is conveyed to the utilization process. Most of these units are equipped with a spring-loaded gate so that any tramp material that enters with the feed can be removed without serious damage to the unit.

FIG. 23 Roller-type mills: (a) roller, bowl type; (b) roller, race type; (c) ball, race type.

In this type of unit, the size reduction forces are applied to a static bed of material rather than to individual particles [14]. This type of size reduction requires more energy than is associated with breakage of particles, such that neither the original particles nor their products are in substantial contact [14]. An excessive amount of fines present in the bed of material will tend to cushion the large particles from the applied size reduction forces and result in a decrease of capacity for a particular unit [14]. However, the bed of material is essential to prevent excessive wear to the grinding elements. Besides an excessive amount of fines, an excessive amount of sticky material, and an excessive feed moisture will also result in decreasing the capacity of these types of units [14].

Other factors that affect roller mill performance are the hardgrove index of the coal, feed rate, air rates, and air temperature. For a given product fineness, the power requirements will decrease as the Hardgrove grindability index (HGI) of the coal increases because the HGI increases for "softer" coals [25]. Also, the gross power consumption per ton of feed decreases as the feed rate increases [26]. For a given classifier setting, an increase in the air rate or an increase in the outlet air temperature will normally result in a decrease in the fineness of the mill output [27]. With other conditions remaining constant, the fineness of the output from a roller mill will increase as the feed rate to the unit is decreased [26].

Roller mills are capable of reducing feed material with a top size of 2 in. to a product that normally contains 70% -200 mesh material. Units are designed to handle material at rates as high as 100 tons per hour, with power requirements of approximately 1000 hp, that produce a product that contains 70% -200 mesh material.

G. Miscellaneous Devices

The size reduction machines discussed in preceding sections are industrial-level (large-throughput) devices employed for coal size reduction. Associated with the increased interest in new areas for coal utilization has been interest in the reduction of the top size of coal into the subsieve range or below. Consequently, various devices used for ultrafine size reduction have been tried on coal, including vibrating mills, jet mills, high-speed hammer mills, and attrition (stirred) mills.

These devices have not become prevalent, but they justify a thumbnail sketch of each.

1. *Vibrating Mill*

The vibrating mill is a cylindrical mill completely filled with grinding media. This mill utilizes transverse vibrations along the longitudinal axis of the mill to cause media/particle impact rather than rotation around the longitudinal axis that is employed for rod and ball mills. The vibrations and subsequent media dropping can be applied at up to 1800 oscillations/min.

2. *Jet Mill*

A jet mill is a size reduction device that has no moving parts and utilizes air, steam, or other gases at pressures of approximately 100 psig as its power source in order to generate particle/particle collisions. The original jet mill utilized opposing jet streams to impact the particles to be treated. The particles of a suitable fineness were removed by upward air classification.

More recent developments have resulted in the use of flat cylindrical (pancake) and oval-shaped (doughnut) grinding chambers with impinging jets of gaseous fluid introduced around the periphery.

3. *High-Speed Hammer Mill*

High-speed hammer mills are size reduction devices that are similar in configuration to the hammer mills discussed in Sec. III.C. The material being reduced in size is struck by the rotor and thrown against a cover. The material is then air-swept into a classifier, from whence oversized material is returned to the mill.

The high-speed hammer mill operates at higher peripheral speeds than do conventional hammer mills, speeds of up to 21,000 ft/min. The high-speed hammer mill produces a finer product with corresponding lower capacities as compared to the units discussed in Sec. III.C.

4. *Attrition Mill*

Attrition mills utilize a paddle wheel or impeller to stir the grinding media and material, which completely fill the cylindrical container, at speeds of 100 to 1500 rpm. These devices are normally incorporated in wet systems.

IV. ENVIRONMENTAL CONSIDERATIONS

Since 1969, the effect of environmental factors has also become a factor in selecting a size reduction device. The two environmental factors that currently have the most direct effect on size reduction devices and systems are fugitive dust and noise.

A. Fugitive Dust

The Federal Coal Mine Health and Safety Act of 1969 includes a section that deals with mandatory health and safety regulations for respirable dust. Title 30, Part 71, of the *Federal Register* deals with the specific standards for respirable dust and quartz in surface installations.

Respirable dust is generally defined as particulate matter less than 5 μm in diameter [28]. For surface facilities, the air quality must be maintained at an average concentration at or below 2 mg of respirable dust per cubic meter [29].

However, should the quartz content of an air sample exceed 5%, the average concentration of respirable dust must be maintained at or below a concentration expressed as milligrams per cubic meter of air. This concentration is determined by dividing the percent of quartz present into the number 10 [29]. Figure 24 illustrates the relationship between the respirable dust concentration and the amount of quartz present in the air.

In breathing airborne dust, the respirable dust portion can penetrate and accumulate in the lungs. After prolonged exposure, one might contract pneumoconisosis, or black lung disease.

Reduction of airborne dust concentrations in order to avert violations of dust standards can be accomplished via prevention or removal techniques, applied either separately or in combination. Preventative controls include appropriate housekeeping techniques and dust suppression. Removal techniques generally employ ventilation followed by dust collection.

B. Noise

The Federal Coal Mine Health and Safety Act of 1969 also contains a section that deals with noise. Title 30, Part 70, Subpart F, pertains to underground facilities and Part 71, Subpart D, deals with surface installations.

Sound is the result of a mechanical-electrical mechanism in which air vibration, in turn, is transmitted through the middle ear to the inner ear, where the mechanical energy is converted into electrical energy in the form of nerve impulses.

Noise can be generally defined as any objectionable or unpleasant sound. Depending upon the intensity and length of exposure, noises can either immediately or eventually result in a measurable permanent hearing loss as a result of impairment of the mechanical-electrical hearing mechanism.

Relative sound intensity is usually quantified through the use of the A-weighted decibel system. The units for this system are A-weighted decibels or dBA [30]. For this system, 0 dBA represents the threshold of hearing

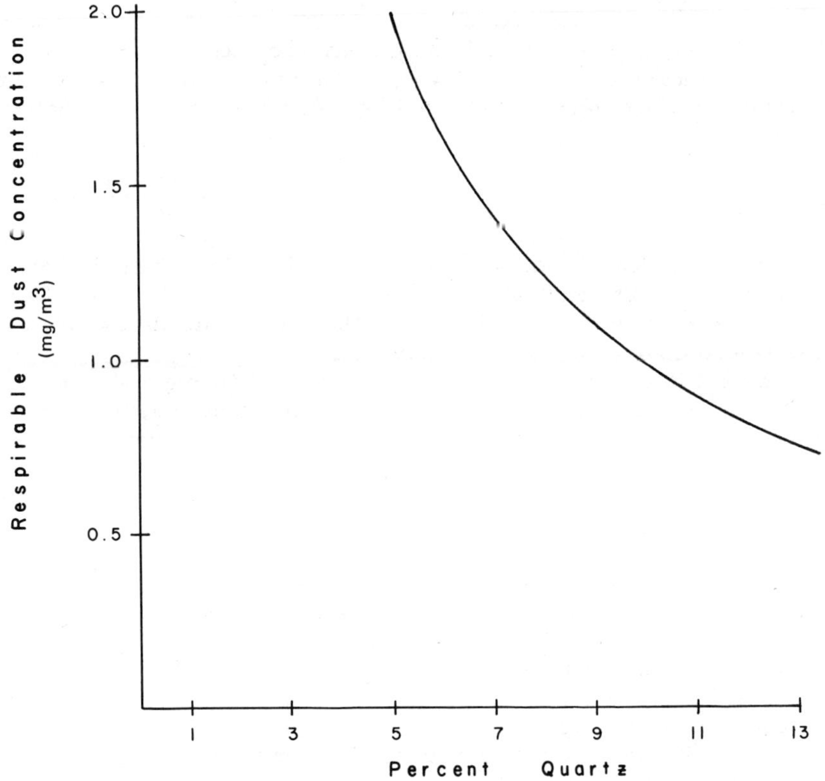

FIG. 24 Respirable dust concentration as a function of the quartz content of an air sample.

TABLE 2 Noise Standard

Exposure duration (hr/day)	Maximum permissible noise level (dBA)
8	90
6	92
4	95
3	97
2	100
1.5	102
1	105
0.5	110
0.25 or less	115

and approximately 140 dBA represent the threshold of pain. On the decibel scale, 1 dBA is the same relative change everywhere on the scale [31].

On the average, the human ear can discern an increase in sound level of approximately 3 dBA, which is actually a doubling of the sound power or intensity [30]. As a result, if two identical sound sources are combined, they do not result in a sound level of twice the individual sound level, rather a 3-dBA increase over the individual sound level. Conversely, the combined sound level from two identical sound sources cannot be cut in half by eliminating one of the sound sources.

Although the actual decibel increase or decrease will vary, the same general relationship holds no matter the number of sound sources or whether the sound level of the sources is different. This relationship must be taken into account when trying to reduce the total sound level in any facility. Tabulated information depicting this relationship is available in published articles [30,32].

Table 2 shows the noise standards contained within the Federal Coal Mine Health and Safety Act of 1969, in which the maximum permissible noise level is a function of the length of exposure [33]. This standard is based on a maximum noise level of 90 dBA for 8 hr of exposure. In addition, the standard halves the permitted exposure time for every 5-dBA increase in noise level, which is a 5-dBA time/intensity trade-off. Currently, there is a proposal in Congress that would provide for a maximum 8-hr exposure of 85 dBA with a time/intensity trade-off value of 3 dBA [34].

An individual is in violation of the noise standard when the individual's noise exposure index (NEI) exceeds unity. The NEI is the ratio of the actual exposure time at a given noise level to the permitted exposure time. If an individual is exposed to multiple noise levels, the NEI is the sum of the individual ratios of actual exposure time at a given noise level to the permitted exposure time [34].

In general, there are two methods of noise control by which the noise exposure index of an individual can be brought into compliance. These are administrative noise controls, in which the work duties or work process of an individual are modified to reduce the time spent in high-noise-level areas, and engineering noise controls, in which either the noise source or the noise receiver, the individual, is isolated [35]. Usually, a combination of administrative and engineering noise control methods can be utilized to achieve the most economical solution.

A noise source is generally isolated through the use of either permanent enclosures or barrier curtains. However, in utilizing such devices, one should provide means of control and inspection from the exterior so that an individual does not have to routinely enter the enclosure and be exposed to the noise levels that are being isolated.

Permanent enclosure can result in noise reductions of as much as 15 to 25 dBA. The addition of absorptive materials in the interior of the enclosure can result in further reductions of from 3 to 4 dBA [35]. Permanent enclosure should include access ways, inspection windows, and interior lighting.

Barrier curtains are made from flexible, high-density materials and generally offer greater flexibility and accessibility than do permanent enclosures. Inspection ports can be provided through the use of plexiglass windows. Barrier curtains can result in noise reductions of as much as 8 to 12 dBA.

Again, the addition of absorptive materials to the interior of the curtains can result in further reductions of from 3 to 4 dBA [35].

An individual can be isolated from a noise source through the use of an enclosure or booth. With this arrangement, an individual must remain within the enclosure for a portion or the entire length of the work shift so that the individual's calculated NEI does not exceed unity. Up to now, ear protection devices such as earmuffs or earplugs have not been considered permanent solutions to the problem of isolating individuals from noise sources.

Portions of the preceding discussion of environmental concerns are not restricted to size reduction devices. For most systems, size reduction devices are not isolated from other units. As a result, any attempt to meet the various environmental standards must be based on a total systems approach rather than an individual-component basis.

V. FUTURE TRENDS

A. Chemical Comminution

Chemical comminution is a size reduction process in which a reagent is contacted with coal, resulting in the selective breakage of the coal by weakening and disrupting the bonding forces at the natural interfaces between the coal and the mineral matter [36]. It is claimed that chemical comminution results in increased liberation at a coarser product size distribution than would be obtainable with mechanical size reduction. This would permit higher product recovery through the use of simplified coal preparation facilities. However, the potentially lower coal preparation expenses must offset increased size reduction expenses which result primarily from reagent storage facilities and reagent recovery systems.

One of the reagents that has been tested is ammonia in the gaseous form or in solution. To date, all information on the ammonia process has resulted from laboratory tests. However, the developers of the process plan to erect and operate a pilot facility in the near future.

The product size distribution of coal treated by this process is largely a function of the concentration of the reagent and the time of exposure. Test results indicate that the process requires approximately 100 lb of reagent per ton of coal, with greater than 99% of the reagent recoverable [36].

B. Coal Slurry Grinding

For this discussion, coal slurry grinding is defined as the wet grinding of coal. Coal slurry grinding can be divided into the general areas of coarse grinding and fine grinding.

Coarse coal slurry grinding yields a product in which approximately 99% of the material is less than 14 mesh in size. Since this approximate product size distribution is usually required for the pipeline transport of coal, water would generally be the liquid used in grinding.

Fine coal slurry grinding would generally produce a product in which approximately 70% of the material passes 200 mesh. A hydrocarbon liquid would normally be used in grinding. The coal slurry product from this type of grinding circuit could be utilized directly in combustion or receive further processing via gasification or liquification.

REFERENCES

1. American Society for Testing and Materials, Description: D431-44, Standard Methods for Designating The Size of Coal from Its Sieve Analysis, *1977 Annual Book of ASTM Standards*, Part 26. Philadelphia, 1977.
2. American Society for Testing and Materials, Description: D409-71, Standard Method of Test for Grindability of Coal by Hardgrove Machine Method, *1975 Annual Book of ASTM Standards*, Part 26. Philadelphia, 1975.
3. M. B. Flavel, Power and the Way It Is Applied Affects Equipment Performance and Production Costs, Proceedings of the SME-AIME Annual Meeting, Mar. 1977.
4. A. F. Taggart, *Handbook of Mineral Dressing*, John Wiley & Sons, New York, 1945.
5. T. Hicks, Crusher Circh Fuel Sizing Cleaning, *Power*, pp. 73-75, July (1951).
6. G. H. Roman, Coal Crushers and Crushing, *Coal Age*, pp. 107-122, Feb. (1967).
7. J. D. McClung, Breaking and Crushing, *Coal Prep.*, AIME, New York, 1968, Chap. 7.
8. R. K. Sirghal, Breaking and Crushing of Coal, *Colliery Guardian*, pp. 81-83, Jan., 1968.
9. D. W. Kestner, Jr., D. S. Confer, and H. B. Charmbury, *The Effect of Crusher Type on the Liberation of Sulfur in Bituminous Coal*, 1962.
10. K. S. Narasimhan and S. R. S. Sastri, How to Estimate Size Distribution of Crushed Product, *Chemical Eng.* 82:77 (June 9, 1975).
11. B. N. Popelnik, V. G. Shkodkin, M. V. Donde, N. K. Ivanov, and P. V. Akulov, Research into Hammer Crusher Performance, *Coke Chem. USSR* 25:46 (1968).
12. E. A. Dolganov, A. M. Shteinberg, E. V. Belyaev, I. M. Lazoviskii, and N. S. Gryaznov, Preparing Coals for Carbonisation in an Air-Swept Crushing Circuit, *Coke Chem. USSR* 28:11 (1971).
13. T. G. Callcott, A Study of the Size Reduction Mechanisms of Swing Hammer Mills, *J. Inst. Fuel* 33:529 (1960).
14. T. G. Callcott, Pulverizing Coals, Proceedings of the Australian I.M.M. 183, 1957, pp. 43-76.
15. C. E. Bird, Report of Tests Carried Out on the Swing Hammer Pulverizer, *J. Chem. Met. Miner. Soc. S. Africa*, pp. 109-117 (1951).
16. M. V. Donde and A. P. Fomin, The Influence of the Design and Operating Conditions of Hammer Crushers on the Grinding of The Charge, *Coke Chem. USSR* 9:11 (1952).
17. V. V. Rakov, The Crushing of Coal Concentrate, *Coke Chem. USSR* 27:10 (1970).
18. M. N. Yampolskii and A. M. Khesin, Use of Vibration in Hammer Crushers, *Coke Chem. USSR* 27:14 (1970).

19. J. W. Leonard and J. D. Clendenin, Some Practical Considerations for The Application of Coal Pulverization at Coke Plants, *Blast Furn. Steel Plant* 49:45 (Jan. 1961).
20. *KVS Rock Talk Manual*, McCormick-Armstrong, Wichita, Kansas, 1974.
21. F. J. Hiorns and L. E. Reed, The Performance of an Air-Swept Tube Mill, *J. Inst. Fuel* 31:383 (1958).
22. Nordberg Grinding Mills, Bulletin 315D, Milwaukee, Wisconsin , 1975.
23. Hardinge Rod Mills, Bulletin 25-D, York, Pa., 1965.
24. G. R. Fryling, Pulverizers, *Combustion Eng.*, Chap. 16, Combustion Engineering, Inc., Stanford, Conn., 1967.
25. A. Fitton and R. Jackson, Some Experiences of Coal Grinding Plant Tests on a Ring-Ball Mill and a Tube Mill, *J. Inst. Fuel* 32:520 (1959).
26. R. C. Ellman and J. W. Belter, Pulverizing Lignite in a North Dakota Power Plant, Bureau of Mines Information Circular IC-8032, pp. 1-31, 1961.
27. F. J. Hiorns, Operational Data on a Ring-Ball Pulverizer, *J. Inst. Fuel* 32:464 (1959).
28. Respirable Dust Control, U.S. Bureau of Mines Information Circular, IC-8753, 1977.
29. Dust Standards, *Fed. Regist.*, Title 30, Parts 70 and 71, Subpart B, 1977.
30. N. P. Chironis, Reducing Noise in Preparation Plants, *Coal Age*, pp. p. 164 (Feb. 1978).
31. B & K Instruments, Inc., Measuring Sound, Pamphlet BR0047, Denmark, 1971.
32. E. E. Ungar and H. K. Sacks, Guidelines to Queting Noise in Preparation Plants, *Coal Min. Process.*, p. 60 (July 1977).
33. Noise Standard, *Fed. Regist.*, Title 30, Part 70 and 71, 1977.
34. T. G. Bobick and D. A. Giardino, The Noise Environment of the Underground Coal Mine, MESA Information Report 1034, 1976.
35. T. G. Bobick, Noise Reduction in Coal Preparation Plants, to be published shortly as a MSHA Informational Report.
36. Will Chemicals Replace Crushing? *Coal Age*, pp. 176-179 (Feb. 1978).

6 Coal Cleaning

D. GLANVILLE WILLIAMS / North American Mining Consultants, Inc., New York, New York

I. INTRODUCTION

Cleaning, as applied to coal, is an omnibus word requiring definition and clarification. It implies that coal is a dirty material requiring treatment to improve it but says nothing of the nature of the dirt. Run-of-mine (ROM) coal is usually a heterogeneous broken material consisting of lumps and particles ranging from 2 ft or more across down to the finest dust. It is frequently wet and often contaminated with pieces of rock or sticky clays. In such a condition the coal is unsuited for use as boiler fuel or cokemaking, which are its two principal uses. On occasion the coal seams being worked may be relatively thick, unfaulted and uniform, free of associated rock partings, and dry. In such cases the ROM coal may require only some breaking or crushing and screening to produce a uniform "pure" coal. Such instances are rare and have become rarer as a result of increased mechanized mining because continuous-coal-mining machines capable of distinguishing between coal (macerals) and rock (minerals), although under development, are not yet available. Thus, although the need for cleaning can sometimes be reduced by careful mining, and many modern mines include means for breaking and crushing oversize lumps within their precincts, the cleaning of ROM coal is essentially a separate operation, carried out in a surface installation that may be close to the mine-mouth, but not necessarily so because economic advantages can sometimes accrue from providing centralized coal-cleaning installations that service a number of mines within a given area, and this practice has been increasing in recent years, although the great majority of coal-cleaning installations are still essentially mine-mouth installations.

A. Objectives of Coal Cleaning

The primary aims of preparing ROM coal for the market and own use are, therefore, the reduction and control of lump size within limits determined by the needs of transportation, handling, and utilization; the removal of extraneous mineral matter, usually referred to as control of ash content; and the control of moisture content within specified limits usually determined by handling and use requirements. This industrial activity is very old, employs methods commonly employed in the even-older activity known as minerals dressing, but is a distinct activity called "coal preparation." Often dubbed "the black art," coal preparation demands knowledge and skills quite different from those otherwise required in coal mining. Its practitioners receive a different education and training, and in most coal-producing countries coal preparation has become recognized as a distinct engineering discipline with its own professional affiliations.

Even where coal has been prepared to meet the applicable specifications for size, ash, and moisture contents, it may still be "dirty" by environmental standards. In this case, the most important and universal contaminant is sulfur. During subsequent combustion the sulfur in coal is almost entirely converted to gaseous products which, unless removed, end up in stack gas emissions. For metallurgical use, in which coal is usually first carbonized to produce coke, very little of the sulfur appears as atmospheric pollution, the bulk of it ending up in the slag and metal produced by smelting. But its presence in the metal is objectionable because it affects metallic properties adversely, and the need for low-sulfur coals for metallurgical use has long preceded the present environmental demands. For reasons that will be made clear later, the conventional methods of preparing coal usually result in some reduction in sulfur content, but this is often incidental and in the nature of a bonus. Only in recent years have deliberate attempts been made to achieve desulfurization on an industrial scale by modification of established coal preparation practices. In part, this has been due to the development, since World War II, of more precise methods for the separation of coal and mineral. But these practices continue to have severe limitations and are highly dependent upon the properties of the coal to be treated. Where low-sulfur coals are a necessity, the deliberate selection of low-sulfur seams for mining remains the most effective solution and has long been the practice followed in producing metallurgical coals. However, given the enormous resurgence in demand for coal as an energy source for production of electric power, and the steady diminution in availability of low-sulfur coal reserves, particularly of bituminous coals having premium value as fuels, the opportunities for selection are limited and becoming more so. In the United States, New Source Performance Standards (NSPS) promulgated in 1978 for regulating the emission of sulfur oxides in stack gases will require removal of 85% of the sulfur oxides present in primary products of combustion. The methods available for achieving this desulfurization (FGD) require the use of scrubbing systems still under development or the use of fluidized combustion employing limestones or dolomites as the bed material to trap the primary sulfur oxides during formation. The latter process has not yet been demonstrated on a commercial scale.

These problems have created renewed interest in developing processes for removing sulfur from coal before combustion, or precombustion cleaning (PCC). The extent to which they will succeed will depend upon the success achieved in overcoming the difficulties inherent in their application, which will be discussed in more detail later, and the costs associated with PCC compared with those of full or partial FGD, or, in the case of fluidized combustion systems, the costs associated with use, regeneration, and disposal of limestone/dolomite.

Current opinions on this topic are mixed. Deurbrouck has recently expressed the following viewpoint [1]

> Since passage of the Clean Air Act of 1970, coal preparation research efforts in the United States have emphasized development of technology for the reduction of sulfur. However, the new sulfur emission standards being promulgated by the U.S. Environmental Protection Agency are so stringent . . . that they preclude the use of physical coal preparation alone as a sulfur control technology capable of producing a compliance coal. Therefore, new research efforts will show more interest in providing maximum recovery of Btu's, while accepting any sulfur reduction as an additional benefit. . . .

Although this view does not rule out the desirability of achieving sulfur reduction by direct chemical or other treatment of coal, it is indicative of the measure of reduction to be expected if such processes are to be viable.

Another angle of attack on the sulfur problem is provided by coal conversion, which involves direct thermal and chemical treatments of coal to produce essentially sulfur-free liquid or gaseous fuels, the sulfur being recovered as a by-product of the conversion process. These are separate technologies, discussed in Chaps. 10 and 11, and form no part of the present discussion.

Before leaving the question of "dirty" coal, even though it may have been "prepared," it is becoming of increasing importance to consider the environmental hazards presented by trace elements. Coal, and its associated mineral matter, have long been known to contain minute quantities of elements such as arsenic, cadmium, germanium, and mercury, and even radionucleides, some of which are believed to be carcinogenic and all of which may otherwise be toxic [2-4]. Fears have been expressed that such trace elements are found in the very finest particles in the combustion products, which, since it is the smallest particles that escape capture in presently available equipment for flue gas particulate capture, become more concentrated in actual particulate emissions. A subsequent, but equally important concern, is that these very finest particles are drawn most deeply into physiological systems and are hence the most difficult to eliminate and potentially the most harmful. The question arises, therefore, as to what extent coal-cleaning processes can be expected to alleviate this problem. This aspect of coal cleaning is considered in more detail later.

The preceding discussion has touched upon the following operations relating to ROM coal: handling, breaking and crushing, screening, reduction and control of ash, reduction and control of moisture, reduction of sulfur, and problems presented by hazardous trace elements. The technology developed for breaking and crushing, screening, and handling in general has been fully discussed in Chaps. 6 and 7 and, although these operations are essential to the overall engineering discipline of coal preparation, they require no further description here except, perhaps, as to certain applications specific to coal-cleaning processes.

The energy content of coal is related to its rank (degree of coalification) and its content of nonfuel components (e.g., minerals and moisture) (see Chap. 1). Sulfur plays a minor role in these relationships. A primary objective of coal cleaning must therefore be to maximize the recovery of energy values in the ROM coal consistent with achieving set specifications for ash, moisture, and sulfur contents. Unfortunately, the close matching of these ideals is rarely achievable. Coal cleaning is an activity of compromise in which, to achieve its desired products, some rejection of usable energy is inevitable. Every coal refuse pile in the world contains some measure of recoverable energy. One major challenge the industry has always faced has been to reduce this loss to the minimum practical. The introduction of new processes aimed at still-cleaner fuels will render even more complex the optimization of production rates, yields, qualities, and costs. But already, certainly as regards electric power generation, it has become necessary to include these exercises of compromise and optimization within the much larger consideration of alternative fuel conversion systems; environmental control before, during, or after combustion; and the relative effects of the associated costs on the price of electricity at the busbar. Control strategies directed toward this end are presently a major concern of every industrialized country, but are of even greater concern within the United States because of (1) the commitment to immensely increased coal utilization for electric power production as required in the National Energy Plan, and (2) the widespread occurrence of high-sulfur bituminous coals in recoverable reserves.

B. Historical Development

The foundations of the modern coal preparation industry were largely laid down in Europe during the quarter century 1915-1940. On reflection from our present vantage point, it is clear that astounding progress was made during this period, particularly during the years 1925-1935, years otherwise notable as a period of financial crisis and severe economic recession. With the exception of hydrocyclones, which were invented and developed by the Dutch after 1945, most of the unit operations in modern coal preparation were further developed, or invented, and brought to high levels of reliability and efficiency during this period. Coal washing by hydraulic methods, jigging, laundering, and classifiers, some of which had been introduced as early as 1830, were fully mechanized and improved. Gravity separation by dense-medium processes, such as the Chance, Barvoys, Tromp, and Loess Processes, were invented; froth flotation was developed to a stage where selective depression (e.g., of pyrites) was possible; oil agglomeration, currently finding renewed interest as a means of recovering coal from ultrafine suspensions, had actually been developed as the Trent Process during the 1914-1918 war. The principles of colloid chemistry were established and applied to the development of efficient, special-purpose flocculants, and effective water clarification was achieved by 1935. However, in many cases the success achieved in mechanization preceded satisfactory elucidation of the basic principles involved, which has been a task accomplished principally since World War II. It may be noted that in spite of this early achievement of high technical and operating efficiencies on a commercial scale, it is still considered necessary or desirable to construct pilot plants containing various assemblies of the unit operations mentioned in order to provide "test beds" for processing various coals or as "demonstration plants" for proving the utility of coal preparation as a means of reducing environmental problems.

Chemical treatment processes to produce clean fuels have not met with the same success. Removal of mineral impurities by acid leaching has been demonstrated and employed on a laboratory scale for special purposes for many years. The range of possibilities for industrial processes of this kind were reviewed as long ago as 1937 [5], but there have been no commercial applications. Chemical treatments specifically designed to remove all or part of the pyritic sulfur, and processes capable of removing pyrite together with substantial proportions of organically combined sulfur have found renewed and intensive interest in recent years, particularly in the United States, and the current status has been reviewed by a number of workers [6-8]. In all the cases reviewed, the process developers face problems of demonstrating successful operation on a commercial scale.

The majority of chemical processes demand a level of technical sophistication that goes far beyond that required in even the most modern of coal preparation plants, and their introduction will require higher levels of operating skills. To cite just one example, the JPL Chlorinolysis Process for desulfurization requires treatment of the coal, suspended in organic liquids, with about 45% by weight of chlorine, the unused chlorine, together with the organic solvent, being recovered and recirculated. A cleaning plant handling 20,000 tons/day of coal will be required to cycle about 9000 tons/day of chlorine. This requirement may be contrasted with the size of chlorine production systems, which range up to 1200 tons/day.

II. REVIEW OF WORLD PRACTICES

Coal cleaning is practiced in every country where coal is produced, but a digest of information summarizing this worldwide activity is not available. In 1975, world production of bituminous coals and lignites was as shown in Table 1. This table does not contain information on anthracites, which is mined in a number of countries and which is usually cleaned for use in the industrial and domestic markets. Historically, the commercial cleaning of anthracite has resulted in the development of a number of highly efficient cleaning systems in Europe and the United States—the Chance Process and classifiers such as the Hydrotator being typical examples of cleaning processes developed in the anthracite mining industry.

A comprehensive statement of world coal cleaning practices therefore cannot be drawn from the information available. Reference to Table 1 shows that more than two-thirds of the world's production of bituminous coals and lignites is produced in nine countries: Australia, Canada, West Germany, India, Poland, South Africa, the United Kingdom, the United States, and the USSR. In each of these countries except West Germany and the United Kingdom, coal production is being rapidly expanded. In the two exceptions, coal production is static, but major reconstruction of the coal-cleaning industry is taking place. Each country has its own diverse coal characteristics, production, market, and utilization requirements, in which a comparatively few flexible methods of coal cleaning are interwoven into a rich fabric of world practice. Before reviewing the individual countries, a few general observations may be drawn. In every country, the need for reduction of ash content is fundamental, but the difficulties of achieving this are diverse and undoubtedly present the

TABLE 1 World Production of Bituminous Coal and Lignite

Country[a]	1975 Millions of short tons	1975 Percent	1976[b] Millions of short tons	1976[b] Percent
North America				
United States	648.4	19.0	665.0	19.0
Canada	27.9	0.8	28.0	0.8
Other	5.7	0.2	6.2	0.2
South America	9.4	0.3	9.7	0.3
Europe				
Bulgaria	30.6	0.9	28.0	0.8
Czechoslovakia	126.9	3.7	129.8	3.7
France	22.3	0.6	22.3	0.6
East Germany	272.4	8.0	272.6	7.8
West Germany	229.4	7.0	239.4	6.8
Greece	19.4	0.6	24.5	0.7
Hungary	27.4	0.8	27.8	0.8
Poland	233.1	7.0	241.0	7.0
Romania	29.8	0.9	29.5	0.8
USSR	689.0	20.2	701.0	20.0
United Kingdom	139.1	4.1	133.6	3.8
Yugoslavia	39.2	1.1	40.6	1.2
Other	25.9	0.7	26.4	0.7
Africa				
South Africa	74.8	2.2	82.2	2.3
Other	5.0	0.1	5.7	0.2
Asia				
China	495.0	14.5	510.0	14.6
India	108.8	3.2	115.6	3.3
Japan	20.8	0.6	20.3	0.6
Other	30.0	0.8	29.0	0.8
Oceania				
Australia	104.9	3.1	110.1	3.1
Other	2.6	—	2.7	—
Total	3417.9	100.0	3501.0	100.0

[a]Countries with greater than 20 million short tons annual production are listed individually. Countries with less than 20 million short tons annual production are grouped under "Other."

[b]Preliminary.

Source: National Coal Association, *International Coal 1977*, Washington, D.C.

greatest difficulties with the finely disseminated mineral matter contents of the Gondwanaland coals of India and South Africa. Concern about moisture content is greatest in countries having very cold winters—Canada, Poland, the United States, and the USSR—which suggests that difficulties encountered in handling frozen coal are more important than is the thermally inert content it represents. Thermal drying is not practiced in any other country considered. Similarly, dry concentration methods, although of limited range of application, are of greatest interest in the coldest winter climates. High sulfur contents are rarely encountered in the coals of the southern hemisphere, are not a major problem in western Europe or western Canada, but are of increasing concern in the United States, Poland, and the USSR. Interest in coal desulfurization is greatest in these three countries.

A. Australia [9]

Australia is the third largest coal-exporting nation in the world following a major expansion into export markets that began in the 1960s. Coal preparation expanded to meet this export trade, which was mainly for metallurgical coals. Until 1950, coal beneficiation consisted mostly of handpicking, crushing, and sizing. During 1977, 57 coal preparation plants were in operation, treating 78.2% of the total ROM coal production of 81,638,000 tons. A summary of plant practices is as follows:

Number of operating plants	57
Types of unit operations	
Jigs	32
Launders	2
Dense-media vessels	13
Dense-media cyclones	27
Water-only cyclones	17
Tables	8
Froth flotation	19
Air concentration	0

Australian coal preparation is effectively based on the adaptation of foreign practices and equipment that have been tailored to its local coals and conditions. Some processes, such as dense-media vessels, fine coal jigs, and wet and dry concentrating tables, have not found wide application. Australian coals generally are low in sulfur. There is consequently little interest in coal desulfurization.

B. Canada

In the 10 years 1967-1977, Canadian coal production tripled from 11.1 to 31.6 million tons/year. During 1977, the production of metallurgical coal, at 14.4 million tons, increased by 12% over 1976. Thermal coal produced, at 17.2 million tons, was 13% higher. The rapid growth in metallurgical coal has been absorbed by the world steel market, principally by Japan, and was produced principally in the western mountain coalfields of Alberta and British Columbia, the remainder being produced in Nova Scotia.

Thermal coal is produced mainly in Alberta and Saskatchewan and consists primarily of low-ash, low-sulfur, high-moisture lignites and subbituminous coals, most of which require little preparation. Smaller quantities of bituminous thermal coals are produced in the Maritime provinces, Nova Scotia and New Brunswick. These generally contain 2.5-8% sulfur and present particular difficulties for coal cleaning.

Canada currently has eight coal preparation plants, ranging in outputs from 1.5 to 6.0 million tons/year. The majority are central plants handling outputs from groups of mines.

The western coals, although low in sulfur (usually less than 0.8%), often contain substantial quantities of intergrown materials, are friable, and are frequently contaminated with intercalated clay bands. High-precision separations, not obtainable with jigs, are often necessary, so that heavy-media systems are chiefly used. Heavy-media vessels and drums are employed for coarse coal and heavy-media hydrocyclones for sizes down to 28 mesh. Froth flotation is employed for cleaning sizes below 28 mesh and is generally efficient except where substantial oxidation has occurred. When oxidized coal is present, a major feature of current practice is the use of compound water cyclones, developed in Canada by the Federal Department of Energy Mines and Resources. These have proved effective for sizes down to 100 mesh, and the frequent occurrence of soft, highly friable coals has made their development especially valuable [10]. Froth flotation is also employed for recovering -100 mesh coals.

In Nova Scotia coal is produced from a group of undersea mines and is cleaned to provide metallurgical and thermal grades at a multistream central coal preparation plant of 4.3 million tons/year capacity. The coals have limited intergrown material, but the small quantities present are associated with much of the high pyrite contents, so that precise control of gravity separation is necessary for satisfactory control of sulfur. This is achieved by a combination of primary and secondary heavy-media hydrocyclones and froth flotation.

C. West Germany

In 1977, the industry produced 84,500,000 tons of salable hard coal and approximately 107,300,000 tons of brown coal. All the hard coal is mined from comparatively thin seams (1 to 4 m), at depths averaging 850 m and the ROM coal is heavily contaminated with dirt, averaging up to 40% [11]. The entire production of ROM coal is therefore processed in fully equipped coal preparation plants. Ash content of the lignite falls in the range 2 to 8% and hence does not require cleaning.

Germany and its neighboring countries France, Belgium, and Holland have always been in the forefront of coal cleaning technology and have developed many cleaning techniques which later became adopted as standard practices worldwide. These have included [12] the plunger-type jig, developed in 1830; the Baum jig, 1892; sand flotation, 1916; pneumatic cleaning, 1927; Barvoys (Sophia Jacoba) dense-medium system, 1933; and the Tromp Process, 1937. It has been the practice, since 1921, of separating dust (-28 mesh) by dedusting or desliming systems and either adding it back as dry material to the washed coarse product or cleaning it by froth flotation. A major improvement in jig washing was made in 1967 with the development of Batac jigs [13], which, during a major rebuilding program carried out during the past 10 years, have replaced all the heavy-media systems formerly employed for coal sizes below 10 mm and 40% of those formerly employed for coarse coal. Since 1970, only systems based on Batac jigs have been installed, and all fine coal dense-medium systems are being phased out. The need for rebuilding or replacing the wash plant, usually within buildings already housing the plant being replaced, was used as an opportunity to move to equipment items having much greater throughput capacities. Wash plant capacities have been increased by as much as 30% without extending the buildings and has resulted in significantly lower labor and maintenance requirements and costs.

Centralized control with manual operation was widely introduced. A number of attempts at full automation have been tried but have frequently encountered failure because of breakdown and the high maintenance costs of measuring probes and control devices [14,15]. It has been necessary to develop new and more robust instrumentation for measuring qualities, yields, and so on, so that, once again, on-line control procedures are being installed [16].

Some problems with high-sulfur coals are encountered, although these are not as widespread or severe as those found in the United States, Poland, and the USSR. Nevertheless, determined attempts to achieve a significant degree of desulfurization by physical cleaning methods to reduce the costs of flue gas desulfurization (FGD) are being made. The new preparation plant at Walsum Mine in the Ruhr district incorporates systems in which a combination of hydrocyclones, concentrating tables, and froth flotation are being employed to achieve enhanced removal of pyritic sulfur [17]. Progress in desulfurization has recently been reviewed by von Gathen [18]. There are no major projects and only limited interest in chemical desulfurization methods.

D. India

In 1977, India's [36] planned ROM coal production capacity was 108,000,000 tons, distributed as follows:

Metallurgical	28,000,000
Steam/electric	30,000,000
Railways	14,000,000
Industrial/domestic/export	36,000,000

Only coal intended for metallurgical use is cleaned by washing, present capacity treating 80 to 85% of ROM coking grades at 14 wash plants of 26 million tons/year capacity. Two plants under construction will raise this capacity to 31.6 million tons/year.

Most Indian coals have very difficult washing characteristics and high inherent ash contents. Much of the prepared coal has ash contents in the range 12 to 18%. For this reason the coals require substantial pretreatment, including close attention to sizing, before washing. Washing equipment used includes Baum jigs and feldspar jigs, heavy-media vessels and cyclones, water-only cyclones, and froth flotation cells. For future plants two general schemes are being favored. For the easier coals, pretreatment will produce raw feed crushed to 75 mm (3 in.) top size and split at 20 mm (1 in.) or 13 mm (1/2 in.). The coarse fraction will be washed in heavy-media baths and the small fraction in air-pulsated jigs or heavy-media cyclones. A prewash employing a jig will remove excessive rock dilution if this is a problem. More difficult coals will be crushed smaller, to 30 mm (1 1/4 in.) or even to 15 mm (5/8 in.) top sizes, deslimed, and cleaned in heavy-media cyclones. Raw slurries (-28 mesh) will be cleaned by flotation or in hydrocyclones. Increasing demand for coking coals will require cleaning of even poorer ROM feed in the future, and it is expected that crushing to even smaller top sizes (1/2 in.) will be necessary, followed by cleaning in heavy-media cyclones. The consequent increases in fines will be very great, and there is much current interest in oil agglomeration as a solution to this problem, one advantage being the elimination of vacuum filters. A large increase in employment of oil agglomeration is foreseen.

Concentrating tables have found little application, undoubtedly because of the severe near-gravity conditions encountered. Little information on the extent of the sulfur problem has been published and there appears to be no current activity in the application of physical and chemical desulfurization processes, although the shift toward finer crushing and fine coal cleaning methods will obviously increase incidental desulfurization.

E. Poland

Coal production has increased very rapidly during the past 25 years [19,20], so that Poland is now the world's fourth largest producer and is second only to the United States in exports. The growth of the coal industry has been as follows:

	Coal production (millions of short tons)					
Salable output	*1950*	*1955*	*1960*	*1965*	*1970*	*1975*
Bituminous	85.8	104.0	114.4	130.9	154.0	189.2
Lignite	5.3	6.6	10.2	24.9	36.0	43.6

Production is concentrated in the Upper Silesian Basin, which produces more than 97% of total coal production.

In 1975, 46% of the total production of salable bituminous coal was cleaned, the types of cleaning processes employed being as follows:

	Percent of total bituminous coal produced	*Percent of coal cleaned*
Jigs	25	54.4
Heavy media	18	39.1
Flotation	3	6.5
	46	100.0

Coal exports in 1975 were 44 million short tons, of which 25%, or 11 million tons, consisted of metallurgical grades. Total coking coal production was approximately 38.5 million short tons.

Current plans for coal preparation provide for a doubling of capacity by the end of the century. The principal problems encountered in preparation include difficulties from clay contamination (montmorillonite, illite) and desulfurization by reduction of pyrites. It is reported that some coals contain up to 4% sulfur, which is not appreciably reduced by existing washing methods. Consistency of product quality is also receiving attention, and blending and homogenization methods are receiving much attention. The pretreatment of coal to prepare suitable feedstocks for gasification and liquefaction has recently received research and development funding.

Problems presented by high-sulfur coals are not as severe or extensive as those of the United States, so that, although the possibilities of improving physical desulfurization during cleaning are being investigated, there is limited interest and no current activity in chemical desulfurization techniques.

F. South Africa

About 50% of South Africa's current production of 75,000,000 tons receives some degree of cleaning at 44 coal washing plants [21]. The latest available data on coal preparation plant statistics (1975) are quoted as follows:

	Tons washed (millions)	*Percent of total tonnage*	*Number of plants*
Jigs	6.4	20	9
Dense-medium bath only	9.8	30	19
Jigs and dense medium	2.1	7	3
Dense medium: bath and centrifugal	11.1	34	11
Dense medium and froth flotation	3.1	9	2

Present coal production is expected to increase to 123 million tons by 1985. About 40% of this increase, principally for export and carbonization coals, will require substantial preparation, so that demand for new coal plant capacity to 1985 is approximately 20 million tons.

It will be observed that 80% of current coal cleaning capacity consists of dense-media system, jigs cleaning 27% of the total.

Similarly to Indian coals, the Gondwanaland coals of South Africa frequently contain very high inherent ash contents, resulting from the widespread occurrence of intergrown material (near-gravity middlings). Effective cleaning therefore requires crushing of ROM to small sizes before washing. This creates severe additional problems because the intergrown material persists down to very fine sizes, so that the resulting slurries are not readily amenable to cleaning by froth flotation. In addition, it is reported that water-only cyclones have not proved effective, and recourse has been made to alternative process developments. Chief among these has been the employment of heavy-media cyclones of enhanced operating efficiencies for treating coal in down to 200 mesh (75 μm), which is close to the limit for centrifugal separation of the denser particles. A research investigation reported by the U.S. Bureau of Mines in 1974 [22] had demonstrated that improved efficiency of cleaning of fine coals in heavy-medium cyclones could be obtained if feed sizes were regulated to exclude material greater than 1/8 in. South African experience confirms this observation and has shown that separation efficiency when fine coal particles are washed alone is much greater than that obtained when the particles are similarly washed in small coal of wider size range, say 1/2 × 0 in. The South African decision to treat -28 mesh material by heavy-media cyclones in conditions where water-only cyclones are ineffective represents a considerable challenge to the froth flotation process. Since the innovation can also be expected to handle successfully the +28-mesh wedge wire oversize, which frequently causes difficulty with froth flotation, the potential advantages become even more marked. It is worth noting that these developments are claimed to be achievable at costs that remain competitive in coking coal export markets.

Going beyond these innovations, other investigations have shown that further improvement in ash contents results from finer grinding of the coal to, say less than 20 μm. For these reasons there is growing interest, as in India, in oil agglomeration processes. Sulfur in coal does not represent as great a problem as the need for reducing ash contents in metallurgical grades and, in any case, sulfur levels are lower than those encountered in the United States. Nevertheless, the techniques under development for ash reduction, which require fine grinding of the coal, present an interesting parallel, and go much beyond, related attempts in the United States and West Germany aimed at reducing pyritic sulfur.

G. United Kingdom

Coal output consists of anthracites and bituminous categories, there being no lignites or brown coals. During recent years coal output, having declined after 1958, has stabilized at about 130 million short tons/year. Future expansion is dependent upon the relative balance that will be struck among coal, oil, and nuclear energy for electric power generation. Demand for metallurgical coals is not expected to alter significantly; coals for conversion to gaseous and liquid products are not expected to be required during this century; and coal exports are not, and are not expected to be significant. Major discoveries, first of natural gas and lately of oil in the British section of the North Sea, and their rapid commercialization, have played an important part in the evolution of this present situation. Development plans for the coal industry are principally directed to replacement of capacity and improvement of productivities. Coal cleaning, and the depth of cleaning employed, is related directly to coal utilization. Jenkinson [23] has reported that "it is common practice in the U.K. to wash only part of the raw coal to produce an acceptable power station fuel of say 16% ash. The part that is treated is the coarse fraction, defined as plus 4 mm or 6 mm."

Current coal preparation capacity [24] totals approximately 73,000 short tons/hr at 182 preparation plants, ranging in individual capacities from 66 to 1430 short tons/hr. Principal cleaning systems used are Baum jigs (126 units), heavy-media vessels (80 units), and froth flotation (89 units). Heavy-media cyclones are being introduced (6 units in operation) and a smaller number of water-only cyclones. Wet concentrating tables have not been popular and only one system is recorded in current operation. Air tables are not used. The National Coal Board has recently released the following information concerning coal preparation planning:

> An important feature of the Plan for Coal is the substantial investment of new coal preparation plants. Forty-seven plants are being built or reconstructed at a cost of just over Ŀ300 m in this 10 year program which started in 1974. Capital approval for 30 projects has already been given, with more than 20 of these either in operation or under construction. 10 more plants are planned to be under construction in 1978/79.
>
> Current capital expenditure on coal preparation plants is averaging Ŀ40 m a year, embracing investment in a wide range of processes and techniques, including the introduction of computer based control systems at 3 selected plants. More than 20,000 tons/hr. of new capacity is being installed in this 10 year period, almost 15,000 tons/hr. of which is provided by Baum Washboxes with just over 5,000 tons/hr. by dense medium processes. The treatment of fine coal minus 0.5 mm has always been an important aspect of coal preparation, and approximately 15% of all new washing capacity includes fines treatment processes mainly froth flotation plants.
>
> Six of the major new plants will prepare coal from the new mine prospects, including the proposed North East Leicestershire coalfield, at a cost of Ŀ120 m. Major new plants for new mines prospects will become an increasingly important feature of the Coal Preparation Plant Program beyond 1985 as Plan 2000 provides a further expansion of output. [24]

A major aim of raw coal pretreatment is dry screening to exclude fines (-4 mm or -6 mm) from the washing process, because a major part of treatment costs are for water clarification and slurry dewatering equipment [23]. This aim is seriously hampered by the ever-increasing amounts of water used for dust suppression, which wets fine coal and clay so that screens are quickly blinded. A novel approach to this problem has been developed and is currently undergoing plant trials. Called the Rotating Probability Screen, it does not make a precise separation at some predetermined size but automatically regulates the amount of material passed for washing and that left untreated for use in power stations. Initial results obtained with units of 100 tons/hr capacity handling difficult and clayey coals have been promising.

Other important areas of current research and development include plant automation and control, including microprocess computer applications for unit operations in cases where full automation is not required or desirable, improvements in mass flow from bunkers, water clarification, and tailings recovery and handling [25].

H. USSR

Targeted coal production for 1978 is reported [26] to be 746 million tons (820 million short tons), production during the past 5 years having been as follows:

Year	Estimated production (ROM) (short tons equivalent)
1973	734,340,000
1974	752,404,000
1975	771,415,000
1976	782,674,000
1977	794,000,000

During the period tabulated, the production of brown coals and lignites increased from 173 to 181 million short tons, and anthracite from 83.4 to 86.9 million short tons, so that the major part of the increasing production is of bituminous and subbituminous categories.

The same source reported that in 1977 coal preparation plants processed 324.5 million short tons of coal and yielded 170.5 million short tons of clean coals for an average yield of 52.5%. This indicates a high mean ash content in ROM coal.

Deurbrouck and Smithson [27] have reported that "coal preparation as practiced and researched in the Union of Soviet Socialist Republics reflects essentially the same goals and long range needs as in the United States. The similarities in basic washing characteristics of the coals, the application of like preparation equipment, quite similar market requirements, and a determination to utilize coal more heavily in the future to supply growing energy demands led to common technological problems in the two countries." They found that about 50% of Soviet anthracite and hard coals are washed, this proportion tending to remain constant as total coal production increases. Types of cleaning systems are as follows:

Jigs	50%
Heavy-media systems	25%
Froth flotation	15%
Launders, air tables, etc.	10%

In keeping with the expansion in coal output, extensive construction of coal preparation is taking place, more than 30 new plants being scheduled to come on-stream during the current Five-Year Plan.

Equipment used is generally of Soviet design and manufacture. Changes in mining practices and market demands are resulting in some changes in practices. Thus, pneumatic separation plants, long popular because of the very low winter temperatures experienced in most coalfields, are encountering increasing difficulty from wetter coal as a result of increased use of dust-suppression methods in the mines. Jigs are employed to treat coals down to 28 mesh size, feldspar jigs being used for the smaller sizes. Three-product separations (multistream cleaning) is carried out in heavy-media vessels. Heavy-media cyclones have recently been introduced. Recent new developments have included a three product dense-media separation, consisting of cylindrical and cyclonic units coupled directly as a two-stage unit. Froth flotation is widely used and has been extensively researched and developed,

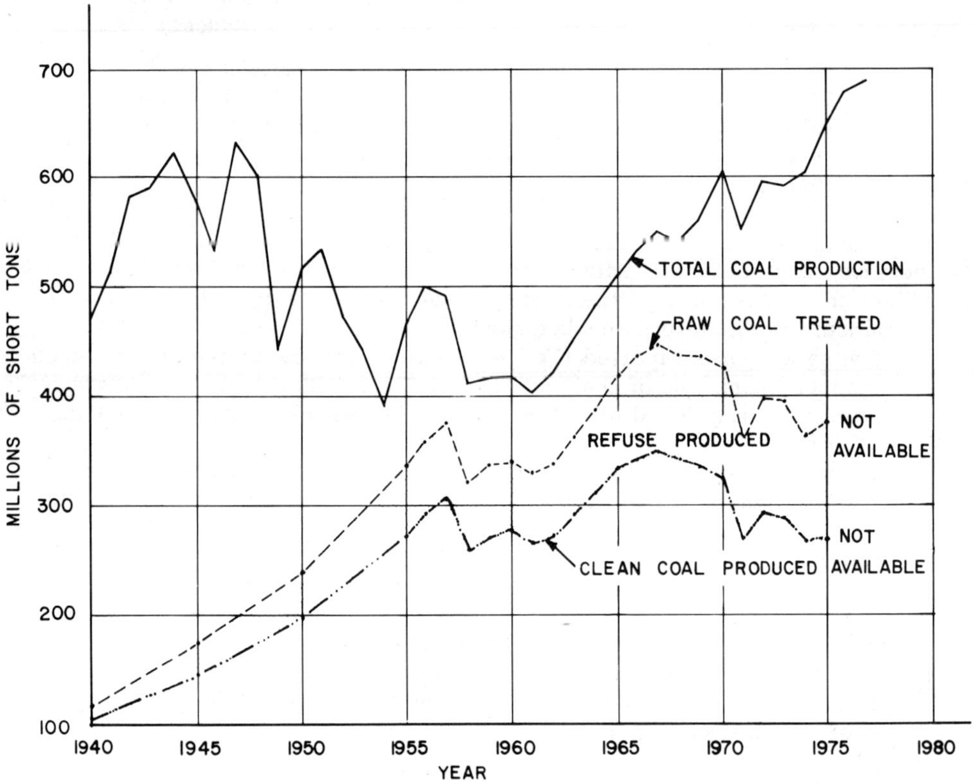

FIG. 1 U.S. production of bituminous coal and lignites.

including its use for achieving desulfurization by selective flotation. It is reported that many coking coals have sulfur contents in the range 1.8 to 2.5%, 80% of which is pyrites so finely disseminated that a 20% reduction by conventional cleaning is regarded as worthwhile. Unlike in the United States, some Russian lignites are reported to contain very high sulfur contents, up to 6%, which is being reduced significantly by washing in meavy-media systems.

Other aspects of coal preparation undergoing development include mechanical dewatering, static thickening techniques, thermal drying, and mathematical approaches to unit and system optimization as a preliminary to automated control of cleaning plant.

I. United States

Coal cleaning in the United States is mainly a post-World War II development, which achieved peak capacity in 1967. Since then it has declined substantially both as to the total tonnage of raw coal handled by treatment plants and as to the yield of clean coal produced. The historical record is presented in Fig. 1 for bituminous coals and lignites. Lignites are produced in the Western

TABLE 2 Mechanical Cleaning of Bituminous Coal in the United States, 1960-1975

	By wet methods							By pneumatic methods	Total mechanically cleaned
Year	Jigs	Concentrating tables	Classifiers	Launders	Dense media	Flotation	Total wet		
1960	136,633	30,741	11,012	7,561	66,251	1,826	255,030	18,139	273,169
1961	133,360	30,158	9,263	6,529	65,148	2,567	247,020	17,691	264,711
1962	136,879	31,859	5,681	5,986	68,565	3,959	252,929	18,704	271,633
1963	142,540	37,492	5,558	5,221	74,177	4,539	269,527	19,935	289,462
1964	145,918	40,878	6,725	6,000	84,159	5,123	288,803	21,400	310,203
1965	151,541	43,197	5,844	4,801	94,636	6,853	306,872	25,384	332,256
1966	156,789	45,427	4,775	4,691	97,301	7,438	316,421	24,205	340,626
1967	161,302	49,529	3,902	4,627	101,043	7,732	328,135	21,268	349,402
1968	159,028	47,268	4,871	4,498	99,497	8,961	324,123	16,804	340,923
1969	155,027	45,328	3,401	4,644	97,636	9,560	315,596	19,163	334,761
1970	140,457	44,058	3,593	5,199	101,593	10,694	305,594	17,855	323,452
1971	115,407	35,656	2,071	4,896	89,764	9,098	256,892	14,506	271,401
1972	130,331	38,232	2,980	5,467	91,058	13,050	281,119	11,710	292,829
1973	132,655	34,935	3,297	5,121	88,203	14,201	278,413	10,505	288,918
1974	129,302	28,869	2,698	3,577	82,283	10,863	257,592	7,557	265,150
1975	N.A.[a]	N.A.	N.A.	N.A.	N.A.	N.A.	260,300	6,700	267,000

[a]N.A., not available.

Source: U.S. Bureau of Mines and National Coal Association.

states of Montana, North Dakota, and Texas. Although of rapidly growing importance to the national energy budget, lignite production is still at comparatively low levels, currently in the region 20 to 25 million tons/year, so that the picture presented in Fig. 1 is essentially that of the bituminous industry. Not shown is anthracite; current production is about 6.5 million tons/year, which contains a very small proportion (less than 0.2 million tons/year) of coal recovered by dredging. The remainder, freshly mined, is all washed and sized for its special markets. Small though the industry may be, it has played an important part over the years in the development of efficient coal preparation systems.

About 42% of current production is cleaned by mechanical methods. The types of mechanical cleaning systems utilized, together with their relative contributions toward the total cleaned coal, are shown in Table 2. The trends illustrated indicate that jigs are the principal cleaning method, accounting for 45 to 50% of total clean coal. Dense-medium systems produced 30 to 35% of the total, whereas the other systems commonly employed for cleaning coarse sizes (e.g., classified and launders) have diminished in importance and now handle less than 3% of total output. Concentrating tables, which have always been more popular and more widely used in the United States than in all other countries, continue to make a substantial contribution to the total output of fine coal. Froth flotation, initially slow to be adopted, is now being more widely used and is producing around 5% of the total production. It may be noted that to a large degree the contribution made by froth flotation consists of fine coal formerly discarded to tailings ponds, and its growth thus represents a significant improvement in recovery efficiency of the industry. Finally, it should be noted that the use of pneumatic cleaning has declined considerably, although recent developments in pneumatic and other dry cleaning devices may act to reverse this trend in the future [28,1]. Other changes that might be expected include the replacement of some heavy-media systems by the newer generation of high-efficiency jigs, and the further replacement of heavy-media vessels by heavy-media cyclones. Wider use of water-only cyclones, particularly of the compound type, may replace some capacity presently provided by feldspar jigs and concentrating tables. These changes will be discussed in more detail later.

In addition to reviewing the types of cleaning systems employed, the present and future structure of the coal preparation industry requires some attention as a result of the overall change in demand and market requirements faced by it. Traditionally, the industry is a mine-mouth activity preparing bituminous coals and anthracites mined by underground methods. It has not been customary to wash subbituminous coals or lignites, although this practice may change in the future and, already, one of the biggest coal preparation plants in the West is cleaning subbituminous coal. In 1975, there were approximately 5100 mines in operation, but only 388 reported cleaning plants [29,30]. The geographical distribution of these plants is as shown in Table 3. The plants are thus seen to be preponderantly in Appalachia, where the mine capacities are generally smaller. For the future, recalling that the future growth of coal cleaning is highly geared to future electric utility requirements, the sizes of cleaning plants relative to the demands of generating plants may become an important relationship in the chain of costs contributing to the busbar cost of electricity. A 1000-MW (electricity) generating station operating

TABLE 3 Distribution of Coal-Cleaning Plants by State

Alabama	21
Colorado	3
Illinois	34
Indiana	11
Kentucky	62
Ohio	19
Pennsylvania	64
Utah	6
Virginia	23
West Virginia	124
All other states	21
	388

at a heat rate of 11,000 Btu/kWhr (η = 0.31) and 85% load factor, burning a bituminous coal having a higher heating value of 12,500 Btu/lb, requires 9000 tons/day over 7 days, or 12,600 tons/day, for a cleaning plant working a 5-day week. Inspection of the sizes of cleaning plants presented in Table 4 shows that only 20% of the present plant could individually support a power-generating plant of this size, which is not large by present standards. Unless a move to larger coal- leaning plants occurs in the future, these power-generating stations will be required to take coal supplies from a multiplicity of sources, a situation already common in the Eastern states. The logistics of large central coal preparation plants handling raw coal outputs from a number of mines within their catchment areas versus a larger number of small self-contained mine-mouth plants is an important consideration in deriving the delivered cost of clean coal and its overall effect on electricity cost at the bus bar. However, although the question of scale on capital and operating costs of coal preparation will be considered below, this wider logistical exercise is beyond the scope of this chapter.

TABLE 4 Distribution of Coal-Cleaning Plants by Daily Capacity

Plant capacity (tons/day)	Number	Percent
>12,000	77	19.8
10,000-12,000	81	20.9
<5,000-10,000	51	13.1
5,000	179	46.2
	388	100.0

TABLE 5 Estimates of Future Demand for Bituminous Coal Production

	East of Mississippi (millions of tons)	West of Mississippi (millions of tons)	Total (millions of tons)
1977 (actual)	522	166	688
1985			
High	610	380	990
Medium	644	472	1116
Low	668	520	1118
1990			
High	660	454	1114
Medium	758	763	1521
Low	806	1050	1856

Source: U.S. Department of Energy, Washington, D.C., June 1978.

The National Energy Plan requires that coal production increases from its present level (688 millions tons in 1977) to 1.0 to 1.2 billion tons by 1985 and to 1.5 billion tons (1.1 to 1.85 range) by 1990. Estimates of future demand are shown in Table 5.

The figures imply production rate increases of about 15 to 20 million tons/year in Eastern coalfields and 25 to 60 million tons/year west of the Mississippi. Much of the increasing Western production will be low-sulfur subbituminous and lignite grades requiring a minimum of coal preparation. Presently, 42% of total production is cleaned by mechanical methods. If present trends of coal preparation are maintained in the future, it is expected that 70 to 75% of the higher-rank bituminous coals will require treatment so that the requirement for new plant capacity will be about 15 to 25 million tons of new capacity per year. If the present practice regarding plant sizes continues, this represents a need for about 20 new plants each year. In 1975, H. L. Lovell [31] estimated that a total of 150 new plants would be required in the period 1976-1985 and that 100 existing plants would require upgrading.

The cost of new capacity varies widely, being dependent on the size of plant depth of cleaning required, coal properties, coal-handling facilities required, expected plant life, and plant location. Units under construction vary from portable "package" plants of throughputs of around 100 to 200 tons/hr to major permanent installations having hourly throughputs of 1200 tons and more. It is customary in the United States to base capital costs on hourly capacity of raw coal handled, since it is raw coal that determines the basic unit sizes. Present costs can vary in the range $3000 to $50,000 per ton of hourly capacity [32,33], hourly capacities being related to annual throughputs according to the number of shifts worked per week.

The cost of new capacity to produce 20 million tons/year of salable coal is conservatively estimated at $200,000,000 to $300,000,000 per year. Investment in the 10-year period 1975-1985 could, therefore, amount to $2 to $3 billion. An estimate prepared by Lovell [31] in 1975 put this cost at $3.4 billion. Estimates prepared by the staff of *Coal Mining and Processing Journal*, also in 1975, put this figure at $4 billion [34].

Certain other factors are of great importance in assessing the future of coal cleaning in the United States. The decreasing yield accompanied by an ever-increasing problem of waste disposal, shown in Fig. 1, is a consequence of greater quantities of impurities in ROM coal resulting from increased mechanization, and until more selective mining machines are developed, this trend will continue to grow. The awkward truth, therefore, is that in relative and absolute terms, coal cleaning has been a declining industry in the United States in very recent times, even though the production of coal has continued to increase and is, at 688.6 million tons in 1977, at its historic greatest level.

Grundy [35] has recently explained that

> . . . the National Energy Plan envisions increasing coal utilization by electric utilities from 450 million tons in 1975 to 860 million tons by 1985, or 100 million tons more than would otherwise occur. . . . The National Energy Plan envisions a four-fold increase in industrial coal consumption by 1985 from 156 million tons in 1977 up to 410 million tons, or 200 million tons greater than would otherwise occur in 1985. The majority of this increase would have to take place in non-metallurgical uses which by 1985, would have to increase from 70 million tons in 1977 to 305 million tons. However, nearly 60 percent of 1976 usage was for coking purposes, a function of steel production. Only 61 million tons were used for other industrial uses. Therefore in order to meet the goal of the National Energy Plan, non-steel uses excluding electric utilities use would have to increase five-fold. . . .

The changes in coal consumption since 1960 are summarized in Fig. 2. A remarkable feature is that a large increase in coal consumption by the electric utilities industry has actually been accompanied by a decrease in coal's share of the total primary energy consumption in electric power generation.

The EPA New Source Performance Standards (NSPS) for fossil-fueled boilers promulgated in 1971 set the limits for sulfur oxides at 1.2 lb/million Btu of fuel consumed. In 1977, Congress revised the basis for NSPS to require "the degree of emission limitation and the percentage reduction achievable through application of the best technological system of continuous emission reduction" by means of setting either or both a maximum emission rate and an enforceable percentage reduction. Congress expressly provided that, in considering new standards, the EPA administrator could allow credits for "any cleaning of the fuel or reduction in the pollution characteristics of the fuel after extraction and prior to combustion. . . ." These amendments to the law and their possible effects have recently been reviewed by Grundy [35]. Two alternative approaches to the NSPS are being considered by the EPA for meeting the percentage removal requirements. One option would require the use of the best available technology for treatment of the entire flue gas stream of all new power plants, no matter what the sulfur content of the coal, so as to reduce the untreated emission level by 85%. A second option would allow use of coal with a lower sulfur content to meet a less stringent percentage removal standard. This alternative would exempt facilities that reduce stack emissions to less than 0.5 lb of sulfur oxides per million Btu from the 85% reduction requirement. These options are usually referred to as the full and partial scrubbing options and their significance for the U.S. coal reserve

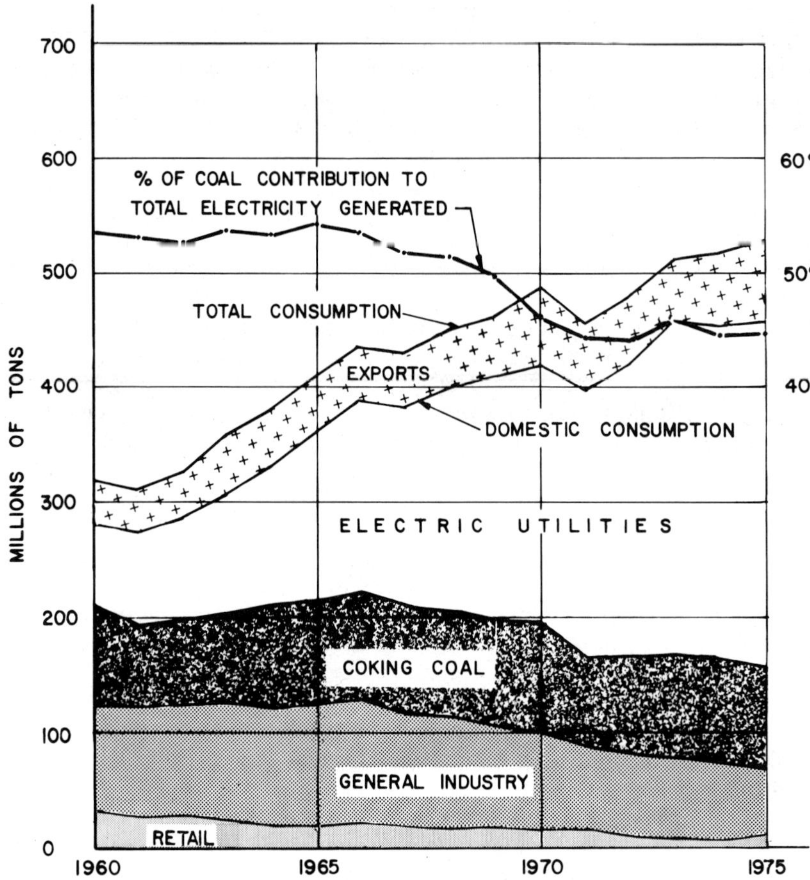

FIG. 2 Consumption by category and percentage contribution by coal to total electricity generated.

base is considered in greater detail in Chap. 7. Their effects on the future of coal cleaning, and more particularly on coal desulfurization, have yet to be worked out. A critical factor will be the relative costs of precombustion desulfurization compared with the other control options (Fig. 3) as a function of electric power costs at the plant bus bar. Such an analysis, already being considered by a number of utilities, takes into account not only fixed and variable costs of preparation, but also the attendant transportation, handling, and remaining cleanup costs at the power plant. Since the market for carbonization coals, the traditional market for deep-cleaned coals, is not expected to grow substantially in the foreseeable future, the future growth of coal cleaning is mainly dependent upon the extent of the contribution it is able to make to the problems faced by the electric utilities and major industrial users.

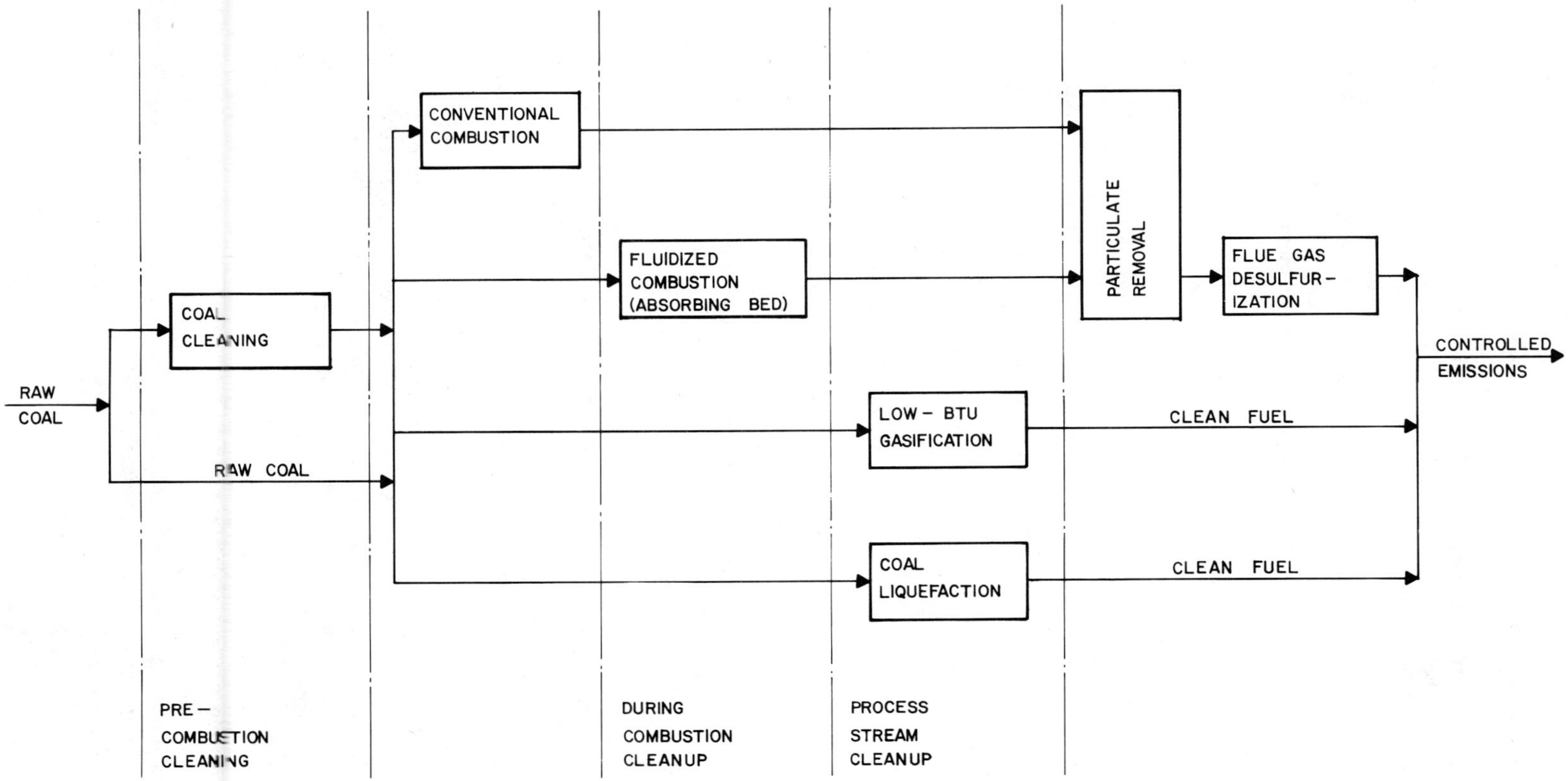

FIG. 3 Current options for controlling flue gas emissions.

III. ELEMENTS OF CLEANING PROCESSES

Although the direct objectives of present coal-cleaning practices are the reduction and control, within predetermined limits, of size, moisture, ash, and, sometimes, sulfur, the properties of coal that have a direct bearing on whether, and how, coal should be cleaned are wider than this list suggests. Thus, the rank of a coal (rank being a complex property descriptive of the nature of the coal substance but not of its impurities) can play a very important role in determining the feasibility and the extent of cleaning.

A. Composition and Rank

Pure coal consists mainly of carbon (C), with lesser quantities of hydrogen, oxygen, nitrogen, and sulfur (H, O, N, S). Compounds of these elements provide the combustible material that constitutes coal's value as a fuel, as measured by the calorific (higher heating) value. These compounds are contained in an organic matrix said to consist of "macerals," to identify them from inorganic materials, which are usually present as impurities assembled in various crystalline matrices termed "minerals." Except for minute quantities of silicon, calcium, and magnesium associated with vegetal material, virtually all other elements found in a coal, other than C, H, O, N, and S, will be associated with the mineral impurities.

When coal is burned, the whole of the macerals present are converted to gaseous combustion products and the mineral impurities present remain as an ignited residue termed "ash." It is important to observe that ash is not the mineral matter originally present in the coal, but the residue remaining after those minerals have been heated to high temperature in air, so that, for example, carbonates initially present will have decomposed and lost carbon dioxide; and pyrites will have oxidized, losing sulfur as sulfur oxides to the products of combustion, leaving iron oxides as a constituent of the ash. Changes resulting from decomposition of mineral matter in this manner do not make any contribution to the calorific value of the coal, usually detract from it, and where calorific values must be measured accurately, require corrections of the measured values. However, these corrections are not large and it is sufficient for practical purposes to take the ash content, which is easily and rapidly determined, as a direct measure of the mineral impurities present in a coal.

In coalification, or the metamorphosis of vegetal materials to coals, which is generally recognized to follow the sequence peat, lignite, subbituminous, bituminous, and anthracite, significant changes in the relative proportions of C, H, O, N, and S occur, the chief of which is the increase in C and the elimination of some H, O, N, and S as large volumes of methane and carbon dioxide are released. Where coals are pyrolyzed, this process is continued by virtually complete elimination of H and O. The carbon residue that remains, however, still contains small quantities of N and S. Typical values for C, H, pyrolyzable matter (volatiles), and residual carbon (fixed) of the various classes of coals are shown in Table 6. This table provides the basis of a system for describing coal rank, which increases as one proceeds down the table. Increasing rank is associated with decreasing volatile matter and increasing calorific value. There are a number of systems of classification

TABLE 6 Coal Properties Related to Coalification

Coal types	Dry mineral-matter-free basis				
	Carbon (%)	Hydrogen (%)	Limits of volatile matter (%)	Fixed carbon (%)	Calorific value (Btu/lb)
Lignites	73-78	5.2-5.6	45-50	50-55	<8,300
Subbituminous	78-82.5	5.2-5.6	40-45	55-60	8,300-11,500
High-volatile bituminous	82.5-87	5.0-5.6	30-40	60-70	11,500-14,000
Medium-volatile bituminous	87-92	4.6-5.2	20-30	70-80	>14,000
Low-volatile bituminous	91-92	4.2-4.6	15-20	80-85	>14,000
Anthracite	95-98	2.9-3.8	5-10	91-95	>14,000

of coals by rank [e.g., ASTM (U.S.) (see Chap. 1, Table 1), NCB (U.K.), DIN (West German), Polish, and ISO (International)]. Most of the systems employ further subdivisions of coal types according to coking, caking, or agglomerating properties exhibited by the coals on heating. In the ISO system of classification, lignites are further subdivided according to whether or not they evolve tars on pyrolysis. One further difference, of great consequence for coal preparation and utilization, is that increasing coalification is a function of increasing pressure and heating of the coal beds, and is accompanied by a large and progressive decrease in bed moisture. Thus, brown coals and lignites frequently contain moisture contents up to 75% (American lignites usually contain about 35%), subbituminous coals up to 25%, whereas those for bituminous coals and anthracites are usually below 5% and can be below 1%. It is necessary to distinguish this bed, or inherent moisture, from moisture increases caused during mining, handling, or preparation operations.

All the properties related to coal rank (i.e., calorific value, coking properties, moisture) have an important bearing on coal preparation, transportation, and utilization. Transportation costs are usually charged on a ton-mile basis which does not distinguish between coal substance and moisture content. Long-range transportation of lignites, more than one-third of which consists of water, can more than triple the initial mine-mouth costs calculated on an energy basis. Coal drying, an essential feature in many modern preparation systems, requires about 1800 to 2500 Btu per pound of water evaporated, which can therefore represent a substantial proportion of the energy content for low-rank coals. These few examples are sufficient to show that coal preparation, its systems, and its costs are an integral part of the production and use of coals. Their effects on costs can be as important as the planning of mine layouts or decisions concerning mining systems and should be an essential element in all mining feasibility studies.

It is now evident that coal cleaning is a process that separates coal macerals and minerals. Cleaning methods are, accordingly, processes that recognize and exploit differences in the properties of macerals and minerals (see Chap. 1, Tables 3 and 4). In general, there are two broad classifications that may be employed: physical properties and chemical properties. Physical properties that may be used include hardness, density, attenuation of radiation (light, x-rays, microwave), magnetic susceptibility, wettability, and zeta potential. Table 7 lists some physical properties of coals, shales, and pyrites and illustrates the differences.

The success achieved by methods designed to separate macerals from minerals depends upon the extent to which these components occur as massive, discrete, loosely associated structures or the extent to which they are intergrown. Some coal seams are virtually pure macerals, so that cleaning is largely a matter of removing extraneous "country" rock that becomes mixed with the coal during mining. In other cases, the coal seams may contain so much intergrown, finely disseminated mineral components that they are uneconomical to mine. In the great majority of cases, seam structures fall between these extremes, the pure coal macerals being diluted with disseminated minerals or interspersed with partings and cleats of intergrown material or clays, or the separation between coal and rock at the floor and roof of the seam may be indistinct. The proportions of these "middlings" materials have an important consequence for the "washability" of coals.

TABLE 7 Some Physical Properties of Raw Coal Components

Property	Coal macerals	Minerals (excluding pyrite)	Pyrite
Color	Black	Grey	Golden
Specific gravity	1.15-1.5	2.2-3.9	4.8-5.0
Reflectance (%)	0.5-7.5	4-8	42-47
Hardness (Mohs' scale)	2.1-2.5	2.0-4.5	6.0-6.5
Wettability (contact angle) (deg)	49-08	11-15	00 75
Magnetic susceptibility (cgs)	-0.5 to -0.6	-20 to +400	+40 to +65
Dielectric constant	2.0-2.5	4.7-7.8	5.2-8.5

The chemical structure of coal is described in Chap. 1. The nitrogen content is almost entirely combined in the maceral components, whereas oxygen and sulfur occur in both organic and inorganic combinations. In the cleaning context, oxygen is unimportant because it is nonpolluting. Sulfur is a special case because it is a serious pollutant and occurs in various forms distributed in both macerals and minerals. As "organic sulfur" it occurs combined entirely in macerals. As pyritic sulfur, it occurs as discrete particles distributed in both macerals and minerals. These two forms usually account for virtually all the sulfur present, but occasionally small quantities of sulfates are found and are associated with the minerals. The bulk of the sulfur present, whether organic or pyritic, and whether associated with the macerals or minerals, reports to the gaseous products of combustion. It should be noted that because pyritic sulfur is associated with both macerals and minerals, determination of the pyritic sulfur content of a raw coal does not itself indicate the extent of desulfurization that will occur on separation of the macerals. For this reason some processes that are effective in removing pyrites from certain coals are ineffective with other coals. The crucial factors in such cases are the relative proportions of pyrites in macerals and minerals, and the size and distribution of the pyrites in the macerals themselves.

B. The Concept of Washability

Washability is a concept that measures and exploits the differences that exist between the densities of coal macerals and the heavier associated minerals as a basis for predicting the yields and qualities of the products obtained for any given partition density.

Density is an absolute physical property measured in terms of mass per unit volume, which is not convenient to measure directly. For practical purposes it is customary to employ the specific gravity of a substance, which is

TABLE 8 Washability Characteristics of a High-Sulfur Bituminous Coal

1	2	3	4	5	6	7	8	9	10	11	12	13	14	15	16	17	18	19	20	21
Size of Coal		Sp. Gr.				Productivity Values			Cumulative Float						Cumulative Sink					
	Size (Wt %)	Fraction	Wt %	Size (Wt %)	% Ash	% S	Ash	S	Wt %	Size (Wt %)	Ash Prod.	% Ash	S Prod.	% S	Wt %	Size (Wt %)	Ash Prod.	% Ash	S Prod.	% S
1½ x3/4 in.	15.7	1.25	2.13	0.34	1.89	1.72	0.64	0.59	2.13	0.34	0.64	1.89	0.59	1.72	100.00	15.7	327.67	20.87	69.78	4.45
		1.25 - 1.30	31.06	4.88	2.97	2.20	14.49	10.74	33.19	5.22	15.13	2.90	11.33	2.17	97.87	15.36	327.03	21.29	69.19	4.51
		1.30 - 1.35	14.47	2.27	6.69	3.44	15.19	7.81	47.66	7.49	30.32	4.05	19.14	2.56	66.81	10.48	312.54	29.82	58.45	5.57
		1.35 - 1.40	11.06	1.74	9.94	4.90	17.30	8.53	58.72	9.23	47.62	5.16	27.67	3.00	52.34	8.21	297.35	36.22	50.64	6.17
		1.40 - 1.50	12.77	2.00	15.01	6.36	30.02	12.72	71.49	11.23	77.64	6.91	40.39	3.60	41.28	6.47	280.05	43.28	42.11	6.51
		1.50 - 1.60	5.11	0.80	23.08	7.88	18.46	6.30	76.60	12.03	96.10	7.99	46.69	3.88	28.51	4.47	250.03	55.94	29.39	6.58
		1.60 - 1.70	3.83	0.60	29.06	8.92	17.44	5.35	80.43	12.63	113.54	8.99	52.04	4.12	23.40	3.67	231.57	63.10	23.09	6.29
		1.70 - 1.80	2.55	0.40	36.80	9.70	14.72	3.88	82.98	13.03	128.26	9.84	55.92	4.29	19.57	3.07	214.13	69.75	17.74	5.78
		1.80 - 2.00	2.55	0.40	46.50	9.68	18.60	3.87	85.53	13.43	146.86	10.94	59.79	4.45	17.02	2.67	199.41	74.69	13.86	5.19
		2.00	14.47	2.27	79.65	4.40	180.81	9.99	100.00	15.7	327.67	20.87	69.78	4.45	14.47	2.27	180.81	79.65	9.99	4.40
3/4 x¼ in.	40.5	1.25	1.96	0.79	2.03	1.38	1.60	1.09	1.96	0.79	1.60	2.03	1.09	1.38	100.00	40.5	687.04	16.96	188.63	4.66
		1.25 - 1.30	23.96	9.70	2.78	2.04	26.97	19.79	25.92	10.49	28.57	2.73	20.88	1.99	98.04	39.71	685.44	17.26	187.54	4.72
		1.30 - 1.35	26.43	10.71	4.78	2.94	51.19	31.49	52.35	21.20	79.76	3.76	52.37	2.47	74.08	30.01	658.47	21.94	167.75	5.59
		1.35 - 1.40	12.19	4.94	9.62	4.92	47.52	24.30	64.54	26.14	127.28	4.87	76.67	2.93	47.65	19.36	607.28	31.47	136.26	7.06
		1.40 - 1.50	11.17	4.52	15.20	6.40	68.70	28.93	75.71	30.66	195.98	6.39	105.60	3.44	35.46	14.36	559.76	38.98	111.96	7.80
		1.50 - 1.60	5.80	2.35	22.40	8.98	52.64	21.10	81.51	33.01	248.62	7.53	126.70	3.84	24.29	9.84	491.06	49.90	83.03	8.44
		1.60 - 1.70	3.58	1.45	28.40	9.24	41.18	13.40	85.09	34.46	289.80	8.41	140.10	4.07	18.49	7.49	438.42	58.53	61.93	8.27
		1.70 - 1.80	2.39	0.97	36.10	9.56	35.02	9.27	87.48	35.43	324.87	9.17	149.37	4.22	14.91	6.04	397.24	65.77	48.53	8.03
		1.80 - 2.00	2.39	0.97	45.08	9.20	43.73	8.92	89.87	36.4	368.55	10.13	158.29	4.35	12.52	5.07	362.22	71.44	39.26	7.74
		2.00	10.13	4.10	77.68	7.40	318.49	30.34	100.00	40.50	687.04	16.96	188.63	4.66	10.13	4.10	318.49	77.68	30.34	7.40
¼ in.x28 M	30.7	1.25	3.15	0.97	2.24	1.64	2.17	1.59	3.15	0.97	2.17	2.24	1.59	1.64	100.00	30.7	605.31	19.72	157.74	5.14
		1.25 - 1.30	35.28	10.83	3.08	2.40	33.36	26.00	38.43	11.80	35.53	3.01	27.59	2.34	96.85	29.73	603.14	20.29	156.15	5.25
		1.30 - 1.35	10.07	3.09	7.74	4.24	23.92	13.10	48.50	14.89	59.45	3.99	40.69	2.73	61.57	18.90	569.78	30.15	130.15	6.89
		1.35 - 1.40	2.54	0.78	8.98	5.02	7.00	3.92	51.04	15.67	66.45	4.24	44.61	2.85	51.50	15.81	545.86	34.53	117.05	7.40
		1.40 - 1.50	16.28	5.00	12.37	6.28	61.85	31.40	67.32	20.67	128.30	6.21	76.01	3.68	48.96	15.03	538.86	35.85	113.13	7.53
		1.50 - 1.60	7.79	2.38	19.78	7.20	47.08	17.14	75.11	23.05	175.38	7.61	93.15	4.04	32.68	10.03	477.01	47.56	81.73	8.15
		1.60 - 1.70	2.64	0.81	25.66	8.76	20.78	7.10	77.75	23.86	196.16	8.22	100.25	4.20	24.89	7.65	429.93	56.20	64.59	8.44
		1.70 - 1.80	3.70	1.14	28.84	8.80	32.88	10.03	81.45	25.00	229.04	9.16	110.28	4.41	22.25	6.84	409.15	59.82	57.49	8.40
		1.80 - 2.00	3.77	1.16	38.74	8.82	44.94	10.23	85.22	26.16	273.98	10.47	120.51	4.61	18.55	5.70	376.27	66.01	47.46	8.33
		2.00	14.78	4.54	72.98	8.20	331.33	37.23	100.00	30.70	605.31	19.72	157.74	5.14	14.78	4.54	331.33	72.98	37.23	8.20
28 M x 0	13.1		100.00	13.1	25.0	4.7														
Composite	100.0	1.25	2.50	2.50	2.13	1.63	5.33	4.00	2.50	2.50	5.33	2.13	4.00	1.63	100.00	100.00	1914.19	19.15	482.54	4.83
		1.25 - 1.30	29.20	29.20	2.98	2.23	87.02	65.12	31.70	31.70	92.35	2.91	69.12	2.18	97.50	97.50	1908.86	19.58	478.54	4.91
		1.30 - 1.35	18.60	18.60	5.57	3.27	103.60	60.82	50.30	50.30	195.95	3.90	129.94	2.58	68.30	68.30	1821.84	26.67	413.42	6.05
		1.35 - 1.40	8.20	8.20	9.58	4.93	78.56	40.43	58.50	58.50	274.51	4.69	170.37	2.91	49.70	49.70	1718.24	34.57	352.60	7.09
		1.40 - 1.50	13.50	13.50	18.79	6.33	253.67	85.46	72.00	72.00	528.18	7.34	255.83	3.55	41.50	41.50	1639.68	39.51	312.17	7.52
		1.50 - 1.60	6.50	6.50	21.49	8.03	139.69	52.20	78.50	78.50	667.87	8.51	308.03	3.92	28.00	28.00	1386.01	49.50	226.71	8.10
		1.60 - 1.70	3.20	3.20	27.56	8.91	88.19	28.51	81.70	81.70	756.06	9.25	336.54	4.12	21.50	21.50	1246.32	57.97	174.51	8.12
		1.70 - 1.80	2.90	2.90	32.46	9.18	94.13	26.62	84.60	84.60	850.19	10.05	363.16	4.29	18.30	18.30	1158.13	63.29	146.00	7.98
		1.80 - 2.00	3.00	3.00	41.94	9.04	125.82	27.12	87.60	87.60	976.01	11.14	390.28	4.46	15.40	15.40	1064.00	69.09	119.38	7.75
		2.00	12.40	12.40	75.66	7.44	938.18	92.26	100.00	100.00	1914.19	19.15	482.54	4.83	12.40	12.40	938.18	75.66	92.26	7.44

its density relative to water at 4°C (sp. gr. = 1.000), and which may be easily and conveniently measured for liquids. Washability data are therefore always reported in terms of mean specific gravities of particles and those of the liquids used to effect separation.

Laboratory washability data are determined using true solutions under quiescent conditions. It is customary to screen the coal sample into convenient size fractions, usually related to the particular application, and to test each size fraction in a series of solutions of incremental specific gravity. The material sinking at any particular specific gravity is recovered and tested at the next-higher gravity level. The material floating is collected and analyzed as representing the particular size and gravity fraction. The analysis carried out depends on the intended coal utilization. Since removal of mineral matter is always the principal objective, ash contents are always determined. Other tests may include calorific value, sulfur, volatile contents, coking proportions, and ash-fusion temperatures. Since it is difficult, even under laboratory conditions, to obtain free, unhindered, separation of ultrafine sizes, it is customary to analyze the -28 mesh (-0.5 mm) material as a single composite fraction. Table 8 shows typical washability data on a high-sulfur coal crushed to a top size of 1 1/2 in. In this table the fractional weights of the sizes selected are shown in the second column. The third and fourth columns show the distribution of each size fraction by specific gravity obtained by the experimental float and sink procedure described above (e.g., for the 1 1/2 × 3/4 in. size fraction, the quantity of material sinking at sp. gr. 1.25 but floating at sp. gr. 1.30 was 31.06% of the size fraction and 29.2% of the composite raw coal sample. The fifth column, "size wt %," is the value in the previous column expressed as a percentage of the total raw coal. Columns 6 and 7 are experimentally determined values of ash and sulfur contents on the corresponding size/specific gravity fractions. The remainder of the table consists of a series of calculated weight mean averages of yield, ash, and sulfur contents for floats and sinks, the column headed "product values" being the accumulators employed. Tabulated data are presented in graphical form called washability curves, the shape of the curves being indicative of the relative ease or difficulty of cleaning. The data in Table 8 are shown plotted in Figs. 4 and 5 for ash and sulfur, respectively. The curves shown are plotted for the composite data shown in Table 8 but may be similarly plotted for any size fraction, as required.

Curve A shows, for any given yield, the maximum ash content of the highest specific gravity fraction comprising that yield. It is based on the assumption that for any given specific gravity fraction, one-half of the material present will have an ash lower and one-half higher than the measured fractional ash content. To obtain the curve, the ash contents shown in column 6 are plotted against the yield for the previous ash content plus half the fractional yield between them, obtained from column 10.

Curve B shows the clean coal yield curve (column 10) plotted against the cumulative float ash content (column 13).

Curve C is a corresponding curve showing refuse yield (column 16) plotted against the cumulative sink ash (column 19).

Curve D shows the relationship between clean coal yield (column 10) and specific gravity (higher figure in column 3). Curves B and D in combination provide the specific gravity cutpoint and yield corresponding to any desired coal quality expressed in terms of its ash content.

Curve E shows the quantity of coal that will float within ±0.1 unit of specific gravity range of any given value of specific gravity, the appropriate values being obtained from columns 3 and 10. The smaller the range of yield, the sharper will be the separation achieved.

The curves illustrating sulfur distributions in Fig. 5 follow similar methods of plotting and are similarly interpreted. The type of information clearly lends itself to electronic data processing and analysis, and these methods are now commonly employed [38].

The data thus presented readily furnish information on the theoretical yields and quality at any predetermined partition specific gravity and are thus an invaluable starting point in the design of coal-cleaning systems. They require additional interpretation when applied to actual systems to allow for such factors as dynamic, rather than quiescent conditions; product yields where more than one recovery stream is employed; size degradation; and the effect of crushing and recycling of intergrown, near-gravity material. With regard to the latter point, it was customary even in recent years to interpret near-gravity data as indicating difficult washing characteristics if the value,

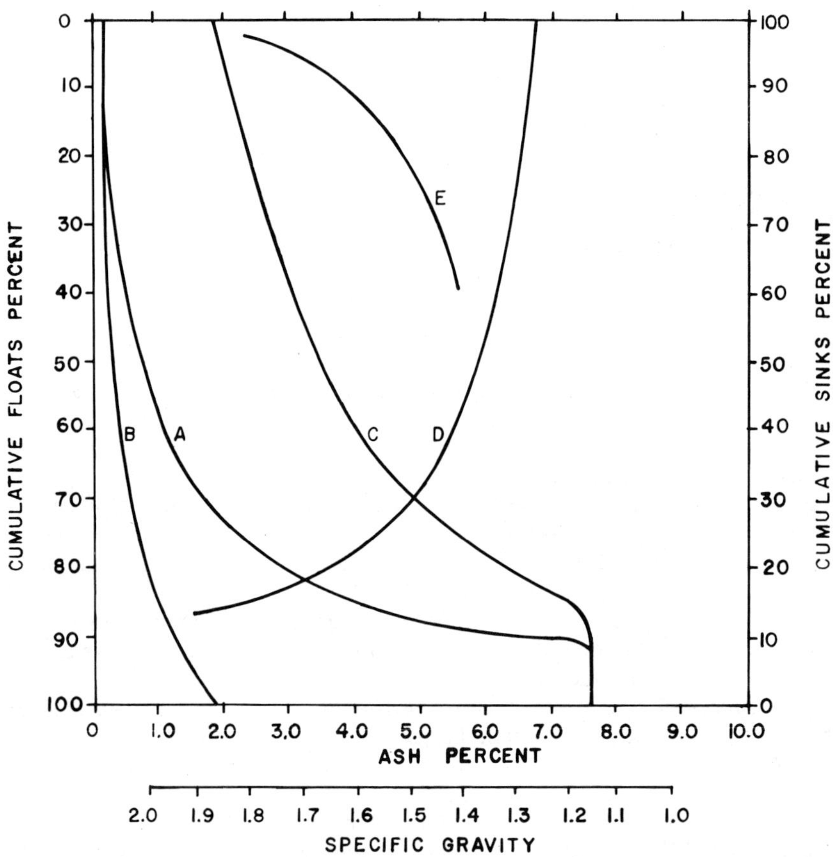

FIG. 4 Washability curves for a high-sulfur coal, showing ash distribution.

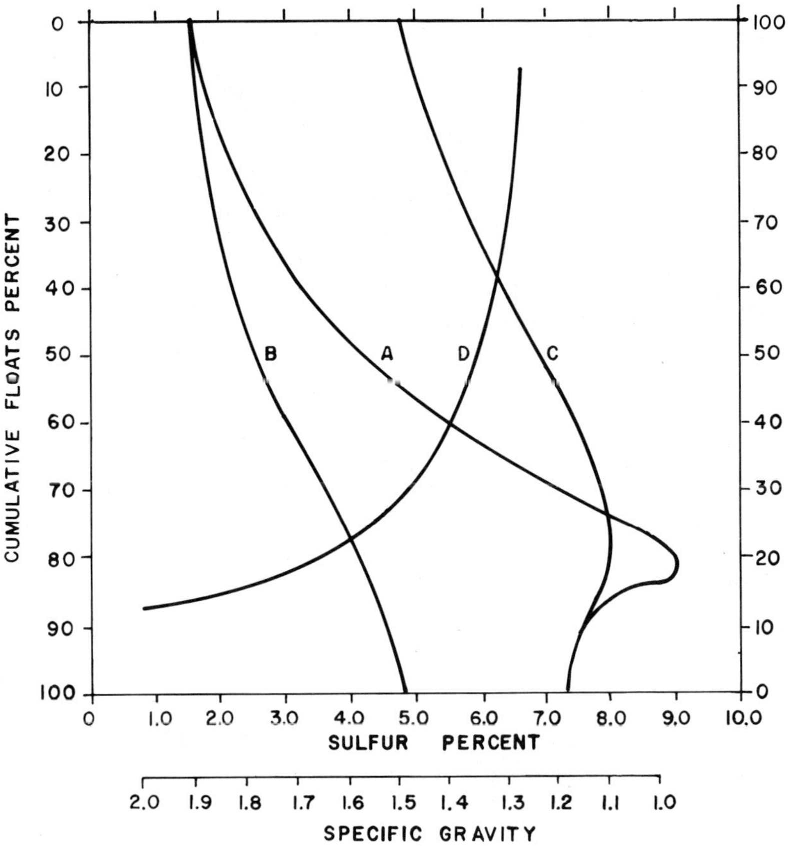

FIG. 5 Washability curves for a high-sulfur coal, showing sulfur distribution.

at any particular partition specific gravity, exceeded ±10%, this difficulty becoming formidable if it exceeded ±25% [39]. Later, Yancey and Geer [40] demonstrated that +0.1 specific gravity yields have little effect on washing efficiencies and that the -0.10 sp. gr. material is the operative factor, so that loss of coal to refuse is a greater problem in operations than is inclusion of refuse in clean product. However, accuracy of separation is not a simple enough concept to have an axiomatic definition and, as will be shown, there are many expressions for it.

Many of the problems of process selection in coal preparation arise from the need to draw comparisons between the predicted performance of alternative systems (e.g., jigs, heavy-media cyclones) that are not possible using these conventional methods of representing washability. In a series of papers published between 1943 and 1953, F. W. Mayer [41,42] described a system of graphical presentation of washability data, referred to as the "M curve," whose geometrical properties were directly analogous to the composition of coal and its associated shale, including intergrown material, although it can in fact be applied to the separations of other materials (Fig. 6). Its essential

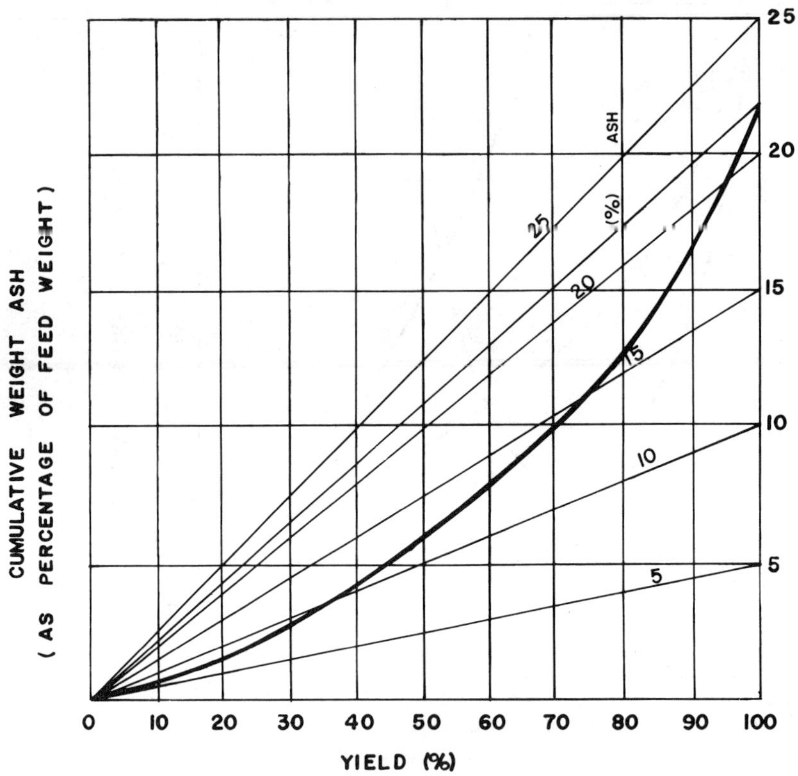

FIG. 6 M curve using normal axis.

feature is that the abscissa always represents the mass of the whole sample and the ordinate the relative mass of some component or other property that can be usefully expressed as a mass ratio of the whole sample. The geometrical properties of the curve are such that the mean slope of the curve to the point of intersection represents the percentage ash content and the actual slope at the point of intersection represents the highest ash material (instantaneous ash) present in the yield. The use of M curve paper ruled as shown in Fig. 6 easily permits the mean ash content corresponding to a given yield to be read by extending a line drawn from the origin through the point of intersection to the right-hand ordinate. However, determination of the instantaneous ash requires measurement of the tangent to the curve at the point of intersection and requires the use of a parallel rule or a special protractor. For some raw coals of difficult washability, the curve becomes too flat for satisfactory measurement of the gradients, in which case the difficulty can be overcome by the use of nonrectilinear coordinates as illustrated in Fig. 7 [43], which shows the same data as Fig. 6.

The utility of the M curve may be better understood by reference to Fig. 8, which shows a three-product separation. A clean coal at 6.5% nominal ash would be represented by a line (OA) having this mean slope and, as indicated, corresponds to an yield of 35%. A middlings fraction of 20% nominal ash content, having a mean slope of this value drawn through A, would cut the yield curve at B equivalent to a yield of 33% (OD - OC). The mean slope of the line joining BE corresponds to a reject ash content of 69% and the yield of refuse will be 28%. The weighted mean ash of the three products, in this example, also equals 28% and corresponds to the raw coal ash shown at point E. The slope of the curve at A corresponds to an instantaneous ash of 10.4%, so that the clean coal contains all particles at this ash content or less. The slope of the curve at B corresponds to an instantaneous ash content of 37%, so that the middlings fraction comprises all material of ash content in the range 10.4 to 37%.

The British National Coal Board carried out a detailed investigation of the properties and utility of the M curve and found that its properties and utility were considerably more extensive than first appeared [44]. Thus, it is not merely a graph relating yield and quality of product but a vector plot in which the horizontal length of a vector represents weight and the gradient represents ash (or other component), so that the washability curve consists of an infinite number of vectorial elements which possess additive characteristics. Fig. 8 shows use of an M curve to indicate a three-product separation. A "dual M curve," representing both ash and size consist of a coal may be constructed

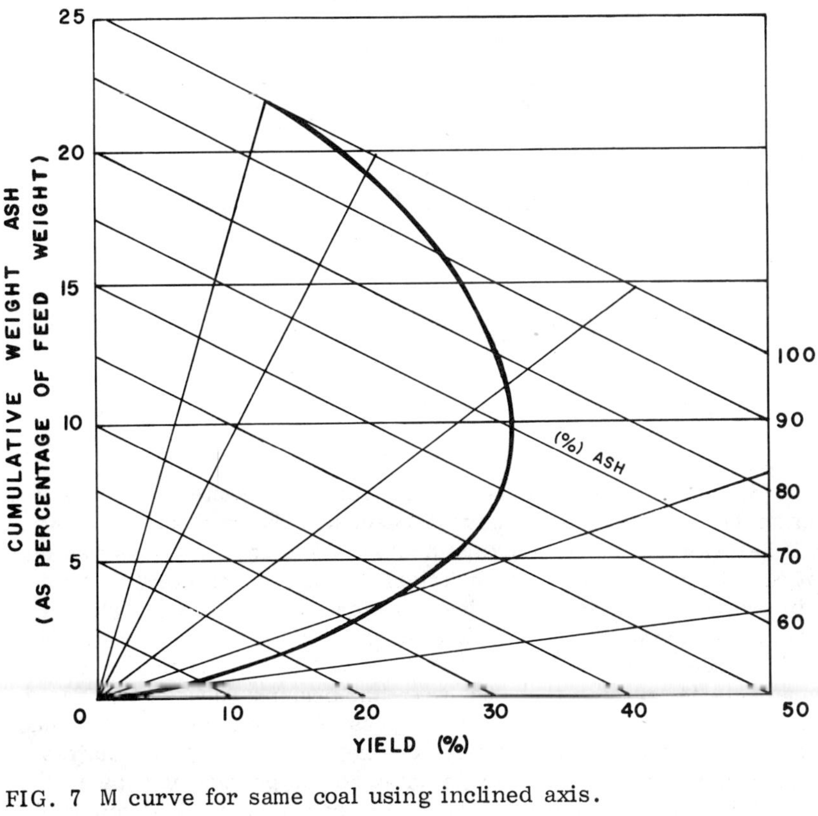

FIG. 7 M curve for same coal using inclined axis.

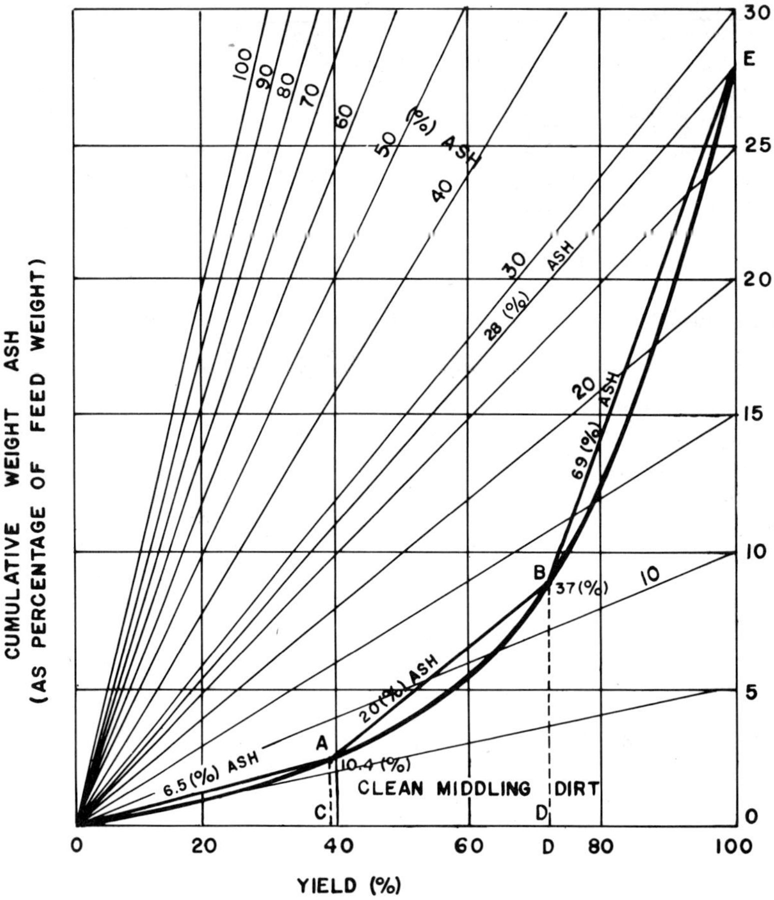

FIG. 8 A three-product separation.

which is a complete definition of the physical state of the coal with regard to ash and size and from which the quantity of any defined fraction, or fractions, may be determined (Fig. 9).

Thus, in Fig. 9, which describes the washability characteristics of a fine coal fraction of 1/4 in. top size, a curve A describes the total sample, curves B, C, and D describe the ash/yield relationship of the coal fractions having top sizes of 32, 60, and 250 mesh, respectively. The curves are displaced to the left in proportion to the quantities of material of these sizes present. Thus, for curve B, 78% of the sample is greater than 250 mesh (65 μm) for curve C, 56% is greater than 60 mesh (250 μm), and so on. It follows that the amount of material greater than 250 mesh but passing 60 mesh is 22%. This type of graphical representation is not possible using conventional washability curves. Separation characteristics, developed from partition curves derived directly from the M curve can be obtained for perfect or imperfect separations which show both the point (gravity) and sharpness of separation

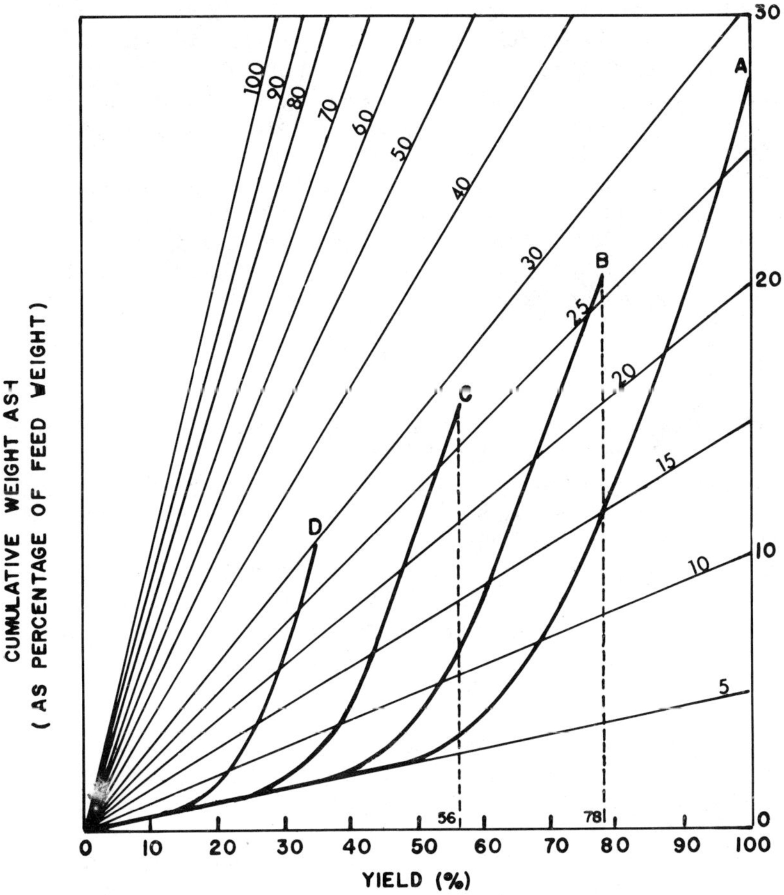

FIG. 9 Dual M curve.

with size, particularly valuable for those processes where partition gravity is dependent upon size. The M curve can be readily applied to the washing of blends of raw coals and for comparing the performance of machines having differing rated throughputs.

In spite of these numerous advantages, the use of the M curve has not found widespread acceptance, and a combination of conventional washability curves and distribution (or partition) curves with their associated estimation of probable error (Ep) continues to be mostly widely employed. These are discussed later in connection with separation efficiency, yield, and recovery.

C. Influence of Particle Size and Shape

The principal driving force separating particles of differing density results from gravitational or centrifugal forces acting on the relative buoyancies of the particles in the fluid, but dynamic effects also introduce factors in which particle sizes and shapes become important. The coal particles and shale particles in a given raw coal exhibit substantial differences in size and shape, which may act to hinder efficient separation, especially when systems are highly loaded. On the other hand, hindered settling is sometimes deliberately utilized to promote separation. Size and shape are difficult to measure or quantify, prismatic ratios alone being insufficient to predict particle behavior, and solution to these problems is often based on empirical observations and practical experience of the equipment designer. At ultrafine sizes, Stokes' law effects increase in importance to a point where they become the determining factors, so that, for example, hydrocyclone performance is subject to sharp cutoff at certain minimum particle sizes.

D. Density Separation Techniques

The specific gravities of particles are true if they are nonporous and apparent if they are porous and are not permeated by the separating fluid. Coal macerals are highly porous materials but the pores have micro dimensions, so that while readily penetrated by say, helium gas, they are, for practical purposes, impenetrable to water. The observed specific gravities of the particles are consistent enough and reproducible enough for all practical purposes.

The specific gravities of separating media may be true, mean, or apparent. If the media is a true, stable liquid or solution, its observed specific gravity is true. Except for laboratory separations, the use of true solutions for coal cleaning has not been successful, usually because of the corrosivity of the concentrated solutions necessary (e.g., Lessing Process, which employs concentrated solutions of calcium chloride). More recently, interest has revived employing heavy organic liquids (e.g., Otisca Process [112]), or solutions of reagents that combine increased density with the capability of effecting chemical desulfurization (Gravichem Process [45]).

If the separating media is a suspension, whether of solids in liquids or immiscible liquids in liquids,* the observed specific gravity will be a mean value subject to stratification. When Tromp introduced magnetite suspensions as heavy media, he deliberately employed this tendency to stratification by arranging to extract middlings from the mid-region of the bath. Such systems require external agitation and recirculation for effective control of separation.

If the separating medium does not employ either solution or suspension of denser material, but relies on a combination of dynamic buoyance and hindered settling effects, its specific gravity is "apparent." Thus, in jigs, the apparent partition specific gravity may often be adjusted to about 1.55, whereas at no time does the specific gravity of the water change from 1.0, the apparent gravity being entirely due to the dynamics of the system.

Finally, as an alternative to separations based on relying on difference in specific gravity under natural gravitational forces, equipment has been developed in which natural gravity is replaced by other forces. This is the basis of cyclonic separations, which rely on centrifugal forces, often 20 to 30 times greater than gravitational forces, first developed in Holland in 1947 [46].

---

*Media of specific gravities as high as 7.0 have been used in laboratory-scale mineral flotation employing suspensions of mercury in tetrabromoethane.

Initially developed as devices for separating solids from fluids, they were consequently modified so that differences in the densities of the solids became an operative factor. Highest efficiencies are achieved where the specific gravity of the fluid is close to that of the desired partition gravity, so that heavy-media cyclones are more efficient than water-only cyclones for coal cleaning.

E. Interfacial Separation Techniques

Separation of coal macerals and minerals are possible by exploiting differences in the surface properties of these components. Froth flotation and oil agglomeration techniques are examples of how separations are achieved. Although differences exist in the surface properties of the macerals themselves, these are generally small in relation to those of the minerals present. The different properties arise because coal is an organic matrix exhibiting nonpolar hydrophobic surface behavior. Minerals, especially clays, have polar, hydrophilic surfaces. Some exceptions to this general rule occur, particularly in the case of pyritic sulfur, which is usually present as pyrite, which exhibits markedly hydrophobic behavior and causes difficulties when present in discrete particles that are selectively floated with coal particles. In such cases, selective flotation procedures may be desirable and such processes are under development [47]. Coal rank has an effect on the extent of hydrophobicity, tending to follow coking characteristics, and being maximum for medium- and low-volatile bituminous grades. The differences are usually referred to as wettability, which can be quantified by measurement of the contact angle of the solid with water. Methods for calculating a floatability index of solid hydrocarbon materials have been suggested which utilize the elemental composition C, H, O, N, S, in which the carbon content is modified to allow for combinations into higher paraffins [48]. However, in the design of process systems these methods are of limited utility, being subject to serious interference, for example in the case of oxidized coals, so that it is customary to measure the floatability of a coal directly by empirical methods in suitably designed laboratory apparatus.

Addition of small quantities of hydrocarbon oils to raw coal pulps causes preferential wetting of the macerals by oil, tending to increase surface-tension forces operating at the coal/water interface. If air is admitted to the system, bubbles attach preferentially to the oiled coal surfaces by acting to reduce the forces at the coal/water/air interface. The bubbles increase the buoyancy of the coal particles, hence causing rapid separation from the heavier, unoiled mineral particles. The clean coal is collected as a froth on the surface of the system and recovered by skimming.

Oil agglomeration operates by preferential "collection" of coal particles by an excess of light hydrocarbon oil added to the pulp. The lower gravity of the coal-in-oil suspension causes it to float and it is separated by decantation or skimming.

The collection and recovery of fine particles by flocculation, although it may be regarded as a surface phenomenon, is not related to the flotation and agglomeration processes. The operative mechanism in flocculation is the neutralization of electric charges carried on the surface of particles, which, being of similar polarity, repel each other and tend to keep the particles in suspension. The corresponding physical property is the zeta potential, which is a measure of the electrokinetic behavior of the particles in applied electric fields. Recent developments in flocculation render it possible to achieve selective flocculation of one kind of particle from suspensions containing mixtures of particles, wherein resides its potential importance for cleaning ultrafine coal particles.

F. Separation Efficiency, Yield, and Recovery

A coal preparation engineer's chief objectives are to obtain the maximum yield of clean coal at or within target quality specification. The target specification usually includes size consist, moisture content, ash, and sulfur. The first two functions are ancillary to the main cleaning processes and outside the scope of this discussion, whose criteria are, therefore, recovery efficiencies and quality in terms of ash* and sulfur contents. Measures of performance are necessary for indicating washing efficiency obtained with given raw coal and for predicting and drawing comparisons between different cleaning operations and equipment. Since the yield and ash and sulfur contents are directly dependent on the washability characteristics of a given raw coal, and since raw coals differ widely in characteristics and are often inherently variable, it is clear that measurement of product yields and ash contents are not themselves adequate measures of performance because observed variations give no indication of the cause of the variation, which might be due to raw coal quality or to process malfunction. Nevertheless, satisfactory measurement of efficiency must somehow depend upon reporting actual results as a function of theoretical results, and of basing it on output/input ratios.

Many attempts have been made to overcome these difficulties during the past 50 years by development of expressions which, relating to commercial washing operations, take into account the sharpness of separation of the process and the washability characteristics of the coal feed at the desired specific gravity cut point. Detailed reviews have been provided by Mitchell and Charmbury (1963) [39] and Geer and Yancey (1968) [50], the latter attempting to draw a classification of methods depending upon whether the performance criteria are directly dependent, indirectly dependent, or essentially independent of the raw coal washability. Table 9 summarizes some of the methods proposed.

Among the dependent criteria, Sarkar and Das [52] have recently suggested a new expression, essentially derived by relating the degree of washing at the actual specific gravity observed with that expected from the washability characteristics at the same specific gravity. The degree of washing is a function of percentage recovery and reduction in ash content, and its relationship to the organic efficiency and ash error expressions is obvious. Its utility derives from the extent to which earlier expressions are combined into a unified value.

Misplaced material, the basis of Anderson's expression, was formerly measured at the single or target specific gravity at which the cleaning unit

---

*Thermal, or Btu efficiency, is dependent on ash content and hence is not a direct control function.

TABLE 9 Methods for Expressing Efficiencies of Cleaning Processes

Category	Originator(s)	Coefficient	Expression
Dependent	Fraser and Yancey [49] (1923)	Recovery efficiency (organic efficiency)	$\text{Efficiency} = \dfrac{\text{yield of washed coal}}{\text{yield of float coal of same ash}} \times 100$
	Anderson [50] (1950)	Separation efficiency	Efficiency = 100 − (weight percent of clean coal in sinks and weight percent of sinks in clean coal)
	Grounds and Needham [51] (1950)	Ash error or yield error	(a) Ash error = ash in clean coal − ash in float coal at equivalent yield (b) Yield error = yield of clean coal − yield of float coal at equivalent ash
	Sarkar and Das [52] (1978)	Washing efficiency	$\text{Efficiency} = \dfrac{\text{recovery \% of clean coal} \times \left(\text{ash \% of raw coal} - \text{ash \% of clean coal}\right)}{\text{recovery \% of float coal} \times \left(\text{ash \% of raw coal} - \text{ash \% of float coal}\right)} \times 100$
Indirectly dependent	Terra [53] (1938)	Ecarte probable (Ep) (probable error)	$\text{Ep} = \dfrac{\text{sp. gr. at 75\% yield} - \text{sp. gr. for 25\% yield}}{2}$ (yields obtained from distribution curve)
	Driessen [54] (1945)	Error area (dimensionless)	Obtained by planimetry from standardized distribution curve
Independent	Belugou and Ulmo [55] (1950)	Imperfection (I)	$I = \dfrac{\text{Ep}}{\text{specific gravity of separation} - 1}$

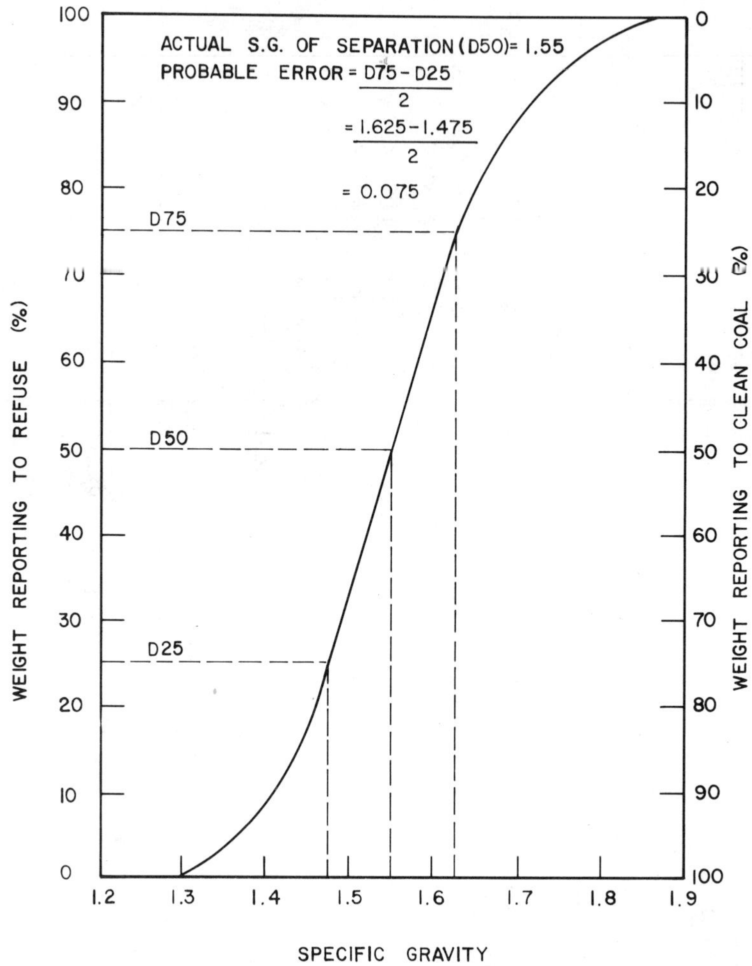

FIG. 10 Distribution curve and probable error.

was set to operate, the clean coal and the refuse being sampled and tested in a solution of the corresponding specific gravity. However, it had been shown many years previously that the distribution of misplaced material is governed by the laws of probability, so that if the proportions of misplaced material are plotted against specific gravity expressed as a difference from the specific gravity of separation, an error distribution curve following a normal distribution is obtained. Tromp, in 1938, used this as the basis for plotting a distribution curve in which percentage recoveries obtained by performing a float and sink analysis on the clean coal and refuse products at given density increments are plotted against the mean specific gravities of those fractions. A typical curve shown is Fig. 10. Now, although there are no recognized standard methods for determining and reporting *washing efficiency*, the International Standards Organization has issued a standard method which requires that a complete float and sink analysis be carried out on clean coal and refuse in order to report *misplaced material*.

If the linear percentage recovery scale of Fig. 10 is replaced by a probability scale, arithmetic or logarithmetic, the characteristic S shape becomes more linear, which is significant because the slope of the line becomes a direct measure of the sharpness of separation. This feature lead Terra [53] to propose the so-called écarté or "probable error," which defines the separating gravity (D_{50}) as the specific gravity corresponding to a partition value of 50% recovery and the probable error (Ep) as half the difference between the value of specific gravity corresponding to partition values of 75% (D_{75}) and 25% (D_{25}). This derivation is shown in Fig. 10 and Table 9. The lower the value of Ep, the sharper is the separation obtained.

Initially, it appeared that probable error had been removed from dependence upon the separating gravity, and hence upon washability characteristics, and Driessen [54] proposed a method of rendering its value dimensionless by introducing an "error area," obtained by uniformly plotting the distribution curves to a scale in which a given length unit represented 2% on the recovery percentage scale and 0.1 unit of specific gravity scale.

Experience, however, showed that there was a relationship between its value and the density of separation, the value tending to decrease as separation density increased. Belugou and Ulmo [55], recognizing this dependence, proposed the use of a concept "imperfection," defined as "the quotient of the probable error divided by the density of separation minus 1.0" (see Table 9) and suggested that this quotient was a constant for any particular cleaning device. Thus, in general, it is able to compensate for the effect of size composition or the particular process considered—jigging, heavy media, classifying—but is affected by large variations in throughput or ash content of the raw coal feed, which cause major alteration in the rate at which refuse has to be handled.

Geer and Yancey [40] have pointed out that sharpness of separation, whether measured by probable error, error area, or imperfection, must not be confused with efficiency of separation. These must be considered together with the actual gravity of separation in conjunction with the washability curve, since only in this way can the amounts of misplaced material, on which efficiency depends, be assessed. Fractional ash contents tend to increase rapidly with mean density of fraction, over the range of specific gravities encountered in practice (Fig. 11). In these circumstances, sharpness of separation has a large bearing on efficiency in conjunction with the near-gravity characteristic of the coal and will clearly be more severe at lower gravities of separation. This condition frequently arises in cleaning coals to meet low-sulfur specifications.

In general, the combination of washability and partition (distribution) curves has found the widest utility for determining process design factors and measurement of plant efficiencies.

G. Control of Moisture Content

The water content of a coal reduces its heating value, can cause severe handling difficulties, increases handling and transportation costs, and reduces yields of carbonization and conversion processes. Its reduction and control are therefore desirable.

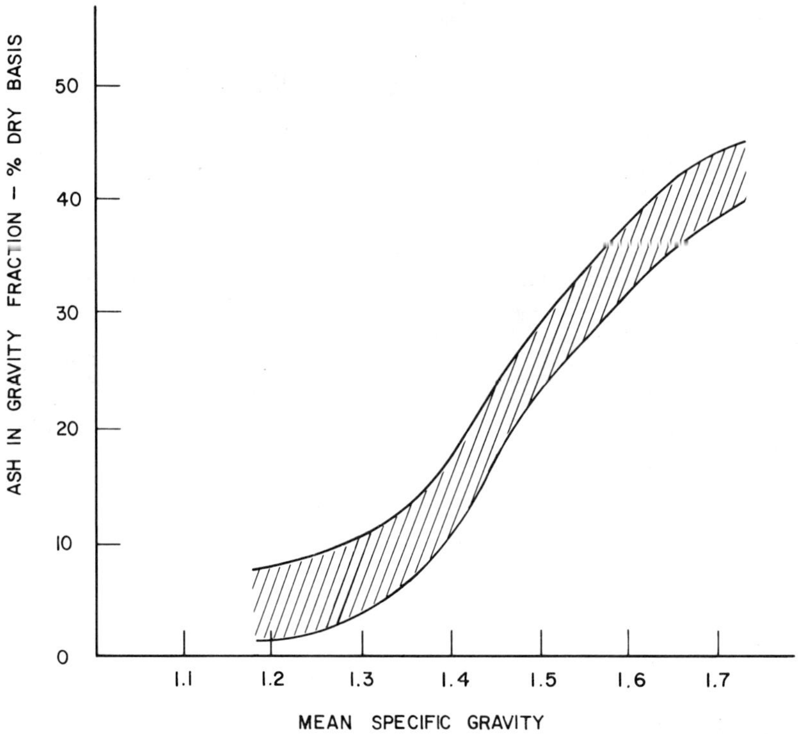

FIG. 11 Approximate relationship between fractional ash contents and specific gravity.

Water occurs in coals in three ways: (1) as inherent moisture contained in the internal pores of the coal substance, including water of hydration and zeolitic water associated with the mineral impurities; (2) as surface moisture wetting the external surfaces of the coal particles' in which adsorption may play a small part; and (3) as free water held by capillary forces in the interstices between the coal particles. The size distribution of the particles plays a very important part in determining the amount of surface and free moisture contents, because both wettable surface and capillarity increase markedly as particle sizes decrease.

Inherent moisture is related to coal rank, as described above, being greatest for low-rank lignites and brown coals and least for high-rank bituminous coals and anthracites. In North America, lignites contain 25 to 40% moisture content when freshly mined. In some other countries, this value can run as high as 75% (Yallourn, Australia). On standing in air, considerable reduction in moisture content can occur but is accompanied by shrinkage and marked structural alteration of the coal. High-rank coals will generally dry, on exposure in air, to moisture contents below 2% with no observable structural changes.

Surface moisture is related to the amount of available surface and the wettability of the coal. Its practical importance is greatest at coal sizes below

1/2 in., since size fractions greater than this are free-draining and the available surface is generally insufficient to hold moisture contents greater than about 5 to 7%. The moisture contents are lower than might be expected from the available surface area because of the low wettability of coal compared with the surfaces of minerals.

For size fractions smaller than 1/2 in., surface effects combine with capillary effects so that the coal forms a matrix capable of holding substantial quantities of free water. The quantities held are highly dependent on particle size. Thus, coal slacks crushed below 1/4 in. will readily contain 12 to 14% moisture even after drainage whereas material below 28 mesh will contain up to 45%. In general, mixtures of coal and water will exhibit solid properties up to 45% moisture content, and liquid suspension properties at higher water contents. Control of moisture can therefore involve the following steps:

1. Concentration and recovery of solids in suspension
2. Dewatering of the concentrated solids
3. Thermal drying of the dewatered solids

The range of methods employed are summarized in Fig. 12.

1. Concentration and Recovery of Solids

For coarse coal, concentration and recovery present few problems because it can be easily achieved by passing the coal suspension over fixed or vibrating screens of suitable aperture, the overflow from the screen decks being fed directly to dewatering devices.

For ultrafine coal the problems of concentration increase so that settling in cones, static thickeners, or accelerated separation in hydrocyclones may be necessary. In this context, it may be noted that froth flotation provides a means of both cleaning and concentrating coal fines. For the finest sizes, flocculation may be required to assist concentration. In Europe, it is finding increasing application for preconditioning of froth flotation pulps and feed pulps to dewatering equipment such as vacuum filters and solid bowl centrifuges.

2. Dewatering of Concentrated Solids

Mechanical methods can be broadly subdivided into methods that effect dewatering by:

1. Repeated mechanical disruption of the water-retaining coal particle matrix (e.g., by vibrating screens)
2. Increasing the normal gravitational force assisting drainage (e.g., by centrifugation)
3. Applying a pressure differential across the coal particle matrix (e.g., as in vacuum filtration)

Coal wettability and water surface tension are interrelated factors that can be altered, for example, by addition of surfactants or merely by heating the system. Water viscosity is also temperature-dependent, particularly up to 160°F (70°C), so that the drainage of free water can be measurably in-

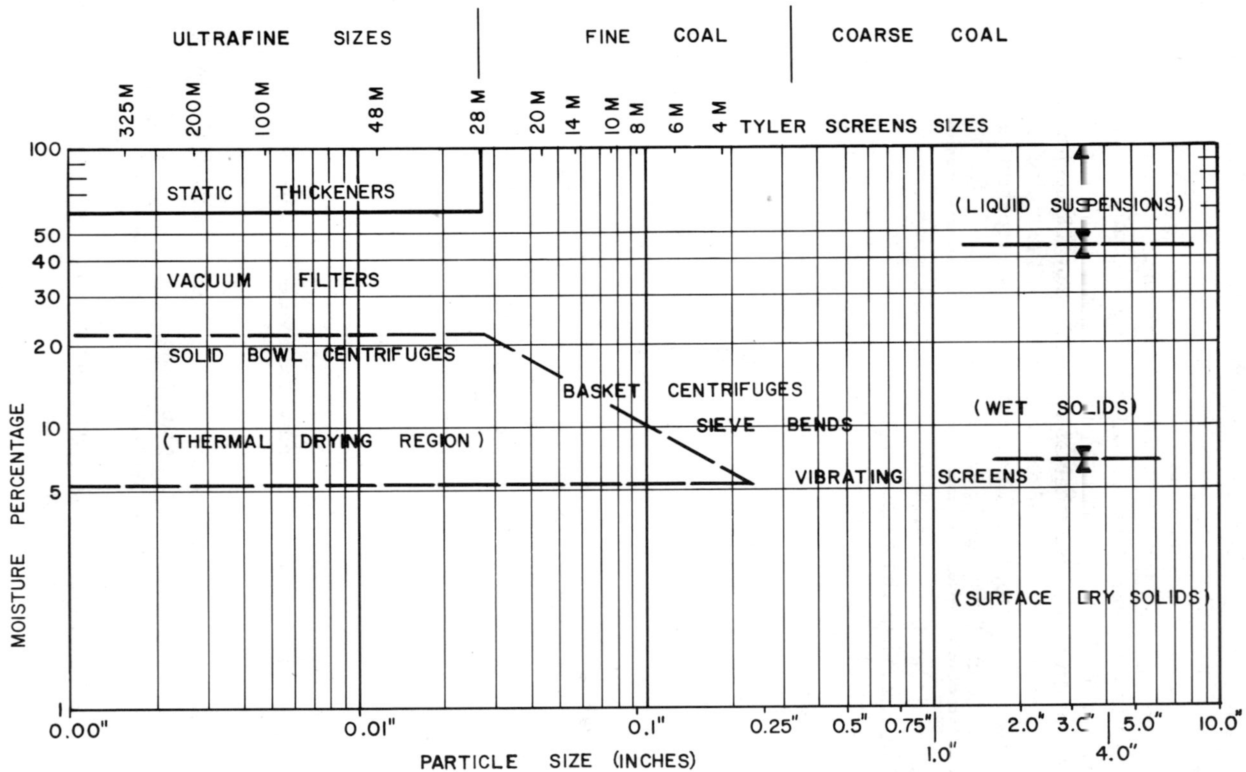

FIG. 12 Typical applications of equipment for moisture control.

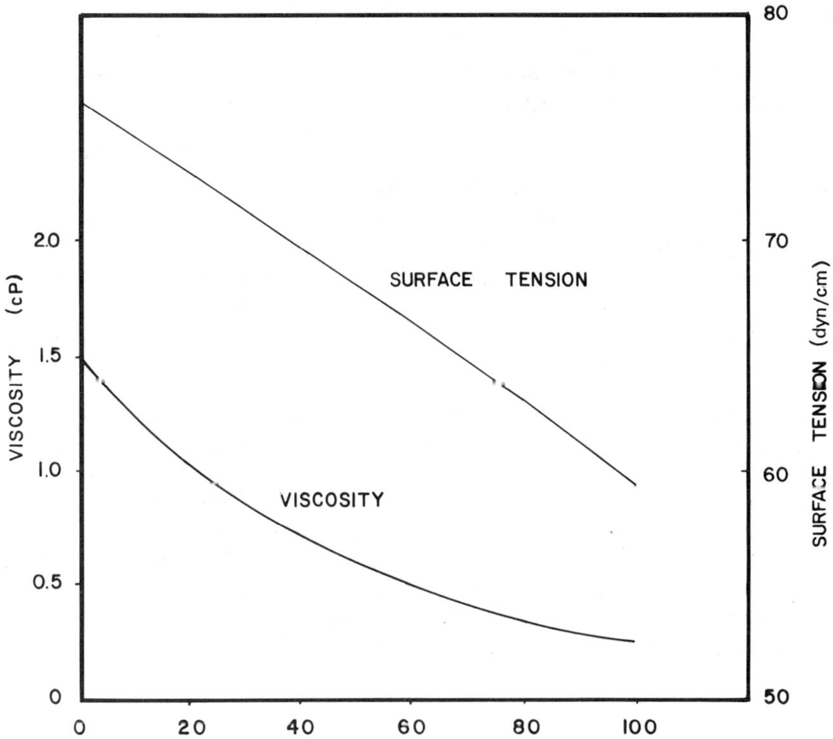

FIG. 13 Properties of water at atmospheric pressure.

creased by raising the temperature of the system. Figure 13 shows the relationship of viscosity and surface tension with temperature.

The addition of surfactants, or raising the system temperature, are special effects that serve to enhance these general methods. Heating of filter cakes with steam during vacuum filtration has found some application but is not widely used. Washing of filter cakes with hot surfactant solution [56] has also been shown to be effective but has not found commercial application.

3. *Thermal Drying of Dewatered Solids*

Thermal drying, particularly of metallurgical coals, has found extensive application in North America, but is little used in most other coal-producing countries except Poland and the USSR. Lignite drying is practiced in West Germany for certain uses (e.g., briquetting), but a different technique is used. There is growing worldwide interest in the development of efficient methods for drying low-rank coals.

Thermal drying entails contacting wet coal particles with hot gases, usually direct products of combustion from oil, gas, or coal burners, under conditions that promote evaporation of the surface moisture without causing thermal degradation or incipient combustion of the coal. Gas/solid contact is promoted by one of the following general methods:

1. Passing hot gases through disturbed beds or falling streams of coal particles
2. Direct fluidization of coal particles with hot gases
3. Suspension of coal particles in rapid, hot gas streams

The design of thermally efficient equipment requires careful application of psychrometric principles and carefully regulated plant operation. Hot gas temperature at entry to the drying zone is the greatest single factor determining the capacity of a coal dryer, so that this should be as high as safety permits. However, the production of dry coal dust atmospheres within drying units creates potentially severe explosion hazards and demands efficient removal of particulate matter from moist gas exit streams. Work in a number of countries has demonstrated that coal dust explosivity is related to the volatile-matter content of the coal. The U.S. Bureau of Mines employs a volatile index which is a ratio of volatile matter to the total ash-free coal substance (i.e., volatile matter plus fixed carbon) in assessing the explosibility of coal dusts. Thus, in general, high-volatile bituminous coals are more hazardous than are medium- or low-volatile coals. As rank decreases, coal reactivity increases and subbituminous coals and lignites ignite in air at appreciably lower temperatures than do bituminous coals. For these reasons the design of satisfactory dryers for low-rank coals has been very difficult to achieve [57], and those units currently in commercial operation for lignites employ indirect steam heating to produce a product containing 10 to 15% moisture [58].

H. Desulfurization

The total sulfur content of a raw coal is distributed throughout the macerals and minerals present. It may occur as elementary sulfur, as sulfates, as sulfides, or in organic combination in the "pure" coal. For these reasons it has been customary to classify the forms of sulfur as inorganic, pyritic, and organic: inorganic sulfur comprising the sulfates, pyritic the sulfides, and organic the remainder, including any elemental sulfur that may be present. Standard methods of analysis have been devised for direct determination of total sulfur, sulfate sulfur, and pyritic sulfur, the organic sulfur being reported as the difference between the total sulfur and the sum of sulfate plus pyritic sulfur contents [59]. In an overwhelming preponderance of cases, it has been observed that inorganic sulfur content is less than 0.1 wt %, and for practical purposes the sulfur content of a coal consists of its pyritic and organic forms. Meyers [6] reviewed selected data worldwide on sulfur forms, which are summarized in Table 10. The table illustrates that there is no direct relationship between the occurrence of organic and pyritic sulfur and that each coal seam requires individual examination to establish this characteristic. This aspect is considered in detail in Chap. 7, where the overall cleaning potential of U.S. coals is considered. The occurrence and forms of sulfur makes possible a variety of approaches to the question of coal desulfurization. At present, coal desulfurization can be achieved on a commercial scale only by means of physical and physicochemical methods, the elements of which have been described earlier and which consist principally of density separation techniques or other techniques that exploit the surface

TABLE 10 Sulfur Forms in Selected Bituminous Coal—Worldwide

Region and country	Location or mine	Sulfur (% w/w)[a]			Ratio pyritic to organic sulfur
		Total	Pyritic	Organic	
Asia					
USSR	Shakhtersky [5]	0.38	0.09	0.29	0.031
China	Taitung [5]	1.19	0.87	0.32	2.7
India	Tipong [5]	3.63	1.59	2.04	0.78
Japan	Miike [5]	2.61	0.81	1.80	0.45
Malaysia	Sarawak [5]	5.32	3.97	1.35	2.9
North America					
United States	Eagle No. 2 [6]	4.29	2.68	1.61	1.7
Canada	Fernie [5]	0.60	0.03	0.57	0.053
Europe					
Germany	— [7]	1.78	0.92	0.76	1.2
United Kingdom	Derbyshire [8]	2.61	1.55	0.87	1.8
Poland	— [5]	0.81	0.30	0.51	0.59
Africa					
South Africa	Transvaal [9]	1.39	0.59	0.70	0.84
Australia	Lower New-castle [10]	0.94	0.15	0.79	0.19
South America					
Brazil	Santa Caterina [11]	1.32	0.80	0.53	1.5

[a]Moisture-free basis, pyrite + sulfate reported as pyrite.

properties of coals and minerals. The basis for these methods has recently been reviewed in detail by von Gathen [18]. They exploit the difference in properties that exist between the various forms of pyritic sulfur—chiefly pyrite and marcasite, but occasionally including galena—and those of the coal macerals. The various forms of pyrite occur as:

1. Epigenetic; mostly coarse crystallites >50 μm
2. Syngenetic; fine crystallites 5 to 50 μm
3. Finely intergrown and ultrafine pyrite; crystallites <5 μm

Deliberate removal of pyrite by mechanical cleaning methods requires equipment and processes that exhibit high efficiency and high precision of density separation. Substantial size reduction of the raw coal may be required, with resulting significant increase in cleaning costs. Table 11 summarizes the more important characteristics of equipment and processes that may be considered. It is generally limited to the removal of pyrite greater

TABLE 11 Cleaning Equipment Effective for Separating Free Pyrites

Process	Equipment	Feed	Effective size range	Effective specific gravity range	Machine capacity (tons/hr)
Dry	Air table	Fine coal	3/4 in.-48 mesh	1.70-2.00	150
Wet	Feldspar jig	Fine coal	1/2 in.-28 mesh	1.45-2.00	100
Hydraulic	Batac jig	Fine coal	3/4 in.-100 mesh	1.45-2.00	700
	Compound water cyclone	Fine coal	2 in.-150 mesh	1.30-1.90	65
Heavy media	Cyclone	Fine coal	1/4 in.-200 mesh	1.30-2.00	100
Interfacial	Froth flotation	Ultrafines	28 mesh-0	-	120
	Oil agglomeration	Ultrafines	100 mesh-0	-	35

than 10 μm. The success of these methods for any coal is therefore dependent upon the size distribution of the pyritic particles because they are considerably harder than coal macerals and less affected by comminution. A petrographic examination of the raw coal can provide useful data on the possibilities of achieving substantial desulfurization by these methods. It should be noted that, although the wettability of coal macerals and pyrites are comparable, the density differences are great, so that froth flotation, even at the roughing stage, can effect substantial desulfurization when the size distribution of the pyrite is appropriate. The double-flotation technique under development by the U.S. Bureau of Mines [47] is designed to improve this process still further by suppressing coal and enhancing pyritic flotation in a second-stage treatment of the rougher froth.

Desulfurization by chemical methods is not yet practiced commercially, but a number of methods are under active research and development in the United States. They can be divided into three general groups according to whether they selectively remove pyritic sulfur, organic sulfur, or both forms. In common with other aspects of the chemical constitution of coal, the combinations in which organic sulfur are bound are highly complex and not fully understood. In general it is known that sulfur replaces oxygen in the coal molecules, so that oxygen decreases as sulfur content increases. The nature of the sulfur linkages has been studied by a number of workers and has identified thiophenols, sulfides, disulfides, and heterocyclic compounds in high-sulfur coals. Of these, heterocyclic sulfur is known to be the least reactive, and hence most difficult to remove, but has been found to account for 70% of the organic sulfur present. Effective desulfurization therefore requires vigorous reaction by the reagents employed. Meyers [6] identified the following criteria as being generally applicable to the reagents used: (1) the reagent must be highly selective to either pyritic or organic sulfur (or both) and not significantly reactive with other coal components; (2) the reagent must be regenerable so that once-through reagent cost is not a major factor; (3) the reagent should be either soluble or volatile in both its unreacted and reacted form so that it can be near totally recovered from the coal matrix; and (4) the reagent should be inexpensive.

IV. PROCESS SYSTEMS FOR COAL PREPARATION

A wide variety of highly efficient equipment is available for the cleaning of coal, even those containing large amounts of intergrown material. They range from portable, quickly erected "package plants" having throughputs of 100 to 400 tons/hr to large, permanent multisteam facilities producing 20,000 tons/day and more of clean coal. Detailed descriptions of the equipment are readily available, as are portfolios of process flow sheets [30]. Manufacturers provide comprehensive information on the design, capacities, performance, and costs of their equipment, which are usually available in a range of sizes to facilitate plant designs tailored to meet the requirements of any coal feedstock. Some major problems remain, relating to more efficient and cheaper methods for the recovery, cleaning, and dewatering of ultrafine coals. These are proving somewhat intractable but are receiving close attention by equipment manufacturers and some research and development organizations.

TABLE 12 Levels of Coal Preparation [32]

Level and brief designation	Scope	Yield weight (%)	Recovery Btu (%)	Reduction potential: Ash	Reduction potential: Sulfur	Work done on raw coal[a]	Typical circuits and equipment used	Refuse	Comments
A, absence	No prep.	100	100	None	None	None—ship ROM coal	None	None	Not general practice
B, breaking	Top size control only	98-100	100	None to minor	None	Crushing to 3 in. or less and removal of coarse refuse	Scalping screen; crusher; rotary breaker	Dry lumps and trash	General practice on all ROM coals
C, coarse	Coarse beneficiation	75-85	90-95	Fair to good	None to minor	Level B followed by dry screen at 3/8 in. and wet beneficiate at +3/8 in. only. Ship 3/8 in. × 0 as is	Same as level B plus vibr. screens; jigs; heavy-media vessels or cyclones; dewatering thickeners; filters	+3/8 in. drained -28 mesh ponded	Used where -3/8-in. fraction fairly clean or much rock present in +3/8-in. fraction
D deliberate	Fine beneficiation	60-80	80-90	Good	Fair	Level B followed by wet screen at 1/4 in. and wet beneficiate +28 mesh discard 28 mesh × 0	Same as level C plus concentrating tables or hydrocyclones. Some thermal drying	+1/4 in. drained +28 mesh dewatered -28 mesh ponded or filtered	Used with coals having good washability characteristics
E elaborate	Very fine beneficiation	60-80	80-90	Good to excellent	Fair to good	Level D plus wet beneficiate 28 mesh × 0	Same as level D plus flotation circuits; thermal drying prevalent	Same as level D except more fines	Used with coals having excellent washability characteristics
F, full	Deep cleaning	60-80	85-90	Clean coal stream: excellent; middling stream: none to fair		Level E after greater than normal size reduction and separation into two streams: clean coal and middlings	Same as level E plus additional size reduction	Substantially same as level E	Two or more washed coal products of different qualities are obtained

[a]Coal sizes shown are typical but will vary somewhat with each coal and process selected.

A. Levels of Coal Preparation

The extent of preparation a coal requires is determined by its condition in the ROM state, by its washability characteristics, and by the market specification and location. All coal preparation incurs costs that can be justified only if the resulting handling, transportation, and utilization receive benefits and costs savings that outweigh the preparation costs. Thus, in the United Kingdom at the present time the bulk of coal-fired power plant can operate without trouble burning thermal coals averaging 16% ash. It is therefore the policy not to clean raw smalls that have lower ash contents. This policy, determined by economics, is suitable for the prevailing British conditions but may not apply, for example, to Appalachian thermal coals used for power generation in Florida.

The stringent market specifications for anthracite and metallurgical coals requires that these fuels be delivered at minimum levels of impurity. Historically, they have always received elaborate preparation. The level of preparation required by thermal coals has been determined primarily by handling and transportation costs and occasionally by technical factors, for example the modification of ash fusion temperatures by removing extraneous minerals. These traditional approaches are now being affected by tougher environmental regulations, so that the future levels of preparation of thermal coals is expected to become more elaborate or, in the language of the industry, deeper cleaning will be required.

General classifications of levels of preparation have been provided by a number of authors, a typical classification by Phillips and De Rienzo [32] being shown in Table 12. A similar classification, together with corresponding flow sheets, has been provided by Zimmerman [33], who defined the levels "to help classify and establish for physical—as opposed to chemical—coal cleaning, in order to determine the extent of pyritic sulfur removal for different U.S. coals. These levels may also be used to classify the degree of ash removal and Btu recovery."

Other authors have recognized fewer levels within the same range of treatment options. In most cases, the levels defined have served as the basis for detailed economic analyses and cost/benefit comparisons of coal cleaning versus other environmental control technologies.

The range of treatments options may be represented on a simplified flow diagram as shown in Fig. 14.

B. Raw Coal Pretreatment

The importance of adequate raw coal pretreatment cannot be overemphasized, because many of the operating problems encountered at cleaning plants can be attributed to inadequate or inefficient pretreatment, which results in excessive quantities of oversize and/or undersize material in plant feeds to the various cleaning units, causing blockages or loss of cleaning efficiency and frequently resulting in plant shutdown. It is often observed that a plant can cope with intermittent periods of overloading better than it can handle inadequate pretreatment.

Pretreatment essentially consists of breaking, crushing, and screening the ROM coal in a manner that provides a uniform raw coal feed of predeter-

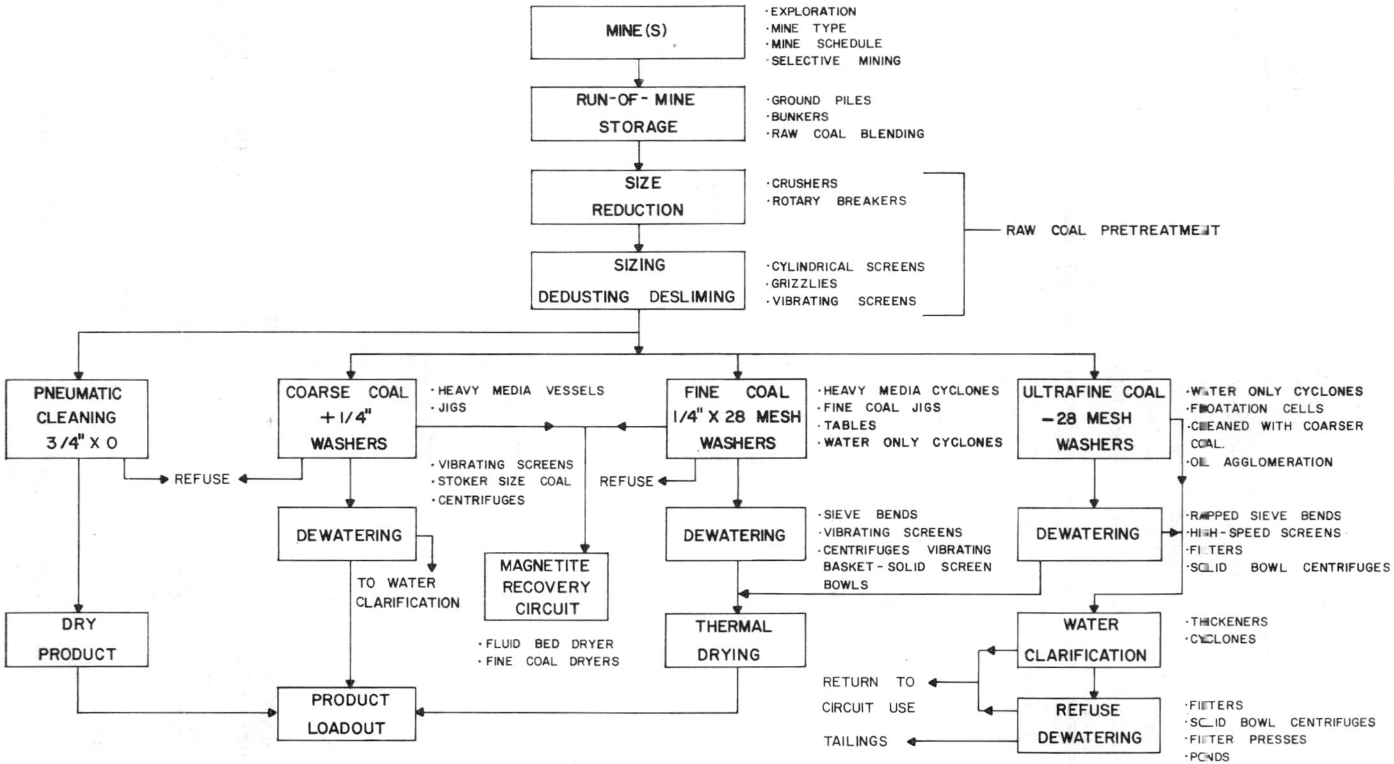

FIG. 14 General coal preparation plant flow sheet: range of options.

mined top size and minimizes the production of ultrafine sizes through excessive crushing or degradation by handling. Breaking is sometimes an effective method for removing much of the extraneous shale, which, being harder than the coal, resists breakage and is removed as oversize material. Breakers and cylindrical screens have also, on occasion, been found effective in removing sticky clays, which tend to roll into balls and become oversized material. The choice of crushers, with regard to the particular coal to be handled, can be of great importance in reducing the production of unwanted ultrafines.

Equipment used for controlling size includes rotating cyclindrical, fixed, and vibrating screens. If the raw coal moisture is not excessive, generally less than about 6%, dedusting is effective in removing a substantial proportion of ultrafines and reducing the loads on the cleaning units. Dedusting, by removing the finer, lighter particles, can function as a partial cleaning device [12], the dry dust product being readded to the clean coal product without excessive increase in overall product ash. Centrifugal dedusters have been widely used in West Germany for many years and have been developed to machines of 8 m (26.2-ft) diameter capable of handling up to 800 tons/hr of raw coal feed, removing a substantial proportion of the -28 mesh material present. Although their use can be expected to become more widespread, the first American installation already having been reported [60], these will be limited by the necessity of processing raw coals of 3 to 5% surface moisture, since predrying of the raw coal is uneconomic.

The final stage of raw coal pretreatment is to screen it into size fractions convenient for handling by the various process streams. Top sizes may be 6 to 8 in. if coarse coal washing by jigs or heavy-media vessels is employed, but 1 1/2 to 2 in. if heavy-media cyclones are used. Although some jig plants are designed to handle an 8 × 0 in. feed, the majority of plants employ coarse and fine coal streams, the coarse coal having a lower limiting size of 1/2 or 1/4 in. and the fine coal, if prewetting or desliming is employed, will have a lower limiting size of 28 mesh (1/2 mm). The 1/2 in. or 1/4 × 28 mesh fine coal is cleaned in fine coal jigs, hydrocyclones (heavy media or water only), or tables. The raw slurry underflow from the desliming screens passes to the ultrafines circuit for treatment and recovery. Ultrafine slurries resulting as screen underflows from dewatering of cleaned coarse and fine coal streams, mostly produced by breakage, are also usually added to the raw slurry before cleaning and recovery of the ultrafines. Where dry cleaning by pneumatic methods is employed, raw coal pretreatment usually consists of reducing the total feed to -3/4 in. by successive scalping and crushing operations.

C. Coarse Coal-Cleaning Techniques

1. *Dry Methods*

Modern dry cleaning methods in the United States, where this technique was developed and has always been most widely used, handle raw coal feeds of up to 2 in. top size, but usually of 3/4 in. top size. As shown in Fig. 14, it is a cleaning system distinct from wet methods and has distinct advantages and disadvantages in application, although its use in combination with wet methods (e.g., heavy-media cyclones) has recently been advocated [28].

Dry cleaning techniques employ a combination of upward air currents and mechanical agitation, the combined cleaning action sometimes being referred

TABLE 13 Performance Data of an Air Table Cleaner (28)

Size (in. and mesh)	2×1	1×$\frac{3}{4}$	$\frac{3}{4}$×$\frac{1}{2}$	$\frac{1}{2}$×$\frac{3}{8}$	$\frac{3}{8}$×$\frac{1}{4}$	$\frac{1}{4}$×8	8×12	14×28	28×48	48×100	100×200	2×0
Screen analysis (%)												
Feed	7.1	22.9	25.1	10.0	10.8	11.0	4.9	2.9	2.0	0.9	0.8	100.0
Clean coal	6.9	22.6	24.1	9.4	10.5	11.6	5.6	3.4	2.2	1.0	0.9	100.0
Refuse	8.3	24.5	30.9	13.3	12.7	7.2	0.8	0.4	0.4	0.4	0.3	100.0
Ash (%)												
Feed	17.8	14.5	14.1	14.1	14.6	14.0	12.5	12.7	13.3	14.6	16.3	14.5
Clean coal	10.5	8.6	7.9	7.8	8.3	9.6	11.6	12.1	12.5	13.3	15.4	9.2
Refuse	55.2	48.5	44.0	42.0	47.3	57.7	53.7	45.7	41.0	56.4	35.7	47.2
Pyritic sulfur (%)												
Feed	6.24	5.44	5.06	4.75	4.63	3.95	2.98	3.20	3.09	3.25	4.12	4.80
Clean coal	2.80	2.67	2.45	2.39	2.52	2.82	2.67	3.10	3.01	3.12	4.05	2.67
Refuse	23.83	21.33	17.74	15.18	15.46	15.19	16.02	8.36	5.71	5.44	5.93	18.04
Total sulfur (%)												
Feed	7.57	6.74	6.47	6.16	5.97	5.09	4.49	4.25	4.15	4.38	5.05	6.11
Clean coal	3.96	3.82	3.61	3.57	3.60	3.93	4.20	4.15	4.07	4.22	4.96	3.82
Refuse	26.01	23.51	20.33	17.60	18.11	16.64	16.97	9.45	7.00	6.91	7.21	20.35
Actual recovery (%)	83.6	85.2	82.9	81.6	83.7	90.9	97.7	98.1	97.2	94.1	95.9	—
Theoretical recovery (%)	85.2	87.8	87.1	87.2	88.3	92.8	98.1	98.7	98.5	97.4	98.3	—
Efficiency (%)	98.1	97.0	95.2	93.6	94.8	98.0	99.6	99.4	98.7	96.6	97.6	—

to as pneumatic oscillation. A variety of machines employing laundering, tabling, or jigging actions were formerly employed, but because of increasing raw ash and moisture contents of ROM coal, laundering and tabling machines encountered increasing difficulties with dirt extraction capacities, so that, by 1965, they had fallen into disuse in the United States. Present dry cleaning capacity is now solely provided by air-pulsated jigs, referred to as air table cleaners. Currently production is about 6.5 million tons or about 2.5% of totally mechanically cleaned coal.

The underlying scientific principles of dry concentration have never been expressed with precision and the technique continues to be an empirical art in which the fractional density composition of a raw coal, provided by conventional washability data, together with some rule-of-thumb methods governing particle shapes, are used for assessment of cleaning potential. Ultimate suitability must be determined by full-scale trials in an available plant or by pilot-plant runs. Experience suggests that it is most effective for coals having optimum yield/ash values at partition specific gravities in the range 1.7 to 1.9. This is well above the range commonly encountered in wet separations, so that dry separation is a complementary rather than a competitive technique. Because the air has a very low density compared with the particles, a medium of sufficient density has to be formed by the raw coal itself, the required mobility for this classification being provided by the mechanical action. It is necessary to prevent short-circuiting of the air flow by suitable distribution of pulsating air from the plenum chamber.

The cleaning action for ash and free pyrites becomes less efficient at lower sizes, and is ineffective below 48 mesh. A substantial proportion, up to 70%, of the -48 mesh fraction is removed by the air flow and must be recovered by cyclones, any dust-laden air from the cyclone discharge receiving final cleaning in air filter baghouses before discharge to the atmosphere. It is reported [61] that this dust does not differ significantly in analysis from fine coal recovered directly, so that its final disposal as clean product or refuse is determined by its ash and sulfur contents. The performance data given in Table 13 are typical for a modern air table handling a high-sulfur coal.

The major advantages of air cleaning are that no water is required, dry clean coal that does not freeze is produced, the need for water treatment is eliminated, and stream pollution is avoided. Capital and operating costs are low compared with wet cleaning methods and maintenance requirements greatly reduced. Its major disadvantage arises from difficult dust control problems, especially within plant precincts, and the method is not generally applicable to all coals.

2. *Wet Methods*

Wet methods for cleaning coarse coal fall into two broad classes: (1) hydraulic (water-only) methods, and (2) heavy-media methods, the latter now almost universally employing suspensions of magnetite in water as the heavy media.

a. Hydraulic Hydraulic equipment employed in coarse coal preparation have included jigs, launders, and various types of classifiers. In 1974, the last year for which full data are available, 41.2% of total bituminous and lignite production in the United States was mechanically cleaned, of which 97.5%

was cleaned by wet methods (see Table 2). Jig cleaned coal accounted for 50.2% of total wet cleaned, a proportion that has shown little change for over 40 years. Launders and classifiers accounted for 2.4% of mechanically cleaned production and have steadily declined in importance. They are unlikely to provide a significant contribution to cleaned coal production in the future, the continuing decrease in raw coal size having resulted in their replacement by wet concentrating tables and froth flotation. The review of worldwide coal preparation practices also shows that jigging is the predominant coal-cleaning method in all the countries considered here except South Africa and Canada.

Jigs. Lovell [62] defines jigging as "a process of particle stratification in which the particle rearrangement results from an alternate expansion and compaction of a bed of particles by a pulsating fluid flow." The bed of particles is carried on a perforated deck through which water passes upward and downward in repeated "pulsion" and "suction" strokes, and at the same time the bed is moved forward through the jig chamber in a lateral direction by the horizontal flow of water. Product separation is achieved by cutting the stratified bed at various heights by subsurface gates to remove shales and middlings and by lifting the clean coal product over a weir on the upstroke at the exit end of the wash box.

Although an explanation of this kind provides a general understanding of jig operation, it should be observed that the phenomenon has proved particularly intractable to rigorous scientific analysis and mathematical expression. A number of diverse opinions exist, which were reviewed by Lovell [62]. They range from a classical concept, which considers individual particle motions, to Mayer's "potential energy theory," which considers the entire particle bed. One aspect of practical importance is that particle densities predominate over particle size and shapes in determining stratification, so that commercial jigs capable of handling a very wide range of raw coal sizes are available. However, coals containing significant amounts of flat, slabby shale sometimes present difficulties, and special refuse extraction designs are necessary to handle these coals successfully.

Jigs may be fully mechanical in operation (plunger or diaphragm type) or pneumatically operated (Baum, Tacub, and Batac types). Plunger or diaphragm jigs have a very long history, going back some 150 years and are still employed for lower throughput capacities up to about 250 tons/hr. In these jigs the differential stroke action is produced by suitable cam action. A diaphragm jig of unusual design in which dual eccentrics operate at different strokes and frequencies was introduced in Europe in 1950 (Wemco-Remer) as a ragging jig for handling fine sizes.

The Baum jig was invented in Germany in 1892 but not introduced to the United States until 1928. Baum washboxes consist of U-shaped cells, one limb of which is open and contains the perforated deck, the other terminating in a closed chamber to and from which air is repeatedly admitted and extracted by rotary values coupled to an air compressor. Differential stroke action is obtained by appropriate rotary valve design. The Tacub jig is a Japanese modification of the Baum design, introduced in 1954, in which the air chambers are located below the perforated deck. It provides generally improved control over differential wave form and uniformity.

The Batac jig, introduced in 1965 in Germany [13], was a modification of the Baum principle developed specifically in response to the need for wider

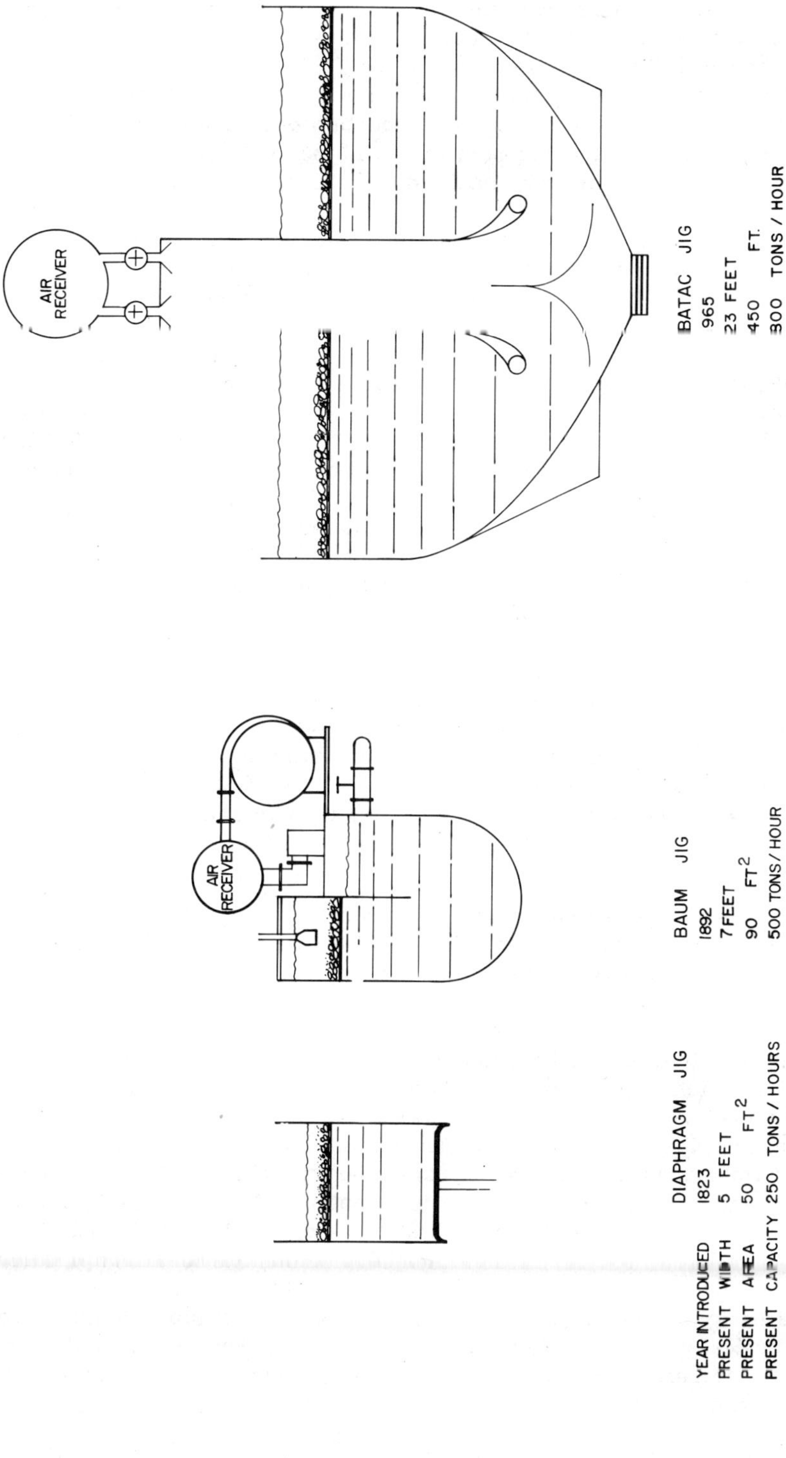

FIG. 15 Evolution of the coal jig.

jigs having increased throughput capacities. Figure 15 depicts the evolutions of the coal jig. One problem encountered in attempts to produce wider Baum jigs were the swirl effects caused, but not controlled, by the pulsations, which interfered with efficient stratification. This effectively limited the width of the washboxes to 7 ft and jigging areas to 180 ft^2 and lead to the introduction of "twin jigs" consisting of parallel jig decks within a single frame, individually pulsated, but sharing common units for certain operations, such as shale and middlings extraction. Development of the Batac system resulted in a double-U cross section, the center limb consisting of the closed air chamber driving the outer deck chambers. Additional improvements were obtained by changes in the design of product discharge mechanisms, sensing probes, and measurements of washbox operation and integrated control. A remarkable improvement in capacity and efficiency resulted. A recent coarse coal installation has an operating width of 13.1 ft, a deck area of 269 ft^2, and a throughput of 500 tons/hr [63]. At the same plant, a fine coal jig has an operating width of 23 ft, 452 ft^2 of working area, and a throughput of 550 tons/hr. The Batac jig has recently been introduced to the United States [64], two fine coal jigs each 16.5 ft wide, having 326 ft^2 of jig area and a rated capacity 600 tons/hr, having been installed at the Greenwich Collieries, Barnsboro, Pennsylvania. Performance data for these jigs are summarized in Table 14.

The results indicate a high order of efficiency of separation, even at ultrafine sizes of 28 × 100 mesh. It is reported that these jigs frequently operated at throughputs of 700 tons/hr with peaks at 750 tons/hr without noticeable difference in performance.

Once set up, the performance of all jigs is largely automatic and they are capable of handling reasonable variations in loading and raw coal composition. However, in common with other cleaning devices, product quality is adversely affected if refuse extraction becomes overloaded, causing variation in bed characteristics. Overextraction of shale and middlings also disrupts smooth bed operation, usually resulting in excessive loss of clean coal in the refuse. Middlings extracted from the washboxes are usually crushed and recycled to the raw coal inlet or to a parallel fine coal jig, so that the process is essentially a two-product separation.

b. Heavy Media At present some 82 million tons, or 31% of all mechanically cleaned coal in the United States, is processed in heavy-media systems. However, significant proportions of this quantity represents fine coal cleaned in heavy-media hydrocyclones.

Dense-medium separations of coarse coal are accomplished by immersing raw coal in a fluid of density intermediate between that of clean coal and refuse, or which may be somewhat lower than that of the clean coal but where separation is assisted by inducing upward currents in the fluid bath. Although many organic liquids and aqueous salt solutions possess densities and viscosities that make them candidates for use as heavy media, and which are true solutions and not suspensions, they have not found large-scale application, primarily on the grounds of higher cost, recoverability, or corrosivity. The Belknap Process is the only example to have found commercial application in the United States. The Otisca Process, currently at the pilot plant stage [112], employs heavy organic liquids as the dense media. An ideal medium is low cost, physically stable and chemically inert, easily recoverable and removable from coal and refuse products, and exhibits low viscosities at the densities of application.

TABLE 14 Batac Jig Performance—Greenwich Collieries, Pennsylvania, June 1976 (64)

Coal feed	3/4 in. × 28 mesh				28 mesh × 100 mesh	
Test number	1	2	5	6	3	4
Separating gravity	1.81	1.85	1.585	1.56	2.00	2.07
Actual probable error (Ep) at separating gravity 1.81	0.97	0.083	0.080	0.072	0.190	0.185
Guaranteed Ep at separating gravity 1.81	0.113	0.118	0.085	0.082	0.254	0.274
Float in refuse (%)	2.50	2.50	6.00	6.00	7.54	10.50
Misplaced in refuse (%)	0.76	0.38	1.92	2.17	1.03	0.99
Sink in clean coal (%)	0.90	0.25	0.50	1.20	3.67	4.30
Misplaced in clean coal (%)	0.63	0.21	0.34	0.77	2.76	3.90
Imperfection (%)	0.120	0.097	0.136	0.129	0.19	0.173
Recent yield (%)	69.6	84.80	67.98	63.85	86.55	90.91
Raw coal ash (%)	30.16	19.82	30.64	32.84	18.16	16.05
Clean coal ash (%)	8.38	8.21	7.50	7.41	8.53	9.33
Refuse ash (%)	80.00	84.27	79.77	77.75	80.18	77.55
Organic efficiency (%)	99.6	99.7	98.07	97.53	98.91	99.14

Suitable minerals that may be employed as suspensions include sand, clay, loess, barytes, pyrites, and magnetite. Of these only sand and magnetite have been widely favored, and now, except where the Chance sand process is still in use, magnetite has become the preponderant medium used worldwide. Its major advantage is that highly efficient methods for its recovery and recirculation have been developed which exploit its magnetic properties.

(1) Chance sand process. The sand processes employ unstable suspensions whose effectiveness can be maintained only by high rates of agitation and recirculation. By this means specific gravity separations up to 1.90 are obtainable, but specific gravities of 1.45 to 1.60 are more commonly employed for bituminous coals. The most successful commercial application of this process has been the Chance Cone Process, introduced after 1917. In this process water is pumped into the cone through its sloping sides to produce

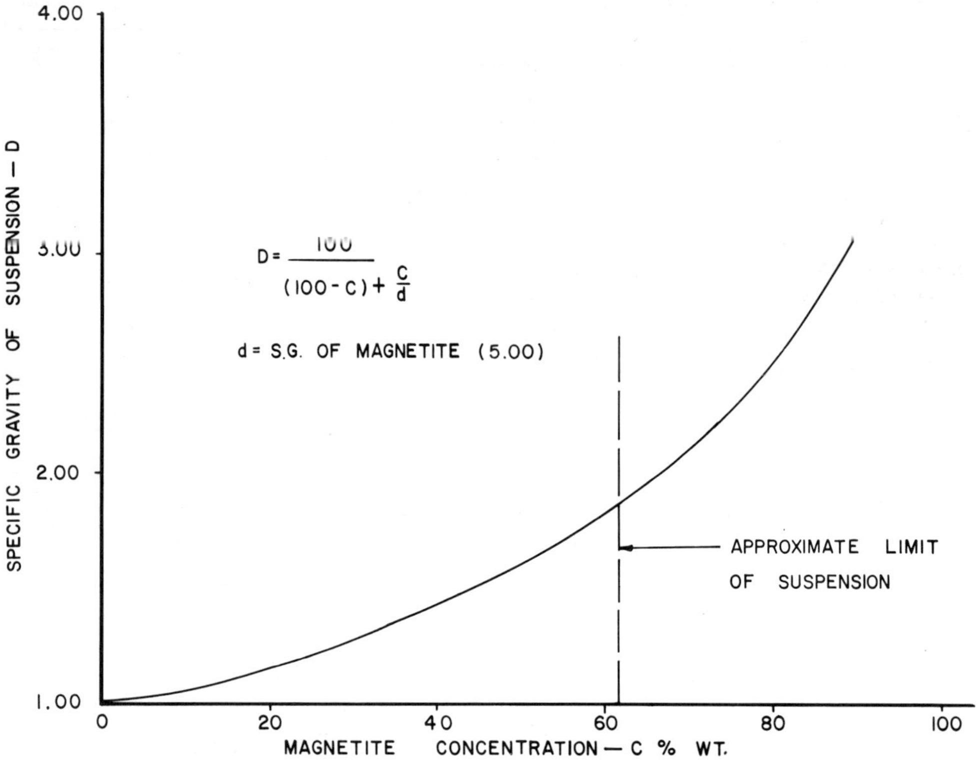

FIG. 16 Typical variation of specific gravity of magnetite suspension with concentration in water.

a rapidly rotating, rising water flow into which an overdense sand suspension is added in quantities sufficient to achieve the desired specific gravity in the mixed flows. Raw coal is added to the top of the cone, clean coal removed by a weir in the surface of the suspension, and sinks removed by a lock hopper in the base of the cone.

(2) Heavy-media vessels. Because densities are additive properties, the apparent specific gravity of a suspension may be easily calculated if the concentration of solids in unit weight of suspension and the true density of the solids are known, as shown in Fig. 16 for magnetite of density 5.00. However, the effective specific gravity of a suspension is strongly dependent on the stability of the suspension, which is, in turn, dependent on the fineness of the suspended particles. In practice, the magnetite is reduced to very fine sizes, usually in ball mills. The viscosity of the suspension increases with increasing fineness of particle size and particle concentration and tends to limiting values beyond 60% concentration. The settling rates may be greatly reduced, and hence the stability of the suspension also improved, by the presence of clays (slimes). Usually, these are present in the raw coal, so that it is necessary to recondition the suspension by bleeding off part of the circulating volume and recovering the magnetite by magnetic drums, rejecting

the slimes or, if they contain much fine coal, passing them to the ultrafines circuit. Similarly, magnetite is recovered from the rinsing waters underflowing the final clean coal and refuse dewatering screens and recycled to use. These operations are highly efficient, so that magnetite losses are commonly only 1 to 2 lb/ton of washed product. In some countries, loess being readily available and usually associated with the coal being mined, is used as the suspended medium, particularly for coals of lower rank. Medium conditioning in these cases is carried out by flocculation and concentration in settling cones. Dense-medium operations have been greatly assisted by the development of reliable instrumental methods for measuring the densities of suspensions, which have enabled heavy media to be automatically monitored and controlled. Heavy-media densities are easily altered by appropriate adjustment of concentration, and in practice a control better than ±0.005 specific gravity is obtainable.

Heavy-media vessels are the most effective means of cleaning coarse coals containing significant amounts of intergrown material. Probable errors (Ep) of 0.05 or better are commonly achieved with this equipment. Heavy-media vessels have demonstrated capability for handling fluctuating feed rates of raw coal, but like all coal cleaning equipment, their operation is adversely affected if their middlings and refuse handling capacities become saturated. Also, product quality deteriorates if the quantities of coal of particle sizes lower than used in design of the equipment becomes appreciable, so that efficient pretreatment of the raw coal, to screen out fine coal, is a necessary prerequisite of efficient operation. Occasionally, difficulties are encountered with friable coals which break down during the handling operations. Difficulties of this kind are best avoided at the process selection and equipment design stages and are usually the result of inadequate information on raw coal properties at the preliminary design stage of a coal cleaning project.

The principles of heavy-media vessel design have been described in some detail by Palowitch and Deurbrouck [65]. In general, since the settling velocities of smaller particles are lower than those of larger particles for a given specific gravity, and because equipment design invariably requires that all particles that sink move through some fixed distance before a clean coal product cut is made, smaller particles of higher specific gravity and hence of ash content will report with the larger, less dense material unless residence times and sinking distances are carefully determined. The need for accurate information on fractional specific gravity distributions of the raw coal are evident.

Efficient heavy-media vessels are available from a number of manufacturers in a variety of types too numerous to review here. They tend to fall into three broad categories: cones, troughs, and drums (wheels). In some machines the raw coal is fed, together with incoming heavy medium, to the surface of the vessel contents. In other types, the raw coal and heavy media is premixed and injected beneath the surface of the vessel contents. Three product separations are easily arranged, and it is a matter of record that the Tromp process, which was the first commercial magnetite process introduced in 1935, was designed for three-product separation.

Heavy-media vessels exhibit a capability of handling high throughputs in units requiring relatively small space compared with jigs, this feature being particularly marked in the case of drums and wheels. Units of throughputs up to 1000 tons/hr are available.

(3) Heavy-media cyclones. Although heavy-media vessels are effective in size ranges of 6 × 1/4 in., and have been used at top sizes to 15 in., the introduction, in recent years, of heavy-media cyclone washers has blurred the distinction between coarse and fine coal-cleaning processes and introduced equipment of intermediate capability for highly efficient separations in the size range 2 in. × 28 mesh. Indeed, the latest experience with these units suggests that a high level of separating efficiency is maintained as low as 200 mesh [21], which is close to the lower limit imposed by Stokes' law effects. The upper particle size is limited by the need for employing an exit cone diameter of certain maximum dimensions which determine the effective cut point for the rejects. Heavy-media cyclones were originally developed for cleaning coal sizes below 3/4 in. and are discussed in greater detail when considering fine coal-cleaning methods.

(4) Fluidized solids as heavy media. The readiness with which fine, dense solids may be fluidized by air has long attracted interest as a dry method of heavy media concentration of coals. To date, the development of a commercially successful system has not been achieved.

D. Fine Coal-Cleaning Techniques

Methods employed for cleaning coal sizes of 1/2 × 1/4 in. × 28 mesh may also be conveniently classified into hydraulic or heavy-media categories. The former include jigs, water-only hydrocyclones, wet concentrating tables, launders, and upward-current washers. The latter consist of heavy-media cyclones. Some authorities consider that water-only cyclones utilize a type of self-generated dense medium composed of the heavier particles present and that they are a type of heavy-media separator. Canadian work, in developing compound water hydrocyclones, exploited hindered settling effects produced by the heavy particles, but this mechanism also occurs in jigging action and may be better considered as a consequence of the dynamics of a suspended particle bed than as a heavy medium. In any case water-only cyclones do not require the provision of a separate circuit for preparation and conditioning of a heavy medium and, in this sense, are properly classed as hydraulic systems.

The relative employments of these systems in the United States is difficult to assess because the available statistical records make no distinction between coarse and fine coal cleaned in jigs and dense-medium systems. Until recently it was believed that wet concentrating tables provided the greatest fine coal cleaning capacity, but there can be little doubt that the introduction of high-capacity Batac jigs and the growing employment of heavy-medium cyclones capable of handling coal sizes to 28 mesh and lower will alter this position. The contributions made by wet concentrating tables, launders, and classifiers are indicated in Table 2.

The performance of fine coal cleaning equipment is a function of particle-size distribution, the efficiency of separation decreasing as particle size decreases. In most cases there are, for each type of equipment, optimal mean sizes at which separating efficiency are maxima. Figure 17 indicates typical performance ranges for the most widely used systems.

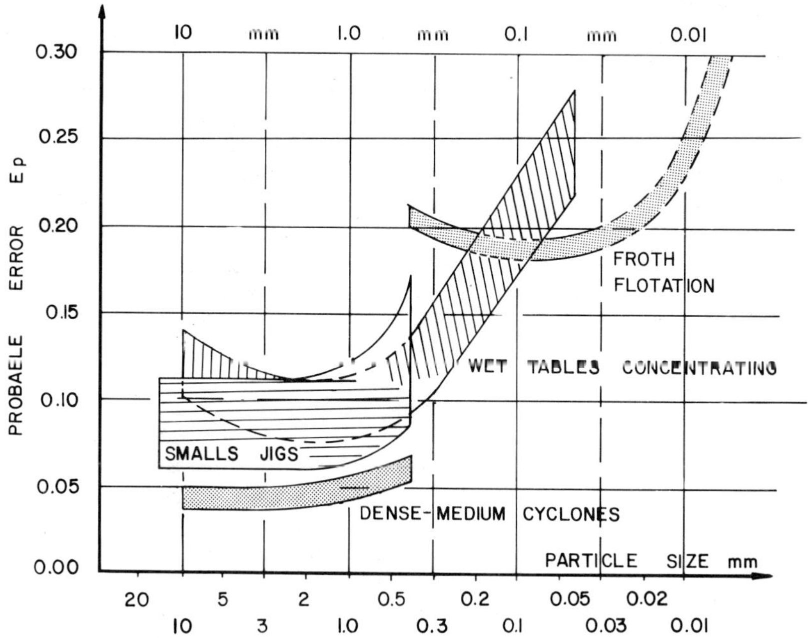

FIG. 17 Efficiency of separation and particle size [18].

1. *Fine Coal Jigs*

The operating principles of the jigging operation have already been described, from which it may be recalled that particle density rather than particle size and shape is the principal factor. The ability of jigs to handle size ranges down to 28 mesh has been noted. However, for a fine coal feed having significant near-gravity proportions, the effect of particle interactions can increase to an extent where efficiency of separation and product quality fall away, and in those cases the deliberate introduction of a hard, uniform, dense material to form a permanent "refuse bed" within the jig has been found effective in improving separations and qualities. A European development, it has been widely adopted there, and has found some application in the United States. Feldspar exhibits properties of density, particle shape, and resistance to degradation that make it particularly useful for this type of ragging application, and it has been used so widely for the purpose that the Baum-type jigs modified in this way are usually called feldspar jigs. Separations are two-product, the sinks passing through the feldspar beds into the hutch compartment of the jig for removal. Feldspar can also be employed in Batac jigs handling fine coal and is reported in use in the final cells of the Batac jigs of the Greenwich Collieries plant [64].

2. *Wet Concentrating Tables*

These units have been used for over 90 years for cleaning 1/4 × 0 in. coal in the United States, but have not found extensive application in other countries, although interest in them, particularly as an efficient method of removing free pyrites, is growing. A new plant at Walsum, West Germany, designed to reduce pyritic sulfur in a thermal coal, employs wet tables in the fine coal cleaning circuit [63].

The separatory principles, stages of development, and operating factors have been described by Deurbrouck and Palowitch [66]. The primary action is stratification of heavy and light particles by repeated agitation of a thin layer of the raw coal feed on a horizontal or lightly inclined table. Initially, this agitation was achieved by bumping, but in modern units it is obtained by cam-operated differential motion. The action not only produces vertical stratification but also causes progression of the solids across the table. At the same time, a transverse flow of water is applied along one longitudinal edge of the table and the surface interrupted by parallel riffles along the direction of solids flow. Lighter particles are preferentially lifted over the riffles so that a net diagonal separation of lighter from heavier particles is obtained. It is believed that hindered settling assists the separation so that the overall process of separation combines stratification and hindered settling. Clean products are removed from the edge opposite the water feed and are progressively classified by size and density, so that smaller coal particles separate with coarser intergrown (middlings) particles as shown in Fig. 18. Efficient operation is dependent on providing uniform feed and water flows and the initial setup of table operation requires appropriate adjustments and balancing of horizontal and longitudinal table inclinations and differential stroke lengths. Stroke speeds are normally in the range 250 to 300 per minute. Separation efficiencies for varying particle sizes are shown in Fig. 17.

Modern designs provide for suspension of tables by cables, which dampen transmission of vibrations to plant main structures, and the use of double and triple decks to improve throughputs per unit area of floor. Feed rates of about 30 tons/hr of 1/4 × 0 in. raw can be handled on triple-deck units requiring 120 ft^2 of floor area.

3. *Hydrocyclones*

The earliest applications of cyclonic cleaning employed heavy media [67,68]. Later, fundamental research on hydrocyclones, especially by Visman in Canada, led to the development of efficient "water-only" units [10].

Writing in 1968, Sokaski et al. [69] wrote that

> . . . inasmuch as the first cyclone plant in this country was built as recently as 1961, it would be premature to assess the role the cyclone will play in American coal preparation practice. . . . At the time the cyclone was developed, the argument was made that it would not be popular in America because, since it could not treat the extreme fines, it necessitates the addition of a flotation circuit. Now, however, many plants, . . . have flotation circuits. . . . many table plants are treating about the same size range that is handled in cyclones. . . . Deterioration in raw coal quality, stricter market specifications, or the two combined, could change a cleaning problem that could be handled by any cleaning device into one that could be handled economically only by a device as efficient as the cyclone.

What these authors foresaw has largely occurred. Cyclones have rapidly developed into the leading technique for cleaning fine coals, and improvements

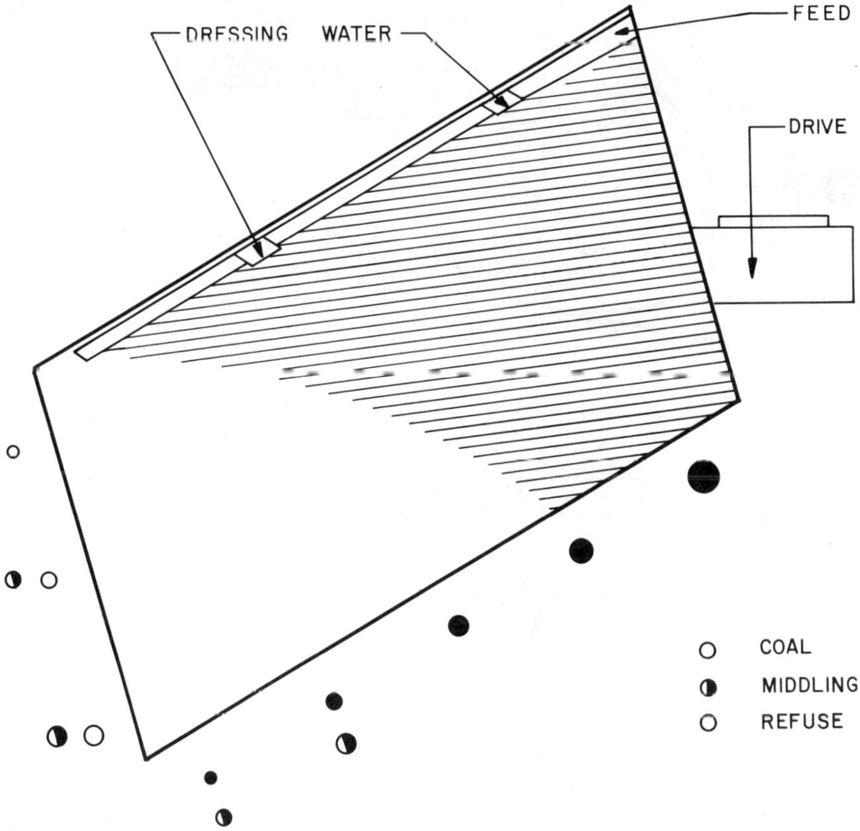

FIG. 18 Distribution of table products by particle size and specific gravity.

in design have extended the top range capable of being handled from 3/4 to 1 3/4 to 2 in., and the lower size limit to 200 mesh [21]. Heavy-medium cyclone systems now straddle intermediate coal particle sizes as cleaning units of incomparable throughput, efficiency, and economy, even for difficult coals containing large amounts of intergrown material. The introduction of compound water cyclones for handling size 1/4 × 0 in. further avoids the need for providing dense-medium preparation and recovery circuits, with additional savings in capital and operating costs. Because of the high precision of separation, cyclones are particularly effective in controlling sulfur contents for coals where the pyritic sulfur is mainly associated with shales and middlings. The design of plant in which rejects from a primary heavy-media cyclone circuit become feed, with or without crushing to liberate intergrown material, to a secondary heavy media cyclones circuit is easily and conveniently arranged.

The elements of cyclonic cleaning are illustrated in Fig. 19. Raw coal and fluid enter tangentially close to the top of the cylindrical section, forming a powerful vertical flow. Forces acting on the particles are proportional to V^2/r, where V is the tangential velocity and r the radius of the cylindrical section and are equal for all the particles in the inlet flow. Separation arises

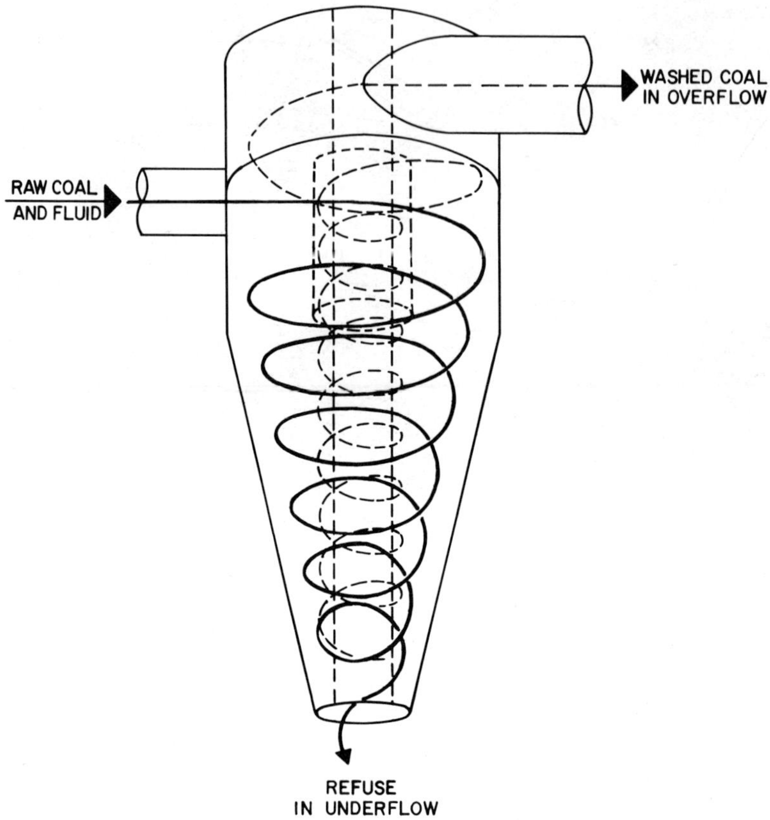

FIG. 19 Density separation in hydrocyclones.

because the other proportionality factor operating is buoyancy, defined as the difference between the mass of any particle and mass of an equivalent volume of displaced fluid. If the fluid is water, this factor is equivalent to specific gravity. The densest particles therefore move to the wall and downward, and exit with the cyclone underflow. Conditions can be arranged so that all the particles will exit this way, as in solids recovery and classifying, but can also be adjusted so that less dense particles remain suspended in the vertical flow and exit from the top of the cyclone. In practice it has been found that this density separation can be controlled with great precision. The force V^2/r is much greater than ordinary gravitational force, so that separations occur under greater density gradients than are achievable in ordinary heavy-media vessels, and the effects of particle sizes are less important. One of the control mechanisms employed is the selection of cone angle and orifice diameter for the underflow, the latter in particular being dependent on particle size and, to some extent, shape. Flat, slabby particles are particularly prone to blockages, so that this control parameter generally limits the top sizes of coal feed.

Many papers describing the theory and optimum design of hydrocyclones have been published relating to their functions as solid/fluid clarifiers, classifiers, and gravity separators, with or without heavy media as fluid. In their functions as classifiers, the control parameter employed being the particular particle diameter at which half the particles report to the overflow and half to the underflow, commonly referred to as d_{50}. This decreases as cyclone diameter decreases. But cyclone throughput also decreases with cyclone diameter and increasing pressure drop of the system, so that optimum design of a hydrocyclone corresponds to a criterion of minimum pressure drop required for a desired separation.

In their functions as gravity separators, the theories of operative mechanisms have been reviewed by Weyher and Lovell [70]. But for practical coal-cleaning applications, separating efficiencies are best expressed in terms of probable error, error area, imperfection, the M curve, and so on, whichever method allows direct comparison with other coal-cleaning devices. By modifying the cone section of the hydrocyclone to include subsections or frusta of successively lower cone angles, Visman [10] was able to generate conditions within a single unit in which the particles passed progressively through regions of hindered settling (as in jigs) to regions of classification. The resulting compound water cyclone exhibited a significant increase in efficiency of separation over conventional single-angle cyclones. A further improvement, resulting in its application to -28 mesh raw coal feed has been described by Picard et al. [71], which is of particular value in cases where the ultrafine coals, because of rank or oxidation, are not susceptible to cleaning by froth flotation.

4. *Launders and Classifiers*

The use of these cleaning devices is steadily decreasing under competition from the systems described above, and they are not regarded as likely candidates for wide employment in future coal-cleaning practices. Launders of the Rheolaveur type have a long history of successful application for cleaning bituminous coals, and upward-current classifiers of the Hydrotator design have a successful record in the American anthracite industry.

E. Ultrafine Coal-Cleaning Techniques

1. *Origin of Ultrafines*

The raw slurry feed requiring cleaning by ultrafine techniques usually has several sources in a modern wash plant. It comprises small particles of size below 28 mesh (595 μm) in the United States or mesh sizes equivalent to 500 μm in other countries. The presence of sizes larger than this are undesirable because they frequently "sand out" during processing, reporting to rejects and occasionally causing blockages. The small particles have their origin in the raw coal and are also formed by attrition during the cleaning of the larger sizes. Depending on the friability of the coal and the mining methods employed, the size fraction below 28 mesh may be as high as 30 to 35% of the raw coal and is, nowadays, rarely lower than 15%. The introduction of increasingly mechanized mining methods has resulted in substantial increases in the ultrafine fraction.

It should be noted that the size range 500 μm to zero is, itself, a wide distribution in which certain types of equipment may function effectively down to some lower limit but be ineffective below that limit. The design of efficient systems for cleaning ultrafines therefore requires reliable information on fractional size distribution below 28 mesh and presents problems of sampling and testing comparable with those for the macro sizes.

Raw slurry forms part of the principal plant water circuit, so that, after separation and recovery of clean ultrafines and removal of high ash slimes by clarification, the supernatant water flow is recirculated, makeup water being added as required. Where cleaning of ultrafines is not carried out, the raw slurry is discharged directly to settling ponds or lagoons, the ultrafines being recovered from the drained ponds for use or discarded according to quality. Growing environmental objections are likely to limit the use of once-through systems of this type in the future.

Alternative processes include oil agglomeration and selective flocculation, both of which operate best for particle sizes below 100 mesh. Oil agglomeration has been introduced on a commercial scale in Europe (Olifloc Process [37]) and in Canada.

2. *Hydrocyclones*

The general principles of hydrocyclones have already been described. Further improvement of the compound water cyclone, involving the use of elongated cylindrical sections, has been demonstrated by Canadian workers so that it can be applied to ultrafines cleaning in the range 28 to 150 mesh [71]. Performance obtained is illustrated in Fig. 20. The method is clearly inapplicable if there is a significant fraction of ultrafines smaller than 150 mesh. One application that has been suggested for this technique is the recovery of coal from old lagoons.

3. *Froth Flotation*

This has long been the most important method of cleaning ultrafines but is generally inapplicable to low-rank coals and encounters difficulties with oxidized bituminous coals. The increasing production of clean coal by froth flotation in the United States is shown in Table 2. A typical flotation circuit is shown in Fig. 21.

Froth flotation is one of the family of adsorbed bubble (adsubble) concentration techniques [72], and its general principles were described earlier. Its application calls for control of raw pulp concentration, adequate dosing with collector and frother reagents, efficient mixing of pulp and air, sufficient retention time for separation of the froth, and an adequate rate of froth collection and removal. Many types of cells have been developed, but the two most widely used in current coal preparation practice are the D-R Denver and Fagergren systems. Both use submerged rotors to mix the pulp and air, the former using low-pressure air, the latter using atmospheric air induced by rotor action. A newer development, the Cyclo-cell, avoids the use of moving parts by mixing and discharging low-pressure air and a high-velocity jet of raw pulp in a submerged vortex chamber.

Other factors that affect flotation efficiency, such as pulp size distribution and water quality, particularly pH and dissolved solids, may be present. In

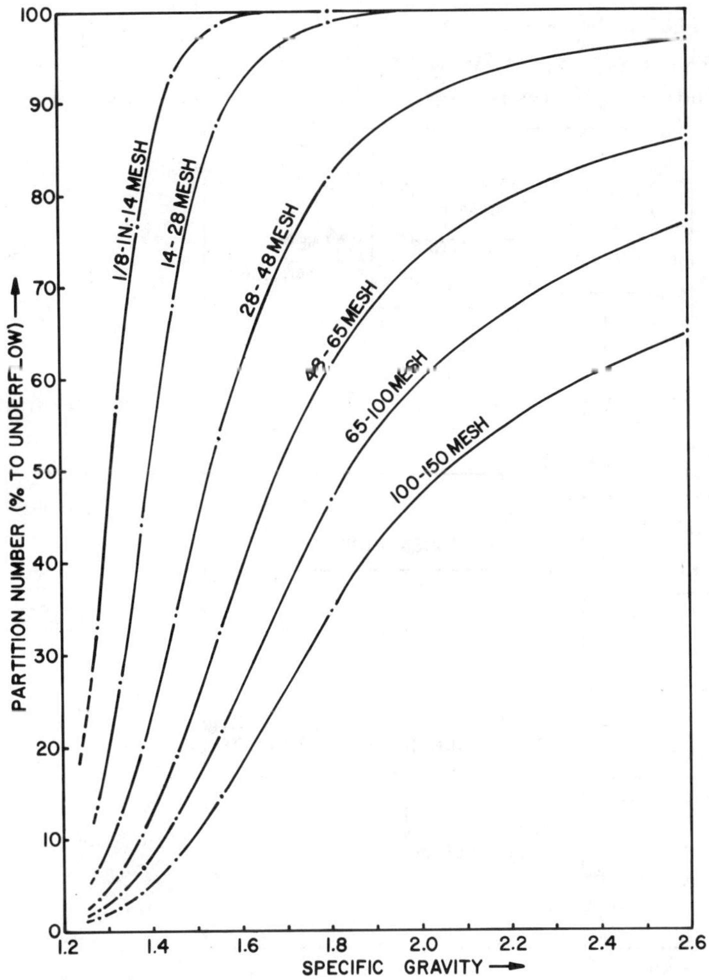

FIG. 20 Average experimental error curves for standard compound water cyclone (CWC-8) [71].

many instances no specific control of feed pulp concentration is employed, the actual concentration resulting from relative rates of raw coal feed and washing water and the proportion of ultrafines in the raw coal. Solids concentration varying in the range 5 to 15% are thus often encountered, so that precise dosing with flotation reagents is impossible. Greater attention to pulp concentration and conditioning is now being applied, especially in Europe. Static thickeners, flocculation, and larger flotation cells are being introduced. Dosing with reagents is directly controlled by pulp flows and concentrations. These improvements result in greater utilization of cell capacity and improved efficiencies of separation [73].

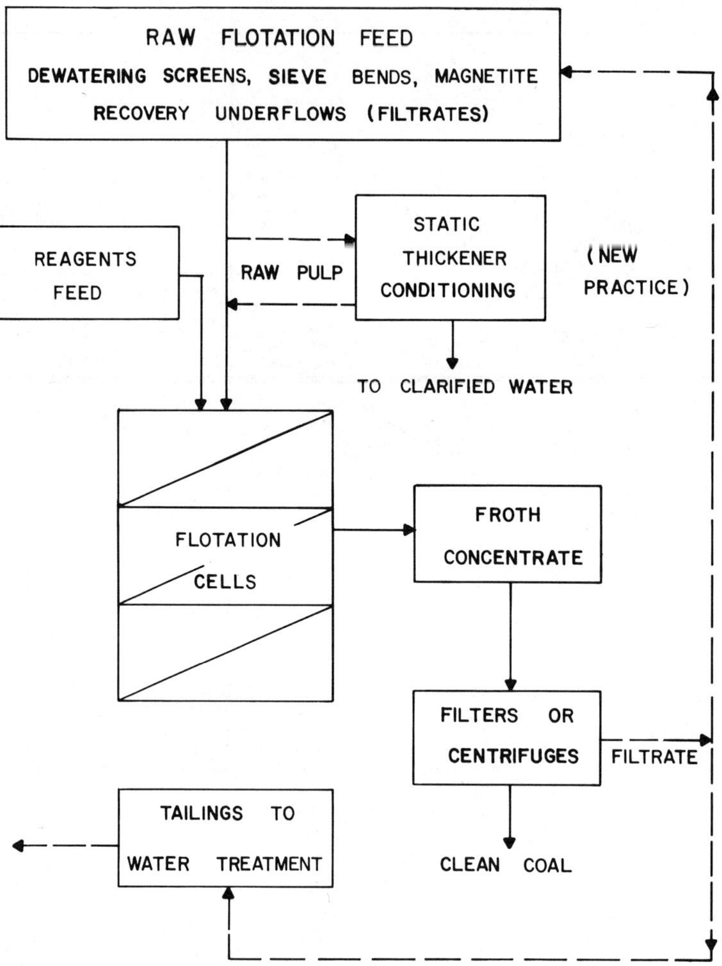

FIG. 21 Typical froth flotation circuit.

The use of two-stage froth flotation, a primary roughing froth being re-flotated to produce a cleaner final product, is commonly encountered in mineral dressing, but has found little application in coal cleaning because the costs involved would rarely be recovered in improved product quality and price. However, a selective second flotation may be of value in separating pyrite, where substantial quantities of this are recovered with a primary froth product [74,75]. The first commercial installation of the Coal-Pyrite Flotation Process has been reported [1].

Cleaned froth is collected, allowed to disintegrate, and the resulting suspension is fed to dewatering devices. Rotary disk or drum vacuum filters or solid-bowl centrifuges are employed. The underflow high-ash tailings from the flotation cells pass to the water treatment plant.

V. ANCILLARY OPERATIONS

It is evident that the basic processes for coal cleaning must be supported by other unit operations which dewater, dry, and store the products; treat water for reuse; and provide means for disposing of refuse and tailings. In addition, environmental regulations require control of fugitive dust, polluting liquid effluents, and excessive noise. All these ancillary operations demand careful engineering design, require substantial capital costs, and often incur heavy operating costs, and no discussion of coal cleaning is complete unless they are considered. The provision of general utilities and services, although an essential cost element, is universal to all industrial operations and does not need special consideration.

A. Dewatering

All the products from wet coal-cleaning processes require dewatering at some stage, even the operation of lagoons and settling ponds being essentially a dewatering operation. As discussed earlier, of the three general types of water retained by coal, only interstitial and surface-held water are available for removal in ordinary plant operations. Interstitial water is readily removed from particle size fractions exceeding 1/4 to 1/8 in., and the tendency of the surfaces of coal macerals to shed water assists in this process. Agitation on dewatering screens is usually sufficient, and many types of efficient vibrating screens are available for this purpose.

As particle sizes decrease, a twofold problem arises. First, separation of solids from the liquid flow in which it is carried, followed by removal of water from wet solids. The former can be efficiently accomplished on sieve bends (Møgensen screens) or shaker screens of 1/2-mm wedge wire slots for sizes generally 1/2 in. × 28 mesh. Final dewatering for this size range may be accompli5hed on high-speed vibrating screens or in basket-type centrifuges. The efficiency of available equipment is such that final moisture contents of 6 to 9% are obtainable. For particles below 28 mesh top size, increasing difficulty is encountered both in recovering solids from the liquid flows and in achieving a satisfactory dewatering of the recovered solids. The difficulty increases markedly as the mean particle size continues to fall, to an extent that may even cause difficulties with the equipment specially designed for handling ultrafine sizes.

Partial solids concentration can be obtained by hydrocyclones acting as particle separators, by froth flotation, or by settling in cones. Final recovery is by means of rotary vacuum filters or by solid-bowl centrifuges. Rotary vacuum filters, both disk or drum types, have been used for many years, the former being generally more popular because of the substantially lower floor-space requirements for a given filter area. Recent increases in machine capacities have pushed filter areas of disk types to 4300 ft^2 per unit, and drum types to 1300 ft^2 per unit. Filtering capacity depends upon complex relationships among pulp concentration, pulp temperature, mean size of particles and size distribution, vacuum applied, and speed of rotation of the disks or drum. These are adjusted in operation to produce maximum cake thicknesses for acceptable discharge moisture contents, which are usually in the range 20 to 25%. Because of the viscosity and surface-tension relationships of water with temperature, improved filtration and lower final moistures are obtainable if pulps are heated or if steam is admitted through the wet cake on the nonsubmerged portions of the filter. Both methods incur additional costs and have been employed in certain cases, but heated filtration has not proved to be of general, economic utility. Similar results obtainable with surfactants at ordinary or elevated temperature are being tested [76]. The use of flocculants to condition the pulp before filtration is becoming more widely used, reagents of the polyelectrolyte type being generally suitable.

Solid-bowl centrifuges are a more recent introduction but are steadily gaining acceptance [77], particularly where the quantities of coal particles smaller than 50 μm do not exceed 20 to 25% of the feed solids. If these values are exceeded, unacceptable "slip" of fine solids to the bowl effluent is usually encountered. These machines, currently operating at 500 × gravity centrifugal forces, produce dewatered solids containing 17 to 22% moisture content. Heating the feed pulp or injecting steam to the bowl may achieve a small improvement on these figures but is not generally economical. Similarly, surfactants may produce an improvement in moisture, as already noted for rotary vacuum filters. Conditioning of the feed with flocculants is a growing practice.

Environmental pressures to reduce discharge of liquid wastes has generated the need for developing efficient dewatering of high-ash slimes from the water clarification circuits. These particles are often clays and so present exceptional difficulties. Rotary filters are rapidly rendered ineffective because of blinding of the filter fabric, and solid-bowl centrifuges have proved more successful. The use of flocculants assists dewatering. Filter presses are also used, particularly in Europe, for dewatering slimes and are considered the most dependable and advantageous equipment for this purpose [78].

Development of improved rotary filtration has generally been concerned with the manufacture of large equipment and greater capacities. Improvement of centrifuges is directed toward attaining higher g forces, the present range of 200 to 800 g being increased to 1500 to 2000 g, with consequent improvements in machine throughputs and product moisture [79].

The use of settling ponds and lagoons, often followed by bucket dredging and draining to recover coal values, is no longer environmentally acceptable, and use of this method will decline.

B. Drying

Moisture is an inert component of a coal which makes no contribution to its fuel value and, in fact, requires heat for its evaporation and release in the products of combustion. In cold climates wet coal freezes in stockpiles and during transit. Subsequent handling becomes difficult for coals having free surface moisture. These problems arise in North America, Poland, and the USSR but are of smaller consequence in other major coal-producing countries, including western Europe, where winters are less severe. Coal freezing can be avoided if dry concentration methods are employed, but these are of limited application. Thermal coal drying has been most widely developed and used in the United States and more recently, in western Canada and has found its widest employment in the preparation of metallurgical coals.

Drying is the evaporation of moisture from a material and takes place in air at ordinary temperatures at rates determined by the relative humidity of the air and the rate of air flow over the material surfaces. For drying coals having moisture contents ranging from 5 to 25% depending on size range, this process is too slow and drying is accelerated by passing hot combustion gases over or through beds of coal to be dried. The hot gases are produced by burning a fuel, sometimes pulverized coal extracted from the coal to be dried, more usually using liquid or gaseous fuels. Fully automatic systems have been developed for producing hot combustion gases from each of these fuels.

Three basic mechanisms are commonly employed for drying bituminous coals: (1) tumbling or agitation of coal beds in streams of the drying gas (e.g., in cascade trays and drum-type dryers); (2) suspending coal particles in high velocity streams of hot combustion gases, followed by separation of dry coal and water-saturated gases; and (3) fluidizing beds of coal with hot gas. In each case efficient methods for removing coal dust from coal exit gas streams are required and form an important part of the overall process requirements, because stringent dust emission standards are imposed. The psychrometric principles underlying the design of efficient equipment were described earlier.

Fluidized-bed dryers have found the widest acceptance for general drying purposes. Suspension-type flash dryers are finding increasing use for the combined drying and preheating of metallurgical coals charged to coke ovens.

Drying, because it incurs substantial capital and operating costs, is applied only to the smaller size ranges, usually 1/2 × 0 in. Final coal moistures are usually specified in the range 5 to 8% for bituminous coals, these levels ensuring ready handling and avoidance of freezing.

Low-rank subbituminous coals and lignites, which can contain 20 to 60% of inherent moisture, are often highly susceptible to spontaneous ignition and combustion and provide very difficult drying and handling problems [80]. Direct drying by contact with hot combustion gases has not always proved suitable, and the development of other methods is a subject of much current investigation. In West Germany, methods for drying brown coals containing up to 60% moisture have been developed which involve indirect steam heating of the coals in tumble dryers [58] to achieve final moisture contents of about 9 to 15%. Throughputs are low and this method is best applied for special purposes (e.g., the production of briquettes). In Australia a new method for thermal dewatering of brown coals without evaporation (thermodynamically more efficient) in which the coal is heated with oil under pressure, is under development but is not yet in commercial use. Its principal utilization is expected to be a pretreatment process of the coal before liquefaction by hydrogenation [81]. Another method for drying power station coals is also being developed at Monash University, but technical details are not available at the present time [82].

C. Refuse and Tailings Disposal

Coal mining and cleaning is an industry that has a substantial impact on the environment, and refuse disposal forms a major part of this problem for which a substantial body of literature exists [83]. Its magnitude is indicated in

Fig. 1, which shows that about 29% coal submitted to mechanical cleaning in the United States requires subsequent disposals as refuse and tailings. Current production is approximately 110 million tons, a figure that will increase as coal production and increasing preparation of thermal coals increases.

Disposal presents a twofold problem. First, unless return of the coal wastes to the mine is possible, large areas of land are required, and fixed and mobile materials handling plant must be provided. Substantial amortization, capital, and operating costs are therefore incurred [84]. Second, land sterilization and spoil-heap drainage control present problems of environmental control which may require extra capital costs and operating expenses. The danger of spontaneous combustion of refuse piles arises where the coal wastes contain sufficient carbonaceous material and where spoil heaps are constructed by inefficient methods that allow ingress of sufficient air to support combustion. Refuse disposal requires the handling of wet granular material of size greater than 28 mesh and of sludges composed of ultrafine refuse. The latter present severe difficulties during handling and if mixed into main spoil heaps can lead to stability problems of the piles, and if disposed of separately usually require additional treatment, including chemical dosing, to achieve acceptable stability [85]. Figure 22, adapted from a summary provided by Hummer [86], presents current methods of refuse removal and optional disposal methods.

D. Water Treatment

Cleaning plants may require up to 2000 gal/ton of feed coal, although volumes of only half this quantity are now required by the latest jig designs, which are the units requiring the greatest quantities of water. If no water treatment and recycling is practiced, this volume represents a net demand for fresh water. However, in all major coal-producing countries, the installation of total recycle closed water systems has become a major aim, one advantage being that the water requirement reduces to that necessary for replacing water lost in the products so that water consumption drops to 18 to 30 gal/ton of feed coal. This net requirement is independent of the types of processes employed for wet cleaning.

Cleaning plants require no special quality of feed water, most surface supplies meeting requirements provided that total dissolved solids are not abnormally high. High concentrations of dissolved salts may interfere with froth flotation.

Unless the coal being washed contains appreciable quantities of soluble salts, cleaning processes do not materially alter water composition. The greatest difficulties are encountered in the presence of iron carbonates and pyrite, particularly where the coal passes through stockpiles which allow some prior oxidation of the iron salts. The contamination is usually indicated by a substantial lowering of pH, from a normal range of 6 to 7.5 to values of 3 or even less, and may necessitate dosing with alkali (caustic soda or lime) to restore neutral conditions. This type of problem is not often encountered and the basic water treatment required is clarification to remove suspended slimes so that the water may be recirculated.

Clarification may be partially achieved using batteries of small-diameter hydrocyclones to remove particles down to about 50 μm. Full clarification

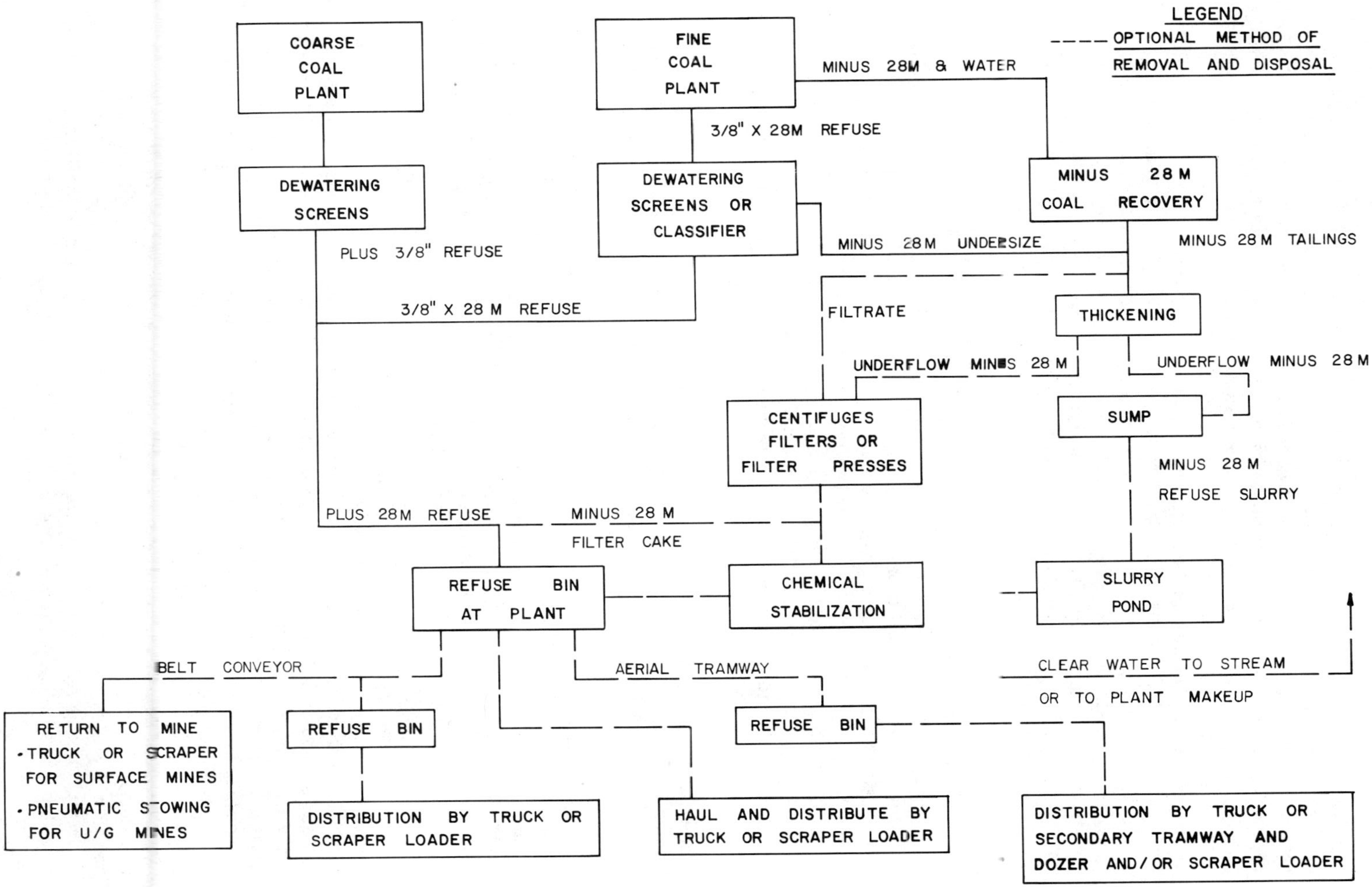

FIG. 22 Plant refuse removal and some optional methods of disposal.

requires settling in cones and static thickeners, in conjunction with chemical dosing to promote flocculation and rapid settling. This process requires long retention times and the provision of corresponding retention volume, which represents a major capital addition to plant costs. Efforts to reduce costs and improve efficiency are being pursued along two broad lines. First, improved flocculants are being developed so that higher settling rates and lower retention volumes can be achieved. The introduction of polyelectrolyte flocculants resulted in major improvements, and many effective reagents are available. However, plant waters are variable so that suitable experiment is usually necessary for selection of reagent and optimum dosing. A major comparative survey and indexing of commercial flocculants has been reported [87]. Second, improvements in the design of settling cones and thickeners are being developed. A major advance has been the successful introduction of the Lamella Thickener [88] which is an inclined, shallow-depth sedimentation device. It performs the same function as conventional gravity thickeners but occupies only a fraction of the ground space. Another major advance is the invention, in the United Kingdom, of the Deep Settling Cone, which thickens suitable sludges to such a high concentration and consistency that the solid may be discharged directly from the base of the cone onto a conveyor for disposal [89]. This advance holds the potential for eliminating sludge dewatering by filters, filter presses, or solid-bowl centrifuges, which will result in significant savings in capital and operating costs.

Many hundred of investigations have been, and are continuing to be, made into various aspects of water treatment for coal cleaning in all the major coal-producing countries. A problem of increasing concern is the quality of runoff for coal storage piles, which often requires collection and treatment before discharge. This subject was discussed in detail at the NCA/BCR Coal Conference at Louisville, Kentucky, October 1977.

E. Dust Control

Dust presents a serious problem in dry cleaning plants and in the pretreatment sections of wet cleaning plants. It also arises during thermal drying, and elaborate provisions for dust containment form an integral part of all thermal drying units. The purposes of dust collection are air pollution control, safety- and health-hazard elimination, reduction of equipment maintenance, and recovery of product values. In addition to the nuisance and health hazards, coal dust clouds can in certain circumstances present serious explosion hazards. Certain bituminous coals and anthracites also continue to release measurable quantities of methane gas for days, and sometimes weeks, after they have been mined, which compounds the explosibility hazard. Bunkers, dryers, and other confined areas through which dry coal passes must therefore be designed with these hazards in mind. Provision of efficient ventilation and restriction of electrical or thermal sources of ignition are required by federal and state regulations.

The main sources of fugitive dust are crushers, breakers, dedusters, dry screens, and transfer points between handling units in pretreatment sections and dry concentrating tables. The sizes of coal particles that become airborne generally fall into the range 1 to 100 μm, particles larger than this depositing close to their origin. It is the usual practice to extract the dust clouds from the various sources through ducts that exhaust to a central air-cleaning unit, although some equipment may include integral dust-suppression devices.

Dust collection equipment used at coal cleaning plants includes dry and wet processes. Common dry units employed include cyclones, dynamic collectors, and baghouses (fabric filters). Wet methods include dynamic, impingement, or centrifugal devices; or gravity, disintegrator, or venturi scrubber systems. A detailed description of these units used in coal preparation has been provided by King [90]. The capital costs of this equipment is a major component to total plant costs, and their operating expenses contribute to overall cleaning costs. Efficiencies are variable, and careful maintenance and operation is necessary to comply with the stringent emission standards.

Other sources of dust nuisance are windblown losses from storage piles and from railcars during loading and transportation. Various chemical treatments are available for sealing strategic stockpiles and are being used for spraying the tops of railcars, but active stockpiles present an intractable problem. One solution used in Germany is to totally enclose the stockpiles and coal-handling machines [95].

F. Noise Control

Coal preparation plants, which invariably contain a variety of rotating, reciprocating, and vibrating machinery in addition to large throughput tonnages of free-falling solids, have long been recognized as presenting a serious noise problem, but only in 1969 was federal legislation introduced to regulate noise levels. It is noteworthy that the third, and most recent, edition of *Coal Preparation* (American Institute of Mining, Metallurgical and Petroleum Engineers, Inc., New York), published in 1968, does not include noise or its control in its subject index. The Code of Federal Regulations [91] sets the maximum permissible sound level to which a worker may be exposed for an 8-hr shift at 90 dbA. Exposure to louder sounds requires specified reduction in the period of permissible exposure calculated according to a "noise exposure coefficient."

The results of a major study to survey and reduce noise at a new coal cleaning plant were described in 1976, together with the measured improvements resulting from machine modifications [92]. The results of other surveys and guidelines for quieting noise in coal preparation plants have since become available [93], and the techniques are being incorporated in the necessary design specifications of new plants.

Most of the noise control measures found effective increase initial capital and replacement costs but do not add significantly to operating costs or effect productivity. They may be classified broadly into measures that (1) suppress noise formation or (2) provide noise containment. Examples of the former include installation of mufflers on air pumps and of rubber screen cloths and resilient screen decks and chute liners. Noise containment requires erection of barriers, either walls or curtains. Their combined employment enables noise control adequate for meeting current operator exposure standards. In some plants some of the methods described have not been successful with regard to noise, cost, or plant productivity, but the reasons for these failures are not clear [94].

VI. PROCESS CONTROL

Fully mechanized and integrated machine operations have long been a feature of coal preparation systems, so that high labor productivity is customary. Most of the present operating labor effort is directed to visual monitoring of operations and housekeeping duties, supported by necessary maintenance duties requiring craft workers. Centralized control is common and has been achieved by suitable plant instrumentation in conjunction with mimic panels and manually operated controls, leaving direct visual monitoring and local control of some ancillary operations for division between a relatively small number of operatives. Many attempts to extend this type of centralized control to full process optimization and automation have been attempted in a number of countries, especially the United Kingdom, West Germany, and the USSR, but until very recently these attempts have not been completely successful. The reasons for failure are various but, in general, can be attributed to lack of reliable methods for rapid, automatic measurement of all the variables required, and to inadequate mathematical modeling and integration of all the unit operations involved. A review of early work on modeling and the various attempts at automation has recently been provided by Walters and Ramani [96].

Earlier, unsuccessful attempts made up until about 1970 were followed by a period of retrenchment for further development of the necessary tools [97]. A coal preparation plant presents an extremely harsh environment. Heavy vibration, dust, moisture, and widely varying ambient temperatures provide a serious hindrance for methods requiring sensitive instrumentation, but the remarkable development of solid-state electronic devices during the past 5 years is having a major impact on reducing these problems, including the possibility of introducing automated local control of certain operations such as heavy-media circuits. Their availability has enabled improved methods to be introduced for sensing and measuring liquid and solid flows: coal analysis, particularly ash, sulfur, and moisture contents; pulp concentrations; and other essential parameters. Improved systems engineering has resulted from such advances as computerized analysis of coal washability data [38], extended mathematical modeling of unit operations, and linear programming of the overall process steps. In the United Kingdom, this process forms an extension of automated mine conveyor systems control by the MINOS system [23]. West German experience has been described in detail and a fully automated coal preparation plant control is currently being put in operation at Haus Aden mine in the Ruhr district [14-16].

VII. ECONOMICS OF COAL CLEANING

In recent times modern wash plants in the United States has been standing idle because the premium on price necessary to cover cleaning costs could not be recovered in a slack coal market. Coal users in a competitive market tend to evaluate their coal purchases by methods that indicate minimum costs in energy terms (i.e., delivered costs per million Btu). A second stage of evaluation is then based on user costs deriving from coal properties. These can be complex and include such items as ash content, ash composition and fusion temperatures, fixed carbon (cokemaking), sulfur, and crushing and pulverizing characteristics. Considerations of these factors may modify the primary evaluation, but as a general rule will justify a premium for cleaning only if substantial cost savings in utilization will result.

However, calorific value is not a primary control function in coal cleaning. It has an approximately linear relationship to ash and moisture contents. But after gross rock dilution in a raw coal has been removed, the yield in terms of thermal efficiency of recovery with ash content becomes nonlinear and an increasing penalty in thermal recovery for unit decreases in ash content becomes the rule. Losses of yield in cleaning represent the biggest item contributing to total cleaning costs. The most immediate effect is on capital costs because the size, and hence the capital cost of a cleaning plant, is based on the throughput capacity for *raw* coal. It is generally accepted that capital costs reduce with increases in plant size, although not all the items comprising a cleaning plant are directly related to the throughput capacities of the cleaning units. For example, the sizes of raw coal storage and handling units, and the sizes of cleaned coal bunkers and loading facilities, are often determined by the needs for strategic stockpiles or the requirements of the transportation system.

Economic analysis is also greatly influenced by the relationship of the coal producer to the coal user. At least three different considerations may arise: (1) cleaned coal is produced for sale under comparatively short term contracts (1 to 3 years) in a competitive market, (2) coal is produced on long-term (7 years and more) supply contracts, and (3) coal production and cleaning forms part of a totally integrated operation in which coal is mined and used by a single industrial undertaking. Types (1) and (2) are most representative of past and present practices for metallurgical and thermal coals. Type (3) may apply following the current reorganization and restructuring of the coal industry and the growth of large coal conversion plants and mine-mouth generating stations.

Additional complexity arises from tougher environmental regulation in which coal cleaning is only one aspect of control technology available for limited emissions of sulfur oxides and particulates, including hazardous trace elements. It has already been noted that environmental regulation relating to the operation of cleaning plants—liquid effluents, fugitive dust, and noise—have resulted in significant increases in capital and operating costs.

The future growth of coal cleaning in the United States will be largely determined by the attitude of the electric utilities industry and the extent to which utilities companies enter into full or joint ownership of the means of coal production. This may result in decisive changes in the way in which economic evaluation of new coal projects, including cleaning, are made. The financial yardsticks applied, until very recent times, for determining the economic worth of coal cleaning were relatively simple measures. The quality, tonnage, and expected market price of raw coal were compared with similar parameters for cleaned coal. Capital and operating costs of coal cleaning were then estimated and added to the basic production costs of ROM coal. The difference between anticipated gross production costs and forecast selling prices became the basis for applying various accounting devices to determine the balance of economic advantage. Usually, a discounted cash flow (DCF) analysis enabled calculation of return on investment (ROI) over periods of about 10 to 15 years for the different options. This exercise was principally of concern to coal-producing companies because they carried the fiscal responsibility for the decision to clean or not to clean. As a general rule of thumb, a decision to proceed with cleaning depended on a DCF-ROI of at least 15% per year, and more commonly, 20%. If the case for cleaning was a foregone necessity, because of the markets' requirements, the analysis was used to calculate a selling price that would produce the necessary ROI. These rules had worked well in an industry that, although still important and substantial, had been declining.

The utilities employ different accounting principles from the CDF-ROI type of evaluation and they employ different funding methods, particularly as regards debt/equity ratios. Traditionally, their payback periods are substantially longer, 30 to 40 years for fossil-fired plant. These differences can result in substantial changes in the fixed capital charges set against new plants, which is important because under current practices these costs form the largest element in coal cleaning costs, being greater than 50% at all levels of preparation. Although a number of cost studies are in progress, the full impact of these changes and their wider importance for coal cleaning cannot be assessed at the present time. In this new situation it may not be correct to consider a "selling price" for cleaned coal. It becomes a "transfer cost" in the chain of operations that ends in the busbar cost of distributed electric power.

The elements of the cost of coal cleaning comprise fixed costs arising from capital charges, and fixed or variable costs arising from plant operation.

Capital charges are determined by the depreciation of costs incurred in land acquisition and preparation; design, procurement, and construction; provision of utilities and transportation and communications facilities; taxes; working capital; and interest charges. They are a fixed-cost burden unaffected by actual plant throughputs.

Operating expenses include salaries, wages, power costs, water costs, and productive supplies, including fuels, refuse, and waste disposal. This category may include fixed costs independent of plant throughput (e.g., salaries) or variable costs tied directly to throughput (e.g., productive supplies).

For a given annual production, capital costs are highly dependent upon raw coal quality as determined by ash content and size consist. The former determines the yield of clean coal, and since plant capacities are dictated by *raw* coal throughputs, this factor has the major influence on capital requirements. As shown in Fig. 1, the average yield of current American coal cleaning plants is 71%. The most expensive items of capital equipment to provide and operate are required for handling fine and ultrafine coal sizes (i.e., froth flotation, vacuum filters, centrifuges, water clarification, and thermal drying). Cleaning costs are therefore highly sensitive to the quantity of sizes smaller than 1/4 in. in the feed.

Capital and operating costs are also sensitive to plant utilization and the system of working. Plant practices vary from single-shift, 5-day-week operation to continuous-shift, 7-day operation. The latter normally allows one to three shifts per week downtime for planned maintenance. It is clear that, for a given annual production, plant sizes and hence capital costs are related to the working practice adopted.

The elements of capital and operating costs for the "traditional" operation of the industry have been described in detail by Zachar and Gilbert [95].

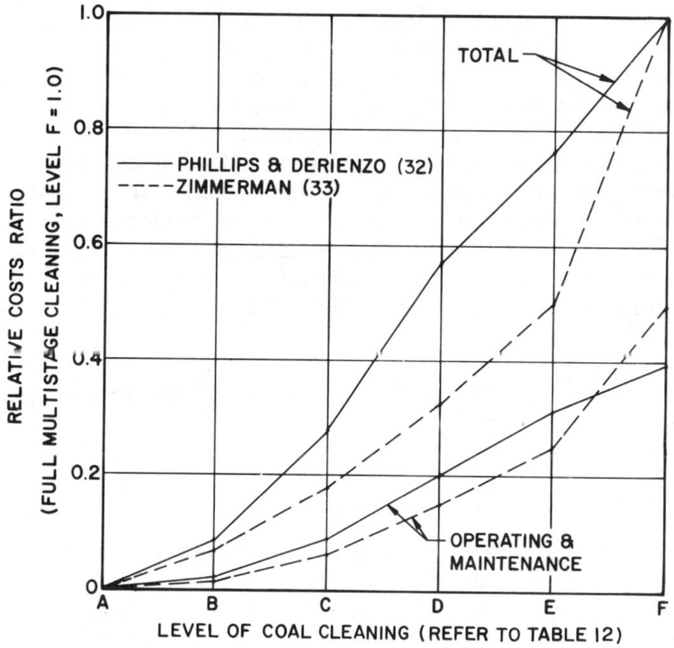

FIG. 23 Comparison of relative coal-cleaning costs.

More recently, several authors have provided generalized cost estimates for the various levels of coal preparation illustrated in Table 12 [32,33]. Direct comparison of the two sets of costs are not possible because of differences in the base data used. In particular, Phillips and DeRienzo [32] report total capital costs including land charges, whereas Zimmerman [33] reports turnkey costs excluding land charges and costs of utilities provided at battery limits. Also, the former include capital charges at 30% on a 10-year straight-line amortization, whereas the latter excludes capital charges. Allowing for these differences, but assuming plant capacities of 2.5 to 3.0 million tons/year of raw coal, the relative preparation costs of the various levels of cleaning are shown in Fig. 23, taking full multistage cleaning, level F, as unity.

The capital costs (1978 dollars) corresponding to the ratios shown range from $6 million (level B) to $50 million (Level F) for a plant handling 3 million tons/year raw coal. Operating costs range from 30 cents/ton (level B) to $7.00 per ton (level F) of clean, dry coal. It may be observed that, for all levels, capital charges account for more than 50% of the total costs of preparation, which are, accordingly, highly sensitive to the rules used for amortization.

VIII. DESULFURIZATION—THE STATUS OF DEVELOPMENT

As described earlier, the forms that sulfur takes in coal may be divided into two broad categories: pyritic and organic. The former, whether distributed in the minerals or within the macerals, usually occurs as discrete inclusions of iron pyrite and more rarely as marcasite. The organic sulfur is always combined in the macerals present. Physical methods of separating coal and shale are unable to reduce organic sulfur but may reduce pyritic sulfur, according to the relative distributions of the pyrite in coal and shale. Deliberate size reduction of the coal and middlings fractions during preparation, in order to liberate additional pyrite, will be successful only in those cases in which the pyrite occurs as inclusions of size greater than about 10 μm. In these cases, physical methods of coal cleaning are effective in separating significant quantities of the pyrite present, but for steam coals the economic justification for this type of desulfurization requires that the degree of desulfurization achieved be compared with the additional costs incurred in cleaning coals crushed to finer size consists and the resulting savings in flue gas desulfurization.

Organic sulfur can only be reduced by chemical treatments that selectively destroy the sulfur compounds present in the coal matrix and render the sulfur soluble for extraction by leaching.

A. Physical Methods

The capability of the common methods of density separation, including interfacial methods, for achieving reduction in pyrite have long been known and exploited as a process incidental to the reduction of ash content. However, because of the much greater costs of cleaning and drying ultrafine coals, and the higher preparation losses inevitably incurred with these methods, deliberate crushing or grinding to enhance sulfur removal has not been practiced. Many studies of the effectiveness of a particular cleaning method have been published and a large body of literature exists. In addition to the common wet and dry cleaning methods described previously, several other physical techniques have been investigated, including magnetic and electrostatic separation. During the past 5 years a systematic review and assessment of these physical methods has been carried out in West Germany and described by Leining and Monostory [98] and von Gathen [18]. The studies have included the increased cleaning costs associated with size reduction, which have then been compared with the costs of flue gas desulfurization. A summary of the principal test results is given in Table 15. It was found that preparation costs, including coal losses resulting from cleaning the finer sizes, related to the reduction of sulfur achieved in a given case and reached equivalence with the costs of flue gas desulfurization for raw coal crushed to 1/8 × 0 in. It was concluded that this represented the economic limit for those cleaning methods shown in Table 11. Taking into account current developments and future cost estimates, it is projected that the future application of physical methods of desulfurization will be limited to (1) general reduction of raw coal size to 10 × 0 mm (approximately 3/8 × 0 in.), and (2) crushing middlings fractions to 3 × 0 mm (1/8 × 0 in.) before rewashing.

The results of detailed American studies of coal cleaning as an option in increased coal utilization and control strategies for sulfur oxides emissions have recently been published [99]. Kilgroe et al. reviewed the status of physical and chemical methods in North America. Table 16 summarizes the status found for physical methods.

TABLE 15 Physical Desulfurization—Summary of Most Important Test Results (98)

Machine	Raw feed size (mm)	Clean coal		Refuse		Pyritic sulfur removal (%)
		Yield (%)	Ash (%)	Yield (%)	Ash (%)	
Wet methods						
Fine coal jig	10–0.5	90.8	7.7	9.2	78.0	53.0
		89.6	14.1	10.4	79.8	32.7
Concentrating tables	6–0	85.2	10.5	14.8	74.9	53.7
		81.3	11.7	18.7	70.0	57.7
		95.2	8.4	4.8	72.0	38.9
Froth flotation	<0.75	80.0	5.1	20.0	60.5	39.0
		67.7	13.1	32.3	70.7	65.5
Hydrocyclones	6–0.6	84.7	11.5	15.3	43.1	42.8
Launder	<3	82.2	14.9	17.8	45.9	60.2
High-intensity magnetic separator	<1	80.0	17.5	20.0	44.0	43.5
		80.0	18.0	20.0	67.0	33.0
Dry methods						
Deduster		62.9	16.8	37.1	38.3	38.4
Electrostatic precipitator		80.0	12.5	20.0	41.0	71.0
		90.0	16.5	10.0	47.0	49.9
High-intensity magnetic drum		80.0	14.5	20.0	74.0	69.0
Dry-flow separator		47.3	29.3	52.7	54.2	81.7

TABLE 16 Status of Physical Cleaning Processes in the United States (99)

Method of separation	Size of typical feed	Status of use
Wet methods		
Washing		
Jigs, launders	4 to 1/4 in.	Commercial
Concentrating table	3/8 in. to 35 mesh	Commercial
	30 mesh	Experimental
Hydrocyclone	-1/4 in.	Commercial
Concentrating spiral	-30 mesh	Commercial and experimental
Dense medium		
Laminar flow	4 to 1/4 in.	Commercial
Cyclone	3/8 in. to 32 mesh	Commercial
Froth flotation		
Conventional	-28 mesh	Commercial
Two-stage vacuum	-28 mesh	Pilot plant
	-14 mesh	Experimental
Oil agglomeration		
Trent	-28 mesh	Experimental
Convertol	-325 mesh	Pilot plant
Spherical agglomeration	-200 mesh	Commercial in Canada
Selective flocculation	40 μm or less	Experimental
Dry methods		
Air concentration		
Jigs, tables, launders	3/8 in. to 48 mesh	Use diminishing
Classifiers	80% -200 mesh	Commercial and experimental
Centrifugal	70% -200 mesh	Experimental

They concluded that physical coal cleaning could be used (1977) to meet a variety of state and federal SO_2 emission regulations, singly or in combination with flue gas desulfurization. The technology was commercially available and provided the lowest-cost control method for some coals. However, it was not applicable to all high-sulfur coals. Uncertainties relating to future costs, and regulatory and institutional factors, were inhibiting its wider application.

In the same report, Hoffmann and Deurbrouck considered engineering/economic analyses of coal preparation compared with SO_2 cleanup processes for 12 different coals to be used at 12 different locations. Four cases were considered for each coal company: (1) physical cleaning followed by stack gas scrubbing with (2) sulfur cleanup exclusively by stack gas scrubbing, these cases being applied to new generating plant at the chosen locations compared with use in existing plant. Physical desulfurization was estimated on a coal top size of 3/8 in. (10 mm) in each case and hence corresponds with the condition specified in the German work for this parameter, but the effect of deliberate crushing of middlings fractions below 1/8 in. (3 mm) and re-washing was not considered. The data indicated that many coals can be beneficiated to remove ash and sulfur at an attractive net cost. It was also found that where coal can be physically cleaned to a level not too far removed from that required to meet emission standards, flue gas desulfurization treating only a portion of the flue gas would provide for environmental satisfaction, so that for many potential situations, the economic advantage of combined physical cleaning and reduced desulfurization was quite significant.

During 1978, the New Source Prevention Standards requiring 85% removal of sulfur for all sources was introduced, which has drastically altered the ground rules on which these studies were conducted. One response to this change has been an assertion that physical desulfurization should be subordinated in favor of obtaining maximum Btu recoveries from coal cleaning [1].

B. Chemical Methods

The chemical removal of sulfur from coal requires treatment of solid coal particles with reagents under physical conditions of temperature and pressure that will allow selective removal of the sulfur without thermal degradation of the coal substance itself. Since bituminous coals start to decompose thermally at temperatures of about 325 to 350°C, the upper limit of permissible reaction temperature is about 300°C.

Although reactions are possible in the suspended gaseous phase, the need for handling solid coal particles throughout usually makes reactions in the suspended liquid phase easier to execute. In these conditions the reactions are essentially at constant volume, so that, by Le Châtelier's principle, elevated pressures are unnecessary. One exception may be where elevated pressures are necessary in order to maintain a normally gaseous reactant in solution in the desired concentration. This is frequently encountered in oxidation reactions in which air or oxygen is one of the reactants. It may be noted that if the sulfur reaction products are gaseous, elevated pressures are liable to impede reaction rates.

Finally, the need for commencing and ending with solid coal particles requires that reactants diffuse into and resultants diffuse out of the particles, so that because these rates are generally slow, reaction times are essentially diffusion-controlled. Reaction time can be reduced by fine grinding of the particles and by stirring the reactor.

It has been demonstrated that sulfur occurs in crystalline ionic form as pyrite, and in various covalent combinations as "organic sulfur."

Pyrite undergoes many chemical reactions, which may be classified into four general classes:

1. Substitution:

$$M + FeS_2 \rightarrow MFe + [S_2]^{2-}$$

2. Acid-base neutralization:

$$4H^+ + [S_2]^{2-} + 2e \rightarrow 2H_2S$$

3. Oxidation:

$$[S_2]^{2-} \xrightarrow{[O]} 2S + O^{2-}$$

TABLE 17 A Summary of the Chemical Desulfurization Processes (1)

Process	TRW [100]	Magnex [101]	Battelle [102]	Ledgemont [103]	Ames [104]	PERC [105]	JPL [106]	KVB [107]	GE [108]	ARCO [109]
Method	Acid leach	Dry chemical pretreatment plus magnetic separation	Caustic leach	Oxygen/ammonia leach	Oxygen/sodium cabonate leach	Air/water leach	Chlorine/water leach	Dry oxidation	Microwave caustic leach	Two-stage chemical oxidation
Reagent(s)	$Fe_2(SO_4)3$, H_2O, O_2, acetone	$Fe(CO)_5$	NaOH, $Ca(OH)_2$, H_2O, CO_2	O_2, H_2O, NH_3, lime	O_2, H_2O, Na_2CO_3, lime	Air, H_2O, lime	Cl_2, H_2O, CH_3CCl_3	O_2, N_2, NO, H_2O, NaOH	H_2O, NaOH	N.A.[a]
Pressure (psia)	15–80	Atmospheric	350–2,500	0–1000	0–1000	500–1500	Atmospheric	15–300	Atmospheric	N.A.
Temperature (°C)	90–135	150–200	220–350	10–230	10–230	150–200	65–95	40–260	N.A.	N.A.
Retention (hr)	1/2–10	1/10–1	1/4–1/2	1/4–24	0–1	0–1	2–4	1	N.A.	N.A.
Ash reduction	0	Up to 40	0	0	0	0	0	0	N.A.	N.A.
Pyritic sulfur reduction (%)	90	90	90	90	90	90	90	90	N.A.	95

Organic sulfur reduction (%)	0	0	Up to 40	Up to 25	Up to 25	Up to 40	Up to 50	Up to 40	N.A.	40
Sulfur product	Iron sulfates, elemental sulfur	Dry sulfurous mineral matter	H_2S, elemental sulfur	Gypsum	Gypsum	Gypsum	H_2SO_4	Gypsum	N.A.	N.A.
Status	Pilot plant, 8 tons/day	Pilot unit, 200 lb/day	Pilot unit, 0.25 ton/year	Inactive	Bench scale	Bench scale design of continuous unit	Bench scale	Inactive	Bench scale	Laboratory

[a]N.A., not available.

Source: From Ref. 1, with permission.

4. Reduction:

$$[S_2]^{2-} \xrightarrow{[H]} H_2S$$

Examples of type 1 have not been reported. Type 2, more familiar as acid leaching, is used for laboratory demineralization of coal but has found no commercial application. Type (3) represents most of the pyritic coal desulfurization processes under development. Type 4 is the hydrodesulfurization reaction by which pyrite is decomposed in coal gasification and liquefaction, discussed elsewhere in the book.

The main types of combination of organic sulfur have been shown to be

1. Heterocyclic: =CH-S-CH=
2. Thioethers: R-S-R'
3. Mercaptans: R-S-H
4. Thiophenols: C-S-H

of which type 1 usually predominates. These are all stable sulfur bonds that require vigorous chemical treatment to produce cleavage and release the sulfur. There is thus only a narrow margin between conditions vigorous enough to release the sulfur but not drastic enough to produce general decomposition of the coal. A few treatments have been devised which are capable of removing up to 40 to 50% of the sulfur present without causing excessive loss of coal.

The chemistry of desulfurization has been comprehensively reviewed by Meyers [6]. A review of the status of the processes under development, all of which are American, has recently been provided by Deurbrouck [1] and Kilgroe et al. [99]. A summary is presented in Table 17. The most highly developed are the TRW [100] and Magnex [101] processes, the latter being, strictly speaking, a magnetic separation in which the paramagnetism of the pyrite is increased by chemical treatment with iron pentacarbonyl, which decomposes leaving an iron film on the surface of the pyrite. It has been successfully demonstrated in a 200-lb/day pilot unit.

The TRW, or Meyers, Process employs aqueous ferric sulfate solutions as leachant, which may be prepared at specific gravities high enough to be used as dense media for physical coal preparation. This forms the basis of the TRW Gravichem Process [45]. The Meyers Process removes only pyrite, and in this respect its utility is limited. The primary leaching stage, which oxidizes pyritic sulfur to free elemental sulfur, has been successfully tested in an 8-ton/day pilot demonstration unit. The second process stage, extraction of free elemental sulfur, has not yet been demonstrated.

The Battelle Hydrothermal Process [102] employs a hot leach solution containing caustic soda and lime which removes pyritic and a substantial proportion, up to 40%, of organic sulfur.

The Ledgemont [103], Ames [104], and Pittsburgh Energy Research Center (PERC) [105] processes are all variations of oxydesulfurization, employing oxygen gas or air as the oxidants in conjunction with alkaline solutions as leachants, and hence operate at high pressures. A continuous bench-scale operation comparison of the three processes was performed during 1980 at the DOE Multi-Use Energy Processes Test Plant at TRW's San Juan Capistrano Test Site in California (DOE contract number DE-AC01-79ET14292). The results are not available as yet.

Work on the KVB Process [107] is presently suspended. The GE Microwave Caustic Leach Process [108] is at bench-scale stage, and the ARCO Process is believed to be still at the laboratory experimental stage.

Bacteria are capable of breaking down pyritic sulfur, their presence being associated with the natural production of sulfurous waters draining from mines and waste piles. A recent report describes the laboratory culture of the active organism *Thiobacillus ferooxidans* and its capability for microbiological desulfurization of coal [111]. The results showed that about two-thirds of the pyritic sulfur in a high-sulfur coal could be removed after 14 days, but the process was ineffective for removing organic sulfur.

In general, all the processes are at too preliminary a stage for reliable estimates of the coal cleaning costs to be made. However, in 1977, a report on technical and cost comparisons for the Magnex, KVB, Battelle, TRW, Ledgemont, and PERC Processes was prepared by the Bechtel Corporation for the U.S. Bureau of Mines [8]. It was based on an assumption that all the processes would remove 90% of the pyritic sulfur present and that the Battelle KVB, and PERC Processes would additionally remove 25% of the organic sulfur. The results indicated that the Magnex Process could be the cheapest, but would not remove organic sulfur. It concluded that the oxydesulfurization processes were the most promising.

Very recently, TRW [112] has demonstrated chemical removal of 80-90% of both the organic and pyritic sulfur from coal via the Gravimelt Process. This method involves the two-step treatment of coal with molten sodium hydroxide and then with dilute sulfuric acid. The first step removes the coal sulfur while the second step removes 95-96% of the mineral matter, providing a coal product with less than 0.8 lb of $SO_2/10^6$ Btu and less than 0.4 lb of ash/10^6 Btu.

ACKNOWLEDGMENTS

I wish to thank my colleagues in North American Mining Consultants, Inc., and the Stone & Webster Group of companies for assistance in the preparation of this chapter. Mr. Karl Heinz Hawner and Dr. Manfred Hampel of Montan-Consulting GmbH, Essen, and Mr. M. D. Barnes of the National Coal Board, U.K., kindly provided information on the coal preparation industry in Europe; Ms. Mary VanBuskirk at the Canada Institute for Science and Technical Information provided invaluable assistance in a worldwide review of literature references since 1970. I am grateful to Mrs. Herma Joel and Mr. A. Berlingeri for their skillful work in preparing the typescript and illustrations.

REFERENCES

1. A. W. Deurbrouck, in Coal Preparation Research in the United States, Paper D1, 4th International Conference on Coal Research, Vancouver, Canada, Oct. 1978.
2. C. T. Ford, Coal Cleaning to Remove Trace Elements Prior to Utilization, 4th Symposium on Coal Utilization, NCA/BCR Conference, Oct. 1978.

3. C. T. Ford and J. F. Boyer, Effect of Coal Cleaning on Fugitive Elements: A Progress Report, Bituminous Coal Research, Inc., Sept. 1978.
4. T. Alexander, New Fears Surround the Shift to Coal, *Fortune Magazine*, Nov. 20, 1978, pp. 50-60.
5. H. Lustmann and R. Lehnert, *Brennst.-Chem.* 18:353-356.
6. R. A. Meyers, *Coal Desulfurization.* Marcel Dekker, Inc., New York, 1977.
7. R. R. Oder, L. Kulapaditharom, A. K. Lee, and E. L. Ekholm, *Min. Congr. J.* 63:42-40 (Aug. 1977).
8. Bechtel Corporation, An Analysis of Chemical Coal Cleaning Processes, U.S. Bureau of Mines Contract No. JO166191, June 1977.
9. G. E. Edwards, *World Coal* 4:45-48 (July 1978).
10. J. Visman, The Cleaning of Highly Friable Coals by Water Cyclones, Transactions of the 4th International Coal Preparation Congress, 1962, pp. 155-172.
11. Montan-Consulting GmbH, private communication.
12. F. Brinkmann, *World Coal* 3(5):33-35 (1977).
13. E. Hoffmann and B. Michaelowski, *Glueckauf* 101:637 (1965).
14. D. Bartelt and R. von Gathen, Improvement of Profitability of West German Collieries by Changes in Coal Preparation, Paper 14E, 6th International Coal Preparation Congress, Paris, 1973.
15. D. Bartelt and W. Rosner, Operational Use of Automation and Process Control in the West German Mining Industry, Paper D.1, 7th International Coal Preparation Congress, Sydney, 1976.
16. K. Whitworth, *World Coal* 2(5):56 (1976).
17. M. Becker, A. Simmering, and H. Ritter, *Aufbereit.-Tech.*, No. 6, p. 265 (1978).
18. R. von Gathen, Possibilities and Limits of Coal Desulfurization, Paper D3, 4th International Conference on Coal Research. Vancouver, Canada, Oct. 1978.
19. B. Stranz, Poland: Mining Industry and Protection of the Environment, Paper II-4, Proceedings of the 9th World Mining Congress, Düsseldorf, West Germany, May 1976.
20. M. Swiss, *World Coal* 4:36-37 (Nov. 1978).
21. D. W. Horsfall, Coal Preparation Development in the Republic of South Africa, Paper D5, 4th International Conference on Coal Research, Vancouver, Canada, Oct. 1978.
22. A. W. Deurbrouck, U.S. Bureau of Mines, Report of Investigations 7982, 1974.
23. D. E. Jenkinson, Coal Preparation R&D in the United Kingdom, Paper D4, 4th Internat.onal Conference on Coal Research, Vancouver, Canada, Oct. 1978.
24. National Coal Board (U.K.), Coal Preparation Plants, 1978.
25. J. Abbott, C. Dell, B. Denison, D. Knott, and N. W. Hill, Coal Preparation Plant Effluent Disposal by Means of Deep Cone Thickeners, 6th International Coal Preparation Congress, Paris, Mar. 1973.
26. *Coal Daily*, Sept. 18, 1978.
27. A. W. Deurbrouck and G. R. Smithson, *World Coal* 3(7):44-47 (1977).

28. R. L. Llewellyn, Pneumatic Cleaning of Fine Coal for Sulfur Reduction, 3rd Symposium on Coal Preparation, NCA/BCR Conference, Oct. 1977.
29. National Coal Association, *Coal Data*. Washington, D.C., 1976.
30. McGraw-Hill, Inc., *Keystone Coal Industry Manual*. New York, 1978.
31. H. L. Lovell, paper presented to panel on SO_2 Control Technologies, U.S. Commerce Technical Advisory Board, Washington, D.C., 1975.
32. P. J. Phillips and P. P. DeRienzo, Steam Coal Preparation Economics, 2nd Symposium on Coal Preparation, NCA/BCR Coal Conference, Kentucky, Oct. 1976.
33. R. E. Zimmermann, *World Coal* 4, pp. 000-000 (May-July 1978).
34. *Coal Min. Process. J.*, p. 76 (Sept. 1975).
35. R. D. Grundy, Impact of Environmental Policies on Outlook for Coal Utilization in the United States, Paper E2, 4th International Conference on Coal Research, Vancouver, Canada, Oct. 1978.
36. G. G. Sarkar, *World Coal* 3(5):48 (1977).
37. B. Bogenschneider, *Publ. Tech. Charbon. Fr.* 3(125) (1977).
38. H. B. Charmbury and H. L. Lovell, *Mechanization* 26 (Apr.-May 1962). 1962).
39. H. H. Lowry, ed., *Chemistry of Coal Utilization*. John Wiley & Sons, Inc., New York, 1963, p. 326.
40. M. R. Geer and H. F. Yancey, in *Coal Preparation*, J. W. Leonard and D. R. Mitchell, eds. AIME, New York, 1968, pp. 18-22, 18-23.
41. F. W. Mayer, *Glueckauf* 79:15 (1943), BCURA Translation No. 41 (U.K.).
42. F. W. Mayer, *Glueckauf* 86:498 (1950); 89:584 (1953).
43. K. Aso, The Modification of the M-Curve, Paper C8, 2nd International Conference on Coal Preparation, Essen, West Germany, 1954.
44. C. C. Dell, National Coal Board, CRE Report No. 1255, United Kingdom, Oct. 1955.
45. TRW Inc., TRW Gravichem Process, Redondo Beach, Calif.
46. F. J. Fontein and C. Dijksman, The Hydrocyclone, Its Applications and Its Explanation, Conference on Recent Developments in Mineral Dressing, London, 1953, pp. 229-257.
47. R. E. Hucko, Beneficiation of Coal by Selective Flocculation, a Laboratory Study, U.S. Bureau of Mines, Report of Investigations 8234, 1977.
48. S. C. Sun, in *Coal Preparation*, J. W. Leonard and D. R. Mitchell, eds. AIME, New York, 1968, pp. 10-66, 10-73.
49. T. Fraser and H. F. Yancey, *Trans. AIME* 69:447-469 (1923).
50. W. W. Anderson, *Trans. AIME* 187:256-264 (1950).
51. A. Grounds and L. W. Needham, The Performance and Efficiency of British Coal Preparation Plants, Paper G4, First International Conference on Coal Preparation, Paris, 1950.
52. G. G. Sarkar and H. P. Das, *World Coal* 4(6):23-25 (1978).
53. A. Terra, Theory of Coal Washing, *Rev. Ind. Mineral.* 425:383-403 (1938).
54. M. G. Driessen, *J. Inst. Fuel (Lond.)* 19:33-45 (1945).
55. P. Behrgou and M. Ulmo, Representation of Washing Results, Paper A3, 1st International Conference on Coal Preparation, Paris, 1950.
56. A. F. Baker, Hot Surfactant as a Dewatering Aid During Filtration, 2nd Symposium on Coal Preparation, NCA/BCR Conf. Kentucky, 1976.

57. P. T. Luckie and E. A. Draeger, *Coal Age* 81(1):106-109 (1976).
58. Verlag Die Braunkohle, *Brown Coal Utilization.* Düsseldorf, West Germany, 1976.
59. American Society for Testing and Materials, ASTM D3177-75; D2492-68, Philadelphia, 1974.
60. R. E. Zimmermann, *World Coal* 3(10):29-31 (1977).
61. J. W. Leonard and D. R. Mitchell, eds., *Coal Preparation,* 3rd ed. AIME, New York, 1968, p. 11-8.
62. H. L. Lovell, in *Coal Preparation,* 3rd ed., J. W. Leonard and D. R. Mitchell, eds. AIME, New York, 1968, p. 9-38.
63. M. Becker, A. Simmering, and H. Ritter, *Aufbereit.-Tech.*, No. 6, pp. 265-270 (1978).
64. C. E. Tackett, Greenwich Collieries. A Variation in the Art of Coal Washing by Jigs, AIME, Preprint No. 77-F-53, Mar. 1977.
65. E. R. Palowitch and A. W. Deurbrouck, in *Coal Preparation,* 3rd ed., J. W. Leonard and D. R. Mitchell, eds. AIME, New York, 1968, p. 9-10.
66. A. W. Deurbrouck and E. R. Palowich, in *Coal Preparation,* 3rd ed., J. W. Leonard and D. R. Mitchell, eds. AIME, New York, 1968, p. 10-2.
67. M. G. Driessen, *J. Inst. Fuel (Lond.)* 19:33-45 (1945).
68. F. J. Fontein and C. Dijksman, The Hydrocyclone, Its Applications and Its Explanation, Conference on Recent Developments in Mineral Dressing, London, 1953, pp. 229-257.
69. M. Sokaski, M. R. Geer, and H. F. Yancey, in *Coal Preparation,* 3rd ed., J. W. Leonard and D. R. Mitchell, eds. AIME, New York, 1968, p. 10-26.
70. L. Weyher and H. L. Lovell, The Response of Parameter Variation in the Hydrocyclone Processing, *Trans. Soc. Min. Eng. AIME,* pp. 333-340 (Dec. 1966).
71. J. L. Picard, M. W. Mikhail, and O. E. Humeniuk, Improved Compound Water Cyclone for Cleaning Coal Slimes, Canada Centre for Mineral and Energy Technology, Report No. ERP/ERL 75-14R (IR).
72. R. H. Perry and C. H. Chilton, eds., *Chemical Engineers Handbook,* 5th ed. McGraw-Hill Book Company, New York, 1973.
73. J. M. Mueller and J. Brockhoff, *Aufbereit.-Tech.* 19(3):129-134 (1978).
74. K. J. Miller, Flotation of Pyrite from Coal, U.S. Patent No. 3,807,557, Apr. 30, 1974.
75. K. J. Miller, U.S. Bureau of Mines, Technical Progress Report 91, May 1975.
76. A. F. Baker, Hot Surfactant as a Dewatering Aid During Filtration, 2nd Symposium on Coal Preparation, NCA/BCR Conference, Kentucky, 1976, p. 175.
77. J. S. Orphanos, Pollution Reduction and Product Recovery by Centrifugal Dewatering, *Trans. ASME* 258:301 (Dec. 1975).
78. B. C. Radford, *Aufbereit.-Tech.* 19(5):265-270 (1978).
79. F. W. Keith, R. T. Moll, and C. O'Donnell, Dewatering Tailings and Coal Fines by Centrifugation, 4th Symposium on Coal Utilization, NCA/BCR Conference, Kentucky, Oct. 1977.
80. P. T. Luckie and E. A. Draeger, *Coal Age* 81(1):106-109 (1976).
81. Monash Process Dries Out Brown Coal, *Electr. Rev.* 201:47 (July 1977).

82. Monash Process Dries Out Brown Coal, *Electr. Rev.* 201:34 (July 1977).
83. V. E. Gleason, A Bibliography on Disposal of Refuse from Coal Mines and Coal Cleaning Plant, Bituminous Coal Research Inc., May 1978.
84. R. E. Brown, T. C. Wilson, and D. L. Thomasson, Economic Evaluation of Coal Refuse Disposal Systems, 3rd Symposium on Coal Preparation, NCB/BCR Conference, Kentucky, Oct. 1977, p. 141.
85. D. C. Hoffman, Calcilox Additive Stabilization of Coal Fines Refuse—A Cost Approach, 3rd Symposium on Coal Preparation, NCA/BCR Conference, Kentucky, Oct. 1977, p. 163.
86. E. D. Hummer, in *Coal Preparation*, J. W. Leonard and D. R. Mitchell, eds. AIME, New York, 1968, p. 16-7.
87. H. A. Hamza and J. L. Picard, Index of Commercial Flocculants—1974, CANMET, Department of Energy, Mines and Resources, Canada.
88. R. L. Cook, Compact Lamella Thickeners in Coal Preparation Plants, 2nd Symposium on Coal Preparation, NCA/BCR Conference, Kentucky, Oct. 1976, p. 117.
89. N. P. Chironis, *Coal Age* 82:126-129 (Oct. 1977).
90. D. T. King, in *Coal Preparation*, J. W. Leonard and D. R. Mitchell, eds. AIME, New York, 1968, p. 14-1.
91. CFR, Title 30, Chapter 1, Subchapter 0, 71-0, July 1977.
92. N. P. Chironis, *Coal Age* 81:97-155 (Jan. 1976).
93. H. K. Sacks and M. N. Rubin, *Noise Control in Coal* Preparation Plants, 3rd Symposium on Coal Preparation, NCA/BCR Conference, Kentucky, Oct. 1977.
94. E. E. Ungar and H. K. Sacks, *Coal Min. Process.* 14:64 (Aug. 1977).
95. H. Hennig, *Aufbereit.-Tech.* 18:259-273 (June 1977).
96. A. D. Walter and R. V. Ramani, A Review of Computer Oriented Models of Coal Preparation Plants, 2nd Symposium on Coal Preparation, NCA/BCR Coal Conference, Kentucky, Oct. 1976.
97. E. Lippek and D. Uhlig, *Neue Bergbautech.* 7:289-294 (Apr. 1977).
98. D. Leininger and F. P. Monostory, *Glueckauf* 111(22):1079-1082 (1975).
99. Coal Cleaning—An Option for Increased Coal Utilization, EPA-600 17-77-130, Aug. 1978.
100. R. A. Meyers, M. J. Santy, W. D. Hart, L. C. McClanathan, and R. A. Orsini, Reactor Test Project for Chemical Removal of Pyritic Sulfur from Coal, Final Report, EPA-600/7-79-013a, b (1979).
101. A. W. Deurbrouck, in Coal Preparation Research in the United States, Paper D1, 4th International Conference on Coal Research, Vancouver, Canada, Oct. 1979, p. D1.13.
102. E. P. Stambaugh, A. Levey, R. D. Glammar, and K. C. Sekhar, Hydrothermal Coal Desulfurization with Combustion Results, Proceedings of 4th National Conference on Energy and the Environment, Oct. 1976, AICHE, New York, pp. 386-394.
103. S. S. Sareen, Sulfur Removal from Coals: Ammonia/Oxygen System, Proceedings of the Symposium on Desulfurization of Coal and Coal Char, ACS, Washington, D.C., Mar. 1977.
104. T. D. Wheelock, C. Y. Tai, and G. V. Graves, Desulfurization of Coal by Oxidation in Alkaline Solutions, 3rd Symposium on Coal Preparation, NCA/BCR Conference, Kentucky, Oct. 1977.

105. S. Friedmann and R. P. Warzinski, *J. Eng. Power, ASME* 99:361-364 (July 1977).
106. J. Kalvinskas, personal communication, Jet Propulsion Laboratory, California Institute of Technology, Pasadena, Calif.
107. U.S. Patent No. 3,909,211, Sept. 30, 1975.
108. Environmental Protection Agency, *Coal Clean. Rev.* 1(1):000 (1977).
109. Environmental Protection Agency, Coal Cleaning—An Option for Increased Coal Utilization, EPA-600/7-77-130, Aug. 1978, p. 81.
110. M. Steinberg, *Fuel* 56(2):227-228 (1977).
111. P. R. Dugan and W. A. Apel, Microbiological Removal of Sulfur from Pulverized Coal Blend, 3rd Symposium on Coal Preparation, NCA/BCR Conference, Kentucky, 1977.
112. R. A. Meyers, W. D. Hart, and L. C. McClanathan, *Chemical Engineering Progress Technical Manual, Coal Desulfurization*, Chemical Engineering Progress, New York, in press.

7 Coal Utilization

RONALD F. AYERS* / Battelle, Columbus Laboratories, Columbus, Ohio

---

*Present affiliation: Allis-Chalmers Corporation, Milwaukee, Wisconsin

I. HISTORICAL USE OF COAL

A. Period Before 1850

Compared with its European counterparts, the coal industry in the United States developed late [1]. By the time the Americas were being colonized, the European coal industry was mature. In Europe, the coal industry began in the twelfth and thirteenth centuries, whereas the American industry did not begin until the eighteenth century. Coal production in Britain had reached about 1 million tons/year by 1600; output in America did not reach that level until the 1830s.

Coal was used sporadically in the Americas before its colonization. Native Americans used coal from outcroppings for ornamental purposes. The Hopis mined coal in what is now Arizona as early as A.D. 1000, for home-heating and firing pottery. Other small amounts were extracted from outcroppings but not on a systematic basis.

The abundance of wood was the greatest factor that inhibited early coal mining ventures. Demand for energy to power steamboats, locomotives, and households was supplied by wood—often obtained directly by the user. Energy supply was a "do it yourself" enterprise, well suited to American pioneer practices.

Associated with the abundance of wood was an early lack of transportation facilities and an ignorance of coal use. Since energy supply was not a truly commercial enterprise, the infrastructure for what would become coal depots did not exist. Also, potential users were suspicious of coal. A report, supposedly dated 1840, told of a businessman who transported coal to Mobile from the Warrier River fields. Unable to sell the coal, he was forced to give it away and to provide assistance on how to light and burn it.

The first recorded commercial shipment of coal in the United States dates back to 1758. However, until the 1830s, the production was small and sporadic. Between 1830 and 1850, the coal industry expanded rapidly both regionally and in volume; and by 1850, coal mining had been established on a commercial basis.

B. Emergence of Coal, 1850-1885

The rise of the coal industry was linked with the industrialization of the United States. Although coal (used for steam power and metalworking) was a vital ingredient to the Industrial Revolution in Britain, the use of wood and charcoal prevailed in the United States until the mid-nineteenth century.

Several factors contributed to the increased importance of coal and decreased importance of wood and charcoal. The spread of blast furnaces was important to the American coal industry. By the 1850s, these furnaces were large consumers of charcoal and coal. The post-Civil War expansion of the railroads further increased the demand for steel and also hastened the demise of charcoal because sufficient quantities were not available near the furnaces. Wood supplied the riverboats, which were a major form of transportation during this period. However, the forests near the waterways became depleted, and a similar situation arose along the rail lines. Thus, coal found ready markets, supplying energy to the railroads and riverboats. The use of coal for home-heating also emerged during this period. It was demonstrated that anthracite could burn cleanly without an artificial draft, and the location of the anthracite fields near East Coast urban centers further spurred its use for home-heating.

The earliest data on coal production and utilization date back to 1851. Production was 8.4 million tons, of which 50% was anthracite. Consumption by users was:

14-25% for pig iron, castings, and wrought iron
75-86% for steam power, mechanical work, and homes

Coal production doubled nearly every decade between 1850 and 1890. However, anthracite production, which equaled bituminous output in 1870, began to decline in relative importance. Bituminous coal use increased in the expanding metallurgical operations, while anthracite was limited to space heating and localized industrial needs.

In summary, the second half of the nineteenth century was a period of rapid growth in coal utilization. The major growth factor was its replacement of charcoal in the metals industry, the major growth industry of the time.

C. Supremacy of Coal, 1885 to World War I

In 1885, coal overtook wood as the primary energy source in the United States. Wood was still the dominant fuel for home-heating, but coal had captured most of the important industrial markets. About 110 million tons of coal was consumed in 1885, with the following distribution:

42% for rail fuel
13% for coke production
45% for other uses

Yet, despite the increasing use of coal, per capita coal consumption did not approach that of Britain until the early 1900s [2].

At that time the United States was in a general state of rapid expansion, and it appeared that coal consumption would increase to astronomical levels. Between 1899 and 1909, gross national production increased by 50%, manufacturing output by 58%, and steel output by 125%. Industrial machinery was becoming more powerful, and railroad mileage continued to increase. The electric utility industry also began to expand around the turn of the century.

In 1907, coal accounted for 78% of the total energy supply. This share has never been exceeded. Production continued to increase through 1918, but at a slower rate. In that year, because of the war, coal production reached a peak of 679 million tons—a total that was not surpassed until 1947. The year 1918 was also the last year that coal contributed more than 75% of the American energy supply.

D. Coal's Declining Importance, Post-World War I to the Present

After World War I, several factors combined to reduce the demand for coal. The Great Depression was important in reducing coal demand, although coal consumption in the boom year of 1929 was already lower than in 1918. Coal continued to supply the expanding electricity generation market; however, increased efficiency of coal use moderated the demand. The most important factor in coal's early growth, the railroad, was also critical in its decline. Railroads ceased rapid expansion and switched to diesel power because oil-burning diesels proved more efficient to operate, despite the higher-per-Btu fuel costs. During this period, transportation growth occurred via automobiles and air travel—hardly markets for coal. Cheap oil and gas found ready markets in the expanding transportation industries.

Other minor factors also contributed to coal's decline. One factor was the loss of the maritime transport market, which declined by 95% between 1920 and 1955. Oil and gas replaced coal as the preferred fuels for residential and commercial energy needs. In 1935, coal accounted for 78% of mineral fuels in these markets; by 1979, the share had dropped to about 0.8%. Coal's share of the industrial energy market also declined, although not so precipituously. Until the 1960s, natural gas captured most of the new industrial energy demand. In the 1960s, coal's fortunes suffered further as firms switched to oil and gas because of environmental considerations.

In the market where interfuel substitution is least feasible—cokemaking—coal experienced only lethargic growth. The amount of coal required to produce a ton of steel decreased from 1.5 tons in 1935 to 0.55 ton in 1978. This decline was the result of greater use of scrap metal, use of injected supplementary fuels in the blast furnace, and less waste in the steel production process.

The only true growth market for coal since World War I has been for electricity generation. In a sense, then, coal's share of energy use in some of the market sectors noted above has not declined as much as the data suggest—instead, it has been replaced in part by coal-fired electricity generation. In

TABLE 1 Coal Consumption in the United States and Exports, 1946-1978 (Thousands of Tons)

Year	Electric power utilities	Railroads (Class I)	Coking coal	Steel and rolling mills	Cement mills	Other manufacturing and mining industries[a]	Retail deliveries to other consumers	Total U.S. consumption	Exports
1946	68,743	110,166	83,288	12,151	6,990	120,364	98,684	500,386	41,197
1947	86,009	109,296	104,800	14,195	7,919	127,015	96,657	545,891	68,667
1948	95,620	94,837	107,306	14,193	8,546	112,612	86,794	519,909	45,930
1949	80,610	68,123	91,236	10,529	7,966	98,685	88,389	445,538	27,842
1950	88,262	60,969	103,845	10,877	7,923	97,904	84,422	454,202	25,468
1951	101,898	54,005	113,448	11,260	8,507	105,408	74,378	468,904	56,722
1952	103,309	37,962	97,614	9,632	7,903	95,476	66,861	418,757	47,643
1953	112,283	27,735	112,874	8,764	8,167	96,999	59,976	426,798	33,760
1954	115,235	17,370	85,391	6,983	7,924	78,359	51,798	363,060	31,041
1955	140,550	15,473	107,377	7,353	8,529	91,110	53,020	423,412	51,277
1956	154,983	12,308	105,913	7,189	9,026	94,772	48,667	432,858	68,553
1957	157,398	8,401	108,020	6,938	8,633	88,566	35,712	413,668	76,446
1958	152,928	3,725	76,580	7,268	8,256	82,327	35,619	366,703	50,291
1959	165,788	2,600	79,181	6,674	8,510	74,365	29,138	366,256	37,253
1960	173,882	2,101	81,015	7,378	8,216	77,432	30,405	380,429	36,541
1961	179,629	b	73,881	7,495	7,615	78,050	27,735	374,405	34,970
1962	190,833	b	74,262	7,319	7,719	79,453	28,188	387,774	38,413
1963	209,038	b	77,633	7,401	8,138	83,467	23,548	409,225	47,078
1964	223,032	b	88,757	7,394	8,679	83,639	19,315	431,116	47,969
1965	242,729	b	94,779	7,466	8,873	86,269	19,048	459,164	50,181
1966	264,202	b	95,892	7,117	9,149	89,941	19,965	486,266	49,302
1967	271,784	b	92,272	6,330	8,922	84,009	17,099	480,416	49,528

(continued)

(Table 1 continued)

Year	Electric power utilities	Railroads (Class I)	Coking coal	Steel and rolling mills	Cement mills	Other manufacturing and mining industries[a]	Retail deliveries to other consumers	Total U.S. consumption	Exports
1968	294,739	b	90,765	5,657	9,391	83,054	15,224	498,830	50,637
1969	308,462	b	92,901	5,560	9,131	78,557	14,666	507,275	56,234
1970	318,921	b	96,009	5,410		83,207[c]	12,072	515,619	70,944
1971	326,280	b	82,809	5,560		68,862[c]	11,351	494,862	56,633
1972	348,612	b	87,272	4,850		67,294[c]	8,748	513,776	55,997
1973	386,879	b	93,634	6,356		60,953[c]	8,200	556,022	52,870
1974	390,068	b	89,747	6,155		57,899[c]	8,840	552,709	59,926
1975	403,249	b	83,272	2,715		59,783	7,282	556,301	65,669
1976	445,750	b	84,324	2,743		57,762	6,900	597,479	59,406
1977	475,700	b	77,300	3,200		57,300	7,000	620,500	53,700
1978	481,200	b	71,400	2,900		55,800	7,700	625,200	40,000
1979[d]	528,800	b	77,100	e		65,900	e	680,900	66,300

[a]Includes bunker fuel.

[b]Data included in other manufacturing and mining industries.

[c]Beginning in 1970, the data listed in these two columns were combined.

[d]Preliminary.

[e]Included with "Other manufacturing."

Source: Ref. 21.

relative terms, coal's share of the electricity generation market has declined since the 1920s; however, total electricity generation increased by over 6000% from 1920 to 1975, providing coal with a 1000% increase in that market.

In sum, the 60-year period from 1918 to 1978 was a period of little growth as well as major structural market changes for coal. Oil and gas replaced large quantities of coal in households, commercial establishments, industry, and transportation. However, coal did hold a major share of electricity generation, a rapidly growing energy source [3]. Table 1 reviews coal consumption for recent years.

E. The Future—A New Coal Age?

Current U.S. energy policy suggests a return to coal. The balance of the chapter considers the prospects for such a return by examining in detail coal's present and future markets.

II. COAL FOR ELECTRICITY GENERATION

Electricity generation is the largest market for coal in the United States. By virtue of the Energy Supply and Environmental Coordination Act of 1974, the only present competition for coal in new power plants is nuclear. That act prohibits oil and natural gas burning in new power plants. Therefore, the amount of coal required for electricity generation will increase in the future. That electricity generation has become a "legislated" market makes it no less interesting from the standpoint of coal marketing. Coal must still compete with nuclear power. In addition, the competition among various coals is intense.

Table 2 provides historical and projected distribution of electric power generation by fuel type. Several trends can be noted from this table. First was rapid growth in natural gas use, followed by a rapid decline. Surprisingly, the growth rate for oil consumption continues at rather high levels. Of course, nuclear power, beginning from a zero base, experiences the highest growth rate. The use of coal has exhibited similar growth rates as those for total generation. Finally, the slowing of electricity demand, from 1971 to 1978, is thought by many to be a permanent state of affairs,* making the projected future growth rate open to some question.

The following sections address coal utilization for electricity generation. The first reviews the market from a regional standpoint. Figure 1 shows the U.S. Bureau of the Census regions used in this discussion. The second discusses the outlook for the future demand for electricity. The last section describes coal procurement in the electricity industry.

A. Geographical Fuel Use

1. New England

New England is a classic case of interfuel competition. The region relied on water power for its early generation needs. When these limited resources were fully exploited, the region turned to fossil fuels imported from other

*United States electricity sales between September 1979 and September 1980 increased 0.8% on an annual basis [33].

TABLE 2 Distribution of Electric Power Generation by Energy Source, 1946-1985

	Electric power output (millions of kWhr)					Percent of total					Average annual percentage increase			
	1946	1959	1971	1978	1985	1946	1959	1971	1978	1985	1946-1959	1959-1971	1971-1978	1978-1985
Hydroelectric and other	78,404	137,782	266,320	289,368	255,989[a]	35.1	19.4	16.5	13.2	8.3	4.7	5.7	1.2	(1.8)
Coal	111,654	378,424	714,756	974,054	1,455,544	50.0	53.0	44.3	44.4	47.3	10.2	6.0	4.5	5.9
Oil	14,082	46,840	218,162	355,619	441,658	6.3	6.6	13.5	16.2	14.4	3.2	14.2	7.3	3.2
Gas	18,820	146,619	375,939	295,298	155,014	8.4	20.6	23.3	13.5	5.0	13.5	8.5	(3.4)	(8.8)
Nuclear	—	188	37,899	278,068	767,916	—	0.0	2.3	12.7	25.0	—	44.0	34.0	15.7
Total	223,178	710,006	1,613,936	2,192,407	3,076,121	100.0	100.0	100.0	100.0	100.0	9.4	7.4	4.4	5.0

[a]Hydroelectric total calculated at minimum, low-water conditions.

Source: Edison Electric Institute; projections for 1985 from same source, Apr. 1979.

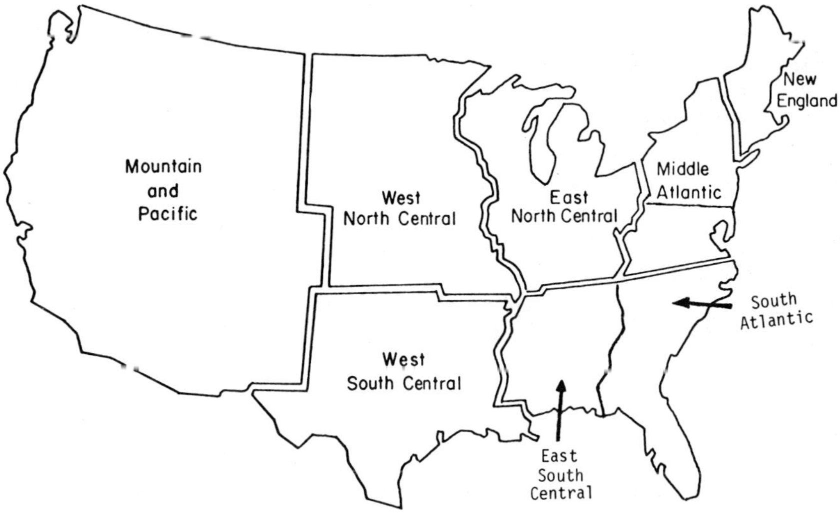

FIG. 1 Geographic regions used in Sec. II.A.

areas. Coal was an early favored fuel but was quickly replaced by oil. As early as 1946, New England used oil for 24.3% of its generation needs [4]. Until the early 1960s, there was intense competition between oil and coal. With the erosion of oil import quotas in the period 1958 to 1966, oil quickly became the preferred fuel. In the year of the oil embargo, oil accounted for 94% of the region's fossil fuels.

2. Middle Atlantic

The Middle Atlantic states are more diverse from a fuel-choice standpoint. The coastal areas of New York City, Philadelphia, and New Jersey experienced the same intense coal/oil competition as New England. In 1973, oil use heavily dominated in these three areas. Stringent environmental regulations in these urban areas continue to favor oil over coal, despite the price differential.

The balance of the Middle Atlantic region, inland Pennsylvania and New York State, are closer to coal sources, and New York has significant water resources. In the absence of the oil embargo-related price rise, oil would probably have captured most of the New York State and eastern Pennsylvania fossil-fuel markets [5].

3. South Atlantic

Fuel choice in the South Atlantic states has always been a location-specific issue. Much of the region is close to coal supplies. However, coastal areas switched heavily to oil in the 1960s and 1970s. Florida, an isolated fuel market, depended almost entirely on oil since the beginning of the electric power industry. The exception is Tampa, where midwestern coal has been barged in via the Mississippi River to Tampa Bay. Additional coal-fired stations will fill future Florida generation requirements. Potential coal sources include Appalachian and imported coal. The infrastructure and experience gained from using imported oil as well as the distance from domestic coal sources makes the coastal region a prime candidate for imported steam coals.

4. *East North Central*

The East North Central region is dominated by coal-producing states. Recently, the region has become the center of competition between Eastern and Western coal. The water transport resources of the region make delivery of western coal relatively cheap. For example, Western coal has been transported into the heart of the Ohio coalfields via the Ohio River system, and western coal will be moving to Detroit via the Great Lakes. Current plans call for 15 to 20% of the 1985 coal burned to be Western coal.

5. *East South Central*

The East South Central region is dominated by a single utility company, the Tennessee Valley Authority (TVA). TVA first used the untapped hydro resources and then began burning coal in large quantities. The only non-coal-burning state is Mississippi, where gas and, more recently, oil are dominant.

6. *West North Central*

The fuel situation in the West North Central region differs by state. In the more westerly states, locally produced gas is used to generate electricity. The Dakotas possess substantial lignite deposits as well as a small population. Other states, notably Minnesota and Missouri, are favorably located for Illinois coal. However, as with the East North Central region, states in this region are in a good position to receive western coal. By 1985, it appears that the region will be burning only Western coal [6,7].

7. *West South Central*

The West South Central region, the location of most of the domestic oil and gas production, has traditionally been heavily gas dependent. The future, however, belongs to coal. The coal competition here is between Western subbituminous coal and locally produced lignite. The Texas Utility Company has operated lignite-fired stations since the 1950s and plans additional units. Arkansas Power and Light has committed itself to local lignite for a mid-1980s station. Based on Battelle studies [8] in the region, it appears that lignite resources are limited and extraction costs are high enough to prohibit region-wide use. Lignite does present a viable choice in local areas and will provide strong competition for Western subbituminous coal.

8. *Mountain and Pacific*

The sparsely populated Mountain region is turning to large local coal resources for future generation needs. At one time, the region used hydro and gas for generation needs. In 1978, coal was the predominant fuel in every state.

The Pacific region can be divided into subregions, composed of California and its northern neighbors, Oregon and Washington. Until recently, the Northern states were heavy exporters of hydro-generated electricity. Washington has one local coal-fired station; Oregon has none. California is in a particularly complex situation. Having outstripped available water resources in the 1950s, the state turned to gas, supplemented by oil. Environmental regulations precluded coal use in the state, so that large mine-mouth complexes were built in Arizona. Future mine-mouth complexes were planned to supplement in-state nuclear plants. However, the defeat of the Kaiparowits Plateau mine-mouth project in Utah meant that California would have to locate future coal-fired plants within the state. As a result, Pacific Gas and Electric of San Francisco has announced plans for the first such in-state plant based on Utah coal [9a].

The Pacific region, together with New England, has been a focus of debate on overall energy futures. Much research and development is being conducted on low-energy-technology options, as opposed to the central energy systems that are presently constructed. As a result, even though this region has planned on large nuclear and Western coal-fired stations, strong opposition places the results of such planning in doubt. Southern California Edison, one of the larger Pacific region utility companies, is currently planning that 30% of its additional 1990 capacity will be from renewable and alternate sources [9b].

B. Future Demand for Electricity

The controversy over energy futures in California points out the fundamental energy question at present: How much energy is enough? In the past, energy was cheap and was substituted for labor and capital in production. As seen in Table 2, electricity demand doubled every 10 years since the 1920s. Today, with much higher energy prices, the future growth of energy and particularly electricity is a controversial subject.

There are several questions concerning the future demand for electricity, including:

Is electricity generation inherently inefficient?
Is the demand for electricity nearing saturation?
Are the environmental impacts of electricity generation too high?

These questions are complex and interrelated but are critical to understanding coal's future role in the electricity generation market.

1. *Is Electricity Generation Inherently Inefficient?*

In modern steam-electric power plants, about 3 Btu of fuel is required to produce 1 Btu of electricity (referred to as the "heat rate"). When distribution losses to the ultimate consumer are added, the overall efficiency declines further. From 1938 to 1978, the heat rate declined by 36% (from 16,500 to 10,492 Btu/kWhr). Although there have been significant improvements in the heat rate at electric power plants, the efficiency of a power plant is low even compared with coal gasification.

The focus of the heat rate/efficiency debate is whether electricity generation is an inherent waste of natural resources. Electricity is certainly indispensable for many applications, such as electronics, lighting, and small appliances. However, it has been estimated that only 4% of the electricity generated is actually consumed for these purposes [10]. The balance is used in applications where other energy systems and fuels could be substituted [10].

A major technological problem with electricity supply is that storage is impractical and expensive. Hence, a utility company must have sufficient active capacity to meet its "peak" demands. During lower demand periods, this equipment sits idle. Much current research and development is being conducted on load leveling and storage to help ameliorate this traditional problem.

Finally, there is the question of capital availability for a greatly expanded electric utility industry. Many companies are now paying huge sums for pollution control equipment that does not add to generating capacity. State regulatory commissions are becoming more suspicious of rate increases. Continued growth will require more and more funds for new plants and equipment. Whether such a high level of investment can be tolerated is open to debate and intimately tied to the question of electricity efficiency.

2. *Is the Demand for Electricity Nearing Saturation?*

As noted previously, the demand for electricity in the 1900s has approximately doubled every decade since 1910-1970. Owing primarily to the above-mentioned increases in efficiency, the *real* (adjusted for inflation) price of electricity actually declined from 1938 until the early 1970s [11]. As a result, electricity found many new uses, including heating, cooling, smelting, and providing mechanical work. The role of electricity in air conditioning has become so pervasive that most utility companies have their peak loads in the summer.

The price elasticity (the quantitative change in demand for a good or service in response to a change in price) of electricity is being closely studied both in the public and private sectors. As shown in Table 2, the annual growth rate for electricity in the 1970s has decreased from that of the 1950s and 1960s. The recession following the 1973 oil embargo artificially depressed demand; nonetheless, the overall growth rate has decreased. The regional growth rates shed even more light on the discussion. Mature areas, such as New England, have slowed considerably, while Texan demand continues to grow, although at a lower-than-historical level. Thus, higher prices as well as a stabilizing population and industrial base could slow electricity demand considerably in the United States.

3. *Are the Environmental Impacts of Electricity Generation Too High?*

The huge modern power plant is a significant source of air and water pollution. The electric utility industry is devoting massive amounts of talent and money to this problem, and there is no doubt that modern plants are much cleaner than were their ancestors. However, the issue is the cumulative environmental impact of a continued doubling of power plants every 10 years or so.

Environmental impacts of coal-fired power production are felt far beyond the site of the plant. Advanced monitoring techniques have traced airborne emissions hundreds of miles. Substantial research and development is being conducted on this problem. Nuclear plants also have environmental effects. The tremendous heat from reactors heats water in surrounding streams, and more recently, the entire question of nuclear waste disposal has been brought under closer scrutiny.

4. *Conclusions*

The preceding discussion was not intended to resolve the debate on whether the demand for electricity will or should continue to grow at high rates. However, it does illuminate some of the contemporary questions concerning energy and electricity futures. Because the future of coal is so closely tied to the fortunes of the electric utility industry, it is important to understand some of the broader issues in that industry.

C. Coal Procurement for Electricity Generation

A complex optimization process is involved in determining fossil-fuel-fired generating plant location. Site selection initially involves a trade-off between two transportation steps: transporting fuel from the production point to the power plant, and transporting power to the consumer. At one time, the cost of transporting power was more expensive than the cost of transporting fuel; therefore, power plants were located close to the load (i.e., usually an urban area). Although generalizations are risky in this type of analysis, it can be broadly stated that today the most economical option is to build power plants close to fuel sources and transmit the power to the load. One reason is that extra-high-voltage transmission has reduced line losses and thus lowered the delivered power costs. Another is that, for many reasons, large power plants are extremely expensive to site in urban areas.

However, not all plants can be built near the source of fuel. Companies with service areas far removed from coal fields do not have this option. Also, there is the problem of obtaining suitable blocks of coal reserves. For example, a modern 1000-MW plant, operating at a 70% plant factor with a heat rate of 9800 Btu/kWhr, would need over 2.6 million tons of 11,500 Btu per pound of coal per year. Assuming a lifetime plant factor of 50% and an average plant yeat rate of 10,000 Btu/kWhr, such a plant would consume over 57.1 million tons of coal in 30 years. If the fuel were from an underground mine with a 50% recovery factor, 114 million tons of reserves would be needed. If a cleaning plant were required to assure consistent coal quality, and cleaning plant losses were 15%, the required reserves would be approximately 134 million tons.

Coal procurement involves large amounts of money and is intimately tied to plant choice (coal versus nuclear) and plant siting. For example, the plant described above would have to spend about $52 million per year on coal (at $20 per ton). Once a coal plant has been selected in preference to a nuclear plant, the utility company must decide how to obtain the coal. This may be done by directly investing in a "captive" mine, securing the coal on a long-term contract, or buying the coal on a "spot" basis.

1. *Long-Term Coal Contracts*

The objective of a long-term contract is to bond buyer and seller together for mutual benefit. In 1978, approximately 79% of the coal delivered to utility companies was obtained under long term contract [12]. This includes captive coal. Research at Battelle [13] concludes that there are no formal industry-wide procedures for identifying potential coal sources other than bidders' lists maintained by the buying utility company. Formal bids are then requested from persons on the list via a competitive negotiation process. These bids are evaluated and screened to less than five potential sources. At that point, the buyer begins to negotiate contract terms with the bidders, assuming that a captive mine or other similar arrangement has been ruled out (see below). These negotiations, often spanning several years, can result in a contract based on a fixed price, cost of production plus profit, or base price plus escalation.

a. Fixed-Price Contracts This type of contract sets a fixed price on a supply of coal over time. These contracts were quite common in the 1950s when a buyers' market prevailed and coal production costs were stable or declining. Such contracts are not relevant to today's coal market but are of historical interest.

b. Cost-of-Production-Plus-Profit Contracts This type of contract reimburses the producer for total costs and pays a profit (or management fee) on a per ton basis. These contracts are generally favored by the coal producer or seller and disliked by the buyer. The advantages for the seller are obvious—increased costs can be passed directly through to the price. However, there can be subtle buyer advantages to this type of contract. Capitalizing on these advantages requires the buyer to become intimately acquainted with the mining operation, to carefully examine large cost items, and to audit frequently.

c. Base-Price-Plus-Escalation Contracts This type of contract begins with some base value per ton, then escalates the base over time to account for rising costs. These escalations are made by applying formulas, indices, and so on, contained in the various clauses of the contract. Table 3 lists various contract clauses according to whether they are favored by the buyer or the seller. It is the most common contract among utility companies and is clearly favored by most utility buyers, the principal reason being that the risks are allegedly borne more equally by the buyer and the seller.

2. *Spot Coal Purchases*

This type of procurement option refers to the purchase of short-term coal supplies on the open market. About 21% of the coal delivered to electric utilities in 1978 was obtained through spot purchases [12]. Utility companies use this method to avoid expenses of unnecessarily large stockpiles, to cover peak coal demands, to take advantage of lower coal prices during certain periods, and to experiment with new coal sources.

The spot coal market is highly competitive and, like most commodity markets, volatile in terms of prices and quantities offered. Some utility companies buy little spot coal, whereas others fill a major portion of their needs from the spot market. Such a decision is usually based on the structure of the coal industry near the power plant.

3. *Captive Mining Operations*

A captive mine may refer to:

Utility ownership and operation of a coal mine and related facilities, directly or through a wholly owned subsidiary;

TABLE 3 Advantages and Disadvantages of Basic Contract Types

Type of contract	Advantages to buyer	Disadvantages to seller	Advantages to seller	Disadvantages to buyer
(1) Cost of production plus profit	(1) Greater buyer risk should mean lower profit margin for seller (2) Buyer has right to audit books and inspect mine; results in some control over expenses	(1) Because of lower risk, accompanying profit could be less (2) Producer must open his books for audit (3) Can encourage laxness in management	(1) Little producer risk (2) Actual costs incurred passed on to buyer (3) Price adjustments made quickly (4) Reimbursed automatically for declining productivity	(1) Increased buyer risk for declining productivity and inflation (2) Few incentives for production costs reduction (3) Cost increases passed through without delay (4) Heavy administrative burden
(2) Base price plus escalation	(1) Less risk to buyer; production costs are separated into components and all costs increases need not be absorved by buyer (2) Productivity losses absorbed by producer (3) Some escalation clauses aid buyer	(1) More risk to seller (2) Producer absorbs productivity losses (3) Escalation clauses may fail to track producer costs accutately	(1) Many escalation clauses protect producer's interest and detail price adjustments (2) Incentives promote good management	(1) Most escalation clauses transfer risk to buyer

Source: Ref. 13.

Utility ownership of reserves and possibly mining equipment with mining done by a mining company; or

Financial involvement, such as a loan or a loan guarantee, with a coal-mining company.

The net effect of captive mining is that the utility has greater control of the mine's reserves and output.

There were about 26 utility companies with captive coal arrangements in 1974 [14]. The amount of coal captively mined increased from 8.9% in 1973 to 12.7% in 1976, and is expected to increase to 18.8% by 1985 [15].

There is some controversy today concerning the continuing growth of captive operations. The utility companies maintain that captive mines provide a reliable source of coal that is sheltered from market-induced price fluctuations and offer bargaining leverage in contract discussions with independent coal producers. On the other hand, several regulatory agencies and consumer groups suggest that captive operations can result in questionable practices. For example, they argue that captive mines are actually a "new business venture" for the utility companies, designed to earn a rate of return for the parent company, that captive mines offer a tax shelter to the parent company that can be outside state regulatory authority, and that captive mines are operated to maximize the profits of parent and subsidiary rather than keep coal costs low.

III. COAL FOR INDUSTRY

As noted earlier, the direct use of coal by industry has been steadily declining since World War I. The most precipitous declines in coal use came after World War II. In 1947, coal supplied industry with about 67% of its energy needs. by 1954, this share dropped to about 40%. Coal's decline in industry had many roots, but the primary cause was the ready availability of oil and gas. Other associated conditions included environmental legislation and the development of industry far from the coal fields.

Following the oil embargo and higher energy prices, the federal government embarked on a policy to encourage increased coal use in industry. Although laws passed in 1974 and 1975 gave authority to the now defunct Federal Energy Administration (FEA) to force industry to switch to coal, this authority was never really used. Instead, FEA concentrated its efforts on large electric power plants [16]. However, under the National Energy Act, the Department of Energy plans to increase its efforts in industrial coal conversion. The National Energy Act will attempt to force such a conversion by using tax incentives and penalties. This will be done on a case-by-case basis and will undoubtedly be a slow process.

Industry is generally reluctant to turn back to coal. The primary reason is the risky financial returns from installing coal-burning equipment and, more important, the necessary environmental controls.

A. Industrial Energy Use

Table 4 compares energy and coal consumption in industry for 1962 and 1976 by type of industry. No industry clearly dominates industrial energy consumption, although the chemical and metals industries account for over 40% of the total.

Despite the large number of industrial energy consumers, their use of energy does not differ significantly. Basically, energy use in industry is limited to process steam (boilers), open-fired applications (e.g., kilns), and process heaters (e.g., pyrolysis units). Dow Chemical Corporation completed a report that documented energy usage in the five largest energy-consuming industries—paper, chemicals, petroleum stone, clay and glass, and metals [17]. The findings show the following distribution within these industries:

Fuel-consuming application	*Total energy consumption (%)*
Process steam	64.6
Open-fired	9.7
Process heater	25.7
Total	100.0

Analysis of energy-consuming patterns in other manufacturing sectors has led Battelle to conclude that U.S. industry generally follows these same energy-consuming patterns, although the boiler percentage is higher [18].

B. Industrial Boiler Markets for Coal

Three major factors determine the industrial boiler coal conversion market: boiler size, fuel design capability, and geographic location.

1. *Boiler Size*

The focus of federal coal conversion legislation has been to determine the range of boiler sizes that could be converted to coal without undue financial hardships. The new National Energy Act appears to define the minimum size as 250 million Btu/hr.

2. *Fuel Design Capability*

In 1974, FEA found that fuel consumption in industrial boilers was 55% natural gas, 23% petroleum, and 22% coal [19]. In reality, as shown in Table 5, 22% of the boilers in the size range 101 to 250 × 1000 lb/hr and 33% of boilers in the size range 251 to 500 × 1000 lb/hr are capable of burning coal. Thus, coal is not being burned in all units designed for coal.

Figure 2 shows why the characteristics of a boiler are crucial to its ability to burn coal. Converting a boiler originally designed for oil or gas requires derating; therefore, it may be cheaper to buy a new boiler designed specifically for coal. According to the American Boiler Manufacturers Association, orders for new boilers designed for coal are finally increasing. However, such orders are of little help to the industrial coal market in the short term.

TABLE 4 Consumption of Energy by SIC Group, 1962 and 1976

SIC group		1962 Total energy use (billions of Btu)	1962 Coal use (billions of Btu)	1962 Coal use (thou- sands of net tons)	1962 Coal use (% of total	1971 Total energy use (billions of Btu)	1971 Coal use (billions of Btu)	1971 Coal use (thou- sands of net tons)	1971 Coal use (% of total
20	Food and kindred products	732,071.5	219,629.3	8,586	30	937,500.0	86,230.6	3,292.1	9
21	Tobacco manufactures	11,671.5	7,008.9	274	60	20,000.0	N.A.[a]	N.A.	N.A.
22	Textile mill products	224,257.6	76,586.3	2,994	34	328,600.0	29,910.9	1,141.9	9
23	Apparel and other fabricated textile products	21,282.0	3,274.1	128	15	56,800.0	1,341.1	51.2	2
24	Lumber and products	56,362.0	3,018.4	118	5	243,800.0	3,020.2	115.3	1
25	Furniture and fixtures	27,725.3	11,485.3	449	41	46,800.0	2,116.5	80.8	4
26	Paper and allied products	881,536.2	375,846.8	14,693	43	1,294,000.0	221,142.8	8,442.5	17
27	Printing and publishing	34,302.7	2,174.3	85	6	85,700.0	N.A.	N.A.	N.A.
28	Chemicals and allied products	1,775,084.0	542,346.6	21,202	31	3,017,100.0	336,626.9	12,851.3	11
29	Petroleum and coal products	1,070,459.0	19,952.4	780	2	1,291,700.0	6,151.0	234.9	<1
30	Rubber products	118,285.3	58,066.5	2,270	49	237,000.0	23,611.9	901.5	10
31	Leather and leather products	20,532.9	7,980.5	312	39	22,400.0	1,207.5	46.1	5

32 Stone, clay, and glass products	896,549.8	258,024.9	10,087	29	1,219,600.0	288,665.7	11,020.5	>4
33 Primary metal products	2,092,138.0	277,337.9	10,842	13	2,380,500.0	135,572.3	5,175.7	6
34 Fabricated metal products	187,623.2	25,861.5	1,011	14	380,600.0	10,309.9	393.6	3
35 Nonelectrical machinery	206,731.7	57,810.9	2,260	28	330,100.0	16,656.8	635.9	5
36 Electrical machinery	152,970.4	36,349.0	1,421	24	233,300.0	13,136.3	501.5	6
37 Transportation equipment	270,935.9	96,001.6	3,753	35	380,300.0	50,339.6	1,921.8	13
38 Instruments and related products	32,616.4	13,837.8	541	42	73,300.0	N.A.	N.A.	N.A.
Total	8,813,130.0	2,092,586.0	81,806	24	12,400,100.0[b]	1,226,044.0	46,806.6	10

Note: These are not inclusive figures but only those for the sectors actually reported in the survey.

[a]N.A., not available or not applicable.

[b]Coal-reporting sectors only.

Source: U.S. Department of Commerce, Survey of Manufacturers, Washington, D.C.

TABLE 5 Description of Industrial Boiler Population by Design Fuel Capability and Coal-Burning Method (Percent Distribution)

	Boiler capacity range (1000 lb steam/hr)	
	101-250	251-500
By fuel		
Oil only	30.0	22.0
Gas only	22.0	22.0
Coal only	18.0	22.0
Oil and gas	26.0	23.0
Oil and coal	0.5	3.0
Gas and coal	0.5	3.0
Other (solid fuels)	3.0	5.0
Total	100.0	100.0
By coal-burning method (for boilers burning coal)		
Spreader	45.0	20.0
Underfeed	15.0	0.0
Overfeed	5.0	0.0
Pulverized	15.0	65.0
Other	20.0	15.0
Total	100.0	100.0

Source: Ref. 20.

3. *Geographic Location*

DOE has collected detailed information on the location of large boilers but refuses to make the data available. In the absence of this valuable information, boiler data have been developed by Battelle for the Environmental Protection Agency in the course of other work [20]. As would be expected, the industrial Midwest (East North Central) has the largest number of industrial coal-capable boilers (34.9%) and New England has the lowest number (1.2%). These are actually boilers designed for solid fuels: coal, lignite, wood chips, and bagasse. The percentages of these boilers located in the other regions are: Middle Atlantic, 16.9%; South Atlantic, 18.1%; West North Central, 7.2%; West South Central, 8.4%; and Mountain and Pacific, 13.3%.

4. *Coal Consumption Potential*

There is a wide range of estimates of the potential coal market in these industrial boilers. Estimates range from 12.5 to 41 million tons. Battelle has estimated the market to be from 16 to 26 million tons, depending on assumptions about environmental regulations.

C. Industrial Process Heater Markets for Coal

The industrial process heater market is typified by the tubed furnaces used in the petroleum industry and the pyrolysis units used in the chemical industry.

Because there is no direct product contact with the combustion gases, characteristics of these units are similar to those of steam boilers.

Industrial process heaters are concentrated in the West South Central region, where 47.7% of the total are located [18]. Of the remaining heaters, 19.5% are in the combined New England, Middle Atlantic, and South Atlantic regions (with only 0.2% in New England). The Mountain and Pacific and East North Central regions both have 13.8%, and the West North Central region has 5.2% [4].

The chemical and petroleum industries are the two major users of process heaters. Usually, a plant in either industry will have both process heaters and industrial steam at a single site. The chemical industry experienced its greatest growth during the past 30 years, during the same period when coal usage was declining. Consequently, chemical experts doubt that there are many in the chemical industry operating process heaters that could be readily converted to coal. Thus, the potential market here is considered slight.

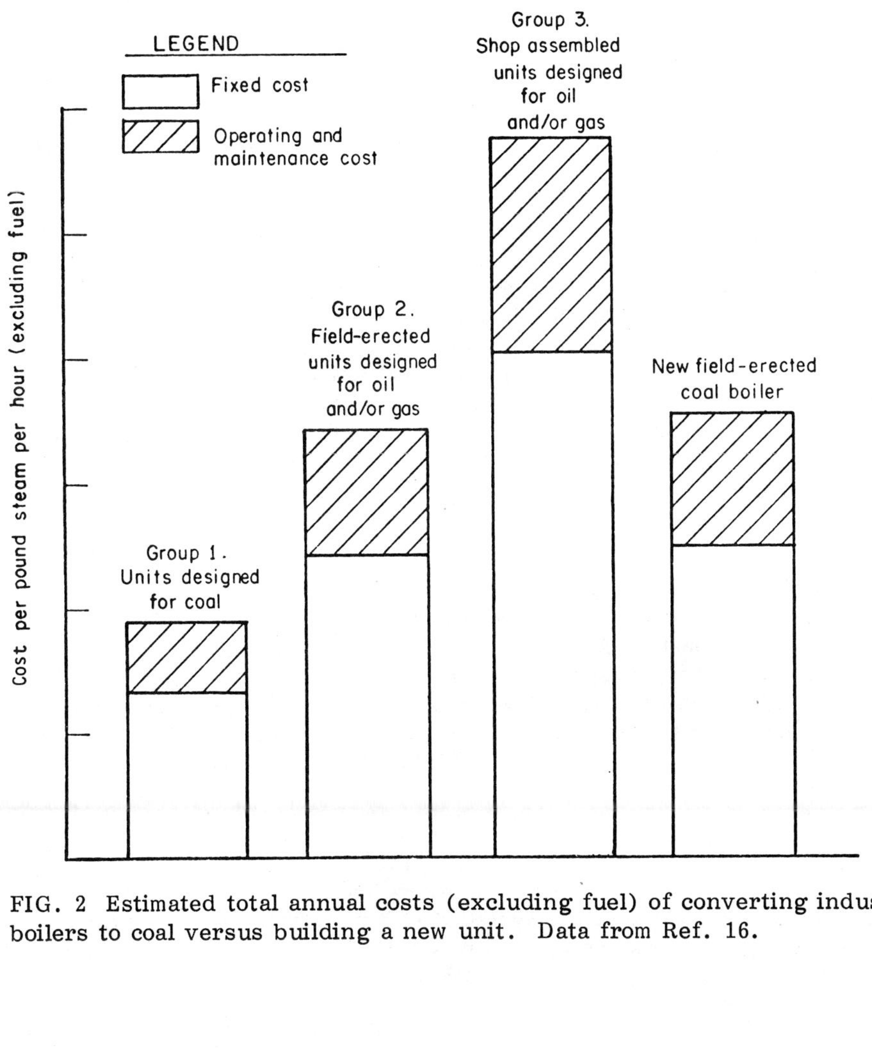

FIG. 2 Estimated total annual costs (excluding fuel) of converting industrial boilers to coal versus building a new unit. Data from Ref. 16.

D. Open-Fired Markets for Coal

Open-fired industrial energy usage encompasses a broad range of processes, including ovens, calciners, driers, and kilns. Unlike boilers and heaters, the product to be heated is usually in direct contact with the flame, requiring a clean fuel. For these units, coal conversion is impractical from a technical standpoint. In others, such as cement and lime calcining, coal is a practical and widely used fuel.

Industrial open-fired coal systems are concentrated in the eastern part of the country, with 67.9% in the four regions east of the Mississippi River [18]. The combined Middle and South Atlantic regions account for 39.2% of the open-fired systems, and the North Central region accounts for 26.6%. For the West North Central, West South Central, and Mountain and Pacific regions, the percentages are 7.0, 12.4, and 12.7%, respectively [21].

The cement and lime industry has been the most active to date in coal conversion. As a result, a "new" coal market in these industries probably totals no more than 5 million tons.

IV. COAL FOR COKEMAKING

The term "coking" refers to a process where a high-rank bituminous coal is heated under a slight negative pressure to drive off the volatile matter (e.g., gases) to leave a substance with a significantly higher carbon content. Steelmakers consume coke in the manufacture of pig iron or, in most cases, "hot metal," a necessary step in steel production. By-products from the processing and manufacture of coke also provide the steelmaker with usable auxiliary fuels or commercial products. The other major use of coke is in foundries, where it is used as a fuel to melt a metallic charge of iron and/or steel scrap, often with some pig iron, to produce molten iron for pouring gray iron, ductile iron, and malleable iron castings. In addition, small quantities of coke are used for ferroalloy production, lead smelting, calcium carbide production, elemental phosphorus production, and other industrial uses. As shown in Fig. 3, these coke uses are minor compared to furnaces or foundries.

A. Blast Furnace Coke

As shown in Fig. 3, the quantity of coal used to produce blast furnace coke far exceeds coke production for any other use. The amount of blast furnace coke required is intimately tied to raw steel production. Thus, when steel production is high, the demand for coking coal is also high.

In 1979, 77 million tons of coal was carbonized in furnace coke plants [22]. This was a decline from about 78 million tons in 1976, and a 9% decline from the record 84 million tons carbonized in 1973. This is not indicative of a reduced need for coking coal—merely a reflection of overall U.S. industrial activity and steel imports.

1. *Coking Coal Supply*

Coking coal consumers in the United States procure their coal through several sources: they produce their own coal from captive or joint-venture mines, they purchase on contract from large and small producers, or they

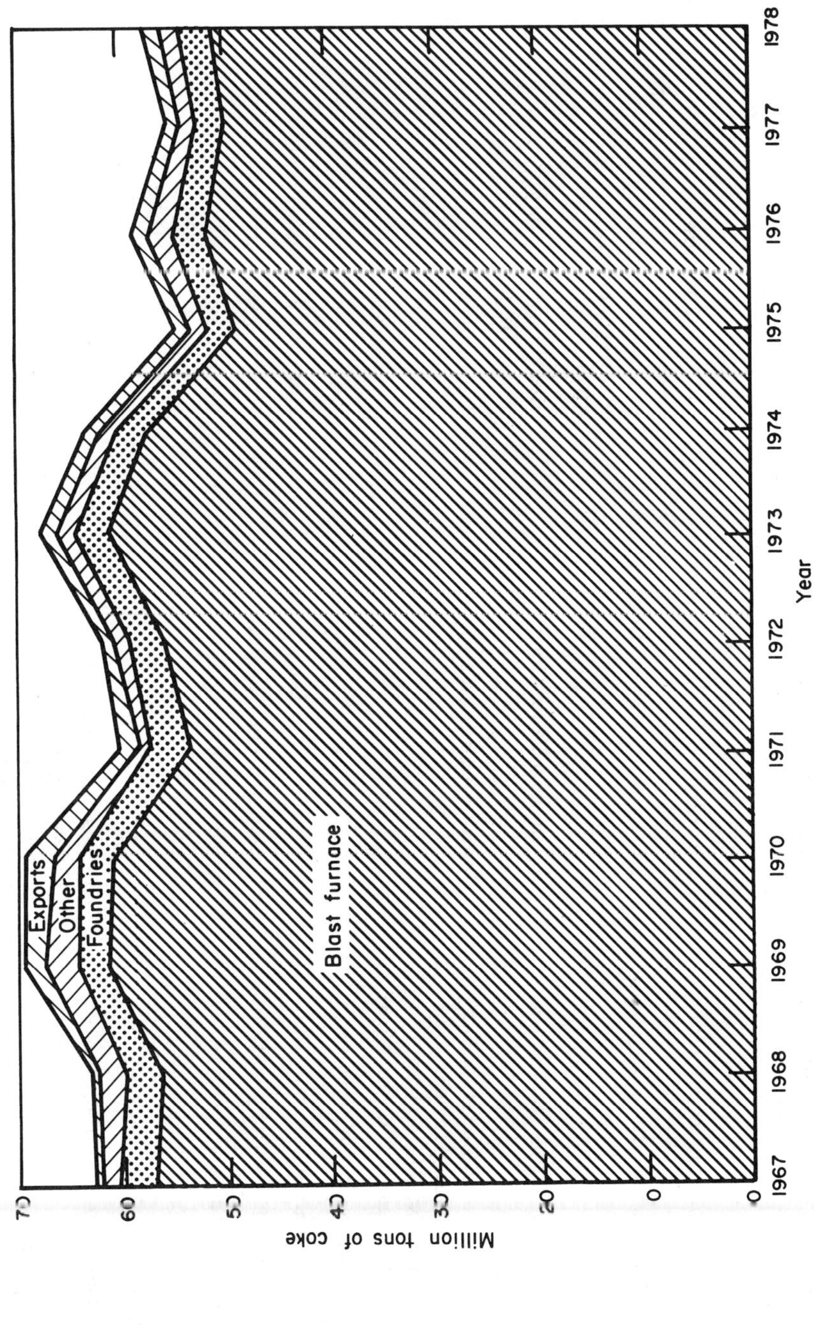

FIG. 3 Historical consumption of coke (including breeze). Data from Ref. 21.

TABLE 6 U.S. Coking Coal Production by Steel Companies, 1979

Operating company	Parent or controlling company	Location	Production (thousands of tons)
United States Steel Corporation	United States Steel Corporation	Alabama, Kentucky, Pennsylvania, Utah	14,979
Bethlehem Mines Corporation	Bethlehem Steel Company	Kentucky, Pennsylvania, West Virginia	12,499
Jones & Laughlin Steel Corporation (including former Youngstown Mines with 2213 thousand tons in 1976)	Jones & Laughlin Steel	Pennsylvania	5,323
Beth-Elkhorn Corporation			
Republic Steel Corporation	Republic Steel Corporation	Kentucky, Pennsylvania	3,342
National Mines Corporation	National Steel Company	Kentucky, Pennsylvania	3,245
Armco Steel Corporation	Armco Steel Corporation	West Virginia	3,898
Buckeye Coal Company			
Olga Coal Corporation			
Mathes Coal Company	Lykes Resources, Incorporated Stelco (Canada) National Steel Consolidation Coal	Pennsylvania	1,455

Inland Steel Coal Company	Inland Steel Company	Illinois	1,936
Kaiser Steel Corporation	Kaiser Steel Corporation	Utah, New Mexico	1,992
Beatrice Pocahontas Company	Republic Steel Corporation Island Creek Coal Company	Virginia	654
CF&I Steel Corporation	CF&I Steel Corporation	Colorado (est.)	600
Harmor Coal Company	Consolidation Coal Company Wheeling Steel Corporation	Pennsylvania	377
Total			50,300

Note: Figures may not add to total because of rounding.

Source: Ref. 27, 1980 ed.

buy spot coal. Table 6 lists captive coal production in 1979. It should be noted that not all of the captive production was necessarily consumed for coking by the producer. A certain amount may have been sold to other cokemakers, and some would not have met specifications and may have been marketed or used as steam coal.

Independent coal producers also produce significant quantities of coking coal. The major independent producers include:

Pittston
Westmoreland Coal
Island Creek Coal
Eastern Associated Coal
Consolidation Coal

These companies produce a wide variety of coals that are often custom-blended for customer specifications. There are also many small producers that market specific coals to cokemakers.

2. *Contracts and Market Procedures*

Prior to the heavy contract commitments made by the Japanese in the late 1960s, coking coal was sold almost exclusively on annual "evergreen" contracts. Essentially, price and quantities were negotiated annually by buyers and sellers. Substantial quantities are still sold under this method; in fact, there appears to be a return to this procedure, although it is more formal than in the past.

Captive production has a major impact on domestic coking coal market procedures. Captive production allows the buyer to be more sophisticated and demanding during contract negotiations. Also, the purchased coal must be compatible with captive coal in order to make good coke.

Prices are usually quoted FOB mine, and the buyer usually pays for transport. Test shipments for a new coal can range from an envelope full to a 500 lb test.

3. *Future Outlook for Furnace Coke*

Future furnace coke demand will depend on the level of pig iron production, which in turn will depend on total steel requirements and on the need for hot metal for basic oxygen furnaces (BOFs). In addition, two major technological developments that are currently under way could greatly alter future furnace coke needs: formed coke and direct reduction of iron ore.

The objective of formed coke processes is to make furnace coke from coals that are now considered "nonmetallurgical." Much of the interest in formed coke centers around Western low-sulfur coals and Illinois nonmetallurgical coals. A secondary objective, which has developed in recent years, is to produce coke by continuous processes in plants that do not have the emission-control problems of conventional slot-oven coke plants. Some of the early evaluations of formed coke in blast furnaces were disappointing, but more recent and more extensive tests have been more encouraging. This does not signal the demise of the coking coal industry but will act to depress the demand for additional future coking coal supplies. This is particularly true in nations, such as Japan, that lack large prime coking coal supplies.

In contrast to formed coke processes, direct reduction processes have been highly publicized for about 20 years. All direct reduction processes are methods to reduce iron ore to metal without the use of the blast furnace. Thus, all direct reduction processes avoid the need for traditional coke in the reduction of iron ore. However, because they do not produce molten pig iron, the "sponge iron" product must be melted and refined in steelmaking furnaces (as is scrap) or, in some versions, charged as a partially reduced material in blast furnaces to decrease their coke rate and increase their productivity. Most reduction processes rely on natural gas, although progress is being made on coal-based processes. Because of this, the U.S. industry has not invested widely in direct reduction. However, the growth could be much larger abroad, particularly in oil- and gas-producing nations. Therefore, this technology presents more of a threat to export coking coal sales than to domestic sales.

B. Foundry Coke

Compared to furnace coke, foundry coke is a relatively minor product, although, as with furnace coke, its demand fluctuates with the level of overall industry activity.

Buyers of foundry coke require large sizes, usually over 4 × 6 in. Other specifications include extremely low ash and sulfur. Foundry coke is produced in merchant ovens that rely heavily on open-market sales. Important merchant foundry coke producers include:

Allied Chemical
Shenago
Alabama By-Products

Foundry coke is also produced captively, although only a small portion of the coke supply comes from captive production.

The technologies that will affect future furnace coke markets also apply to foundry coke markets but to a lesser extent. In addition, other factors that could decrease the needs for foundry coke include electric cupola melting and the replacement of iron castings. Electric melting replaces the traditional cupola in foundries. Foundries have found electric melting to have lower capital costs and lower environmental problems than do cupolas. Several years ago it appeared that electric melting would eventually replace cupola melting; however, recent cupola additions suggest that complete replacement is not taking place. On the other hand, the replacement or iron castings appears to be increasing. The entire iron castings industry is constantly threatened by lighter-weight materials. The principal United States market for iron castings is the automobile industry; therefore, this threat seems well founded in light of the desire for more-fuel-efficient automobiles. The ferrous iron castings industry has been in a decline, which is expected to continue.

C. Market Outlook Summary

Taken in total, the outlook for increased coking coal production for domestic markets is not one of growth. Increased coke demand caused by population growth will be offset by technological or product changes.

V. OTHER MARKETS FOR COAL

A. Chemical Production

Each ton of coal coked produces something on the order of 1000 lb of by-products. In the early days of the coke industry, simple ovens or "beehives" were used. The gases and other by-products simply escaped through vents in the roof of the oven. However, it soon became apparent that those escaping by-products could be sold or used at the plant. Thus, the modern "slot oven" was developed to capture the by-products from coking. There are still a few beehive ovens in use, mostly in Virginia. However, these are insignificant from a statistical viewpoint.

The by-products from coking are basically gases, oils, and tars. Coke plants are the ancestors of the modern oil refinery/chemical complex. In fact, several steel companies have large chemical divisions that evolved from coking generations. The list of products that can be derived from coke oven by-products reads like the familiar "tree" used in high school chemistry classes. Important by-products and their derivatives include:

Coke oven gas
Coke tar (source of acids, creosote oil, naphthalene, pyridine bases, and hard or soft pitch
Crude light oil (source of benzene, toluene, xylenes, and solvent naphtha)
Ammonia

Any of these by-products can in turn be converted to industrial or chemical products, such as fertilizer from ammonia.

Coke oven gas is a low-heating-value gas (500 Btu/ft^3) that is widely used within the integrated steel plant. About 33% of the coke oven gas is used for fueling "soaking pits" when ingots are heated before further working [23]. The next largest use (about 32%) is for underfiring the coke ovens. Other uses include fuel for steel-melting furnaces, ovens, and boilers. A small, declining percentage is sold and/or wasted.

Coal tar, condensed from coke oven off-gases, is refined to yield simple products such as creosote. Some tar is injected into the blast furnace as a supplementary fuel.

Light oils and ammonia are among the more valuable by-products of coking. However, coke ovens account for only a small, declining portion of U.S. consumption; the major sources are oil refineries.

In conclusion, the coal industry was the first large-scale source of many of the carbon-based products and fuels used in modern industrial societies. This contribution could be greatly increased in the future with the construction of new coal gasification/liquefaction plants. This potential new use of coal will be examined in a subsequent section.

B. Residential and Commercial Coal Use

Coal was once a dominant fuel for heating homes and commercial establishments. As late as 1940, coal deliveries to retail dealers for household and commercial use was about 20% of total coal shipments [1]. By 1978, this had declined to less than 1% of the coal market. It is estimated that in 1979, bituminous coal and anthracite directly supplied about 0.8% of energy needs in these sectors, although indirect coal use, such as for electricity generation, is not included here [24].

It is possible that there will be some incremental increase in coal use in these sectors. Rural markets have developed as well as for supplemented heating needs in many homes. However, this market is small, highly seasonal, and statistically insignificant.

C. Railroads and Mine Use

As noted earlier, railroads were the major single coal consumer until the 1940s. In 1976, the railroads accounted for 0.04% of the coal market [24]. This market will not increase to any substantial degree.

A small amount of coal is often used at the mine for steam generation, for example at cleaning plants. However, this use could actually decline as a result of environmental regulations.

VI. SYNTHETICS FROM COAL

The preceding discussions of present and future coal uses require an examination of the prospects for the development of new coal uses. The major considerations are economics and the desire to employ the technology [25].

A. Background

The major research on synthetic fuels was performed in Germany before and during World War II. Here, of course, economics was not an issue. Following the war, oil imports replaced the wartime technology in Germany.

The only commercial synthetic fuel plants in the world today are in South Africa. As in Germany, economics are not an issue because South Africa maintains some measure of energy autarchy.

In the United States, interest in synthetic fuels from coal was minimal until the oil embargo. Since that time, much basic and applied research has been conducted. The first demonstration plants are now being tested. Commercial-size plants could be built on a limited scale in the next few years.

B. Gasification Markets

Basically, two types of gasification schemes have been envisioned: production of either high- or low-heating-valued gases. Despite common processing techniques, the products are sufficiently differentiated that each represents a different marketing problem for the owner and operator of a plant.

Gasification plants producing gas with a heating value equivalent to natural gas (referred to as "pipeline quality") present minimum marketing problems. Plants would be very large and would probably locate near the source of coal in order to minimize transport costs. The gas would merely enter existing natural gas lines as a supplemental fuel source. Under present conditions the synthetic gas would be more costly than natural gas. The cost impact on the consumer would presumably be minimized by "rolling in" the price of the synthetic plant gas into the total rate base for all gas. Thus, the consumer would not have to pay the full gas cost initially; instead, it would be prorated according to the amount of synthetic gas entering the system. The gas would be suitable for all classes of consumers.

The other types of gasification products are low- or intermediate-heating-value gas. The product is cheaper to produce because it does not involve the costly methanation step necessary to upgrade the product to a pipeline-quality basis. The gas is more expensive to pipe on a delivered-heating-value basis and requires modifications in combustion equipment.

Lower-heating-value gas would serve industrial or large commercial users. Unlike the residential consumer, such companies would have the financial and technical capability to burn such gas. Undoubtedly a dedicated pipeline would have to be built to distribute the gas. Customers would have to be clustered geographically to be serviced economically. The gas would have to be priced at a level that would make it attractive enough for the customer to alter his combustion system. The ideal customers for lower-heating-value gas would be companies that employ large fuel-using processes that are accustomed to using diverse fuels. The greatest potentials are in the metals and, particularly, steel industries. Other industries include stone, glass and clay, chemicals, and pulp and paper.

C. Liquefaction Markets

The term "liquefaction" encompasses many processes that produce a liquid or solid fuel. From a marketing standpoint, a liquid fuel could penetrate many more markets than could solid or gaseous fuels. A liquefaction process geared to the production of synthetic crude oil could essentially substitute traditional oil production. The synthetic crude could be easily stored (a problem with gasification) and converted to products in a traditional refinery. Liquid fuels could directly replace and/or supplement supplies of gasoline, heating oil, jet fuel, or kerosene.

Liquefaction processes yielding solid fuels would present the traditional coal marketing problems. Liquids and gases are much more convenient, even to large-scale users. Synthetic solid fuels would have to compete with naturally low-sulfur coal as well as liquid and gaseous fuels. Thus, synthetic solid fuels would be in direct competition with coal. Since coal's largest market is electricity generation, the fuel would have to be at least equal in cost to scrubbing technology. Certainly, there is a market for a clean fuel in older power plants or industry, but this market is highly fragmented and would be difficult to penetrate. In sum, unless a "clean" solid fuel could demonstrate comparable economics to scrubbers, the market prospects are dim.

VII. TRANSPORT OF COAL

Transportation costs are of critical importance to the coal industry. The delivered price of coal, like all bulk commodities, can be more influenced by transport than by production costs. For example, the price of a Powder River Wyoming coal, delivered to Texas, may be 80% transport charges [8]. Since coal is usually sold FOB mine, transport charges are closely monitored by buyer and seller. Neither party can seriously influence transport rates, but each searches for options to lower costs. Sellers can trim FOB mine price margins, while buyers have been known to form transport subsidiaries or switch planned plant locations to minimize transport rates.

The coal industry is the largest single user of railroad services and the second largest user of barge and lake carriers (Table 7). Many carriers depend almost entirely on the coal industry for their continued existence. Coal accounted for 20.2% of total rail carloadings in 1977 [25]. By comparison, the second largest commodity hauled by rail was chemicals, accounting for only 6.1% of 1977 carloadings [25].

A. Rail

The railroads generally offer three major types of service to the coal industry: multiple-carload, trainload, and unit train rates. Another type is single-carload rates; they apply to low-volume movements that are usually from smaller mines. East Coast export rates are usually single-car rates.

Multiple-carload rates may be for as few as 5 or 10 cars or as many as 100 (or may be related to tonnage rather than to number of cars) and are usually significantly lower than single-carload rates. The multiple-carload

TABLE 7 Shipment of Coal from the Mine (Thousands of Net Tons)

Year	Loaded at mine for shipment by rail	Loaded at mine for shipment by water	Trucked to final destination	Used at mine-mouth generating plants	Used at mine[a]	Total production
1933	293,258	13,021	15,463	N.A.[b]	11,888	333,630
1961	293,546	46,348	51,044	N.A.	12,039	402,977
1971	382,201	58,817	60,323	47,727	3,124	552,192
1973	397,158	60,604	57,268	64,424	4,284	591,738
1975	481,148	69,094	79,365	73,543	8,288	648,438
1976	431,051	69,675	89,834	79,182	8,994	678,685
1977	407,304	65,969	122,294	83,240	12,536	691,344

[a]Includes coal used at mine for power and heat, made into beehive coke at mine, used by mine employees, used for all other purposes at mine, and shipped by slurry pipeline.

[b]N.A., not available or not applicable.

Source: Ref. 20.

rates may include reduced loading and unloading time allowances and annual volume minimums. Large minimum-volume or car requirements are called trainload rates and move from origin to destination as one unit. Minimum annual volumes, if applicable, range from 50,000 tons to several million tons. A higher penalty rate is assessed if the annual volume requirements are not met.

Unit train rates are usually the lowest rates possible. However, in some cases, rail carrier management will not negotiate unit train rates because they feel that unit trains are not operationally desirable. Frequently, electric utilities provide the hopper cars (and sometimes locomotives) for use in train service. Unit trains differ from trainload shipments in that unit trains use dedicated equipment and operate on a regular schedule. Unit train operation is more common in the South and West than in the northeastern United States. This is because coal shipments in the Northeast usually involve several rail carriers and because the higher density and poor roadbed conditions reduce the benefits of run-through unit trains.

B. Barge

Barge transportation is usually a lower-cost alternative than rail for coal shipments. Many power plants and coke ovens are located on inland waterways to take advantage of this alternative. In general, if a mine is reasonably near a navigable waterway and the delivery location is on or near a waterway, barge transportation will be used. Trucks or conveyors are used for shipping coal short distances to a barge terminal (up to 10 miles for conveyor and 30 or so miles for truck). Rail transportation to a river terminal of 100 miles or more distant may be used, depending upon the relative benefits of barge transportation. The economics are affected by the circuitry of the waterway route and the nature of the waterway. For instance, barge tows of 30,000 tons or more are possible on the Lower Mississippi River, where there are no locks to pass through (this compares with a maximum unit train or trainload size of about 10,000 tons). However, the size and number of locks to be encountered tend to reduce economic two size on other waterways. The Ohio River has been partially modernized and there are fewer and larger locks to be encountered. This is important because tows must be broken up and "double-locked" if they cannot fit in the lock. The Upper Mississippi River has many smaller locks and is not suitable for large tows. There are only seven locks between the mouth of the Illinois River and the Lower Mississippi. However, a modernization program has been delayed and this has become a bottleneck area. Although the Missouri River is "authorized" for a 9-ft draft, the same as most the U.S. inland waterways system, only about 7 ft of draft is actually available. The Missouri River is therefore not economical for coal transportation. The Gulf Intercoastal Waterway also has size limitations. In addition, ocean-going vessels are needed to reach much of Florida.

C. Lake Carrier

Until recently, coal transportation on the Great Lakes was primarily a backhaul for the iron ore shipments. This consisted of steam coal going to upper Great Lakes ports. Recently, Detroit Edison began to receive Western coal via Lake Superior. Illinois coal is also exported to Canada through Chicago.

Lake carrier shipments are a very low cost transportation option. However, the Great Lakes are closed to traffic for almost 3 months each year, so stockpiling and/or direct rail service is needed for the winter months. In addition, the rail carriers have been reluctant to grant low trainload rates to Great Lakes ports, because they prefer to provide the entire movement. Nevertheless, shipment by lake carrier is likely to increase in importance.

D. Trucks

Truck transportation of coal is expensive and suitable only for short distances. Truck is used for shipment from mine to nearby utilities, railheads, and barge terminals. Because investment in trucks is low in comparison to other modes, many different business arrangements prevail. The buyer or the seller can own the trucks. More often, truck transportation is provided by transport companies or by independent truckers. An independent trucker may either subcontract to a larger trucking company or contract directly with the buyer (or seller).

Short-term increases in coal traffic are often moved by truck. Also, poor rail service in many portions of Appalachia has encouraged greater use of trucks. Unfortunately, increased trucking of coal has resulted in severe highway damage in many regions of the United States.

Truck movements are also used extensively to haul small quantities of coke to foundries. In this instance, the buyer often elects truck to minimize breakage of the fragile coke. This is of minor significance, however, compared to rail shipments.

E. Conveyors

Conveyors of up to 10 miles are used to ship coal to power plants and river terminals. Conveyors have the advantage of low labor costs and high operator control. However, reliability problems have limited their use in some instances. If a conveyor breaks down, there is often no alternative transportation available. A large, constant volume of coal over a short distance is required for the economic feasibility of conveyor systems.

F. Slurry Pipelines

The first major U.S. coal slurry pipeline was built from Cadiz, Ohio, to Cleveland by the Consolidation Coal Company, a distance of 108 miles. This pipeline was operationally successful. Its introduction forced the railroads to offer trainload rates, at which time the pipeline was abandoned. The only currently operating coal slurry pipeline in the United States is the 273-mile-long Black Mesa pipeline in Arizona. This pipeline has been an economic success.

The success of Black Mesa has created much interest in slurry pipelines. However, to be economically viable, the following conditions should exist: (1) long distance, (2) high volumes, point to point; (3) no adequate existing rail service; and (4) long-term (20-year) commitment.

One of the chief advantages of the slurry pipeline is that, although it is capital-intensive, operating costs will not increase as rapidly as the general rate of inflation (because the capital will be depreciated). If environmental problems and the eminent domain question can be settled, it seems likely that some slurry pipelines will be constructed. They will probably never move a major portion of the U.S. coal industry output, however, and will likely be confined mainly to the Western states with scattered Eastern applications.

G. Transport for Export

Overseas coal exports are shipped from East and Gulf Coast ports. Hampton Roads, Virginia, and Mobile, Alabama, are major exporting ports. Almost all coal shipments to these ports go by rail, although there are some barge shipments to Mobile. Coal exports from New Orleans are generally shipped in by barge. However, no coal transfer facilities are presently available at New Orleans, so direct transfer from barge to ship is necessary. Canadian exports consist of the previously mentioned lake carrier shipments out of Chicago and all-rail shipments from Appalachian fields to eastern Canada.

H. Advanced Transport Modes

Opportunities for advances in bulk transportation technologies are minimal. However, as demonstrated by slurry pipelines, new technology can be employed successfully in the bulk transport industry. Two methods currently being explored are pneumatic and containerized transport. A pneumatic pipeline concept, Tube-express, has been developed by a subsidiary of the Transco Companies. In brief, covered gondolas of coal can be moved pneumatically in a sealed tube. This technique shows promise as a substitute for conveyors, trucks, or rail in relatively short haul applications. Akin to Tube-express are other schemes to containerize coal shipments. A company in Australia is examining this technology as a means of delivering a highly uniform product sheltered from the weather.

VIII. WORLD COAL RESERVES AND PRODUCTION

Coal reserves are widely dispersed worldwide. Reserves have been developed on every continent, and undoubtedly much coal remains to be discovered in unexplored areas, such as central Africa and Latin America. The world's reserves of coal far exceed those of any other fossil fuel and are believed sufficient to support a massive increase in consumption well into the next century.

The World Energy Conferences' *Survey of Energy Resources* estimates the world's total resources of all ranks of coal to be 11,500 billion metric tons, of which about 740 billion tons are deemed economically recoverable. Five regions hold 96% of the known reserves. North America has 31%, the USSR and other nations in eastern Europe 26%, western Europe 17%, China 15%, and Australia 6%. Table 8 summarizes reserves by nation.

Total world production of all types of coal in 1977 was 3400 million metric tons. Of this total, anthracite and bituminous (i.e., "hard" coal) types represented 2500 million metric tons or about 70%. The leading producers

TABLE 8 Estimated World Coal Reserves[a] (Millions of Net Tons)

	Recoverable reserves[b]	Percent of world total	Total reserves[c]	Percent of world total	Total resources[d]	Percent of world total
USSR	150,576	23.1	301,151	19.2	6,298,158	53.1
United States	200,379	30.7	400,758	25.6	3,223,709	27.2
Canada	6,103	0.9	9,958	0.6	119,906	1.0
China	88,185	13.5	330,693	21.1	1,102,310	9.3
India	12,765	2.0	25,529	1.6	91,466	0.8
Rest of Asia	6,589	1.0	19,109	1.2	27,645	0.2
West Germany	43,620	6.7	109,702	7.0	315,426	2.7
United Kingdom	4,267	0.7	108,993	7.0	179,472	1.5
Poland	24,956	3.8	42,851	2.7	66,803	0.6
Rest of Europe	66,903	10.3	90,980	5.8	107,976	0.9
South Africa	11,667	1.8	26,702	1.7	48,875	0.4
Rest of Africa	5,560	0.9	6,688	0.4	15,989	0.1
Australia[e]	26,814	4.1	81,946	5.2	218,879	1.8
Rest of Oceania	213	—	396	—	1,202	—
Latin America	3,089	0.5	10,129	0.7	36,295	0.3
World total	651,676	100.0	1,565,582	100.0	11,854,114	100.0

Note: Estimates of U.S. coal reserves in this survey do not agree with other domestic data; among other reasons, criteria used for thickness of seam and depth of overburden are not those employed by the U.S. Bureau of Mines or the U.S. Geological Survey.

[a]Excludes peat.

[b]Amount of reserves in place that can be recovered under present local economic conditions using available technology.

[c]The portion of total resources that has been carefully measured and assessed as being exploitable under local economic conditions and available technology.

[d]Total amount available that can be successfully exploited and used within the foreseeable future.

[e]Does not include additional resources for Queensland.

Source: Ref. 29.

TABLE 9 World Production of Bituminous Coal and Lignite (Thousands of Net Tons)

Country[a]	1972	1974	1976	1977	1978
North America					
Canada					
Bituminous	17,426	19,419	22,926	25,000	28,000
Lignite	3,283	3,842	5,156	6,000	6,000
Greenland					
Bituminous	4	—	—	N.A.*	N.A.
Mexico					
Bituminous	3,598	5,694	6,228	6,600	N.A.
United States					
Bituminous	574,387	585,504	639,365	663,000	629,000
Lignite	10,999	15,496	25,635	28,000	36,000
South America					
Argentina					
Bituminous	765	690	678	421	409
Brazil					
Bituminous (marketable)	2,722	3,436(r)	3,200(e)	2,207	N.A.
Chile					
Bituminous (marketable)	1,472	1,554	1,285	1,300	N.A.
Columbia					
Bituminous[c](e)	3,500	3,472(r)	3,600(e)	4,070	N.A.
Peru					
Bituminous	83	2	2	1	
Venezuela					
Bituminous	44	63	98	106	N.A.
Europe					
Albania					
Lignite[d]	843	940(r)	940	N.A.	N.A.
Austria					
Lignite[f]	4,139	4,001	3,543	—	N.A.
Belgium					
Bituminous	8,316	6,694	6,737	7,791	7,264
Bulgaria					
Bituminous	252	209	220(e)	319	308
Lignite[d]	29,644	26,453	27,748	27,379	27,500
Czechoslovakia					
Bituminous	30,782	30,745(r)	31,158	31,000	31,000
Lignite	93,620	91,260(r)	98,628	100,000	105,000
France					
Bituminous	23,455	18,650	18,814	28,700	26,100
Lignite	3,245	3,087	3,489		
Germany, East					
Bituminous	898	655	505	<500	<500
Lignite[d]	273,870	268,368	272,140	280,000	279,000

Country[a]	1972	1974	1976	1977	1978
Germany, West					
Bituminous[g]	104,161	96,443	91,124	93,000	93,000
Lignite[d]	121,712	138,940	148,299	135,000	136,000
Greece					
Lignite	12,259	15,552(r)	24,512	26,200	N.A.
Hungary					
Bituminous	3,309	3,537	3,235	7,700	7,700
Lignite[d]	24,439	24,859	24,606	24,783	23,650
Italy					
Bituminous	166	4	1	4,000	N.A.
Lignite	1,539	2,160(r)	2,235		
Poland					
Bituminous	166,115	178,579	197,646	205,000	212,000
Lignite[d]	42,131	43,899	43,331	45,000	45,000
Romania					
Bituminous[g]	7,288	7,812(r)	7,780(e)	7,766	7,700
Lignite[d]	18,958	21,813(r)	21,730(e)	21,590	23,650
Spain					
Bituminous	8,820	8,147(r)	7,710	12,320	13,464
Lignite	3,359	3,177(r)	4,562	6,050	6,600
Svalbard (Spitz-bergen)					
Bituminous[h]	502	481	572	N.A.	N.A.
USSR[i]					
Bituminous	467,437	493,878	511,000	527,000	530,000
Lignite[d]	171,651	177,076	180,000	184,000	186,000
United Kingdom					
Bituminous	130,660	118,681(r)	133,604(e)	132,000	133,000
Yugoslavia					
Bituminous	660	662(r)	646	600	500
Lignite[d]	33,446	36,355	39,969	41,000	43,000
Africa					
Algeria					
Bituminous[c]	13	14(r)	3	1	N.A.
Botswana					
Bituminous	—	36	247	200	N.A.
Mozambique					
Bituminous	370	469(r)	660(e)	400	N.A.
Nigeria					
Bituminous	376	306	350(e)	330	N.A.
Rhodesia, Southern					
Bituminous	3,045	3,080(r)	3,226	2,750	N.A.

(continued)

Table 9 (continued)

Country	1972	1974	1976	1977	1978
South Africa					
Bituminous (marketable)	62,946	71,232	82,233	91,000	97,000
Swaziland					
Bituminous	139	154(r)	170	100	N.A.
Tanzania					
Bituminous	3	2	1(e)	1	N.A.
Zaire					
Bituminous	141	105	110	100	N.A.
Zambia					
Bituminous	1,033	892	864	880	N.A.
Asia					
Afghanistan					
Bituminous[k]	150	169	181	220	N.A.
Burma					
Bituminous	23	19	23	20	N.A.
China					
Bituminous and lignite(e)	420,000	475,000	510,000	606,000	681,000
India					
Bituminous	82,421	92,706	111,324	111,000	112,000
Lignite	3,381	3,355	4,260	4,000	4,000
Indonesia					
Bituminous	197	172	202	220	N.A.
Iran					
Bituminous	1,102	1,323	1,100	990	N.A.
Japan					
Bituminous	29,249	22,211	20,236	17,000	N.A.
Lignite	108	84	58		
Korea, North					
Bituminous	N.A.	28,380	34,320	35,200	37,000
Lignite	7,400	8,700	8,800	8,800	9,200
Mongolia					
Bituminous	117	151	220	N.A.	N.A.
Lignite	2,367	2,576	2,921		
Pakistan					
Bituminous and lignite[m]	1,379	1,653	1,213	1,100	N.A.
Philippines					
Bituminous	43	56	123	330	N.A.
Taiwan					
Bituminous	4,313	3,235	3,567	3,190	N.A.
Thailand					
Lignite	380	534	750	220	N.A.
Turkey[n]					
Bituminous	5,116	5,185(r)	4,295	8,140	N.A.
Lignite	5,151	5,933(r)	6,411		

Country	1972	1974	1976	1977	1978
Oceania					
Australia					
Bituminous	65,796	69,482(r)	83,000	86,000	88,000
Lignite	26,121	30,096(r)	34,104	32,000	36,000
New Zealand					
Bituminous	2,237	2,669	2,554	3,000	2,000
Lignite	168	159	187	N.A.	N.A.
Total bituminous	1,833,052	1,867,377(r)	2,004,621	N.A.	N.A.
Total lignite (including peat)	886,813	920,015(r)	985,214	N.A.	N.A.
Total mixed grades[o]	421,379	476,653	511,213	N.A.	N.A.
Total all grades	3,141,244	3,264,045(r)	3,501,046	N.A.	N.A.

Note: (e), estimate; (p), preliminary; (r), revised.

[a]In addition to the countries listed, Ecuador produces coal, but output was less than 500 tons annually in the years covered by this table.

[b]Less than 1000 net tons.

[c]May include a small amount of anthracite.

[d]Includes material reported in national sources as brown coal.

[f]Available sources report only lignite production; a small amount of bituminous coal may also be produced.

[g]Official sources report the aggregate of anthracite and bituminous; distribution to these separate grades is estimated from reported total.

[h]Output from Norwegian-controlled portion only; output of that portion of Svalbard controlled by USSR is presumably included in the total output recorded for that country.

[i]Run-of-mine output.

[j]Sales for year ended August 31 of that stated.

[k]Year beginning March 21 of that stated.

[l]Includes low-ranked coals, including some lignite.

[m]Year ending June 30 of that stated.

[n]Includes private sales.

[o]Bituminous plus lignite for China and Pakistan.

*N.A., not available or not applicable.

Source: Refs. 29, 30, 31.

of hard coal are the industrialized nations. As shown in Table 9, more than 75% of the world coal output is produced by five nations: the United States, the USSR, Poland, the United Kingdom, and West Germany. Among the developing nations, India is the only major producer. The United States accounts for more than half of all the coal mined and consumed by the non-Communist countries. The USSR and China each produce roughly as much as does the United States.

World production of brown coal and lignite has been increasing recently. East Germany is the most important producer, with 250 million tons. The next most important producers are the USSR, West Germany, and the United States. Because of their low heating value, brown coal and lignite are not statistically important in meeting world energy needs.

IX. WORLD COAL TRADE

The first recorded export shipment of coal took place in 1325. A ship from France loaded with wheat landed at Newcastle-on-Tyne and took back a load of "charboun de meer," or sea coal [27]. France remained an important buyer of British coals despite periodic wars between the nations. Great Britain remained the most important coal exporter until World War II. British exports peaked in 1923 at 79.4 million tons.

U.S. overseas coal exports began on a regular basis in 1897 (Canada was actually the first foreign buyer, before 1897). The first U.S. overseas shipments were to Latin America for transport and industrial use. Exports increased slowly until World War I, when U.S. coal helped to fill the void created by war-damaged European production. In 1920, the United States exported over 22.3 million tons of coal (versus 5.7 million tons in 1917).

U.S. coal exports declined rapidly in the mid-1920s and, with the exception of a surge in 1926 caused by a British miners' strike, never exceeded 20 million tons again until 1947. Several factors caused this export decline, including a rebuilt European coal industry, the Great Depression, and the growth of oil and gas use.

The end of World War II created another bonanza for U.S. overseas coal exports. Again, U.S. coal replaced the output from European coal industries. However, the high quality and cost-competitiveness of U.S. coal was exploited by U.S. exporters and the post-World War I decline was not repeated, and U.S. exports have dominated world coal trade until very recently.

A. Types of Coal Traded

The predominant type of coal traded on a worldwide basis is coking coal, a necessary ingredient to steelmaking. Buyers can afford relatively high transport charges to produce a strong, high-quality coke. Also, continued exports have been encouraged because specific coals are required for blending purposes. For example, coking coal from the Pocahontas seams in Appalachia are a basic ingredient in many coke blends. Technological advances in coking coal blending in the 1960s (led by the Japanese) has decreased reliance on specific coal types. Concurrently, the development of new coal deposits has increased the types of coking coal available on the world market.

Large-scale international trade in noncoking coal is a relatively recent event, dating mainly from the oil embargo. In 1977, about 60 million tons of noncoking coal was traded. By 1985, this figure could about double [28a, b]. Most of the noncoking coal is imported by western Europe for electricity generation. Japan is also increasing its imports of noncoking coal, which were about 1.8 million tons in 1976 [29].

B. Major Exporters

The large markets for world coal and the wide geographic dispersion of coal resources have encouraged several nations to increase coal exports. This section highlights a few of the more important exporting nations.

1. *United States*

As noted previously, the United States has been the world's largest coal exporter. Close examination of Table 10 shows that the United States has been losing its dominant export position. In 1960, the United States accounted for about 33.5% of world coal trade. By 1973, this had declined to 28.4%. More recent data shows further erosion.

Many reasons are given for the declining world coal trade role of the United States: (1) the major coal importer, Japan, decided to diversify its coal sources; Japanese interests fostered the development of an export industry in several nations, notably Australia and Canada; (2) sophisticated blending techniques cut requirements for premium low-volatile U.S. coking coals; (3) labor, environmental, and regulatory problems raised the price of U.S.

TABLE 10 Coal Exports by Nation of Origin (Thousands of Metric Tons)

	1960	1970	1973	1976	1978
United States	34,456	65,075	48,613	53,881	39,434
Poland	17,497	26,374	35,954	38,822	44,091
Australia	1,584	18,320	25,800	30,418	41,790
USSR	12,300	24,499	25,000	26,890	N.A.
West Germany	17,974	15,906	13,856	12,801	20,565
Canada	774	3,987	10,933	11,759	15,055
United Kingdom	5,547	3,191	2,693	1,435	2,497
Czechoslovakia	2,195	2,790	2,931	N.A.[a]	N.A.
Others	10,436	8,938	5,286	N.A.	N.A.
Total	102,703	109,000	171,000	N.A.	N.A.

[a]N.A., not available.

Source: Ref. 29.

coal vis-à-vis other suppliers; and (4) buyers complained of poor treatment during periods of high coal demand.

Nonetheless, the United States should retain its position as a major coal-exporting nation in the future. U.S. coal remains the world standard, and not all of it can be replaced from other sources. U.S. coal exporters have been very aggressive in adopting blending techniques to deliver a consistent, high-quality product. Foreign investment in U.S. coal mining has increased recently. Finally, U.S. coal reserves are large enough to permit a sizable increase in export capacity without affecting domestic buyers.

2. *Australia*

The Australian coal export industry is presently tied to Japan. In 1979-1980, 67.1% of Australian coal exports went to Japan [29]. In comparison, only 24.2% of U.S. exports went to Japan [29]. Australia has been diversifying its markets in recent years.

As shown in Table 10, the Australian coal export industry has grown rapidly since 1960. The coal industry is characterized by large surface mines operating in shallow, thick seams. Much foreign capital is involved, including U.S., European, and Japanese companies. The coal lies near large, modern, deep-water ports developed mainly for coal exports. Some observers believe that Australian coal exports will exceed those of the United States in the 1980s.

The Australians have become aggressive coal marketers, yet there are some problems. Australia's geographic isolation from non-Asian markets presents a transport cost disadvantage when attempting to diversify markets. In addition, China could present future regional competition. The labor situation is uncertain as well as the institutional arrangements between local and federal governments.

3. *Canada*

Canada is in the unique position of being both a major coal exporter and a coal importer. The locus of Cdnadian coal consumption is Ontario and Quebec—historical importers of nearby U.S. coal. Most of Canada's low-cost coals are in the distant western provinces (Alberta and British Columbia). There is a continuing public debate concerning shipping Canadian coal east, although transport charges are high and shipments have been minimal to date. The Canadian export industry, like that of Australia, is heavily tied to Japan. In 1979, over 75% of Canadian exports were destined for Japan [29]. Canadian coal producers are understandably anxious to open new markets.

4. *Poland and South Africa*

Poland and South Africa are relative newcomers to international coal trade, both specializing in noncoking coal. Both have highly developed coal industries and a strong, government-encouraged desire to derive foreign currency from coal exports.

Poland has about 65 highly mechanized mines. About 80% of Polish coal production is consumed internally. Of the exports, 70% is noncoking coal. Poland's largest market is the USSR, with about 25% of the exports, followed by other western European nations. Poland has exported small quantities of steam coal to Florida.

South Africa imports coking coal and exports noncoking coal. Its principal coking coal suppliers are the United States and Poland. South Africa's principal noncoking coal market is Europe, notably France. In addition, South African coal was burned in Massachusetts during the oil embargo.

5. *Other Coal Exporters*

The balance of the coal exporters are characterized either by underdeveloped or declining coal industries. India is a good example of the former, whereas West Germany characterizes the latter.

West German coal production has been declining despite government initiatives to stabilize the industry. Falling domestic coal demand has paralleled the production declines. German mines are modern and efficient, but mining conditions are poor relative to other nations. Currently mined seams are thin and deep, the result of centuries of coal production.

India has been attempting to develop a viable coal export industry for several years. Much like South Africa and Poland, coal exports are seen as a means of earning hard foreign currency. However, the capital and infrastructure requirements are high and much of the equipment would have to be imported.

The USSR is the second largest coal producer in the world after the United States. Soviet exports are primarily to Eastern Bloc nations, notably Bulgaria and East Germany. Soviet exports have grown very slowly and the future is highly uncertain.

China is the newest entry into coal export circles. China was the world's second largest coal producer in 1979 [29]. Domestic coal consumption is very high and two-thirds of China's energy needs are derived from coal. Previous coal exports were nominal, but it appears that China could become a major exporter to Japan. Chinese coal exports are principally noncoking.

Once the world's largest coal exporter, the United Kingdom is presently a net coal importer. The situation is similar to West Germany, a mature industry producing from thin, deep seams. The future of the British coal industry is closely tied to the success of North Sea oil production.

C. Major Importers

The steel industries of various nations are the chief coal importers. Power companies are increasing their demands for noncoking coal imports, although not as much as was expected in the mid-1970s.

1. *Japan*

As shown on Table 11, Japan is by far the world's largest coal importer. In 1973, Japan accounted for about one-third of all coal imports. As noted in a previous section, the Japanese desire to expand coal sources led to the development of coal-exporting industries in Australia, Canada, and probably in the future, China. Japanese coal buyers have traditionally favored long term (10 years or longer) contracting arrangements. In several Canadian and Australian mines, the Japanese have essentially contracted for the entire mine output—often financing part of the mine. The Japanese were the first buyers of American coal who desired contracts longer than 1 year.

TABLE 11 Coal Imports by Selected Nations

	1960	1970	1973	1978
Japan	8,292	50,172	56,854	56,355
Canada	12,299	17,125	14,959	14,330
France	10,112	13,722	12,499	18,403
Netherlands	0,808	5,180	3,200	5,545
Italy	9,739	12,637	11,705	12,936
West Germany	6,705	9,719	7,021	7,236
Belgium-Luxembourg	4,148	7,694	7,486	7,724
USSR	4,776	7,100	10,000	N.A.
Others	39,824	45,725	47,342	N.A.
Total	102,763	169,080	171,066	N.A.

[a]N.A., not available.

Source: Ref. 29.

Japan is also a coal producer, supplying about 20% of its needs [29]. However, the coal deposits have been mined extensively for many years and production has steadily declined. Increased Japanese coal demand will be met by imports.

2. *European Community*

European steelmakers are the second largest buyers of imported coking coal. Some nations (e.g., France) import coal to supplement domestic coal production. Others (e.g., Italy) rely completely on coal imports for coking needs.

Unlike the Japanese, European buyers are more willing to accept short-term contracts or spot purchases for coal needs. This willingness is undoubtedly tied to the availability of regional coals. However, the Europeans have purchased coking coal mines in the United States for future needs. As mentioned previously, the Europeans have been active in importing noncoking coal for electricity generation. Principal suppliers are Poland and South Africa.

3. *Latin America*

Aside from Mexico and Brazil, the need for coking coal in most nations is small. The entire continent has not been explored for coal, and there will probably be new coal discoveries in the future. Many nations in Latin America have coal industries on a small scale. However, much of the Latin American coal is not of coking quality, so that it must be supplemented by imports. The United States is by far the most important supplier.

4. *Africa and Asia*

Aside from a few exceptions (South Africa, India) the coal requirements in these regions are small. Again, much land area has yet to be explored for coal. A few nations (e.g., Saudi Arabia) have ambitious plans for a modern steel industry which could create future export markets.

APPENDIX: WORLD COAL PRODUCTION, RESERVES, AND CONSUMPTION TABLES FOR 1977-1979*

Data on world coal production and reserves for the complete years 1977 and 1978 and the first two quarters of 1979 are presented in metric ton units in Table 1, short ton units in Table 2, and British thermal units in Table 3. Coal consumption data are presented in Tables 4 to 6.

The term "hard coal," as used in the tables, means both anthracite and bituminous coal. "Brown coal," as used outside the United States, generally means low-Btu coal in gradations from subbituminous through lignite. For "hard coal equivalent," an average heat value of 7000 kcal (net calorific value) per kilogram (or 7 tcal per 1000 metric tons) has been used. On a Btu (British thermal unit) basis, this relationship is the same as an equivalent of 27.78 million Btu per metric ton (1 Btu = 252 calories).

For coal production, trade, and inventories, primary governmental sources were used wherever available (see Table 7). The European Economic Community (EEC) and the United Nations (UN) were used where primary data were unavailable. The United Nations was also the source for data on Eastern European countries.

The U.S. Central Intelligence Agency supplied the data for coal reserves as of January 1, 1978. Quarterly reserves are computed by subtracting quarterly production from the annual beginning-year estimates, thus not accounting for new reserve additions during the period.

---

*Prepared by J. Bruce Carruthers II, Business Information Display, Inc., San Diego, California.

TABLE 1 Coal Production/Reserves (Thousand Metric Tons Hard Coal Equivalent)

Trends: % Change—Latest vs. Prev. Q	YTD	4 Qs	Tot Yr		Year-to-Date 1979	Year-to-Date 1978	Current 9 Quarters: 1979 Q2	1979 Q1	1978 Q4	1978 Q3	1978 Q2	1978 Q1	1977 Q4	1977 Q3	1977 Q2	Total Years 1978	Total Years 1977
				PRODUCTION—Hard Coal													
108.5	116.2	58.9	−2.9	Argentina	253	117	171	82	141	113	37	80	88	114	72	371	382
5.5	2.9	2.6	5.6	Australia	38426	37344	19727	18699	20684	19479	20339	17006	19354	19877	19620	77507	73431
−16.4	−4.0	−4.2	−6.8	Belgium	3398	3541	1547	1851	1807	1242	1712	1829	1796	1391	1874	6590	7068
na	na	na	na	Brazil	na	na	na	na	na	na	na	na	501	501	501	na	2002
17.6	14.8	19.8	12.0	Canada	8749	7624	4729	4020	4222	4192	4019	3604	3469	3236	3810	16038	14318
na	12.4	na	12.4	China*	617958	549964	na	na	154476	154476	154476	154476	137484	137484	137484	617958	549964
na	na	na	1.1	Czechoslovakia	na	10479	na	5487	5292	4816	5136	5343	5318	4863	5032	20588	20371
−7.0	−6.0	−7.2	−7.5	France	10331	10995	4979	5353	5110	3533	5435	5561	5512	3941	5893	19689	21293
−11.5	5.0	1.1	−1.3	Germany, West	47742	45455	22408	25334	23559	21088	21141	24314	23333	22576	20632	90098	91304
na	na	na	−.6	India	na	35125	na	20824	16958	16293	15983	19142	16991	16011	15867	68376	68782
na	na	na	4.4	Japan	na	9638	na	4505	4589	4823	4633	5006	4860	4418	4347	19050	18245
−2.2	2.5	3.5	4.8	Korea, South	7767	7577	3841	3926	4213	3686	3839	3738	3907	3653	3596	15476	14770
69.1	2.8	−3.4	−7.5	New Zealand	183	178	115	68	81	102	105	73	91	109	101	360	389
−18.7	−40.4	−16.1	.8	Norway	136	228	61	75	72	93	104	124	102	30	110	393	390
na	na	na	3.5	Poland	na	62398	na	33166	30740	30710	31018	31381	30618	29311	29492	123847	119662
na	na	na	−5.7	Portugal	na	56	na	29	30	31	27	29	31	32	30	116	123
na	na	na	na	Romania	na	na	na	na	na	na	na	na	895	895	895	na	3578
na	4.9	na	5.9	Spain	3920	3738	na	na	2127	1942	2380	2066	2142	1838	2218	8514	8042
na	na	na	−.2	U.S.S.R. (includes lignite)	na	295065	na	na	151301	146368	145545	149519	148834	147190	146368	592732	593690
−8.2	−.3	.8	.8	United Kingdom	61289	61484	29335	31954	34013	26139	30201	31283	33028	25925	30636	121677	120666
13.8	34.0	20.1	−5.3	United States (includes lignite)	311773	232659	165971	145801	165250	146302	161206	71452	139413	146981	154699	544212	574686
na	na	na	na	Venezuela	na	23	na	na	na	15	23	24	24	24	24	na	96
na	na	na	−7.7	Yugoslavia	na	140	na	62	64	59	69	71	75	66	67	262	284
				PRODUCTION—Brown Coal/Lignite													
21.3	−4.7	5.2	5.9	Australia	5294	5557	2902	2392	2643	3069	2981	2577	2227	2674	2880	11268	10637
na	na	na	−2.3	Austria	na	664	na	300	316	329	320	344	361	341	312	1309	1340
na	na	na	na	Bulgaria	na	3516	na	na	na	na	1745	1771	1849	1777	1831	na	7104
−13.5	8.0	6.1	.2	Canada	4211	3900	1954	2258	2234	1742	1633	2267	2051	1763	1803	7876	7861
na	na	na	2.2	Czechoslovakia	na	16923	na	8678	9043	8067	8067	8856	8578	7883	8160	34033	33296
na	na	na	−.2	Germany, East	na	36611	na	na	18429	17353	17564	19047	18721	17449	17581	72393	72517
−16.1	10.5	6.5	.5	Germany, West	17679	15999	8067	9612	9462	8081	7417	8582	9263	7796	7505	33542	33369
na	na	na	−5.2	Greece	na	2589	na	1252	1251	1232	1236	1352	1322	1263	1235	5071	5348
−13.1	.1	.6	.9	Hungary	3688	3685	1715	1973	2057	1722	1755	1930	1975	1763	1709	7464	7401
−1.6	−13.9	−7.4	−12.2	New Zealand	512	595	254	258	346	367	323	273	356	370	335	1308	1490
na	na	na	.8	Poland	na	5960	na	2621	3083	2672	2831	3129	2930	2778	2726	11715	11622
na	na	na	na	Romania	na	na	na	na	na	na	na	na	1248	1248	1248	na	4992
na	53.5	na	54.4	Spain	2240	1459	na	na	1392	1154	791	1002	757	748	799	4340	2811
na	na	na	na	Thailand	na	79	na	na	na	76	79	0	81	69	60	na	275
na	na	na	1.2	Yugoslavia	na	5956	na	3425	3375	2944	2822	3134	3205	3135	2775	12275	12130
				RESERVES—Hard Coal (million mt)**													
.0	.0	.0	.0	Brazil (1)	8100	8100	8100	8100	8100	8100	8100	8100	8100	8100	8100	8100	8100
−.1	−.2	−.2	−.2	Canada	9373	9392	9373	9378	9383	9388	9392	9397	9400	9404	9407	9400	9416
na	na	na	−.6	China*	na	98585	na	na	98276	98430	98585	98739	98894	99031	99169	98894	99512
−.1	−.3	−.3	−.3	Germany, West	34260	34352	34260	34283	34308	34331	34352	34374	34398	34421	34442	34398	34489
na	na	na	−.2	India	na	33663	na	33608	33629	33646	33663	33679	33698	33715	33731	33698	33767
.0	.0	.0	.0	Indonesia (1)	1400	1400	1400	1400	1400	1400	1400	1400	1400	1400	1400	1400	1400
.0	.0	.0	.0	Iran (1)	200	200	200	200	200	200	200	200	200	200	200	200	200
.0	.0	.0	.0	Mexico (1)	900	900	900	900	900	900	900	900	900	900	900	900	900
.0	.0	.0	.0	Nigeria (1)	100	100	100	100	100	100	100	100	100	100	100	100	100
na	na	na	−.8	Poland	na	21702	na	21554	21606	21654	21702	21750	21799	21847	21893	21799	21985
na	na	na	−.7	Romania	na	na	na	na	na	na	na	400	401	402	403	400	403
na	na	na	−.2	U.S.S.R. (total coal)	na	255727	na	na	255469	255600	255727	255853	255983	256111	256238	255983	256498
−.1	−.3	−.3	−.3	United Kingdom	44814	44936	44814	44843	44875	44909	44935	44966	44997	45030	45056	44997	45118
−.1	−.3	−.3	−.3	United States	177739	178363	177739	177905	178051	178217	178363	178524	178596	178735	178882	178596	179170
na	na	na	−10.1	Venezuela	na	945	na	na	na	928	945	972	1000	1028	1056	1000	1112
na	na	na	−3.2	Yugoslavia	na	8360	na	8174	8236	8300	8360	8428	8500	8574	8640	8500	8783

*Quarterly averages based on annual data. **Proven reserves derived from estimates at year-end 1977.

(1) Not accounting for year-to-date production.

TABLE 2 Coal Production/Reserves (Thousand Short Tons Hard Coal Equivalent)

Trends: % Change—Latest vs. Prev. Q	YTD	4 Qs	Tot Yr		Year-to-Date 1979	1978	Current 9 Quarters 1979 Q2	1979 Q1	1978 Q4	1978 Q3	1978 Q2	1978 Q1	1977 Q4	1977 Q3	1977 Q2	Total Years 1978	1977
				PRODUCTION—Hard Coal													
108.9	115.5	58.5	−2.9	Argentina	278	129	188	90	156	124	41	88	97	125	79	409	421
5.5	2.9	2.6	5.6	Australia	42357	41165	21746	20612	22800	21472	22419	18745	21334	21910	21627	85436	80943
−16.4	−4.0	−4.2	−6.8	Belgium	3745	3903	1705	2040	1992	1369	1887	2016	1980	1533	2066	7264	7791
na	na	na	na	Brazil	na	na	na	na	na	na	na	na	552	552	552	na	2207
17.6	14.8	19.8	12.0	Canada	9645	8404	5213	4432	4654	4621	4431	3973	3823	3567	4200	17679	15783
na	12.4	na	12.4	China*	681178	606229	na	na	170279	170279	170279	170279	151549	151549	151549	681178	606229
na	na	na	1.1	Czechoslovakia	na	11551	na	6048	5834	5309	5662	5890	5862	5360	5547	22694	22455
−7.0	−6.0	−7.2	−7.5	France	11388	12120	5488	5900	5632	3894	5991	6130	6076	4344	6495	21703	23471
−11.6	5.0	1.1	−1.3	Germany, West	52626	50105	24700	27926	25970	23245	23303	26802	25721	24885	22742	99315	100644
na	na	na	−.6	India	na	38719	na	22954	18693	17959	17618	21100	18730	17649	17490	75372	75818
na	na	na	4.4	Japan	na	10624	na	4966	5058	5316	5107	5518	5357	4870	4791	20999	20111
−2.2	2.5	3.5	4.8	Korea, South	8562	8352	4234	4328	4645	4063	4232	4120	4307	4027	3964	17059	16281
68.0	3.1	−3.1	−7.5	New Zealand	202	196	126	75	89	112	115	81	100	120	112	397	429
−19.3	−40.2	−16.1	.9	Norway	150	251	67	83	79	103	114	137	112	33	121	434	430
na	na	na	3.5	Poland	na	68782	na	36560	33885	33851	34191	34591	33751	32310	32509	136518	131904
na	na	na	−5.2	Portugal	na	62	na	32	33	34	30	32	35	35	33	128	135
na	na	na	na	Romania	na	na	na	na	na	na	na	na	986	986	986	na	3944
na	4.9	na	5.9	Spain	4321	4121	na	na	2344	2140	2624	2277	2361	2026	2445	9385	8865
na	na	na	−.2	U.S.S.R. (includes lignite)	na	325252	na	na	166780	161342	160435	164816	164061	162248	161342	653372	654428
−8.2	−.3	.8	.8	United Kingdom	67559	67774	32336	35223	37492	28814	33291	34483	36407	28578	33770	134125	133011
13.8	34.0	20.1	−5.3	United States (includes lignite)	343669	256461	182951	160718	182156	161269	177699	78762	153676	162018	170525	599887	633479
na	na	na	na	Venezuela	na	25	na	na	na	17	25	26	26	26	27	na	106
na	na	na	−7.7	Yugoslavia	na	154	na	69	70	65	76	78	83	72	74	289	313
				PRODUCTION—Brown Coal/Lignite													
21.4	−4.8	5.2	5.9	Australia	5835	6126	3199	2636	2913	3383	3286	2840	2455	2947	3175	12421	11725
na	na	na	−2.3	Austria	na	732	na	330	348	363	353	379	398	376	344	1443	1477
na	na	na	na	Bulgaria	na	3876	na	na	na	na	1924	1953	2038	1959	2018	na	7831
−13.5	8.0	6.1	.2	Canada	4642	4299	2154	2489	2462	1921	1800	2499	2261	1943	1987	8682	8665
na	na	na	2.2	Czechoslovakia	na	18654	na	9566	9968	8893	8892	9762	9455	8690	8994	37515	36703
na	na	na	−.2	Germany, East	na	40356	na	na	20315	19128	19361	20996	20636	19234	19380	79800	79936
−16.1	10.5	6.5	.5	Germany, West	19488	17635	8892	10596	10430	8908	8176	9460	10211	8594	8272	36974	36782
na	na	na	−5.2	Greece	na	2854	na	1380	1378	1358	1363	1491	1457	1392	1361	5589	5895
−13.1	.1	.6	.9	Hungary	4065	4062	1890	2175	2268	1898	1935	2128	2177	1943	1884	8228	8158
−2.1	−14.0	−7.4	−12.2	New Zealand	564	656	279	285	381	404	356	300	392	408	369	1441	1642
na	na	na	.8	Poland	na	6569	na	2889	3399	2946	3121	3449	3229	3062	3005	12913	12811
na	na	na	na	Romania	na	na	na	na	na	na	na	na	1376	1376	1376	na	5503
na	53.4	na	54.4	Spain	2469	1609	na	na	1534	1273	872	1105	834	825	881	4784	3099
na	na	na	na	Thailand	na	87	na	na	na	84	87	0	89	76	67	na	303
na	na	na	1.2	Yugoslavia	na	6555	na	3775	3720	3245	3110	3455	3533	3456	3059	13531	13371
				RESERVES—Hard Coal (million st)**													
.0	.0	.0	.0	Brazil (1)	8928	8928	8928	8928	8928	8928	8928	8928	8928	8928	8928	8928	8928
−.1	−.2	−.2	−.2	Canada	10332	10353	10332	10338	10343	10348	10353	10358	10362	10366	10370	10362	10379
na	na	na	−.6	China*	na	108670	na	na	108329	108500	108670	108840	109011	109162	109314	109011	109692
−.1	−.3	−.3	−.3	Germany, West	37765	37866	37765	37790	37818	37843	37866	37890	37917	37942	37965	37917	38017
na	na	na	−.2	India	na	37107	na	37046	37069	37088	37107	37124	37145	37164	37182	37145	37221
.0	.0	.0	.0	Indonesia (1)	1543	1543	1543	1543	1543	1543	1543	1543	1543	1543	1543	1543	1543
.0	.0	.0	.0	Iran (1)	220	220	220	220	220	220	220	220	220	220	220	220	220
.0	.0	.0	.0	Mexico (1)	992	992	992	992	992	992	992	992	992	992	992	992	992
.0	.0	.0	.0	Nigeria (1)	110	110	110	110	110	110	110	110	110	110	110	110	110
na	na	na	−.8	Poland	na	23922	na	23759	23816	23869	23922	23975	24029	24082	24132	24029	24234
na	na	na	−.7	Romania	na	na	na	na	na	na	na	441	442	443	444	441	444
na	na	na	−.2	U.S.S.R. (total coal)	na	281889	na	na	281605	281749	281889	282028	282172	282313	282453	282172	282739
−.1	−.3	−.3	−.3	United Kingdom	49399	49533	49399	49431	49466	49504	49532	49566	49601	49637	49666	49601	49734
−.1	−.3	−.3	−.3	United States	195923	196610	195923	196106	196267	196449	196610	196788	196867	197021	197183	196867	197500
na	na	na	−10.1	Venezuela	na	1042	na	na	na	1023	1042	1071	1102	1133	1164	1102	1226
na	na	na	−3.2	Yugoslavia	na	9215	na	9011	9079	9150	9215	9291	9369	9452	9524	9369	9682

*Quarterly averages based on annual data. **Proven reserves derived from estimates at year-end 1977.

(1) Not accounting for year-to-date production.

TABLE 3 Coal Production/Reserves (Net Quadrillion Btu Heat Equivalent)

Trends: % Change—Latest vs. Prev. Q	YTD	4 Qs	Tot Yr		Year-to-Date 1979	1978	Current 9 Quarters 1979 Q2	Q1	1978 Q4	Q3	Q2	Q1	1977 Q4	Q3	Q2	Total Years 1978	1977
				PRODUCTION—Hard Coal													
150.0	*133.3*	*55.6*	*-9.1*	Argentina	.007	.003	.005	.002	.004	.003	.001	.002	.002	.003	.002	.010	.011
5.6	*2.9*	*2.6*	*5.5*	Australia	1.067	1.037	.548	.519	.575	.541	.565	.472	.538	.552	.545	2.153	2.040
-15.7	*-4.1*	*-4.3*	*-6.6*	Belgium	.094	.098	.043	.051	.050	.034	.048	.051	.050	.039	.052	.183	.196
na	*na*	*na*	*na*	Brazil	na	na	na	na	na	na	na	na	.014	.014	.014	na	.056
17.0	*14.6*	*19.8*	*11.8*	Canada	.243	.212	.131	.112	.117	.116	.112	.100	.096	.090	.106	.445	.398
na	*12.4*	*na*	*12.4*	China*	17.166	15.277	na	na	4.291	4.291	4.291	4.291	3.819	3.819	3.819	17.166	15.277
na	*na*	*na*	*1.1*	Czechoslovakia	na	.291	na	.152	.147	.134	.143	.148	.148	.135	.140	.572	.566
-7.4	*-5.9*	*-7.2*	*-7.4*	France	.287	.305	.138	.149	.142	.098	.151	.154	.153	.109	.164	.547	.591
-11.6	*5.0*	*1.1*	*-1.3*	Germany, West	1.326	1.263	.622	.704	.654	.586	.587	.675	.648	.627	.573	2.503	2.536
na	*na*	*na*	*-.6*	India	na	.976	na	.578	.471	.453	.444	.532	.472	.445	.441	1.899	1.911
na	*na*	*na*	*4.3*	Japan	na	.268	na	.125	.127	.134	.129	.139	.135	.123	.121	.529	.507
-1.8	*2.9*	*3.6*	*4.9*	Korea, South	.216	.210	.107	.109	.117	.102	.107	.104	.109	.101	.100	.430	.410
50.0	*.0*	*.0*	*-9.1*	New Zealand	.005	.005	.003	.002	.002	.003	.003	.002	.003	.003	.003	.010	.011
.0	*-33.3*	*-20.0*	*.0*	Norway	.004	.006	.002	.002	.002	.003	.003	.003	.003	.001	.003	.011	.011
na	*na*	*na*	*3.5*	Poland	na	1.733	na	.921	.854	.853	.862	.872	.851	.814	.819	3.440	3.324
na	*na*	*na*	*.0*	Portugal	na	.002	na	.001	.001	.001	.001	.001	.001	.001	.001	.003	.003
na	*na*	*na*	*na*	Romania	na	na	na	na	na	na	na	na	.025	.025	.025	na	.099
na	*4.8*	*na*	*6.3*	Spain	.109	.104	na	na	.059	.054	.066	.057	.059	.051	.062	.237	.223
na	*na*	*na*	*-.2*	U.S.S.R. (includes lignite)	na	8.196	na	na	4.203	4.066	4.043	4.153	4.134	4.089	4.066	16.465	16.491
-8.2	*-.4*	*.8*	*.8*	United Kingdom	1.702	1.708	.815	.888	.945	.726	.839	.869	.917	.720	.851	3.380	3.352
13.8	*34.0*	*20.1*	*-5.3*	United States (includes lignite)	8.660	6.463	4.610	4.050	4.590	4.064	4.478	1.985	3.873	4.083	4.297	15.117	15.964
na	*na*	*na*	*na*	Venezuela	na	.001	na	na	na	.000	.001	.001	.001	.001	.001	na	.003
na	*na*	*na*	*-12.5*	Yugoslavia	na	.004	na	.002	.002	.002	.002	.002	.002	.002	.002	.007	.008
				PRODUCTION—Brown Coal/Lignite													
22.7	*-4.5*	*5.5*	*6.1*	Australia	.147	.154	.081	.066	.073	.085	.083	.072	.062	.074	.080	.313	.295
na	*na*	*na*	*-2.7*	Austria	na	.018	na	.008	.009	.009	.009	.010	.010	.009	.009	.036	.037
na	*na*	*na*	*na*	Bulgaria	na	.098	na	na	na	na	.048	.049	.051	.049	.051	na	.197
-14.3	*8.3*	*6.1*	*.5*	Canada	.117	.108	.054	.063	.062	.048	.045	.063	.057	.049	.050	.219	.218
na	*na*	*na*	*2.2*	Czechoslovakia	na	.470	na	.241	.251	.224	.224	.246	.238	.219	.227	.945	.925
na	*na*	*na*	*-.1*	Germany, East	na	1.017	na	na	.512	.482	.488	.529	.520	.485	.488	2.011	2.014
-16.1	*10.6*	*6.5*	*.5*	Germany, West	.491	.444	.224	.267	.263	.224	.206	.238	.257	.217	.208	.932	.927
na	*na*	*na*	*-5.4*	Greece	na	.072	na	.035	.035	.034	.034	.038	.037	.035	.034	.141	.149
-12.7	*.0*	*.5*	*.5*	Hungary	.102	.102	.048	.055	.057	.048	.049	.054	.055	.049	.047	.207	.206
.0	*-17.6*	*-8.1*	*-12.2*	New Zealand	.014	.017	.007	.007	.010	.010	.009	.008	.010	.010	.009	.036	.041
na	*na*	*na*	*.6*	Poland	na	.166	na	.073	.086	.074	.079	.087	.081	.077	.076	.325	.323
na	*na*	*na*	*na*	Romania	na	na	na	na	na	na	na	na	.035	.035	.035	na	.139
na	*51.2*	*na*	*55.1*	Spain	.062	.041	na	na	.039	.032	.022	.028	.021	.021	.022	.121	.078
na	*na*	*na*	*na*	Thailand	na	.002	na	na	na	.002	.002	0	.002	.002	.002	na	.008
na	*na*	*na*	*1.2*	Yugoslavia	na	.165	na	.095	.094	.082	.078	.087	.089	.087	.077	.341	.337
				RESERVES—Hard Coal**													
.0	*.0*	*.0*	*.0*	Brazil (1)	224.99	224.99	224.99	224.99	224.99	224.99	224.99	224.99	224.99	224.99	224.99	224.99	224.99
-.1	*-.2*	*-.2*	*-.2*	Canada	260.36	260.89	260.36	260.51	260.64	260.76	260.89	261.02	261.12	261.22	261.32	261.12	261.54
na	*na*	*na*	*-.6*	China*	na	2738.46	na	na	2729.88	2734.17	2738.46	2742.75	2747.04	2750.86	2754.68	2747.04	2764.21
-.1	*-.3*	*-.3*	*-.3*	Germany, West	951.66	954.22	951.66	952.30	952.99	953.63	954.22	954.83	955.49	956.13	956.72	955.49	958.02
na	*na*	*na*	*-.2*	India	na	935.08	na	933.55	934.13	934.61	935.08	935.52	936.05	936.52	936.97	936.05	937.97
.0	*.0*	*.0*	*.0*	Indonesia (1)	38.89	38.89	38.89	38.89	38.89	38.89	38.89	38.89	38.89	38.89	38.89	38.89	38.89
.0	*.0*	*.0*	*.0*	Iran (1)	5.56	5.56	5.56	5.56	5.56	5.56	5.56	5.56	5.56	5.56	5.56	5.56	5.56
.0	*.0*	*.0*	*.0*	Mexico (1)	25.00	25.00	25.00	25.00	25.00	25.00	25.00	25.00	25.00	25.00	25.00	25.00	25.00
.0	*.0*	*.0*	*.0*	Nigeria (1)	2.78	2.78	2.78	2.78	2.78	2.78	2.78	2.78	2.78	2.78	2.78	2.78	2.78
na	*na*	*na*	*-.8*	Poland	na	602.82	na	598.71	600.16	601.49	602.82	604.16	605.52	606.85	608.13	605.52	610.68
na	*na*	*na*	*-.7*	Romania	na	na	na	na	na	na	na	11.11	11.14	11.17	11.19	11.11	11.19
na	*na*	*na*	*-.2*	U.S.S.R. (total coal)	na	7103.54	na	na	7096.37	7100.01	7103.54	7107.04	7110.65	7114.21	7117.73	7110.65	7124.96
-.1	*-.3*	*-.3*	*-.3*	United Kingdom	1244.83	1248.22	1244.83	1245.64	1246.53	1247.47	1248.19	1249.05	1249.92	1250.83	1251.55	1249.92	1253.28
-.1	*-.3*	*-.3*	*-.3*	United States	4937.22	4954.53	4937.22	4941.83	4945.88	4950.47	4954.53	4959.01	4961.00	4964.87	4968.95	4961.00	4976.96
na	*na*	*na*	*-10.1*	Venezuela	na	26.25	na	na	na	25.78	26.25	27.00	27.78	28.55	29.33	27.78	30.89
na	*na*	*na*	*-3.2*	Yugoslavia	na	232.21	na	227.07	228.79	230.57	232.21	234.12	236.10	238.18	240.01	236.10	243.98

*Quarterlv averages based on annual data. **Proven reserves derived from estimates at year-end 1977.

(1) Not accounting for year-to-date production.

TABLE 4 Coal Consumption (Thousand Metric Tons Hard Coal Equivalent)

Trends: % Change—Latest vs. Prev. Q	YTD	4 Qs	Tot Yr		Year-to-Date 1979	1978	Current 9 Quarters 1979 Q2	Q1	1978 Q4	Q3	Q2	Q1	1977 Q4	Q3	Q2	Total Years 1978	1977
				COAL REQUIREMENTS													
.1	12.5	7.5	4.1	Belgium	8084	7185	4045	4039	4022	3242	3568	3617	3644	3446	3527	14450	13886
223.9	30.4	15.24	4.3	Canada	11886	9117	9081	2804	9263	5996	6285	2832	6984	7454	5739	24375	23375
			12.0	China												457970	438970
19.4	15.7	20.8	11.8	Denmark	2926	2530	1592	1333	2158	1412	1302	1227	1306	1540	1522	6099	5457
−1.1	2.4	1.7	2.1	France	23990	23421	11930	12061	12010	9179	11739	11683	11401	9591	11433	44708	43803
−4.7	13.5	11.0	6.4	Germany, West	64328	56692	31392	32937	33462	27554	27835	28858	29865	26315	26338	117677	110571
.0	.0	−4.8	−6.8	Ireland	384	384	192	192	192	192	192	192	225	198	153	823	883
−.3	−17.0	−8.9	−.6	Italy	5464	6583	2727	2736	3325	2959	3254	3329	3153	3156	3015	12858	12941
na	na	na	−14.3	Japan*	na	26064	na	na	13032	13032	13032	13032	15192	15192	15192	52124	60813
−13.6	11.7	7.2	−.1	Netherlands	2693	2411	1248	1445	1411	1110	1443	968	1386	1067	1269	4930	4934
na	19.3	na	na	Spain	8052	6751	na	na	na	na	na	na	na	na	na	na	15171
60.5	33.8	12.1	−9.7	Sweden	943	705	581	362	717	533	399	306	554	697	na	1955	2165
			7.5	Switzerland												360	335
1.9	3.5	1.5	−1.7	United Kingdom	62981	60873	31781	31201	32743	25889	29636	31238	32255	26731	30722	119546	121634
−10.8	11.9	9.0	2.1	United States	278340	248665	131212	147127	136585	134310	114301	134360	121171	133995	117704	519559	509032
				EEC CONSUMPTION BY POWER PLANTS													
−18.0	5.7	8.5	12.4	Belgium	1773	1677	799	974	951	679	768	909	797	661	675	3347	2977
−25.9	26.6	23.8	8.5	Denmark	2569	2030	1093	1476	1187	930	977	1054	1098	657	831	4154	3829
−26.2	13.3	15.8	17.4	France	7312	6456	3104	4208	4344	2797	2904	3552	3748	2283	2195	13602	11588
−22.3	8.6	6.7	4.0	Germany, West	32012	29484	13999	18014	17497	13233	13616	15869	16735	12569	12468	60062	57735
−15.3	.2	−2.3	−3.5	Ireland	485	484	222	262	245	251	195	289	268	252	231	981	1017
9.7	55.7	38.5	24.5	Italy	1615	1037	845	770	722	481	441	596	640	359	351	2229	1791
−19.8	12.6	9.5	1.7	United Kingdom	38418	34113	17093	21325	18327	13525	15093	19020	16848	13199	15048	65984	64852

*Quarterly averages based on annual data.

TABLE 5 Coal Consumption (Thousand Short Tons)

Trends					Year-to-Date		Current 9 Quarters									Total Years	
% Change—Latest vs. Prev.							1979		1978				1977				
Q	YTD	4 Qs	Tot Yr		1979	1978	Q2	Q1	Q4	Q3	Q2	Q1	Q4	Q3	Q2	1978	1977
				COAL REQUIREMENTS													
.2	12.5	7.5	4.1	Belgium	8911	7920	4459	4452	4434	3574	3934	3987	4017	3799	3888	15928	15307
223.8	30.4	15.2	9.9	Canada	13102	10050	10010	3091	10210	6609	6928	3121	7699	8217	6326	26869	24448
			12.0	China												504823	450810
19.4	15.7	20.8	11.8	Denmark	3225	2788	1755	1470	2379	1556	1436	1353	1440	1698	1678	6723	6015
−1.1	2.4	1.7	2.1	France	26444	25817	13150	13294	13239	10118	12940	12878	12567	10572	12603	49282	48284
−4.7	13.5	11.0	6.4	Germany, West	70909	62492	34603	36306	36886	30373	30683	31810	32921	29008	29033	129716	121883
.0	.0	−4.8	−6.8	Ireland	423	423	212	212	212	212	212	212	248	218	169	907	973
−.3	−17.0	−8.9	−.6	Italy	6022	7256	3006	3016	3665	3261	3587	3669	3475	3478	3323	14174	14265
na	na	na	−14.3	Japan*	na	28730	na	na	14365	14365	14365	14365	16746	16746	16746	57457	67034
−13.6	11.7	7.2	−.1	Netherlands	2968	2658	1376	1592	1556	1224	1591	1067	1527	1176	1399	5435	5439
na	19.3	na	na	Spain	8876	7441	na	na	na	na	na	na	na	na	na	na	16724
60.4	33.7	12.1	−9.7	Sweden	1039	777	640	399	790	588	440	337	611	768	na	2155	2386
			7.6	Switzerland												397	369
1.9	3.5	1.5	−1.7	United Kingdom	69425	67101	35032	34393	36093	28538	32668	34433	35555	29466	33865	131776	134078
−10.8	11.9	9.0	2.1	United States	306815	274105	144635	162179	150559	148051	125994	148106	133568	147704	129746	572712	561109
				EEC CONSUMPTION BY POWER PLANTS													
−17.9	5.7	8.5	12.4	Belgium	1954	1849	881	1073	1049	748	847	1002	878	729	744	3690	3282
−25.9	26.5	23.8	8.5	Denmark	2832	2238	1205	1627	1309	1026	1077	1161	1211	724	916	4579	4220
−26.2	13.3	15.8	17.4	France	8060	7116	3422	4638	4788	3084	3201	3916	4131	2517	2420	14994	12774
−22.3	8.6	6.7	4.0	Germany, West	35288	32501	15431	19857	19287	14586	15009	17492	18447	13855	13743	66207	63642
−15.2	.2	−2.3	−3.6	Ireland	534	533	245	289	270	277	215	319	296	278	254	1081	1121
9.7	55.7	38.5	24.5	Italy	1780	1143	931	849	796	530	487	656	705	396	387	2457	1974
−19.8	12.6	9.5	1.7	United Kingdom	42348	37603	18841	23507	20202	14909	16638	20966	18572	14550	16587	72735	71487

*Quarterly averages basedon annual data.

TABLE 6 Coal Consumption (Net Quadrillion Btu Heat Equivalent)

Trends: % Change—Latest vs. Prev. Q	YTD	4 Qs	Tot Yr		Year-to-Date 1979	Year-to-Date 1978	Current 9 Quarters: 1979 Q2	1979 Q1	1978 Q4	1978 Q3	1978 Q2	1978 Q1	1977 Q4	1977 Q3	1977 Q2	Total Years 1978	Total Years 1977
				COAL REQUIREMENTS													
.2	12.5	7.5	4.0	Belgium	.2245	.1995	.1123	.1121	.1117	.0900	.0991	.1004	.1012	.0957	.0979	.4013	.3857
223.7	30.4	15.2	4.3	Canada	.3301	.2532	.2522	.0779	.2572	.1665	.1745	.0786	.1940	.2070	.1594	.6770	.6493
			12.0	China												12.7214	11.3603
19.5	15.7	20.8	11.8	Denmark	.0812	.0702	.0442	.0370	.0599	.0392	.0361	.0340	.0362	.0427	.0422	.1694	.1515
−1.1	2.4	1.7	2.1	France	.6663	.6505	.3313	.3350	.3336	.2549	.3260	.3245	.3166	.2664	.3175	1.2418	1.2167
−4.7	13.5	11.0	6.4	Germany, West	1.7868	1.5747	.8719	.9149	.9295	.7653	.7731	.8016	.8295	.7309	.7316	3.2688	3.0714
.0	.0	−4.8	−6.9	Ireland	.0106	.0106	.0053	.0053	.0053	.0053	.0053	.0053	.0062	.0054	.0042	.0228	.0245
−.4	−17.0	−8.9	−.6	Italy	.1517	.1828	.0757	.0760	.0923	.0821	.0903	.0924	.0875	.0876	.0837	.3571	.3594
na	na	na	−14.3	Japan*	na	.7240	na	na	.3620	.3620	.3620	.3620	.4220	.4220	.4220	1.4479	1.6892
−13.7	11.8	7.2	−.1	Netherlands	.0748	.0669	.0346	.0401	.0392	.0308	.0400	.0268	.0384	.0296	.0352	.1369	.1370
na	19.3	na	na	Spain	.2236	.1875	na	na	na	na	na	na	na	na	na	na	.4214
61.0	33.8	12.1	−9.7	Sweden	.0261	.0195	.0161	.0100	.0199	.0148	.0110	.0084	.0153	.0193	na	.0543	.0601
			6.5	Switzerland												.0099	.0093
1.9	3.5	1.5	−1.7	United Kingdom	1.7494	1.6909	.8828	.8666	.9095	.7191	.8232	.8677	.8959	.7425	.8533	3.3207	3.3787
−10.8	11.9	9.0	2.1	United States	7.7316	6.9073	3.6447	4.0868	3.7940	3.7308	3.1750	3.7322	3.3658	3.7221	3.2695	14.4322	14.1398
				EEC CONSUMPTION BY POWER PLANTS													
−18.1	5.8	8.6	12.3	Belgium	.0492	.0465	.0221	.0270	.0264	.0188	.0213	.0252	.0221	.0183	.0187	.0929	.0827
−25.9	26.6	23.8	8.5	Denmark	.0713	.0563	.0303	.0409	.0329	.0258	.0271	.0292	.0305	.0182	.0230	.1153	.1063
−26.2	13.3	15.8	17.4	France	.2031	.1793	.0862	.1168	.1206	.0777	.0806	.0986	.1040	.0634	.0609	.3778	.3218
−22.3	8.6	6.7	4.0	Germany, West	.8892	.8190	.3888	.5003	.4860	.3675	.3782	.4407	.4648	.3491	.3463	1.6684	1.6037
−15.3	.0	−2.3	−3.5	Ireland	.0134	.0134	.0061	.0072	.0068	.0069	.0054	.0080	.0074	.0069	.0064	.0272	.0282
9.9	55.6	38.5	24.5	Italy	.0448	.0288	.0234	.0213	.0200	.0133	.0122	.0165	.0177	.0099	.0097	.0619	.0497
−19.9	12.6	9.5	1.7	United Kingdom	1.0671	.9475	.4747	.5923	.5090	.3756	.4192	.5283	.4680	.3666	.4179	1.8328	1.8014

*Quarterly averages based on annual data.

Source: Data reproduced from *World Energy Industry*, Vol. 1, No. 1, with permission of the publisher, Business Information Display, Inc., San Diego, Calif.

TABLE 7 List of Main Sources

National sources
- Algeria
  - Sonatrach
- Argentina
  - Banco Central de la Republica Argentina
  - Ministry of Economy
- Australia
  - Australian Bureau of Statistics
  - Australian Information Service
  - Australian Institute of Petroleum
- Austria
  - Osterreichisches Statistisches Zentralamt
- Belgium
  - Institute National de Statistique
- Brazil
  - Banco Central do Brazil
  - Petrobras
- Canada
  - Bank of Canada
  - Department of Energy, Mines and Resources
  - Statistics Canada
- Colombia
  - Colombia Information Service
  - Empresa Colombiana de Petroleos
- Denmark
  - Danmarks Statistik
- Finland
  - Bank of Finland
  - Tilastokeskus
  - Board of Customs
- France
  - Ministere de L'Industrie
  - Comite Professionnel du Petrole
  - Institut National de la Statistique et des Etudes Economiques (INSEE)
- Germany, West
  - Statistisches Bundesamt Wiesbaden
- Greece
  - National Statistical Service of Greece
- Indonesia
  - Biro Pusat Statistik
  - Directorate General of Oil and Natural Gas
  - Pertamina
- Italy
  - Instituto Centrale di Statistica
  - Union Petrolifera
- Japan
  - Ministry of International Trade and Industry
  - Industrial Bank of Japan
  - Petroleum Association of Japan
  - Japan Tariff Association
- Korea, South
  - Bank of Korea
  - National Bureau of Statistics
- Mexico
  - Banco de Mexico
  - Direccion General de Estadistica
- New Zealand
  - Department of Statistics

Norway
  Statistik Sentralbyra
Portugal
  Instituto Nacional de Estatistica
  Direccao-General de Energia
Singapore
  Department of Statistics
  Singapore Trade Office
Spain
  Spanish Economic News Service
  Instituto Nacional de Estadistica
  Ministerio de Industria y Energia
Sweden
  Statistiska Centralbyran

Switzerland
  Eidgenossisches Amt fur Energiewirtschaft
  Eidgenossische Oberzolldirektion
  Union Petroliere
Thailand
  Bangkok Bank Ltd.
Trinidad and Tobago
  Central Statistical Office
  Ministry of Energy and Energy-Based Industries
United Kingdom
  Department of Energy
United States
  Department of Energy
Venezuela
  Banco Central de Venezuela

General sources (applying to more than one country)
  Intergovernmental organizations
    European Economic Community (EEC)
    International Monetary Fund (IMF)
    Organization for Economic Cooperation and Development (OECD)
    Organization of Arab Petroleum Exporting Countries (OAPEC)
    Organization of Petroleum Exporting Countries (OPEC)
    United Nations (UN)
  Trade associations
    American Nuclear Society
    American Petroleum Institute
  General journals
    Oil and Gas Journal (crude oil and natural gas reserves at beginning year)
    Petroleum Intelligence Weekly (crude oil prices in $US/barrel)
  Other organizations
    United States Central Intelligence Agency (CIA)

REFERENCES

1. S. H. Shur and B. C. Netschert, *Energy in the American Economy, 1850-1975.* Resources for the Future. John Hopkins Press, Baltimore, Md., 1960, 774pp.
2. J. Darmstadter, *Energy in the World Economy.* Resources for the Future. John Hopkins Press, Baltimore, Md., 1971, 876pp.
3. R. L. Gordon, *U.S. Coal and The Electric Power Industry.* Resources for the Future. John Hopkins Press, Baltimore, Md., 1975, 210pp.
4. National Coal Association, *Steam-Electr. Plant Factors*, various issues, Washington, D.C.
5. R. F. Ayers, Oil Consumption in Electricity Generation, M.S. thesis in mineral economics, Pennsylvania State University, 1973, 89pp.
6. National Electric Reliability Council, Fossil and Nuclear Fuel for Electric Utility Generation Requirements and Constraints, 1977.
7. Edison Electric Institute, 1978 Annual Electric Power Survey, Washington, D.C., Apr. 1978, 46pp.
8. R. F. Ayers and R. A. Winslow, Coal Market Study for Gulf Coast Lignites to 2035, Society of Mining Engineers, AIME, New York, Feb. 1979.
9. (a) Coal Emission Technology Raises Doubts in Calif., *Energy Users News*, Nov. 13, 1978, p. 21.
9. (b) Leff, Laurel, "Southern California Edison Raises Commitment to Alternative Energy in 'Major Change'," *Wall Street Journal*, October 20, 1980.
10. A. B. Lovins, Energy Strategy: The Road Not Taken? *Foreign Affairs* 55(1):65 (1976).
11. U.S. Department of Energy, Typical Electric Bills—January 1, 1978, Energy Information Administration, Washington, D.C., Aug. 1978, 191pp.
12. U.S. Department of Energy, Annual Summary of Costs and Quality of Electric Utility Plant Fuels, Federal Energy Regulatory Commission, Washington, D.C., Sept. 1978, 200pp.
13. R. F. Ayers, W. A. Gordon, and R. A. Winslow, Technical Economic Research on Ohio Electric Utility Coal Procurement Practices and Prices, Battelle's Columbus Division report to Touche Ross and Company for the Office of Consumer's Counsel (Ohio), Columbus, Ohio, Oct. 1978, 91pp.
14. Federal Power Commission, Electric Utilities Captive Coal Operations, Bureau of Power, Washington, D.C., June 1977, 40pp.
15. Battelle Columbus Laboratories, report to private client, Columbus, Ohio, 1979.
16. R. F. Ayers, Coal—New Markets, New Prices, prepared for *Coal Week* by Battelle-Columbus Laboratories, McGraw-Hill, Inc., New York, 1977, 71pp.
17. R. W. Barnes, Evaluation of New Energy Sources for Process Heat, Dow Chemical USA for National Science Foundation, PB245-604, Washington, D.C., 1975.
18. Battelle-Columbus Laboratories, report to private client, Columbus, Ohio, 1978.
19. W. H. Jackson, The Impact of Coal on the Industrial Economy, *Tappi J.* 59:(June 1976).
20. D. W. Locklin et al., Design Trends and Operating Problems in Combustion Modification of Industrial Boilers, Battelle's Columbus Laboratories for EPA, Grant No. R-802403, Apr. 1974.

21. U.S. Department of Interior, Bureau of Mines, Coal, Cement, Lime, *Minerals Yearbook*, various issues, Washington, D.C.
22. American Iron and Steel Institute, Annual Statistical Report, various issues.
23. Battelle's Columbus Laboratories, Energy Use by the Steel Industry in North America, report to private clients, Columbus, Ohio, 1971.
24. U.S. Department of Interior, Bureau of Mines, Bituminous Coal and Lignite Distribution—Calendar Year 1976, Washington, D.C., 1977.
25. R. F. Ayers, Evaluating Market Prospects for "Clean" Coal, Energy and Environment, Proceedings of the Fifth National Conference, American Institute of Chemical Engineers, Dayton, Ohio, 1978, pp. 285-291.
26. Association of American Railroads, *Yearbook of Railroad Facts*, Economics and Finance Department, 1978, p. 26.
27. McGraw-Hill, Inc., *Keystone Coal Industry Manual*. New York, 1980, p. 116.
28. (a) Battelle Memorial Institute, report to private clients, Columbus, Ohio, and Frankfurt, Germany, 1976.
28. (b) C. L. Wilson, *Future Coal Prospects: Country and Regional Assessments*, World Coal Study. MIT Press: Cambridge, Massachusetts, 1980.
29. National Coal Association, *International Coal*, Washington, D.C., various issues.
30. U.S. Department of Energy, *Annual Report to Congress*, Energy Information Administration, Government Printing Office. Washington, D.C., 1980, 204 pp.
31. Central Intelligence Agency, *Handbook of Economic Statistics*, Washington, D.C., August 1979.
32. U.S. Department of Energy, *Monthly Energy Review*, December, 1980.

8 Coal Combustion

ARTHUR LEVY, RICHARD E. BARRETT, ROBERT D. GIAMMAR, and HERBERT R. HAZARD / Battelle, Columbus Laboratories, Columbus, Ohio

I. INTRODUCTION

Coal combustion is a complex science and art which cannot be characterized by a set of simple principles because of the variety of physical and chemical properties of coal. As a consequence it must be considered in terms that relate to overall system design. The user is interested not only in the amount of energy available from burning the coal, but must also address other integral aspects, such as fuel handling, ash removal, emissions, and environmental control techniques.

The purpose of this chapter is to provide the coal user with:

A brief explanation of coal properties and how they are used
Coal combustion principles
Coal-firing procedures and operation
Emission characteristics and environmental control techniques
Economics of coal use

Because of the highly variable nature of coal and the wide variety of equipment for firing coal, there is a vast number of technical papers and articles, as well as books, describing specific coal combustion problems and/or solutions. Hence, this chapter is limited to providing an overview of the nature of coal, its characteristics and use, and the problems associated with firing coal in today's energy- and environmentally conscious world.

II. COAL PROPERTIES

Coal properties can provide an indication of the performance of coal burning under specific operating conditions. There are a number of analyses that are available to determine coal properties. Also, their are certain empirical relationships that are convenient descriptors of coals. A brief description of important coal analyses and empirical relationships and their use is given below.

A. Coal Rank

As discussed in an earlier chapter, the American Society of Testing and Materials' (ASTM's) coal rank is an empirical method for classifying coals by certain properties, including apparent age, fixed carbon, volatile matter, inherent moisture, oxygen, and agglomerating and nonweathering indices. Thus, no single property completely defines coal rank. Coal rank is generally used only to identify broad types of coal. In addition to ASTM coal rank, other criteria used for classification of coal by rank have been proposed [1], but none has received widespread use.

B. Coal Analyses

In identifying the characteristics of coals that determine their classification and suitability for given applications, proximate and ultimate analyses are customarily used to report the various components of the coal. Proximate analysis involves determining the moisture, volatile matter, and ash proportions (by weight) and calculating the fixed carbon by difference (ASTM D3172). Proximate analysis may be reported in several ways: as-received, moisture-free, and ash (or mineral matter)-free. The as-received basis is used for combustion calculations, the dry and ash-free basis for classification purposes.

The ultimate analysis involves determining coal composition in weight of carbon, hydrogen, nitrogen, and sulfur; oxygen is determined by difference. This analysis is used in detailed fuel studies and in heat-balance computations. Table 1 lists typical proximate and ultimate analyses of various coals representative of major coal reserves in the United States. Brief descriptions of some of the more common analyses are given below.

1. *Volatile Matter*

Volatile matter is that portion of the coal driven off as gas or vapor when the coal is heated according to a standardized temperature test (ASTM D3175). It consists of a variety of organic gases, generally resulting from distillation and decomposition. Volatile products given off by coals when heated have higher hydrogen/carbon ratios than the remaining material.

2. *Fixed Carbon*

Fixed carbon is the combustible residue left after the volatile matter is driven off. In general, the fixed carbon represents that portion of the fuel that must be burned in the solid state.

3. *Moisture*

Total moisture content of a sample is customarily determined by adding the moisture loss obtained when air-drying the sample and the measured moisture content of the dried sample. Moisture does not represent all the water present in coal; water of decomposition (combined water) and of hydration are not given off under standardized test conditions.

4. *Ash*

Ash is the noncombustible residue remaining after complete coal combustion and is the final form of the mineral matter present in coal. Ash in coal is determined by ASTM D3174.

5. *Sulfur*

Sulfur is found in coal as iron pyrites, sulfates, and in organic compounds. It is undesirable because the sulfur oxides formed when it burns contribute to air pollution, and sulfur compounds contribute to combustion-system corrosion and deposits. Sulfur in coal is determined by ASTM D3177.

6. *Nitrogen*

Nitrogen is found in coal molecules and is also an undesirable coal constituent, because the nitrogen oxides that are formed when coal burns contribute to air pollution. Nitrogen in coal is determined by ASTM D3179.

TABLE 1 Typical Proximate and Ultimate Analyses and Heating Values of Coal Representative of Major Coal Reserves in the United States

Coal type	Ultimate analysis (dry) (wt %)						Proximate analysis (as received) (wt %)				Heating value (as received) (Btu/lb)
	C	H	S	N	Ash	O	Volatile matter	Fixed carbon	Moisture	Ash	
Lignite	65.7	4.5	1.0	1.2	9.2	18.4	31.4	25.9	35.5	7.2	7,100
Western subbituminous	62.6	4.0	1.0	1.0	13.6	17.8	36.6	42.8	8.1	12.5	9,400
Illinois no. 6	70.0	4.9	3.8	1.4	9.2	10.7	41.1	39.6	11.2	8.1	11,300
Eastern bituminous (high-sulfur)	70.4	4.6	4.6	1.4	10.5	8.5	37.0	46.4	6.9	9.7	11,700
Eastern bituminous (low-sulfur)	79.9	5.5	1.3	1.5	5.4	6.4	36.9	53.9	4.0	5.2	14,000

7. *Ash-Fusion Temperatures*

Ash-fusion temperatures are a set of temperatures that characterize the behavior of ash as it is heated. These temperatures are determined by heating cones of ground, pressed ash in both oxidizing and reducing atmospheres according to ASTM D1857. When coal ash is heated to high temperatures it becomes soft and sticky and, finally, fluid. These temperatures give an indication of where slagging may occur within a boiler. The initial deformation temperature is related to the temperature at which coal begins to fuse and become soft. The softening temperature is related to the temperature at which the coal ash shows an accelerated tendency to fuse and agglomerate in large masses. The fluid temperature is related to the temperature at which the coal ash becomes fluid and flows in streams. These temperatures are also affected by the furnace atmosphere; a reducing atmosphere generally gives lower ash-fusion temperatures than does an oxidizing atmosphere. The ash-fusion temperature determination is an empirical laboratory procedure, and thus it is sometimes difficult to make general statements as to its appropriateness for all types of applications.

8. *Grindability Index*

The grindability index indicates the ease of pulverizing a coal in comparison to a reference coal. This index is helpful in estimating mill capacity. The two most common methods for determining this index are the Hardgrove Grindability Method (ASTM D409) and Ball Mill Grindability Method. Coals with a low index are more difficult to pulverize.

9. *Free Swelling Index*

The free swelling index (ASTM D720) gives a measure of the extent of swelling of a coal and its tendency to agglomerate when heated rapidly. Coals with a high free swelling index are referred to as caking coals, whereas those with a low index are referred to as free-burning coals.

10. *Burning Profile*

The burning profile of a coal, obtained by thermal gravimetric analysis, is a plot of the rate of weight loss when a coal sample is heated at a fixed rate. The burning profile of a solid fuel offers an indication of ignition and burning characteristics by comparing it with other burning profiles of fuels with known performance. Figure 1 presents burning profiles for several coals.

11. *Ash Analysis*

Ash analyses give percentages of inorganic oxides present in an ash sample. Ash analyses are used for evaluation of the corrosion, slagging, and fouling potential of coal ash. The ash constituents of interest are silica (SiO_2), alumina (Al_2O_3), titania (TiO_2), ferric oxide (Fe_2O_3), lime (CaO), magnesia (MgO), potassium oxide (K_2O), sodium oxide (Na_2O), and sulfur trioxide (SO_3). An indication of ash behavior can be estimated from the relative percentages of each constituent (see the discussion on slagging in Sec. IV.D).

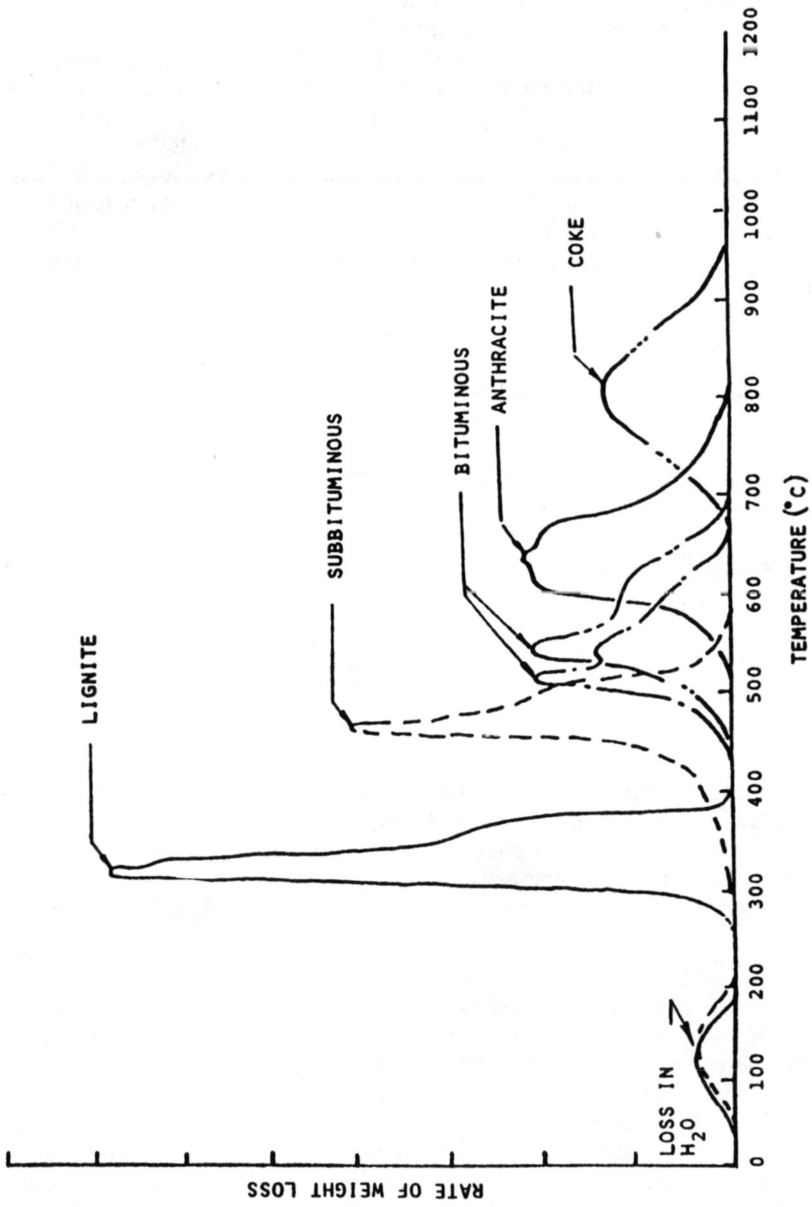

FIG. 1 TGA-derived burning profile indices used for coal-fired boiler designs [2,3].

III. COAL COMBUSTION PRINCIPLES

Coal, for purposes of thermodynamic and kinetic considerations, is a black solid made up of C, H, N, S, O, and mineral matter (ash). When a coal molecule is oxidized (burned) in air, it produces (1) heat and combustion products (CO_2 and water vapor), and (2) pollutants. The thermodynamic and kinetic aspects of coal are discussed from both these points of view.

Let us examine a typical coal, assigning it a composition (wt %) as follows:

C	H	N	S	O	Ash
68	4	1	4	8	15

On a molecular scale this coal molecule may be described as $CH_{0.7}N_{0.012}S_{0.022}O_{0.088}$, which if completely combusted would produce

CO_2	1 mol
H_2O	0.35 mol
NO	0.012 mol
SO_2	0.022 mol

On a mass basis 1 lb of this coal would evolve 10,775 Btu of heat:

$$68\%\ C \times 14{,}093\ \text{Btu/lb} = 9{,}583\ \text{Btu/lb}$$

$$4\%\ H \times 0.5 \times 51{,}623\ \text{Btu/lb} = 1{,}032\ \text{Btu/lb}$$

$$4\%\ S \times 3{,}980\ \text{Btu/lb} = 159\ \text{Btu/lb}$$

$$10{,}774\ \text{Btu/lb}$$

A. Thermodynamics

Thermodynamically, the principal equilibrium of concern from a combustion or gasification point of view is the CO/CO_2 equilibrium. Because of the present critical concern regarding pollutant emissions, however, NO/air, NO/NO_2, and SO_2/SO_3 equilibria are also of importance with respect to the burning of coal.

1. *CO/CO_2 Equilibrium*

Reactions of carbon are presented in Table 2 and their associated equilibrium constants are presented in logarithmic form in Table 3, where

$$K_p = \frac{(X)^x(Y)^y(Z)^z \cdots}{(A)^a(B)^b(C)^c \cdots}$$

$$A^aB^bC^c \cdots \rightarrow X^xY^yZ^z \cdots$$

TABLE 2 Heats of Reaction (4)

C (solid)	ΔH (kcal)
Reacting with O_2	
(1) $C + O_2 = CO_2$	-94.0
(2) $C + \frac{1}{2}O_2 = CO$	-26.6
(3) $CO + \frac{1}{2}O_2 = CO_2$	-67.4
Reacting with CO_2	
(4) $C + CO_2 = 2CO$	+40.8
Reacting with H_2O (gas)	
(5) $C + H_2O = CO + H_2$	+31.1
(6) $CO + H_2O = CO_2 + H_2$	-9.7
(4) $C + CO_2 = 2CO$	+40.8
(7) $C + 2H_2 = CH_4$	-17.9
Reacting with H_2	
(7) $C + 2H_2 = CH_4$	-17.9

TABLE 3 Equilibrium Constants for Gas/Carbon and Associated Reactions (4)

Temperature (K)	Reaction 1	Reaction 2	Reaction 3	Reaction 4	Reaction 5	Reaction 6	Reaction 7
300	+68.67	+23.93	+44.74	-20.81	-15.86	+4.95	+8.82
400	+51.54	+19.13	+32.41	-13.28	-10.11	+3.17	+5.49
500	+41.26	+16.26	+25.00	-8.74	-6.63	+2.11	+3.43
600	+34.40	+14.34	+20.06	-5.72	-4.29	+1.43	+2.00
700	+29.50	+12.96	+16.54	-3.58	-2.62	+0.96	+0.95
800	+25.83	+11.93	+13.89	-1.97	-1.36	+0.61	+0.15
900	+22.97	+11.13	+11.84	-0.71	-0.37	+0.34	-0.49
1000	+20.68	+10.48	+10.20	+0.28	+0.42	+0.14	-1.01
1100	+18.80	+9.94	+8.86	+1.08	+1.06	-0.02	-1.43
1200	+17.24	+9.50	+7.74	+1.76	+1.60	-0.16	-1.79
1300	+15.92	+9.12	+6.80	+2.32	+2.06	-0.26	-2.10
1400	+14.78	+8.79	+5.99	+2.80	+2.44	-0.36	-2.36
4000	+5.14	+5.84	-0.70	—	—	—	—

The ratio of CO/CO_2 has been shown to be controlled [5] by

$$\frac{CO}{CO_2} = 10^{3.4} e^{-12,400/RT}$$

This relationship is of limited value to coal combustion engineers because equilibrium is rarely achieved. Additionally, mixing of air and fuel is not perfect and temperatures are not constant. Hence, CO emissions from coal combustion systems are measured rather than estimated from chemical thermodynamics.

2. SO_2 / SO_3 *Equilibrium*

The sulfur oxides emitted from coal combustion are a function of the sulfur and ash composition of the coal. If we go back to our "typical" coal above, the ash content (15 wt %) will be made up of SiO_2, Al_2O_3, TiO_2, Fe_2O_3, CaO, MgO, K_2O, and Na_2O. The last four oxides, the alkaline oxides, which may make up anywhere from 5 to 30% of the ash, will capture as much as 60% of the sulfur based on the alkaline content. The remaining sulfur in the coal will be emitted as SO_2 and SO_3. The ratio of SO_2/SO_3 is a function of the excess air and the temperature of the system as presented in Table 4.

3. *NO Equilibrium*

NO_x ($NO + NO_2$) emissions in combustion are significantly more complex than are SO_x emissions. Whereas all the sulfur in coal can be considered as oxidized to SO_2 or SO_3, including the portion removed as alkali sulfate in the ash, the nitrogen in the coal is oxidized by more complex chemistry and is not greatly influenced by thermodynamics.

TABLE 4 Theoretical Conversion of SO_2 to SO_3 in Flue Gases at Different Temperatures and Excess Air Levels

Temperature		$\frac{P_{SO_3}}{P_{SO_2} + P_{SO_3}} \times 100$ for excess air			
°F	°C	10%	5%	1%	0.1%
800	427	97.1	96.1	91.7	77.9
980	527	81.2	75.8	58.7	31.1
1460	627	47.0	39.1	22.6	8.5
1340	727	20.0	15.3	7.6	2.55
1520	827	8.2	6.1	2.87	0.93
1700	927	3.7	2.7	1.25	0.40
1880	1027	1.34	1.33	0.61	0.20
2060	1127	1.01	0.73	0.33	0.11
2240	1227	0.60	0.43	0.20	0.06

Note: P, partial pressure of species indicated.

On the other hand, however, some of the nitrogen in the air used for combustion may be fixed ($N_2 + O_2 \rightarrow 2NO$) by the high temperatures of the combustion process. The fixation is controlled by kinetic and thermodynamics principles. The major NO_x emissions from coal combustion are related to fuel-bound nitrogen (some 80%); thus, nitrogen fixation equilibria are of only minor concern.

Thermodynamic equilibria are of practical significance mainly in flame temperature calculations. Theoretically, one considers many more species than those indicated in the preceding discussion. The primary combustion products CO_2, H_2O, NO, and SO_2 contribute to many atomic species (H, O, OH) and molecular species (CH_2, NH_3, HCN, C_2N_2, SO_3), which do not substantially affect the total heat release because of their negligible composition. In rich flame processes, such as in staged combustion practices, one might calculate substantial amounts of the nitrogen-containing species (NH_3, HCN, C_2N_2), yet their actual levels are extremely low in practice.

Fields et al. [6] introduce the concept of thermal equilibrium, which in contrast to thermodynamic equilibrium is of major significance in analyzing real systems. In the second or so available for the coal particle, thermodynamic equilibrium is not attained, but there does exist a thermal equilibrium or a steady state wherein the rate of heat generated by the fuel/air input rate equals the rate of heat removal, where the latter is made up of heat losses to the combustor surroundings plus enthalpy flux to the combustion products. Considerations of the rate of heat generated leads us to kinetic considerations.

B. Kinetics

1. *Coal Combustion*

Kinetic processes that control the burning of coal particles are complex, encompassing both heterogeneous and homogeneous processes. In a much oversimplified manner, there are at least four major processes:

1. Transfer of oxidant molecules to the particle surface from the gas phase
2. Devolatization and diffusion of gaseous hydrocarbons out of the coal particle
3. Chemical reaction
4. Transfer of products into the gas stream

In the practical scheme of things, the principal role of kinetics is aiding in the calculation of combustion intensity, burnout time (reaction time), and reaction volume. On the basis of the above-cited major steps for controlling the combustion of coal particles, the combustion process might be defined in terms of R_t, the mass rate of consumption of carbon per unit area of surface:

$$R_t = \frac{1}{1/R_{mt} + 1/R_{chem}}$$

in other words, combustion intensity or total burnup time will be mass- and chemically controlled. It is outside the scope of this discussion to examine this concept further. One can state in a simplified fashion [7] that

TABLE 5 Ranges of Reaction Times for Hydrocarbon Fossil Fuels at Normal Pressure [8]

Fuel	Conditions of reaction	Time (msec)
Solids (coal)	Normal pulverized coal combustion	10^3 (1 sec)
	Possible limit in explosion flame	10^2
Fuel oils	Heavy fuel oils (carbon-forming)	10^2
	Light fuel oils: large drops	10
	Light fuel oils: small drops	1
Gases	Premixed flames (industrial)	1
	Adiabatic flat flame combustion	10^{-1}
	Detonation or shock combustion	10^{-2}

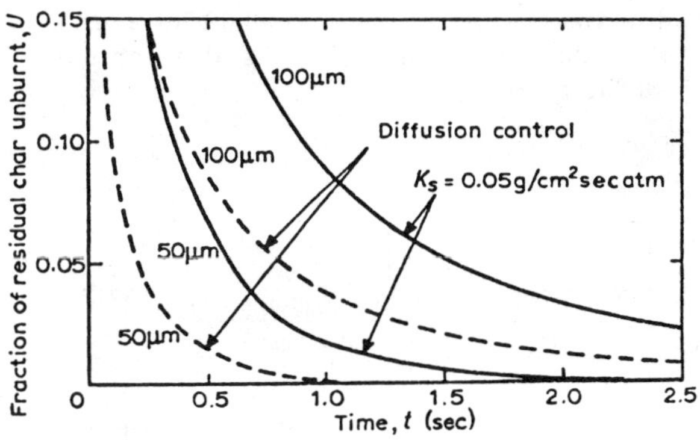

FIG. 2 Burnout times [9].

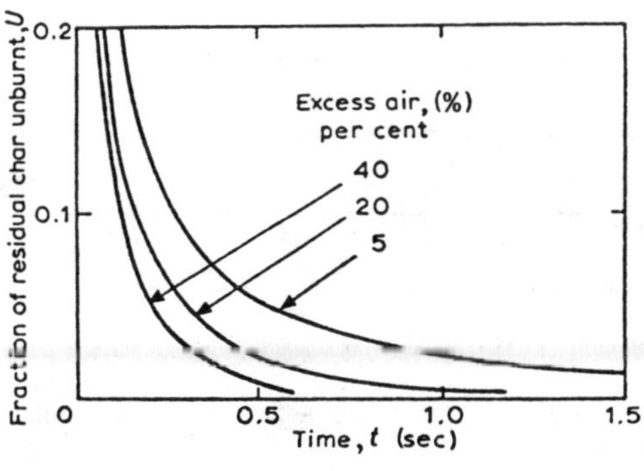

FIG. 3 Effect of excess air on burnout times [9].

Combustion intensity (C.I.) =

$$\frac{\text{mass throughput of reactants}}{\text{reactor volume}} \times \frac{\text{heat release}}{\text{unit mass of reactants}} = \frac{W}{V_c} Q_m$$

Table 5 presents ranges of reaction times calculated from the Rosin equation as applied by Essenhigh [8].

Figures 2 and 3 illustrate burnout times based on the work of Field et al. [9] on the combustion rates of pulverized coal.

Present kinetic theory is hard-pressed to calculate burnout times with an accuracy better than one to two orders of magnitude. In essence, if coal particles are in the large range, >100 μm, diffusion control is dominant, whereas for smaller particles, chemical control appears dominant.

2. *Pollutant Kinetics*

a. SO_2-SO_3 The conversion of SO_2 to SO_3 is kinetically controlled in combustion processes, and as a rule of thumb 1 to 2% of the SO_2 produced in the combustion process is converted to SO_3. The SO_3 produced in combustion is above the thermodynamic equilibrium values and is controlled by excess air and reaction kinetics. Figure 4 illustrates the effect of sulfur content on the percent conversion fuel sulfur to SO_3.

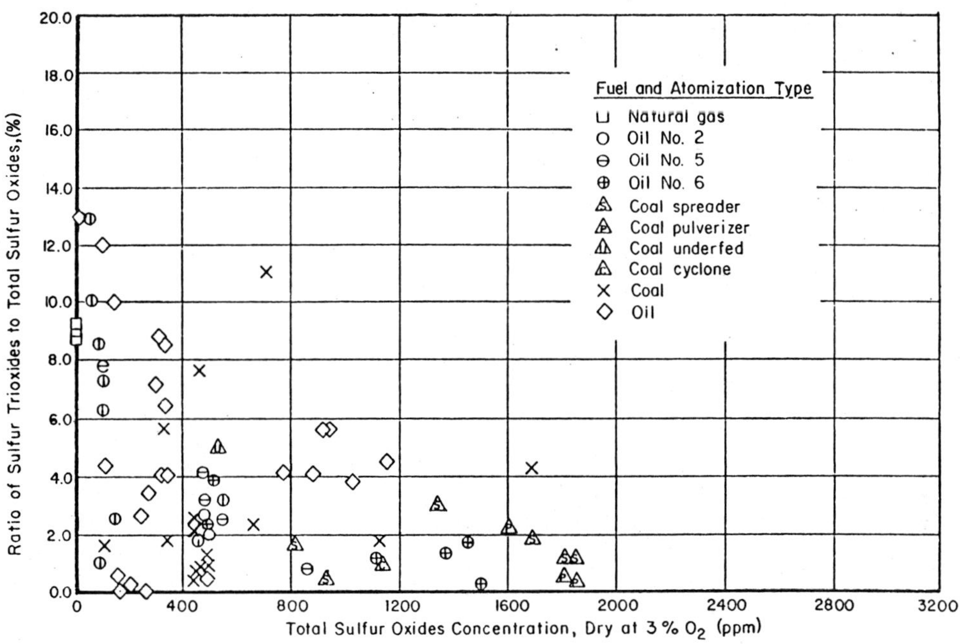

FIG. 4 Ratio of sulfur trioxides to total sulfur oxides as a function of total sulfur oxides measured (10).

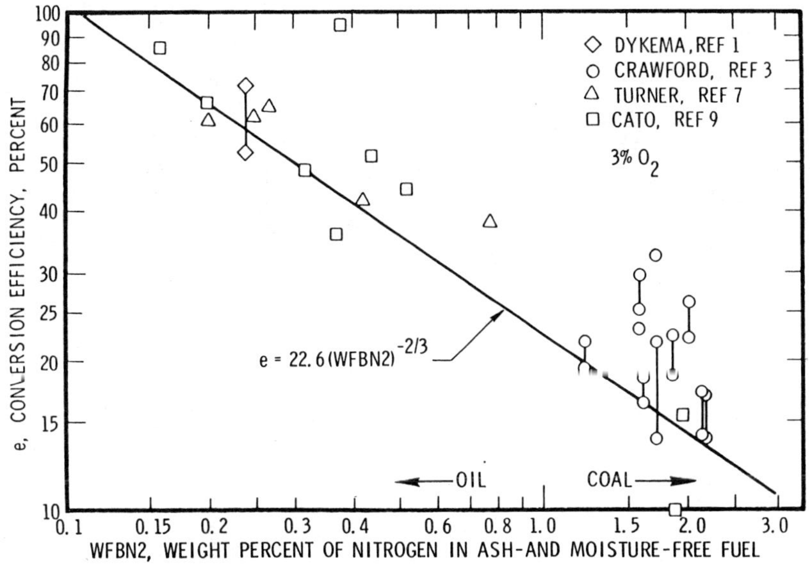

FIG. 5 Variation of the efficiency of conversion of fuel-bound nitrogen to NO_x with the concentration of nitrogen in the fuel (11).

b. Fuel Nitrogen—NO The conversion of fuel-bound nitrogen, present as high as 2% in some coals, to nitric oxide is a chemically and kinetically controlled process. Whereas all fuel sulfur is converted to SO_2 and/or SO_3, less than half the fuel nitrogen is converted to NO. Figure 5 illustrates the effect of fuel nitrogen on NO yields. Analogous to SO_3, the percent conversion falls off with increasing fuel-nitrogen content.

C. Coal Combustion Calculations

Calculations of the quantity of air required for combustion and the quantity of flue gas products generated during combustion are frequently needed for sizing system components, as input to efficiency calculations, and for emissions calculations. The calculation of other values, such as excess air and theoretical CO_2, are useful when evaluating combustion system performance.

Coal combustion calculations are generally made on a weight basis. However, coal combustion calculations can frequently be simplified by using molecular weights as the basis for the calculations. Molecular weights may be expressed in any weight units; in English units, the pound molecular weight or pound mole is commonly used.

The data required for performing coal combustion calculations include fuel analysis (C, H, O, S, N, and ash content) and flue gas analysis (CO_2, CO, and O_2).

TABLE 6 Combustion Constants

Number	Substance	Formula	Molecular weight[a]	lb/ft^3 [b]	Ft^3/lb[b]	Sp. gr. air = 1.000[b]	Heat of combustion[c] Btu/ft^3 Gross	Btu/ft^3 Net[d]	Btu/lb Gross	Btu/lb Net[d]
1	Carbon	C	12.01	—	—	—	—	—	14,093	14,093
2	Hydrogen	H_2	2.016	0.005327	187.723	0.06959	325.0	275.0	61,100	51,623
3	Oxygen	O_2	32.000	0.08461	11.819	1.1053	—	—	—	—
4	Nitrogen (atm)	N_2	28.016	0.07439[e]	13.443[e]	0.9718[e]	—	—	—	—
5	Carbon monoxide	CO	28.01	0.07404	13.506	0.9672	321.8	321.8	4,347	4,347
6	Carbon dioxide	CO_2	44.01	0.1170	8.548	1.5282	—	—	—	—
.										
.										
.										
29	Sulfur	S	32.06	—	—	—	—	—	3,938	3,983
.										
.										
.										
30	Hydrogen sulfide	H_2S	34.076	0.09109[e]	10.979[e]	1.1898[e]	647	596	7,100	6,545
31	Sulfur dioxide	SO_2	64.06	0.1733	5.770	2.264	—	—	—	—
32	Water vapor	H_2O	18.016	0.04758[e]	21.017[e]	0.6215[e]	—	—	—	—
33	Air	—	28.9	0.07655	13.063	1.0000	—	—	—	—

Number	Substance	Ft^3/ft^3 of combustible: Required for combustion O_2	N_2	Air	Flue products CO_2	H_2O	N_2	Lb/lb of combustible: Required for combustion O_2	N_2	Air	Flue products CO_2	H_2O	N_2	Experimental error in heat of combustion (% + or −)
1	Carbon	—	—	—	—	—	—	2.664	8.863	11.527	3.664	—	8.863	0.012
2	Hydrogen	0.5	1.882	2.382	—	1.0	1.882	7.937	26.407	34.344	—	8.937	26.407	0.015
3	Oxygen	—	—	—	—	—	—	—	—	—	—	—	—	—
4	Nitrogen (atm)	—	—	—	—	—	—	—	—	—	—	—	—	—
5	Carbon monoxide	0.5	1.882	2.382	1.0	—	1.882	0.571	1.900	2.471	1.571	—	1.900	0.045

6	Carbon dioxide	—	—	—	—	—	—	—	—	—	—	—	—	—
.														
.														
.														
29	Sulfur	—	—	—	—	—	—	0.998	3.287	4.285	1.998		3.287	0.071
.														
.														
.														
30	Hydrogen sulfide	1.5	5.646	7.146	SO_2 1.0	1.0	5.646	1.409	4.688	6.097	1.880	0.529	4.688	0.30
31	Sulfur dioxide	—	—	—	—	—	—	—	—	—	—	—	—	—
32	Water vapor	—	—	—	—	—	—	—	—	—	—	—	—	—
33	Air	—	—	—	—	—	—	—	—	—	—	—	—	—

Notes: All gas volumes corrected to 60°F and 30 in. Hg. dry. For gases saturated with water at 60°F, 1.73% of the Btu value must be deducted.

[a]Calculated from atomic weights given in *J. Am. Chem. Soc.*, February 1937.

[b]Densities calculated from values given in grams per liter at 0°C and 760 mm in the International Critical Tables allowing for the known deviations from the gas laws. Where the coefficient of expansion was not available, the assumed value was taken as 0.0037 per °C. Compare this with 0.003662 which is the coefficient for a perfect gas. Where no densities were available, the volume of the mol was taken as 22.4115 liters.

[c]Converted to mean Btu/lb (1/180 of the heat/lb of water from 32 to 212°F) from data by Frederick D. Rossini, National Bureau of Standards, letter of April 10, 1937, except as noted.

[d]Deduction from gross to net heating value determined by deducting 18.919 But/lb mol of water in the products of combustion. Osborne, Stimson, and Ginnings, *Mech. Eng.*, p. 163 (March 1935), and Osborne, Stimson, and Flock, National Bureau of Standards Research Paper 209.

[e]Denotes that either the density or the coefficient of expansion has been assumed. Some of the materials cannot exist as gases at 60°F and 30 in. Hg pressure, in which case the values are theoretical ones given for ease of calculation of gas problems. Under the actual concentrations in which these materials are present, their partial pressure is low enough to keep them as gases.

Source: Reprinted from *Fuel Flue Gases*, 1941 ed., courtesy of American Gas Association [12].

TABLE 7 Example of Combustion [13]

Weight method: The values in the example given below are generally developed from the data below. These calcuiations determine the quantities and the products of combustion air involved per pound of fuel.

Example 1

Weight method: solid fuel--standard air[a]
(complete combustion assumed)
See below for multipliers

Ultimate analysis (lb/lb fuel as fired)			Required for combustion (lb/lb fuel at 100% total air) O₂	Dry air
C	0.728	× 2.66 and × 11.53	1.936	8.394
H_2	0.048	× 7.94 and × 34.34	0.381	1.648
O_2	0.062		—	—
N_2	0.015		—	—
S	0.022	× 1.00 and × 4.29	0.022	0.094
H_2O	0.035		—	—
Ash	0.090		—	—
Sum	1.000		2.339	10.136
Less O_2 in fuel (deduct)			-0.062	-0.268*
Required (at 100% total air)			2.277	9.868

*Air equivalent of O_2 in fuel.

Mole method: The mole method, as exemplified below, is not quite as direct as the weight method for determining air and flue gas quantities, but is more flexible. It gives the flue gas analysis by only a simple additional step.

Example 2

Mole method: solid fuel--standard air
(complete combustion assumed)
See below for mol. wt. and multipliers

Ultimate analysis (lb/100 lb fuel as fired)		Mol/100 lb fuel		Required for combustion (mol/100 lb fuel at 100% total air) O_2	Dry air
C	72.8 ÷	12=6.07	×1.00 and ×4.76	6.07	28.89
H_2	4.8 ÷	2.016=2.38	×0.5 and ×2.38	1.19	5.66
O_2	6.2 ÷	32=0.19		—	—
N_2	1.5 ÷	28=0.05		—	—
S	2.2 ÷	32=0.07	×1.00 and ×4.76	0.07	0.33
H_2O	3.5 ÷	18=0.19		—	—
Ash	9.0	— —		—	—
Sum	100.0	8.95		7.33	34.88
Less O_2 in fuel (deduct)				-0.19	-0.90*
Required (at 100% total air)				7.14	33.98

*Air equivalent of O_2 in fuel = 0.19 × 4.76 = 0.90 mol air/0.19 mol O_2.

	Required for combustion (lb/lb fuel at 125% total air) O_2	Dry air
O_2 and air × 125/100, total	2.846	12.335
Excess air = 12.335 - 9.868	—	2.467
Excess O_2 = 2.846 - 2.277	0.569	—

Products of combustion

			lb/lb Fuel at 125% total air
CO_2	0.728	×3.66	2.664
H_2O	0.048	×8.94+0.035 +0.013×12.335	0.624
SO_2	0.022	×2.00	0.044
O_2	(excess)		0.569
N_2	12.335	×0.7685+0.015	9.629
Weight, wet			13.530
Weight, dry = 13.530 - 0.624			12.906

	Required for combustion (mol/100 lb fuel at 125% total air) O_2	Dry air
O_2 and air × 125/100, total	8.93	42.48
Excess air = 42.48 - 33.98	—	8.50
Excess O_2 = 8.93 - 7.14	1.79	—

Products of combustion

			Mol/100 lb fuel	% by vol, dry basis
CO_2	6.07	×1	6.07	14.30
H_2O	2.38	×1+0.19+0.89*	3.46	—
SO_2	0.07	×1	0.07	0.17
N_2	42.48	×0.79	33.56	80.91
O_2	(excess)		1.79	4.32
Wet			44.95	
Dry = 44.95 - 3.46			41.49	100.00

*= mol H_2O in air
= (42.48 × 29 × 0.013) ÷ 18 = 0.89.

[a] Air at 60% relative humidity and 80°F dry bulb, or 0.013 lb of moisture/lb dry air.

1. *Combustion Reactions*

The three most prevalent elements in hydrocarbon fuels are combusted (oxidized) by the following reactions:

$$C + O_2 \rightarrow CO_2$$

$$2H_2 + O_2 \rightarrow 2H_2O$$

$$S + O_2 \rightarrow SO_2$$

These reactions are readily expressed in molecular, mole, or weight units. Since these oxidations occur with air (79% N_2, 21% O_2 for practical purposes) the preceding reactions become

$$C + O_2 + 3.76N_2 \rightarrow CO_2 + 3.76N_2$$

$$2H_2 + O_2 + 3.76N_2 \rightarrow 2H_2O + 3.76N_2$$

$$S + O_2 + 3.76N_2 \rightarrow SO_2 + 3.76N_2$$

Table 6 summarizes many values related to the use of these equations.

2. *Detailed Combustion Calculations*

Combustion calculations can be made by performing the step-by-step analysis associated with balancing the foregoing reactions, or they may be made using simplified formulas. Performing the step-by-step analysis may give a better understanding of the combustion process, but it can become laborious because of the need for making many routine calculations.

Table 7 presents the step-by-step calculation procedures for making coal combustion calculations. The following sections give formulas for quickly calculating important values.

a. Air Required for Combustion Stoichiometric or theoretical air is the exact quantity of air required to provide sufficient oxygen for complete combustion of a fuel.

The weight of dry air required to supply a given quantity of oxygen is 4.32 times the weight of the oxygen required. Oxygen contained in the fuel, except that in ash, should be deducted from the amount of oxygen required for combustion, because this oxygen is already combined with fuel components. Also, water vapor is always present in atmospheric air, but generally is neglected in coal combustion calculations (i.e., calculations are based on "dry air" requirements).

From the relationships presented above, the weight of dry air required for stoichiometric combustion of any hydrocarbon fuel can be calculated by

Theoretical combustion air (lb dry air/lb fuel)

$$= 0.0144(8C + 24H + 3S - 3O) \qquad (1)$$

where C, H, S, and O are the weight percentages of carbon, hydrogen, sulfur, and oxygen in the fuel, respectively.

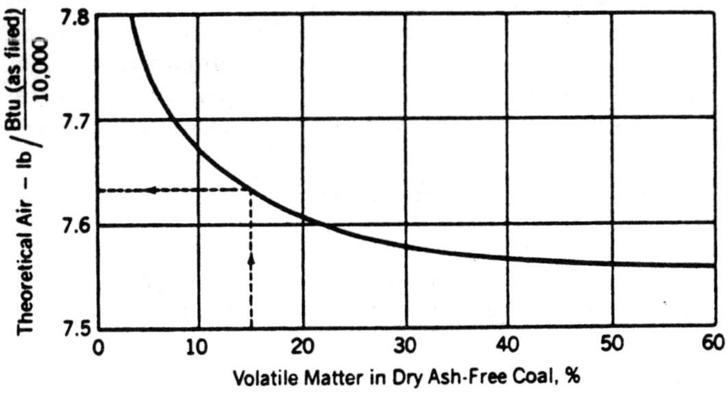

FIG. 6 Theoretical air in lb/10,000 Btu heat value of coal with a range of volatile (14).

For coal combustion, theoretical air is frequently expressed in terms of pounds per 10,000 Btu. Figure 6 can be used to estimate this value for coals with a wide range of volatile contents.

Excess air, in addition to that theoretically required for combustion, must be supplied to practical combustion systems to ensure that complete combustion is achieved.

$$\text{Excess air (\%)} = \frac{\text{air supplied} - \text{theoretical air}}{\text{theoretical air}} \times 100 \tag{2}$$

The excess air level at which a combustion process operates significantly affects its overall efficiency. Too much excess air dilutes flue gases excessively, lowering the heat-transfer temperature and increasing the sensible flue gas loss. Conversely, too low a level of excess air may cause incomplete combustion and loss of unburned combustible gases from the equipment.

General practice is to supply from 5 to 50% excess air, the exact amount depending on the type of fuel burned, combustion equipment, and other factors.

The amount of dry air supplied per pound of fuel burned may be obtained from Eq. (3), which has reasonable precision for most solid and liquid fuels.

$$\text{Dry air supplied (lb/lb of fuel)} = \frac{3.04N_2}{CO_2 + CO} \tag{3}$$

This value, together with the value for theoretical air, may be used in Eq. (2) to determine excess air.

Because excess air calculations are almost invariably made from flue gas analysis results, and theoretical air requirements are not always known, another convenient method for calculating excess air is

$$\text{Excess air (\%)} = \frac{100(O_2 - CO/2)}{0.264N_2 - (O_2 - CO/2)} \tag{4}$$

b. Quantity of Flue Gas Produced The weight of dry flue gas produced per pound of fuel burned is required in heat loss and efficiency calculations. This weight is equal to the sum of the weights of (1) fuel (minus ash retained in the furnace), (2) air theoretically required for combustion, and (3) excess air. For solid fuels, this weight, determined from the flue gas analysis, is

$$\text{Dry flue gas (lb/lb of fuel)} = \frac{11\ CO_2 + 8O_2 + 7(CO + N_2)}{3(CO_2 + CO)} \times C \tag{5}$$

where values for CO_2, CO, and N_2 are percentages, by volume, from the flue gas analysis dry, and C is the weight of carbon burner per pound of fuel, corrected for carbon loss in ash.

c. Theoretical CO_2 The theoretical CO_2 (also ultimate CO_2 or maximum CO_2) concentration attainable in products of combustion of a hydrocarbon fuel with air is the CO_2 concentration obtained when the fuel is completely burned to CO_2 and H_2O. Theoretical CO_2 varies with the carbon/hydrocarbon ratio of the fuel. Comparison of actual CO_2 values to theoretical CO_2 is a means of quickly estimating excess air values. For combustion with excess air present, theoretical CO_2 values may be calculated from a flue gas analysis:

$$\text{Theoretical } CO_2\ (\%) = \frac{\%\ CO_2 \text{ in flue gas sample}}{1 - (\%\ O_2 \text{ in flue gas sample}/21)} \tag{6}$$

In practice, desirable CO_2 values depend on the excess air and hence on the fuel, firing method, and other considerations.

The theoretical CO_2 value can also be computed from the formula

$$\text{Theoretical } CO_2\ (\%) = \frac{31.3C + 11.5S}{1.504C + 3.55H_2 + 0.56S + 0.13N_2 - 0.45O_2} \tag{7}$$

where chemical symbols other than CO_2 represent weight percentages of each material in the ultimate as-fired analysis of the fuel.

Equations (6) and (7) both assume complete combustion.

IV. COAL-FIRING PROCEDURES AND OPERATIONS

The selection of the most appropriate burning equipment for a specific application is dependent on factors which include economics, operating characteristics, type of coal, and environmental standards. The recent trend has been to fire coal in stoker-boilers with a capacity up to 200,000 lb of steam per hour and in pulverized-coal boilers above that capacity. The cyclone furnace has been used in utility applications and could satisfactorily fire low-ash-fusion coals. However, cyclones have not been built recently, as their NO_x emissions exceed federal standards. Conversely, fluidized-bed combustion has received much attention, as its provides for in situ SO_x control. In addition, as a means to extend the supplies of oil, the firing of coal mixed with oil (coal/oil mixture, COM) is being considered for boilers initially designed for oil firing.

A. Combustion Equipment

1. Stoker-Boilers

Stokers of various types are designed to mechanically feed coal uniformly onto a grate or tuyeres within a furnace, to supply combustion air to the fuel bed, and to remove ash from the zone of combustion. Although mechanical stoker technology is nearly 200 years old, it has been steadily improved, so that the modern stoker is still considered to be an important element in the industrial combustion of coal. The modern stoker-boiler system incorporates controls to coordinate air and fuel supply with changing loads, dust-collecting equipment to minimize emissions, and, in many cases, fly carbon return systems to increase efficiency.

Stokers are classified according to the method of feeding fuel to the furnace and by the type of grate. The prominent stoker types are:

1. Spreader stoker
2. Chain-grate or traveling-grate stoker
3. Vibrating-grate stoker
4. Underfeed stoker

The major features of each are summarized in Table 8. Generally, the type of stoker selected is based on boiler capacity, responsiveness to load swings, and the properties of coal to be fired. Coal properties that most significantly affect stoker performance include:

1. Moisture content
2. Percentage of coal fines
3. Friability
4. Heat content
5. Ash-fusion temperature
6. Mineral constituents
7. Volatile matter/fixed carbon ratio
8. Caking tendency

The coal properties most desired for firing of each type of stoker are discussed below in the brief description of the stoker types.

a. Spreader Stokers Spreader stokers utilize a combination of suspension burning and grate burning. As illustrated in Fig. 7, coal is continually projected into the furnace above a burning bed of fuel. The coal fines are burned in suspension; the larger particles fall to the grate, where they are burned in a thin, fast-burning fuel bed. Because ignition is almost instantaneous upon an increase in firing rate, this method of firing provides good flexibility to most load fluctuations, and the spreader stoker is favored over other types of stokers in many industrial applications.

Spreader stokers are designed to burn approximately 50% of the fuel in suspension; thus, much higher particulate loadings are generated by spreader stokers than is usual from other types of stokers. Thus, dust collectors are required to reduce particulate emissions.

TABLE 8 Characteristics of Various Types of Stokers

Stoker type and subclass	Typical capacity range (pph steam)[a]	Maximum burning rate (Btu/hr/ft^2)	Characteristics
Spreader			Capable of burning a wide range of coals, best ability to follow fluctuating loads, high fly ash carryover, low load smoke
Stationary and dumping grate	20,000–80,000	450,000	
Traveling grate	100,000–400,000	750,000	
Vibrating grate	20,000–100,000	400,000	
Underfeed			
Single or double retort	20,000–30,000	400,000	Capable of burning caking coals and a wide range of coals (including anthracite), high maintenance, low fly ash carryover, suitable for continuous-load operation
Multiple retort	30,000–500,000	600,000	
Chain grate and traveling grate	20,000–100,000	500,000	Characteristics similar to vibrating-grate stokers except these stokers experience difficulty in burning strongly caking coals
Vibrating grate	30,000–150,000	400,000	Low maintenance, low fly ash carryover, capable of burning wide variety of weakly caking coals, smokeless operation over entire range

[a]pph = lb steam/hr; 1 pph ≅ 1000 Btu/hr.

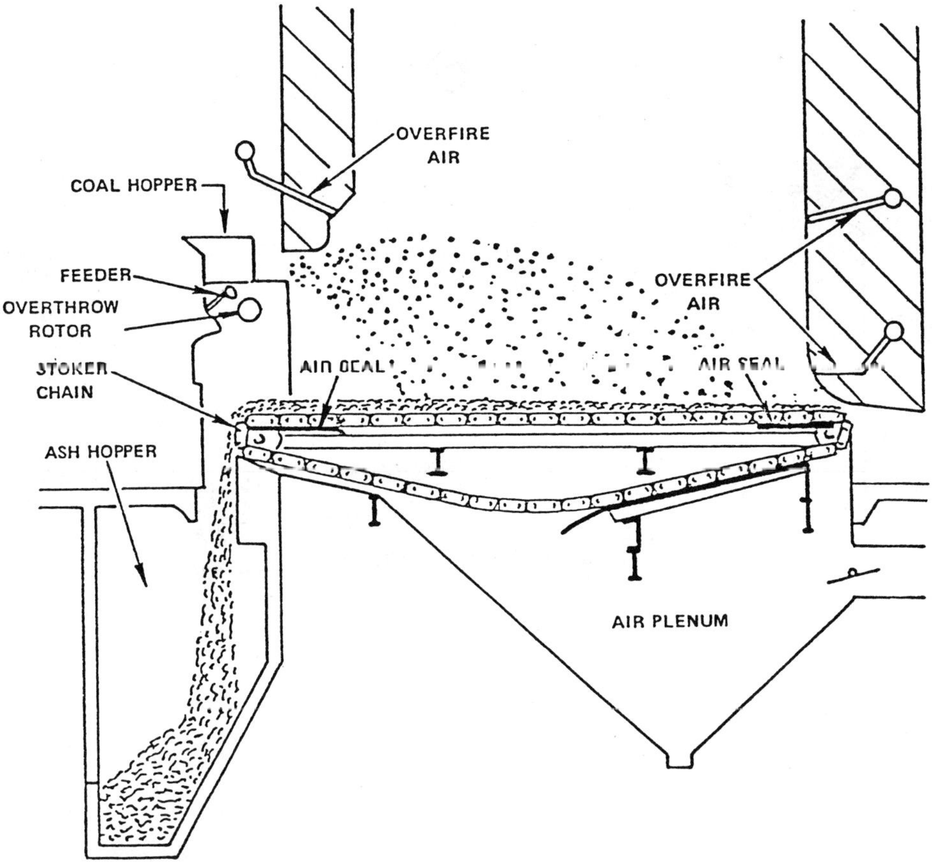

FIG. 7 Spreader stoker, traveling-grate type.

To minimize carbon loss, fly carbon reinjection systems are sometimes utilized to return to the furnace, for complete burnout some of the fly ash collected. This process increases furnace dust emissions and hence can be used only with very highly efficient dust collectors.

Spreader stokers can use stationary, dump, vibrating, or traveling grates. For high-efficiency operation, a continuous ash-discharging grate is necessary. All spreader stokers, in particular those with traveling grates, are able to fire fuels with a wide range of burning characteristics. Spreader stokers can be fired with coals having caking tendencies because the rapid surface heating of the coal in suspension destroys the caking tendency. High-moisture, free-burning bituminous and lignite coals are commonly burned, while coke breeze can be burned in a mixture with a high-volatile coal. However, anthracite, because of its low-volatile content, is not a suitable fuel for spreader-stoker firing.

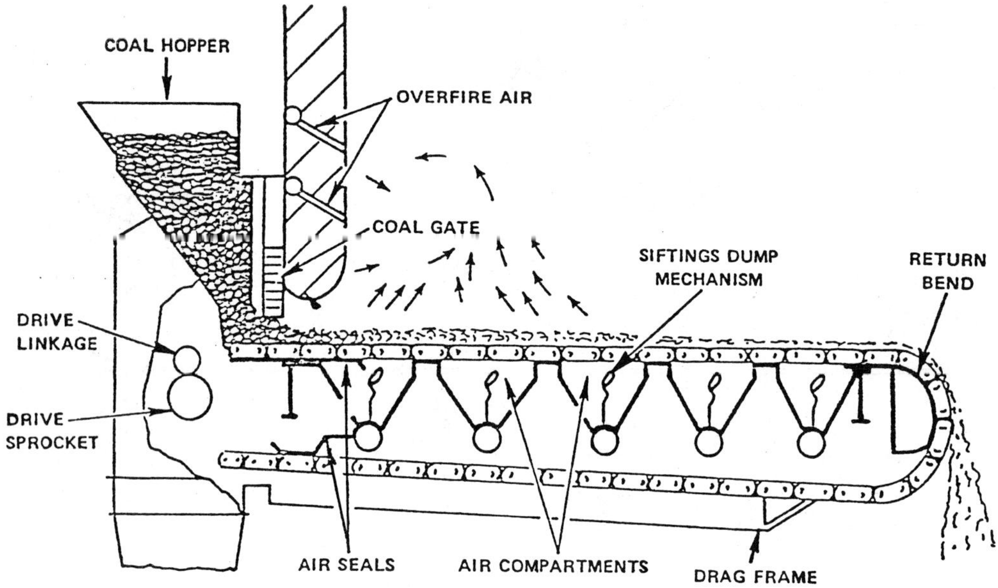

FIG. 8 Chain-grate overfeed stoker.

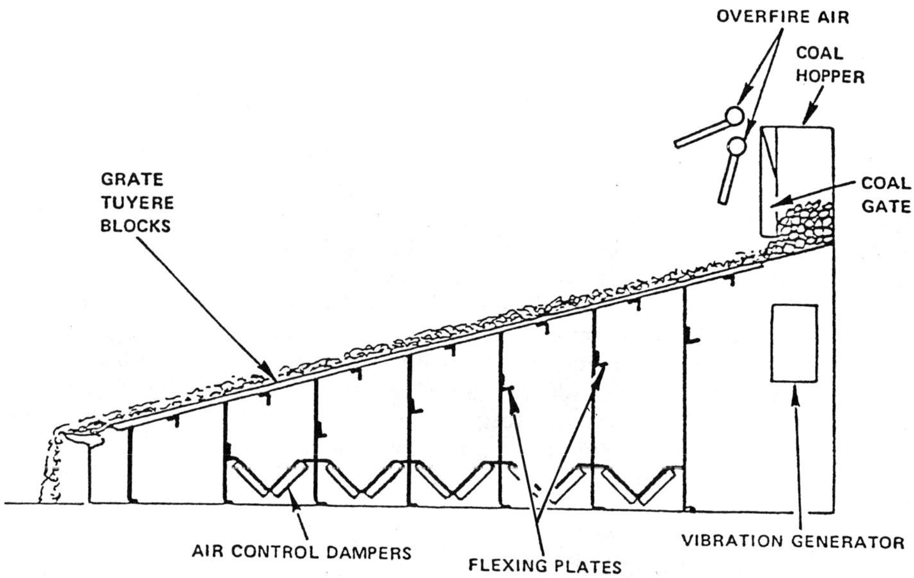

FIG. 9 Vibrating-grate stoker.

b. Chain-grate or Traveling-grate Stokers Chain-grate or traveling-grate overfeed stokers are characterized by the coal being fed onto a grate at one end of the furnace, the coal burning as the grate moves to carry the coal across the furnace, and the ash being dumped at the opposite end of the furnace. The coal-feed hopper runs along the entire coal-feed end of the furnace. A coal gate is used to control the rate at which coal is fed into the furnace by controlling the thickness of the fuel bed. The coal entering the furnace is heated by radiation from the furnace gases or from a hot refractory arch. As volatile matter is driven off by this rapid radiative heating, ignition occurs. The fuel continues to burn as it moves across the furnace, and the fuel bed becomes progressively thinner. At the far end of the furnace, where the combustion of the coal is completed, the ash is discharged into the ash pit.

In 1954, a chain-grate stoker was developed with overfire air jets that eliminated the need of a front arch for burning volatiles [15]. As shown in Fig. 8, the stoker was zoned or sectionalized and was equipped with individual zone dampers to control the pressure and quantity of air delivered to the various sections. The chain-grate and traveling-grate stokers can burn a variety of fuels, including peat, lignite, subsituminous coal, free-burning bituminous coal, anthracite coal, and coke—provided that the fuel is sized properly. However, strongly caking bituminous coals have a tendency to mat and prevent proper air distribution to the fuel bed. Also, a bed of strongly caking coal may not be responsive to rapidly changing loads.

c. Vibrating-grate Stokers The vibrating-grate stoker, as shown in Fig. 9, is similar to the chain-grate stoker in that both are overfeed, mass-burning, continuous-ash-discharge units. However, in the vibrating-grate stoker, the sloping grate is supported on vertical plates that are periodically vibrated in a rectilinear motion, causing the fuel to move from the hopper through an adjustable gate into the active combustion zone. Air is supplied to the stoker through laterally exposed areas beneath the stoker (formed by the individual flexing of the grate support plates). Ash is discharged into a shallow or basement ash pit. The grates are water-cooled and are connected to the boiler circulating system. The rate of coal feed and fuel-bed movement are controlled by the frequency and duration of the vibrating cycles and regulated by automatic combustion controls that proportion the air supply to optimize heat-release rates. Typically, the grate is vibrated about every 90 sec for durations of 2 to 3 sec, but this is dependent on the type of coal and boiler operation.

The vibrating-grate stoker has found favor because of its simplicity, inherently low fly ash carryover, low maintenance, wide turndown (10:1), and adaptability to multiple-fuel firing.

The water-cooled vibrating-grate stoker is suitable for burning a wide range of bituminous and lignite coals. Because of the gentle agitation and compaction caused by the vibratory actions, a uniform fuel bed without blowholes and thin spots can be maintained while firing caking coals. The uniformity of air distribution and resultant fuel-bed conditions produce good response to load swings and smokeless operation over the entire load range. Fly ash emission is probably greater than from the traveling grate because of the intermittent agitation of the fuel bed.

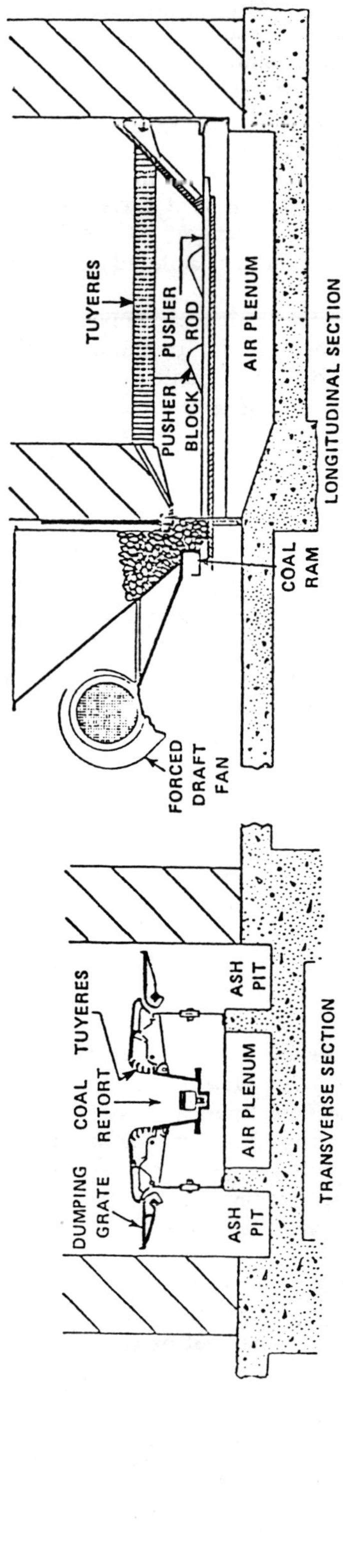

FIG. 10 Horizontal underfeed stoker with single retort.

d. Underfeed Stokers Underfeed stokers, as their name implies, introduce the raw coal into a retort beneath the burning fuel bed. The underfeed stokers are classified as horizontal feed and gravity feed. In the horizontal type, coal travels within the furnace in a retort which is parallel with the floor; in the gravity-feed type the retort is inclined by 25 deg. Generally, the horizontal-feed stokers are designed with single or double retorts, and, rarely, with triple retorts, while gravity-feed stokers are designed with multiple retorts.

In the horizontal stoker, as shown in Fig. 10, coal is fed to the retort by a screw (for the smaller stokers) or a ram (for the larger units). Once the retort is filled, the coal is forced upward and spills over the retort to both form and feed the fuel bed. Air is supplied through tuyeres at each side of the retort and through air ports in the side grates. As the coal moves upward within the retort, heat is conducted downward from the active bed above, and the volatile gases are distilled off and burned. The rising fuel bed then ignites from contact with the burning bed. The incoming raw coal displaces the fuel bed gradually over the tuyeres and side grates. The coal is completely burned by the time it reaches the dump grates at both sides, from which the ash and clinker are dropped into pits. Overfire air is used to provide additional combustion air to the flame zone directly above the bed to prevent smoking, especially at low loads.

The gravity-feed units are similar in operating principle. These underfeed stokers consist of sloping multiple retorts and have rear ash discharge. Coal is fed into each retort, where it is moved slowly to the rear while simultaneously being forced upward over the retorts.

Either type of underfeed stoker can burn a wide range of coal, although the horizontal type is better suited for free-burning bituminous coal. In general, these units can burn caking coal, provided there is not an excess amount of fines. The ash-softening temperature is an important factor in selecting coals as the possibility of excessive clinkering increases with lower ash-softening temperatures. Because combustion occurs in the fuel bed, these stokers respond slowly to load change.

2. *Pulverized-Coal Firing*

For pulverized-coal-fired boilers, coal is pulverized to a fine size (typically 70 to 80% - 200 mesh) in a pulverizer or mill; then the pulverized fuel is carried to the burners in an air stream and is blown into the furnace, where it is burned in suspension. Most of the ash remaining after combustion of the fuel remains airborne and is carried from the furnace by the flue gas stream. Some ash deposits on walls, where it agglomerates and may sinter or fuse. Ash that falls to the bottom of the furnace is removed via an ash hopper.

Most modern pulverized-coal-fired boilers contain little refractory and thus have cold walls; maintaining fuel ignition can be a problem. Thus, many pulverized-coal-fired boilers utilize preheated combustion air at temperatures of 400 to 600°F. Preheated air is also used in the pulverizer, to evaporate surface moisture and maintain satisfactory pulverizer operation.

There are several methods of firing pulverized coal, which are related to the location and position of the burner(s) in the furnace. The burners may be installed (1) horizontally for front, rear, or side wall firing; (2) vertically for firing through the roof of the furnace; (3) both in front and rear walls for opposed firing; and (4) in the corners of the furnace for tangential firing. The type of firing method used primarily depends on boiler size and

FIG. 11 B & W circular burners for pulverized coal [16].

type and the coal type. Figure 11 shows a typical pulverized-coal-fired burner. Figure 12a and b illustrates some typical burner-boiler types.

Pulverized-coal-fired boilers can fire a wide range of coals. Although bituminous coals are by far the most common fuels for pulverized coal units, everything from anthracite to lignite has been fired in utility-size pulverized-coal units.

The most important operating considerations in selecting fuel for a pulverized-coal-fired unit are:

1. Ignition and flame stability
2. Effect of ash properties, including ash handling and deposits

Ignition can be maintained by firing fuels with a minimum of about 20% volatile matter, together with the use of preheated combustion air. This includes most U.S. coals except anthracite.

Ash properties of coals affect their ash handling and deposit characteristics. Ash deposits and slagging are more of a problem in pulverize coal firing than in stoker firing. Most modern pulverized-coal-fired boilers have dry-bottom furnaces; that is, the ash is intended to be removed as a dry solid. (Wet-bottom or slag-tap pulverized-coal-fired boilers remove the ash as a flowing slag.) For dry-bottom furnaces, it is necessary that the ash-fusion temperature of the fuel be sufficiently high to avoid the ash becoming a running slag. Thus, specification for the purchase of coals for such units generally include a minimum ash-fusion temperature. (Conversely, specifications for coals for slag-tap furnaces include maximum ash-fusion temperature limits.)

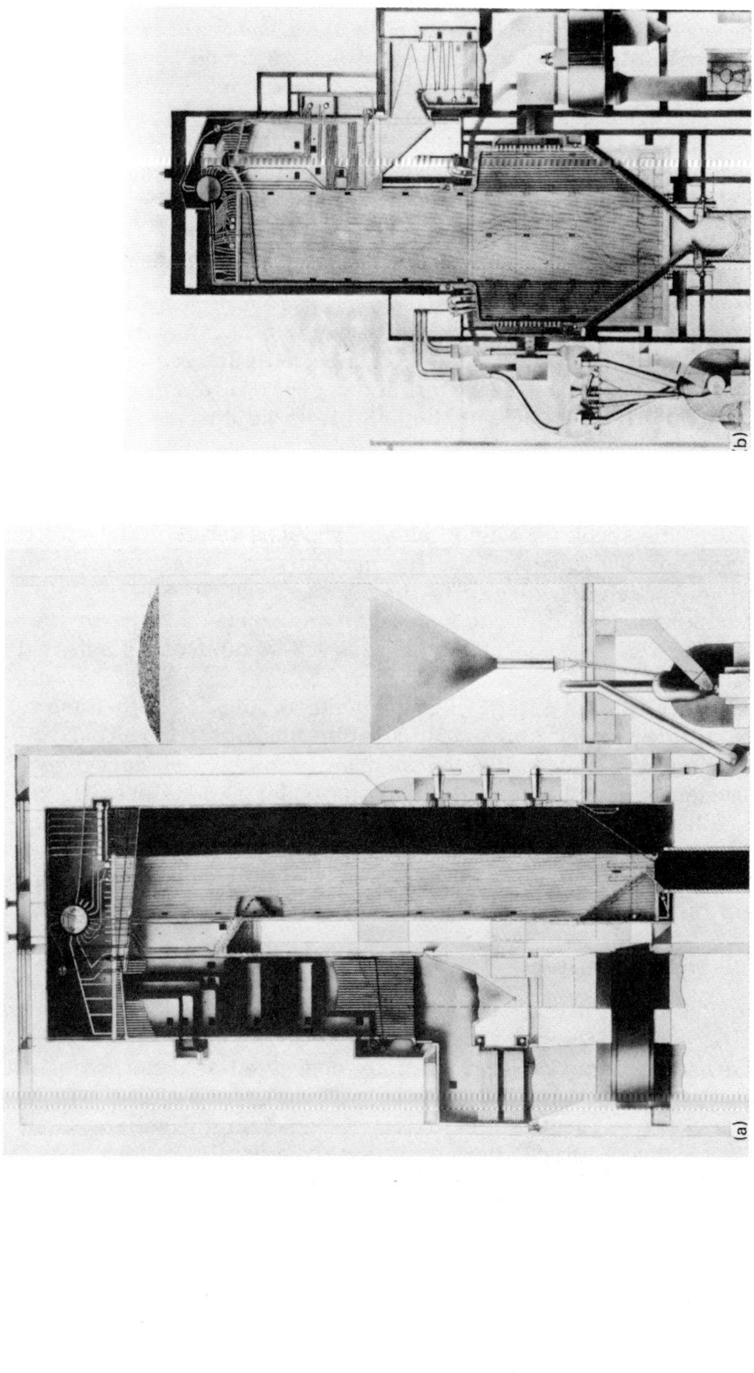

FIG. 12 (a) Hopper-bottom horizontal-fired pulverized coal furnace. Superheater section of separated furnace; natural circulation; differential fired steam generator: capacity, 1,850,000-lb of steam per hour. (b) Down-fired pulverized coal furnace installation (Holtwood Steam Electric Station). (Courtesy, Foster Wheeler Corp.)

The second ash-related consideration is the accumulation of deposits on the furnace walls and/or convection surfaces. Pulverized-coal-fired furnaces generally include wall blowers and soot blowers to remove deposits via a stream of compressed air or steam. However, the presence of certain minerals or elements in the coal ash can cause the deposits to sinter or fuse and thus reduce the effectiveness of wall and soot blowers. Iron and alkalies (sodium and potassium) are especially important ash constituents when considering furnace wall deposits and convection tube deposits, respectively.

The minimum size of pulverized coal fired boilers has varied, mostly as related to costs of operation and maintenance of boilers firing competing fuels. Pulverized-coal-fired boilers are more expensive to construct and maintain than are oil- and gas-fired boilers. Because of the relatively higher increases in the price of oil and gas in comparison to coal, consideration is now being given to firing pulverize coal in boilers with a capacity of less than 200,000 lb of steam per hour.

3. *Cyclone Furnace*

The cyclone furnace is a water-cooled horizontal furnace in which crushed coal (approximately 4 mesh) is fired and heat is released at high rates. A cyclone furnace is shown in Fig. 13. The water-cooled surfaces are covered with refractory so that gas temperatures of about 3000°F are achieved. This temperature is sufficiently high to melt the ash into a liquid slag which forms on the walls of the furnace.

The crushed coal is conveyed into the burner by the primary air (20% of the total air), which enters the burner tangentially, imparting a whirling motion to the coal. The secondary air is also introduced tangentially into the furnace at a velocity of 300 ft/sec. This further imparts a whirling action to the coal particles, which are thrown to the furnace wall by centrifugal force. These particles are held in the slag layer and burned. Most of the ash is retained in the slag layer, which minimizes the amount of fly ash that is carried over into the boiler.

The cyclone offers the advantage of being able to burn low-ash-fusion coals that create problems when burned in conventional pulverized-coal burners. In addition, the cyclone has an inherently low fly ash carryover and minimizes erosion and fouling problems in the boiler and reduces the size of a particulate collector. A major disadvantage of the cyclone, which has seriously restricted its use, is the relatively high levels of NO_X that are generated. The high NO_X levels can be attributed to high heat-release rates and the resulting high gas temperatures.

4. *Fluidized-Bed Combustion*

The concept of burning a fuel in a fluidized bed has a long history; however, not until the early 1960s did fluidized-bed combustion (FBC) receive attention as a method of firing boilers. Initially conceived as a method of using low-grade fuels, the fluidized-bed combustion process gained added attention when European workers suggested that fluidized-bed firing of boilers could reduce the capital cost of utility boilers and, for some configurations, possibly increase the overall power-plant cycle efficiency. FBC is still in the research and development stage but has the potential for widespread application. By introducing limestone into the bed at Ca/S ratios of about 3:1, significant

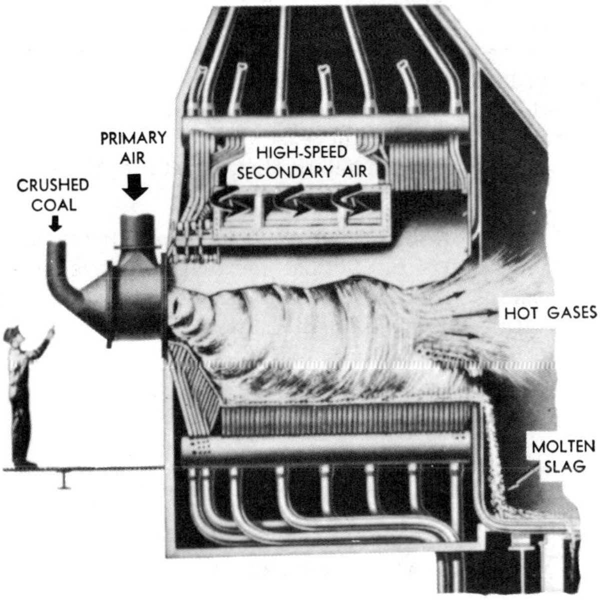

FIG. 13 Cyclone furnace, in the form of a horizontal cylinder, completely water-cooled by connection to the main boiler circulation. All combustion gases leave through the reentrant throat at the rear. Molten slag drains from the bottom at the rear through a small opening into the adjacent boiler furnace [17].

reductions in SO_2 emissions have been experienced, on the order of 80% and above. NO_x emissions from FBC also appear to be less than in pulverized coal firing. Research is continuing to develop FBC systems with costs and thermal efficiencies competitive or better than alternative coal utilization technologies.

FBC requires efficient mixing and contact of bed particles with combustion gases to promote optimum combustion and heat transfer and to minimize hot spots and stagnant areas. This is accomplished by proper design of air-distributor and solids-feeding system. In addition, the design of bed internals must not impair mixing.

Some advantages attributed to FBC over conventional pulverized coal combustion with emission controls include:

1. In situ removal of SO_2
2. Ability to handle wide range of fuels
3. Low NO_x emission
4. Increased availability of steam generator
5. Higher bed-to-tube heat transfer rate
6. Higher volumetric heat-release rate in the bed
7. Lower capital and operating costs

There are three basic designs of fluidized-bed combustion systems, although there are variations in each of these designs. The conventional atmospheric fluidized-bed design is applicable to commercial/industrial and utility boilers as well as to process heaters. The multisolid atmospheric fluidized-bed design has the same application as the conventional design and has the potential for increased load response, turndown, and compactness. The pressurized fluidized-bed design is oriented to a combined-cycle power-generating system and has the potential for increased compactness, improved limestone utilization, and higher system efficiency than do atmospheric systems.

a. Conventional Atmospheric FBC A simplified diagram of an atmospheric fluidized-bed combustor is shown in Fig. 14. Coal and limestone are introduced to the fluidized bed, where combustion takes place. Limestone serves as a fluidizing solid as well as a sulfur dioxide sorbent. Spent limestone is withdrawn from the bed and is either discarded as solid waste or reused after regeneration. A heat-transfer surface can be immersed in the bed to effectively remove the combustion heat. Ranges of the typical parameters of importance are as follows:

Bed temperature	1400-1800°F
Bed material	Limestone, dolomite, or other inert solid (alumina, zirconia, etc.)
Superficial gas velocity	3-12 ft/sed
Static bed height	1.5-7.0 ft
Ca/S ratio	2.0-4.0

b. Pressurized FBC Fluidized-bed combustors can be operated at elevated pressures. Pressurized FBC are designed to be one of the components in a combined-cycle power-generating system. The main difference between an atmospheric and a pressurized FBC is the size reduction of the pressurized system. As an example, for similar fludizing velocity, the area of an atmospheric FBC would be 10 times as great as the area of a bed for operation at 10 atm pressure. Because the pressurized FBC will be used in a combined cycle, it requires a means to remove particulate efficiently at high pressure and temperature.

c. Multisolid FBC Basically, the multisolid FBC can be regarded as a recirculating or entrained fluidized bed superimposed on a conventional dense bed. Two distinct regions are formed. The bottom zone is a dense-phase fluid-bed region consisting of a relatively coarse, high-specific-gravity bed material. Above this dense-phase region is an entrained-phase fluidized-bed zone, which comprises relatively fine particles of the same or another bed material. Under operating conditions, the coarse, high-specific-gravity material remains in the bottom of the combustor (forming a dense-phase fluidized bed) while allowing the fine bed material to pass through the dense-phase region to form an entrained-phase region. This fine bed material is recycled and returned to the bottom of the dense-phase region.

The superficial gas velocity of the multisolid FBC is about 30 to 40 ft/sec with heat-release rates about two to three times those of conventional systems. Wide system operation flexibility can be obtained by operating in a conventional dense fluidization mode at much lower velocities.

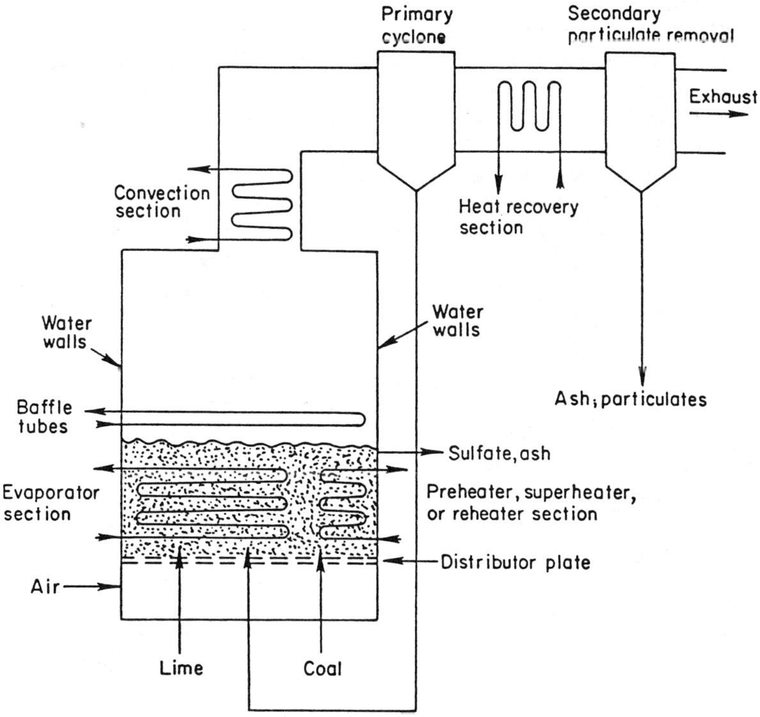

FIG. 14 Fluidized-bed combustion boiler.

5. *Coal/Oil Mixtures*

Slurries of coal in oil, coal-in-oil fuels, oil/coal suspensions, colloidal fuel, and coal/oil mixtures are all terms used over the past century to describe a mixture of solid particles of coal (generally in the size range of 70% passing through -200 mesh) and a liquid hydrocarbon fuel, usually a crude or residual fuel oil. The latter term, coal/oil mixture (COM), is one that is currently being used. Interest in such fuels has been sporadic, with much emphasis during World War II when supplies of petroleum were limited and alternative methods were sought to extend available sources of fuel oil. The present situation resulting from high prices for imported petroleum and shrinking local production and reserves of crude oil has again aroused interest in using COMs in the United States. The advantages of utilizing COMs are threefold:

1. Extending our oil supplies
2. Utilization of our vast coal reserves
3. Exploitation of the advantages and capability for handling and burning a fluid fuel

COMs for steam generation and for industrial heating have many advantages over coal alone, mainly because as a fluid fuel, they are easier to transport, to meter and control, and easier to burn than coal.

Combustion Considerations Coal/oil mixtures pose no unique problems in combustion [18,19]. Little experience is available on field tests of COMs for extended periods of time to indicate long-term operational problems. However, some comments can be directed toward potential problems.

There is a potential loss in efficiency when burning COMs not normally encountered with burning oil alone. Unburned carbon, attributed primarily to coal particles in the fly ash, can be expected but, as with pulverized coal, carbon losses should be controllable with proper burner-furnace design.

Ash in the coal may cause some problems in firing COMs in boilers not designed for coal firing. A typical COM will contain more than 30 times as much ash as the original fuel oil on a comparable Btu basis. The characteristics of coal ash vary widely, but with this high burden of ash in the furnace compared with oil, radical differences can be expected in the fouling of heat-receiving surfaces such as wall tubes, convective tube banks, economizers, and air heaters. Although removal of ash from such surfaces is seldom necessary with oil firing, it is likely to be imperative when burning COM. Similarly, refractory surfaces in oil-fired boilers are rarely a major source of trouble, but they can be expected to fail rapidly under attack of molten ash from the coal when burning COMs.

The high ash burden with COMs will pose problems with emissions of particulates not present when burning fuel oil properly. The amount of ash retained in the boiler is likely to vary, depending upon the furnace design, the firing rate, and the ash composition. Electrostatic precipitators or baghouses will probably be needed to control particulate to acceptable levels.

B. Coal Handling and Ash Removal

Coal handling and ash removal are important elements in coal combustion systems. The specific equipment that is used for these operations will depend on the type of coal-burning equipment, the overall capacity of the coal burners, and the application for which the coal is burned.

1. *Coal Handling*

Coal is delivered to a site by either rail or truck. The ease of onloading the coal is dependent on the moisture content of the coal. Dry coal flows freely, but if too dry, high winds can blow the coal fines away as dust. Conversely, high-moisture coal that is transported during periods of freezing temperatures may be frozen into a solid mass, creating serious unloading problems. Oil sprays can be used to minimize the dust problems, and calcium chloride and rock salt can be used to minimize some of the freezing problems.

Unloading of coal is generally accomplished using mechanical equipment. For small plants, portable belt and screw conveyers are used for onloading, filling bunkers, and transporting coal inside to mechanical equipment. For larger plants the coal-handling facilities are more elaborate, with coal storage silos, screw conveyors, and elevators.

Coal for stokers is generally supplied with the correct size distribution. Coal for pulverized firing requires grinding in pulverizers to a size consist of 80% passing through -200 mesh. Coal for FBC requires some size reduction (about 1/4 in. top size), which can be accomplished in crushers.

2. *Ash Handling*

The mineral matter in coal presents ash removal problems. In stokers the majority of the ash remains on the grate and is discharged into a hopper. Another source of ash is that which drops through the holes in the grate into a hopper below. A third source of ash is that collected in particulate control device. The ash from all these sources can be removed pneumatically through an ash-piping system that connects to an ash silo.

In pulverized-coal firing, the method of ash handling is somewhat dependent on whether the system is dry bottom (most common) or wet bottom. In dry-bottom furnaces about 80% of the ash is emitted as fly ash. This fly ash is generally collected in electrostatic precipitators and fabric-filter baghouses and then dropped into collection hoppers. From the hopper the dry fly ash is pneumatically conveyed to either an intermediate storage bin or a hydro exhauster, where it is mixed with water to form a slurry. The remaining 20% of the ash remains in the furnace as bottom ash or in the economizer. These sources of ash are collected and stored in hoppers located beneath the furnace and economizer. This ash is generally hydraulically conveyed to the disposal site.

In wet-bottom furnaces over 50% of the generated ash is retained. The fly ash that is emitted is collected and disposed of in the same manner as that of a dry-bottom furnace. However, in a wet-bottom ash-removal system, the ash is collected from the furnace in a pool or water to isolate it thermally from the hot furnace gases. It is then hydraulically or mechanically conveyed away for disposal.

C. Heat-Transfer Considerations

In the design of coal-fired equipment, the dominating factor in equipment design is usually heat transfer rather than the combustion characteristics of the coal. This is true in most industrial applications, such as steam boilers, rotary kilns, and reverberatory furnaces. For this discussion, a typical large steam generator for utility service will be used as an example to show how coal properties and heat-transfer requirements are integrated in boiler design and operation.

Figure 15 shows a cross section of a typical large pulverized-coal-fired steam generator for utility service. A typical unit of this type matches a 600-MW generator and burns about 225 tons of coal per hour. Coal is burned in the furnace, where combustion products are cooled to about 2200°F, following which they pass through the superheaters and reheaters, the economizer, and the air heater, in sequence. For a typical steam generator operating at 2400 psi, 1000°F/1000°F, about 40% of the fuel energy is absorbed in the furnace walls for steam evaporation, about 40% is used for steam superheating and reheating, and remaining 20% is absorbed in the economizer and air preheater. Radiant superheater panels in the upper furnace aid in achieving this distribution while avoiding excessive gas temperatures at the inlet to the superheater convection surfaces.

The design and operation of the boiler furnace is critical to satisfactory performance and operation of the steam generator, as the furnace serves as both the combustion space and the largest heat exchanger. The size and

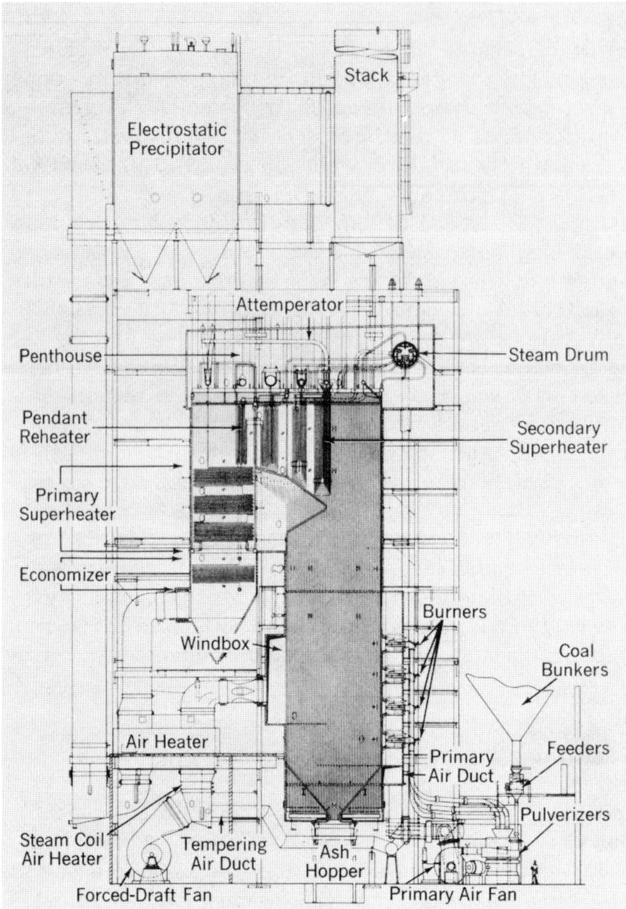

FIG. 15 Utility-type boiler for pulverized coal [20].

shape of the furnace influence flame temperature, the characteristics of ash deposits on the walls, and NO_X emissions. As a generalization, the furnace plan area per unit fuel burned (e.g., Btu/ft^2) is the design parameter that influences peak flame temperatures, which occur just above the top burners. A larger plan area per unit fuel burner increases the wall cooling area per unit of heat released, thus reducing both the heat flux per unit wall area and the flame temperature. A larger plan area is required for coals having low ash fusion temperatures, and a smaller plan area is acceptable for high-ash-fusion-temperature coals. The heat release per unit plan area ranges from 1.8 to 2.5×10^6 Btu/ft^2 for Eastern bituminous coals, from 1.7 to 2.0×10^6 Btu/ft^2 for Midwestern coals, and may be considerably lower for Western coals.

With a plan area established, the furnace height determines the extent to which the gas is cooled by radiation as it flows upward through the furnace, and hence establishes furnace exit temperature. The required furnace exit temperature depends upon the coal ash characteristics, in that the gas temperature must be low enough to prevent formation of sticky ash deposits on the superheater (fouling). Furnace exit temperatures are usually in the range 2000 to 2300°F for Eastern and Midwestern coals, although some Western coals and lignites require lower temperatures because of their extremely severe superheater fouling characteristics.

It is evident that the heat-transfer rates to furnace walls will vary with the thickness of ash covering on the walls, the emissivity of the ash, and the thermal conductivity of the ash. These characteristics vary with each coal and are not usually known when the boiler is designed. Thus, it is usual to select furnace size and shape based on previous experience with a coal having a similar ash analysis. Once in operation, the extent of deposits on furnace walls are controlled by wall blowers, which use jets of steam or air to clean the wall surrounding each blower. A large boiler for a Midwestern coal may have 100 wall blowers, a steam or air supply to operate two blowers simultaneously, and an automatic sequencer that cycles all blowers, in pairs, on a continuous basis. Such active wall cleaning permits use of a smaller furnace than would be possible otherwise, but increases operating cost.

In summary, the design of the furnace for a large steam generator is based almost entirely on heat-transfer requirements, and the volume required for heat transfer is always sufficient for complete combustion. Such design is a rather complex subject and is treated in a comprehensive manner in the book *Steam* [21].

Figure 16 shows a small industrial boiler with a spreader stoker for firing both coal and wood. The heat-transfer requirements of small boilers are somewhat different from those of large boilers, in that surface/volume ratios of small furnaces are much higher than for large furnaces, and full water cooling of the furnace may cause combustion problems. Small furnaces are often of refractory construction, with widely spaced boiler tubes in front of the refractory to provide a suitable flame temperature; or patches of refractory-covered water tubes may be used in the burner region to assure adequate flame temperature at the burner for flame stability.

Burner design for other industrial processes is similarily dominated by heat-transfer considerations. Burners for firing rotary kilns, for example, introduce coal and primary air as a high-velocity jet surrounded by an annular jet of secondary air, with very slow mixing between the two. This results in a long flame that may extend as far as 100 ft from the burner. Conversely, where very high temperature is required, combustion can be completed in a small space by using high burner pressure drop with intense mixing. It is quite feasible to complete combustion of coal within 100 msec, although the usual large boiler furnace provides a residence time of several seconds, to meet heat-transfer requirements. The Babcock and Wilcox cyclone furnace, for example, operates with heat-release rates on the order of 500,000 Btu/ft^3-hr, compared with usual rates below 20,000 Btu/ft^3-hr for large open furnaces. The heat transfer from combustion products in the cyclone furnace is very small, with resulting gas temperatures above 3200°F and NO_x levels above 1000 ppm.

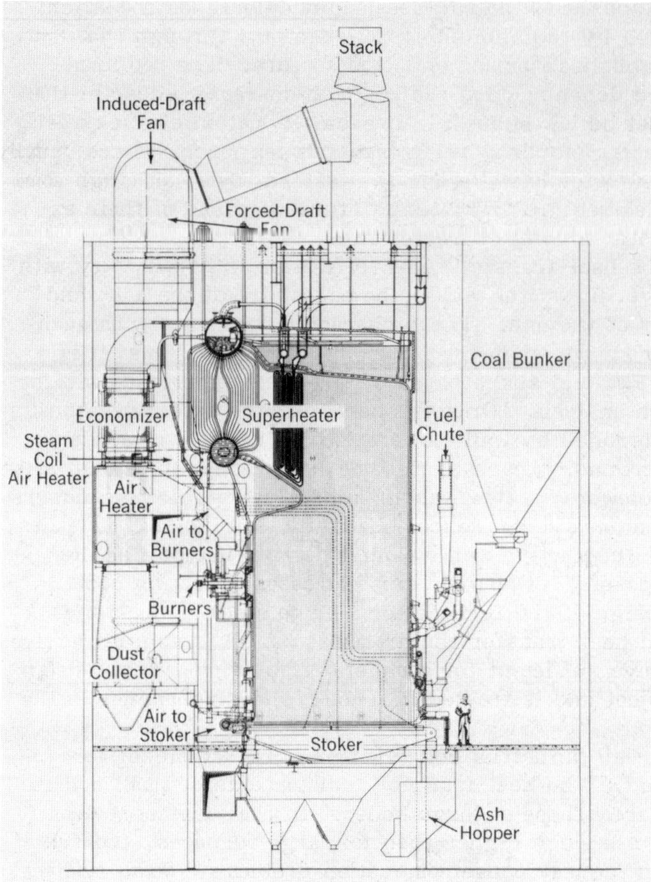

FIG. 16 Industrial boiler with spreader stoker for firing coal and wood [22].

D. Corrosion, Fouling, and Slagging

The presence of SO_2, SO_3, and ash in boiler furnaces creates an environment that is detrimental to furnace surfaces and to surfaces downstream of the furnace. The major problems caused by the presence of sulfur oxides and ash in combustion gases are:

Deposits
Corrosion
Erosion

Deposits are a detriment in boilers because they:

1. Act as an insulating layer on heat-transfer surfaces. This reduces the rates of heat transfer and affects the heat balance for the boiler.

2. Plug gas passages and increase pressure drop.
3. Create environments that are conducive to corrosion.

Corrosion and erosion are detrimental because they produce metal wastage, lead to boiler tube failures and outages, and shorten boiler life.

Deposit and corrosion problems are generally more severe in pulverized-coal-fired utility units than in other furnaces; the following discussion emphasizes this equipment.

1. Furnace Deposits

All pulverized-coal-fired boiler furnaces accumulate deposits on the water tubes that make up the furnace walls. Figure 17 shows areas frequently subject to deposits, corrosion, and erosion in utility boiler furnaces. The resulting deposit layer has a relatively low thermal conductivity and acts as an insulating layer, reducing heat transfer from the flame to the wall tubes and hence to the

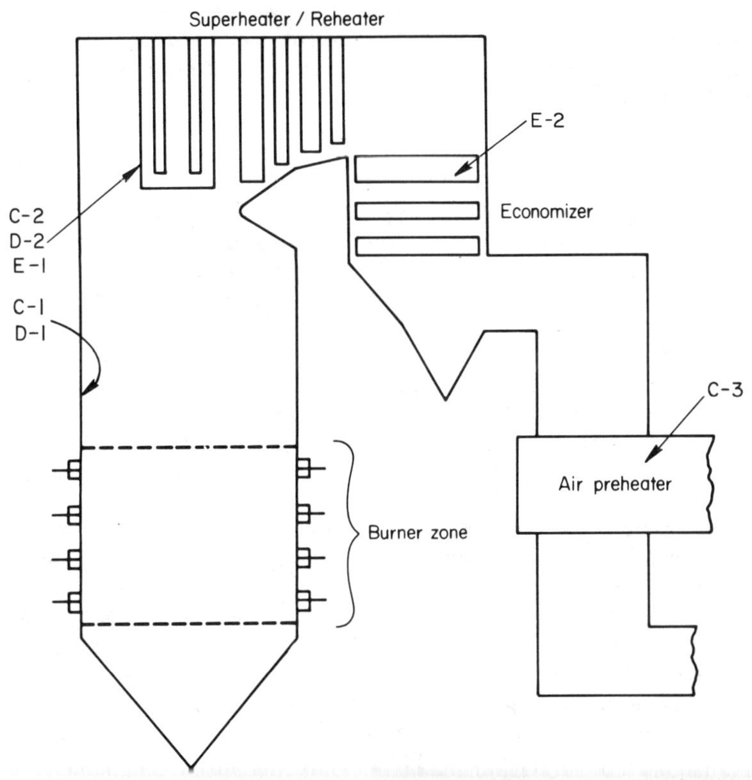

FIG. 17 Utility boiler. D-1, furnace deposits; D-2, convection pass deposits—superheater and/or reheater; C-1, wall-tube corrosion; C-2, superheater/reheater corrosion; C-3, air preheater corrosion; E-1, superheater/reheater erosion; E-2, economizer erosion.

water. As a consequence of the reduced heat transfer to the water walls covered by deposits: (1) less steam is generated in the water tubes, (2) the flame is not cooled as much as for a clean furnace, and (3) the furnace exit gas temperature is increased. The higher exit-gas temperature means that the superheater/reheater sees a higher gas temperature and more heat is removed from the furnace gases in the superheater/reheater. The combination of less steam being generated in the water walls and more heat being absorbed in the superheater/reheater could result in excessively high steam temperatures that might be damaging to the turbine if these effects differ from design conditions. Boiler designers compensate for expected ash deposits by providing a suitable furnace volume (i.e., providing added heat-transfer surface in the furnace to compensate for the reduced heat-transfer rates).

Once the boiler is constructed, furnace size is fixed. Thereafter, boiler operators have two means of controlling furnace heat absorption. First, wall blowers are installed on furnace walls to blow off the accumulated deposits. Wall blowers are retractable nozzles that spray a stream of compressed air or steam at the water walls. Their effective range is about 8 ft, so that some boilers require over 100 wall blowers. Second, desuperheater water, or attemperator water, may be sprayed into the secondary superheater inlet to control superheat temperatures to design values. However, desuperheating water must be extremely clean, as any solids present may be deposited in the superheater or in the turbine. Only limited supplies of desuperheating water are available to the boiler operator and, once the limit of desuperheating water usage is reached, the firing rate must be reduced to control steam temperature. Thus, successful operation of the wall blowers becomes critically important in maintaining a boiler on-line.

The nature of the coal fired can strongly influence the water-wall deposit problems that a boiler experiences. (The other factors are boiler furnace design and boiler operation.) Certain coals have high tendencies for slagging, that is, the formation of adherent deposits that are difficult to remove with wall blowers.

It is useful to consider deposits with regard to their physical nature. Dry deposits which are accumulations of separate particles of ash can easily be removed by wall blowers and offer no problem. Generally, soft, sintered deposits also can be removed by wall blowers. It is the occurrence of hard, sintered deposits and of plastic deposits that is most troublesome, as these may not respond to wall-blower operation. Rather than flaking or peeling off the wall when soot blowers operate, plastic deposits may, depending on their viscosity, be relatively unaffected by soot blowers or may be pushed aside but not blown off the water wall.

Flowing molten deposits can be observed in some boiler furnaces operating under high load. Flowing deposits occur as a molten layer over cooler plastic and sintered deposits (Fig. 18). Flowing deposits are self-limiting in thickness and hence are usually no more troublesome than plastic deposits. However, flowing deposits may become plastic in cooler regions of the furnace and offer severe problems in those regions. Thus, it is the factors that lead to the formation of molten (plastic or flowing) deposits that are the most troublesome.

Several indicators exist for evaluating (1) the tendency of coal ash to form molten deposits, or (2) the thickness of accumulated deposit layers that

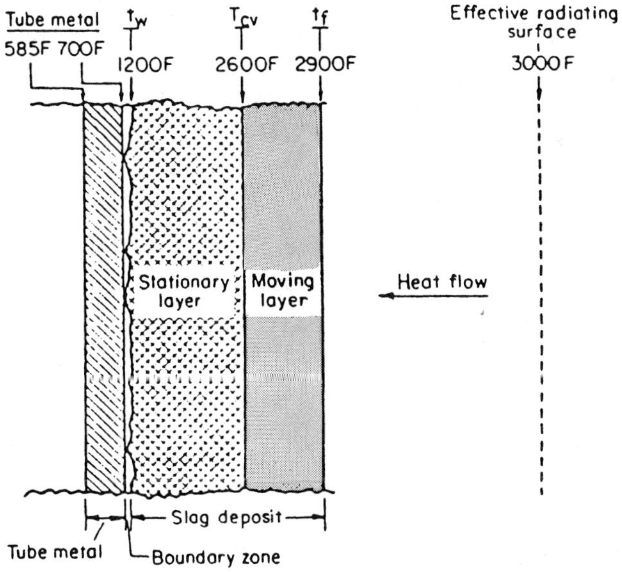

FIG. 18 Temperature distribution in a slag deposit [23].

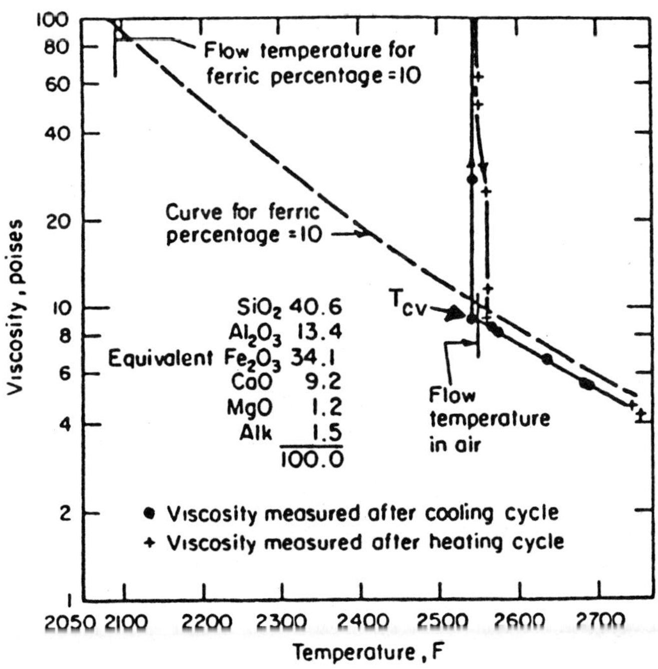

FIG. 19 Viscosity-temperature plot of a typical slag having a definite liquidus temperature [24].

become self-limiting owing to the presence of running slag. These indicators include:

1. Temperature of critical viscosity (T_{cv})
2. Ash viscosity
3. Ash-softening temperatures
4. Slagging factors

Figure 19 is a viscosity/temperature curve for a coal ash slag. The T_{cv} point is shown on the figure. Temperature of critical viscosity is that point at which, as an ash is cooled, a break occurs in the viscosity/temperature curve. The break signifies the transition of the ash from a Newtonian fluid to a plastic material and is probably associated with the crystallization of solid phases. T_{cv} and viscosity/temperature curves may be determined experimentally, but more often they are calculated from empirical relationships; Fig. 20 shows the relationship between viscosity at 2600°F and silica percentage.

Ash-softening temperatures are determined from ASTM tests and are defined as the temperatures at which a cone of ash is deformed when heated slowly. Ash-softening temperatures are limited in accuracy to 100 to 150°F.

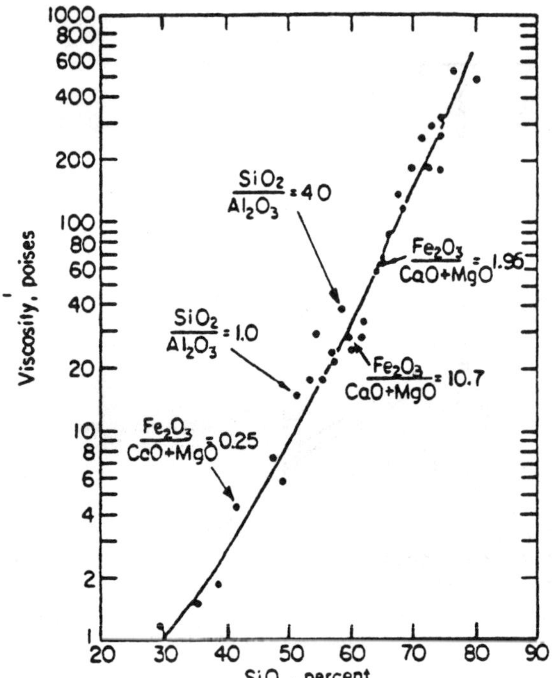

FIG. 20 Viscosity of coal ash slags at 2600°F in air (plotted against SiO_2 content on the basis of omitting Al_2O_3 and making SiO_2 + equivalent Fe_2O_3 + CaO + MgO = 100) [24].

T_{CV}, viscosity, and ash-softening temperatures may be used as indicators of the tendency of a coal ash to cause waterwall deposit problems. Obviously, coals with low T_{CV} and ash-softening temperatures tend to form molten deposits at lower temperatures than coals with high T_{CV} and ash-softening temperatures. Such data can be useful in anticipating deposit problems where the more molten fluid condition exists. Unfortunately, the criteria cited above are of only limited value for predicting problems associated with the plastic state of ash at the low-temperature end of the molten range.

Considering the composition of coal ash or ash deposits, SiO_2 and Al_2O_3 are frequently the most abundant compounds present. Both of these compounds have high melting points and are used in the manufacture of high-temperature refractories. However, most other compounds present in abundance in coal ash (CaO, MgO, Fe_2O_3, Na_2O, and K_2O) act as fluxes and interact with the SiO_2 and Al_2O_3 to produce low-melting-temperature compounds or glasses.

The remaining deposit criterion, the slagging factor of Attig and Duzy [26], evaluates the slagging potential of coals as a function of the base/acid ratio of coal ash times the sulfur content of the coal.* For high-SiO_2 and -Al_2O_3 coal ashes, the base/acid ratio is equivalent to a flux/refractory ratio. Although the slagging factor apparently is used as a basis for furnace design, some boiler operators express reservations about its usefulness.

Table 9 presents some ash characteristics of typical U.S. coals, showing various ash-softening temperatures and some slagging factors [25].

2. *Superheater/Reheater Deposits*

The other locations of troublesome ash deposits in pulverized-coal-fired boiler furnaces are the convection sections—the first closely spaced superheater and reheater sections. For pulverized-coal-fired furnaces, 80 to 90% of the ash in the coal is carried from the furnace as entrained fly ash in the flue gas stream. Some of the fly ash is deposited on the convection tubes, which project into the gas stream. Ash deposits on convection passes that cannot be easily removed by soot blowers are generally referred to as fouling.

Fouling on convection-pass surfaces serves to insulate the convection tubes from the hot gas stream and reduces heat transfer, thus upsetting the heat balance for the furnace. The boiler operator may not be able to obtain the design superheater temperature if convection passes are fouled.

A second problem created by convection-pass fouling is increased gas-side pressure loss through the unit. As deposits accumulate and the gas passages become plugged, the forced and induced draft fans may become unable to provide design gas flow rates. With combustion air flow reduced, the firing rate must also be reduced.

Finally, convection-pass deposits may provide an environment that permits the formation of corrosive compounds, compounds that are not stable except in the protected environment of the deposit. (This is discussed in detail in subsequent paragraphs.)

As contrasted with furnace deposits (slagging), convection-pass deposits (fouling) are more often loose or sintered deposits and, except for tubes

---

*Slagging factor [14], $Rs = \dfrac{CaO + MgO + Fe_2O_3 + Na_2O + K_2O}{SiO_2 + Al_2O_3 + TiO_2} \times S$

TABLE 9 Ash Characteristics of Typical U.S. Coals [25]

	Rank							
	Low-volatile bituminous	High-volatile bituminous				Sub-bituminous	Lignite	Lignite
Seam	Pocahontas No. 3	No. 9	Pittsburgh	No. 6				
Location	West Virginia	Ohio	W. Virginia	Illinois	Utah	Wyoming	Texas	No. Dakota
Ash (dry basis)(%)	12.3	14.10	10.87	17.36	6.6	6.6	12.8	11.2
Sulfur (dry basis)(%)	0.7	3.30	3.53	4.17	0.5	1.0	1.1	0.8
Analysis of ash (wt %)								
SiO_2	60.0	47.27	37.64	47.52	48.0	24.0	41.8	28.4
Al_2O_3	30.0	22.96	20.11	17.87	11.5	20.0	13.6	11.0
TiO_2	1.6	1.00	0.81	0.78	0.6	0.7	1.5	0.4
Fe_2O_3	4.0	22.81	29.28	20.13	7.0	11.0	6.6	14.0
CaO	0.6	1.30	4.25	5.75	25.0	26.0	17.6	18.0
MgO	0.6	0.85	1.25	1.02	4.0	4.0	2.5	5.0
Na_2O	0.5	0.28	0.80	0.36	1.2	0.2	0.6	3.6
K_2O	1.5	1.97	1.60	1.77	0.2	0.5	0.1	0.7
Total	98.8	98.44	95.74	95.20	97.5	86.4	84.3	81.1
Ash fusibility								
Initial deformation temperature (IT) (°F)								
Reducing	2900+	2030	2030	2000	2060	1990	1975	2170
Oxidizing	2900+	2420	2265	2300	2120	2190	2070	2170
Softening temperature (°F)								
Reducing		2450	2175	2160		2180	2130	2200
Oxidizing		2605	2385	2430		2220	2190	2210
Hemispherical temperature (HT) (°F)								
Reducing		2480	2225	2180	2140	2250	2150	2230
Oxidizing		2620	2450	2450	2220	2240	2210	2230

Fluid temperature (°F)								
Reducing		2620	2370	23.20	2250	2290	2240	2310
Oxidizing		2670	2540	2610	2460	2300	2290	2370
Type of ash	Eastern	Eastern	Eastern	Eastern	Lignite	Lignite	Lignite	Lignite
Base/acid ratio (B/A) $Fe_2O_3 + CaO + MgO + Na_2O + K_2O/SiO_2 + Al_2O_3 - TiO_2$	0.079	0.382	0.635	0.439	0.622	0.933	0.482	1.04
Silica ratio $SiO_2 \times 100/SiO_2 + Fe_2O_3 + CaO + MgO$	92	65	52	64	57	37	61	43
Dolomite % $(CaO + MgO)$ $100/Fe_2O_3 + CaO + MgO + Na_2O + K_2O$	17	8	17	23	78	72	73	56
Slagging potential for Eastern coals (B/A × % S)	Low: 0.06	Medium: 1.26	High: 2.24	Medium: 1.83	Low: 0.31	Medium: 0.93	Low: 0.53	Medium: 0.83
Slagging potential for lignitic ash max HT + 4 IT/5	—	High: 2148	High: 2114	Severe: 2090	Severe: 2092	Severe: 2042	Severe: 2022	High: 2182
Fouling potential for eastern coals (B/A × % Na_2O)	Low: 0.04	Low: 0.11	Medium: 0.51	Low: 0.16	High: 0.75	Low: 0.19	Medium: 0.29	Severe: 3.74
Fouling potential for lignite ash (% Na_2O)	Low-medium: 0.5	Low-medium: 0.28	Low-medium: 0.80	Low-medium: 0.36	Low-medium: 1.2	Low-medium: 0.2	Low-medium: 0.6	High: 3.6

exposed to high-temperature furnace radiation, are not likely to be molten. If the deposits are loose or mildly sintered, operation of soot blowers usually removes the deposit. Soot blowers are long, retractable, rotating versions of wall blowers. As the soot blowers penetrate the gas passage, the rotating jet of compressed air or steam blows the deposits off the tubes.

Especially troublesome fouling is experienced when the nature of the coal ash generates strongly sintered deposits and when furnace slagging raises the temperature of gas passing over the convection surfaces.

The coal component that is most critical to fouling problems is sodium. Sodium compounds are volatilized in the high-temperature flame environment; as the flue gas approaches the cooler convection tubes, these compounds condense on the tubes. The condensed sodium compounds serve as a binder to retain pieces of fly ash and for the formation of strongly sintered deposits.

Several methods have been developed for evaluating fouling tendencies of coals. These techniques include:

1. Ash sintering strength
2. Ash bonding strength
3. Fouling factors

The first two methods are more appropriately considered as measures of the ease with which convection-pass deposits may be removed. The last probably relates to both deposit accumulation tendencies and ease of removal.

Barnhart and Williams [27] proposed a technique for predicting relative fouling problems by determining the crushing strength of a pellet made from the coal ash. Higher crushing strengths are indicative of stronger (and more difficult to remove) deposits.

Combustion Engineering and Pennsylvania State University have investigated fouling by measuring the shear force required to remove a deposit from a metal surface [28,29]. Again, higher shear forces are required to remove more strongly bonded deposits.

Attig and Duzy [26] proposed a fouling factor which is similar to their slagging factor but recognizes the importance of sodium to fouling. Their fouling factor is defined as the base/acid ratio times the sodium content of the coal ash.*

3. *Wall-Tube Corrosion*

Gas-side corrosion of wall tubes has been observed, mostly on wet-bottom units in the 1940s [30-32]. The observed wall-tube corrosion was attributed to pyrosulfates ($Na_2S_2O_7$ and $K_2S_2O_7$) or alkali iron trisulfates [$Na_3Fe(SO_4)_3$ and $K_3Fe(SO_4)_3$]. The solution to this problem was to reduce SO_2 levels over the deposits to the point that the pyrosulfate and/or trisulfates would not be stable, even in protected zones beneath deposits.

Wall-tube corrosion of coal-fired boilers has not been a major problem in recent years. However, the use of two-stage combustion techniques may

*Determining the sodium content of coal ash by ASTM procedures (at 800°F) may result in the loss of some sodium. Low-temperature ashing of coal at not over 400°F is gaining acceptance as the preferred technique for ashing for alkali analysis.

generate reducing conditions locally and may create an environment conducive to sulfide corrosion.

4. *Superheater/Reheater Corrosion*

During the period ending about 1960, superheat/reheat temperatures were slowly pushed to higher and higher values. The benefit gained by using higher superheat temperatures was increased plant thermal efficiency. About 1960, superheat temperatures had reached 1100°F. About the same time, severe superheater tube corrosion was detected on boilers operating at these high superheat temperatures.

As a result of the efforts of a number of researchers, the severe superheater corrosion problem on coal-fired units was determined to be the result of the presence of molten alkali iron trisulfates on the tube surfaces. Mixtures of $Na_3Fe\ (SO_4)$ and $K_3Fe\ (SO_4)_3$ were determined to have melting points as low as 1030°F. However, the SO_3 concentration in the gas stream (about 30 to 60 ppm) was well below that required for these compounds to be stable at 1000°F or above (250 ppm at 1000°F).

The mechanism that finally explained how the alkali iron trisulfates could be present on superheater tubes was that the presence of deposits permitted locally high SO_3 concentrations to occur. Such reasoning was supported by the fact that the trisulfates were found only beneath layers of deposit, never on exposed tubes. Also, both Fe_2O_3 and fly ash were shown to be catalysts for the formation of SO_3.

5. *Air Preheater Corrosion*

As utilities have attempted to recover more heat from furnace exhaust gas, the temperatures at which air preheaters operate has been lowered. As air preheater temperatures have been reduced, some units have experienced severe air preheater corrosion. Most air preheater corrosion on coal-fired power plants has been related to attack on preheater surfaces by condensed H_2SO_4, formed from SO_3 and H_2O.

It has been determined that SO_3 is produced in a boiler furnace by two mechanisms; by reactions in the flames, and by catalytic oxidation on fly ash and Fe_2O_3-covered surfaces. Sulfur trioxide in the flue gas can be reduced by (1) reducing fuel-sulfur levels, or (2) reducing excess air levels to very low values. Neither method has been widely adopted in the United States; rather, most U.S. effort has involved use of more-corrosion-resistant air preheater materials, including coatings, or accepting operation at higher air preheater temperatures.

As consideration is given to firing refuse in coal-fired utility plants, the potential for HCl corrosion of air preheaters must be considered.

6. *Erosion*

Erosion of convection tube surfaces (superheaters, reheaters, and economizers) has been related to boiler design but not to coal ash characteristics. Specifically, gas velocities over 50 ft/sec tend to produce unacceptable erosion of convection tubes. Thus, boiler designers attempt to stay below this value when designing new units. If erosion is unacceptable, some convection tube surface may be removed from a boiler to reduce gas velocities.

Although not specifically defined, it is likely that convection-pass erosion is related to selected coal ash variables. For example, increased levels of ash in coal would be suspected of increasing erosion rates. Also, the nature of the ash particles (SiO_2, CaO, etc.) may influence their erosive potential.

V. ENVIRONMENTAL CONTROL TECHNIQUES

The combustion of coal results in the emission of a variety of pollutants into the air. These pollutants may be classified as:

1. Particulates
   a. Fly ash
   b. Carbon (coal, coke, or soot)
2. Gases
   a. SO_2 and SO_3
   b. NO_X
   c. CO
3. Organics
   a. Hydrocarbons
   b. POM (polycyclic organic matter)
4. Trace elements

The order of this list reflects the general history of concern over emissions from coal combustion. Emissions of particulates were the first to be of concern and to be regulated, because they are highly visible. Although there was earlier concern over SO_2 emissions, it was during the 1960s that emissions of SO_2 and NO_X became a serious issue. Finally, during the 1970s, emissions of POMs and trace elements have become of concern, at least partially because of their possible carcinogenic effects.

Emissions from the combustion of coal originate from:

1. Inorganic compounds or elements present in the coal
2. Incomplete combustion of coal
3. Combustion air (for thermal NO_X only)

Fly ash, sulfur oxides, and trace elements are emissions that originate from inorganic compounds or elements that are present in the coal. (Coal contains inherent ash that is intricately bound within the coal structure, and extraneous ash or noncoal matter that is mined together with the coal.) Sulfur is present in coal as organic and inorganic compounds; inorganic sulfur is mainly pyrites, FeS_2. Incomplete combustion of coal will produce emissions of carbon as coke or soot particles, carbon monoxide and hydrocarbon gases, and POM compounds.

A. Gas Emissions

1. *SO_x Emissions*

A variety of techniques have been developed for reducing SO_X emissions from the use of high-sulfur coals. These include:

1. Cleaning the coal prior to combustion
   a. Physical cleaning
   b. Chemical cleaning
2. Reacting the SO_x with particulate during or immediately after the combustion process
   a. Coal/limestone fluidized-bed combustion
3. Capturing the SO_x after it leaves the combustor
   a. Dry processes
   b. Wet scrubbers

a. Coal Cleaning Processes that can be used to reduce SO_x emissions by cleaning coal prior to combustion include:

1. Physical coal cleaning
2. Chemical coal cleaning
3. Liquefaction
4. Gasification

Physical coal-cleaning processes are limited to processes that employ physical separation processes—such as density or magnetism. Chemical cleaning processes separate coal from ash and sulfur by dissolving either the coal or the impurities. The more extreme coal-cleaning processes include coal liquefaction and coal gasification. These processes are discussed in Chap. 6.

b. SO_x Control During Combustion

(1) SO_2 control. As indicated earlier, it has generally been assumed here that the coal combustion process could not be used to control SO_2 for the widely used types of combustion equipment. Hence, for pulverized coal firing, some 95% of the fuel sulfur is emitted as SO_2 and only 5% is retained in the ash; for stoker firing of coal on beds, the amount of sulfur retained in the ash may reach 25% for Eastern coals and 50% or more for Western coals. Recent work with high-calcium Western coals and with coal/limestone pellets has shown that 50% or more the fuel sulfur can be retained in high-calcium ash [33].

(2) SO_3 control. The SO_3 concentration in combustion products can be reduced by firing at low excess air content, or by using MgO-based additives to react with the SO_3 or to coat potentially catalytic surfaces with a non-catalytic layer.

(3) Fluidized-bed combustion. The fluidized-bed combustion of coal, discussed earlier, offers a means for in-furnace control of SO_2. When finely ground limestone is added to the bed of a coal fluidized-bed combustor, up to about 85% of the fuel sulfur can be captured as calcium sulfate.

c. Flue Gas Controls A wide variety of processes have been developed to remove SO_2 from flue gas, including wet scrubbing processes, dry processes, and catalytic oxidation/reaction processes.

(1) Wet scrubbing. Wet scrubbing processes for removing SO_2 are the principal SO_2 control processes presently used in power stations, primarily because other processes have not been commercially demonstrated.

Flue gas scrubbing entails a variety of processes. There are at least seven categories of flue gas scrubbing that have been demonstrated at 20 MW or greater:

1. Wet limestone scrubbing
2. Wet lime scrubbing
3. Single-alkali scrubbing (Na_2CO_3 or NaOH scrubbing without regeneration)
4. Double-alkali scrubbing (sodium alkali scrubbing with regeneration)
5. Chiyoda Process (sulfuric acid scrubbing with limestone neutralization
6. MgO scrubbing (using coke for regeneration)
7. Wellman-Lord Process (sodium-alkali scrubbing with thermal regeneration and sulfur recovery)

There are several basic ways of grouping the wet processes. For example, there are throwaway versus recovery processes, or clear liquor scrubbing versus slurry scrubbing processes. Throwaway processes include wet limestone, wet lime, single-alkali, double-alkali, and Chiyoda; recovery processes include MgO scrubbing and Wellman-Lord. Within the category of throwaway processes, wet limestone and lime scrubbing are slurry scrubbing processes, whereas the single-alkali, double-alkali, and Chiyoda processes involve clear liquor scrubbing. The MgO process involves slurry scrubbing, whereas the Wellman-Lord Process involves clear liquor scrubbing.

Throwaway processes have the advantage of freeing the process user from having to market a by-product, whereas recovery processes make it possible to recover part of the process operating cost through the sale of a by-product. The recovery processes lend themselves to both the manufacture of sulfuric acid and/or the recovery of elemental sulfur. The MgO process produces sulfuric acid, whereas the Wellman-Lord Process recovers elemental sulfur.

The throwaway processes create a solid-waste disposal problem and a greater water pollution problem than do the recovery processes, but the recovery processes are in general more complex to operate. The clear liquor scrubbing processes eliminate the problem of scrubber scaling, permit the use of more efficient scrubbers operating at smaller liquid/gas ratios, and generally have a higher efficiency for SO_2 removal than do the slurry scrubbing processes. On the other hand, the slurry processes have accumulated more power-plant operating experience and in general are less complex than are the clear liquor processes. All current wet scrubbing processes suffer from the disadvantage that stack gas reheat is needed to increase plume buoyancy.

Throwaway processes The biggest advantage of the wet limestone or lime scrubbing process is its relative simplicity and the use of a relatively inexpensive source of alkali. However, to date, the limestone process has been plagued by operational problems, mainly scaling in the scrubber and mist eliminator plugging.

The single-alkali process is relatively simple and holds the promise of high SO_2 removal efficiency at low liquid/gas ratios in the scrubber. However, it suffers from the major disadvantages that Na_2CO_3 is an expensive alkali source and water-soluble Na_2SO_4 produced in the process must be disposed of in an environmentally acceptable manner.

The double-alkali process attempts to circumvent scrubber scaling and mist eliminator plugging by putting in a second liquor loop that utilizes soluble alkali (either Na^+ or NH_4^+ ion) as the scrubbing liquor. The scrubbing liquor is then neutralized with lime or limestone external to the scrubber circuit, and every effort is made to keep Ca^{2+} ion out of the scrubbing circuit. Later versions incorporate a third loop using an ion-exchange resin to ensure that there will be no scaling in the scrubber.

Similar to the double-alkali process, the Chiyoda Process tries to prevent scrubber scaling and mist eliminator plugging by eliminating Ca^{2+} ion from the scrubber circuit. The scrubber liquor in this case is dilute sulfuric acid that is neutralized external to the scrubber with limestone. Dry gypsum is produced and thus minimizes the waste disposal problem common to throw-away processes. The Chiyoda Process is rather insensitive to the reactivity of the limestone used as the raw material. The greatest disadvantage of the process is that it scrubs an acid gas with an H_2SO_4 solution. Thus, very high liquid/gas ratios must be used, which lead to high power requirements, very large scrubbers and oxidizer towers, and, in general, a large-volume scrubber circuit.

Recovery processes The greatest advantage of the MgO scrubbing process is that there are no major waste streams to be disposed of and the acid produced is of marketable quality. A disadvantage is that it is difficult to separate the fly ash from the recovery liquor because of the presence of insoluble magnesium salts.

The Wellman-Lord Process can recover either elemental sulfur or sulfuric acid. In the former case, elemental sulfur is the most environmentally acceptable by-product. Also, the process economics are relatively insensitive to the marketing or disposal of sulfur. In the Wellman-Lord Process, in which sulfur is recovered, a reducing agent is required to convert stripped SO_2 to elemental sulfur. To permit regeneration of the sodium sulfite scrubbing liquor, fly ash must be kept out of the regeneration system. This necessitates a fly ash collection device that has a high efficiency and is extremely reliable. The process also requires a purge stream to remove sulfate from the system and has rather high steam requirements because of the evaporation of water in the evaporator/crystallizer.

(2) Dry processes. In older dry processes, finely ground limestone was injected into boiler furnaces upstream of the superheater. The limestone was calcined and reacted with the SO_2 as the limestone and flue gas passed through the convection passes. The reacted limestone was collected in the particulate control system. The major reasons why this process did not find favor were (1) fouling of convection passes by the high-calcium ash, and (2) the low utilization of limestone (around 15 to 20%).

After being set aside for 7 to 10 years in favor of wet scrubbing processes, dry limestone SO_2 control processes are again a subject of interest. The increasing cost of scrubbers and scrubber operation and the questionable reliability of wet scrubbers have served as the impetus for a return to interest in dry processes. Another factor favoring dry processes is the possible need for baghouses to meet particulate emission limits. The dry processes of current interest introduce the limestone downstream of the economizer as a fine mist in a spray dryer. By the time the reacted limestone reaches the baghouse, the water has evaporated.

(3) Other processes. Another process, the Monsanto Cat-Ox Process, oxidizes SO_2 to SO_3 at about 85°F on catalytic surfaces such as vanadium. To prevent contamination of the catalyst, the SO_3 thus formed can be captured by condensing H_2SO_4 from the flue gas. The Cat-Ox Process was tried in a 109-MW demonstration at the Illinois Power Company's Wood River Station in the early 1970s. Although successful in reducing SO_2 emissions by over 90% for short periods (24 hr), a wide variety of problems were encountered such that only 27 days of full operation were achieved in the first 2 years of service. No further orders have been placed for this type of process for full scale units.

The Foster Wheeler-Char Absorption Process involves dry adsorption of SO_2 on char followed by thermal regeneration to produce a relatively concentrated stream of SO_2. Foster Wheeler developed a proprietary off-gas treatment system called Resox for the conversion of SO to elemental sulfur. The reaction takes place in a reactor vessel filled with crushed coal. The SO_2 stream is reduced to gaseous elemental sulfur, and the liberated oxygen combines with a portion of the coal to form carbon dioxide. The gases leaving the reactor enter a sulfur condenser, where the sulfur is condensed and carried off in the molten state.

The Shell-Copper Oxide Process, which is licensed in the United States through Universal Oil Products Corporation, concentrates the SO_2 by dry reaction with supported copper oxide pellets and regeneration with hydrogen. The SO_2 concentration in the regenerated off-gas is increased by removing water and impurities in an absorption/stripping system. The concentrated SO_2 is then reduced to elemental sulfur in a Claus plant.

2. NO_x Emissions

Nitrogen oxides are produced during combustion by (1) thermal fixation of nitrogen from combustion air, and (2) the conversion of organic nitrogen compounds in the fuel to NO_x. For conventional combustors, thermal fixation becomes significant at temperatures above 2800°F; fuel nitrogen conversion takes place at much lower temperature and is the main source of NO_x in coal combustion.

The fraction of fuel nitrogen converted to NO_x during combustion varies with the fuel-nitrogen content of the fuel (Fig. 5). Under normal combustion conditions, some 25 to 40% of the organic nitrogen in coal is converted to NO_x. With coal nitrogen content ranging from 0.7 to 2.8%, this level of conversion could easily account for all the NO_x output permitted by federal New Source Performance Standards. Thus, it has become necessary to develop coal combustion techniques specifically oriented toward minimizing the conversion of fuel nitrogen to NO_x.

In pulverized coal-fired utility units without provision for NO_x control, the lowest NO_x levels are obtained with wall or corner fired dry bottom furnaces, and the highest NO_x levels are obtained in hot slag-tap and cyclone furnaces (Table 10). Furnace shape, furnace volume, heat release per unit boiler surface in the burner zone, and the air preheat temperature all significantly affect NO_x emission levels. Techniques that have been developed or proposed for decreasing NO_x emissions include the following:

Thermal NO_x	*Fuel nitrogen NO_x*
Flue gas recirculation	Biased or two stage combustion
Water addition	Low-excess-air combustion
Biased or two-stage combustion in combination with heat removal	

For coal firing, thermal NO_x control (by reducing flame temperature) has met with only limited success, since the NO_x problem is primarily a fuel-nitrogen problem. Hence, most current efforts are directed toward staged combustion and excess air control or flue gas treatment.

Two-stage combustion consists of burning the fuel with only a portion of the stoichiometric air (about 70 to 90%) in the first stage; the remaining air is added and combustion is completed in a second stage. Two-stage combustion reduces both thermal NO_x, by lowering peak temperatures, and fuel NO_x, by maintaining reducing conditions during the early burning. A variation of two-stage combustion is biased firing of multiburner furnaces wherein some burners are run fuel rich and others admit only air to the furnace. Biased firing achieves a form of two-stage combustion but does not require two separate combustion zones. Another form of two-state combustion is achieved by designing burners to delay mixing secondary air with the fuel.

For some combustion systems, NO_x may be reduced slightly by firing the system at a low excess-air level. However, firing at low excess air without smoke requires good mixing of air and fuel, and the high turbulence may increase flame temperatures and thermal NO_x. Therefore, decreasing excess air to decrease NO_x emissions has only limited application.

a. Pulverized Coal Combustion About 80% of the NO_x in pulverized coal combustion is fuel NO_x [34]. NO_x emissions from pulverized, coal-fired boilers can be reduced significantly by biased firing or staged combustion. Table 11 shows the NO_x emission reduction potential of several control techniques. Table 12 reports the lowest NO_x emission values obtained for various firing methods (as of 1977). The use of biased firing and delayed mixing burners

TABLE 10 Emission of Nitrogen Oxides

Type of unit	NO_x (ppm)
Pulverized coal	
Corner firing	375-550
Front-wall firing, dry bottom	500-825
Front-wall firing, wet bottom	825-1500
Horizontal opposed firing	400-1000
Cyclone	900-1500
Stoker	
Spreader stoker	250-400
Commercial underfeed	150-300
Residential underfeed	100-300

TABLE 11 Combustion Modification Assessment Summary [35]

Control method	Basic mechanism of control	Single-face-fired Reduction potential (%) Typ.	Single-face-fired Max.	Horizontally opposed fired Reduction potential (%) Typ.	Horizontally opposed fired Max.	Tangentially fired Reduction potential (%) Typ.	Tangentially fired Max.
Low excess air	Increase fuel/air ratio at burners	0–15	15	0–15	15	0–10	10
Biased firing	Increase fuel/air ratio to majority of burners	5	7	5	8	—	—
Burners out of service[a]	Increase fuel/air ratio to active burners	30	35	25	35	30	45
Overfire air[a]	Increase fuel/air ratio to burners	15	30	30	58	30	35
Flue gas recirculation	Reduce peak flame temperature	—	—	14	17	—	—
New burners	Controlled diffusion flame	—	—	30	60	—	—

Control method	Primary limitations	Limiting factors in effectiveness	Retrofit limitations	Energy penalties	Secondary pollutants	Operational and maintenance considerations	Impact on auxiliary equipment
Low excess air	Slagging, smoke, flame instability	Carbon carryover	Combustion controls and instrumentation	May increase slightly	No effect	Additional comb. controls, increased maintenance	None

Biased firing	Degree of biasing, wall slagging, and corrosion	Carbon carryover, flame stability, smoke	Pulverizer and fan capacity, flexibility in coal feed system	None	No effect	Additional combustion controls, increased maintenance	None
Burners out of service	Degree of staging, wall slagging, and corrosion	Carbon carryover, flame stability, smoke	Pulverizer and fan capacity, flexibility of coal feed system	None	No effect	Additional combustion controls, increased maintenance	None
Overfire air	Degree of staging, wall slagging, and corrosion	Carbon carryover, flame stability, smoke	Furnace configuration, fan capacity	None	No effect	Additional combustion controls, increased maintenance	None
Flue gas recirculation	Much less cost effective than staging	Flame stability	Unit layout, fan capacity	0.5% due to auxiliary fan loadings	No effect	Fan vibration, increased maintenance	Increased fan loadings
New burners	Cost	Windbox configuration, flame stability	Furnace and windbox configuration, fan capacity	None	No effect	Minimal	None reported

[a]Burners out of service and overfire air are other versions of biased firing.

TABLE 12 Status of Coal-Fired Utility Boiler NO_x Control Technology for New Units and Retrofit Applications [35]

	Lowest attainable NOx emission levels			
	New units		Retrofit	
	ppm at 3% O_2	lb/10^6 Btu	ppm at 3% O_2	lb/10^6 Btu
Wall-fired				
Single-face-fired	300-350	0.45	400-500	0.6
Horizontally opposed	300-350	0.45	400-600	0.6
Tangential	250-300	0.35	250-350	0.4

has allowed NO_x emissions as low as 0.2 lb/10^6 Btu to be demonstrated, although not in utility boilers. Since biased firing and staged combustion depend upon maintaining zones of reducing atmosphere, some alteration in furnace slagging and fly-ash characteristics may occur. In addition, if reducing zones occur along walls, severe wall-tube corrosion may result.

b. Cyclone Furnace Firing At the time of this writing, about 100 cyclone-fired boilers are in service, fired by about 500 cyclone units. These are particularly appropriate where high-ash, highly abrasive coal of low grindability and low ash-fusion temperature must be burned. As NO_x emission level for cyclone furnaces using bituminous coals is high, in the range 800 to 1000 ppm, NO_x emission regulations effectively prohibit new cyclone furnaces.

c. Stoker-Fired Furnaces There appears to be limited opportunity to utilize concepts such as biased firing or staged combustion for stoker firing.

NO_x emission from stokers are generally low, representing conversion to NO_x of only 10 to 20% of the fuel nitrogen, as contrasted to the 25 to 40% conversion for uncontrolled pulverized coal combustion [36]. The low NOx emissions from stoker-fired boilers are attributable to the low combustion temperatures (2200 to 2400°F versus 2800 to 3000°F in pulverized coal units), and to the partially burning of coal in a reducing environment, at least locally in the bed.

3. *CO and Gaseous Hydrocarbon Emissions*

Emissions of CO and hydrocarbons, products of incomplete combustion, are not a problem for large coal-burning equipment but can be a problem for smaller equipment. The residence time for combustion and the mixing of air and fuel is sufficient in power-station boilers that, at the excess air levels generally used, CO and hydrocarbon emissions are low. For example, few coal-fired power-plant boiler furnaces emit more than 100 ppm of CO.

Small coal-burning equipment, specifically stokers, can emit large quantities of CO and hydrocarbons. For bed burning, at certain levels within the bed the combustion reactions amount to gasification of the coal. If good mixing of combustion air with the CO- and hydrocarbon-rich gasified coal is not achieved, and temperatures are not sufficient for CO and hydrocarbon oxidation, some CO and hydrocarbons may escape the furnace. The use of over-fired air (air jets above the bed) is common to ensure good mixing and thus oxidation of CO and hydrocarbon gases.

TABLE 13 Combustible Gases Emitted by Bituminous Coal with Various Firing Methods [37]

	Pulverized-coal-fired	Cyclone	Spreader stoker	Chain-grate stoker	Underfeed stoker	Hand-fired
Single-atom Hydrocarbons	1-10	—	6-9	5	36-120	730
Carbon monoxide	5-17	—	29	510	160-1100	3500
Formaldehyde	0.1-0.25	0.17	0.06-0.22	0.14	0.21-0.38	—

Note: lb/10^9 Btu input—multiply by 0.430 to obtain ng/J.

Table 13 lists some mean values for CO and hydrocarbon emissions from various sizes of boilers.

B. Particulate Emissions

The combustion of coal produces particulate emissions of two types, fly ash emissions, which are the result of small ash particles being entrained in the flue gas, and carbon particle emissions, which result from incomplete combustion of the coal.

Both fly ash and carbon emissions from coal combustion are strongly related to the type of combustion systems in which the coal is fired and how the combustor is operated. For example, if coal is fired on relatively undisturbed beds, for example on overfeed or underfeed stoker grates, little ash is entrained as fly ash. Conversely, in pulverized coal firing, most of the ash is entrained in the flue gas as fly ash. Considering carbon emissions, the quantity of carbon emitted as smoke is affected by residence time, air/fuel mixing, combustion temperatures, and excess air.

Table 14 gives estimated particle sizes for particulate emissions from various types of combustion equipment.

TABLE 14 Particle Size of Particulate Emissions from Coal Firing [38]

	Weight Percent Less Than Stated Size			
Particle size (μm)	Pulverized-coal-fired furnace	Cyclone furnace	Spreader-stoker-fired furnace	Stoker-fired (other than spreader)
10	30	76	10	7
20	50	83	20	15
40	70	90	37	26
60	80	92	47	36
80	85	94	54	43
100	90	95	60	50
200	96	97	—	66

Once the fly ash or carbon particles become entrained in the flue gas stream, the air pollution control problem becomes one of separating the particles from the gas stream. The major types of devices for accomplishing this are cyclones, electrostatic precipitators, scrubbers, and baghouses.

1. Cyclones

Cyclones are devices used to separate particles from a gas stream by centrifugal force. In these devices the dust-laden gas stream is caused to swirl rapidly as it enters the cyclone. The gas is then allowed to exit the cyclone along its center line. Because of the high swirl, centrifugal force causes the particles to be thrown outward, toward the wall of the cyclone. Upon reaching the wall of the cyclone, the particles fall to the bottom of the cyclone and are collected. Only small (low-mass) particles are able to follow the gas stream and escape the cyclone. The efficiency of a cyclone collector can be increased by increasing the swirl velocity, but this requires a higher pressure drop across the cyclone. The usual pressure drop for boiler applications is 2 to 4 in. Hg.

The efficiency of cyclone collectors decreases with particle size. Considering modern air pollution control requirements, conventional cyclone collectors are not capable of meeting air emission standards for most large coal-fired systems.

Advantages of the cyclone are low capital cost and reliability (due to simplicity). Operating costs, which consist primarily of the power associated with pressure loss, are very low.

2. Electrostatic Precipitators

The most common particulate control device for large coal-fired installations is the electrostatic precipitator (Fig. 21). In an electrostatic precipitator, the particles become charged when they enter an electrical field, causing the particles to migrate to a surface of opposite charge. As the particles strike the collector plate, they are discharged; simultaneously, they are agglomerated with other particles and are either held to the collector surface or fall to hoppers as larger particles. Rapping of precipitator surfaces is usually required to dislodge accumulated particles.

The efficiency of electrostatic precipitators can be adversely affected by several factors:

1. Excessive gas velocity.
2. Insufficient collector plate area.
3. Inadequate field strength.
4. Poor gas distribution, which overloads portions of the precipitator.
5. High resistivity of fly ash; this causes the accumulated fly ash layer to act as an insulator and prevents draining the charge from newly arriving particles—thus, a charged layer accumulates and repels other particles.

The first four deficiencies can be overcome by proper design practices. Thus, as the ratio of collection surface area to gas velocity is increased, collection efficiency is increased. Similarly, collection efficiency improves for particles greater than 1 μm [40].

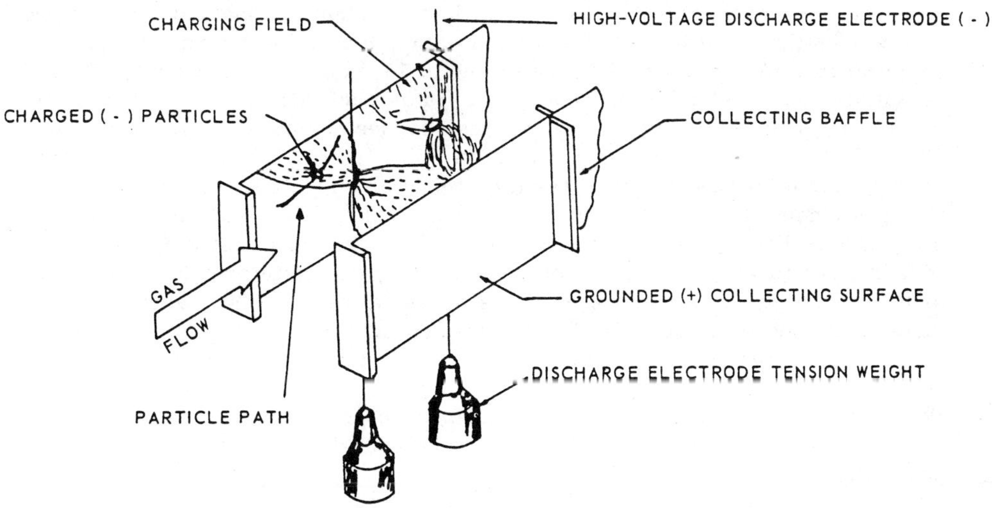

FIG. 21 Operating principle of an electrostatic precipitator [39].

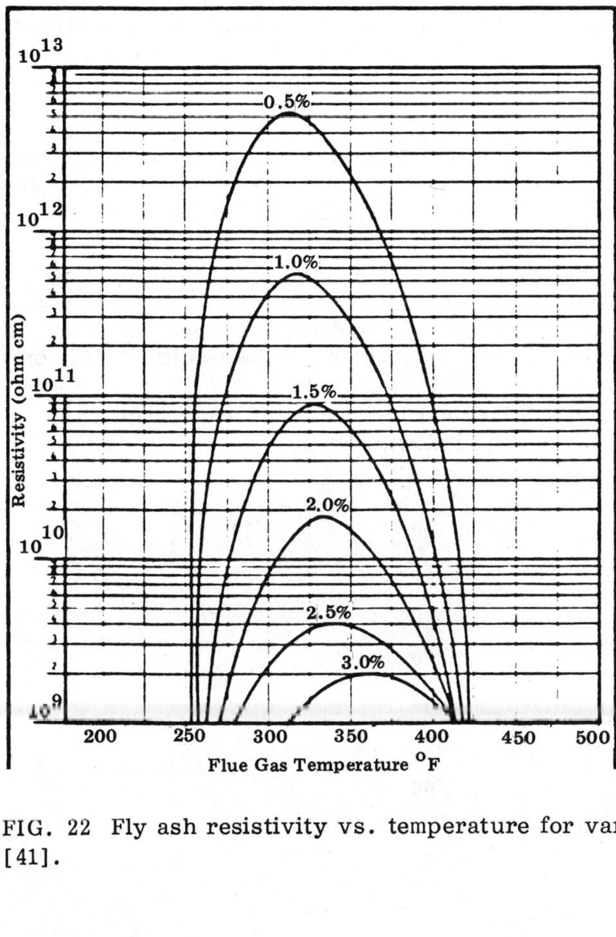

FIG. 22 Fly ash resistivity vs. temperature for various sulfur content fuels [41].

The last factor, fly ash resistivity, is dependent on particle temperature and composition. Figure 22 shows the relationship of fly ash resistivity to temperature. A fly ash resistivity of 10^8 to 10^{10} ohms/cm is considered to be satisfactory for achieving good precipitator performance. For high-sulfur coals, the preferred range of fly ash resistivities occurs at temperatures of 275 to 350°F. Thus, for precipitators on units firing high-sulfur coals, the precipitator is located between the air preheater and the stack, where the flue gas temperature is close to its minimum values. For units firing low-sulfur, low alkali coals, the preferred range of fly ash resistivity occurs at higher temperatures. Thus, "hot" electrostatic precipitators, located upstream of the air preheater where the flue gas temperature is higher, above 600°F, have been used on some units firing these coals.

A second means of achieving acceptable fly ash resistivity with low-sulfur coals is by the addition of additives upstream of the precipitator. The additives alter the fly ash resistivity sufficiently to permit successful precipitator performance. Among the additives used successfully are SO_3, H_2SO_4, and ammonia.

3. *Scrubbers*

Scrubbers have been applied as particulate control devices to a few large coal-fired power plants. One reason for the limited interest in scrubbers as particulate control devices relates to the problem of maintaining in operation equipment that must pump and handle dispersions of fly ash in water.

4. *Baghouses*

As a result of more stringent particulate emission regulations, baghouses are receiving more attention as possible particulate control devices for coal-fired combustors. Baghouses (Fig. 23) are efficient particulate control devices but, formerly, operating cost (pressure drop) and the capital cost and maintenance cost (for bag replacement) generally eliminated baghouses from consideration. However, as emission regulations become more stringent, other control devices may not be capable of achieving necessary emission control or may be more costly. Hence, baghouses are becoming a viable option for some coal-fired boilers.

One important consideration when considering baghouse collectors is the bag material. Bag materials have the following operating-temperature limits (°F) [43]:

Cotton and paper	180
Wool	210
Polyamide (nylon)	220
Polyester (Dacron)	280
Nomex	450
Asbestos	500
Tetrafluoroethylene	500

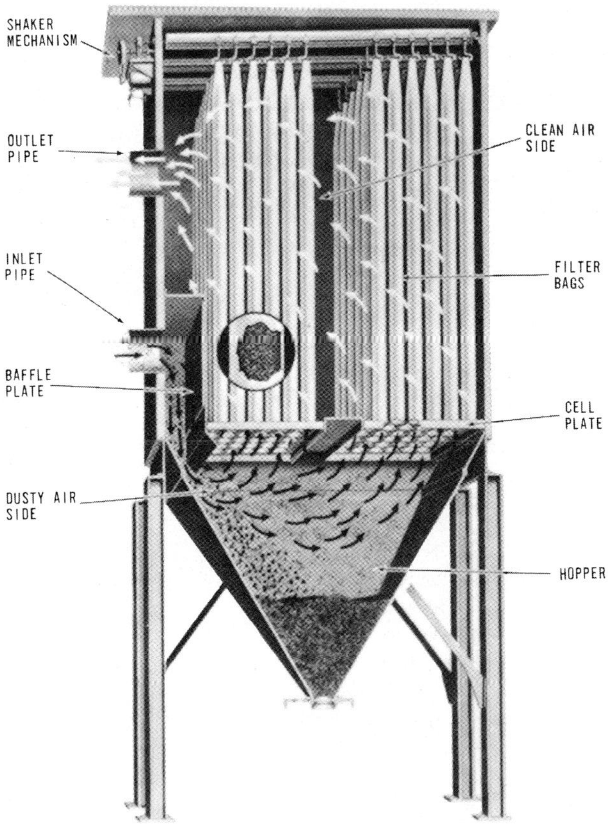

FIG. 23 Typical simple fabric-filter baghouse design [42].

The problem becomes one of selecting the most economical bag material that will withstand the temperatures and stack gas constituents to be encountered and give acceptable life.

Unlike other particulate collection devices, baghouses must be oversized for the flow rate to be encountered. This results from the fact that flow is shut off to sections of the baghouse when bags in the section are rapped to shake loose accumulated deposits.

C. POM Emissions

Polycyclic organic matter (POM) is included within the categories of particulate and/or gaseous hydrocarbons; however, it is considered separately because of its potentially carcinogenic nature. The following list of some POM compounds shows their relative carcinogenic ratings, based on a report by the National Academy of Science [44].

Strongly carcinogenic

Benz(a)pyrene
Dibenz(a,h)pyrene
Dibenz(a,i)pyrene
Dibenz(a,h)anthracene
7,12-Dimethylbenz(a)anthracene
Benz(c)phenanthrene
Dibenz(c,g)carbazole
Benz(b)fluoranthene
Benz(j)fluoranthene
3-Methylchloranthrene
Benz(j)aceanthrylene
Dibenz(b,def)chrysene
Dibenz(a,j)acridine
Dibenz(a,h)acridine

Carcinogenic

Benz(a)anthracene
Dibenz(a,j)anthracene
Dibenz(a,c)anthracene
Dibenz(a,g)fluorene
Indeno(1,2,3-cd)pyrene
Dibenz(def,p)chrysene

Uncertain or weakly carcinogenic

Dibenz(a,h)fluorene
Dibenz(a,c)fluorene
Dibenz(a,l)pyrene
Chrysene
Dibenz(c,h)acridine
Benz(a)carbazole
Dibenz(a,g)carbazole
Dibenz(a,i)carbazole

Not carcinogenic

Naphthacene
Benz(ghi)perylene
Anthracene
Phenanthrene
Anthanthrene
Pyrene
Coronene
Perylene
Ben(e)pyrene
Fluorene

Many of the POM species emitted by combustion systems fall within the non-carcinogenic species. However, concern exists over even low levels of emissions of such carcinogenic POM species as benzo(a)pyrene.

The exact nature of events leading to POM emissions from coal combustion is not known. Coal itself is a highly complex organic species and contains many ring-structure organics that, if released, would constitute POM species. Therefore, the incomplete combustion of coal will release POM compounds. However, POM compounds can also be formed as a result of reactions during the cooling of flue gas from combustion temperature to temperature less than 1000°F. See Table 15.

VI. ECONOMICS OF COAL UTILIZATION

Electric power generation consumes about 55% of all coal mined; firing of industrial boilers for process heat and space heating consumes about 20% of coal production, and direct-heating applications such as kilns and steel production account for about 23%. Each of these uses has different economics. This discussion considers the economics of firing boilers for electric power generation and industrial steam, which account for an aggregate of about 75% of all coal used in the United States.

TABLE 15 Polynuclear Hydrocarbons Emitted by Bituminous Coal with Various Firing Methods [45,46]

	Pulverized-coal-fired	Cyclone	Spreader stoker[a]		Chain-grate stoker	Underfeed stoker[a]			Hand-fired
			A	B		C	D	E	
Benzo(a)pyrene	0.03–0.13	0.49	0.04	0.057	0.082	22	0.26	8.4	880
Pyrene	0.20–0.40	2.25	0.23	1.30	0.86	35	3.70	17	1320
Benzo(e)pyrene	0–0.58	0.87	0.13	0.770	0.290	17	0.510	11.9	220
Perylene	0–0.15	0.07	—	—	—	3.5	—	—	132
Benzo(ghi)perylene	0.02–1.42	0.44	—	—	—	9.9	—	1.28	660
Anthanthrene	0–0.02	—	—	—	—	0.64	—	—	198
Coronene	0–0.15	0.01	0.02	0.057	—	0.73	—	2.64	66
Anthracene	—	—	—	—	—	1.9	—	—	880
Phenanthrene	—	—	—	—	—	22	2.2	64	2200
Fluoranthene	—	0.17	0.11	0.790	1.50	83.9	7.1	103	2200
Benz(a)anthracene	—	—	—	—	—	8.6	—	1.23	—

Note: Lb/10^{12} Btu input (multiply by 0.430 to obtain μg/MJ).

[a]Letter designations refer to different units.

TABLE 16 Relative Costs of Coal, Oil, and Natural Gas, January 1979 [47]

Utility or location	Coal	Jet A	No. 2 oil	No. 6 oil	Natural gas
Alabama Power	140-186	—	297-314	—	225
Arizona Public Service	38-94	—	274-313	207-291	140-252
Baltimore Gas and Electric	—	315	290-302	236-243	172
Cleveland Electric Illumination	115-175	—	299-317	247-250	—
Cincinnati Gas and Electric	87-193	334	287-293	—	220
Commonwealth Edison, Chicago	162	314-353	306-326	297-432	200-244
Florida Power and Light	—	—	290-305	180-241	72-215
Central Texas Power	—	—	268	230	70-241

Note: Range of contract prices for major utility purchasers (cents/10^6 Btu)

A. Relative Costs of Coal, Oil, and Natural Gas

The costs of fuels purchased by public-utility generating stations are published regularly for each utility, and provide a basis for comparing relative costs of fuels to large users in different areas. Utilities often purchase fuel on long-term contracts, resulting in bargain prices in times of rapidly escalating fuel costs. Other fuel, purchased on spot sales in times of fuel scarcity, reflects higher-than-normal costs.

Table 16 shows some selected utility fuel purchase costs for January 1979 as reported to the FERC [47]. It is evident that fuel costs vary widely for each utility and for different locations. For example, the cost of coal varied from $0.38 per million Btu at the mine-mouth plant at Four Corners, to $1.93 for deep-mine coal in Cincinnati. The price of No. 2 oil, used for gas turbine peaking plants, varied from $2.68 to $3.26; jet A aircraft kerosene can be substituted for No. 2 oil in gas turbines when available, and ranged in cost from $3.14 to $3.53. No. 6 fuel oil, used widely for boiler fuel in coastal cities, ranged from $1.80 to $4.32, with most falling between $2.07 and $2.91. The cost of natural gas varied most widely, ranging from $0.70 to $2.52, reflecting different contract time periods. The wide differences in reported costs illustrate the difficulty of discussing cost ratios among fuels.

The cost of coal varies widely depending upon source and quality. Low-sulfur coal can meet 1979 EPA emission requirements without the use of stack-gas scrubbers and is therefore cost-effective at almost any price. Thus, large tonnages of coal are shipped from Montana to Chicago and even into the Ohio and West Virginia mining areas. Mining costs for Western strip mines are around \$6.00 per ton (\$0.30 per 10^6 Btu), but the delivered cost in Chicago is \$1.62 per 10^6 Btu. The cost of Appalachian coal ranges from about \$30 per ton (\$1.20 per 10^6 Btu) to about \$50 per ton (\$2.00 per 10^6 Btu). High-sulfur strip-mine coals, available at low cost in Illinois and Ohio, do not meet SO_2 emission requirements, and their use in the future will probably require stack gas scrubbers. Low-sulfur Eastern coal that meets emission requirements is frequently washed to remove ash and sulfur, thus increasing its cost. Because this coal is usually from deep mines, it has a higher initial cost than does strip-mined coal.

B. Electric Power Generation

Because large-scale storage of electrical energy is not feasible, electric utility power stations generate electricity on demand. This requires sufficient generating capacity to meet (1) the base load used continuously through the year, (2) cycling loads that extend for 8 to 12 hr through each day, and (3) peak loads that occur for only a few hundred hours per hear. Peak loads may include a summer peak on the hottest days, a winter peak during the coldest days, and daily peaks of a few hours. The plant factor (defined as the total annual electrical output divided by the total annual output for a year of operation at rated capacity) is the usual basis for describing annual load factor. Typical values range from about 70% for base-load plants to about 10% for peaking plants.

A base-load plant is usually designed for the highest practical efficiency, at high investment cost and with minimum fuel and operating cost per kilowatt-hour of electric power. A typical steam plant for base-load service would be a reheat plant with steam at 2400 or 3600 psi pressure and 1000°F superheat and reheat temperatures. New base-load plants would normally be constructed with a cooling tower and with a stack gas scrubber for control of SO_2 emission. The overall thermal efficiency of an older plant without scrubbers is in the range 38 to 40%, but the efficiency of new environmentally acceptable plants is about 32% for the same steam conditions. The 1979 capital cost of a base-load plant is about \$640 per kilowatt.

Typical peak-load plants operate for less than 1000 hr/year, with capacity factors between 0 and 20%. The overriding factor in the cost of peaking power is investment cost, because capital cost is distributed over a small number of operating hours; thermal efficiency and fuel cost are much less important than for base-load plants. Gas turbines, burning No. 2 distillate oil or natural gas, are widely used as peaking plants because they can be installed for less than \$150 per kilowatt. At the present time there appears to be no type of coal-fired plant having suitably low capital cost for peaking service. In the future, coal-derived liquid fuels or pipeline synthetic natural gas would appear to be the most economical fuels for utility peaking service, assuming reduced availability of natural gas and oil.

Midload plants carry the cycling load between the base-load units and the peaking units, often coming up to full load at 7 A.M. and returning to a no-load condition at the end of the day. Usual plant factors range from 20 to 70%. Many midload plants are new steam plants designed for cycling service, which requires a relatively simple cycle and moderate capital cost because of the relatively low capacity factor expected. Because the plant factor is low, the economics of stack-gas scrubbing are not attractive. It would appear that cycling plants would find it most economic to burn low-sulfur coal, at a higher fuel cost, rather than carry the high capital cost of a scrubber installation for a few hours of operation each day. Future EPA rules for new-plant construction may provide little opportunity for such economic choices, however, as it is currently being proposed that all new plants install scrubbers regardless of class of service.

1. *Costs of Base-Load Power Plants*

The economics of coal utilization for both electric power generation and industrial steam and power generation are changing rapidly with inflation, with changing coal mining regulations that influence coal cost, and with changing environmental regulations that influence the capital and operating costs of coal-fired systems. Thus, to make sense of any cost estimate for energy production from coal, it is essential to know the exact period covered by the estimate, the rules pertaining to environmental requirements, and the cost structure for that period. As an example, a recent cost study by Ebasco Services showed that the capital cost of a large utility steam plant for the generation of electricity had increased from $122 per kilowatt of electricity for a plant completed in 1968, to $639 per kilowatt for a plant completed in 1978 [48]. Of the cost increase of 420% in 10 years, $120 per kilowatt was due to inflation and $397 was due to regulatory changes. These changes include the cost of stack gas scrubbers to remove SO_2, the cost of more efficient particulate collectors, the cost of cooling towers, and, of even greater importance, the increased cost of interest and escalation incurred because of a greatly increased construction period brought about by the need for environmental-impact statements, public hearings, and hundreds of permits from numerous agencies. Any delays from actions by intervenors who may wish to delay or stop construction also adds to costs.

TABLE 17 Cost Comparison of Boiler-House Equipment for Coal- and Oil-Fired 1000-MWE Utility Boilers [50]

	Coal	Oil
Steam-generating equipment	$30,005,000	$25,379,500
Draft system	10,512,500	10,404,000
Fossil-fuel equipment	11,714,000	7,123,500
Ash-handling system	2,320,500	1,010,000
Instruments and controls	2,275,000	2,275,000
Miscellaneous suspense items	710,000	710,000
Total boiler plant equipment	$57,537,000	$46,902,000
Total plant cost	$174,000,000	$158,000,000

A comprehensive study of power-plant construction costs was carried out in 1971 by United Engineers and Constructors, for the U.S. Atomic Energy Commission, to provide definitive cost comparisons of nuclear plants with plants fired with coal, oil, and gas. The results were published in eight reports [49]. All costs were based on the construction of a 1000-MW plant at a Midwest location with completion scheduled for 1971. Estimated costs for fossil-fueled plants were \$174 per kilowatt for coal, \$158 per kilowatt for oil, and \$112 per kilowatt for gas firing. The costs of the boiler-house equipment are itemized in Table 17. These costs do not include coal storage and handling equipment outside the boiler house. The higher costs for the coal-fired plant reflect the more expensive boilers, coal storage and handling, coal pulverizers, and ash-handling systems. They do not include provisions for stack gas scrubbing, which was not required then, and none of the costs include interest and escalation during construction, which can add 50% to total capitalized cost.

2. *Costs of Future Base-Load Plants*

It has been recognized that it would be desirable to improve the efficiency of environmentally acceptable electric-power-generating plants from the present value of about 32%. The U.S. Department of Energy Energy Conversion Alternatives Study (ECAS), carried out in 1974, reviewed in great detail, on the basis of common ground rules, the characteristics and costs of many of the advanced power cycles that have been proposed [51]. After the first screening phase, a more detailed study was made of the seven most economic cycles, and performance and costs for these cycles were compared with those for the present standard steam cycle.

Figure 24 shows overall efficiency, coal pile to bus bar, plotted against cost of electricity for several types of advanced power cycles. The most attractive cycles are those plotted below and to the right of the reference steam cycle. The dimensions of the oval areas plotted represent the range of values determined for cycle variations in Phase 1 studies. The solid points represent optimized systems in the Phase 2 studies.

Figure 25 shows the distribution of capital costs for the alternative power systems of Fig. 24, as determined in the Phase 2 study. The Reference Steam Cycle is a 3500 psi/1000°F supercritical, single-reheat steam cycle with cooling towers and stack-gas scrubbing. Costs of all systems are based on starting construction in 1974, and the interest during construction and escalation varies with the length of the construction period. From Fig. 25 it is evident that some cycles promised lower capital cost than does the reference steam cycle.

Figure 26 shows the distribution of capital, fuel, and operating costs for the same cycles. It is evident that capital cost dominates all these cycles, with the exception of the combined cycle with water-cooled gas turbine fired by a coal-derived liquid fuel, for which fuel cost is the dominant factor. The gas turbine would appear to be most suitable for the lower load factors because of its lower capital costs. The combined cycle with an open-cycle gas turbine and steam cycle fired with low-Btu gas shows costs very similar to the reference steam cycle. However, if the gas-turbine-inlet temperature is increased to the range 2600 to 3000°F, significant cost reductions occur.

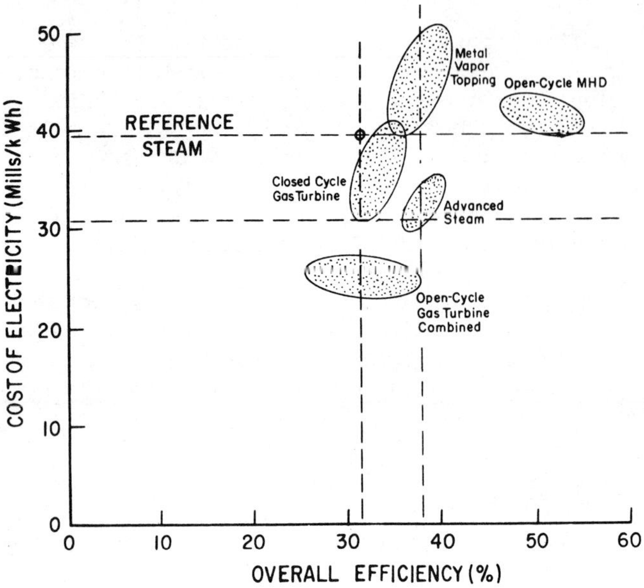

FIG. 24 Efficiency and cost of electricity for ECAS advanced power systems [51].

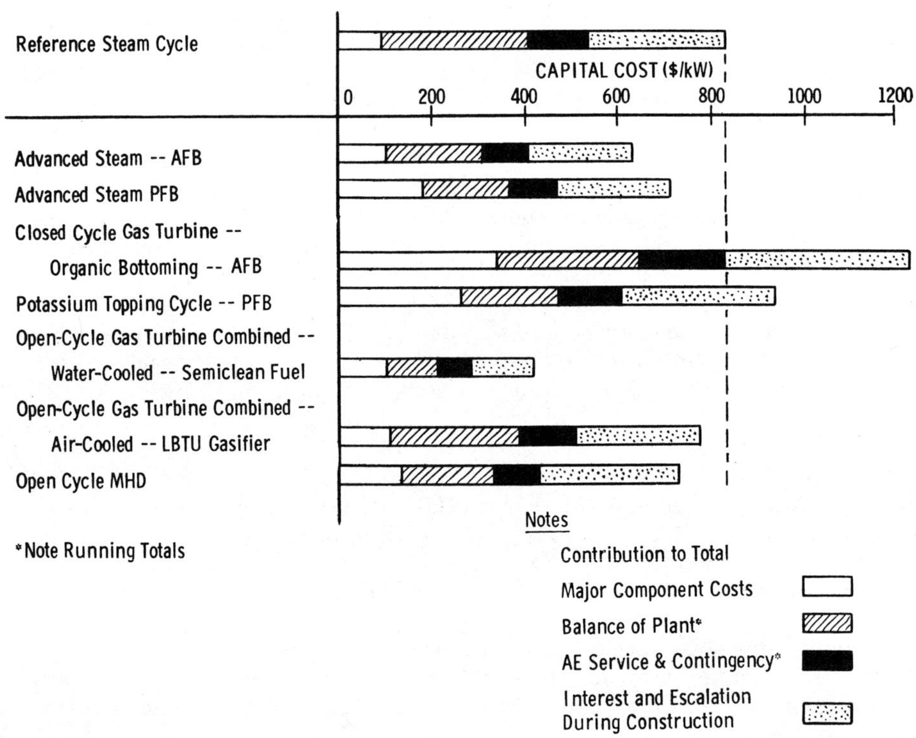

FIG. 25 Comparison of capital cost for ECAS advanced power systems [51].

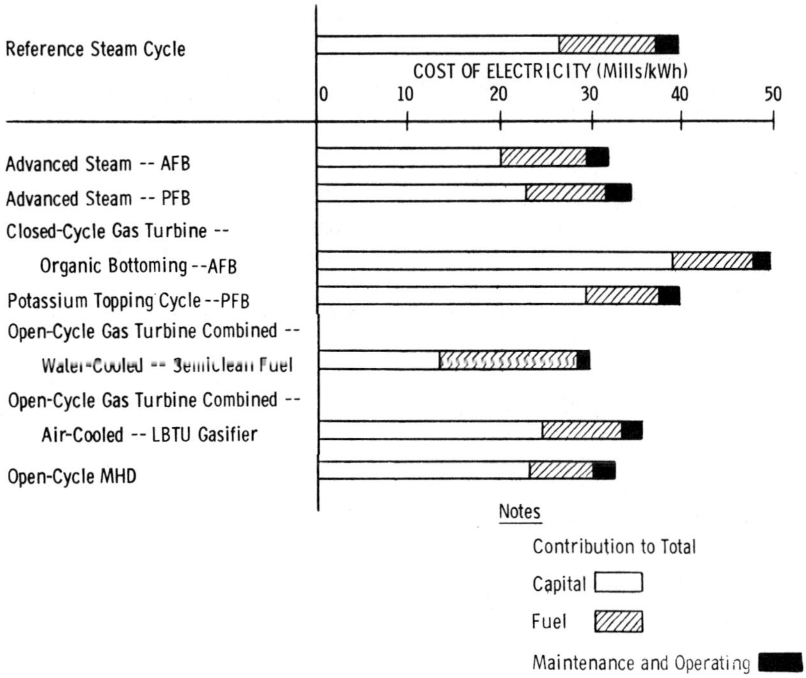

FIG. 26 Components of electricity costs for ECAS advanced power systems [51].

It will be noted that the capital cost of the Reference Steam Cycle plant for the ECAS study is about $820 per kilowatt of electricity, compared with only $174 for the WASH-1230 study. For a direct comparison it is necessary to deduct the cost of interest and escalation during construction from the ECAS costs, adjust for completion at the same date, and adjust costs to meet environmental requirements.

C. Industrial Steam Generation

Industrial steam plants are major users of fossil fuels, accounting for about 60% of all industrial energy consumption. However, at the present time this sector is dominated by oil and gas, which provide much lower steam costs for small installations and those that have lower capacity factors. Coal-fired industrial boilers tend to be larger units operated at relatively high load factors. In 1968, 23% of industrial steam fuel was coal.

With current uncertain supplies of imported oil and declining supplies of natural gas, there is more incentive to install coal fired industrial boilers to assure future operation. In most industrial plants, the cost of steam is a small part of total plant operating expense, and the financial cost of a plant shutdown because of a fuel shortage far outweighs any possible variations in the cost of steam with different fuels. In addition, it appears probable that future costs of oil and gas will increase faster than the cost of coal. Thus, the cost of a coal-fired boiler plant can be justified on the basis of assuring a reliable steam supply and, possibly, on the basis of a saving in life-cycle costs from fuel savings. However, the capital cost of a field-erected coal-fired boiler is several times greater than that for a packaged oil-fired boiler, and it may take many years to pay out the difference by savings in fuel cost.

Figure 27 shows comparative capital costs in 1978 for coal- and oil-fired industrial boilers [51]. Coal fired boilers for steam capacity below 200,000 lb/hr are usually stoker-fired, whereas it is economic to fire larger boilers with pulverized coal if the load factor is high. The capital cost of stoker-fired boilers ranges from about \$60 per pound of steam per hour for a 50,000-lb/hr unit, to about \$40 per pound of steam per hour for a 200,000-lb/hr unit [45]. The comparable capital cost of a pulverized-coal-fired boiler ranges from about \$50 per pound of steam per hour for a 300,000-lb/hr unit to about \$45 per pound of steam per hour for a 500,000-lb/hr unit.

Table 18 shows comparative operating costs for oil-fired and coal-fired boilers. The operating cost of the pulverized-coal-fired boiler is lower than that for a stoker boiler because it can burn lower grades of coal and is more efficient. The requirement for operation at high excess air, and the high carbon loss, lower the combustion efficiency of stoker-fired boilers. Under present economic conditions a pulverized-coal-fired boiler would pay out the difference in capital cost over a stoker-fired boiler in a few years if both operate at high load factors. For low load factors, the lower cost of the stoker would make it preferable.

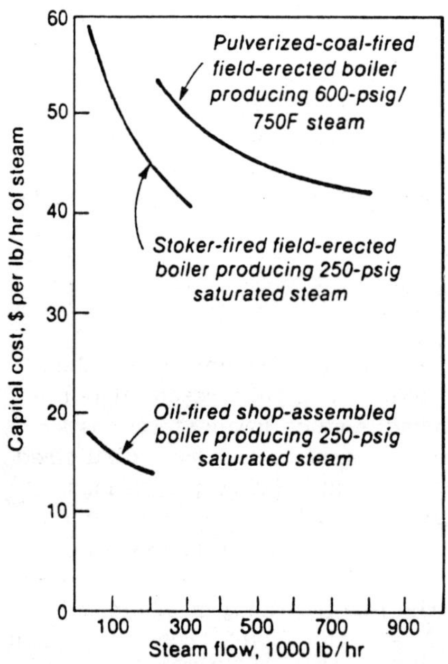

FIG. 27 Capital costs of industrial boilers [50] based on 1968 cost data.

TABLE 18 Industrial Boiler Annual Operating Costs [50]

Operating requirements and costs	Type of steam generator		
	Oil-fired	Stoker-fired	Pulverized-coal-fired
Equivalent full-power operating hours per year	6000	5400	5700
Fuel consumption (gal/yr for oil, tons/yr for coal)	12,000,000	86,400	79,800
Fuel cost (cents/gal for oil, $/ton for coal)	35	42	35
Total fuel cost ($/yr)	4,200,000	3,628,000	2,793,000
Electric-power consumption (kWhr/yr)	3,000,000	3,240,000	8,550,000
Electric-power cost [$/yr (assumes 3 cents/kWhr)]	90,000	97,200	256,500
Maintenance cost [$/yr (assumes repairs cost 2-1/2% of capital investment per year)]	70,000	220,000	265,000
Personnel per shift	1	2.25	2.75
Personnel cost [$/yr (assumes $25,000/yr per person)]	100,000	225,000	275,000
Ash removal (tons/yr)	—	8640	7980
Ash-removal cost [$/yr (assumes $2/ton)]	—	17,280	15,960
Total operating cost ($/yr)	4,460,000	4,187,480	3,605,460

Note: Based on 1978 cost data.

The level of environmental control required at a specific location affects the cost of industrial boilers. All coal-fired boilers require dust collectors to meet particulate emission limits. High-efficiency cyclone collectors currently suffice for stoker-fired boilers, but more expensive electrostatic precipitators or baghouses are required for pulverized-coal-fired units. The cost of an industrial electrostatic precipitator is about $5 per pound of steam per hour [4].

Present New Source Performance Standards limit emission of SO_2 from boilers having a heat input greater than 250 million Btu/hr, equivalent to about 250,000 lb of steam per hour, to 1.2 lb SO_2 per million Btu. This limit can be met by firing low-sulfur coal or by use of stack-gas scrubbing. It is estimated that the capital cost of an industrial stack-gas scrubber is about $5 to $10 per pound of steam per hour. There is also a significant cost of operation and of disposal of scrubber sludge. Under present conditions it appears most economical to avoid scrubber costs by burning low-sulfur coal. However, the design of a new plant should provide space for a future scrubber as insurance against future cost increases for low-sulfur coal or changes in environmental requirements.

It will be noted that the cost of a large industrial boiler, at $40 per pound of steam per hour, is equivalent to about $400 per kilowatt of electrical capacity. For comparison, a utility boiler 10 times as large would cost only $120 per kilowatt of electrical capacity.

VII. FUTURE TRENDS

To attempt to define future trends in the combustion technology or extent of utilization of coal is rather precarious. Future trends depend on (1) energy policy, (2) economics, (3) environmental regulations, and (4) technology. In certain respects, these four factors may be listed in terms of decreasing priority. Unquestionably, energy policy, and for that matter foreign policy, will play a significant role in providing the impetus for coal utilization. As restrictions on oil imports grow more severe and as concern for achieving a more favorable balance of payments increase, there will be more and more pressure to burn coal. The U.S. government's energy projections predict utilization of some 850 million tons of coal per year by 1985. Today, some 70% of U.S. utilities burn coal. Industry, on the other hand, still burns oil in preference to coal.

It is estimated that from 100 to 150 million tons of coal annually will be burned by industry by 1985 [52]. Industry will be burning coal for process steam, space heating, and self-generation of electricity. Currently, the coals used by industry are selected on the basis of cost and compliance with environmental regulations. One can expect to see some relaxation in environmental regulations, primarily because of economic pressures, but the clean burning of coal will remain a major issue.

Industry will be burning coal in pulverized coal furnaces and in stokers. One can expect to see moves toward smaller pulverized-coal units and larger stoker units. Industry will attempt to avoid flue gas scrubbing if at all possible, primarily for economic and operational reasons. Also for these reasons, we will see industry giving serious attention to fluidized-bed combustors. At this writing FBC is still in the "demonstration mode," but in the next 10 years there will probably be a significant move to FBC. Will it have a capacity limitation? Probably, but experience and economics will provide the basis for maximum FBC limitations.

The quality of coal available will also influence how industry applies coal combustion in the future. One can expect to see more interest in burning low-quality coals, such as 6000 Btu/lb lignites, especially in fluidized-bed units, in the future. In this vein, refuse-derived fuels (RDF) will be burned with coal, probably on stokers and fluidized beds, and possibly even in pulverized-coal units. Coal/oil mixtures (COM) may also gain in favor, at least in an interim period wherein industries burning oil want to delay or cannot afford (economically or technically) to convert to coal combustion.

Combustion furnaces and burners will not be changed significantly in the future. As NO_x controls are applied, there will be more emphasis on low-NO_x burners with delayed mixing and/or staged combustion, and on larger furnace volumes. This will have some effect, but not a major one, on the design of coal combustion furnaces.

In essence, then, the crystal ball shows that utilities and industry will burn more coal in the future. For the most part, however, they will burn it the way they are burning it today. Coal combustion is a rather efficient process. Coal is dirty and harder to handle than gas or oil, but it remains the country's number 1 alternative fuel.

REFERENCES

1. H. H. Lowry, ed., *Chemistry of Coal Utilization, Supplementary Volume.* John Wiley & Sons, Inc., New York, 1963, pp. 29-34.
2. C. L. Wagoner and E. C. Winegartner, Further Development of the Burning Profile, *J. Power Eng.*, p. 119 (Apr. 1973).
3. R. W. Bryers, K. Biswas, and T. E. Taylor, Fuel Evaluation Using Differential Thermal Techniques, 68th Annual Meeting of the AIChE, Los Angeles, Calif., Nov. 1975.
4. H. H. Lowry, ed., *Chemistry of Coal Utilization, Supplementary Volume.* John Wiley & Sons, Inc., New York, 1963, pp. 754-755.
5. J. R. Arthur, *Nature* 157:732 (1946).
6. M. A. Field et al., *Combustion of Pulverized Coal.* BCURA, Leatherhead, U.K., 1967.
7. M. F. R. Mulcahy and I. W. Smith, *Rev. Pure Appl. Chem.* 19:81 (1969).
8. H. B. Palmer and J. M. Beer, *Combustion Technology: Some Modern Developments.* Academic Press, Inc., New York, 1974, p. 389.
9. M. A. Field et al., *Combustion of Pulverized Coal.* BCURA, Leatherhead, U.K., 1967, pp. 225-226.
10. D. K. Sommer, Report of Information on Sulfur Emissions from Power Plant Boilers, FEA Report No. Y-8833, Washington, D.C., Sept. 1975.
11. O. W. Dykema, Analysis of Gas-, Oil-, and Coal-Fired Utility Boiler Test Data, Proceedings of the Stationary Source Combustion Symposium, Atlanta, Ga., June 1976, EPA-600/2-76-152C, Vol. 3, p. 115.
12. American Gas Association, *Fuel Flue Gases*, 1941 ed.
13. The Babcock and Wilcox Co., *Steam, Its Generation and Use.* New York, 1978, pp. 6.10-6.11.
14. The Babcock & Wilcox Co., *Steam, Its Generation and Use.* New York, 1978, p. 6.17.
15. The Babcock & Wilcox Co., *Steam, Its Generation and Use*, 37th ed. New York, 1964, pp. 16-2, 16-3.
16. The Babcock & Wilcox Co., *Steam, Its Generation and Use*, 37th ed. New York, 1963, p. 17-17.
17. The Babcock & Wilcox Co., *Steam, Its Generation and Use*, 37th ed. New York, 1963, p. 28-2.
18. Proceedings of the Coal-Oil Mixture Combustion Technology Exchange Workshop, Washington, D.C., October 29, 1976, prepared by the Mitre Corporation/METREIC Division, McLean, Va., Conf. 761019 M 77-8, Feb. 1977.

19. C. B. Foster, ERDA's Coal-Oil Mixture Combustion Program, ASME 77-WA/FU-4, Winter Annual Meeting, Atlanta, Ga., Nov. 27-Dec. 2, 1977.
20. The Babcock & Wilcox Co., *Steam, Its Generation and Use,* 37th ed. New York, 1963, p. 22-6.
21. The Babcock & Wilcox Co., *Steam, Its Generation and Use.* New York, 1972, Chaps. 12-16.
22. The Babcock & Wilcox Co., *Steam, Its Generation and Use.* New York, 1972, p. 27-5.
23. W. T. Reid and P. Cohen, *Trans. ASME* 66:685 (1944).
24. P. Nicholls and W. T. Reid, *Trans. ASME* 62:141 (1940).
25. D. R. Hardesty and J. H. Pohl, The Combustion of Pulverized Coal—An Assessment of Research Needs. I. Background, Justification and Mineral Matter, SAND 78-8804, Jan. 1979, p. 37.
26. R. C. Attig and A. F. Duzy, Proceedings of the American Power Conference, Vol. 31, 1969, p. 290.
27. U. H. Barnhart and P. C. Williams, *Trans. ASME* 78:1229-1236 (1956).
28. R. C. Borio, G. W. Goetz, and A. A. Levasseur, ASME Paper No. 77-WA/Fu-6, 1977.
29. A. K. Meza, K. Shoji, and L. G. Austin, ASME Paper No. 78-WA/CD-1, 1978.
30. R. C. Corey, B. J. Cross, and W. T. Reid, *Trans. ASME* 67(4):289 (1945).
31. R. C. Corey, H. A. Grabowski and B. J. Cross, *Trans. ASME* 71(8): 951 (1949).
32. W. T. Reid, R. C. Corey, and B. J. Cross, *Trans. ASME* 67(4):179 (1945).
33. K. L. Maloney, P. K. Engel, and S. S. Cherry, Sulfur Retention in Ash, EPA-600/7-78-153b, Nov. 1978.
34. D. W. Pershing, G. C. England, M. P. Hays, and G. Flament, Control of NO_X Emissions from Liquid Fuel-Fired Industrial Boilers, 86th National AIChE Meeting, Houston, Tex., Apr. 1-5, 1979.
35. Environmental Control Implications of Generating Electric Power from Coal, Argonne National Lab. Report ANL/ECT-3, Appendix D, p. 20.
36. R. D. Giammar, D. R. Hopper, P. R. Webb, and E. Rhadakrishnan, Evaluation of Emissions and Central Technology for Industrial Stoker Boilers, Proceedings of the Third Stationary Source Combustion Symposium, EPA-600/7-79-050a, Vol. 1, Feb. 1979, pp. 3-34.
37. W. S. Smith and C. W. Druler, Atmospheric Emissions from Coal Combustion—An Inventory Guide, Publ. 999-AP-24, Public Health Service, U.S. Department of Health, Education and Welfare, Cincinnati, Ohio, 1966.
38. Air Quality Criteria for Particulate Matter, HEW Publ. AP-49, Jan. 1969, p. 27.
39. Control Techniques for Particulate Air Pollutants, HEW Publ. AP-54, Jan. 1969, p. 85.
40. S. Oglesby, Jr., and G. B. Nickols, Electrostatic precipitation, in *Air Pollution,* 3rd ed., A. Stern, ed. Academic Press, Inc., New York, 1977, p. 213.
41. E. W. Stenby, J. L. York, and K. S. Campbell, Particulate Removal When Western Coal is Burned, *Pollut. Control Clean Energy AIChE Ser.* 72(156):244.

42. Control Techniques for Particulate Air Pollutants, HEW Pub. AP-51, Jan. 1969, p. 103.
43. K. Jinoya and T. C. Orr, Jr., in *Air Pollution*, Vol. 4, 3rd ed., A. Stern, ed. Academic Press, Inc., New York, 1977, p. 153.
44. Particulate Polycyclic Organic Matter, NAS-NRI, 1972, p. 4.
45. R. D. Hangebrauck, D. J. Von Lehmden, and J. E. Meeker, Sources of Polynuclear Hydrocarbons in the Atmosphere, National Air Pollution Control Administration, HEW Publ. No. 999-AP-33, 1967.
46. W. S. Smith and G. W. Gruber, Atmospheric Emissions from Coal Combustion—An Inventory Guide, Public Health Service, HEW Publ. No. 999-AP-24, 1967.
47. *Energy Week*, May 7, 1979.
48. *Fortune*, May 7, 1979, pp. 114-120.
49. United Engineers and Contractors, Inc., for the U.S. AEC, 1000-MWE Central Station Power Plants Investment Cost Study, AEC Report No. WASH-1230, Vol. 3, Coal Fired Fossil Plant; Vol. 4, Oil Fired Fossil Plant, Washington, D.C., June 1972.
50. B. Dwight Coffin, Estimating Capital and Operating Costs for Industrial Steam Plants, *Power*, pp. 106-111 (Apr. 1979).
51. General Electric Company, Study of Advanced Energy Conversion Techniques for Utility Applications Using Coal or Coal-Derived Fuels, Feb. 17, 1976.
52. R. L. Major, The Outlook for Coal Through 1985, Proceedings of the Engineering Foundation Conference on Clean Combustion of Coal, EPA-600/7-78-073, Apr. 1978, p. 323.

9 Coal Carbonization

MICHAEL O. HOLOWATY, RICHARD G. PHELPS, WILLIAM DuBROFF, and
GEORGE L. LANDSLY / Inland Steel Company, East Chicago, Indiana

I. EARLY DEVELOPMENTS

The term "coke" brings an immediate reference to the blast furnace and thus to a technology that has been one of mankind's greatest achievements—the extraction of Metal from oxide ores.

The "invention" of the iron blast furnace reduction process occurred over a period of many centuries during the era of unwritten history. It could have started with an accidental discovery of iron-ore reduction in a tribal fire. In any case, the forge-type charcoal reduction process was well known in almost all areas of the ancient world around the time of Christ [1]. Gradually, the iron production increased as the market demanded and by the early Middle Ages, a shaft blast furnace known in Germany as "stuckofen" was well established and ironmaking was a respected trade and not a sporadic part-time occupation. The fuel for these furnaces was provided by wood charcoal and, consequently, the vitality of the iron and steel industry was controlled by the availability and cost of wood.

The rapid industrial development of western Europe in the eighteenth century, particularly that of England, was severely hampered by the supply of char-grade timber, which was consumed at an enormous rate and literally devastated the woods of the English Midlands. The solution to the blast furnace fuel problem so desperately sought by England in order to recapture the primary iron production market from the United States, appeared to be mineral coal, which was abundantly available.

The key to the utilization of the structurally strongest and chemically richest coal—anthracite—was found in preheating the blast to the tuyeres, which thus far had been operated at ambient temperature only. This step proved to be the salvation of the English steel industry, but it also gave a boost to the American steel industry, which had at its disposal the extensive anthracite reserves of eastern Pennsylvania [2].

The use of coke came about apparently by accident in the early nineteenth century at about the same time in both England and the United States. Here, small amounts of coke generated by underground mine fires were charged together with bituminous coal or anthracite into blast furnaces in western Pennsylvania. F. H. Oliphant of Fayette County, Pennsylvania, is credited with having operated the first coke-burdened furnace, near Johnstown, although this bit of history is questionable.

The method of coke production was initially the same as for the production of charcoal—stockpiling coal in round heaps, known as milers; igniting the piles; and then covering the sides with clay-type soil. The smoke generated by the partial combustion of coal tars and gases was soon a major problem near residential areas, and the coking process itself could not be controlled because of the elements—rain, wind, and ice. Nevertheless, this new modification of the charring-coking kiln laid the foundation for the modern coke blast furnace: the beehive process.

II. THE BEEHIVE COKING PROCESS

The beehive process is based on self-heating of the coking coal charge in a dome-shaped circular oven by means of partial combustion of the coal, particularly its volatile by-products. The beehive design was conceived in Britain and is really a logical outgrowth of the charcoal kiln process, in which timber was stacked in a conical pile and covered by dust and soil which was packed tightly. The pile was provided with openings for air access, which could be adjusted to maintain the temperature needed for the process.

A. Construction of Ovens

The cross section of a typical beehive oven is shown in Fig. 1. The name "beehive" is literally descriptive of the form of the beehive coke oven. Beehive coke oven plants constructed in the past have followed three general arrangements: (1) the bank system, in which the ovens are built in single rows against a bank of earth, natural or artificial, thus making it necessary to build only one retaining wall along the front of the ovens; (2) the single-block system, which consists of a single row of ovens with retaining walls at both the front and back; and (3) the double-block system, in which the ovens, in a double row, are built back or staggered with a retaining wall extending along the front of each row. Figure 2 illustrates the double-block system. The ovens are usually 10 to 15 ft in diameter and up to 10 ft high.

B. Charging

Beehive ovens are charged as soon as practicable after drawing in order that stored-up heat from the previous charge will be sufficient to start the new coking charge. New ovens must be heated up gradually to the coking temperature by wood and coal fires, after which small charges of coal are coked until the ovens reach normal working conditions. With the oven in readiness for charging, the door is partially bricked up and the charge of 5 to 12 tons is dropped through the trunnel head from the larry car above, leaving the coal in a cone-shaped pile in the oven. To secure uniform coking of the coal, this pile must be leveled so that the coal will lie in a bed of uniform depth over the entire bottom of the oven. This leveling may be done by machine or by hand. After leveling the coal, the door opening is then bricked up to within about 1-1/2 in. of the top.

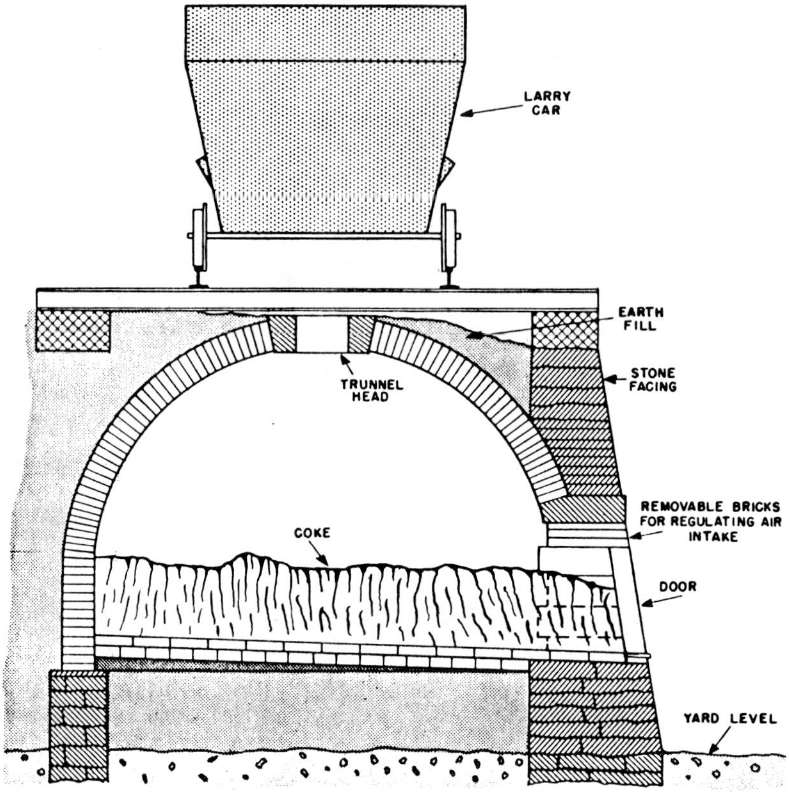

FIG. 1 Beehive oven, cross section. (U.S. Steel.)

The coking process begins very soon after leveling is completed, as the ovens retain enough heat in the brick of the walls and the loam backing to start liberation of the volatile matter from the coal. As more heat is absorbed by the coal charge, the temperature of the oven soon reaches the "kindling" (ignition) points of the combustible gases, which, in the presence of the air admitted to the oven, ignite with a slight explosion at first and then continue to burn quietly in the crown of the ovens or as small candlelike flames at the surface of the coking mass, thus supplying heat to continue the process.

C. Coking Process

Coking proceeds from the top of the coal downward, so that the coking time depends mainly upon the depth of the coal. The generation of gas thus rapidly approaches a maximum, which is maintained for a period, then declines to practically nothing. The burning of volatile matter during this period must be regulated by gradually closing up the opening at the top of the door for admission of air. This regulation is necessary to maintain the temperature at a maximum, and conserve coke, as an excess of air at the beginning of the coking period tends to cool the oven and later consumes the carbon of the coke. The yield is also reduced by improper leveling. If the coal is not of uniform depth in the oven, the thin areas coke through before the thick, and some of the coke of the thin sections is consumed by combustion while the coking of the thick portions is being completed. On the other hand, if the process is stopped when the thin areas have coked through, there will be a loss due to uncoked butts in the thick areas. In the coking of bituminous coals in beehive ovens, coking proceeds downward from the top of the charge in which the coal, at increasing depths, passes through a plastic state as the temperature rises. This produces expansion and contraction of the charge, with the result that the cake is crisscrossed by a great number of irregular vertical fissures, thus giving it a long columnar structure. These very irregular columns extend from the top to the bottom of the cake. This struc-

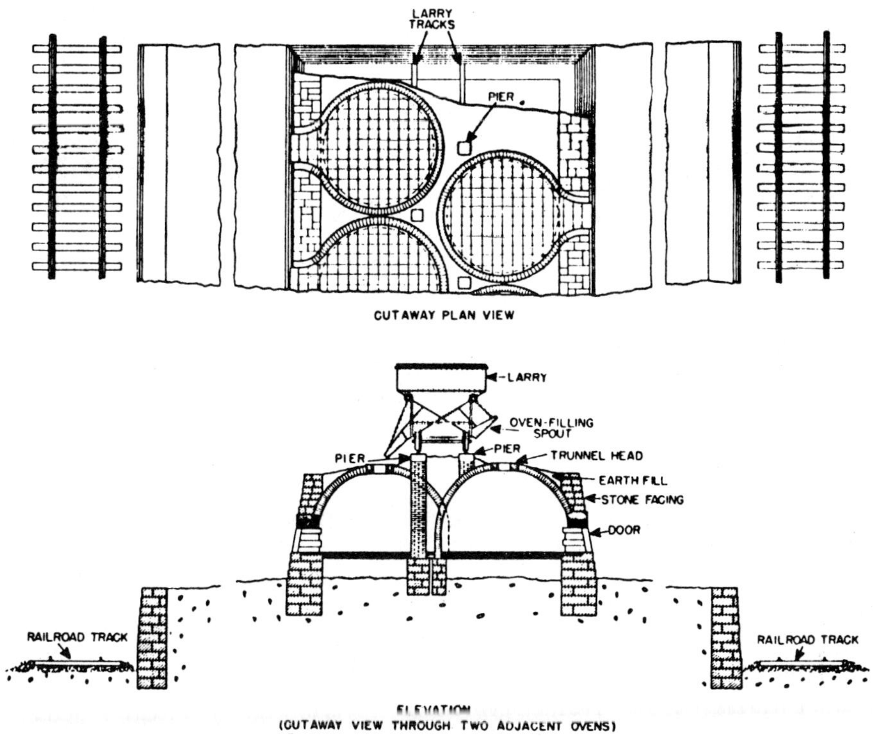

FIG. 2 Beehive oven, double-run layout. (U.S. Steel.)

ture affords a means by which beehive coke can be distinguished from by-product coke.

The by-products of coal carbonization released in the beehive oven are not recovered; therefore, their smoking combustion is characteristic of this process. The history of the coke industry in North America starts with the beehive oven. Their number increased as the hot blast process of ironmaking gained acceptance. The first census of the industry taken in 1900 showed the existence of 57,000 such ovens in the United States and an estimated 4,000 in Canada. The rise and decline of this industry in the United States is shown in Table 1.

The basic weaknesses of the beehive coking process were as follows:

1. Failure to recover valuable by-products, including energy in the form of gas
2. Low product yield, about 50% maximum
3. Extreme air pollution

Other past developments of nonrecovery ovens included the rectangular Mitchell oven, which was heated through a side flue, and the Curran Knowles oven. Neither of these achieved great popularity, although they possessed many advantages over the beehive oven.

The most recent entry into the field of horizontal ovens is the PACTI coke process [3]. It is claimed that this equipment will generate substantial gas and waste-heat credits and will meet current pollution control requirements.

A sketch of the PACTI process is in Fig. 3. Two ovens approximately 9 × 15 ft in cross section are operated in pairs with a common combustion chamber to consume particulates and the contaminants. The coal is charged to a depth of 3-1/2 ft and is heated from both top and bottom. Coking time is 42 to 48 hr. Eight to 10 acres would be required for a 1000-ton/day plant.

TABLE 1 Number of Beehive Coke Plants and Number of Beehive Coke Ovens in the United States, 1900-1970

Year	Number of Plants	Number of ovens
1900	387	57,499
1910	548	100,362
1920	347	75,298
1930	121	23,907
1940	93	15,150
1950	125	17,708
1960	43	7,583
1970	2	520

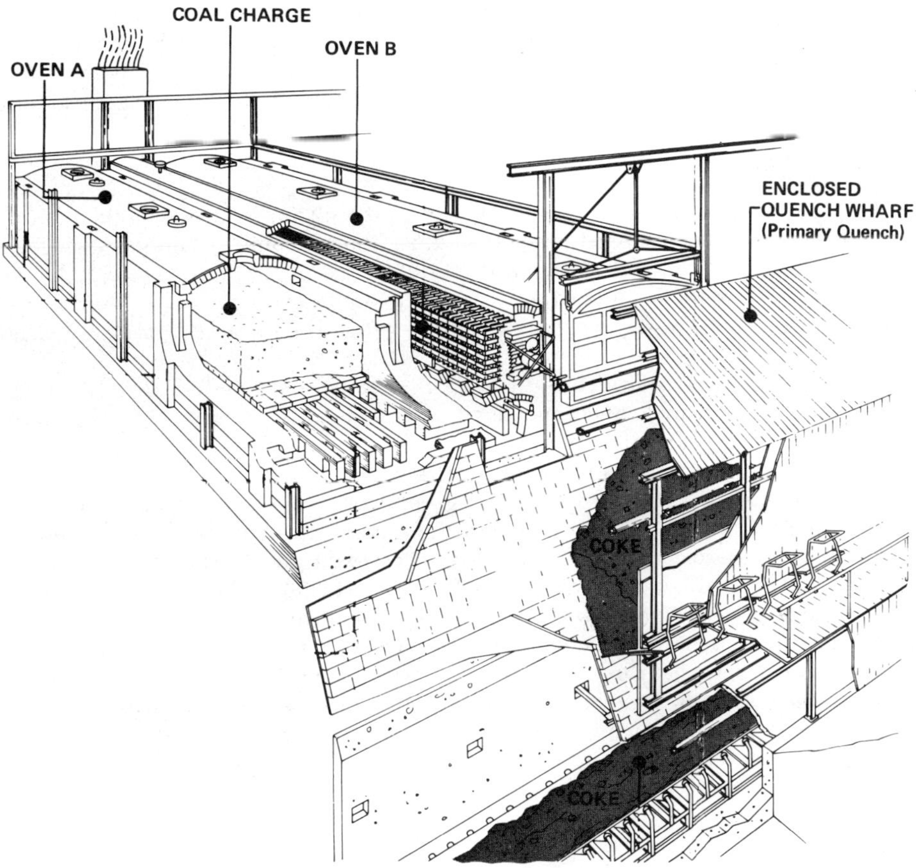

FIG. 3 Pacti coking process.

III. BY-PRODUCT OVEN

A. European Development

The production of manufactured gas from coal in retorts was started on a commercial basis in France, England, Germany, and the United States in the early 1800s. Tar, ammonia liquor, and gas were recovered in this operation. These ventures, however, were only remotely related to the manufacture of coke [4].

Prior to, and for fully 50 years after industrial installations for making manufactured gas, repeated efforts were made to develop a single facility that could produce coke similar in quality to that made in beehive ovens and also provide for recovery of products such as those obtained in gas-producing retorts. Attention was being directed similarly during this period to the design of an oven that utilized a principle of coking different from those then practiced.

In 1856, in Commentry, France, Knab built a group of retort ovens with the principal purpose of producing gas, tar, and ammonia. The gas from the ovens was freed of tar and ammonia in suitable chemical apparatus and then burned in heating flues under the ovens to provide the heat for carbonizing the coal. The Knab ovens were 23 ft long, 6 ft 6 3/4 in. high, and 39 1/2 in. wide [4].

Prior to this time, however, an oven that embodied vertical flues in its walls was developed in France and used both in that country and in the Ruhr district of Germany for producing coke. The purpose of that design was to exclude air from the coking chamber to the greatest degree possible during the carbonization of coal. The volatile gases were drawn by stack draft through orifices in the top of the coking chamber into vertical flues in the oven walls. Gas from the ovens and air were mixed and burned in a downward direction in these flues and passed to the stack through sole flues under the oven. Known as the Francois oven when built in France and as the Rexroth oven when used in Germany (hence the combination name Francois-Rexroth), this so-called waste-heat oven was 26 ft long, 5 ft high, and 35 in. wide.

Both the Francois-Rexroth 5nd the Knab ovens were departures from the then contemporary practice in that air was excluded from the coking chamber and external heat was supplied for carbonization [5].

B. Development of Vertical-Flue Ovens

The course that was pursued later in improving oven performance led to two separate and distinct developments.

Carves of England, using the Knab principle, developed an oven in 1862 (Knab-Carves design) that embodied horizontal heating flu0s in the oven walls. This oven was an improvement over the Knab design, providing for greater yield of coke and requiring less time for carbonization. The principal contribution by Carves, however, was the addition of an exhauster in the chemical-recovery apparatus. In the Carves oven system, gas was drawn from the oven by the exhauster, pumped through the chemical-recovery apparatus, and returned to the ovens for use as a fuel, the present method of recovering coal chemicals being based on this same principle. Because of the inclusion of horizontal flues in the oven walls, the Carves system can claim the distinction of being the originator of the horizontal-flue system of oven heating [4].

Almost simultaneously with the Carves development, Coppee in Belgium, using the basic principles of the Francois-Rexroth oven, introduced the innovation of reducing the width of the oven and providing additional vertical heating flues in the oven walls. The ovens built by Coppee at the time were 30 ft long, 3 ft 7 in. high, and 18 in. wide. In these ovens, stack draft was used to draw the gases of combustion from the vertical downdraft flues through waste-heat boilers, recovering the sensible heat as steam. The ovens built by Coppee embodied 28 vertical flues, and one of the chief contributions of this builder was the control of combustion by dampering the air admitted to the flues. Since the ovens built by Coppee conformed more nearly to present-day conventional-width ovens and also contained a substantial number of vertical flues in the oven walls, this oven may, in some respects, be considered the true predecessor of the modern coke oven.

Ovens of the Coppee design were built in the Ruhr district in Germany by Dr. C. Otto and Company, who added further refinements to the earlier Francois-Coppee design [6]. Attempts at converting these ovens from the waste-heat type to recovery-type by-product ovens have been credited to this builder. In the ovens of Otto design, the volatile products from the coal were removed from an opening in the top of the oven and the gas, after passing through the chemical-recovery apparatus, was returned to the flues in the walls of the ovens. The flow of combustion gases in this oven was different from that of the early Coppee design, with products of combustion rising and descending in alternate flues, thus forming the basis for the later ovens of "hairpin flue" design.

C. Horizontal-Flue Ovens

Evolution of the horizontal-flue oven was also proceeding, but improvements were largely in the direction of by-product recovery rather than in coke oven design. The basic design of the Carves-Knab oven was improved later by Heussner in Germany and Simon in England. Heussner, in 1881, developed the Heussner-Carves oven along more conventional lines by reducing the width. Simon, in England, also revised the Carves oven design and, in 1883, erected a plant in Durham, England, embodying recuperators for the transfer of heat from waste gases to incoming air. A reduction of coking time to 48 hr was the principal contribution of the Simon-Carves oven.

In 1885, Louis Semet, chief engineer of the Brussels Gas Works, and his brother-in-law, Ernest Solvay, built twenty-five 4-ton ovens near Mons, Belgium. The Semet-Solvay design (Fig. 4) was based on the Carves principle of horizontal flues. Sovay's interest involved supplying the lime kilns with coke and the soda process with ammonia. It is of interest that the ovens of horizontal-flue design did not embody regenerators, such facilities being included 10 years later.

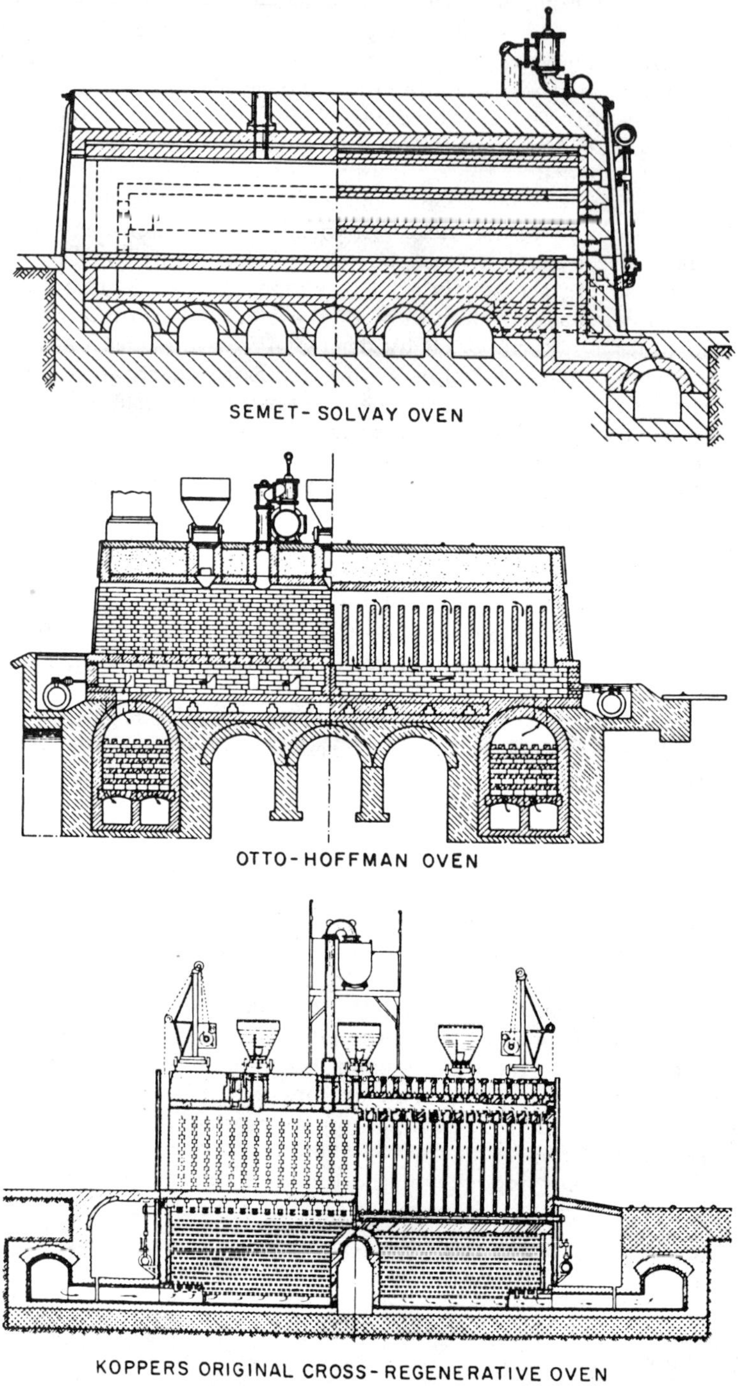

FIG. 4 By-product oven, early designs. (U.S. Steel.)

The Rothberg oven, a half-divided horizontal-flue oven, was developed in 1902; the first of these ovens in the United States began operation in 1903.

D. Improvements in Heating Facilities

Despite the fact that the ovens of Coppee, Knab, Carves, Heussner, Semet-Solvay, Otto, and others represented appreciable progress compared with the simple beehive system, such facilities were lacking in heat economy. Temperatures in the ovens were not high enough to produce an acceptable metallurgical coke, notwithstanding return to the oven for heating purposes of all or nearly all of the volatile gases from the coal. These ovens left much to be desired in uniformity of temperature. Their coke product did not compare favorably in quality with that produced in beehive ovens. As a result of inferior quality and lack of uniformity of the coke produced, the use of recovery-type by-product ovens for producing metallurgical coke was viewed with skepticism until further improvements were made in both design and operation.

From an economical standpoint, the beehive oven was cheap to build, could be operated with a minimum of help, and was capable of being shut down or restarted with little or no difficulty. The recovery-type oven, however, was costly to build and difficult as well as expensive to operate, because of the size of the units and the need for skilled personnel. Consequently, the incentives for building recovery-type ovens were lacking in the steel industry until the economical advantages of the ability of the integrated steel plant to use the excess gas had been clearly established. The industry then began widespread adoption of this process.

E. The By-Product Oven in the United States

Although reasonable success had been achieved abroad with the by-product oven, it was not used in the United States prior to 1890. It is pertinent to recall that the first ovens of the by-product type installed in the United States were not for the production of metallurgical coke. The distinction for launching the industry must be credited to the Solvay Process Company, which was interested in establishing the soda-ash industry in this country. The plant that was built in Syracuse, New York, and placed in operation in 1893, consisted of 12 Semet-Solvay ovens.

Until 1900, only nine plants with 1081 by-product ovens had been constructed in the United States. From 1900 through 1910, however, there was a period of rapid industrial expansion, particularly in the iron and steel industries, and the number of by-product coke plants more than tripled while the number of by-product ovens increased fourfold. The need for coal tar dyes and explosives in World War I added impetus to this explosion, and by 1919 the by-product coke industry had 59 plants with over 9000 ovens. The construction of by-product ovens for gas utilities in the 1920s further expanded the industry and by 1930 there were 89 plants with 12,771 ovens in existence.

Because of the world-wide economic depression, construction of by-product ovens was halted during the 1930s, and by 1940 there were fewer ovens in existence than during the early 1930s. However, World War II started a new construction period, and between 1940 and 1950, 2359 ovens were added to the industry total. The increase in total ovens, however, only partially indicates the vast construction program of this period, because many of the ovens constructed were replacements rebuilt ovens originally built during World War I. Actually, over 5000 ovens were constructed during this period.

The construction of by-product ovens continued after 1950, and in 1958 the number of ovens reached an all-time peak of 16,244. Although ovens increased in number because of the expansion of furnace plants, a number of by-product coke plants were abandoned as markets for coke oven gas were lost to natural gas which became available in many areas through the extension of gas pipelines after World War II. Declining coke markets since 1958 caused further reductions in plants and ovens, and at the close of 1966 there were 66 by-product coke plants with 13,720 ovens in the United States. This number decreased to 64 plants at the end of 1968, and 56 plants in 1975.

1. *General Description*

The by-product coke oven is a long-narrow refractory structure closed by removable doors on each end. Charging ports in the roof are sealed by metal lids and one or more openings are provided to remove the gas generated in the oven during the distillation process. Ovens are arranged in batteries of 30 to 100 ovens side by side with heating flues in between.

Oven heights increased from 10 to 11 ft in the 1920s to 12 to 13 ft in the 1940s. Five- and 6-m batteries were built during the 1960s. Some Japanese batteries have gone to 7 and 7.5 m with plans for an 8-m unit.

Oven lengths were increased from about 37 ft to 50 ft over the same period. Oven widths are expressed as averages since the walls taper, with the widest dimension on the coke side to facilitate pushing the oven. Average oven widths vary from about 14 to 18 in. Oven volumes (to the coal line) range from 500 ft^3 for a 10-ft oven 38 ft long to 1390 ft^3 for a 6-m oven 50 ft long. Since coking time is a function of width, increased volume is directly proportional to increased productivity. Individual coke ovens are assembled into batteries which operate as a unit. Heating flues are in the walls between the ovens and the entire brickwork structure is supported by steel buckstays (upright structural columns), which in turn are held together by adjustable tie rods.

Changes in oven and machinery design have been evolutionary in nature. The basic process encompasses a refractory chamber with provision for charging the coal through the top and withdrawal of the gases generated through ductwork. Removable doors at each end of the oven permit pushing the charge of completed coke into a receiver. Gas is burned in flues enclosed in the walls between adjacent ovens so that the combustion system is completely separate from the gas generated by destructive distillation of the coal. Air is excluded from the coking chamber to the maximum extent possible to avoid combustion of the coal or gas.

Some early oven doors were of the guillotine type, which could be raised or lowered vertically with a simple hoist. These were replaced when machinery was developed to install and remove doors horizontally with the door in a vertical position. All the early doors were sealed with a refractory clay/water mixture known as luting material. This is installed by hand, sometimes by a man riding a small elevator to reach the top of the doors.

Today, most oven doors are self-sealing. A thin metal strip on the door is pressed against the metal door jamb by screw or spring pressure exerted against door latches. Sturdier and more complex door machinery was needed to accommodate this change.

Improvement in the control of coke oven gas discharge evolved through the availability of better gas exhausters, coupled with more sophisticated pressure control devices to maintain uniform pressure at the ovens.

Coal is transported by conveyor or elevator to a bunker above the battery. A charging or larry car travels on top of the battery and under the coal bunker. Individual hoppers on the larry car have discharge chutes that mate with the three or four charging holes on each oven.

The larry car is positioned over an empty oven and the coal is charged in a prescribed sequence. Since the coal dropping by gravity forms peaks, it must be leveled by a bar on the pusher machine. This leveling bar is inserted through a small "chuck" door on the pusher side. After leveling, all lids and doors are replaced and the coking process continues until most of the volatile matter is removed. This process may take as little as 13 hr or as much as 48 hr, depending upon the heating rate and type of coke being produced (Fig. 5).

When the coking process is complete, the doors on each end of the oven are removed by the pusher machine on one side and the door machine on the coke or discharge side. A guide is used to contain the coke on its way from the oven to the quench car. The block of coke, perhaps 18 in. thick by 20 ft high and 50 ft long, is pushed from the oven by a ram on the pusher machine. This ram travels through the oven and into the guide to push all the coke into the quench car, which may travel during the operation to spread the coke evenly (Fig. 5).

The incandescent coke will burn as soon as exposed to air, so it must be quenched immediately. The usual practice is to spray several thousand gallons of water onto the coke in a quench station. Another system not yet in use in the United States is to dry-quench or cool the coke with inert gas and recover the heat as steam in a waste-heat boiler.

2. *Battery and Machinery*

Ovens and combustion flues, which must withstand temperatures in excess of 2400°F, are made of silica brick. While silica brick has a high rate of expansion at lower temperatures, it is relatively stable in the operating range of a coke battery. Bringing a battery up to temperature through critical expansion zones is a slow and painstaking process. Once a battery is at operating temperature, it is seldom or never cooled down, to avoid the structural damage that may occur with contraction and subsequent expansion. Many batteries are never cooled during their entire life of 25 to 30 years (Fig. 6).

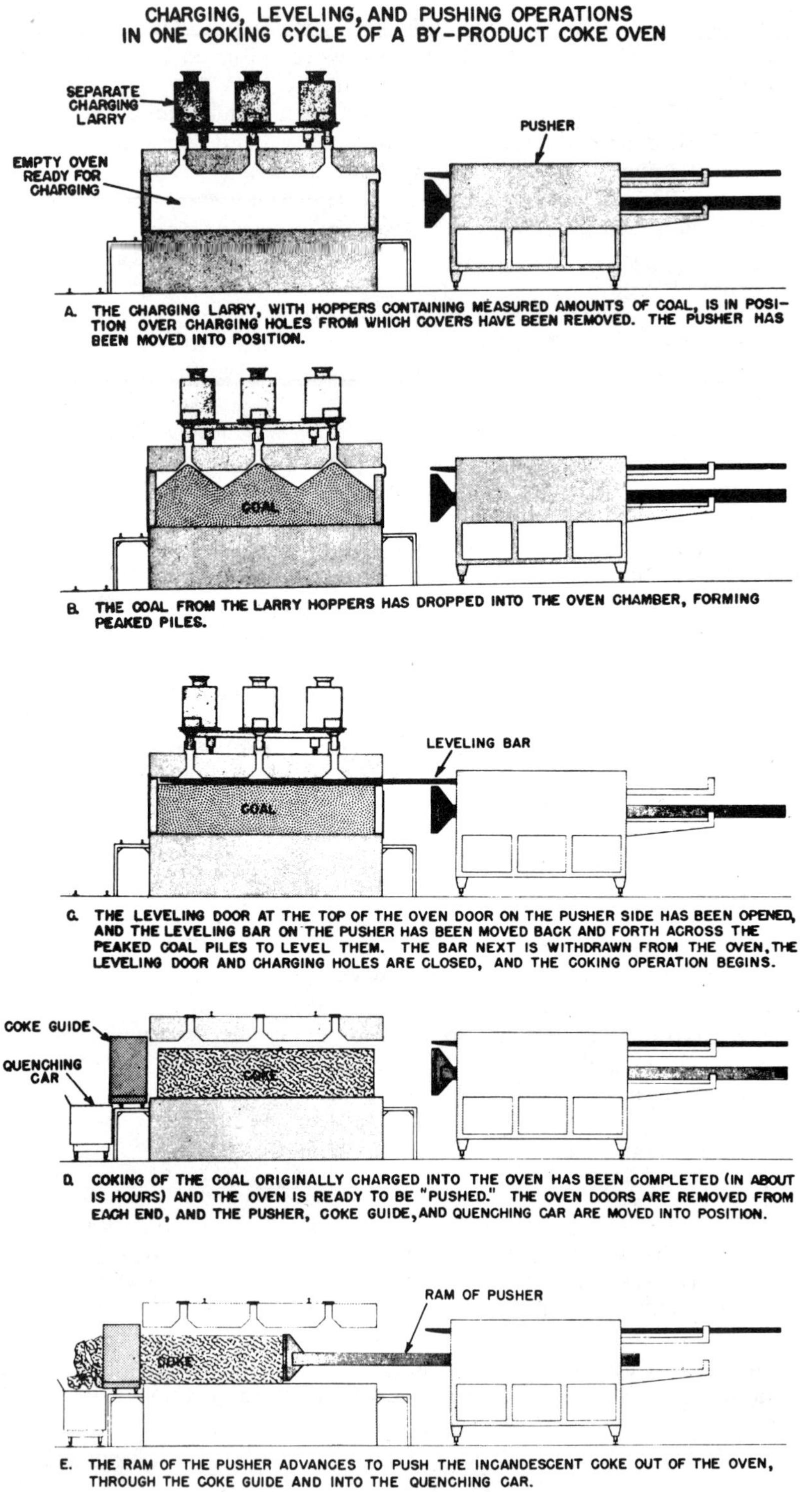

FIG. 5 Coke oven operating sequence. (U.S. Steel.)

FIG. 6 Coke oven design, cross section. (U.S. Steel.)

The battery is erected on a heavy pad which may be piled, depending upon soil conditions. A medium-size battery would contain 20,000 tons of refractory plus 3000 to 5000 tons of steelwork and machinery. The buckstays are usually heavy section I or H beams which extend from the foundations to the top of the battery. They are held together by tie rods which are usually spring-loaded at top and bottom. The battery top, which is exposed to the weather, has a heavy layer of insulation and paving brick. The charging hole castings, gas offtakes, and larry car rails are set in this refractory.

The regenerators, which extract heat from the exhaust combustion gas and then preheat the air (and sometimes low-heating-value gas), are composed of numerous fireclay brick shapes with high surface exposure to the gas or air passing through them. The regenerators are located below the oven floor level.

A bench on the pusher side below the oven floor level permits workers to reach the ovens and doors to perform the necessary cleaning and maintenance work. A similar bench on the coke side supports the door machine and guide as well as providing access to the ovens. The bench level is high enough to permit the coke to fall into the quench car.

Both the door machine and pusher are equipped with mechanisms to remove and replace the oven doors. This rather complex machinery relieves a latch-ing arm pressure either by loosening screws or compressing springs. The door is then lifted and pulled away from the ovens. On many machines it is then rotated 90° to clear the bench and facilitate cleaning. Consistent and thorough cleaning of coke oven seals and jambs is essential to prevent leakage.

The newer door and pusher machines are equipped with automatic door and jamb cleaners. These usually consist of mechanical scrapers which are forced against the sealing surface of the door. A separate head is positioned against the jamb surface. The scrapers are cycled through up and down strokes or by continuous travel on a chain drive. These mechanical cleaners require a great deal of maintenance effort to be effective. Newly designed high-pressure water cleaning machinery shows great promise as a more effective solution.

The guide, which conveys the pushed coke across the bench to the hot car, may be an integral part of the door machine or be towed by it on a separate trailer car. The guide is a vertical chute the height of the oven. It can be extended toward the oven to latch onto the furnace structure or be retracted for traveling along the bench.

In addition to the door-handling machinery, the pusher is equipped with a ram for pushing coke through the oven and a leveling bar to rake the coal charge to a uniform height.

The larry car, which conveys the coal from the coal bunker into the oven, has three or four hoppers depending upon the number of oven charging ports. The hoppers are equipped with measuring sleeves or rings to assure loading to a predetermined volume. The car is equipped with scales or rests on a track scale so that coal weight can be recorded. The hoppers may be lined with stainless steel to facilitate free flowing of the coal, and vibrators may also be used.

Discharge through a slide gate at the bottom of each hopper may be by gravity into the oven or through a mechanical feeder. Screw or table feeders are the two types employed. A boot or discharge sleeve must fit closely to the charging hole casting to prevent the escape of fumes and the spillage of coal. Automatic equipment may be used to sequence the dropping of coal instead of relying on the operator's judgment, since proper procedure is very important in emission control.

The quench car travels below the coke side bench so that it can receive the coke as it emerges through the guide. Formerly, all cars had to travel during the push to spread the coke over the bed of the car. More recently, "one-spot" cars have been built which can receive the entire push without traveling. In either case the car carries the coke to the quench station where it is cooled with water. Removal is accomplished by tilting the car and dumping the coke on an inclined, refractory brick wharf, from which it is discharged onto a conveyor belt.

3. *Heating*

The objective of a coke oven is to heat the coal charge as uniformly as possible. Since the coke side of the oven is wider than the pusher side to facilitate pushing, varying heat must be applied, as the oven walls taper uniformly across the oven.

Many variations of heating systems have been developed over the years. All modern designs utilize vertical flues between the adjacent oven walls. There are, however, two basic methods of introducing combustion gas to the system.

1. A gun flue battery, in which the gas is fed from each side into a horizontal sole flue and then distributed into the vertical flues.
2. The underjet battery, which requires a basement under the regenerator, furnishes gas to individual flue burners through piping. Burners can be removed for cleaning or changing from the basement.

Complex and sophisticated heating systems have been developed by various suppliers of coke ovens. All of them employ regenerators which recover and store waste heat and then preheat air for combustion when the battery gas flow is reversed. When low-Btu blast furnace gas is being fired, it is also passed through regenerators to increase its heating potential.

In all systems, combustion occurs in one flue, which operates at relatively high temperature compared to the flue that returns the products of combustion to the regenerators and eventually out the stack. Since the combustion flue would eventually reach the point of refractory damage, the gas flow is reversed every 20 to 30 min. Air is then pulled through a hot regenerator and the relatively cool waste-gas flue becomes the combustion flue.

The taller ovens present a more difficult problem in vertical heat uniformity. This is overcome by introducing either gas or air at different levels to achieve differential combustion and more even temperature distribution.

Maintenance of the heating system is vital to good battery operation. If burners become plugged through deposits from the gas, uneven heating occurs, with dark streaks of uncoked coal and excessive emissions occurring. A 70-oven battery would have over 2000 burners, so it is not easy to keep them all in good condition.

Waste gases, after heating the regenerators, pass through a waste-heat tunnel and out a stack. Draft is controlled by an automatic valve.

4. *Coke Oven Gas System*

The gas evolved from the distillation of coal contains tar, ammonia, naphthalene, and hundreds of hydrocarbons included in the light oil category. As the gas passes out of the oven through a vertical standpipe, around a return bend or gooseneck into a collecting main, it is sprayed and cooled with flushing liquor from 1100 to 1300°F to 150 to 200°F. Some of the tar condenses at this time and flows to the bottom of the main, where it is carried away by the flushing liquor.

A coke oven battery operates on a very delicate pressure balance. The collecting main is maintained at a slight positive pressure of 6 to 12 mm H_2O. Gas evolution in the oven generates higher pressures, which force the gas into the collecting main. As the gases leave the collecting main, they pass a control valve which maintains the slight pressure. Beyond that valve is the suction main, which is held at several hundred millimeters of negative pressure by gas exhausters. The pressure balance is important to the operation. Excessive pressure in the oven will cause emissions to leak through the door seals, charging ports, and many miscellaneous cracks and openings. Negative pressure will pull air through these same areas and cause combustion within the oven resulting in refractory damage.

Coke oven gas is high in hydrogen and methane, with a heating value of 500 to 550 Btu/ft^3. Volumes generated range from 10,000 to 12,500 ft^3/ton of coal carbonized. The various by-products are removed as the gas temperature is decreased.

The suction main carries the gas to the primary coolers. Tar and flushing liquor flow along the bottom of this main and are diverted to tar decanters for separation of the water and tar. In the primary coolers the gas is cooled to 65 to 90°F by sprays of flushing liquor. Most of the remaining tar condenses out at this point and is recovered.

The exhausters, which are large turbines powered by either steam or electricity, not only pull gas from the ovens but also discharge it at sufficient pressure to operate the by-product facilities. The compression to 10 to 12 in. H_2O also increases the temperature 5 to 15°F. The tar mist remaining in the gas stream is recovered by wet electrostatic tar precipitators.

Ammonia is recovered from both the liquid and gas streams. An ammonia still readily separates the free ammonia from the liquid. The fixed salts must be treated with lime or caustic in a "lime leg." The gaseous ammonia may be absorbed with sulfuric acid to produce ammonium sulfate, processed to anhydrous ammonia, or destroyed by combustion.

Naphthalene must be removed from the gas stream to avoid plugging of lines and valves in subsequent gas usages. It may be scrubbed from the gas with petroleum-based wash oil and then separated by distillation. Another removal method is to reabsorb it in tar. Naphthalene may be extracted in crystalline form as a pure substance. In many cases, it is absorbed back into the tar for burning or further processing in a by-product plant.

The light oils, consisting primarily of benzene, toluene, and xylol, may be left in the gas and burned with it or extracted for subsequent distillation. Extraction is accomplished by scrubbing with a petroleum-based wash oil in packed or spray towers. The light oils are then separated from the wash oil by distillation. Many coke plants sell this light oil to refiners rather than refining it further. Figure 7 shows the coke plant and by-product operation in the form of a diagram.

Environmental requirements are mandating sulfur removal from the gas before burning. Two principal methods of sulfur removal are being used.

a. Liquid Absorption Processes These processes involve absorption of acid gases into a recirculating solution, desorption of these gases from solution, and conversion of the H_2S into sulfuric acid or elemental sulfur. Processes in this group are:

Vacuum Carbonate Process
Sulfiban Process
Carl Still Process

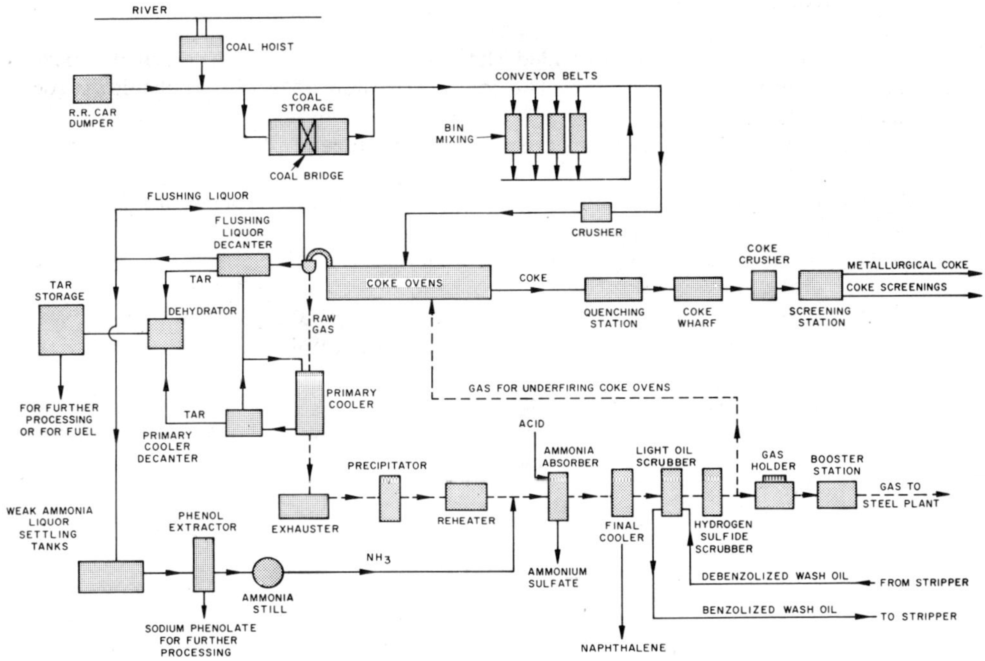

FIG. 7 Flow sheet for by-product plant.

b. Wet Oxidative Processes These processes do not desorb the acid gas once it is absorbed. The H_2S is oxidized directly to elemental sulfur and filtered from the solution. The oxidizing agent is then regenerated by air oxidation. Processes in this group are:

Stretford Process
Takaha Process
Fumas Process
Giammarco Vetrocoke Process

Clean coke oven gas is a valuable fuel. The steel mills use it for firing various reheating furnaces in the plant. With natural gas in restricted supply, many new coke oven batteries will be underfired with blast furnace gas in order to free the 40 to 50% of the coke oven gas used for this purpose for other mill usage. Merchant coke producers will usually sell their surplus gas to a utility or another commercial operation.

F. Environmental Problems

Coke oven operations have caused the most serious environmental problem facing the steel industry since the gross emission of oxide steelmaking fumes was controlled.

Coke oven emissions come from several different phases of the operation:

1. Charging—improperly charged ovens can emit voluminous clouds of coal dust and smoke.
2. During coking—miscellaneous leakage from charging lids, standpipes, and doors can contribute substantially to air pollution.
3. Pushing—poorly coked coal will discharge large volumes of dense, black smoke. There is discharge of particulate matter from a well-coked push because of the thermal effect of convection, which carries particulate matter into the atmosphere.
4. Quenching—the quench station generates very large clouds of steam, which may carry particulates as well as some chemicals dissolved in the water.

1. *Charging*

Charging emissions can be controlled if the pressure at all open ports is maintained at a slightly negative level, but this is not readily accomplished. The negative pressure may be induced by the venturi effect of steam (or high-pressure liquid) jets in the oven offtake area. These jets must have sufficient capacity to overcome the positive pressure in the collecting main and absorb leakage into the oven.

Assuming that sufficient aspiration is available, the next problem is to assure that free passage of gas to the evacuation point is maintained. Since coal must be dropped at three or four points along the oven length, it is necessary to have a gas offtake at each end. This may be accomplished by having permanent offtakes and collecting mains at each end (double collecting main battery) or utilizing a jumper pipe to an adjacent oven. The draft on this oven is then utilized to exhaust gases from the end of the charged oven farthest from the single collecting main.

When coal is dropped in the proper sequence, a gas passage at the top of the oven is always open to one end or the other. This method of charging was developed by the British and is known as "stage charging." When executed properly with good equipment, there is practically no visible charging emission.

Many foreign and U.S. batteries built during the late 1960s and early 1970s were equipped with scrubber larry cars. Each charging port was evacuated by a high-energy fan and the effluent scrubbed before discharge. This equipment performed much better than conventional charging at that time but will probably not meet the standards of the 1980s.

A more recent innovation is the pipeline charging of preheated coal. In this system the oven is completely sealed during the charging process. This has the potential for an almost complete elimination of charging emissions.

2. Topside Seals and Doors

Control of relatively minor emissions that escape from charging hole lids, offtake piping, and collecting mains can generally be accomplished by good maintenance and careful workmanship on an hour-to-hour basis.

The sealing of coke oven doors represents a mechanical design problem as well as a requirement for thorough maintenance and cleaning. The conventional self-sealing doors that have been used for many years will not meet the stringent emission standards under consideration. This is particularly true of the higher 5- and 6-m ovens. New designs of sealing mechanisms have been developed and are being tried experimentally in an effort to solve the problem.

3. Pushing Emissions

Control of pushing emissions is a major problem for coke oven operators. Government regulations require that pushing be accomplished in a closed system. Several approaches have been tested. They are as follows:

1. Sheds enclosing the coke side with fume cleaning equipment
2. Land-based cleaning equipment with traveling hoods
3. Mobile units with cleaning equipment

All of these facilities are very costly, cumbersome, and difficult to maintain. In addition, they use excessive amounts of energy to prevent very small amounts of particulate matter from reaching the atmosphere. In some instances the pollution generated at the power plant far outweighs the collected material, giving a negative pollution effect.

4. Quench Station Emissions

The only control measures for quench station emissions adopted in the United States has been the use of wooden baffles in the tower and the elimination of contaminated water from the quenching system. The baffles trap a great deal of particulate matter and the clean water quench minimizes the contaminants in the steam plume.

The steam plume can be eliminated through the use of a dry-quench system, principally in operation in the USSR and Japan. The incandescent coke is dumped into a tower, where it meets an ascending flow of inert gas. As the coke is cooled, the heated gas is passed through a waste-heat boiler generating approximately 1000 lb of steam per ton of coke. It has not been tried in the United States principally because of the high investment cost.

IV. NOVEL TECHNIQUES OF BY-PRODUCT COKING

The shortage of high-grade coking coal and the constantly increasing cost of these coals all over the world necessitated a critical review of ways and means to utilize lower-grade (marginal) coking coals or even noncoking coals in coking blends.

Studies carried out in practically every country with a coking industry established over a period of many years that the coal preparation is one of the most critical factors in the production of metallurgical coke. The term "coal preparation" includes the thoroughness of coal blending, coal sizing, degree of coal oxidation, and the moisture level of the individual coals and of the blend. Usually, the coal has to attain the optimum grind or particle size distribution and also the proper moisture, both of which affect the density of coal in the oven. The density and the coal agglomerating properties determine the strength of coke expressed as stability.

The density can also be increased by the addition of oil, which frequently is considerably better than addition of water.

A. Coal Charge Briquetting

Since the agglomerating properties of coal cannot be improved in the raw coal blends, some ingenious practices have been developed to aid in and therefore improve coke properties of the specific coal or coal blends. One such practice was developed in the steelmaking area of Saarbrucken, Germany, whose coals are only marginally coking. In this practice, the local coal blend is finely ground to almost 100% - 2 mm and then at the optimum moisture level (3 to 4%) the blend is densified outside the oven in an open steel chamber by stamping. The stamped cake of coal is pushed by a ram from the pusher side into the oven, the door is closed and sealed, and the coking process is under way in the conventional manner. This practice increases the density from 720 to approximately 850 kg/m^3, which makes feasible production of high-quality coke from poorly coking coals. The stamping practice is limited to ovens approximately 36 m high, because higher-coal cakes show an increasing tendency of falling apart when pushed into the oven.

Since the Saarbrucken steel plant was in need of large 20-ft-high ovens, it was decided to densify the charge by briquetting. The present practice is to briquette the raw coal blend to approximately 1 1/2-in. size in roll presses located above the larry cars. The green briquettes thus produced are not very stable, yet they are strong enough to make up approximately 50% of the charge into the oven. The balance is made of fines, although most of these are considerably larger than the original finely ground coal. The density achieved by the briquetting practice is almost identical to that obtained by stamping.

The practice of partial briquetting the oven was also adopted recently by the Japanese, although in the modified version. Nippon Steel, Sumitomo, and Nippon Kokan have developed their own modifications of the practice, one of which is shown in Fig. 8.

In this practice, the coking coals are thoroughly blended and a fraction of this coking blend moves to the mixer, in which noncoking coals and a bituminous binder substance are also thoroughly blended in preparation for the briquetting process. The briquetted fraction usually ranges from 25 to 35% of the total coal blend. The unbriquetted blend of coals and the briquettes are mixed and delivered to the larry car [7].

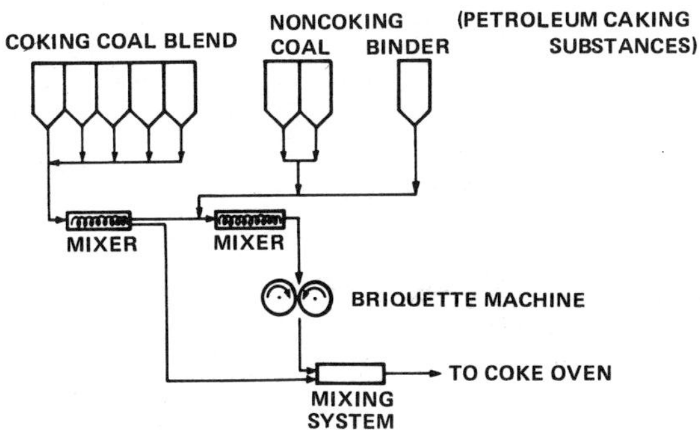

FIG. 8 Partial briquetting practice. (Sumitomo.)

The Japanese coke makers claim that this practice improves the stability by several points and permits them to reduce or eliminate a percentage of important coking coals, which are quite expensive.

B. Coal Preheating Processes

From the extensive research and test work carried out on conventional by-product batteries, it became evident that, in general terms, increased pulverization improves stability whereas increased moisture is detrimental to stability. Therefore, the predrying step of the preheating practice is expected to increase coke stability. The other major anticipated effect of drying and preheating the coal blends is to shorten the coking time, and this is also at least partially responsible for reducing and narrowing the size range of the coke produced.

Therefore, from known relationships, one has the following expectations about preheat coking as compared to conventional coking:

1. Higher stability of coke produced from the same blend or the same stability of coke produced from blends containing a higher proportion of marginally coking coal
2. Shorter coking times and somewhat smaller coke size

The preheating practice is of special value to coking coal blends containing large percentages of Midwestern coals, which are characterized by high inherent moisture contents.

A broad background of test coke oven and commercial coke oven data shows that coking time is related to total blend moisture and level of preheat temperatures as shown in Fig. 9. For Midwestern (Illinois) coal blends, a large segment of the reduced coking time is achieved by drying and preheating the coal blend to about 250°F, as seen in this figure. However, to achieve

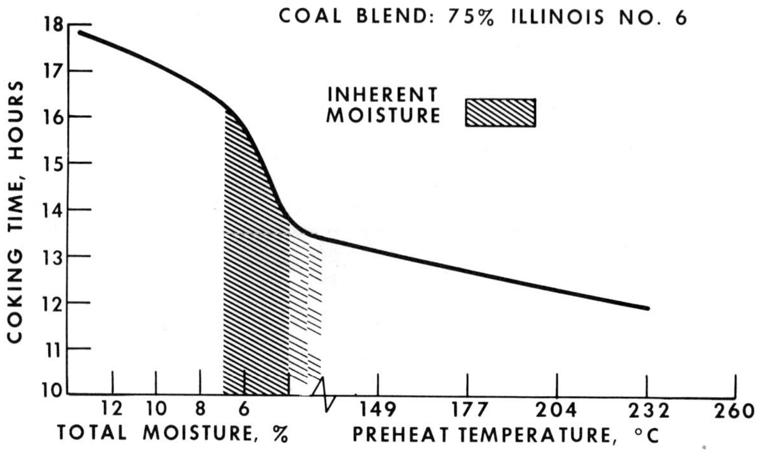

FIG. 9 Effect of moisture and preheat temperature on coking time.

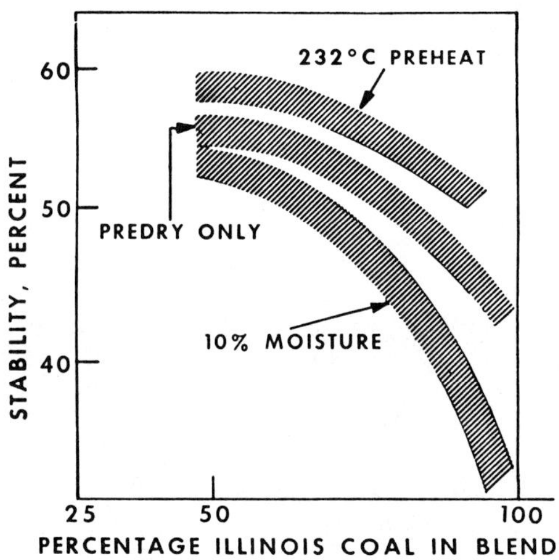

FIG. 10 Effect of blend composition and conditioning on coke.

high stability levels in these coal blends, it is necessary to reach a significant preheat temperature, as shown in Fig. 10.

This critical step of preparing the coal for charging into the oven can be achieved in several different ways. However, the two processes reduced to practice are the one-stage preheat of the Coaltek-Cerchar design and the two-stage preheat process of the Rosin-Bergbauforschung design. The difference between these two systems is in the passage of hot gas, which moves countercurrently to the flow of coal in the (BFF) precarbon system and cocurrently in the Cerchar. Also, the temperature differential between coal and gas is considerably higher in the Cerchar system, since both drying and preheating steps have to be carried out in the very short contact time of only a few seconds.

The two systems, Coaltek-Cerchar and Rosin, differ in still another respect. The coaltek system uses pipelines to convey the coal into the ovens with steam being used as the conveying medium. In this respect, the Rosin system can be divided into two subsystems. The BBF uses an endless Redler conveyor to deliver the coal to two charging legs, which feed the blend into the oven from above. The Simon-Carves method uses an insulated hot larry car to deliver the preheated coal charge to the oven. Thus, both systems use a gravity charging approach.

The flow sheets of the coal moving through the Cerchar-type preheater and the Rosin system are shown in Figs. 11 and 12 [9]. There are, at the

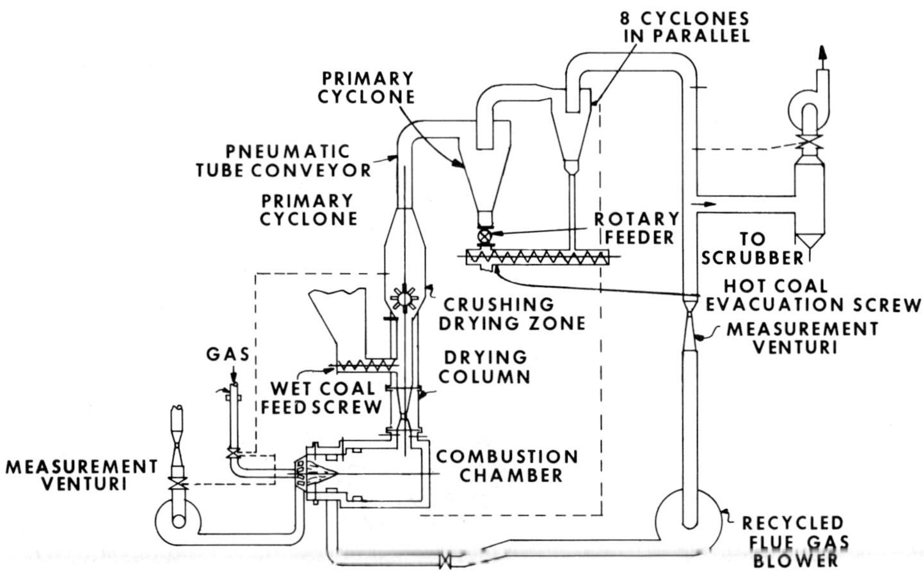

FIG. 11 Flow sheet for Coaltek coal preheating and pipelining process.

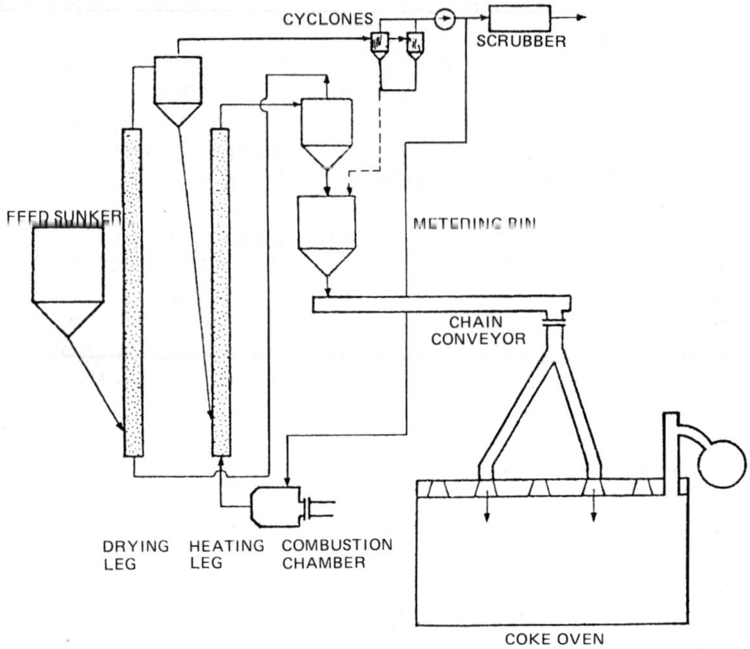

FIG. 12 Flow sheet for precarbon process.

present time (1979), five installations of the Coaltek design and two of the Rosin design in operation.

The design of the battery itself and its operation remains unchanged regardless of the coal preheat system used. The preheater section is really an attachment to the battery which can be located next to the battery or at some distance away. However, the operation of a Coaltek-type battery is quite different from that of a conventional battery.

A typical coal charging operation proceeds as follows. The charging bin receives a predetermined weight of coal, usually 25,400 to 27,200 kg; the weight is determined by load cells supporting the bin. When the oven to receive the charge has been sealed, the diverter valve to this oven is opened, thus closing the main pipeline beyond the valve. The charging bin can then be pressurized with superheated steam (315°C), which forces the coal into the pipeline system. The pressure varies from 0.8 bar for the nearest oven (No. 1) to 2.4 bars for the most distant one (No. 77). The accelerator jets in the pipeline are also steam-pressurized to prevent settling of coal in the pipeline and to provide the coal mass with sufficient kinetic energy to move it through the entire pipeline charging system. The charging operation continues for 10 to 14 min [10].

The original design provided for transfer steam bleed-off once the coal entered the final downward curve leading to the inlet port. This was expected to reduce the amount of steam entering the oven and thus reduce the amount of coal fines carryover. However, for vertical charging the bleed-off was found to be unnecessary.

The design called for the coal charge in the oven to be 300 to 350 mm below the roof. In practice the level varies considerably, and frequently the valleys are 0.9 or 1.2 m deep. This is obviously undesirable, because it lowers coke production and promotes carbon buildup on the roof. Coking time was established at 13 hr, after which time the charge can be pushed.

Thus, in general terms, the main advantages of the coal preheating practice are as follows:

1. Ability to use coal blends containing high percentages of lower rank coals
2. Improved stability and size uniformity of coke
3. Increased pollution controls of the battery
4. Ability to automate the plant operation

V. STATUS OF FORMCOKE TECHNOLOGY

The formcoke process concept and efforts to implement it date back to the nineteenth century. With the exception of small-scale laboratory experiments and one or two small pilot plants, not much happened in this field until the 1930s. The first recorded commercial production of low-temperature coke took place in November 1933 near the Champion coal preparation plant of the Pittsburgh Coal Company, located approximately 20 miles west of Pittsburgh, Pennsylvania. Eventually, two more production units were added in 1934 and 1936, and the three-kiln unit averaged in production about 6500 tons/day of low-temperature coke known as Disco fuel.

The Disco process was actually a very simple one. Medium-range-fluidity coal or coal blend estimated at 500 to 1000 Gieseler dial divisions per minute ground to -7 mm was first dried and preheated to about 370°C and then fed into an externally heated kiln together with undersize coke fines. This mass was gradually heated to approximately 475°C by hot gases passing countercurrently in the annular space between the outer surface of the kiln, made of mild steel, and the stationary outer shell.

The gradual heating of the kiln charge brought about an initial balling and in the subsequent semifluid state agglomeration and carbonization took place over a temperature range 25 to 30°C. The product of the Disco kiln had the appearance of spheroids ranging in nominal diameter from 50 to 200 millimeters. The final Disco product contained on the average 17% volatile matter, but its 2.1% sulfur level was relatively high.

The Disco coking process yielded also about 110 m^3 of gas with a heating value of 3500 to 4000 kcal/m^3 and some 65 liters of low-temperature tar and oil per ton of coal charged. The gas was used as process fuel and the tar was sold on the chemical market.

The primary application of the Disco product was as low-cost smokeless fuel. The only commercial application was a 200-ton lot of +100-mm Disco material used in a cast iron foundry and a nominal 2500-ton lot used as emergency fuel in the blast furnace plant of Jackson Iron Company in Jackson, Ohio, in 1942 and 1943. Both reported satisfactory operation but gave no details. The Disco plant was phased out shortly after World War II.

Since 1945, numerous processes have been developed for producing formcoke that is suitable for blast furnace use. Experiments with different types of formcoke processes have taken place in Japan, Germany, the Netherlands, France, Poland, the USSR, South Africa, and the United States. During the last 10 years, many experimental and full-scale blast furnace tests have been carried out with different types of formcoke.

Although many processes have been proposed and numerous patents granted, only a few have been developed to the pilot-plant stage. The six most significant of these processes, which have actually produced enough coke for blast furnace trials, are listed below, and then are described in sufficient detail to define their basic characteristics:

1. Eschweiler Bergwerkverein (Ancit), Germany
2. Bergbauforschung-Lurgi (BFL), Germany
3. Houilleres du Bassin du Nord (HBN), France
4. Sapozhnikov, Ukrainian Institute of Coal Research, USSR
5. Didier-Keihan Rentan-Sumitomo Metal (DKS), Japan
6. FMC Corporation (FMC), United States

One other process, the Consol-BNR Coke Pellet Process, progressed to the demonstration-plant stage; however, the work was suspended before any significant blast furnace trial could be run. Pertinent features of these processes are given in Table 2.

A. Ancit Process

The Ancit Process (Fig. 13) is a modification of a Dutch State Mines Process for the production of domestic smokeless fuel briquettes [11]. This process utilizes a two-coal, hot briquetting technique for producing formcoke. A noncoking, low-volatile, anthracite coal is heated rapidly to about 1100°F in a horizontal retort. The hot gas from this step is then utilized to preheat the coking coal component to about 550°F. The two components are then mixed in a 75% noncoking coal/25% coking coal ratio, with a resulting mix temperature of 800 to 850°F. The unit is briquetted at this temperature and the briquettes are transported to an insulated holding bin, where they harden while slowly cooling, thus improving their physical properties.

Work on the Ancit Process was initiated in 1950, and in 1952 a 5-ton/hr pilot plant was built near Limburg, Holland. The Eschweiler Coal Company acquired the process in 1967, and in 1971 a 275 ton/day demonstration plant was built in Alsdorf, West Germany. The Ancit formcoke is primarily being used as smokeless fuel; however, several tests lasting 3 to 4 days each were run on a small blast furnace at Thy-Marcinelle, Belgium. The results were discouraging because of strong dusting since the product is not resistant to abrasion and its strength is relatively low.

TABLE 2 Pertinent Features of Various Formcoke Processes

Process	Required	Thermal pre-treatment	Requires binder	Forming	Thermal pre-treatment
Ancit	25% coking, 75% noncoking	Yes	No	Hot briquetting at 850°F	No
BFL	35% coking, 65% noncoking	No	No	Hot briquetting at 850°F	Optional
HBN	85% noncoking, 15% coking	No	Yes	Cold briquetting with supplemental binder	Yes
Sapozhnikov	100% weakly coking coal	Yes	No	Hot briquetting	Yes
DKS	80% noncoking, 20% coking	No	Yes	Cold briquetting with supplemental binder	Yes
Consol-BNR	100% coking	Yes	Yes	Balling at 900°F	Yes
FMC	100% noncoking[a]	Yes	Yes	Cold briquetting with internally generated binder	Yes

[a]Coking coals can be used, but are not required.

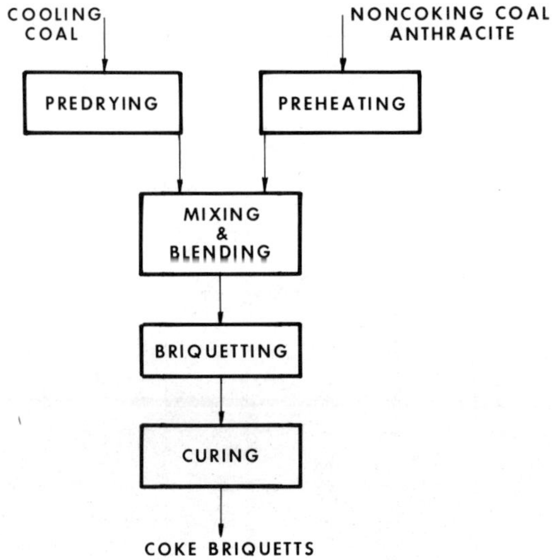

FIG. 13 Ancit formcoking process.

B. BFL Process

The BFL Process (Fig. 14) is similar to the Ancit Process in its basic form. It is a two-component system using 65% noncoking coal, which is heat-dried. The two coals are mixed and then briquetted at an average temperature of 850°F. The green briquettes can be used directly or calcined at 1450 to 1650°F in a vertical shaft coker.

The BFL Process has been extensively developed on progressively larger units, starting with a bench-scale unit in 1960 [12]. The first pilot plant, with a capacity of 100 tons/day, was erected in 1964, and by 1967, this pilot plant has produced formcoke in both the green and calcined briquette form which was used for two blast furnace trials at Rheinhausen in 1970. On the basis of these tests, a decision was made to build a 300-ton/day demonstration plant in order to produce large enough quantities of formcoke briquettes for testing in different blast furnaces and under various conditions. The demonstration plant was built at Prosper, West Germany, in 1976, but after many operating and mechanical problems, the plant was permanently shut down in 1978. A larger BFL Process plant with a designed capacity of over 600 tons/day was built for British Steel Corporation at Scunthorpe, England, in 1977. This plant is currently in the startup phase and is experiencing many of the problems that ultimately resulted in the shutdown of the Prosper plant.

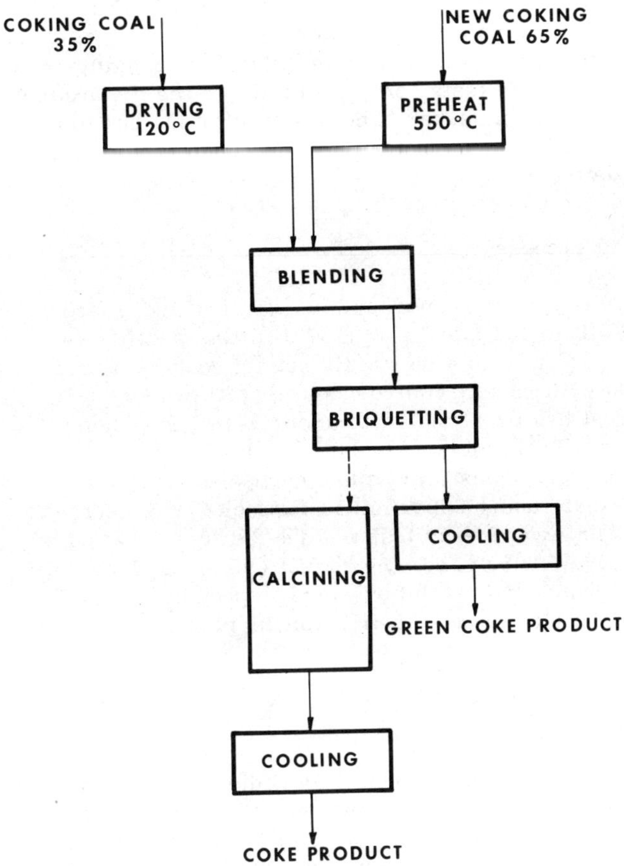

FIG. 14 Bergbauforschung formcoking process.

C. HBN Process

The HBN Process requires a blend of 85% low-volatile noncoking coal and 15% coking coal. The coals are briquetted with the addition of tar or pitch binder. Then the green briquettes are charged into a shaft coker, where they are carbonized at about 1850°F in a closely controlled heating cycle. The briquettes are cooled in the lower zone of the coker, where the recovered heat is recirculated to the high-temperature zone in a recirculating gas stream.

HBN formcoke produced at a 100-ton/day pilot plant at Douai, France, has been tested in blast furnaces during four separate experimental programs. The duration of each of the test programs was quite extensive, ranging from 29 to 58 days. However, during the tests, only about 25% of the by-product oven coke was replaced with HBN formcoke. The design of a 200,000 ton/year HBN plant was completed in 1977, but no orders for construction of such a plant have been placed to date.

D. Sapozhnikov Formcoking Process

The Sapozhnikov Formcoke Process was developed at the Ukrainian Institute for Coal Research in the USSR in the late 1960s [14]. In this process, a mixture of crushed medium- or high-volatile weekly coking coals is heated in a series of fluidized beds with closely controlled temperature increases between beds. After the coal has been heated to about 950°F, it is hot-briquetted without the addition of binder. The hot briquettes are then heated to about 1850°F in vertical ovens to complete carbonization.

Only two blast furnace tests using Sapozhnikov formcoke have been carried out on a small blast furnace, and each test was limited to a duration of 3 days. The formcoke replaced 100% of the by-product oven coke with good operating results. Private discussions with Soviet coke specialists in 1977 indicated that they had no plans to replace their 8-ton/hr pilot plant with a larger unit.

E. DKS Process

The DKS Process was developed as a joint venture by Didier Engineering GmbH, Keihan Rentan Kogyo Co., Ltd., and Sumitomo Metals Industries, Ltd. [15]. This process requires the blending of 70 to 90% noncoking coal, up to 20% coking coal, and 10% pitch binder. This mixture is heated to and briquetted at about 200°F. The briquettes are then carbonized in a vertical slot oven with either a bottom dump floor or an inclined bottom with a discharge door at the lower end. The coking process is not significantly different from the by-product oven process except that the product is in the form of briquettes of uniform size and shape. One necessary precaution is to ensure that the coal blend is within a critical fluidity range; if the fluidity is too high, the coal will all agglomerate together (coke) and the individual briquette form will be lost.

DKS formcoke was used experimentally in six different blast furnaces in Japan in trials of 3 to 21 days' duration. The results indicated that the DKS formcoke can be used without adverse effects as a replacement for up to 50% of the by-product oven coke on large blast furnaces. At the present time, the 10-oven DKS Process experimental battery has been shut down, and no plans have been announced to build a commercial-size installation.

F. Consol-BNR Process

The Consol-BNR Process (Fig. 15) involves preheating a coking coal to its softening point, mixing it with small amounts of preheated char and pitch, forming pellets in a rotating drum, and calcining these pellets at a high temperature to produce the final coke product. Part of the char is obtained from low-temperature carbonization of coal, and part is recycled from elsewhere in the system. This process relies on some of the inherent coking properties of the coal to act as a binder. However, the pitch produced in the low-temperature carbonization step is also used as a binder to assist weakly coking coals and at the same time increase coke yield. A 450-ton/day demonstration plant was constructed at Bethlehem Steel's Sparrows Point, Maryland, plant; however, because of technical difficulties, work on this project has been suspended.

G. FMC Formcoking Process

The FMC Formcoking Process (Fig. 16) was developed by the FMC Corporation [16]. FMC's primary objective when developing the process was to produce coke suitable for their phosphorus furnaces. The early work led to the

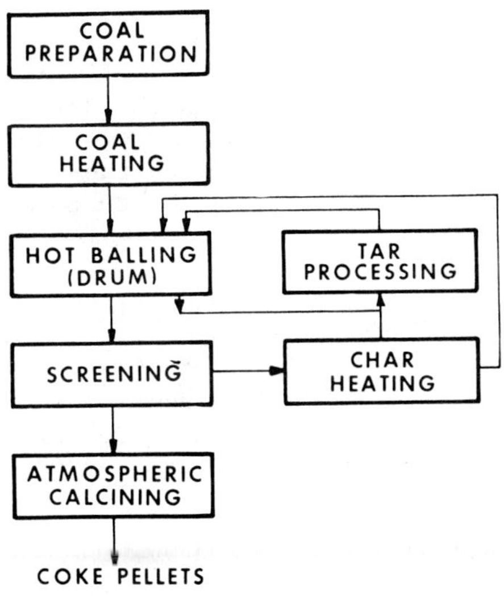

FIG. 15 Consol-BNR process.

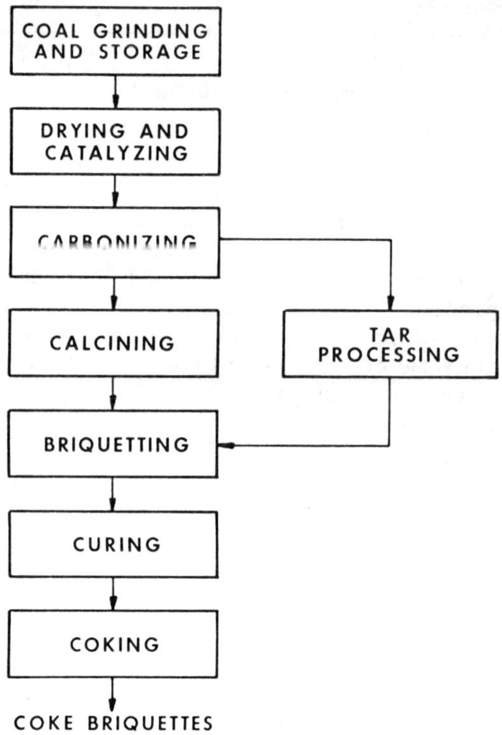

FIG. 16 FMC formcoking process.

building of a demonstration plant at Kemmerer, Wyoming, in 1960 as a joint venture between FMC and United States Steel Corporation. The FMC Coke Process was designed to use a single coal as the raw material, in conjunction with a binder produced from the tar generated during the carbonization step. The coals best suited for this process are low-rank, high-volatile, subbituminous and bituminous coals that are mined in states west of the Mississippi. However, good coking coals, including medium-volatile bituminous coals, have been successfully processed. The Kemmerer plant has been running successfully and continuously since 1966.

The first step in the FMC Coke Process is the devolatization of the finely ground coal to produce a calcinate containing about 3% volatile matter. This is accomplished in a series of three fluid-bed reactors, from which tar is also recovered and processed into binder. The calcinate produced is mixed with an appropriate amount of binder and briquetted. The green briquettes are cured at about 450°F in an oven and coked at about 1600°F in a combination shaft coker and cooler to form a hardened briquette.

Several blast furnace tests have been carried out with formcoke from the FMC Coke Process. In 1962, FMC formcoke was tested on the experimental blast furnace of United States Steel Corporation in Universal, Pennsylvania. Subsequent tests included one at Armco Steel Corporation in 1967 and a 20,000-net-ton test of FMC formcoke briquettes at Inland's No. 5 blast furnace in 1973 [17].

Based on published information, the blast furnace testing of the product of the FMC Coke Process has been more extensive than that of any other formcoke process. The five major blast furnace tests conducted to date are summarized in Table 3.

It appears that the future of the formcoke concept is highly promising because it offers advantages in the field of coal availability, environmental control, and the uniformity of product size and quality.

The next step is to build a large commercial-size formcoke plant and learn to produce good-quality formcoke which would assure lowest coke rate and highest iron production on any given blast furnaces.

VI. COAL AND COKE CHARACTERISTICS

A. Brief Review of Coking Coal Reserves of the United States

It is estimated that carbon accounts for only about 0.04% of the bulk of the earth and only about 0.1% of the earth's crush. Only a very small portion of this, about 0.02%, occurs in a form that is capable of reacting with oxygen [18]. Consequently, the total carbon that is both available for reaction with

TABLE 3 Summary of Blast Furnace Tests with FMC Formcoke

Year	Test site	Coal source	Furnace burden	Quantity tested (tons)	Briquette Size (in.)
1962	U.S. Steel, Universal, Pa.	Elkol	Self-fluxing sinter	2,000	1 1/4
1967	Armco Steel Hamilton, Ohio [5,6,15]	Elkol	95% reserve pellets, 5% reclaimed slag	13,000	2 × 1 3/4
1970	Inland Steel, East Chicago, Ind.	Illinois No. 6	50% pellets, 50% sinter	120	1 1/2
1972	British Steel, Cardiff, Wales [11]	Elkol	50% pellets, 50% sinter	3,000	1 1/2
1973	Inland Steel, East Chicago, Ind. [25]	Elkol	70% pellets, 15% coarse ore, 15% screened sinter	20,000	2

TABLE 4 Coal Reserve Base of the United States by Province, State, Rank, and Method of Mining (10 Short Tons)

	Lignite		Subbituminous		Bituminous		Anthracite		
	Surface	Underground	Surface	Underground	Surface	Underground	Surface	Underground	Total
Eastern province									
Eastern Kentucky					3,500	9,500			13,000
Maryland					150	900			1,050
Northern Alabama					1,200	1,800			3,000
Ohio					3,600	17,400			21,000
Pennsylvania					1,100	22,600	100	7,200	31,000
Tennessee					300	700			1,000
Virginia					700	3,000			3,700
West Virginia					5,000	34,500			39,500
Total					15,550	90,400	100	7,200	105,950
Gulf province									
Louisiana	ISD[a]								ISD
Mississippi	ISD								ISD
Southern Alabama	ISD								ISD
Southeastern Arkansas	ISD								ISD
Southeastern Texas	1,300								1,300
Total	1,300								1,300
Interior province									
Illinois					12,000	53,500			65,500
Indiana					1,600	9,000			10,600
Iowa					ISD	2,900			2,900
Kansas					1,400	ISD			1,400
Michigan					ISD	100			100

Missouri					3,400	6,100			9,500
Northwestern Arkansas					300	400			700
Northwestern Texas					2,000	ISD			2,000
Oklahoma					400	900			1,300
Western Kentucky					3,900	8,700			12,600
Total					25,000	81,600			106,600
Pacific province									
Alaska	300	ISD	5,900	4,250	1,200	ISD			11,650
Washington			100	ISD	400	1,400			1,900
Total	300		6,000	4,250	1,600	1,400			13,550
Rocky Mountain and Northern Great Plains provinces									
Arizona			350	ISD					350
Colorado			ISD	4,700	900	9,200			14,800
Montana	7,100	ISD	35,500	64,500	ISD	1,400			108,500
New Mexico			2,000	600	300	1,500			4,400
North Dakota	16,000	ISD							16,000
South Dakota	400	ISD							400
Utah					300	3,800			4,100
Wyoming			23,800	25,000	ISD	4,500			53,300
Total	23,500		61,650	94,800	1,500	20,400			201,850
Grand total	25,100		67,650	99,050	43,650	193,800	100	7,200	429,250

[a]ISD, insufficient data; although economical deposits are known to exist, the extent has not been determined.

oxygen and within our reach (i.e., the earth's crust) is in a planet-wide concentration of only 0.2 part per million; in this context, carbon is a trace element. It is fortunate that natural processes have concentrated much of this "reactive" carbon in fossil fuels, *most* of it in the form of coal.

The total world resources of coal (anthracite, bituminous, subbituminous, lignite, and brown coal) are approximately 1.1×10^{13} tons, about 27% of this being in the United States [19]. About 3.9×10^{11} tons of this 27% is classified as reserves (i.e., suitable for mining by existing methods) [20]. Table 4 illustrates the distribution of the reserves of the United States in terms of rank, method of mining, and location [20,21].

With respect to the coking industry, only bituminous coal is considered suitable for the production of metallurgical coke. If one assumes mining recovery rates of 50% for underground mining and 80% for surface mining, about 1.2×10^{11} tons of bituminous coal is currently recoverable. The reserves of this recoverable bituminous coal are distributed very unequally; over 50% occurs in just three states—Illinois, West Virginia, and Kentucky (see Fig. 17). Of course, not all of this coal is desirable for coke production from the standpoint of ash and sulfur content; however, there are techniques available to counter the adverse effects of ash and sulfur on the production of good-quality steel. In addition, many high-volatile bituminous coals are of insufficient rank to produce good-quality coke by conventional coking techniques. However, recent advancements in cokemaking technology have enabled steel companies to use coals long thought to be unsuitable for cokemaking. For example, it was once thought that Illinois coals were unfit for the production of blast furnace coke because of their high ash and sulfur contents and lower rank. Now, however, Illinois coal that meets required ash and sulfur guidelines can be used to produce acceptable coke by blending it properly with higher-rank coals.

The supply of bituminous coal in the United States should last for approximately 200 years at the current production rate of 6.0×10^{8} tons/year. Government proposals to double production by 1985 would reduce this to a 100- to 140-year supply. This is a far cry from the oft-quoted figures of 1000 to 1400 years of bituminous coal that were used in the recent past. Additional coal reserves will be proven in the future, so it is doubtful that the current estimates will be accurate in a few years. However, with advancing technologies in steel and coke production (i.e., coal preheating, formcoking, hot-metal desulfurization, etc.) widening the range of coals that are suitable for the production of coke, and with the increasing use of subbituminous coal and lignite for energy generation, the United States should have more than enough coal to last until new technologies for producing steel and energy without coal come of age.

B. Coal Properties and Their Effect on Plant Operations and Coke Characteristics

Coke quality and coke plant operations are affected by the properties of the coals in a variety of ways. Of the many properties or characteristics of coal, the following are considered important in cokemaking; chemistry, petrography, rheology, hardness, moisture, and size.

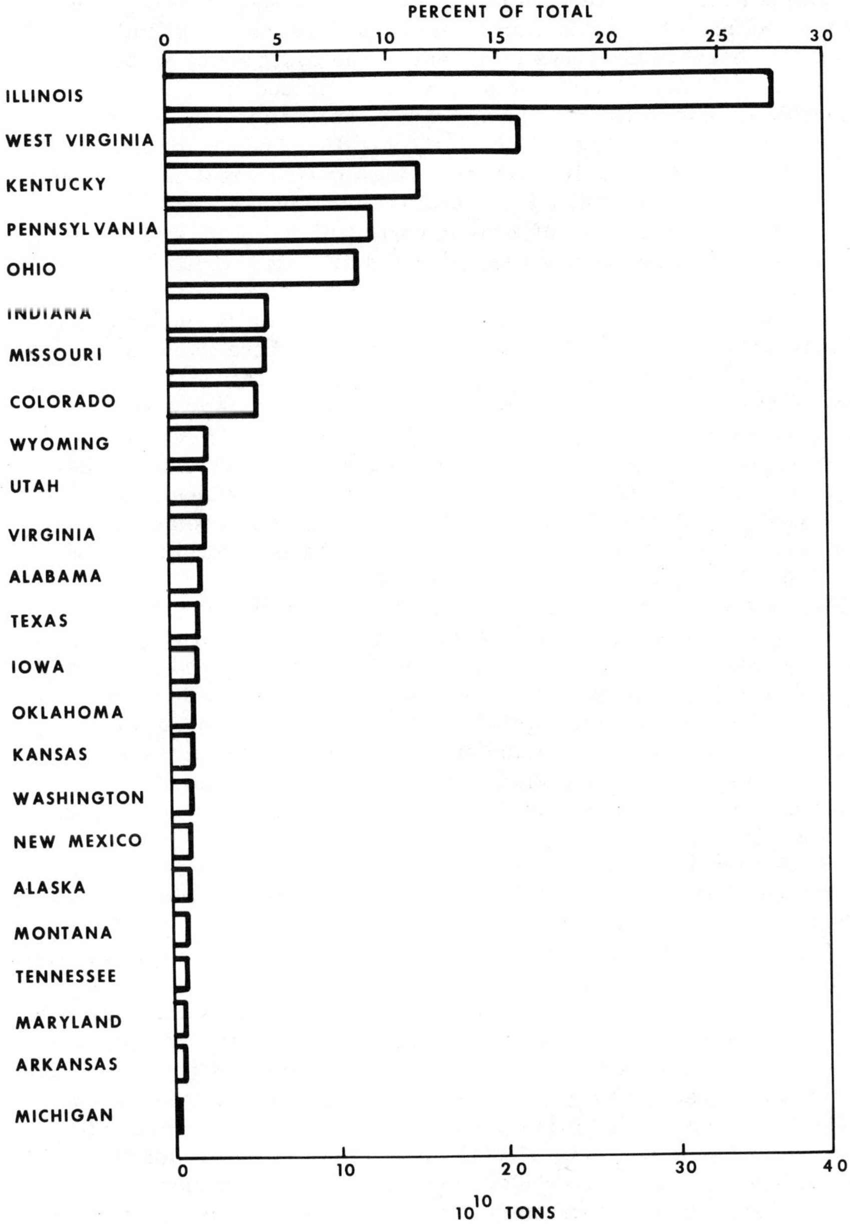

FIG. 17 Probable recoverable bituminous coal reserves of the United States.

The chemistry of coal can be expressed in terms of its elemental composition, which is defined as its ultimate analysis or in empirical terms by a prescribed method which is defined as its proximate analysis. The ultimate analysis is specified when evaluating coals for combustion, gasification, and liquifaction processes, but is less important in the cokemaking processes. The proximate analysis, which is determined by a procedure that uses a heating profile similar to the carbonization process, defines the amount of "volatile matter" driven from coal, the "ash" residue, and the balance, which is termed "fixed carbon." Sulfur content is also given, which is equivalent to the elemental sulfur determined in the ultimate analysis.

The volatile-matter content of a given coal blend will give an indication of the yield of coke to be expected from a blend of coals. Generally, the coking coal blend is made up of varying proportions of high- and medium- or low-volatile coals. All the ash contained in the coal blend is retained in the coke. Therefore, for a higher-volatile-matter blend, there would be a lower yield of coke of higher ash content. A large portion of the organic sulfur is driven off from the coal during carbonization and appears in the by-products. On the other hand, inorganic sulfur is essentially retained in the coke. The effect that the chemistry of the coal has in the coking process can be summarized by saying that the volatile matter determines to a great extent the yields of coke and by-product, ash in the coal is retained to the coke, and sulfur divided between coke and by-products according to the proportions of inorganic and organic forms.

Coal petrography has become widely used in the world for predicting coke quality based on coal analysis. Work by Schapiro and Gray [22] led to a system for predicting coke stability based on petrographic entities and reflectance of coal. With this system, an optimum blend of coals could be selected to produce desired coke quality. Basically, coal is made up of reactive and inert components for which at any given reflectance there exists an optimum ratio of these components to produce maximum coke strength.

When most bituminous coals are heated, they soften, become fluid, and finally solidify into a spongelike coherent solid. It is this flow property that determines whether a particular coal has the potential to make coke. The study of the rheology of coal includes fluidity, plasticity, swelling, expansion-contraction, and dilatation.

The fluid or swelling property of coal determines whether a coal will make coke. The temperatures at which the softening and solidification occurs are important in that all coals in a blend should have similar softening ranges to permit an interaction of coals in the fluid state during coking. During the coking process, the coal mass softens forming a plastic envelope. This envelope contains water vapor and gas, which exerts pressure on the coke oven walls. The magnitude of this pressure is dependent on the strength of the envelope. Pressure in excess of 13.8 kPa is considered unsafe for the structural integrity of the coke oven walls. Another important property is the expansion and subsequent contraction of the coal mass as it is converted to coke. In order for the coke to be pushed from the oven, a certain amount of contraction of the coke is necessary. The consequence of not having contraction at all, or only marginal contraction, is that the coke sticks in the oven or it requires very high power requirements for pushing.

Coking coal blends are pulverized to permit the intimate mixing of the coking components in the blend. Generally, as the percent passing 3.2 mm increases, coke stability increases for constant moisture in the blend. However, an excessively high degree of pulverization results in a decrease in coke apparent specific gravity, which is also an inverse measure of porosity. Coal pulverization is affected by moisture and the hardness of the coal as measured by the Hardgrove test. Coal with too high moisture content reduces the efficiency of most pulverizing equipment. Different coals usually have different hardness values. Therefore, in preparing a blend for coking, each coal should be pulverized separately according to its needs and then blended. Bulk density of pulverized coal is affected by moisture and particle size. Thus, it can be seen that moisture, hardness, and size are interrelated. Low bulk density can affect productivity by decreasing the amount of coke produced per oven, which in turn requires more time for charging and pushing to meet production demands.

C. Coal and Coke Properties and Their Determination

The American Society for Testing and Materials (ASTM) has defined many testing procedures applicable to coke and coal (see Tables 5A and 5B). This section will point out some of the more important test procedures and reference several of the tests performed in other countries.

Of primary importance in the testing of coal and coke is the need for obtaining a representative sample, which is often a very difficult task. Generally, the entire quantity of material is incremently sampled such that equal portions comprise the gross sample. This gross sample is then either mechanically or manually reduced to the amount required for the particular analysis.

Tests for coal and coke characteristics can be divided into two major classes, chemical and physical. Each class is discussed below.

Considering chemical tests first, there is the ultimate analysis and proximate analysis for both coal and coke. The ultimate analysis includes the determination of carbon, hydrogen, oxygen, sulfur, nitrogen, and ash. Usually, a proximate analysis is sufficient for meeting plant needs and consists of moisture, volatile matter, ash, fixed carbon, and generally, sulfur.

Another chemical property, reactivity, which refers to the ability of coke to react with carbon dioxide, is measured [24-29]; however, the ASTM does not provide a standard reactivity test procedure.

With respect to the physical tests, coals are classified into two major classes, banded coal and nonbanded coal. This division is done on the basis of origin, composition, and certain readily observed features. The lithologic components of these coal classes include vitrain (shiny black bands), fusain (material resembling charcoal), and attrital coal (material composed of plant microfragments). In addition, the macerals (organic components of coal distinguishable by microscopic examination) are classified. ASTM procedures describe the preparation of samples and methods for the identification of these organic components.

Other measured coal properties include free swelling index, fluidity, expansion characteristics, size consist, strength, and grindability. The

TABLE 5A ASTM Testing Procedures Applicable to Coke

Designation	Title
General	
D121-72	Standard Devinitions of Terms Relating to Coal and Coke
Sampling	
D346-75	Standard Method of Collection and Preparation of Coke Samples for Laboratory Analysis
Chemical	
D1857-68 (reapproved 1974)	Fusibility of Coal and Coke Ash
D2361-66 (reapproved 1972)	Standard Method of Test for Chlorine in Coal
D2795-69 (reapproved 1974)	Analysis of Coal and Coke Ash
D3172-73	Standard Method for Proximate Analysis of Coal and Coke
D3173-73	Standard Method of Test for Moisture in the Analysis Sample of Coal and Coke
D3174-73	Ash in the Analysis Sample of Coal and Coke
D3175-73	Volatile Matter in the Analysis Sample of Coal and Coke
D3177-75	Total Sulfur in the Analysis Sample of Coal and Coke
D3178-73	Carbon and Hydrogen in the Analysis Sample of Coal and Coke
D3179-73	Nitrogen in the Analysis Sample of Coal and Coke
Physical	
D167-73	Specific Gravity and Porosity of Lump Coke
D292-29 (reapproved 1972)	Standard Method of Test for Cubic Foot Weight of Coke
D293-69 (reapproved 1974)	Standard Method of Sieve Analysis of Coke
D294-64 (reapproved 1972)	Standard Method of Tumbler Test for Coke
D323-70	Specification for Perforated-Plate Sieves for Testing Purposes
D2490-70	Standard Method of Tumbler Test for Small Coke
D3038-72	Standard Method of Drop Shatter Test for Coke
D3402-75	Standard Methods of Tumbler Test for Coke
E11-70	Specification for Wire-Cloth Sieves for Testing Purposes

TABLE 5B ASTM Testing Procedures Applicable to Coal

Designation	Title
General	
D121-72	Standard Definitions of Terms Relating to Coal and Coke
D388-66 (reapproved 1972)	Specifications for Classification of Coals by Rank
Sampling	
D2013-72	Preparing Coal Samples for Analysis
D2234-72	Collection of a Gross Sample of Coal
Chemical	
D407-44 (reapproved 1969)	Definition of the Terms Gross Calorific Value and Net Calorific Value of Solid and Liquid Fuels
D1412-74	Test for Equilibrium Moisture of Coal at 96 to 97 Percent Relative Humidity and 30°C
D1756-62 (reapproved 1974)	Test for Carbon Dioxide in Coal
D1757-62 (reapproved 1974)	Test for Sulfur in Coal Ash
D1857-68 (reapproved 1974)	Test for Fusibility of Coal and Coke
D2015-66 (reapproved 1972)	Test for Gross Calorific Value of Solid Fuel by the Adiabatic Bomb Calorimeter
D2361-66 (reapproved 1972)	Test for Chlorine in Coal
D2492-68 (reapproved 1974)	Test for Forms of Sulfur in Coal
D2795-69 (reapproved 1974)	Test for Analysis of Coal and Coke Ash
D2961-74	Total Moisture in Coal Reduced to Number 8 Top Sieve Size (Limited Purpose Method)
D3172-73	Proximate Analysis of Coal and Coke
D3173-73	Test for Moisture in the Analysis Sample of Coal and Coke
D3175-73	Test for Volatile Matter in the Analysis Sample of Coal and Coke
D3176-74	Ultimate Analysis of Coal and Coke
D3177-75	Test for Total Sulfur in the Analysis Sample of Coal and Coke
D3178-73	Test for Carbon and Hydrogen in the Analysis Sample of Coal and Coke
D3179-73	Test for Nitrogen in the Analysis Sample of Coal and Coke
D3286-73	Test for Gross Calorific Value of Solid Fuel by the Isothermal-Jacket Bomb Calorimeter
D3302-74	Test for Total Moisture in Coal

(continued)

(Table 5B continued)

Designation	Title
Physical	
D197-30 (reapproved 1971)	Sampling and Fineness Test of Pulverized Coal
D291-60 (reapproved 1975)	Test for Cubic Foot Weight of Crushed Bituminous Coal
D311-30 (reapproved 1969)	Test for Sieve Analysis of Crushed Bituminous Coal
D409-71	Test for Grindability of Coal by the Hardgrove Machine Method
D410-38 (reapproved 1969)	Sieve Analysis of Coal
D431-44 (reapproved 1969)	Designating the Size of Coal from its Sieve Analysis
D440-49 (reapproved 1975)	Drop Shatter Test for Coal
D720-67 (reapproved 1972)	Test for Free Swelling Index of Coal
D1812-69 (reapproved 1974)	Test for Plastic Properties of Coal by the Gieseler Plastometer
D2014-71	Test for Expansion or Contraction of Coal by the Sole Heated Oven
D2639-74	Test for Plastic Properties of Coal by the Constant Torque Gieseler Plastometer
D2796-72	Definition of Terms Relating to Lithologic Classes and Physical Components of Coal
D2797-72	Preparing Coal Samples for Microscopical Analysis by Reflected Light
D2798-72	Microscopical Determination of the Reflectance of the Organic Components in a Polished Specimen of Coal
D2799-72	Microscopical Determination of Volume Percent of Physical Components of Coal
E11-70	Specification for Wire-Cloth Sieves for Testing Purposes
E323-70	Specification for Perforated-Plate Sieves for Testing Purposes

free swelling index is used to indicate the caking property and is measured by comparing the volume changes in a sample after having been heated under controlled conditions. Fluidity is a measure of the plastic behavior of a coal while it is being heated at a constant rate. Significant indices of this measure include softening temperature, temperature of maximum fluidity (i.e., when the stirrer immersed in the heated coal sample is rotating at its maximum rate), maximum fluidity (maximum stirrer rotation), and solidification temperature (the temperature at which the stirrer stops). One way that expansion (or contraction) of a coal can be measured is by heating only the bottom surface of a sample while a piston applies a constant force to the top surface. The location of the piston upon reaching the conclusion of the heating cycle indicates the degree of expansion (or contraction). Size consists are determined by sieving and strength is determined by a drop shatter test. Ease of pulverization of coal can also be measured by the Hardgrove grindability test.

Physical properties of coke can be categorized into three groups: (1) the static properties considered for a lump of coke (porosity), (2) the static properties of bulk coke (specific gravity, bulk density, and size), and (3) the mechanical properties (shatter, stability and hardness, and high-temperature strength).

Coke pore structure includes the pore on the external surface of the lump, the internal microscopic pores, and the "submicroscopic" pores, whose diameters are only several nonameters [30-32]. The calculation of porosity can be made with a knowledge of the apparent specific gravity and the true specific gravity.

$$\%\ \text{Porosity} = 100 - 100 \times \frac{\text{apparent specific gravity}}{\text{true specific gravity}}$$

Apparent specific gravity is the average weight per unit volume (relative to water) of individual lumps of coke, and so includes the effect of porosity. True specific gravity is the average weight per unit volume (relative to water) of individual particles of coke, and is measured on a crushed sample to eliminate the effect of porosity. Bulk density of coke is the weight per unit volume, which is affected by both the apparent specific gravity of the individual lumps and the voids between lumps. Two aspects of coke size are meaningful, size distribution and average size. Size distribution is reported as the series of percentages by weight of coke that passes through one sieve and remains on another sieve with finer apertures. Average size is defined as the weighted average of the products of the average size of each size range and the fraction present in that range.

The usual method for measuring coke strength (resistance to abrasion and impact) is by measuring the size degration that occurs in a tumbler drum. Table 6 shows the parameters for tests used in different countries [33]. As can be seen in the table, the parameters vary considerably. These different drum tests were devised to maximize the measurement in degradation between various cokes. In one study, a large number of degradation tests were performed to ascertain the best conditions for one particular tumbler [34]. Figure 18 shows the degration curves obtained in this study for five typical

TABLE 6 Summary of Strength Test Methods in General Use

Country	Test	Drum diameter length	Rotation of drum (rpm × min)	Coke size (mm)	Sample weight (kg)	Sieve type	Screen sizes used (mm)	Strength indices	Index values most desirable for blast furnace use
Japan	JIS drum test	1500 mm, 1500 mm	15 × 2	>50	10	Square	15	+15 mm strength index	+15 mm; index, >92
United States	ASTM tumbler	36 in., 18 in.	24 × 58	>50	10	Square	25, 6	+25 mm (stability), +6 mm (hardness)	+25 mm, >40; +6 mm, >65
Germany and France	DIN Micum test	1000 mm, 1000 mm[a]	25 × 4	>50	50	Round	40, 10	+40 mm, −10 mm	+40 mm, >67; −10 mm, <10
USSR	Gost	2000 mm, 800 mm	15 × 10	>50	410	—	25	+25 mm (Sundgren index)	+25 mm; index, >280 kg
Britain	B.S. 1016 Part 13, Micum test	1000 mm, 500 mm	25 × 4	>20	25	Round	40 mm, <10 mm	3 mm	>40 mm, −75; <10 mm, −8

Note: Except for the Gost apparatus, all of the drums above have angles on their internal cylindrical surfaces. In the case of the Gost drum, the cylindrical surface is formed of steel rods 25 mm in diameter spaced 25 mm apart.

[a]DIN is 990 mm in diameter and 900 mm in length.

Source: British Carbonization Research Association.

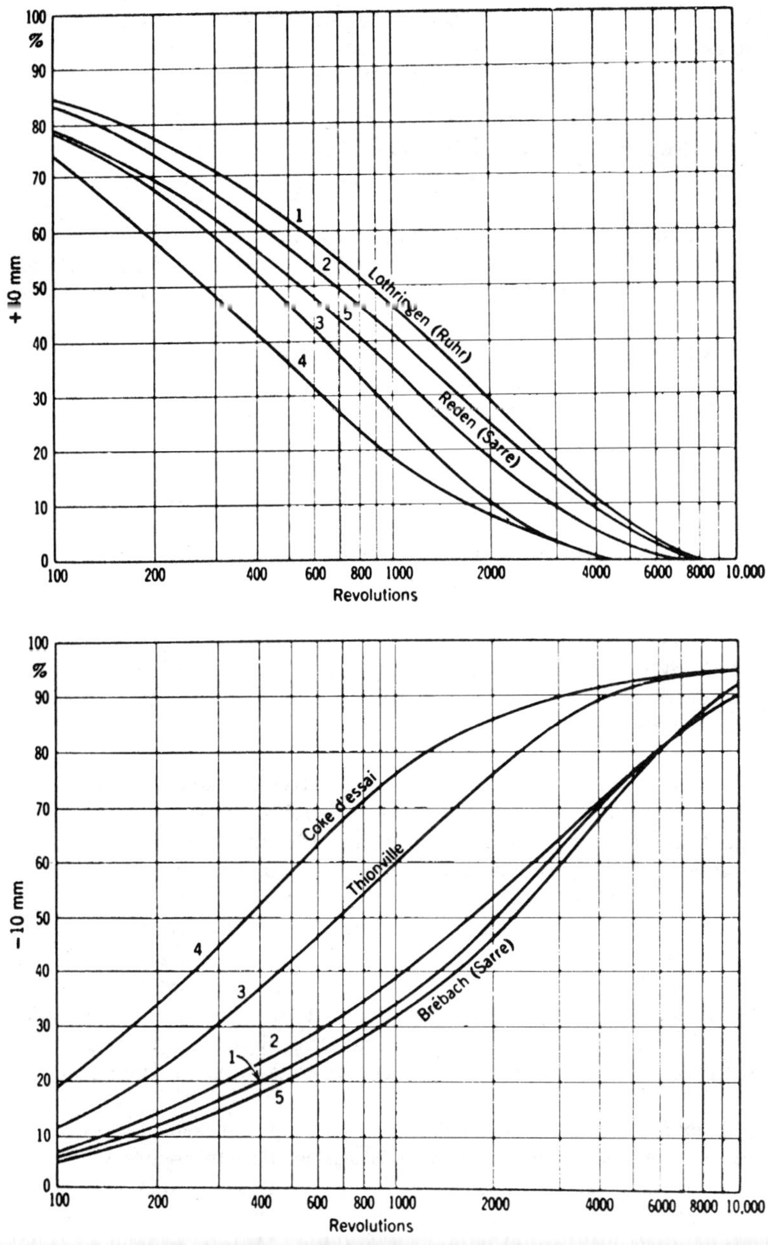

FIG. 18 Typical degradation curves of five cokes as a function of a number of revolutions of a Micum drum.

cokes from France and Germany [35]. These results point out the difficulty of selecting the optimum number of drum revolutions to determine whether or not a difference in coke strength exists. The usefulness of tumbler tests at ambient temperatures has been questioned by blast furnace operators. Consequently, several investigators [36,37] have conducted high-temperature tests of cokes. The results of these investigations are inconclusive.

D. Effect of Coke Properties on Blast Furnace Operation and Performance

There are at present no reliable quantitative relationships regarding the effect of coke properties on blast furnace operation. In spite of this, there is qualitative and, in some instances, semiquantitative evidence which indicates that coke properties can significantly affect blast furnace operation. This is, of course, consistent with the fact that coke is a critical factor in terms of the chemical reactions within the blast furnace as well as in the maintenance of acceptable fluid dynamic conditions. Specifically, coke must provide the following chemical and physical functions:

1. Chemical—Combine with oxygen to provide the following while simultaneously minimizing the introduction of sulfur and slag components which adversely affect blast furnace efficiency:
   a. The heat required for reducing iron oxide and melting slag and iron
   b. The CO required to reduce iron oxide
2. Physical—Provide adequate permeability within the burden so that a countercurrent movement of hot reducing gases and molten slag and iron is maintained

From the combustion point of view, coke is relatively inert and requires relatively high temperatures to initiate a reaction with oxygen that is sufficient to provide the heat necessary for reducing iron oxide and for melting slag and iron. To a good first approximation, this occurs in the immediate vicinity of the blast furnace tuyeres, where heated air (greater than 1000°C) reacts with hot coke to produce a temperature sufficient for blast furnace operation (greater than 1800°C). Thus, to provide the maximum benefit as a fuel, the amount of contained carbon in the coke should be maximized and hence the ash content minimized. After taking this constraint into account, there are no limitations on coke as a fuel source other than the obvious criterion that the bulk of it survives the passage through the furnace and arrives at the tuyeres.

The principal carbon-bearing gas produced during the combustion of coke at the tuyeres is carbon monoxide, which subsequently reacts with iron oxide to produce iron and carbon dioxide. If this reaction occurs above 1000°C, the resulting carbon dioxide can react endothermically with coke (carbon solution) to form additional carbon monoxide. However, below 1000°C, coke is relatively inert to carbon dioxide, and hence the carbon dioxide produced below 1000°C reports to the top gas. Of these two possibilities, it is more desirable to encourage the reaction between carbon monoxide and iron oxide below 1000°C since this will reduce the coke rate and hence increase blast furnace productivity.

In terms of coke properties, the implication of the desirability to reduce iron oxide below 1000°C is that the coke be relatively unreactive in this temperature range so that the temperature is not reduced by the reaction between coke and carbon dioxide. Also, it is important to note that increased blast temperature increases the production of carbon monoxide at the tuyeres, and hence increases the probability for reaction with iron oxide and consequent generation of carbon dioxide, which subsequently reacts endothermically with coke in the lower portion of the furnace. High blast temperatures are, however, desirable in terms of reducing the heat input that must be supplied by the reaction of carbon and oxygen. There is, therefore, a balance required between coke reactivity and blast temperature in order to simultaneously provide sufficient carbon monoxide for reducing purposes and to minimize coke consumption. In most blast furnaces today, burden materials are relatively easy to reduce. This fact, in conjunction with high blast temperatures, results in coke rates that are marginal with respect to providing sufficient carbon monoxide during combustion to reduce the burdens below 1000°C. Consequently, some carbon solution above 1000°C is required to supplement the generation of carbon monoxide by combustion.

Although the reactivity of coke can be measured by a number of techniques, it is not clear that any of the measurements relate directly to blast furnace performance. Specifically, throughout the world, coke with varying reactivities, as measured by existing reactivity tests, ranging from relatively unreactive to nearly completely reactive, have been used successfully in existing blast furnaces operated under standard accepted operating practices. This is not particularly surprising when one considers the factors that contribute to the reactivity of coke. First, the carbon forms found in coke vary from amorphous carbon to graphite. It is well established that graphite is significantly less reactive than amorphous carbon. Consequently, although the kinetics of graphitization are a function of the carbonization conditions and the nature of the coals employed, the thermal conditions in the blast furnace will increase the degree of graphitization and therefore reduce the reactivity of all cokes; the more reactive coke will tend to be more affected and therefore the difference in reactivity between cokes will be reduced. At the present time, it is not certain how much graphitization occurs in the range of time/temperature/atmosphere conditions found in existing blast furnaces, but it is known that the reactivity of all cokes is increased significantly by the presence of alkali, which tends to concentrate in the lower portion of the furnace. This effect would be pronounced with the lower-reactive coke and results in reducing the difference between the reactivities of all cokes. The increase in reactivity due to alkali is so great (5 to 10 times) that it is most probable that the beneficial effects of graphitization in the furnace are, in large part, negated and all cokes are rendered highly reactive. Consequently, with respect to blast furnace performance, coke reactivity is not likely to be a significant property of coke, although this property has been used successfully to monitor coke plant operations, where essentially every process variable can affect the degree of carbonization and graphitization.

The physical nature of the burden changes as it descends in the blast furnace. Initially, it is charged in alternating layers. This practice increases the permeability in the furnace relative to forming one layer consisting of a uniform mixture of all burden materials. As the burden descends, all the burden materials, with the exception of the coke, soften and melt. Under these conditions, it is imperative that the coke retain adequate strength and size in order to ensure a high level of permeability consistent with production requirements. To effect this condition, it is necessary that in the region where molten slag and/or iron is present, the coke size and size distribution be such that the liquid is properly supported by the gas flow, so that it neither fails to descend with the coke nor fills the voids in the coke layer. Specifically, in qualitative terms, the proper conditions are achieved with coke that has a large average size and a narrow size distribution. These conditions suggest the necessity for charging large coke with an initial small size range of sufficient strength, so that size degradation is minimized. It has been demonstrated on operating furnaces that coke stability and size are directly related to productivity, which suggests that low-temperature properties are correlated with high-temperature properties and undoubtedly are useful guides. However, the uncertainty in the data, and the effect of specific practices on the sensitivity of the correlations, indicate that additional information, perhaps under actual and/or simulated blast furnace conditions, is required before correlations between coke properties and blast furnace performance are sufficiently well defined so that the basic controlling properties of coke can be identified.

VII. ECONOMICS OF COKEMAKING

A. Introduction

The decision to invest in cokemaking facilities is troublesome today to most corporate managements, not only because of the large magnitude of the capital investment but also because of the uncertainty regarding future regulatory impositions and developing alternative technologies. As a result of uncertainty, most investments in coke producing facilities tend to be viewed as defensive-type investments rather than discretionary-type investments predicated solely on an attractive return on investment.

In other words, the coke plant complex comprises about 15% of the total investment per shipped ton required for an integrated steel plant. To perpetuate the earning power of an existing integrated plant, either merchant coke must be acquired or a replacement facility must be installed whenever coke capacity is retired. To refrain from providing the necessary coke would mean that the incremental cost of the coke, including approximate capital charges, is greater than the incremental profit generated by operating the balance of the fully integrated increment of capacity. Although the situation is unlikely, the fact that most integrated steel companies have invested interests in raw materials makes this situation extremely remote, based on normal economic situations.

Consequently, management is faced with the difficult decision as to how coke should be provided. This is not an uncommon predicament because it takes almost three-fourths of a ton of coke to support 1 ton of shipments from an integrated steel plant. Further, as of mid-1978, about 70% of the 12,500 slot ovens operating in North America were over 20 years old. The useful life of a slot oven battery varies considerably based on operating practices, productivity, types of coals charged, and maintenance, but experience shows that on average a coke oven battery must normally be rebuilt after 20 years of operation and that such rebuilding perpetuates the battery life another 10 years. The age profile of the existing facilities suggests that a considerable amount of cokemaking capacity must be replaced domestically within the next 5 to 10 years. In addition, growth in worldwide steel demand will further accelerate the need for new cokemaking capacity.

As stated previously, when an integrated steelmaker is faced with the decision to replace a given unit of cokemaking capacity, a choice of two alternatives is available to the decision maker: he can attempt to secure a line of supply on the open market or he can make an investment in his own cokemaking capacity. The first alternative is generally considered a short-term solution, because the risk associated with an outside supplier of such a critical material to the integrated plant is enormous. Further, most steel plants depend on the by-product fuels from a coke oven battery to support subsequent heating and rolling processes. When the second alternative is considered, the decision maker is faced with selecting a facility based on established or emerging technologies.

By-product conventional slot oven batteries involve a very mature proven technology which probably involves the most difficult environmental and health problems of any of the processes in an integrated steel plant. Many worthwhile improvements have been incorporated into the conventional process over the years, but the development of pipeline charging of preheated coal is one of the most significant because it provides the potential to minimize charging emissions and to utilize marginal coking coals. Therefore, in subsequent portions of this discussion, the economical trade-offs between conventional slot oven batteries and preheated coal-pipeline charged batteries will be reviewed.

An emerging new technology known as formcoke offers the decision maker a continuous process. It should be noted that the formcoke process is a process with limited commercial demonstration for use in an integrated steel plant. Domestically, a 250-ton/day plant is operated successfully by FMC and this plant produces formcoke for a phosphorus plant. Formcoke has been tested successfully as a partial replacement for by-product oven coke for short periods of time on medium-size blast furnaces. However, the process must undergo a successful demonstration of a large scale-up in capacity to be compatible with most steel plants and also demonstrate the wholesale replacement of conventional coke in the blast furnace. Further, the practical use of the low-Btu by-product gas from an air-operated FMC formcoke plant must be demonstrated in an existing steel plant complex. For most decision makers, the uncertainty inherent in the formcoke decision is too great to warrant a commitment by an individual company. Subsequent discussion on the economics of this process will be very general in nature because of the lack of demonstrated results.

B. Return on Investment in Cokemaking Facilities

In general terms, the return on investment in cokemaking facilities can be determined by considering the difference between the net cost of producing coke in-house and the cost of obtaining coke by other means, such as by outside purchases on the open market. The selection of the best in-house facility normally relates the incremental difference of the investment dollars in alternate facilities versus the incremental benefits derived from considering one type of facility relative to another facility. However, by comparing each type of facility to an established purchased price, the relative attractiveness of each alternate can be measured against the "base case" of purchased coke.

When a benefit would accrue to the decision maker from a market price of coke lower than the in-house cost of producing coke, the apparent choice would be to avoid the investment decision and rely on outside purchases to fulfill demand. However, it is unlikely that such a market price would be representative of the long-term cost of coke. First, the continuing unattractiveness of the investment decision relative to replacement capacity would cause supply/demand pressures that would distort the current market price equilibrium. Second, the investment decision necessitates a long lead time to implement, in that permit application and construction activities now require from 2 to 4 years. Therefore, a decision not to invest has the inherent risk of the vagaries of the merchant coke market for a prolonged period. Therefore, in defining the comparison base, it is important to consider the viability of any spot price of purchased coke.

C. Investment Alternatives in Cokemaking

Current experience indicates that it requires anywhere from \$200 to \$250 of investment to obtain an annual ton of capacity. This applies to conventional as well as preheat facilities. As an example, a battery capable of producing 2500 tons/day would entail an investment in the range \$150 to \$200 million. Such an outlay would include a "grass-roots" plant with the following breakdown in facilities by percent of total investment:

Facilities	*Percent of total*
Coal and coke handling	15
Battery ovens with oven machinery	35
Preheating and pipeline charging	15

Coal chemicals recovery	20
Pollution control—air and water	10
Auxiliary buildings and yard utilities	5
Total	100

This breakdown represents a battery with coal preheat and pipeline charging facilities. A similar breakdown for a conventional battery would eliminate the percentage allocated to preheat facilities and would add a comparable percentage to the category of battery ovens with oven machinery. The investment difference between a conventional battery and preheat pipeline charged battery favors the conventional battery by a very small amount (5%). That is, the added investment cost of the preheat units and pipeline is almost equivalent to the cost of the additional ovens and larry cars needed by the conventional battery. Note that a conventional battery requires a larger number of ovens because of the longer coking time required with wet coal. From the investment standpoint, other parameters besides capital cost should be considered in making the decision between conventional and preheat batteries.

In the case of a formcoke facility, the current estimates for an annual ton of capacity range widely from \$200 to \$300 because of the uncertainty in the process scale-up and the process modifications needed to make it compatible with an integrated steel plant's requirements. The formcoke approach, which has a number of advantages, which will be discussed later, has to be proven on a commercial scale both from an operation standpoint and from the standpoint of product utilization in the blast furnace. At a given output, the cost of producing coke is most sensitive to the cost of coal and to the value of the by-products. Experience has demonstrated that the use of coal preheating can significantly improve the coking qualities of marginally coking coals. This has permitted increasing the proportion of the less expensive, more plentiful high-volatile coals in a blend for a preheat battery vis-à-vis a conventional battery. On the other hand, experience has indicated that the processing costs of a preheat battery are somewhat higher. By-product credits are comparable for preheat and conventional batteries.

Cost projections for the formcoke plant are again handicapped by the lack of a large-capacity steel plant installation. However, if successful, it is reasonable to assume that the raw material cost would be significantly lower because it is anticipated that even noncoking coals can be used to make high-quality coke briquettes. Operating cost projections vary significantly depending on assumptions relative to unit availability, repair and maintenance efforts, and supplies and utilities requirements. However, the process, as known today, produces inferior-grade by-products that may have limited value in an integrated steel plant.

Table 7 displays the relative cost index for conventional, preheat, and formcoke installations. Based on the cost comparison, the decision as to the selection of which process to use does not seem difficult since it appears that each alternative has comparable costs. This is not the case, however, and

TABLE 7 Comparison Index—Full-Cost-per-Ton-of-Coke Basis

	Conventional battery	Preheat heat battery	Formcoke facility
Capital charge	100	105	100
Raw material	200	180	150
Operating cost	65	75	85
By-product credit	(65)	(65)	(40)
Total cost index	300	295	295

the decision maker must search for significant differences. First, the real estate requirements are different and could alter the relative investment cost for any specific replacement facility. Second, the coal blend may not change relative to the facility because the company probably has established sources. That is, it should be understood that the raw material cost index (mainly coal) is based on the market price for varying blends of medium- and high-volatile coals. Individual company evaluations must take into consideration the relative reserves of each kind of coal that the company holds or controls and the ability to obtain market price if the coal is not used internally. Third, the operating cost of the coke plant for the integrated steel plant cannot be divorced from that of the blast furnace operation. The quality of the coke produced, both physically and chemically, must be evaluated to determine if one process makes significantly different-quality coke from the given coal blend. For example, a small improvement in coke stability can have significant economic benefits to the blast furnace in the form of higher productivity and lower coke rate. Finally, the quality and quantity of the by-products may be more important than the equivalent Btu value that was assumed in the index determination. If an integrated steel plant has developed a complex network of coke oven gas users, the volume and quality of the by-product gases may significantly affect the operations of other in-plant processes. Availability and cost of alternative fuels, including any investment, must be included in the economic analysis.

In conclusion, the economics of cokemaking are not simple because cokemaking is an integral and basic part of the sequential nature of steel manufacturing using virgin iron materials. The ability to completely isolate a cokemaking decision from the balance of an integrated steel company's operations is remote. This complexity requires that the analyst have a thorough understanding of all aspects of the company's operation.

ACKNOWLEDGMENTS

The authors have relied on a number of Inland people who provided vital assistance in writing and editing the material contained in this chapter. Particular thanks go to Messrs. Jerome J. Knoerzer, Dennis D. Kaegi, and Charles A. Osterman, all of the Research Department; and Mr. Thomas M. Mohr of the Facility Planning Department.

REFERENCES

1. R. J. Forbes, *Metallurgy in Antiquity*, Heinman Publishers, Leiden, The Netherlands, 1950.
2. R. A. Mott, History of cokemaking, in *Coke Oven Managers Association Yearbook*. Nexborough, Yorkshire, England, 1936.
3. J. F. Ciarimboli and W. F. Goscinski, Nonrecovery Cokemaking—Past, Present and Future, Pennsylvania Coke Technology, Inc., AISE Meeting, Chicago, Oct. 1978.
4. J. Armstrong, *Carbonization Technology and Engineering*. Lippincott Publishers, London, 1929.
5. Diendonne, Mémoire sur la fabrication du coke, Forbach et Hirschbach Annales des Mines, Paris, 1859-5 series.
6. F. M. Ress, Geschichte der Kokereitechnik, Glueckauf (1957).
7. K. Sugasawa, Y. Suwami, T. Kanoh, and M. Tsuyuguchi, The Utilization of Non-Coking Coals for Metallurgical Coke Manufacture, Agglomeration 77, SME-AIME, Vol. 2, Atlanta, Ga., 1977, pp. 955-969.
8. C. E. McMorris, Inland's Preheat Pipeline Charged Coke Oven Battery, *Ironmaking Proc. Metall. Soc., AIME* 34:330-338 (1975).
9. K. G. Beck, The Precarbon Process—Blast Furnace Coke from Preheated Coal, in *Coke Oven Managers Association Yearbook*. Nexborough, Yorkshire, England, 1936.
10. M. O. Holowaty and N. A. Robins, Coal Preheating and Pipeline Charging at Indiana Harbor, *Iron Steel Int.*, pp. 285-290 (Oct. 1978).
11. W. Hermann and R. Schmit, The ANCIT Process, Agglomeration 77, SME-AIME, Vol. 2, Atlanta, Ga., 1977, pp. 970-995.
12. W. Peters, New Techniques for the Production of Blast Furnace Coke from the Aspects of Enlarging the Coking Coal Base, Panel Discussion: Technological Challenge in Use of Non-Coking Coals, International Iron and Steel Institute Meeting, Osaka, Japan, 1976.
13. J. A. Cordier and R. Rochas, HBNPC Formcoke Tests in USINOR Blast Furnaces, Agglomeration 77, SME-AIME, Vol. 2, 1977, pp. 996-1011.
14. J. E. Barker, The Coking Industry in the USSR: Recent Development and Current Trends, *J. Inst. Fuel* 40:387-396 (Sept. 1967).
15. D. Wagner et al., DKG Formed Coke Process Operation Results and Recent Developments, *Ironmaking Proc. Metall. Soc. AIME*, 36:507-515 (1977).

16. J. Work, The FMC Coke Process, *J. Met.* 18:635-642 (May 1966).
17. P. K. Strangway and M. O. Holowaty, Blast Furnace Test with 20,000 Net Tone of FMC Formcoke at Inland's No. 5 Blast Furnace, Agglomeration 77, SME-AIME, Vol. 2, Atlanta, Ga., 1977, pp. 1012-1050.
18. D. W. van Krevelen and J. Schuyer, *Coal Science.* Elsevier Publishing Co., New York, 1957, p. 3.
19. Work Energy Conference—Survey of Energy Resources, 1974.
20. P. Averitt, Coal Resources in the United States, January 1, 1974, U.S. Geological Survey, Bulletin 1412, 1975, p. 131.
21. G. W. Land, Western Coal, Meeting of the Missouri Valley Electric Association, Kansas City, Mo., Apr. 7-9, 1976.
22. N. Schapiro and R. J. Gray, Petrographic Classification Applicable to Coals of All Ranks, *Proc. Ill. Min. Inst.* 68:83-97 (1960).
23. R. O. Wenger and V. A. Neubaum, The Effects of Coke Plant Operating Variables on the Quality of Metallurgical Coke, *Ironmaking Proc. Metall. Soc. AIME* 28:170-179 (1969).
24. H. Echterhoff, K. G. Beck, and W. Peters, Size Distribution, Strength and Reactivity of Coke and How They Are Affected by the Coking Process, Proceedings of the Blast Furnace, Coke Oven and Raw Materials Conference, 1961, pp. 403-418.
25. N. Schapiro and R. J. Gray, Relation of Coke Structure to Reactivity, *Blast Furn. Steel Plant* 51:273-280 (Apr. 1963).
26. L. D. Miller, R. W. Bouman, and M. C. Chang, Factors Affecting Coke Reactivity in Blast Furnaces, *Blast Furn. Steel Plant* 53:381-388 (May 1965).
27. T. A. Daly and C. F. Budge, Reactivity of Chars from New Zealand Coals, *Fuel* 53:8-11 (Jan. 1974).
28. Y. K. Rao and B. P. Jalon, A Study of the Rates of Carbon-Carbon Dioxide Reaction in Temperature Range of 839° to 1050°C, *Metall. Trans.* 3:2465-2477 (Sept. 1972).
29. M. T. Mackowsky, Nouvelles méthodes d'évaluation de la qualité de coke métallurgique, La Documentation Métallurgique, No. Special SK 1, Feb. 15, 1956, pp. 79-90.
30. R. R. Thompso, A. F. Mantione, and R. P. Aikman, Improvement of Coke Uniformity Through Measurements of Coke Reactivity, *Blast Furn. Steel Plant* 59:161-164 (Mar. 1971).
31. C. Arnu, Congr. Assoc. Tech. Gas., Paris, 1959, pp. 281-314.
32. A. Dahme, *Ing. Arch.* 21:346-51 (1953).
33. H. Joh, Discussion on Measurement of Coke Strength by Japanese Industrial Standard (JIS) Method, *Blast Furn. Steel Plant* 54:610-617 (July 1966).
34. C. G. Thibaut, *Qualités des Cokes Métallurgiques*, 2nd ed. CESSID, Metz, France, 1960, pp. 1-84.
35. C. G. Thibaut, Solid Products of Carbonization, in *Chemistry of Coal Utilization*, H. H. Lowry, ed. John Wiley & Sons, Inc., New York, 1963, pp. 494-538.

36. M. W. Lightner, Burdening the Blast Furnace—Ore, Coke Furnace Operation, *Steel* 136(6):116, 118, 120, 122, 128, 131; 136(7):92, 95-96, 101; 136(8):82, 84, 87; 136(9):96, 98, 102 (1955).
37. M. O. Holowaty and C. M. Squarcy, Blast Furnace Coke Tested at High Temperature, *J. Met.* 9:577-581 (1955).

10 Coal Gasification

WILLIAM W. BODLE and JACK HUEBLER / Institute of Gas Technology, Chicago, Illinois

I. INTRODUCTION

Gaseous fuels have been produced from coal for residential, commercial, and industrial users for more than 100 years. After a waning interest due to the advent of plentiful natural gas in the post-World War II era, coal gasification is again receiving attention.

Manufactured gas is classified by heating value.* High-Btu gas is synonymous with substitute natural gas- and pipeline-quality gas having a heating value in the range 970 to 1000 Btu per standard cubic foot (scf). Its composition is mainly methane, with very small amounts of carbon monoxide (CO), carbon dioxide (CO_2), and nitrogen (N_2). High-Btu gas is generally interchangeable with natural gas and can be manufactured from coal on a large scale in centralized plants and added to the natural gas pipeline system.

Medium-Btu gas has a calorific value between 270 and 600 Btu/scf. At the lower end of this range the gas consists primarily of CO and hydrogen (H_2) with a small amount of CO_2. At the higher end of the range the heating value is increased by inclusion of methane (CH_4) or other hydrocarbons. Medium-Btu gas can be very useful in the manufacturing industries because it burns rapidly and produces a flame temperature as high or higher than natural gas. This gas, however, cannot be put into the natural gas distribution network because it is not interchangeable with natural gas and because of its high carbon monoxide content. These objections can be overcome in certain ways, but the main use for such gas appears to be for industrial complexes or large manufacturing plants.

Low-Btu gas will normally have a heating value in the range 90 to 150 Btu/scf. The combustible components consist of CO and H_2 diluted by CO_2 and N_2. This gas has a low flame temperature unless the combustion air is strongly preheated. Its use is relatively limited.

The two principal routes for converting coal to gas are shown in Fig. 1. They differ in the method of supplying heat for gasification. In the first case (top route in Fig. 1), heat is supplied by combustion with air directly in the process, so that air nitrogen is mixed into the product gas. The ultimate product is low-Btu gas.

Alternatively, heat may be supplied by other means that do not introduce nitrogen into the product gas. This is shown in the lower lines of Fig. 1, where oxygen is used, heat is supplied through indirect heat transfer from hot gases or solids that may or may not take part in the gasification reactions, or a hydrogasification route is followed where gasification heat is supplied by the exothermic reaction of hydrogen with carbon and carbon oxides to form methane. These are the routes for production of medium-Btu gas, high-Btu gas (SNG) hydrogen, and synthesis gas. In addition to air or oxygen, steam or water are generally introduced as supplementary gasification agents, as moderators to exothermic heats, or to supply the quantity of hydrogen required in the final product.

---

*Different heating values are due to different compositions in respect to methane (1012 Btu/scf), carbon monoxide (322 Btu/scf), hydrogen (325 Btu/scf), and the noncombustible constituents such as nitrogen and carbon dioxide.

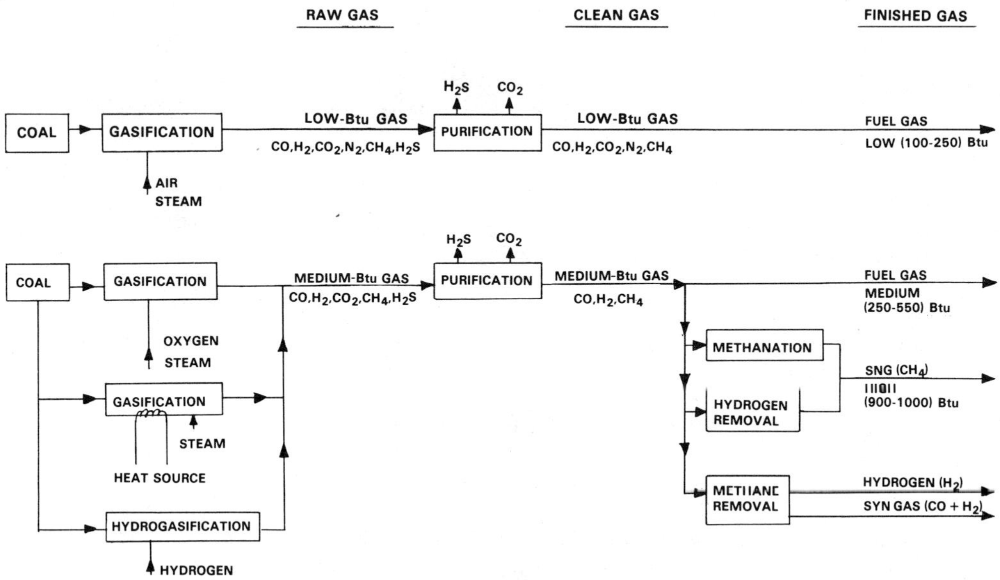

FIG. 1 Coal-to-gas process steps.

Raw gas from coal gasification produced at 250 to 3000°F and containing all the elemental constituents of the coal feed (principally carbon, hydrogen, oxygen, sulfur, and nitrogen) may be used directly if environmental regulations permit this usage. For example, coal containing a low-sulfur content might be used directly, but, more likely, the gas will have to be treated for at least the removal of sulfur compounds and solid particulates either for environmental considerations or for the benefit of downstream equipment. CO_2 removal at this stage will depend on the amount allowable to produce the desired product and will occur as a natural consequence of sulfur removal.

Clean medium-Btu gas can be used directly as fuel or through further processing can be upgraded to SNG, hydrogen, or syngas as the finished product.

II. BASIC CONSIDERATIONS

Coal contains several constituents that must be removed or substantially altered in order to obtain an acceptable fuel gas. Heat must be added and removed at various process stages. Finally, since coal is a solid, special techniques must be employed to feed it into a reactor, to contact it with reacting gases, and to remove ash. The three sections to follow present the chemical, thermal, and mechanical considerations that are basic to all coal gasification processes.

A. Chemistry of Coal Gasification

Coal is chemically a very complex solid substance that primarily consists of carbon (C) with relatively small amounts of hydrogen (H), oxygen (O), nitrogen (N), sulfur (S), and ash. To transform it to fuel gas requires conversion of the C to CO at one extreme or methane (CH_4) or other volatile hydrocarbons at the other extreme. Two of the oldest examples are still practiced in the steel industry and can serve as examples of established technology.

Coke oven gas is produced by heating coal to a moderate temperature so that the C, H_2, and O_2 of the coal are converted to liquids and gas mixtures containing H_2, CO, CH_4, and illuminants (unsaturated higher hydrocarbons). The gas typically has a heating value of 550 Btu/scf and is therefore a medium-Btu gas. The major constituents are, on a volume basis, H_2 (55%) and CH_4 (28%). The majority of the coal is left as coke (mostly C) which contains the ash. Only about 20% of the coal's heating value appears as gas. In the steel mill, coke is the principal product because it is needed in the blast furnace. The gas is a by-product that is a valuable plant fuel. The gas can be the desired product elsewhere if use can be found for the coke and the liquids that are also produced. The temperature and mode of carbonization can be varied to produce more liquids than gas if desired, as in the COGAS Process.

The coke is burned with a deficiency of air in the blast furnace to produce the heat required to make and melt iron.

$$C + 1/2\ O_2 + 2N_2 \rightarrow CO + 2N_2$$

Some of the C is also used to reduce the iron oxide to iron. This partial combustion produces blast furnace gas having a heating value of about 90 Btu/scf (a low-Btu gas) that contains, on a volume basis, CO (27%), a small amount of H_2 (1%), highly diluted with N_2 (60%), and CO_2 (12%). Gas of this sort can be produced in many other types of apparatus by burning coal in a deficiency of air. In the blast furnace, the ash and sulfur are easily disposed of in the slag. This disposal constitutes a major problem in other low-Btu gas processes, although the processes that have the specific purpose of converting coal to low-Btu gas can produce a better gas than blast furnace gas.

There are a large variety of producer gas generators that partially oxidize coal with air. Steam (H_2O) is usually added to the air to moderate the excessive temperature increases caused by the exothermic C oxidation. Steam reacts with C to form CO and H_2, which absorbs heat (endothermic reaction). A typical gas might contain 4% CO_2, 28% CO, 14% H_2, 3% CH_4, and 51% N_2. The heating value is typically 175 Btu/scf.

Simple inspection of the foregoing gas analysis shows that elimination of the N_2 would double the heating value to 350 Btu/scf and produce a medium-Btu gas. In past days, this was accomplished in "blue gas" or "water gas" sets. Coal was heated to incandescence by burning part of it with air. The air was turned off and steam blown through the hot coal forming CO and H_2, thus producing an N_2-free gas:

$$C + H_2O \rightarrow CO + H_2$$

More modern processes employ O_2 instead of air in the producer-type process, thus eliminating the N_2 and increasing the heating value.

In all the foregoing processes, coal was gasified primarily by conversion to CO by reaction with O_2 or by H_2O. Carbon can also be gasified by reaction with hydrogen to form methane:

$$C + 2H_2 \rightarrow CH_4$$

This hydrogasification was initially studied by Dent in England and later highly developed in the 1950s by the Institute of Gas Technology. The use of pure H_2 was abandoned because of its high cost and uncontrollable exothermicity in reaction with carbon. The process is periodically reactivated from time to time, but with little success. The hydrogasification reaction is, however, a key element in many processes that produce high-Btu gas.

In addition to hydrogasification of C to CH_4, CH_4 can also be formed by the catalytic combination of CO and H_2:

$$CO + 3H_2 \rightarrow CH_4 + H_2O$$

In this reaction the products are CH_4 and H_2O. The reaction is commonly referred to as methanation. Most high-Btu gas processes employ a combination of direct hydrogasification and indirect methanation.

B. Thermal Considerations

All coal gasification processes require the input of heat at high temperature levels. Many processes, even though the products they produce are the same, vary in the manner in which this heat is introduced. Basically, the methods employed are combustion of C with O_2 contained in air, combustion of C with O_2 separated from air, and by indirect means with a heat carrier heated in a separate step by combustion of carbon with air:

$$C + O_2 + N_2 \rightarrow CO_2 \pm N_2 \quad \text{or} \quad C + O_2 \rightarrow CO_2$$

The direct use of air results in the production of low-Btu gas diluted with nitrogen. High- and medium-Btu gas generally require an air separation plant or an "external" heat-input step. Exceptions to this are the medium-gas processes that employ pyrolysis, such as in the coke oven.

Much of the earlier work in process development was highly influenced by a desire to eliminate the expensive air separation plant. Improvements in O_2 production have reduced much of this incentive and the majority of the modern processes for medium- and high-Btu gas are directly fired with relatively pure O_2 using H_2O as a temperature control with the added benefit of internal H_2 production from the H_2O:

$$2C + 1/2\ O_2 + H_2O \rightarrow 2CO + H_2$$

Several examples of indirect heat addition (basically using air) are described later. Although none of these have reached the commercial stage, HYGAS Electrothermal, HYGAS Steam-Iron, and the CO_2 Acceptor Processes have been technically proved at the large-pilot-plant scale.

C. Mechanical Factors

The extensive variety of coal gasification processes that are being developed are all influenced to a large extent by the mechanics of:

Feeding solid coal into reactors (some at very high pressure)
Contacting the coal with reactant gases
Removing solid or liquid ash
Collecting and handling of fine partially reacted dust (carried out with the gaseous products)

1. Solids Feeding

There are a large number of solids feeding devices that have been developed over the years for low-pressure gasification applications. The special problems involved in coal gasification processes presently under development arise from the high pressures involved. Low-pressure (15 psig and below) processes use feeding devices such as screws and star valves (see Fig. 2). These are often used in newer high-pressure processes for internal transfer of solids between stages within a process vessel.

Special attention will be paid to the two leading methods of feeding into and out of high-pressure vessels: the lock hopper and the slurry feeder.

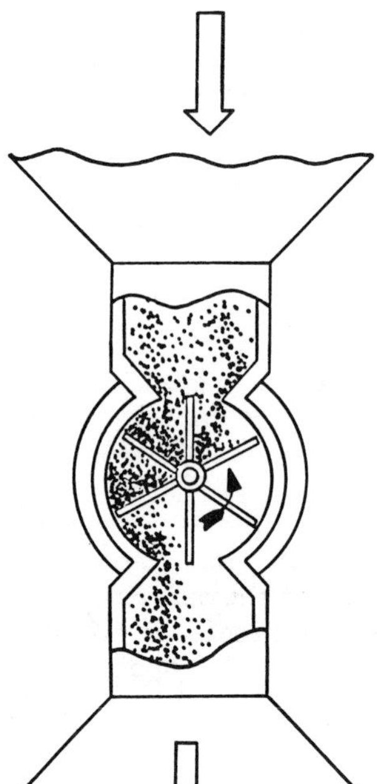

FIG. 2 Star feeder.

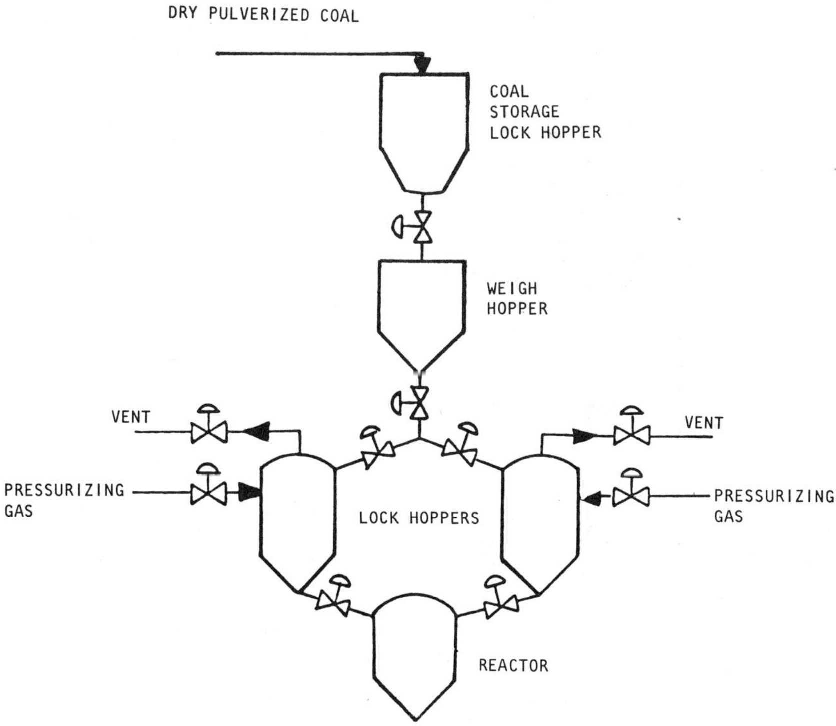

FIG. 3 Continuous lock hopper system.

Coal at atmospheric pressure is dumped from a storage tank into a vessel (the lock hopper; see Fig. 3) open at the top and sealed with a valve at the bottom. A valve at the top is then closed and the vessel is pressurized by admitting cold gas into the vessel until the pressure is equal to reactor pressure. The bottom valve is then opened and the coal introduced into the reactor. The coal usually falls by gravity into the reactor if the process is downflow, or into an intermediate hopper from which it can be screwed or blown into the reactor, depending on the circumstances. When the lock hopper is empty, the bottom valve is closed, the top valve is opened, and the cycle is repeated. The gas in the hopper is generally lost during this operation. If continuous feed to the reactor is desired, two parallel lock hoppers are employed, as shown in Fig. 3. Two or more hoppers can be used in series to save compression energy and gas but this has not been done. Lock hoppers have not been used above about 400 psi.

Slurry feed systems have been used commercially to convey coal from a mine to a power plant. In addition, they have been used at pressures up to 1100 psi in a large pilot plant. The system (Fig. 4) consists of an atmospheric pressure hopper constantly refilled with liquid (water or light oil) and finely ground coal. The coal constitutes about 50% by weight of the mixture. The coal is kept in suspension (slurry) by suitable agitation.

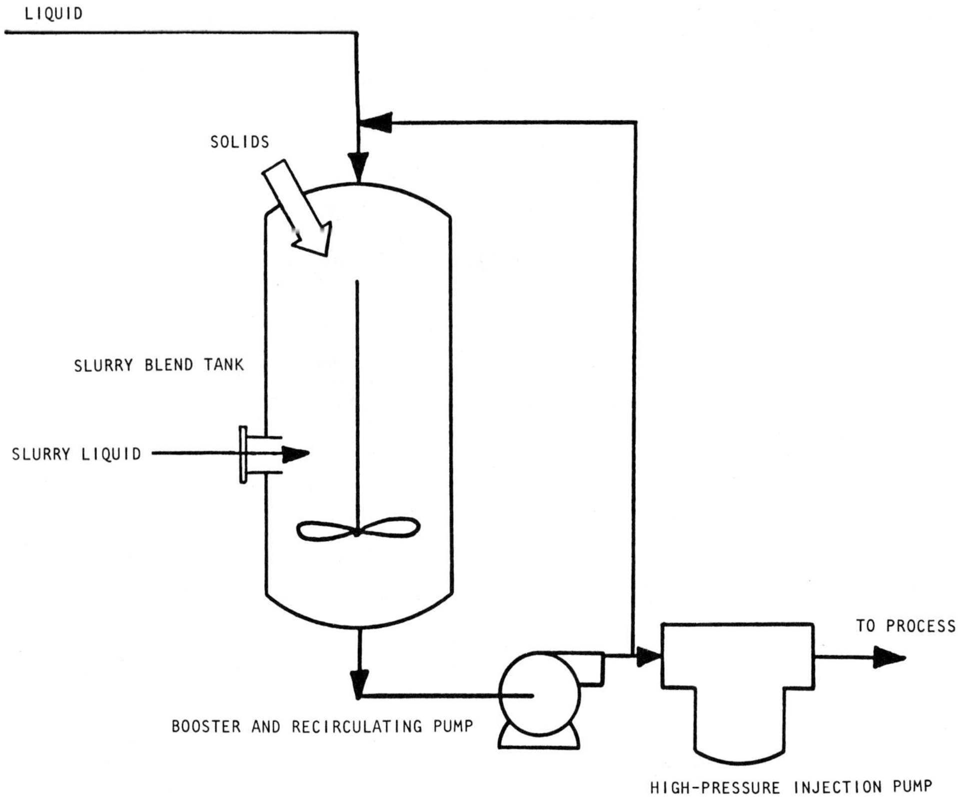

FIG. 4 Slurry feed system.

The slurry is pumped constantly at atmospheric pressure by a centrifugal pump through a pipe loop that passes the inlet to a reciprocating pump and is then returned to the slurry vessel. When the reactor is activated, the reciprocating pump (a conventional oil field mud pump) compresses the slurry to somewhat above reactor pressure. The slurry then feeds through a pipe into a drying zone in the reactor system where the water or oil is vaporized before the dry coal enters the reaction vessel. This liquid is recovered from the gas produced in the reactor and reused.

Pressure feeding devices such as extruding a mud paste and improved star valves are under investigation, but none are near the stage where they can be used.

2. *Contacting for Reaction*

Contacting the solid coal with reactant gases to accomplish the required gasification is a second major mechanical problem. There are three types of contacting methods: a descending bed of solids with upflowing gas, a fluidized bed of solids, and entrained solids.

The descending-bed-of-solids system is often referred to as a moving or fixed bed (Fig. 5). Lumps of coal (usually 1/8 to 1 in. in diameter) are laid down at the top of a vessel while reactant gases (assume O_2 and H_2O) are introduced at the bottom of the vessel and flow at relatively low velocity upward through the interstices between the coal lumps.

As the coal descends it is reacted first by devolatilization using the sensible heat from the rising gas, then hydrogenated by the hydrogen in the reactant gas, and finally burned to an ash. The reactions are, therefore, carried out in a countercurrent fashion.

The fluidized-bed system requires coal to be ground to about 8 mesh or less (Fig. 6). The reactant gases are introduced through a perforated deck near the bottom of the vessel. The volume rate of gas flow is such that its velocity (1 to 2 ft/sec) is high enough to suspend the solids but not high enough to blow them out of the top of the vessel. The result is a violently "boiling" bed of solids having very intimate contact with the upward-flowing gas. This gives a very uniform temperature distribution. The solid flows rapidly and repeatedly from bottom to top and back again, while the gas flows rather uniformly upward. The reactor is said to be completely backmixed; no countercurrency of reaction sequence is possible. If a degree of countercurrency is desired, two or more fluid-bed stages are placed one above the other. Reaction rates are faster than in the moving bed because of the intimate contact between gas and solids and the increased solids surface area due to the smaller particle size.

The entrained flow reactor (Fig. 7) uses a still finer grind of coal (80% through 200 mesh), fine enough that it can be conveyed pneumatically by the reactant gases. Velocity of the mixture must be about 20 ft/sec or higher, depending upon the fineness of the coal. In this case, there is little or no mixing of the solids and gases, except when the gas initially meets the solids.

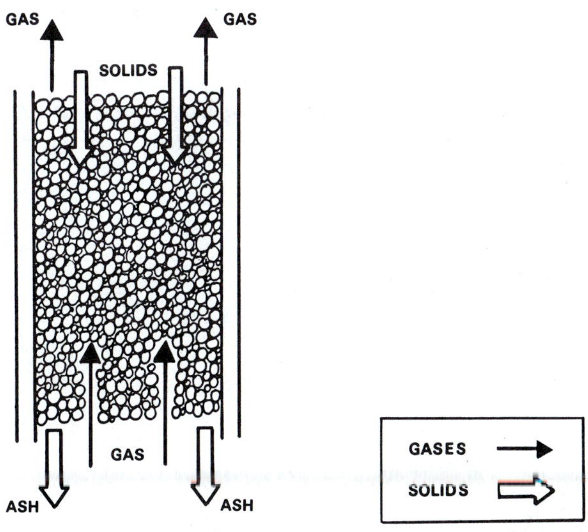

FIG. 5 Countercurrent descending-bed contactor.

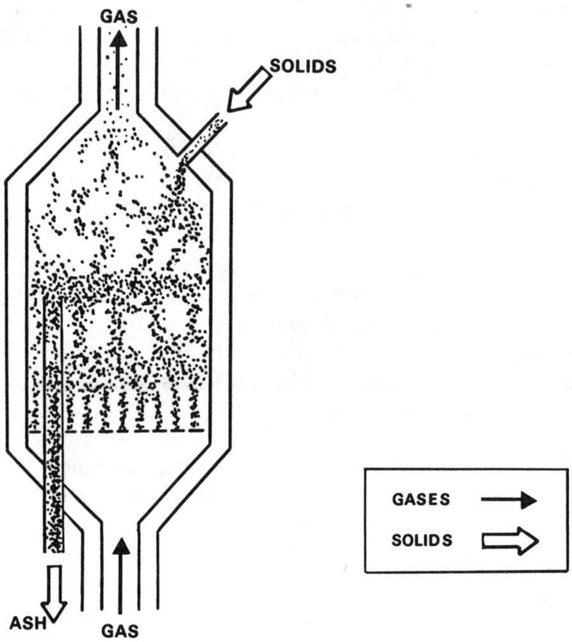

FIG. 6 Fluidized-bed contactor.

The reactions occur in a cocurrent fashion. This type of reactor is only used for very rapid reactions and usually for either combustion in O_2 or the initial reaction of fresh coal and H_2.

3. *Ash Removal*

Ash is withdrawn from reactors in essentially the same way that coal is introduced: as a slurry or in lock hoppers. If the ash is molten, it is quenched in water to break it up before discharge.

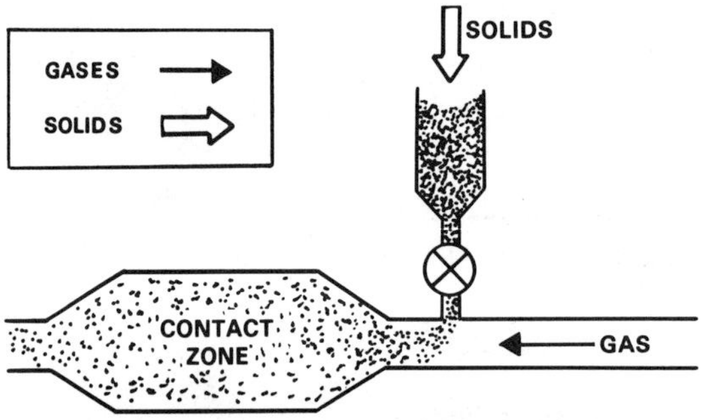

FIG. 7 Entrained flow contactor.

III. GASIFIER DESCRIPTIONS

A. Descending-Bed Gasifiers

Coal is fed to the top of a descending-bed gasifier and moves downward countercurrent to a stream of gas that originates by introduction of air (or O_2) and steam at the bottom of the gasifier. Coal feed is generally cold or at moderate temperature. Steam and air may or may not be preheated. Downward movement of coal, generally at a rate of about 1 to 10 ft/hr, is regulated by the control of ash removal from the bottom of the gasifier. A bed depth of about 10 ft is maintained so that the total residence time of coal through the gasifier is 1 to 10 hr, with the length of time generally decreasing with increasing pressure of operation in the range atmospheric to 500 psi.

As coal descends through the gasifier it is heated by ascending gas from the combustion zone at the bottom of the vessel. This leads to temperature distributions for the coal throughout the length of the gasifier, as shown in Fig. 8. The precise values of temperatures will depend on various conditions such as gasifier pressure and coal reactivity. The relative proportions of steam and air introduced are very important because the steam is used to regulate the maximum temperature reached in the gasifier. This determines whether the ash will be in a dry or molten state. Production of dry ash is accomplished by adding enough steam so that endothermic steam decomposition reactions are adequate to offset exothermic combustion reactions in the combustion zone to the extent that the ash-fusion temperature is not reached. On the other hand, if a molten ash (slagging) operation is desired, the steam rate is reduced considerably.

The temperature changes of the coal as it descends through the gasifier cause changes to take place in the coal, as depicted in Fig. 9. The zones are not distinct as shown but will blend more or less gradually into each other. The drying zone at the top of the gasifier effects evaporation of water. When the temperature rises above 600°F volatile matter consisting

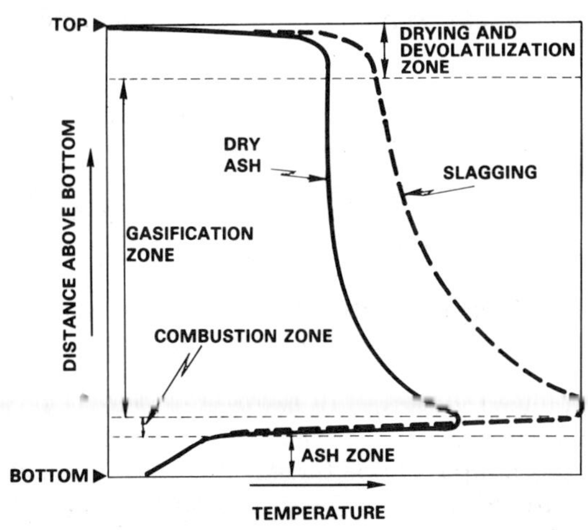

FIG. 8 Temperature distribution in a descending-bed coal gasifier.

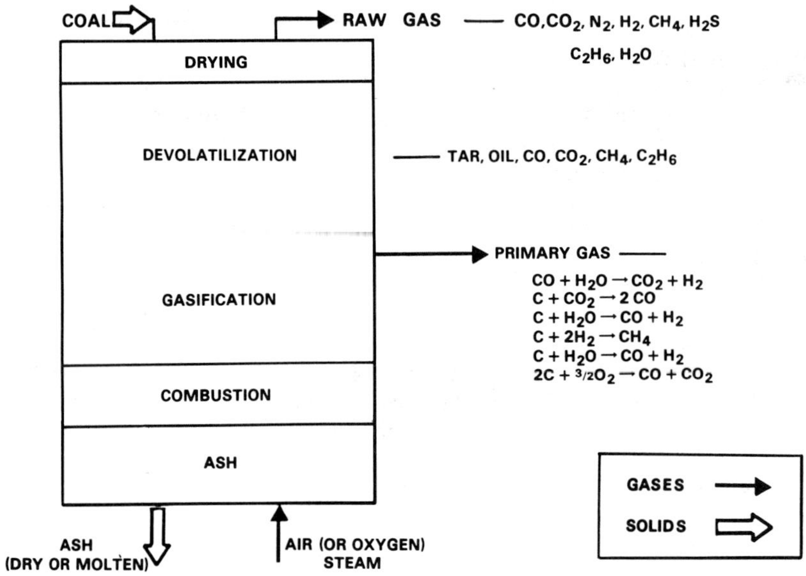

FIG. 9 Zones in descending-bed coal gasifier.

of oils, tars, and gases starts to evolve. This is followed by a region where gasification of the char resulting from devolatilization takes place by reaction with gases and steam moving up from the combustion zone. In the combustion zone, the remaining char is burned and reacted with feedsteam. Finally, the ash is cooled by entering air and steam.

Because of the countercurrent contacting of gas and feed coal at the top of the gasifier, gases are cooled and exit the gasifier at temperatures in the range 500 to 1000°F. This is too low a temperature to effect any appreciable reaction between gas or steam and the tars and oils evolved during devolatilization. Consequently, the raw product gas contains appreciable quantities of these tars and oils; unless these are removed, they will cause problems because of condensation in downstream utilization of the gas. To facilitate this cleanup some descending-bed gasifiers are operated in a two-stage mode whereby a portion of the gas is removed directly from the gasification zone (primary gas of Fig. 9) and only the remainder, sufficient to heat the incoming coal, moves through the devolatilization and drying zones. This provides one stream of gas free of tars and oils. Cleanup of the top gas is also easier because of the reduced volume.

In addition to alternatives of dry or molten ash and one- or two-stage operation, descending-bed gasifiers may be designed for essentially atmospheric or elevated pressure operation, for oxygen or air injection, and with or without provision to handle agglomerating coals. Additionally, various gasifier designs generally have their own specific method of coal feed and ash removal. Table 1 summarizes the unique characteristics of the principal descending-bed gasifiers presently in existence. These gasifiers are described in more detail in the following paragraphs.

1. *Dry Ash: Single-Stage*

a. GEGAS (General Electric Co., Schenectady, New York) [1] This descending-bed gasifier is designed to handle high-caking and swelling coals by provision of three stirrer paddles on a shaft that enters through the top of the gasifier and can move vertically to provide stirring from the very top down to the vicinity of the grate (see Fig. 10). The 5-ft-OD × 24-ft-high experimental unit shown has a 285 grade C steel shell lined with two layers of low-iron castable refractory, leaving an open shaft of 36 in. ID. There is a 3 3/4-in.-thick hard castable on top of an 8 3/4-in.-thick insulating layer. The experimental gasifier is designed for up to 350 psi operating pressure.

Coal feed to maintain a 4- to 12-ft bed depth is continuous by means of a horizontal auger. The auger is fed in 9-ft^3 batches from a coal lock hopper. Fines in the coal are screened out and blended with by-product tar in an

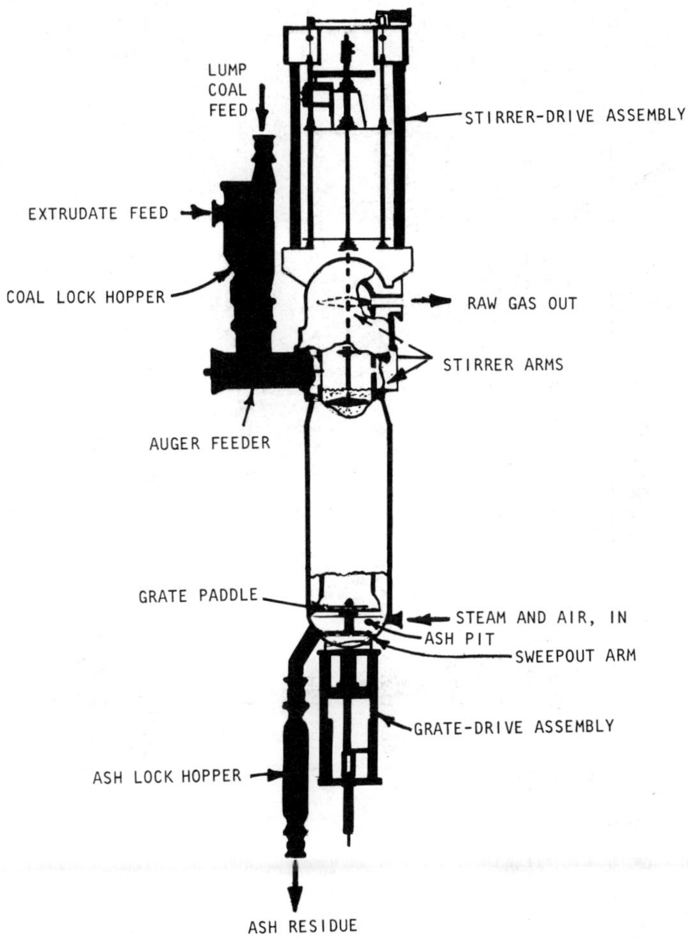

FIG. 10 GEGAS gasifier, 36 in. ID [1].

TABLE 1 Characteristics of Descending-Bed Gasifiers

Number	Operating pressure	Oxygen or air	Coal feed	Ash removal	Agglomeration prevention	Status	References
			Dry ash, single-stage				
GEGAS	To 500 psig	Air	Auger	Lock hopper	Stirrer paddles	5-ft-diameter test unit	1
Lurgi	To 450 psi	Oxygen to air	Lock hopper	Lock hopper	Rotating blades	Commercial	2-9
MERC	To 105 psi	Air	Lock hopper	Lock hopper	Spiraling stirrer	42-in-diameter test unit	10
Riley-Morgan	40 in. H_2O	Air	Lock hopper	Rotating ash pan	Agitator in rotating bed	10.5-ft test unit	11,12
Wellman-Galusha	10 in. H_2O	Air	Valves	Revolving grate	Spiraling arms	Commercial	13,14
Wilputte	Atm	Air		Water-sealed ash flow	Rotating arm	Commercial	15,17

			Dry ash, two-stage				
ATC/Wellman (Incandescent)	Atm	Air	Drum	Rotating grate	None	Commercial	18
FW/STOIC	Atm	Air	Rotary feeder	Water-sealed rotating ash pan	None	Commercial	19,20
Ruhr-100	1500 psi	Oxygen	Lock hopper		Stirrer blades	3-7-ton/hr test unit	21
Woodall-Duckham	40 in. H_2O	Air or oxygen	Lock hopper	Rotating grate	None	Commercial	22,23
			Slagging				
BGC/Lurgi	To 400 psi	Oxygen	Lock hopper	Lock hopper	Stirrer	6-ft-diameter test unit	24,25
GFERC	To 400 psi	Oxygen	Lock hopper	Lock hopper	Stirrer	16-in.-diameter test unit	26

TABLE 2 Operation of GEGAS Experimental Gasifier at 225 psig with Pittsburgh No. 8 Coal [1]

Feed quantities (lb/hr)	
Coal	1320
Air	3470
Steam	594
Ash quench water	5
	5385
Product quantities (lb/hr)	
Dry Gas	4790
Water	271
Dust	16
Residue	190
Tar and oil	58
	5320
Raw gas composition (vol %)	
CO	22.1
H_2	15.8
CH_4	3.0
CO_2	6.2
N_2	45.7
O_2	0.1
H_2O	7.2
	100.0

extruder that is being developed separately. The extruder feeds directly to the reactor at pressure without a lock hopper. At the bottom of the gasifier, a grate is provided with a rotating agitator paddle that can also be moved vertically. A rotating sweep-out arm on the same shaft pushes ash from the pit into the ash lock hopper.

Preheated (400°F) air enters through the center and over the edges of the grate. Sufficient steam is mixed with the air to keep the maximum bed temperature near or below the ash-fusion point. Owing to the deep bed agitation provided by the stirrer, excess steam is reduced so that the steam/air weight ratio required can be less than 0.2, which is only 30% of that normally required without agitation. Thermocouples are installed along two sides of the experimental gasifier wall for measuring bed temperatures at 1-ft intervals.

Crushed run-of-mine Pittsburgh No. 8 and Illinois No. 6 coals have been run in the experimental unit at 300 psig for periods of up to 40 hr. Fines carryover is about 1% of coal feed at the achieved designed rate of 1 ton/hr. Pertinent feed and product rates of 225-psig operation are shown in Table 2.

b. Lurgi (Lurgi Mineraloltechnik GmbH, Frankfurt-Main) The Lurgi gasifier (Fig. 11) is designed for operation at pressures up to 450 psig. Coal is fed through a lock hopper mounted on top of the gasifier. A rotating distributor provides uniform distribution of coal across the bed. When processing caking coals, blades attached to the distributor rotate within the bed to break up agglomerates. A revolving grate supports the bed at the bottom and serves as a distributor for steam and oxygen. Residue solid is removed at the bottom of the gasifier through an ash lock hopper. The entire gasifier vessel is surrounded by a water jacket in which high-pressure steam is generated.

Crude gas leaves from the top of the gasifier at a temperature between 550 and 1100°F, depending on the type of coal fed. The composition of gas also depends on the type of coal feed and is notable for the relatively high methane content as contrasted to gases produced at lower pressures or higher temperatures. Typical compositions of raw gas from various types of coals are shown in Table 3. Also shown are oxygen, steam, and water requirements and by-product production rate (hydrocarbons, liquids, and ammonia). Raw gases, such as shown, can be treated for removal of CO_2 and hydrocarbons to provide synthesis gas for ammonia, methanol, and synthetic oil production, or with CO_2 removal to provide a gas suitable for methanation to SNG. In any of these cases the synthesis takes place at high pressure, which puts the Lurgi gasification process at an advantage over low-pressure gasifiers.

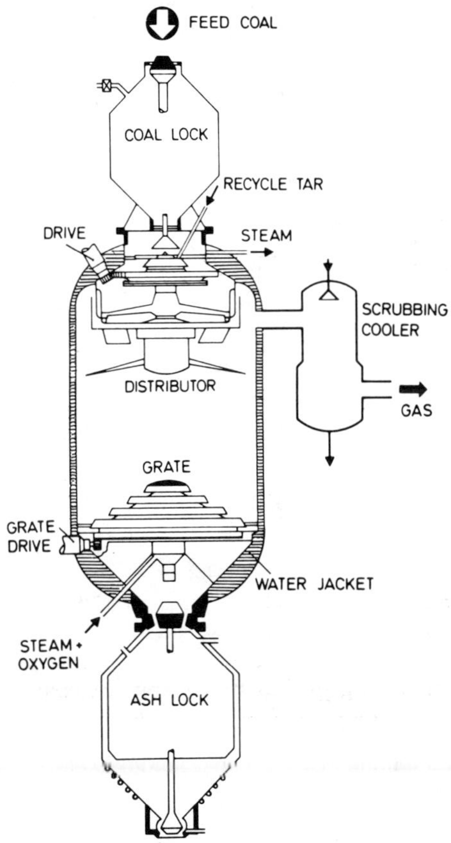

FIG. 11 Lurgi pressure gasifier [2].

TABLE 3 Consumption and Production Figures for Lurgi Gasifiers [3]

Coal type	Lignite	Subbituminous coal	Low-volatile coal
Coal analysis (wt%)			
Ash	3	27	22
Moisture	18	8	3
Related to ash- and moisture-free basis (wt%)			
Volatiles	59	39	9.7
Tar	12	15	2.5
Normal calorific value (kcal/kg)	6471	7430	8440
Consumption and production figures per 1000 n-m^3 raw gas (related to gasifier outlet)			
Gas analysis (vol %)			
CO_2	31.9	28.2	26.5
CnHm	0.5	0.3	0.1
CO	17.4	20.6	21.4
H_2	36.4	39.6	43.5
CH_4	13.5	10.5	8.0
N_2	0.3	0.8	0.5
	100.0	100.0	100.0
Consumption figures			
Coal (ash- and moisture-free) (kg)	657	465	350
Oxygen (N-m^3)	107	140	150
Steam (kg)	600	660	697
Feedwater (kg)	80	160	200
Temperature at gasifier outlet			
°C	290	490	550
°F	553		1020
Producer gas still contains:			
Water vapor (kg)	539	540	525
Tar, oil, light naphtha, etc. (kg)	59	44	4
NH_3 (kg)	6	5	0.4
H_2S and organic sulfur	Depending on sulfur content of coal		

TABLE 4 Lurgi Coal Gasification Plants [4-7]

Date	Location	Gasifiers Number	Gasifiers Diameter (ft)	Service	Capacity (10^6 scf/day)
1936	Hirschfelde, Germany	2	—	Town gas	1
1940	Böhlen, Germany	5	8.3	Town gas	9
1943	Böhlen, Germany	5	8.3	Town gas	10
1944	Most, Czechoslovakia	3	8.3	Town gas	7.5
1949	Most, Czechoslovakia	3	8.3	Town gas	9
1954	Sasolburg, South Africa	10	12.2	Synthesis gas	130
1955	Dorsten, Germany	6	8.75	Town gas	55
1956	Melbourne, Australia	6	8.75	Town gas	22
1957	Daud Khel', Pakistan	2	8.75	Synthesis gas	4.1
1960	Westfield, Scotland	4	8.75	Town gas	25
1963	Coleshill, England	5	8.75	Town gas	45
1963	Naju, Korea	3	10.5	Synthesis gas	15
1966	Sasolburg, South Africa	3	12.2	Synthesis gas	60
1969	Lünen, Germany	5		Fuel gas	160
1978	Sasolburg, South Africa	3	13.1	Synthesis gas	—
1978[a]	Sasolburg, South Africa	1	16	Synthesis gas	—
1979[a]	Secunda, South Africa	36	13.1	Synthesis gas	1500
1982[a]	Taiyuan, China	4	13.1	Synthesis gas	100
1983[a]	Secunda, South Africa	36	13.1	Synthesis gas	1500

[a]Under construction.

Table 4 lists all Lurgi gasifiers that have been or are being installed (late 1978). The only ones presently in operation are the 16 units at Sasolburg [5] and the 5 units at Essen (Lünen) [6]. The former produce synthesis gas for oil production and the latter a fuel gas that fires a combined cycle electric generating plant. The Essen plant is the first Lurgi unit to use air rather than oxygen. As to the three plants under construction, the Secunda, South Africa, plants will produce gas from coal for the manufacture of 2 million metric tons of liquid fuel each and the Taiyuan, China, plant will produce 1000 metric tons/day of ammonia.

The rated production capacity of a given-size Lurgi gasifier has shown a continual increase with time (see Table 4). For example, the ten 12-ft-diameter gasifiers installed at Sasolburg in 1954 were rated at 13 million

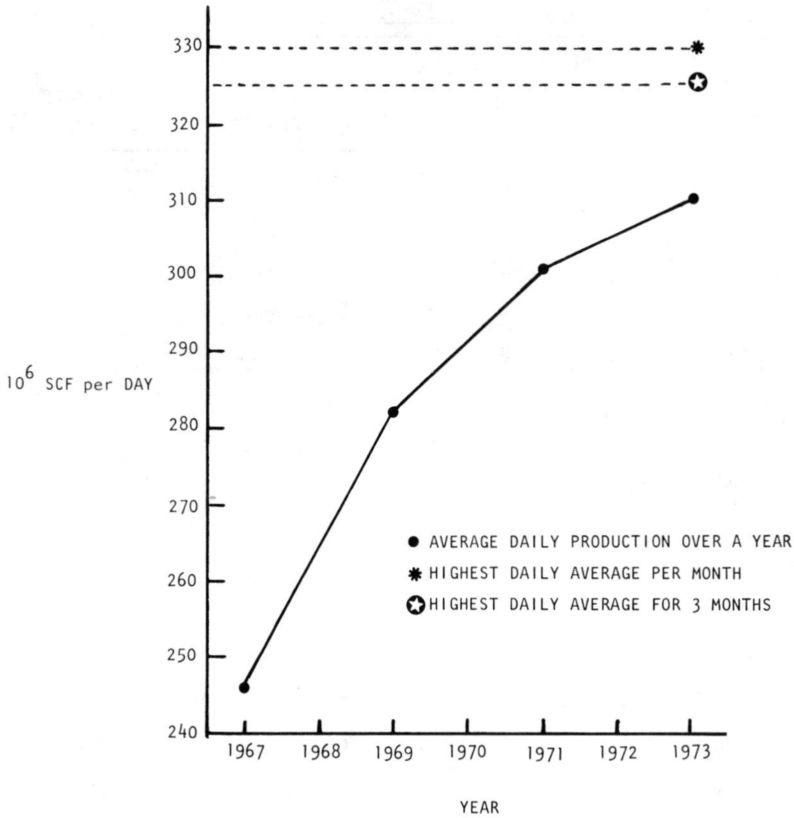

FIG. 12 Raw gas production at Sasol I [5].

scf/day, while the three installed in 1966 were rated at 20 million each. In 1973, all 13 produced an average of 310 million scf/day for an average of about 24 million scf/day each (Fig. 12). This continual increase is attributed to continually improved gasifier operation.

The latest design of the Lurgi gasifier achieves a raw gas production of 55 million scf/day and more. Three gasifiers of this type are already in operation at Sasol, and 76 gasifiers of this type have been installed or will be installed in plants presently under construction.

Recent attention has been directed toward the use of Lurgi gasifiers for plants to produce SNG in the United States. The status of four such projects is shown in Table 5. A $225 million project [9] to start construction in 1979 of a Lurgi gasifier system to produce fuel gas for a combined-cycle electric generating plant has been postponed.

c. MERC (Morgantown Energy Research Center, Morgantown, West Virginia) Coal feed at up to 2 tons/hr to the 42-in.-ID stirred-bed MERC reactor (Fig. 13) is through lock hoppers for up to 300 psi operating pressure. A nominal 79-in. bed height is maintained. The water-cooled stirrer (Fig. 14)

TABLE 5 Status of Commercial and Demonstration High-Btu Coal Gasification Projects [8] (as of August 1979)

Controlling Company	Site	Process	Coal feed (tons/day)	Peak output $M(10^6\ ft^3/day)$	Status
Commercial projects					
American Natural Resources Co., Tenneco, Peoples Gas Co., Columbia Transmission Corp. & Transcontinental Gas Pipeline Corp.	Beulah-Hazen area, Mercer County, N.D.	Lurgi gasification with methanation	12,328	137.5	The plant will be built in two phases. The first phase is scheduled to be operational by the end of 1983 and will be half the size of a full commercial plant. Plant costs are estimated at $890 million (1978 dollars) with another $83 million for a share of the mine. Output from the plant is projected to cost $5.66/1000 ft^3, not including transmission and distribution costs. The project is awaiting a final FERC decision.
El Paso Natural Gas Company	Four Corners area, N.M.	Lurgi gasification with methanation	7,600	72	El Paso Natural Gas Company plans to construct and operate a quarter-size plant on the Navajo Indian Reservation. A joint partnership with Rhurgas AG of West Germany is under consideration. A new application is expected to be filed in 1979.
Panhandle Eastern Pipeline Company	Eastern Wyoming	Lurgi gasification with methanation	13,850	137.5	Plans for a half-size plant remain in a holding stage. No filing has yet been made with FERC.
Natural Gas Pipeline Co. of America	Dunn County, N.D.	Lurgi gasification with methanation	30,000	270	Phase 1 engineering design has been completed. No filing has been made with FERC. Further action on the project has been deferred.

has a normal horizontal rotation of 1/2 rpm and a vertical movement of 6 ft maximum with a 15-min cycle. All ranks of run-of-mine coal can be processed. Air, mixed with superheated (1200°F) steam to give a 400 to 450°F mix temperature, is introduced through a bottom grate.

Early experimental runs at 15 to 30 psig using 30,000 scf/hr of air and steam/coal ratios of 0.3 to 0.7 were unsuccessful because of excessive tar deposits. These deposits occurred because the overhead gas temperature was below 800°F. Subsequent runs at 75 to 105 psig with air rates of 50,000 to 65,000 scf/hr were better but still gave wide fluctuations in gas quality and bed temperature because of intermittent feeding.

Recent efforts have been centered on modernization of the gasifier and installation of a full-flow gas-cleaning system. Upgrading activities included mechanical, thermal, and control instrumentation improvements. Continuous actual ash and bed temperature data are now available through sensors mounted

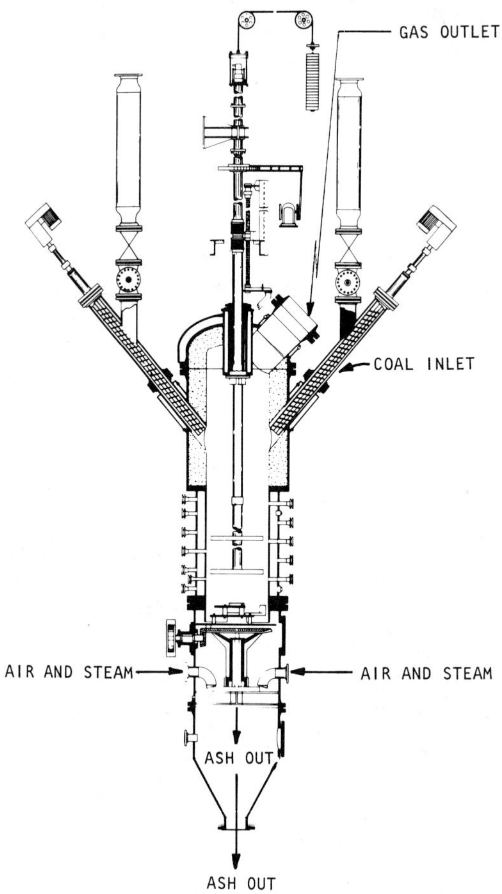

FIG. 13 MERC gasifier [10].

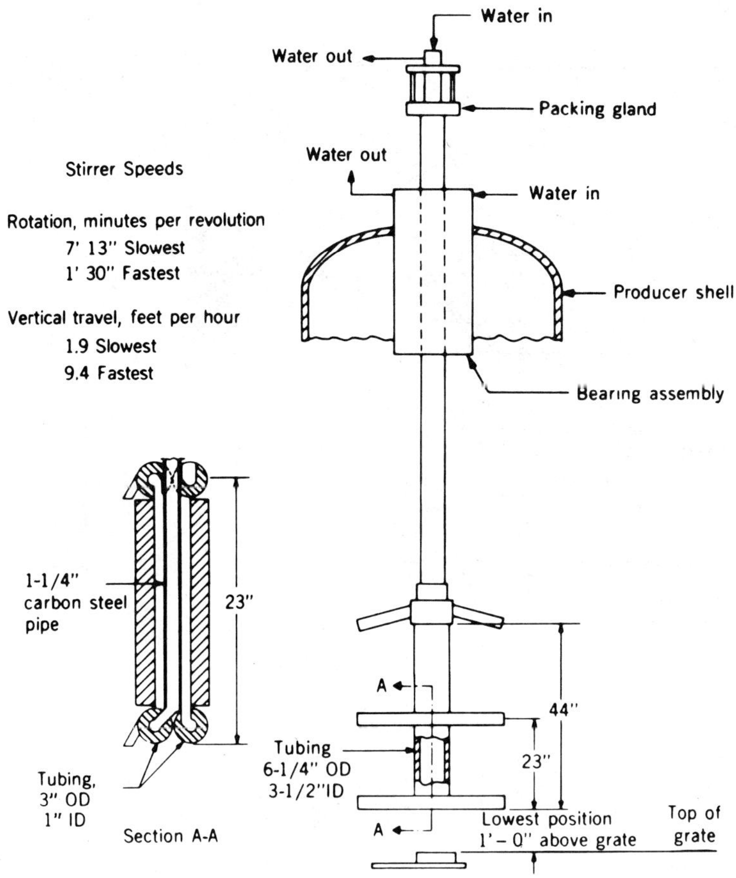

FIG. 14 Stirrer arrangement [10].

on the rotating grate and stirrer. A revised nuclear-level detector system provides data on bed level, bed density, and void formation. Utilizing these data to control the coal feed and ash removal rates has resulted in more uniform gas quality. Near future plans call for a high degree of automation for gasifier operation.

d. Riley-Morgan (Riley Stoker Corporation, Worcester, Massachusetts) The Riley-Morgan gasifier chamber (Fig. 15) is a refractory-lined cylinder that rotates slowly in a water seal. Coal feeds at the top through a lock hopper, which allows operation up to 10 in. H_2O pressure. Entrance of the coal is through a metering feeder across a radius of the revolving bed, which ensures even fuel distribution. Pivoting leveler arms serve as agitators and also sense the position of the bed top, which controls the ashing sequence. Ability to operate at various fuel bed heights allows variation in particle heating rate. This, in conjunction with the agitation function, makes operation with caking fuels possible. Access to the fuel bed, while operating, is accomplished by

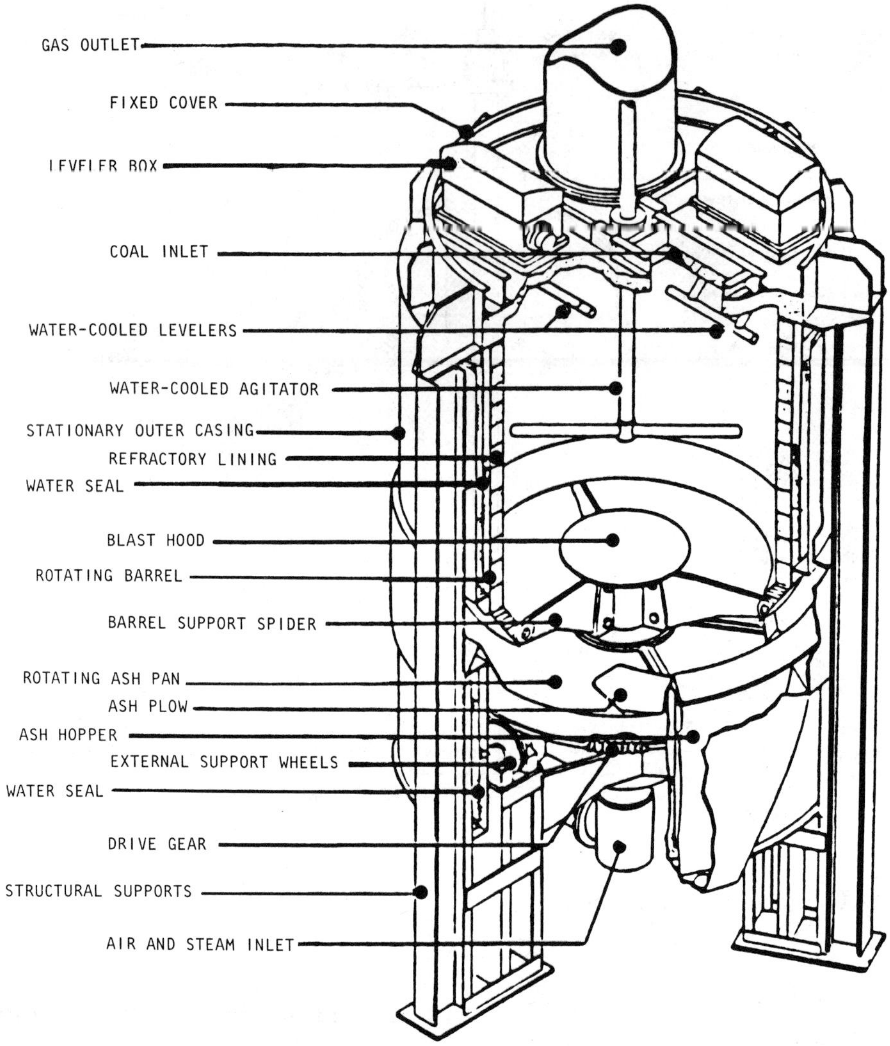

FIG. 15 Riley-Morgan gasifier [11].

means of pressurized ports that effectively eliminate fugitive emissions. Air and steam are distributed across the bottom of the bed by a blast hood, eliminating the need for a grate. Ashes are removed from the bottom of the bed through a water seal by means of a helical plow in the rotating ash pan. Gas leaves at 600 to 1200°F (depending on fuel) from an opening in the fixed cover of the gasifier.

Gasification tests have been carried out in a 10.5-ft-diameter prototype unit, using anthracite and both fine- and coarse-graded bituminous coals. Analyses of coals tested are shown in Table 6. Gas analyses and gasification efficiencies are shown in Table 7, and a material balance for one type of coal is shown in Table 8. Table 9 shows the disposition of sulfur and nitrogen in the gas from gasification of two bituminous coals. About 6000 lb/hr of the sized, high-volatile A bituminous coal could be gasified in the 10.5-ft-diameter unit.

Although no Riley-Morgan gasifier has been operated commercially, the Morgan gasifier, its forerunner, was employed until 1941 by over 9000 producers throughout the world. Units installed in 1933 and 1948 are currently in operation in South Africa. Testing with Western coals has been initiated.

e. Wellman Galusha (McDowell Wellman Co., Cleveland, Ohio) This atmospheric-pressure, air- or oxygen-blown gasifier (Fig. 16) is fed by a two-compartment system. The upper coal storage compartment feeds inter-

TABLE 6 Coal Analysis [12]

	Anthracite	High-volatile A bituminous	Medium-volatile bituminous
Proximate analysis (wt %)			
Moisture	376	3.2	7.1
Ash	8.0	4.7	5.0
Volatile matter	3.6	36.6	21.4
Fixed carbon	84.8	55.5	66.5
	100.0	100.0	100.0
Ultimate analysis (dry basis) (wt %)			
Carbon	86.6	80.8	85.2
Hydrogen	2.0	5.4	4.7
Oxygen and Nitrogen	0.8	0.6	0.7
Sulfur	0.8	0.6	0.7
Ash	8.3	4.8	5.4
	100.0	100.0	100.0
Heating value (as received) (Btu/lb)	11,430	13,916	13,830
Ash-softening temperature (°F) (reducing H = 1/2 W)	2,700	2,510	2,500
Free swelling index	0	4.5	8.5
Size analysis (wt %)			
+1 1/4 in.	16.2	54.9	39.7
-1 1/4 + 1 in.	23.0	38.0	33.5
-1 + 3/4 in.	28.2	5.4	9.4
-3/4 + 1/2 in.	18.2	1.7	3.5
-1/2 + 3/8 in.	1.7		2.3
-3/8 + 1/4 in.	0.9		2.1
-1/4 in.	10.8		9.5

TABLE 7 Gas Analysis [12]

	Anthracite	High-volatile A bituminous	Medium-volatile bituminous
Average dry gas composition (vol %)			
CO	22.70	21.98	23.50
CO_2	9.10	7.53	7.30
H_2	16.60	14.25	16.40
CH_4	0.25	2.24	1.70
Higher hydrocarbons	—	0.92	0.35
H_2S	0.09	0.08	0.12
N_2	51.26	53.00	50.63
	100.00	100.00	100.00
Higher heating value (Btu/scf)	130.00	166.00	153.00
Inlet blast air/steam (lb/lb)	4.81	6.00	5.30
Cold gas efficiency (%)[a]	80.60	69.20	64.30
Cold gas + tar/oil efficiency (%)[b]	80.60	79.80	70.80

[a] $\text{Cold gas} = \dfrac{\text{heat of combustion gas produced}}{\text{higher heating value of unit coal gasified}}$.

[b] $\text{Cold gas + tar/oil} = \dfrac{\text{heat of combustion of gas, tars, and oils}}{\text{higher heating value of unit coal gasified}}$.

TABLE 8 Material Balance HVA Bituminous Coal [12]

	Lb/lb coal feed	Percent
Input		
Coal	1.000	23.2
Air	2.741	63.7
Steam	0.565	13.1
Total	4.306	100.0
Output		
Dry gas	3.787	87.9
Water vapor	0.322	7.5
Tars	0.037	0.9
Light oils	0.040	0.9
Dust recovered	0.007	0.2
Ash	0.113	2.6
Total	4.306	100.0

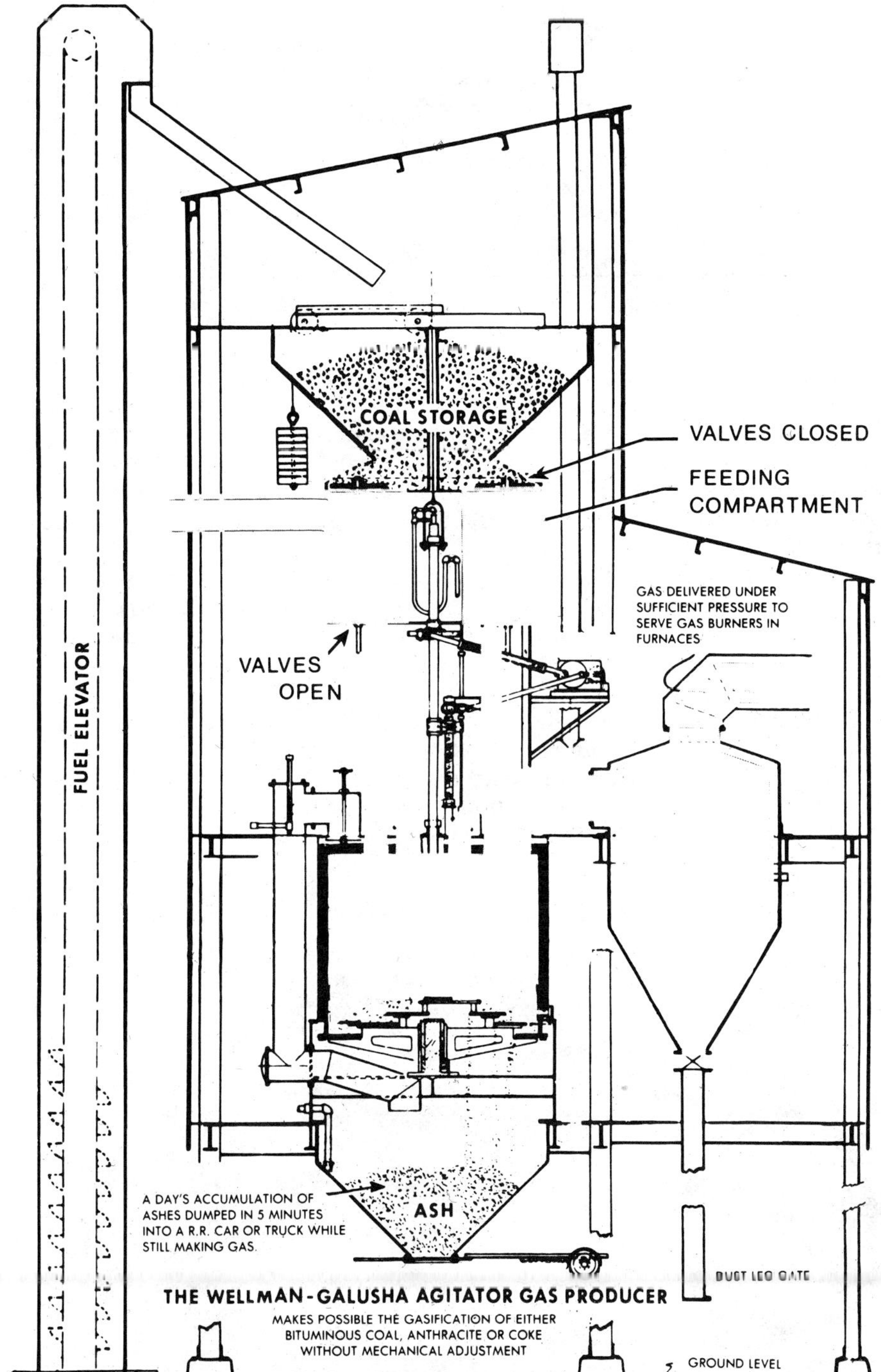

FIG. 16 Wellman-Galusha agitator-type gas producer [13].

TABLE 9 Trace Components Measured in Raw Product Gas [12]

Coal	HVA bituminous	Medium-volatile bituminous
Component (ppm)		
H_2S	1634	1213
COS	60	50
NH_3	190	113
HCN	107	129

mittently into the feeding compartment. The feeding compartment feeds the gasifier section continuously except for brief periods when the feeding compartment is being loaded from storage. On small units (up to 60 in. ID) ash is removed by shaker grates, whereas on larger units ash is removed continuously from the bottom of the gasifier into an ash hopper section by a revolving grate. The grate (Fig. 17) is constructed of flat circular steel plates set one above the other with overlapping edges. The grate, revolving eccentrically within the gasifier, forces the ash through the space between the grate plates and then falls down into the ash hopper. The smaller units are brick-lined, whereas the larger units are unlined and water-jacketed. Combustion air provided by a fan passes over the warm (180°F) jacket water to evaporate it and provide the necessary steam for gasification without requiring an auxiliary boiler.

Product gas leaves from above the fuel bed at a pressure of 5 to 10 in. H_2O and temperatures in the range 800 to 1000°F for bituminous coal and in the range 250 to 400°F for anthracite and lignite. The Wellman-Galusha gasifier includes two design configurations: the "standard type" for gasification of noncaking coals and the "agitator type" for gasification of any other types

FIG. 17 Revolving grates [13].

of coals. The latter type (Fig. 16) uses a slowly revolving horizontal arm, which also spirals vertically a short distance below the surface of the fuel bed to retard channeling and maintain a uniform bed. Use of the agitator not only provides for operation with caking coals with free swelling indexes as high as 9, but can also increase the capacity of the gasifier by about 25% with other coals. An ash-softening temperature of above 2200°F is desirable, but lower values can be handled with some reduction in coal throughput and efficiency. Capacities of various standard-size gasifiers are listed in Table 10. Typical gas compositions are shown in Table 11. One pound of good bituminous coal will make 64 ft^3 of gas and requires 43 ft^3 of air for gasification (Table 11).

Table 12 lists the Wellman-Galusha gasifiers that have been installed or are being erected.

f. Wilputte Producer (Wilputte Corp., Murray Hill, New Jersey) Coal, crushed to less than 4 in. in size and containing up to 10% fines, is fed to a 10-ft-long, 4-in.-diameter brick-lined Wilputte reactor (Fig. 18). A rotating rabble arm makes it possible to handle all classes of coal, including caking varieties. Operation is at atmospheric pressure. Moist air or oxygen is introduced at the bottom of the gasifier through a rotating grate. A stationary ash plow removes ash through a dry ash hopper. Coal residence time in the gasifier is about 2 hr as it moves downward at a velocity of about 4 ft/hr. Product gas leaving the top of the reactor has a superficial velocity of about 3 ft/sec. An air-blown 10-ft-long 4-in.-diameter gasifier will produce 7.2 million scf/day of 165 Btu gas from 60 tons/day of bituminous coal or 7.8 million scf/day of 150 Btu gas from 60 tons/day of anthracite coal. If oxygen is used, the product gas will be 295 to 310 Btu.

Table 13 lists Wilputte gasifiers that have been constructed. Operating characteristics of the installation (Holston Defense Corp., Kingsport, Tennessee), which has been in operation since 1942, have been described in detail [17].

2. *Dry Ash: Two-Stage*

a. ATC/Wellman (Incandescent) (Applied Technology Corp., Houston, Texas, and Wellman Incandescent Ltd., England) The two-stage ATC/Wellman gasifier (Fig. 19) is fed by a drum feeder. Relatively cool gas containing tar and volatile matter is removed from the top of the reactor. Most of the hot gas generated in the lower portion exits through the bottom gas off-take except for a controlled portion that is allowed to rise through the upper section to heat and devolatilize the incoming coal. At the bottom of the gasifier, stationary plows remove ash from a rotating grate through a water seal and into an ash pan. The air and steam mixture entering at the base of the gasifier is distributed across the coal bed by the grate. The lower section of the gasifier is surrounded by a water jacket that provides part of the steam required in gasification.

Applications to date have generally used bituminous coals with a free swelling index between 1 and 3 and ash-fusion temperature above 2200°F. Preferred coal sizes are 2 to 3 in., 1 1/2 to 2 1/2 in., or 3/4 to 1 1/2 in., with a maximum of 10% fines 5/16 to 2 in. and 15% less than 5/16 in.

TABLE 10 Normal Capacities of Wellman-Galusha Gas Producers [13] (Air-Steam Blast)

	Producer size inside diameter	Grate area (ft^2)	Lb/hr coke-charcoal 3/16-5/16 in.	Lb/hr coke-charcoal 5/16-9/16 in.	Lb/hr coke-charcoal 1/4-5/8 in.	Lb/hr bituminous coal 1 1/4-2 in.
A. Brick-lined producers,	18 in.	1.77	—	27	27	—
with shaker grates,	22 in.	2.65	—	40	40	—
for anthracite, char-	24 in.	3.14	—	47	47	—
coal, coke gasification	36 in.	7.07	—	106	106	—
	48 in.	12.05	—	189	189	—
	60 in.	19.6	—	295	295	—
B. Water-jacketed auto-	42 ft	9.62	192	220	241	—
matic producers with	6 ft 6 in. standard	33.18	664	763	830	—
revolving grates for	6 ft 6 in. agitator	33.18	830	956	1037	2900
anthracite, bituminous	8 ft 0 in.	50.27	1000	1150	1250	—
coal, coke gasification	10 ft 0 in. standard	78.54	1600	1840	2000	—
	10 ft 0 in. agitator	78.54	2000	2300	2500	7000

Note: Capacities will increase with size of fuel or with oxygen blast.

TABLE 11 Typical Gas Compositions (Btu/ft^3 HHV) [13]

	Bituminous	Anthracite
CO	25.00	27.1
CH_4	2.27[a]	0.5[a]
Illuminants	0.57	—
H_2	14.49	16.6
CO_2	4.69	5.0
N_2	52.68	50.8
Total	166.7	146.2

[a]Methane increases with increasing volatile content.

TABLE 12 Wellman-Galusha Gas Producer Installations [14]

Period	Number of installations or operator	Number of units	Location
1896-1920	Numerous	Numerous	
1920-1930	106	291	
1930-1937	24	57	
1937-1950	95	146	
1950-1979	45	56	
1950	Hazleton Brick Co.	2	Hazleton, Pa.
1950	Cie Royale Asturienne	4	Auby, France
1950	New Jersey Zinc Co.	1	Palmerton, Pa.
1950	Export	1	Pakistan
1950	Battelle Memorial Inst.	1	Columbus, Ohio
1951	Monsanto Chemical Co.	1	Seattle, Wash.
1951	Delmar Chemical Co.	1	Elkton, Md.
1951	Binghamton Brick Co.	1	Binghamton, N.Y.
1951	Clark Kilns Prop. Ltd.	1	Sidney, N.S.W.
1951	Glen-Gery Shale Brick Co.	1	Watsontown, Pa.
1951	Glen-Gery Shale Brick Co.	1	Middletown, Pa.
1951	Int. Graphite	1	Niagara Falls, N.Y.
1952	Manganese Chemical Corp.	1	Minneapolis, Minn.
1953	University of the Philippines	1	Manila, the Philippines
1953	Export	1	Yugoslavia
1953	Alcoa	1	Messina, N.Y.
1953	Union Carbine	1	Niagara Falls, N.Y.
1954	Glen-Gery Shale Brick Co.	1	York, Pa.
1955	National Lime & Stone Co.	2	Carey, Ohio
1955	Hidro-Nitrol Espanola	2	Monzon, Spain
1955	Taiwan Fertilizer Co.	4	Nankong, Taiwan
1955	Hazelton Brick Co.	1	Hazleton, Pa.
1955	Banco Minero de Peru	1	Lima, Peru
1956	Gypsum Lime and Alabastine	1	Beechville, Ontario

(continued)

(Table 12 continued)

Period	Number of installations or operator	Number of units	Location
1957	Glen-Gery Shale Brick Co.	1	Shoemakersville, Pa.
1957	Glen-Gery Shale Brick Co.	1	Ephrata, Pa.
1957	Chemical Lime, Ltd.	1	India
1958	Steel Co. of Canada	1	Beechville, Ontario
1959	Mississippi Lime Co.	1	Ste. Genevieve, Miss.
1960	M. A. Hanna Co.	1	Cooley, Minn.
1960	Glen-Gery Shale Brick Co.	1	Reading, Pa.
1961	Nopco Chemical Co.	1	Linden, N.J.
1961	National Lead Co.	1	South Amboy, N.J.
1961	India Supply Mission	1	Jamshedjrur, India
1962	Cabot Titanium Corp.	1	Ashtabula, Ohio
1962	Allied Chemical Canada, Ltd.	1	Corunna, Ontario
1963	Olin-Mathieson Co.	1	Ashtabula, Ohio
1966	U.S. Bureau of Mines	1	Morgantown, West Va.
1973	Riley-Stoker Corp.	1	Worcester, Mass.
1974	Glen-Gery Shale Brick Co.	1	Reading, Pa.
1977	Pike County Fiscal Court	2	Pikeville, Ky.
1977	Howmet Aluminum Co.	1	Lancaster, Pa.
1978	Blen-Gery Shale Brick Co.	1	New Concord, Pa.
1978	CAN-DO	2	Hazleton, Pa.
1979	Universal Oil Products	1	Tucson, Ariz.

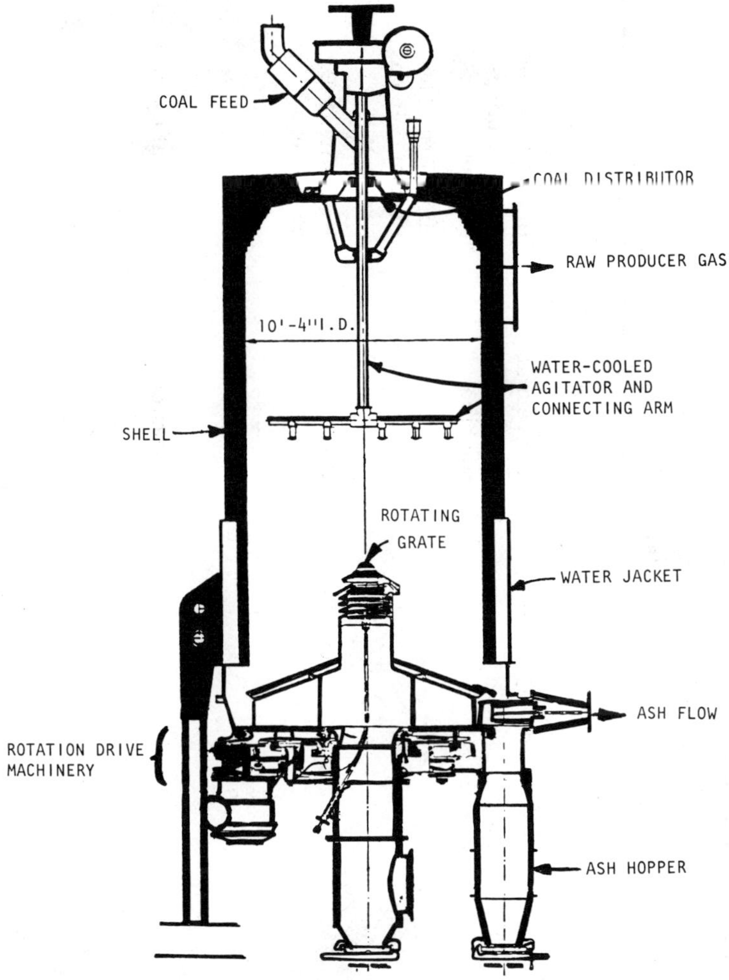

FIG. 18 Wilputte gasifier [15].

TABLE 13 Partial Listing of Commercial Wilputte Gasifiers [16]

Company/location	Capacity: Number of units	Capacity: Gasifier diameter (ft)	Capacity: Heat rate (10^9 Btu/day)	Coal type	Application	Data and Comments
Perth Amboy, N.H., municipality	1	6.6	—	—	—	1918
Pennsylvania Gas and Water	1	—	—	Converted water-gas machine	High-Btu (1000 Btu/ft^3) oil-gas machines	
Paccal, Australia	2	—	—	—	High-Btu (1000 Btu/ft^3) oil-gas machines	
Semet Solvay Ironton, Ohio	2	9.3	0.8 total	Coke	Underfiring BP coke ovens	1930
Staten Island Shipbuilding, Staten Island, N.Y.	1 1	6 9	—	—	—	1912
Tennessee Eastman, Kingsport, Tenn.	2	9.2	—	Bituminous	Chemical operations	1933; 36 tons coal/day
Virginia Carolina Chemicals, Charleston, S.C.	1	9.2	—	Coke	Furnaces	1933; 1600 lb coke/hr

Eastern Rolling Mills, Baltimore, Md.	1 2	5.5 9.2	— 0.2 each	Anthracite No. 1 buckwheat anthracite	Heating rolls Normalizing furnace	1917 1925
Jefferey Mfg. Co., Columbus, Ohio	1	10.5	0.4	Bituminous	Heat-treating gas engines	1919
Allis Chalmers of Canada	4	1-4.5 2-5.0 1-5.5	—	Anthracite	—	1913-1917
A. O. Smith Milwaukee, Wis.	1	9	—	—	—	1912
Citizens Light, Heat & Power, Cangy, Minn.	1	9	—	Buckwheat anthracite	Utility station	1917
Dupont, Carney's Point, N.J.	3	9.2	0.3 each	No. 1 buckwheat anthracite	Chemical operations	1927-1930
Ford Motor Company of Canada, Walkersville, Ont.	3	9.2	—	Bituminous	Heat treating	1926; 1500 lb coal/hr each
General Electric Erie, Pa.	1	5.5	—	—	—	1917
Holston Defense Corporation, Kingsport, Tenn.	12	9.1	—	Bituminous	Furnaces	1945; 4 operating
International Harvester	3	3.5 4.0 5.0	—	—	Test use	1910-1911

(continued)

(Table 13 continued)

Company/location	Capacity: Number of units	Capacity: Gasifier diameter (ft)	Capacity: Heat rate (10^9 Btu/day)	Coal type	Application	Data and Comments
Johns Hopkin University, Baltimore, Md.	1	3.5	—	—	Test use	1915
Kellogg Company, Battlecreek, Mich.	2 (Plus 7 other 9 ft 0 in. size)	9.5	0.2-0.3 each	Nos. 1, 2, 3 buckwheat anthracite	Roasting food products	1927-1930
Long Island Lighting Co.	1	—	—	—	High-Btu (1000 Btu/ft^3) oil-gas machines	

Note: Data for plants sold from 1934 to 1944 are not fully available.

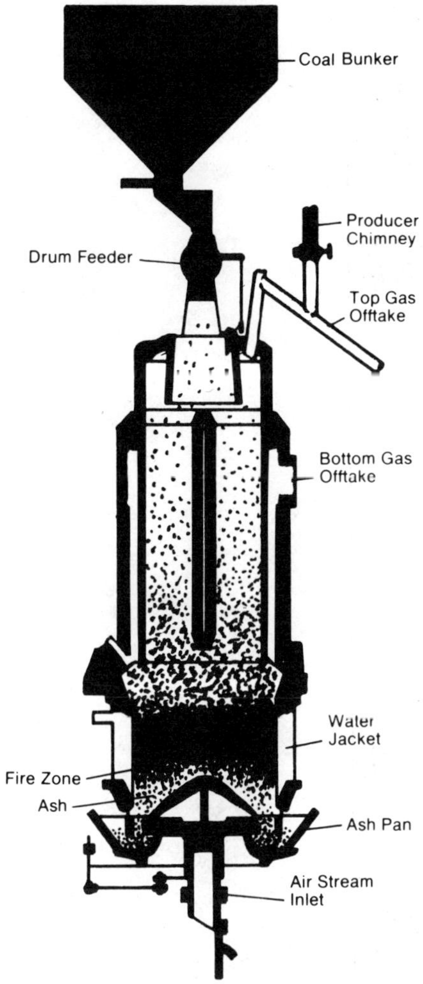

FIG. 19 ATC/Wellman two-stage coal gasifier [18].

Raw gases from the gasifier contain tar (in the top gas) and coal dust (in the bottom gas). These may be separately cleaned of large oil particles and dust to produce a hot raw gas at about 700°F. Further cleaning may be undertaken to produce a hot detarred gas or a cold, clean gas. Capacity for coal feed, gas production rates, and thermal efficiencies for various standard-size gasifiers are shown in Table 14.

Table 15 lists industrial gasifiers that have been installed.

b. FW-Stoic (Foster Wheeler Energy Corp., Livingston, New Jersey, and Stoic Combustion Pty. Ltd., Johannesburg, South Africa) Coal (preferably sized to 1/2 × 1 1/2 in. or 1 1/2 × 3 in. with a free swelling index below 2.7

TABLE 14 Characteristics of Wellman/Incandescent Two-Stage Gasifiers [16]

Gasifier diameter (ft)	Coal feed rate (tons/day)	Thermal output (10^9 Btu/day)			Thermal efficiency (%)[a]		
		Hot raw gas	Hot detarred gas	Cold clean gas	Hot raw gas[b]	Hot detarred gas[b]	Cold clean gas
4.50	13.92	0.30	0.28	0.25	89.8	83.3	76.2
5.50	20.40	0.44	0.41	0.37	90.2	83.8	76.0
6.50±	29.40	0.64	0.59	0.54	90.1	84.0	75.9
8.50	51.90	1.12	1.05	0.95	90.2	84.0	75.9
10.00	71.40	1.54	1.45	1.30	90.1	84.9	76.1
10.75	82.56	1.79	1.67	1.51	90.5	84.3	76.3
12.00	103.20	2.23	2.08	1.88	90.1	84.0	76.0

[a]Based on 12,000 Btu/lb coal.

[b]Includes sensible heat in gas.

and an ash-fusion temperature in reducing atmosphere less than 220°F) is fed by a rotary feeder or by automatic opening and closing valves to the top of this two-stage gasifier (Fig. 20). The steel shell of the gasifier is refractory-lined in the upper section and water-jacketed in the lower section. The upper section is unique in that it is divided into four equal-sized segments through which coal flows downward and which have hollow walls through which bottom gases flow upward to preheat the coal. Additionally, the upper section refractory is tapered outward toward the bottom to allow for coal swelling. Ash at the bottom of the gasifier is removed wet through a water-sealed ash pan that rotates with the grate. Ash removal is adjusted to regulate the position of the 4- to-10-in.-deep fire zone. Air and steam are introduced through the grate. Most of the gasifier steam is generated in the water jacket that surrounds the lower fire zone of the gasifier.

Generally, about one-half of the bottom gas generated in and directly above the fire zone passes up through the refractory walls, transferring heat to descending coal, and exits at 1100 to 1200°F. This gas is tar-free because it is generated near the bottom of the gasifier from devolatilized

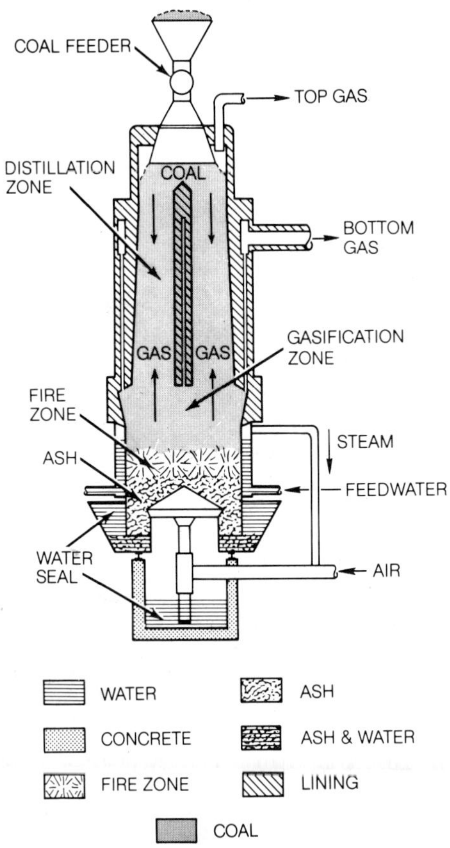

FIG. 20 FW-STOIC two-stage gasifier [19].

TABLE 15 Commercial Wellman/Incandescent Gasifiers [16]

Plant owner[a]	Number of operating gasifiers	Gasifier diameter (ft)	Heat rate (10^9 Btu/day)	Coal type	Scope of system[b]	Application	Year completed
SCAW Metals Ltd.	1	10.00	0.7	Bituminous	Single-stage	Rolling mill and	1963
Stewarts & Lloyds	1	8.50	1.0	Bituminous	Two-stage hot raw	Miscellaneous furnaces	1964
Cullinan Refractories	2	6.50	0.6	Bituminous	Two-stage hot raw	Tunnel kilns	1964
Cullinan Refractories	1	8.50	1.0	Bituminous	Two-stage hot raw	Basic refractory kilns	1965
Union Steel Corp.	4	10.00	4.8	Bituminous	Two-stage (three hot raw, one cold clean)	Steelmaking furnaces and miscellaneous	1965
Southern Cross Steel	1	10.00	1.2	Bituminous	Two-stage cold clean	Stainless steel reheating and heat-treatment furnaces	1966
Consolidated Glass	2	10.00	2.7	Bituminous	Two-stage hot raw	Glassmaking furnace	1967
Union Steel Corp.	1	10.00	0.7	Anthracite	Single-stage	Steelmaking furnaces and miscellaneous	1968
SCAW Metals Ltd.	2	10.75	2.6	Bituminous	Two-stage (one cold clean, one hot detarred)	Rolling mill and miscellaneous	1968

Highveld Steel	4	10.00	5.4	Bituminous	Two-stage cold clean	Rolling mill furnaces	1968
Lurgi, Groolfontein	1	6.50	0.6	Bituminous	Two-stage hot raw	Rotary zinc kiln	1970
SCAW Metals Ltd.	1	10.75	1.5	Bituminous	Two-stage hot detarred	Rolling mill and miscellaneous	1970
Cullinan Refractories	1	8.50	1.0	Bituminous	Two-stage hot raw	Refractory brick kilns	1973
SAICCOR	1	10.00	1.2	Bituminous	Two-stage cold clean	Pulp and paper drying	1973
Johannesburg Consolidated Inv. Co.	1	6.50	0.66	Bituminous	Two-stage hot raw	Antimony concentrate	1975
SCAW Metals Ltd.	1	10.75	1.5	Bituminous	Two-stage hot detarred	Steelworks, reheat annealing furnaces, etc.	1975
Southern Cross Steel	1	10.00	1.2	Bituminous	Two-stage cold clean	Stainless steel reheat furnaces, etc.	1976
Highveld Steel	2	10.00	2.3	Bituminous	Two-stage cold clean	Steel mill reheat furnaces	1976
Caterpillar Tractor Co., York, Pa.	1	10.00	2.5	Bituminous	Two-stage cold clean	Heat treating	1978
Land O'Lakes, Minnesota	1	10.00	1.3	Bituminous	Two-stage hot raw	Boilers and whey drying	1979

[a]All plants are located in South Africa, except Caterpillar Tractor Co. and Land O'Lakes.

[b]All products are low-Btu gas.

TABLE 16 Commercial FW-Stoic Gasifiers [16]

	Company/location		
	Driefontein, South Africa	Lydenberg, South Africa	University of Minnesota, Duluth, Minn.
Capacity			
Number of units	1	1	1
Size, diameter (ft)	10	8.5	10
Heat rate (10^9 Btu/day)	1.44	0.96	1.44
Coal type	Bituminous	Bituminous	Subbituminous lignite
Scope of system	Low Btu gas, hot detarred	Low-Btu gas, cold clean	Low-Btu gas, hot detarred
Application	Brick kiln	Metal drying	Boiler fuel
Status	1973	1976	1978

char. Top gas, which is a mixture from devolatilization of feed coal and from the remainder of the bottom gas that passes upward against feed coal, leaves the gasifier at 250°F and may contain as much as 10 to 15% of the coal energy in tar.

Stoic gasifiers of the type described have been successfully employed commercially in South Africa (see Table 16). Different sizes of gasifiers with their rated output are shown in Table 17. Some projected figures pertinent to a 10-ft-diameter gasifier recently installed at Duluth, Minnesota, for operations with Elkol, Wyoming, coal, are shown in Table 18. Eventually, various other types of coal will be evaluated at that installation.

c. Ruhr 100 (Ruhrgas AG, Essen, West Germany) The design of this gasifier (Fig. 21) is adapted from the Lurgi pressure gasifier. A number of modifications are incorporated for process improvement.

TABLE 17 Nominal Coal Feed Rates for FW-Stoic Gasifiers [16]

Gasifier diameter (ft)	10^6 Btu/hr, maximum	Approximate coal rate (tons/day)[a]
6.50	22.5	31.2
8.50	40.0	52.8
10.00	60.0	72.0
12.50	90.0	108.0

[a]Based on a coal having a HHV of about 12,000 Btu/lb with the gasifier producing hot detarred gas.

TABLE 18 Estimated Performance of FW-Stoic Gasifier, University of Minnesota at Duluth [20]

Coal	
Feed quantity	3 tons/hr
Type	Elkol, Wyoming, subbituminous
Proximate analysis	
Volatiles	32.69 wt %
Fixed carbon	43.19 wt %
Ash	5.37 wt %
Moisture	18.75 wt %
	100.00
Combined gas[a]	
Analysis	
CO_2	3.69 vol %
CO	24.11 vol %
H_2	13.11 vol %
N_2	40.41 vol %
H_2S	0.04 vol %
CH_4	1.92 vol %
H_2O	16.58 vol %
HHV	166 Btu/scf (dry basis)
Delivery temperature	750°F
Delivery pressure	27 in. H_2O
Total heat content	54,700,000 Btu/hr
Tar oil	
Physical properties	
Specific gravity	1.0855
Viscosity (centistokes)	
122°F	555.4
210°F	21.35
Pour point (°F)	80
Flash point, C.O.C. (°F)	315
HHV (Btu/gal)	148,000
Chemical composition	
Carbon	85.92 wt %
Hydrogen	7.92 wt %
Oxygen	4.18 wt %
Nitrogen	1.05 wt %
Sulfur	0.22 wt %
Ash	0.11 wt %
Moisture	0.60 wt %
	100.00
Total heat content	3,000,000 Btu/hr
Overall gasification efficiency	
93.5%, including 750!F combined gas[a] and tar	
Utility requirements	
Electricity	100 kW
Boiler feedwater	2400 lb/hr
Service water	500 lb/hr
Auxiliary steam	1200 lb/hr

[a]Detarred top gas plus bottom gas.

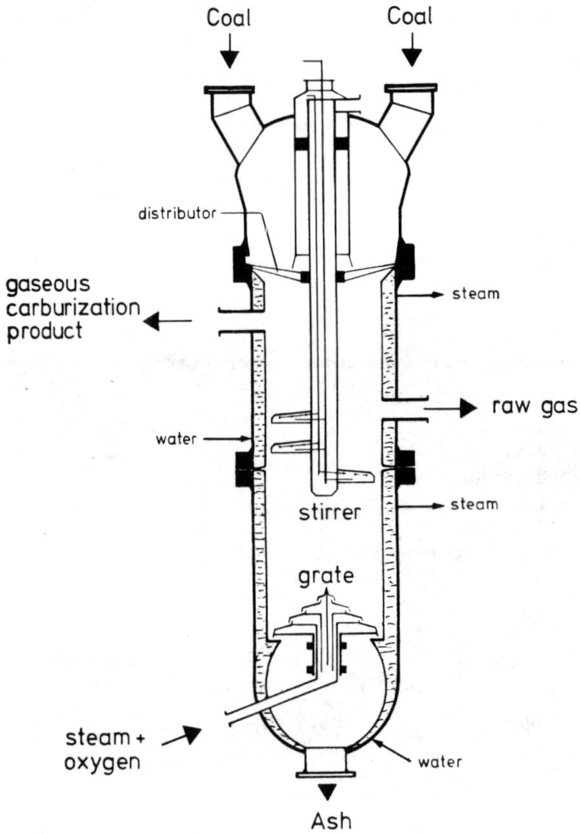

FIG. 21 Ruhr 100 two-stage gasifier [21].

Operating pressure is increased to 100 atm from the normal 30 atm maximum. Higher pressure is expected to increase the capacity of a given-size gasifier, reduce tar formation by 75%, and reduce production of phenols. Methane formation in the gasifier is expected to about double, so that, in SNG production, 70% of the total methane produced will come directly from the gasifier. The exothermic reaction heat of this methane formation will also reduce the quantity of oxygen normally required for gasification.

An essential new feature of the Ruhr 100 design is the provision of separate gas outlets for carbonization products and for raw gas from the gasification zone. Temperatures in the carbonization zone can thereby be controlled by adjusting the flow rate of hot gases from below. Also, if the coal contains a high proportion of fines, the quantity in the raw gas can be reduced considerably by the reduction of upper gas flow rate.

Stirrer blades mounted in the upper part of the gasification zone are provided to prevent caking of coal.

A pilot plant at Dorsten, designed to handle 3 to 7 tons/hr of coal with lock-hopper feed, is expected to be ready for operation in April 1979. This is a joint endeavor of Ruhrgas AG, Ruhrkohle AG, and Steag AG. Plans call for a 5-hr test period. During the first year, operation will be at pressures of 25 to 40 atm, which will be increased to 100 atm as soon as possible.

d. Woodall-Duckham [Woodall-Duckham (U.S.A.) Limited, Pittsburgh, Pennsylvania] This two-stage gasifier consists of a refractory-lined retort (distillation) section atop a water-jacketed gasification and combustion section (Fig. 22). Coal feed into the top through a lock hopper is controlled by the level of coal in the gasifier. Air (or oxygen) and steam are distributed at the bottom through a rotating grate that also removes the ash to twin ash receivers. Operating pressures are 40 to 60 in. H_2O for air and steam entering and 30 to 40 in. for product gas out. Coal feed sized 3/8 to 1 in., 1/2 to 1 1/2 in., or 3/4 to 2 in. must be double-screened to remove fines. Predrying is not required. An ash-fusion temperature of 2050°F can be tolerated, but above 2200°F is preferred. Similarly, a free swelling index of 1 1/2 is minimum but 2 1/2 to 3 is desirable.

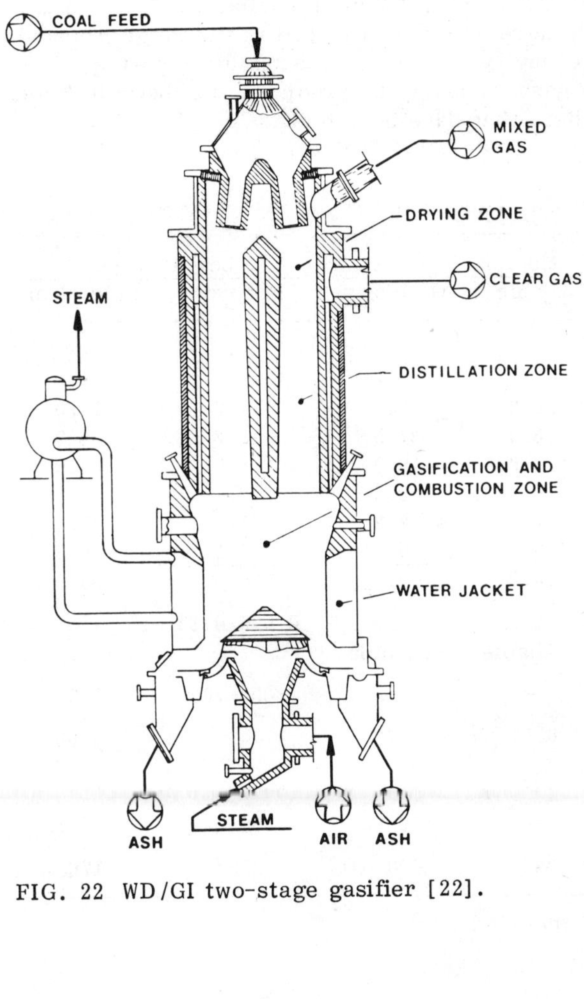

FIG. 22 WD/GI two-stage gasifier [22].

Two sizes of gasifiers are available. A 10-ft-diameter model, which will handle 3 to 3.5 tons/hr of feed coal, has 17-ft-high retort and 8-ft-high water-jacketed sections, respectively. The 12-ft-diameter model will handle 3.5 to 4.5 tons of coal feed per hour.

In air-blown operation top gas leaves the gasifier at about 250°F. This gas contains tars and oils. Bottom gas, which leaves at 900 to 1500°F, must be cleaned of dust. The top gas can be cleaned in an oil-separation cyclone and mixed with bottom gas from a dust cyclone to give a hot raw gas at about 700°F that can be distributed in unlined but insulated mains up to 1000 ft long without condensation of liquids. This hot raw gas will have an effective heating value (taking credit for sensible heat content) of 200 to 210 Btu/scf, with overall gasification efficiency being 88 to 92%.

Alternatively, a cold, clean, oil-free gas can be produced by treating the top gas through electrostatic precipitators and mixing with dedusted bottom gas. This gives a mix gas of approximately 176 Btu/scf at about 120°F. Here, the gasification efficiency is 74 to 76%.

Medium-Btu gas can be produced by using either oxygen or a cyclic operation with air.

An air-blown operation requires about 2.3 lb of air and 0.2 lb of steam per pound of coal. All the steam required is produced in the water-jacketed section. In some cases, there may be an excess of steam for export.

Properties of the various gases that may be produced are shown in Table 19. Table 20 is a list of gasifiers that have been built.

TABLE 19 Product Gas Types Available from WD Two-Stage Process [23]

	Medium-Btu gas		Low-Btu gas	
	Cyclic	O_2-blown	Hot raw	Cold clean
Composition (moisture-, tar-, tar-oil-free) (mol %)				
H_2	52.2	38.4	17.0	17.0
CO	28.5	37.5	28.2	28.2
CO_2	8.0	18.0	4.5	4.5
CH_4	6.5	3.5	2.7	2.7
N_2	4.2	2.2	47.1	47.1
Others (H_2S, COS, NH_3)	0.6	0.4	0.5	0.5
	100.0	100.0	100.0	100.0
Tar and tar oil (lb/lb coal)	Not available	Not available	0.075-0.12	—
Temperature (°F)	—	—	600-700	120
Sp. gr. (air = 1)	0.52	0.73	—	0.83
Wobbe No.	465	334	—	193
Thermal efficiency (%)	—	89-93	88-92	74-76
Btu/ft^3	330	290	210[a]	176

[a]Effective, with sensible heat included.

TABLE 20 Commercial Woodall-Duckham Gasifiers [16]

Company/location	Number of units	Gasifier diameter (ft)	Type	Status
		Industrial Fuel Gas Plants		
Weldless Steel Tube Co., Wedensfield, England	2	10	Bituminous	Nonoperational
Ziar Aluminum Works, Czechoslovakia	7	10	Bituminous	Nonoperational
Chomutov Tube Works, Czechoslovakia	14	10	Lignite	Operational
Istanbul Gas Utility, Turkey	1	8.5	Lignite	Operational
Australian Consolidated Industries, Ltd., Sydney, Australia	4	10	Bituminous	Nonoperational
Melbourne Gas Works, Melbourne, Australia	2	10	Bituminous brown	Nonoperational
Elgin Fireclay Ltd., Springs, South Africa	1	8.5	Bituminous	Operational
Vaal Potteries Ltd., Meyerton, South Africa	1	8.5	Bituminous	Operational
Union Steel Corp., Johannesburg, South Africa	2	10	Bituminous	Operational
Stewarts & Lloyds Steelworks, South Africa	3	10	Bituminous	Operational
Masonite, Escault, South Africa	3	10	Bituminous	Operational
SAAPI, Mandini, South Africa	2	10	Bituminous	Operational
Rand Water Board, Vereeniging, South Africa	1	8.5	Bituminous	Operational
Driefontein, Carltonville, South Africa	2	10	Bituminous	Operational

(continued)

(Table 20 continued)

Company/location	Number of units	Gasifier diameter (ft)	Type	Status
Vereeniging Refractories, South Africa	2	10	Bituminous	Operational
General Refractory, Hitchins, Ky.	1	10	Bituminous	Starting anticipated 1979
	Public Utility Gas Plants			
St. Poelten, Austria	2	6		
Rome, Italy	5	8.5		
Trieste, Italy	2	10		
Como, Italy	1	6		
Genoa, Italy	4	12		
Vierzon, France	2	8.5		
Kensal Green, England	1	10		
Ulm, West Germany	2	8.5		
Zagabria, Yugoslavia	1	6		
Prague, Czechoslovakia	6	8.5		
Warsaw, Poland	3	12		
Posen, Poland	3	10		
Thom, Poland	2	6		
Tokyo, Japan	5	12		

	Number of units	Gasifier diameter (ft)	Coal type	Scope of system
	Synthesis and Water Gas Plants			
OSW Fertilizer Plant, Linz, Austria	4	8.5	Unknown	Medium-Btu gas
Vetrocoke, Porto Marghera, Italy	2	6	Bituminous	Medium-Btu cyclic gas[a]
Montecatini, Crotone, Italy	2	8.5	Unknown	Medium-Btu gas

(continued)

(Table 20 continued)

Company/location	Number of units	Gasifier diameter (ft)	Coal type	Scope of system
Montecatini, St. Giuseppe di Cairo, Italy	2	10	Unknown	Medium-Btu gas
I.M.A.D., Naples, Italy	2	6	Bituminous	Medium-Btu cyclic gas
State Works, Semtin, Czechoslovakia	4	6	Coke[b]	Medium-Btu cyclic gas
D. Swarovski Co., Wattens, Austria	2	6	Coke[b]	Medium-Btu cyclic gas
Edison S.P.A., Milan, Italy	1	10	Coke[b]	Medium-Btu cyclic gas
Marconi S.P.A., Aquila, Italy	1	6	Coke[b]	Medium-Btu cyclic gas
Public Utility, Paris, France	3	14	Coke[b]	Medium-Btu cyclic gas
Public Utility, Fuerth, West Germany	1	6	Coke[b]	Medium-Btu cyclic gas

[a]Cyclic gas by alternate blasting of superheated steam and air separately.

[b]When coke is used as feed, the top retort zone is not required and the gasifier in reality becomes a single-stage system.

3. *Descending-Bed Slagging Gasifiers*

a. BGC/Lurgi (British Gas Corporation, London, and Lurgi Mineraloltechnik GmbH, Frankfurt, Germany) The slagging BGC/Lurgi gasifier (Fig. 23) is designed for pressure operation (350 to 400 psig). Coal feed is through a lock hopper. A coal distributor and stirrer are provided for moving coal through the upper part of the gasifier. A steam/oxygen mixture is introduced through tuyeres near the base of the gasifier. Molten slag, at about 2700°F, runs through a slag tap hole at the gasifier bottom and drops into a slag quench vessel, where it is solidified in quench water. The solid slag is then removed from the system through a slag lock hopper. The gasifier section is refractory-lined and water-jacketed. Gas produced leaves near the top of the gasifier at about 1100°F.

In 1974, an existing 6-ft-diameter conventional dry-ash Lurgi gasifier, at the Westfield (Scotland) Development Center of the British Gas Corporation, was modified for slagging operation. A test program on slagging gasification was then sponsored by a consortium of 14 U.S. industrial firms. Successful conclusion of this program was followed immediately by a Technical Support Program, on a broader range of coals, funded by the U.S. Department of Energy (DOE) and sponsored by a different consortium of U.S. firms.

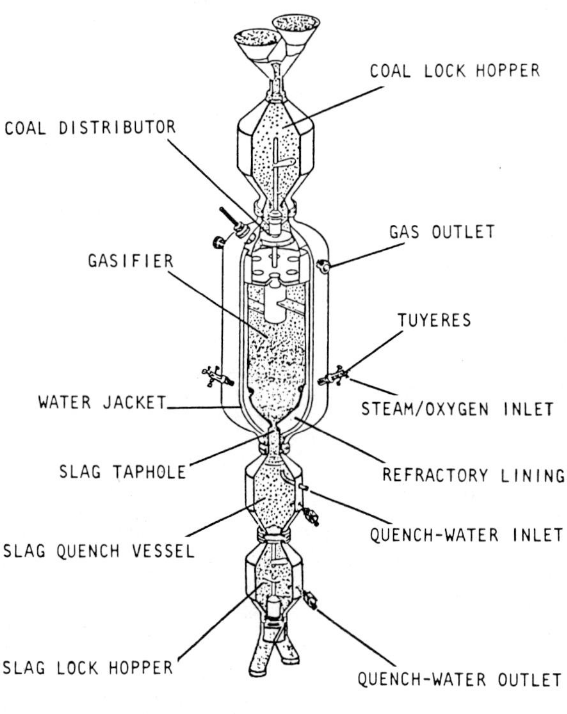

FIG. 23 Descending-bed slagging gasifier [24].

The consortium noted above, headed by Continental Oil Company, has been engaged by DOE since mid-1977 to design a demonstration plant employing BGC/Lurgi slagging gasifiers for producing SNG. As presently envisioned, this demonstration system incorporates one 10-ft-diameter operating gasifier with an additional spare to produce sufficient gas for about 20 million scf/day of SNG. Fifteen runs were completed during the Technical Support Program on Ohio No. 9, Pittsburgh No. 8, and Scottish Frances coals, and metallurgical coke. Table 21 lists the properties of the coals tested.

Limestone and blast furnace slag were added as flux to control gasifier slag viscosity. Noncaking Scottish Frances coal gave no problems and was run continuously for as long as 22 days. However, problems were encountered with the caking U.S. coals, especially Ohio No. 9, a coal that was not washed and varied considerably in the ash content as fed to the gasifier. Modification of the gasifier internals eventually led to satisfactory operation with Pittsburgh No. 8 coal in June 1978. This test lasted for only 4 days because of limited contracted time and availability of feed coal. Comparison of this test run with a test of Pittsburgh No. 8 in a larger dry-ash Lurgi generator and with a test of Frances coal is shown in Table 22. Both the oxygen and steam rates required in the slagger tests are significantly lower than in the dry-ash unit, but the relative capacity (fuel rate) is about six times as high.

TABLE 21 Properties of Test Coals [24]

	Frances	Ohio No. 9	Pittsburgh No. 8
Source	Scotland	Dexter City, Ohio	Champion No. 1, Pa.
Washed	Yes	No	Yes
Ultimate analysis (dry basis) (%)		[a]	
Hydrogen	5.0	4.2	5.2
Carbon	71.2	58.0	78.3
Nitrogen	1.3	0.9	1.3
Sulfur	0.5	4.8	1.7
Chlorine	0.3	—	—
Ash	4.8	25.3	7.3
Gieseler max. fluidity (Dial divisions per minute (DDM))	0	150-3000	8300
Free swelling index	0-1	3.5-5	8

[a]Typical of first shipment.

TABLE 22 Westfield Operating Data [24]

Gasifier	Dry bottom	Slagger	
Shaft cross section (ft^2)	66	28.2	28.2
	Coal inspections		
Coal	Pittsburgh No. 8	Pittsburgh No. 8	Frances
Size consist	1 1/2 × 0 in.	1 1/4 × 1/4 in.	1 × 5/8 in.
Free swelling index	7.0	8.0	0
Proximate analysis (dry basis) (%)			
Volatile	39.4	40.8	35.9
Fixed carbon	52.6	51.9	59.3
Ash	8.0	7.3	4.8
Ash fluid point (°F)	(2500)[a]	2680	2630
Ash iron content (% Fe_2O_3)	16.1	19.1	13.2
	Results		
Operating conditions			
Pressure (psig)	364	350	350
Rates (ton/ton MAF coal)			
Oxygen	0.70	0.56	0.54
Steam	5.10	0.41	0.41
Fuel rate (lb MAF/hr/ft^2)	145	870	852
Yields per ton MAF coal			
Tar + oils (lb)	100	NA	255
Liquor (lb)	5718	NA	400
Raw gas (Mscf, dry)	84	72	70
Gas gross HV (Btu/ft^3)	286	356	352
SNG equivalent (Mscf)	21.4	20.3	20.4
Run duration (hr)	48	88	524
Post-run inspection	—	Good	Good

[a]Not an experimental measurement; typical for coal from Marion Co., W.Va.

TABLE 23 Comparison of Raw Gas Compositions [25]

Crude gas composition (vol %)	Lurgi	Slagging
CO_2	24.6	2.6
CnHm	1.1	0.4
CO	24.6	60.6
H_2	39.8	27.8
CH_4	8.7	7.6
N_2	1.2	1.0
Total	100.0	100.0
Calorific value (Btu/ft^3)	309	370

The slagging gasifier has both a much lower CO_2 content and H_2/CO ratio than dry-ash units because of the lower steam rate and higher temperature of operation (Table 23).

b. GFETC (Grand Fords Energy Technology Center of the U.S. Department of Energy, Grand Fords, North Dakota) The primary objective in operating this small (15 5/8-in. ID) GFETC gasifier (Fig. 24) has been to acquire environmental assessment data on a variety of noncaking Western coals. Operating pressure is between 80 and 400 psi. It is presently (late 1978) being modified by provision of a bed stirrer to enable the handling of caking coals.

In a test program started in 1976 and continued through September 1978, operation of the gasifier was primarily concerned with developing materials of construction and fabrication techniques to provide for a water-cooled hearth plate that allows for continuous slag removal. A total of 66 tests were undertaken, 17 of which were completed on schedule; 10 were terminated by nongasifier-associated events and 39 were terminated by gasifier-related problems.

B. Fluidized-Bed Gasifiers

Only one fluidized-bed process (Winkler) has been commercialized. Others, however, are in advanced stages of development and one may become a commercial reality in the 1980s.

1. *Dry Ash Gasifiers*

a. Winkler (Davy Powergas, Inc., Rheinische Braunkohlenwerke, West Germany) In the Winkler (Fig. 25) Process, crushed coal, typically 0/0 × 0 in., is gasified in a fluidized bed. The gasifier is cylindrical with a conical bottom. Oxygen or air and steam are introduced into the bottom, conical

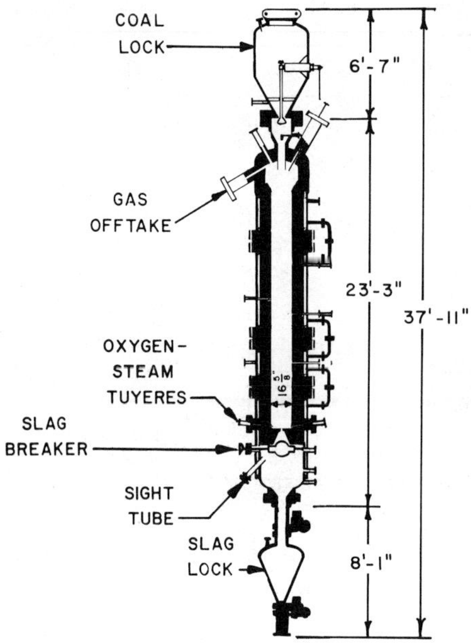

FIG. 24 Original CFETC slagging gasifier (pre-September 1978 configuration) [26].

portion, of the gasifier. The coal is fed via lock hoppers and a screw feeder into the lower cylindrical portion of the gasifier. Heavy ash particles are removed from the bottom of the gasifier by a water-cooled screw conveyor. The ash is then conveyed to a disposal hopper.

Entrained char and ash, together with gas, is subjected to further gasification in the large space above the fluid bed where additional oxygen or air and steam is introduced. Depending on the ash-fusion temperature and coal reactivity, the gasification temperature ranges from 1500 to 2000°F. The hot gas passes up through a radiant water-wall boiler section of the gasifier, where the gas is cooled (200 to 400°F) prior to leaving the gasifier. The gas is then passed through a waste-heat boiler to recover the sensible heat and then through a dry cyclone. Solid particles are removed in both the waste-heat boiler and the cyclone. Conventionally, gas is further cooled and cleaned by wet scrubbing, and if required an electrostatic precipitator can be included in the gas treatment stream.

The best performance of the Winkler Process is obtained by the use of reactive coals, such as lignite and noncaking subbituminous coals; however, noncaking bituminous coals have also been gasified. Careful consideration must be given to the coal ash-fusion temperature, coal reactivity, and size consist when designing the gasifier. The Winkler Process is fully commercial for production of low- and intermediate-Btu gas when operated at near-atmospheric pressure and at low pressure. Davy Powergas, Inc., however, will give normal process guarantees for pressure operation up to 4 atm. In addition, they have considered a joint venture to build an airblown, 8-ton/hr

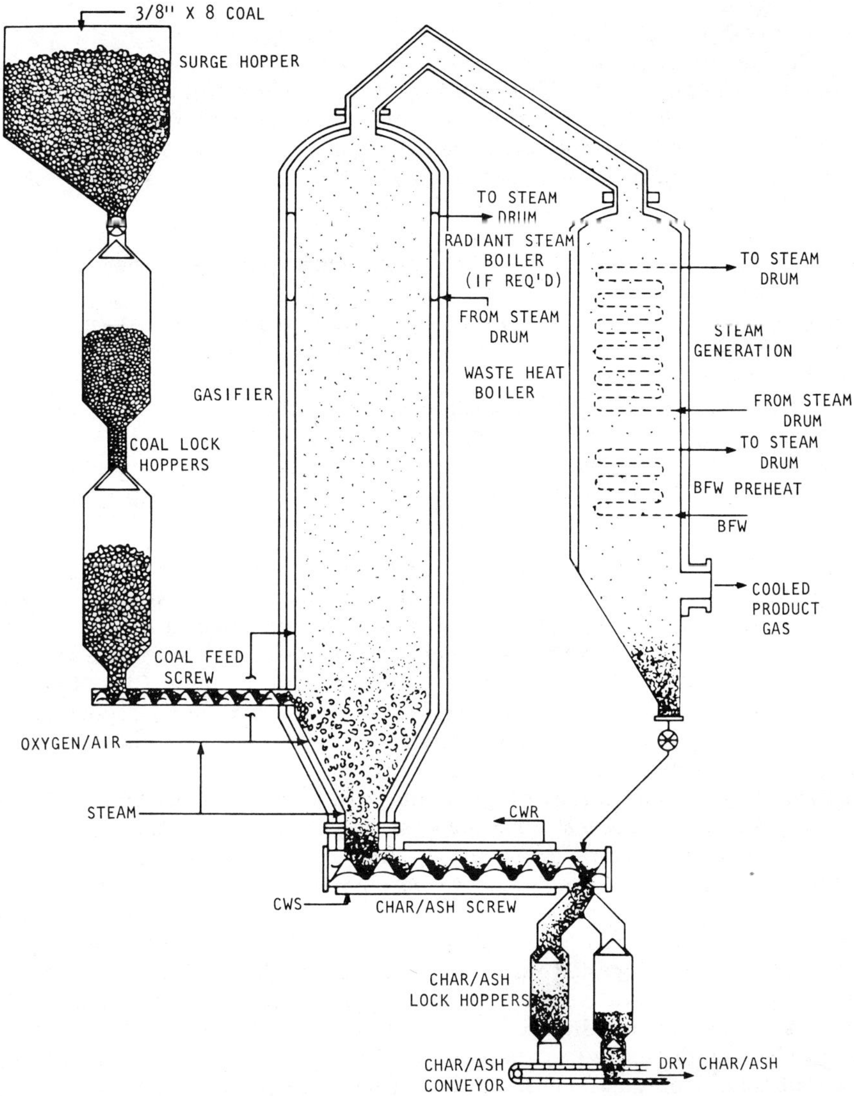

FIG. 25 Davy "Winkler" fluid-bed gasifier [27].

Winkler gasifier that will be designed to operate at pressures of up to 220 psia and will allow the use of a wide range of feedstocks, including high-sulfur coke, crude oil, residual oil, and char.

The largest gasifier built to date is 18 ft in diameter and some 100 ft in height. One 18-ft-diameter gasifier is capable of gasifying up to 700 short tons/day of coal using air as the gasifying agent, and up to 1100 short tons/day of coal using oxygen as the gasifying agent.

Another development of the Winkler Process is that of Rheinische Braunkohlenwerke (Rheinbraun). This is a high temperature/high pressure version of the conventional Winkler Process wherein limestone or dolomite is mixed with the feed coal to raise the ash-fusion temperature of low-ash-fusion-temperature coals. This allows higher gasification temperatures to be maintained. Rheinbraun calls the process the "H-T Winkler Process." In contrast to the foregoing development, this high-pressure process will use oxygen rather than air as the gasifying agent. Rheinbraun plans a 24-metric ton/day pilot plant operating at 10 bars to be commissioned in mid-1978. This pilot plant will primarily gasify lignite to produce a CO/H_2 mixture that can be used as a reducing gas for direct reduction of ores. The gas can also be methanated to produce SNG.

The first commercial Winkler gasifier was put into operation at Leuna-Werke, Merseburg, Germany, in 1926. Presently, 70 gasifiers at 24 plant locations (including two pilot plants) have been designed, constructed, and commissioned (see Table 24). Typical gas analyses for lignite feed for near-atmospheric operation are shown in Table 25.

b. CO_2 Acceptor Process (Conoco Coal Development Co., Library, Pennsylvania) The CO_2 Acceptor Process, a unique fluidized-bed system to convert lignite or subbituminous coal to pipeline gas, was developed at a large-scale pilot plant under contract with the Office of Coal Research and the American Gas Association. This 40-ton/day pilot plant, located in Rapid City, South Dakota, was successful under subcontract with Stearns-Roger Corporation. The plant has been mothballed and further development to demonstrate and commercialize is problematic at this time.

Figure 26 is a process flow diagram and Fig. 27 depicts primary reactions for the CO_2 Acceptor Process. The process requires two reactors at an operating pressure of 150 psig (11 atm). The devolatilization and gasification steps are combined in one reactor, with heat generation in the second.

The CO_2 Acceptor Process functions when heat for the endothermic steam-carbon gasification reaction is supplied by the exothermic reaction of CaO and CO_2. The process gained its name from the CaO "accepting" the CO_2. The acceptor reaction is reversed in the second reactor, the regenerator, with heat supplied by burning residual char from the gasifier with air. Lignite or subbituminous coal is fed to the bottom of the gasifier, where, after rapid hydrodevolatilization, gasification of fixed carbon with steam occurs. The gasifier temperature is in the range 1490 to 1550°F. Heat for the gasification reactions is supplied particularly by the sensible heat of the hot acceptor but principally by the exothermic CO_2 acceptor reaction:

$$CaO + CO_2 = CaCO_3 \quad \Delta H = -76{,}200 \text{ Btu/lb mol } (77°F)$$

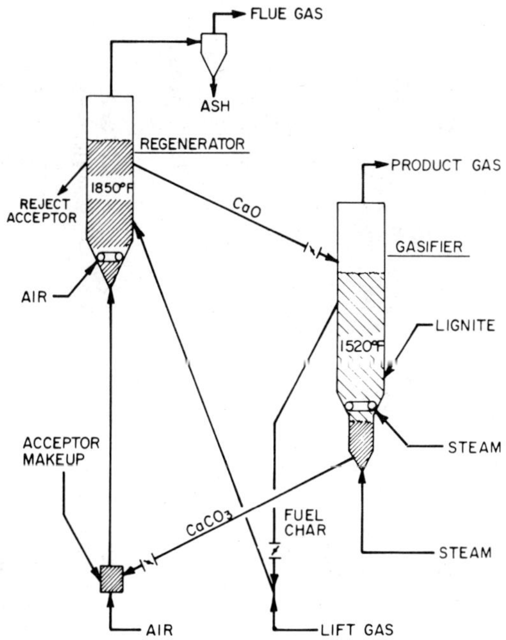

FIG. 26 CO_2-Acceptor Process diagram [28].

The CO_2 acceptor reaction is reversed in the regenerator at about 1850°F from the heat of combustion of residual char from the gasifier. Ash is removed from the regenerator by elutriation and collected via an external cyclone and lock hopper system. Seals between the gasifier and regenerator are maintained by purged standlegs of solids.

Some of the circulating inventory is purposely withdrawn from the gasifier and replaced with fresh stone makeup because the acceptor loses chemical reactivity as it circulates repeatedly between the reactors. This makeup is added to the acceptor entering the regenerator.

	HEAT OF REACTION BTU/MOL at 298.16°K
$C + H_2O(g) \rightarrow CO + H_2$	56,500 Endothermic
$CO + H_2O(g) \rightarrow CO_2 + H_2$	17,700 Exothermic
$CaO + CO_2 \rightarrow CaCO_3$	76,200 Exothermic
$CaCO_3 \rightarrow CaO + CO_2$	76,200 Endothermic
$2C + H_2 + H_2O(g) \rightarrow CH_4 + CO$	24,300 Endothermic

FIG. 27 Pertinent reactions in CO_2-Acceptor Process [29].

TABLE 24 Davy "Winkler" Coal and Coke, Fluid-Bed Gasification Plants [27] (as of January 24, 19"9)

Plant number	Plant	Product	Capacity per gasifier (1000 scf/hr)	Number of gasifiers	Operating dates
1	BASF, Ludwighsaven, West Germany	Pilot plant	75	1	1925–58
2	Leuna-Werke, Merseburg, East Germany	Fuel gas and synthesis gas for MeOH and NH_3	3730 1870	4 1	1926–70
3	BRABAG, Bohlen, East Germany	Hydrogen	1120	3	1938-present
4	BRABAG, Magdeburg, East Germany	Hydrogen	1230	3	1938–45
5	Yahagi, Japan	Ammonia	330	1	1937–60
6	Dai-Nihonyinzo-Hiryo, Japan	Ammonia	520	2	1937–59
7	Nippon Tar, Japan	Ammonia	520	2	1937–60
8	Toyo Koatsu, Japan	Ammonia	750	2	1938–69
9	Fushum, Mandschukko,[a] Japan	Syngas for F.T. fuel	750	4	1939–?
10	BRABAG, Zeit, East Germany	Hydrogen	840	3	1941-present
11	Treibstoffwerke, Brux (now Most), Czechoslovakia	Hydrogen	1120 1200[a]	5 2	1943–73 4954–73
12	Salawad, USSR[b]	Medium-Btu gas	860	7	?-present
13	Baschkirien, USSR[b]	Medium-Btu gas	860	4	?-present

14	Kombinat, Schwarze Pumpa, East Germany[b]	Low-Btu gas	2400	6	?-present
15	Dimitroffgrad, Bulgaria[b]	Medium-Btu gas	670	4	1951-present
16	Stara Zagora, Bulgaria[b]	Medium-Btu gas	1120	5	1962-present
17	Fabrika Azotnih, Jendinjenja, Gorazde, Yugoslavia	Ammonia	260	1	1952-present
18	Calvo Sotelo I, Puertollano, Spain	Ammonia	350	1	1956-70
19	Calvo Sotelo II	Ammonia	350	1	1959-70
20	UKV, Wesseling I West Germany	Synthesis gas for MeOH and NH_3	630	1	1958-67
21	UKV, Wesseling II, West Germany	Synthesis gas for MeOH and NH_3	630	1	1962-67
22	Azot Sanayii TAS, Kutahya, Turkey	Ammonia	450	2	1959-present
23	Meyveli Lignite Corp., India	Ammonia	745	3	1961-present
24	UK-Wesseling HTW,[a] West Germany	Demonstration plant	60	1	1978-present
24	(including pilot plants)			70	(including pilot plants)
11	In operation at present			39	

[a]Replica Winkler plants built by others.

[b]Plants in the Soviet Bloc that were built as replicas of existing plants.

TABLE 25 Gas Analyses from Winkler Gasifier [27]

Case	O_2-blown lignite	Air-blown lignite
Coal ultimate analysis (wt %)		
C	58.9	56.3
H	4.5	4.3
O	21.8	15.0
N	0.9	0.7
S	0.4	3.0
Ash	4.5	12.7
Moisture	9.0	8.0
Total	100.0	100.0
HHV (Btu/lb)	9800	9890
Gas composition (dry basis) (vol %)		
H_2	35.1	12.6
CO	48.2	22.5
CO_2	13.8	7.7
CH_4	1.8	0.7
N_2 + A	0.9	55.7
H_2S + COs	0.2	0.8
Total	100.0	100.0
HHV (Btu/scf	288	125
Scf of dry gas/lb coal	24.8	47.1

The acceptor has three principal advantages:

1. The acceptor is heated and calcined in a separate reactor where air can supply the oxygen for combustion without diluting the product gas with nitrogen. This eliminates the need for an oxygen plant.
2. The required circulation rate is lower than in other heat-carrier processes, because most of the heat is supplied chemically. Figure 28 is a temperature/enthalpy diagram which shows the distribution of sensible heat and heat of reaction as supplied by 1 lb mol of dolomite acceptor with 40% activity and 1870/1500°F temperature differential between the gasifier and regenerator. In this example, only 24% of the heat supplied is sensible heat, the remainder is heat of reaction. Other conditions of acceptor activity and reaction temperatures will alter these proportions somewhat.
3. The acceptor reacts with both H_2S and CO_2, the principal impurities in the gasifier product gas, thus minimizing the gas cleanup requirements.

Operation of the pilot plant successfully confirmed the feasibility of the process. Runs of adequate duration at self-sustained, steady-state operating conditions were accomplished to provide scale-up data.

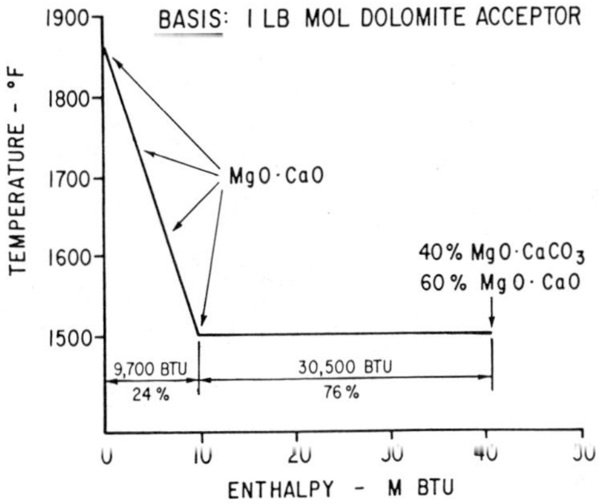

FIG. 28 Heat supply of acceptor at 40% activity [29].

c. HYGAS (Institute of Gas Technology, Chicago, Illinois/Gas Research Institute/Department of Energy) The IGT HYGAS Process for the hydrogasification of coal is tailored to maximize the direct production of methane in the hydrogasification reactor by the use of high temperature (1200 to 1700°F) and high pressure (1000 psig). The process can use caking bituminous coals as well as noncaking lignite and subbituminous coals. Crushed coal is fed in an oil slurry to the gasifier, where, in a hydrogen-rich atmosphere, two-thirds of the methane for the final product gas is produced. The gas produced in the hydrogasifier is purified and raised to pipeline quality by catalytic methanation. The hydrogasified char is used for hydrogen production. Three processes are available for the production of hydrogen from coal char: steam-oxygen, steam-iron, and electrothermal systems.

Methane can be produced directly by the hydrogenation of coal or indirectly by reacting steam with carbon to form CO and H_2, which can then be methanated. The HYGAS Process is tailored toward the improvement of thermal efficiency by maximizing direct methane formation. High temperature is needed to obtain reasonable reaction rates; high pressure is used to increase equilibrium methane yield. The most active fraction of coal is hydrogasified to form methane and the less active fraction is used to generate hydrogen for hydrogasification. The raw gas from the hydrogasifier contains carbon monoxide and hydrogen. These are converted to methane through catalytic cleanup methanation to boost the heating value of the gas and to reduce the carbon monoxide content to the pipeline standard of less than 0.1%. Of the total methane formed, 65 to 70% is produced by the direct route in the hydrogasifier. This key feature contributes significantly to the process's high overall thermal efficiency of over 70%.

Diagrams for steam-oxygen, steam-iron, and electrothermal versions of HYGAS are presented in Figs. 29, 30, and 31, respectively. All versions have been successfully operated in a large pilot plant using 3-ton/hr bituminous, subbituminous, and lignite coals. The steam-oxygen version has been operated in the completely self-sustained mode for several days at a time with each type of coal.

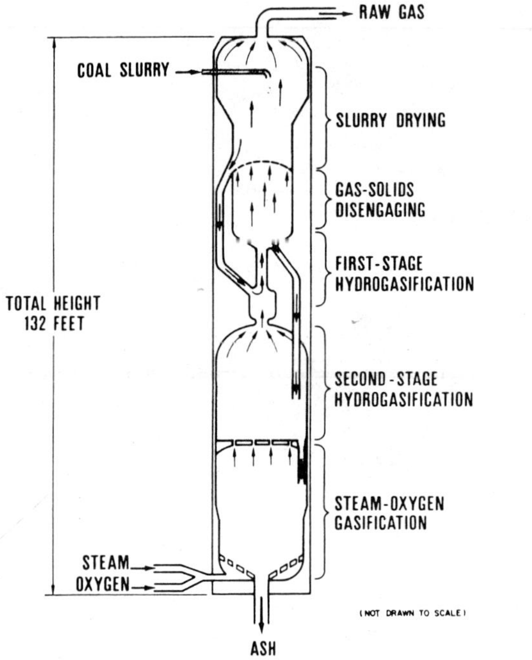

FIG. 29 HYGAS with steam-oxygen [30].

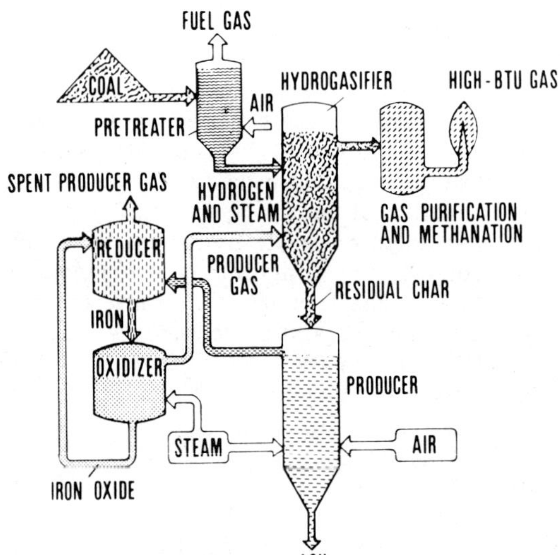

FIG. 30 HYGAS with steam-iron [30].

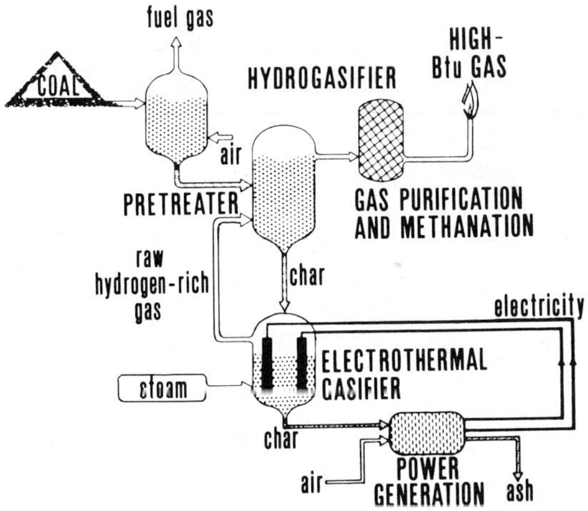

FIG. 31 HYGAS with electrothermal [30].

The coal is dried and crushed to a size consistency of less than 8 mesh, with a minimum of material finer than 100 mesh. Bituminous coal (subbituminous and lignite bypass this step) is pretreated in an air-blown fluid bed at 650 to 750°F. The resulting slight oxidation (about 10%) eliminates the tendency of the coal to cake.

The dry, prepared coal passes on weigh belts into slurry tanks containing a light oil produced in the process. The slurry is kept in constant circulation at low pressure through a pipe loop past the suction side of a conventional oil-field reciprocating mud pump and back to the slurry tanks to prevent solids settling. The mud pump pressurizes the slurry to the gasifier reactor pressure of 1000 psig. The coarse coal particles and low-viscosity oil, both of which make it easy for the coal to settle, cause this slurry to be more difficult to handle than others reported in the literature. In the pilot plant, IGT successfully pumped slurry with concentrations of 45% by weight of solids. A heavy-duty centrifugal pump keeps the slurry in constant motion. This slurry feeding is unique to the HYGAS Process.

In the pilot plant, the hydrogasifier reactor vessel, with four internally connected contact stages, is 135 ft high and has a 5.5-ft ID. The slurry is sprayed into the top stage, which is a fluidized drying bed. The sensible heat in the gaseous reaction products from the later stages vaporizes the oil and heats the dry coal to about 600°F.

The first stage of hydrogasification occurs when coal flows by gravity from the drying bed into a vertical-lift-line reactor. Hot gases (1700°F) from the reaction section below provide the lifting force and the heat to raise the solids temperature to 1200°F. Hydrogen then reacts with about 20% of the coal to produce methane. At the top of the lift line, the gas and coal disengage, and the gas moves up to vaporize the oil in the slurry-drying stage.

The remainder of the partially reacted coal then flows by gravity to the second-stage hydrogasifier. Here the solids are heated in a fluidized bed

TABLE 26 Gas Composition Leaving the Hydrogasification Reactor (Oil-Free) [30]

Component	Steam-oxygen	Steam-iron	Electrothermal
CO	18.0	7.4	21.3
CO_2	18.5	7.1	14.4
H_2	22.0	22.5	24.2
H_2O	24.4	32.9	17.1
CH_4	14.1	26.2	19.9
C_2H_6	0.5	1.0	0.8
H_2S	0.9	1.5	1.3
Other	0.8	1.4	1.0
Total	100.0	100.0	100.0

to about 1700°F and further gasified by the steam and hydrogen-rich gas rising from the steam-oxygen gasification stage below. (Alternatively, the hydrogen-rich gas may flow from the steam-iron reactor of the electrothermal generator.) In this second stage of the reactor, the exothermic hydrogen reactor produces methane, and the endothermic steam reaction produces carbon monoxide and hydrogen. This produces a balanced and controlled temperature so that the heat of methanation, which otherwise would be wasted, is used to produce hydrogen.

The char then flows by gravity to the hydrogen-producer stage, where, depending on the process being used, the char undergoes different degrees of additional gasification. The steam-oxygen gasifier used as the hydrogen producer in the pilot plant is directly below the second-stage hydrogasifier. The steam and high-purity oxygen introduced into the gasifier convert char into hydrogen and carbon oxides at 1850°F. Dry ash is discharged from this stage as a water slurry.

The composition of the gas leaving the slurry drier at the top of the hydrogasification reactor depends on the type of hydrogen producer, as shown in Table 26. In addition to these major components, the gas contains slurry oil, coal dust, and trace constituents such as ammonia and hydrogen cyanide.

The gas is cleaned, purified, and its composition adjusted by water gas shift prior to methanation. As it enters the methanation section, the purified gas would typically have the composition shown in Table 27. The methanation step has two purposes: to raise the heating value of the gas to near that of methane, and to reduce the carbon monoxide concentrations to the requisite 0.1% or less.

(1) Steam-oxygen. The steam-oxygen gasifier (Fig. 29) produces a hydrogen-rich gas by reacting the hot char from the hydrogasifier with steam and oxygen in a high-pressure fluidized bed. An operating temperature

of 1800 to 1900°F is maintained by controlling the quantities of steam and oxygen.

Multiple-feed gas cones inject premixed steam and oxygen at the bottom of the bed. Nonslagging conditions are maintained by a high fluidizing velocity at the reactor bottom and by the gas inlet design, which avoids stagnant areas and promotes rapid initial contact of oxygen, steam, and char. Ash is discharged through the space between the cones.

The hydrogen content of the gas from this gasifier is the lowest of the three HYGAS hydrogen systems. The mechanical simplicity of this system is partially offset by the cost of separating oxygen from air, removing additional carbon dioxide, and by the need for a larger hydrogasifier reactor.

(2) Steam-iron. The steam-iron version of the HYGAS Process (Fig. 30) is complex compared with the other versions. It is, however, both potentially superior in efficiency and in the cost of the gas produced. Spent char from the hydrogasifier is fed directly to the producer, in which the char reacts with air and steam to generate a gas capable of reducing iron oxide to iron. The hot reducing gas is fed directly to the reducer, where it contacts a recirculating stream of iron oxide and reduces it to iron. The iron is contacted with steam in the oxidizer and reoxidized to iron oxide, producing a quite pure hydrogen-steam mixture. The very low carbon oxide content of the gas results in a very low carbon oxide content in the raw gas from the hydrogasifier. This, combined with the very high hydrogen content, permits a small hydrogasifier reactor size and much reduced scrubbing requirements.

(3) Electrothermal. The electrothermal method for producing hydrogen-rich gas (Fig. 31) functions when excess char from the HYGAS hydrogasification reactor reacts with steam. Heat is furnished to the highly endothermic steam-carbon reaction by direct-current heating of a fluidized bed of char. Excess char from the electrothermal reactor is used to produce the required electric power. In the years since the development of this method, the cost of electric power has risen sharply, resulting in this method being the least economical of the three. It is, however, a highly viable method of producing hydrogen from coal or char.

A commercial plant is under preliminary design with a view toward launching a large-scale demonstration project in the near future.

TABLE 27 Gas Composition Entering the Methanator [30]

Component	Steam-oxygen	Steam-iron	Electrothermal
CO	18.0	12.8	16.8
CO_2	2.0	2.0	2.0
H_2	54.0	38.5	50.5
CH_4	25.0	45.0	29.5
C_2H_6	1.0	1.7	1.2
Total	100.00	100.0	100.0

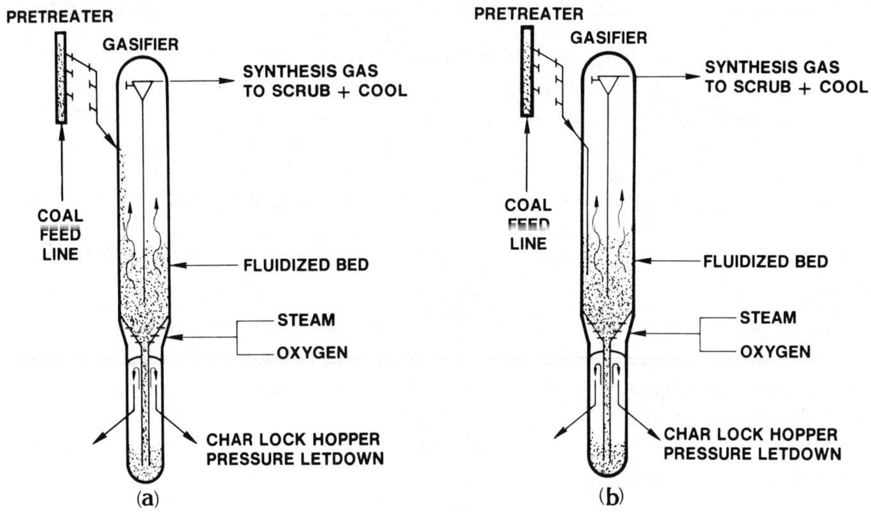

FIG. 32 Synthane fluid-bed gasifier: (a) free-fall feed; (b) deep-bed injection [31].

d. The Synthane Process (Pittsburgh Energy Technology Center of the U.S. Department of Energy) The Bureau of Mines is developing the Synthane Process (Fig. 32), a fluidized-bed gasifier that is superficially similar to the HYGAS-oxygen system. The Synthane Process employs the pretreatment of bituminous coal in free-fall or in a fluid bed with oxygen and steam at the top of the reactor, rather than externally as in the HYGAS Process. The gas produced is, therefore, part of the product. The reactor design is simpler, being one compartment in which the oxygen-steam combustion, hydrogasification, and pretreatment occur.

The advantages claimed by the Synthane Process are as follows:

1. Caking coals can be used directly, as can a wide range of other coals, including lignite.
2. External coal pretreatment is avoided. Raw coal with a minimum of oxygen and steam is fed directly to the gasifier feed pipes; hence, costly external pretreatment with air or oxygen is avoided and all the coal hydrocarbon precursors are retained.
3. Maximum methane is produced in the gasifier. By using the Bureau's concept of operating at relatively low temperature, more than half of the ultimate methane is made directly in the gasifier.
4. Minimum oxygen is required. By providing for maximum methane, minimum oxygen is needed. Also, part (as much as one-third) of the coal is converted to a char, all of which is burned with air to provide power for the process.
5. Modern fluid-bed technology is utilized. Advances in process engineering and design are incorporated for investment savings and increase in efficiency.

6. Costs of compression are minimized because the process operates at high pressure (1000 psi).
7. Lower oxygen plant investment is possible. Substantial technological progress in manufacturing oxygen has been made in the last few years. Oxygen plants of larger size are now possible and are significantly cheaper.
8. Smaller shift purification and methanator units are required. By maximizing methane in the gasifier, the amount of CO + H_2 is minimized with savings in the subsequent shift, purification, and methanation units.
9. The purified product gas contains 20 to 30% methane; the balance is H_2 + CO. The sizes of all process vessels and lines downstream from the gasifier are reduced as much as 30 to 40%; the final methane reactor will be reduced in size about 30% from the size it would need to be if only H_2 + CO were in the feed gas.
10. An improved methanator is used. The Bureau's tube wall and the hot gas recycle system are improvements over previous processes.
11. Technological success is rated high. All elements of the process have a substantial background in coal technology. In particular, the steam-oxygen reaction with coal, including reaction in a fluidized bed, has been well established. The process does not have serious unknown or untested steps.
12. Co-product manufacture is possible: (a) CH_4, (b) CH_4 and H_2, or (c) CH_4 + H_2 + CO. This flexibility can be attractive in providing for the most economical manufacture.

The results obtained in small-scale equipment with four types of coals have widely different properties. Their chemical analyses are given in Table 28.

A 3-ton/hr pilot unit was first operated in July 1976. The operation was terminated at the end of 1978. The results of runs made with subbituminous and bituminous coals are presented in Table 29.

e. COGAS (COGAS Development Company, Princeton, New Jersey) The COGAS Process consists of two distinct steps, both of which are carried out in fluidized beds. The coal is first pyrolyzed in a series of steps during which the temperature is gradually raised to form a liquid product. The char left from pyrolysis is in part gasified and in part burned with air to generate heat for production of hydrogen-rich gas and for sensible heat.

Specifically, unwashed, sized coal is crushed, screened, and pressurized. The coal is transported to pyrolysis, where it is dried and heated in stages to drive off volatile material. The residue char then flows to the gasifier, where it reacts with steam in a fluidized bed to produce synthesis gas. The synthesis gas is circulated back through pyrolysis, where it is mixed with other products of pyrolysis. This mixture of synthesis gas and oil vapors is sent to oil recovery.

Heat for the gasification reaction and for the pyrolysis stage is supplied primarily by the combustion of char fines generated in the process. An oxygen plant is not required in the COGAS Process.

The mixture of synthesis gas and oil vapors is quenched in the oil recovery stage, and the gas and raw oil are separated. The raw oil is sent

TABLE 28 Synthane Gasification Tests with Various Coals [32]

Type of coal	Pittsburgh seam		Illinois No. 6		Lignite	Montana	Goals
Free swelling index	8-8.5	8-8.5	4.5	4.5	N.C.	N.C.	
Coal feed rate							
lb/hr	10.5	10.6	8.7	10.4	21.2	9.7	
lb/hr-ft^3 gasifier	24	22	19	21	47	22	
Steam/coal ratio (scf/lb)	28	23	34	28	14	26	
O_2/coal ratio (scf/lb)	7.6	4.9	7.6	3.9	3.4	3.5	
Bed height (in.)	95	84	84	86	77	72	
Temperature (°C)							
Pretreater	420	430	420	410	440	410	
Gasifier, av.	907	926	887	904	829	834	
Gasifier, max.	975	950	925	950	870	910	
Carbon conversion (%)	67	70	72	74	77	71	65
Product gas/lb coal (MAF)							
H + CO + CH_L (scf)	17	17	18	16	17	16	17
CH_4 (scf)	3.9	4.3	3.7	4.2	3.6	3.0	4.7
Gas analysis (%)							
H_2	34	28	36	31	31	38	
CO	19	25	19	20	25	11	
CH_4	16	18	14	17	14	11	
CO_2	32	29	31	32	30	40	
% CH_4 (CO_2-N_2-free)	23	25	21	26	21	19	
Tar (% of feed)	2.9	6.0	3.4	6.0	1.4	3.3	

TABLE 29 Synthane Gasifier Data [33]

Run number	Subbituminous		Bituminous	
	1-DB-0/7	1-DB-0/8	2-DB-0/1	2-DB-0/2
Data period (hr)	6		31.0	9.5
Ground coal feed (tons/hr)	2.15	2.5	1.9[a]	1.4[a]
Gasifier O_2/ground coal (scf/lb)[b]	3.96	3.96	3.0	3.4
Gasifier steam/ground coal (lb/lb)	1.48	1.43	2.4	2.2
Gasifier bed height (ft)	25.2	23.5	15.5	15.7
Gasifier bed density (lb/ft^3)	6.2	5.6	9.2	8.1
Gasifier bed temperature (°F)	1505	1540	1455	1412
Pretreater O_2/ground coal (scf/lb)[b]	0.62	0.83	0.83	0.89
Pretreater steam/ground coal (lb/lb)	0.44	0.38	1.4	1.6
Pretreater bed density (lb/ft^3)	27.7	26.6	14.3	9.9
Pretreater temperature (°F)	700	700	848	870
Equivalent CH_4/MAF coal (scf/lb)[c]	3.43	3.59	2.3	1.2
Carbon conversion (char analysis) (%)	64.9	78.5	46.8	41.6
Carbon conversion (gas analysis) (%)	69.9	80.1	34.0	56.2
Raw gas production (scf/hr) (dry)	117,900[d]	152,000[d]	66,600[e]	37,360[e]
Dry gas analysis (mol %)				
CO_2	50.9	47.2	35.4	49.0
CO	9.0	10.7	5.8	7.6
H_2	29.8	32.1	21.3	15.3
CH_4	9.9	9.6	10.0	7.1
C_2H_6	0.2	0.1	0.6	0.5
H_2S	0.2	0.1	1.3	0.6
N_2	0.0	0.3	17.8	19.5

[a]Moisture-free coal feed by heat balance.

[b]Standard conditions, 60°F, 1 atm.

[c]Equivalent CH_4/MAF coal = (CH_4 + 2 × C_2H_6)/moisture- and ash-free coal.

[d]Includes transport, Petrocarb, and purge CO_2 for Run 1-DB-0.

[e]Includes about 26,700 scf/hr of transport, Petrocarb, and purge CO_2 for Run 2-DB-0/1 and 7230 for Run 2-DB-0/2.

to oil treatment and the synthesis gas goes through purification, methanation, CO_2 removal, compression, and drying. The 950-Btu/ft^3 product is then sent to market.

A portion of the purified synthesis gas is diverted and sent to a hydrogen plant. The raw oil is hydrotreated in an ebullating bed reactor with hydrogen produced from purified synthesis gas. Further processing yields naphtha and fuel oil products.

The hydrogen sulfide removed from the synthesis gas is recovered as sulfur in a conventional Claus plant.

The flue gas generated by the combustion of char fines in the gasification area is sent to the heat and power recovery area, where it is cooled by producing steam, cleaned, and then expanded through turbines to produce shaft horsepower for the compression of combustion air and for the generation of electricity.

Finally, the flue gas passes through an SO_2 removal process before being discharged to the atmosphere. Coal pyrolysis is an old art which has been significantly advanced by the COED Process developed by FMC Corporation under funding by the Office of Coal Research (OCR). This contract work stemmed from a fluidized-bed pyrolysis process that FMC had previously developed in order to produce formcoke from noncoking Western coals for its elemental phosphorus plant in Pocatello, Idaho. Sponsorship by OCR extended this development into the production of oil with the successful operation of a 36-ton/day pilot plant in Princeton, New Jersey.

This plant processed over 20,000 tons of coal of seven different types from lignite to high-volatile A bituminous. Results of these pilot plant tests are summarized in Table 30.

Since in the COGAS Process, pyrolysis is conducted in fluidized beds in the absence of free oxygen, agglomeration must be avoided by operating each stage at a temperature below the coal-softening point. As volatiles are removed, the softening temperature is raised and the coal can be advanced to a progressively higher temperature stage.

The char from pyrolysis is fed to the gasifier which is designed to use air for combustion, achieve complete char utilization by producing a carbon-free ash, and operate at moderate pressure.

The gasifier is a fluidized-bed unit in which steam serves as the fluidizing medium and reacts with the char to produce synthesis gas (hydrogen and carbon monoxide) with the absorption of heat. The heat for the gasification reactions is supplied from a slagging cyclone combustor, in which a portion of the char—the fines from the system—are burned with air. A recirculating stream of char from the gasifier is used as the heat-transfer medium. This char is heated in a transport lift tube by flue gas from the slagging cyclone combustor, and returns to mix with the gasifier bed. The ash is discharged as molten slag from the combustor via a water quench system.

The Illinois Coal Gasification Group (ICGG) is a partnership of subsidiaries of five major utilities:

1. Northern Illinois Gas Company
2. The Peoples Gas Light and Coke Company
3. Central Illinois Light Company
4. Central Illinois Public Service Company
5. North Shore Gas Company

TABLE 30 Summary of Pyrolysis Pilot-Plant Operations [37] June 1970-November 1974

Coal								
State	Colorado	Wyoming	Illinois	Illinois	North Dakota	Utah	Western Kentucky	West Virginia
Seam	Somerset	Monarch	No. 6	No. 6	—	Hiawatha	No. 9 and 14	Pittsburgh
Mine	Bear	Big Horn	Peabody No. 10	Sahara No. 6	Glen Harold	King	Colonial	Ireland
Rank	hvBb	Sub B	HvCb	HvBb	Lignite	hvBb	hvBb	hvAb
Longest period of continuous oper. (days)	30	16	22	17	14	15	14	23
Coal processed (tons)	3500	1000	6500	290	880	1680	1380	750
Pyrolysis staging temperature (°F)								
Stage 1	540-550	310-525	350-550	550	300-350	310-350	360-535	400-450
Stage 2	850-860	720-885	770-840	810	690-720	810-850	740-810	720-730
Stage 3	1000-1010	940-1050	960-1100	1050	920-990	965-1000	865-950	830-850
Stage 4	1450-1540	1410-1470	1400-1620	1530	1235-1300	1400-1475	1210-1300	1200-1220
Flows								
Coal to stage 1 (lb/hr (wet)	2040-2380	1710-2600	1410-3050	1880	2000-2630	1700-2750	1325-2800	1200-1500
Fluidizing Gas to stage 4	O_2/N_2	O_2/steam	O_2/steam	O_2/N_2/steam	O_2/steam	O_2/steam	O_2/steam	O_2/steam
Product output per ton of coal (as received, wet)								
Oil (bbl)	0.70-0.79	0.43-0.54	0.74-1.01	0.88	0.16-0.19	0.83-1.05	0.63-0.87	0.80-0.85
Char (lb)	960-1050	640-880	980-1160	1200	685-760	980-1150	940-1250	1225-1375
Gas (N_2-, CO_2-, H_2O-, H_2S-free) (scf)	6710-8340	11600-14510	4260-11490	12800	6060-8930	6990-9080	3500-4200	3300-3800
Liquor (gal)	16-32	14-19.2	16-110	23	88-103	13-44	39-71	160-170
Coal moisture (as received (wt %)	8.1-10.9	14.7-22.6	10.2-15.9	5.5	36.8-41.1	6.3-7.6	7.8-13.9	1.6-3.2

They have agreed to design and build both a COGAS demonstration plant and a commercial gasifier.

At this stage of the project, definitive design of the demonstration plant is only partially complete; however, the conceptual commercial design is finished. The demonstration plant will process 2300 tons/day of a mixture of Illinois Nos. 5 and 6 coals (the only other inputs required are approximately 1.7 million gallons of water per day, 600 kW of purchased power, chemicals, and catalysts). With 2300 tons/day of the Illinois Nos. 5 and 6 coal blend, the demonstration plant will produce approximately 24 million ft^3/day of synthetic pipeline gas (SPG), 1570 barrels of fuel oil, 390 barrels of naphtha, 57 tons of sulfur, and 4 tons of ammonia. The gas produced represents about 65% of the total Btu yield of the process. In addition, the process will be tested for shorter periods using a Western subbituminous coal and a highly caking Eastern coal to demonstrate the flexibility of the process for a wide range of coal feeds.

f. Catalytic Coal Gasification (Exxon Research and Engineering Co., Baytown, Texas) Exxon Research and Engineering Company has been pursuing coal gasification in general since the late 1960s and catalytic coal gasification (CCG) since 1971. It has been known for some time that alkali metal salts of weak acids (e.g., K_2CO_3, Na_2CO_3, K_2S, Na_2S) catalyze the steam gasification of coal. Exxon Research confirmed that at catalyst concentrations on the order of 10 to 20 wt % K_2CO_3 on Illinois bituminous coal, commercially acceptable gasification rates could be obtained at 1300°F (700°C) versus about 1700°F (925°C) without catalyst. Relatively high catalyst loadings are required to achieve acceptable rates. The high cost of catalyst necessitates catalyst recovery systems for economic balance.

This first part of the concept was well known by researchers in the field. The second part of the concept was the discovery that this same catalyst system also promoted the methanation reaction.

The third element of the process concept relates to the thermal balances for the individual reaction steps and for the overall steam gasification to methane and carbon dioxide reaction. The gasification reaction is highly endothermic, requiring the addition of 64 kcal of heat for each 2 mol of carbon.

The shift reaction is slightly exothermic, releasing 8 kcal of heat for each mol of CO shifted. By contrast, the methanation reaction is highly exothermic, releasing 54 kcal for each mol of methane formed. The overall reaction shown is slightly endothermic for graphite and slightly exothermic for real coals. In the catalytic coal gasification concept all three reactions are carried out in one vessel, avoiding the need for heat input via air or oxygen. Using the catalyst, commercially acceptable rates are claimed at 1300°F. At this temperature the catalyst equilibrates both the shift and methanation reactions. The net products are forced to CH_4 and CO_2 by recycling the CO and H_2. The embodiment of this concept in a process is illustrated in Fig. 33.

Feed coal is crushed to -8 mesh, a size distribution suitable for fluidization. Catalyst is then added as an aqueous solution, and the feed is dried and fed through lock hoppers to a fluid-bed gasifier. The gasifier operates at 1300°F (700°C) and 500 psi (3.5 MPa). The fluidization gas is a mixture of steam and recycled CO and H_2 which has been preheated as shown. Unreacted steam is condensed and CO_2 and H_2S removed by conventional acid

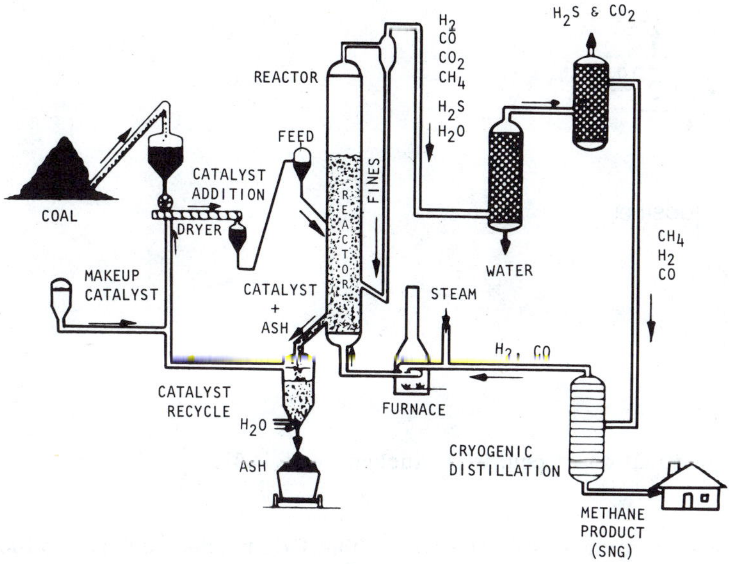

FIG. 33 Exxon catalytic gasification process [38].

gas treating (physical solvent). The product methane is separated cryogenically from the recycle CO and H_2. Char containing the catalyst, unreacted carbon, and coal minerals is removed continuously. Most of the catalyst is recovered and recycled.

Unique to this catalytic gasification system are the catalyst addition/recovery block, the fluid-bed gasifier, and the cryogenic distillation section.

The process has just emerged from the laboratory phase. Development is being initiated on a larger scale (100 lb/hr).

g. Nuclear Heat Coal Gasification In conventional gasification processes only 60 to 70% of the coal is converted to gas, while the remainder must be burned to provide the required energy. A project is under way in Germany to use process heat from a helium-cooled high-temperature nuclear reactor (HTR) as a source of energy, thus transforming the total coal to gas. The advantages claimed for this process compared with existing processes are: a savings of coal because more gas can be produced from less coal; lower pollutant emissions because the HTR is used for the production of steam and electricity instead of a coal-fired boiler; and in countries where coal prices are higher than the costs expected for the heat from HTRs, production costs are less for the gas.

Two methods of using nuclear heat have been proposed. The first is as a direct substitution for oxygen in a fluid-bed gasifier as shown in Fig. 34. Here, steam reacts directly with coal to produce CO, CO_2, and CH_4. The heat is supplied indirectly by helium flowing through tubes buried in the coal bed. The helium is also indirectly heated in an HTR.

The second method employs hot helium from the HTR to furnish heat for a steam-methane reformer (Fig. 35). The H_2 produced, after shifting and

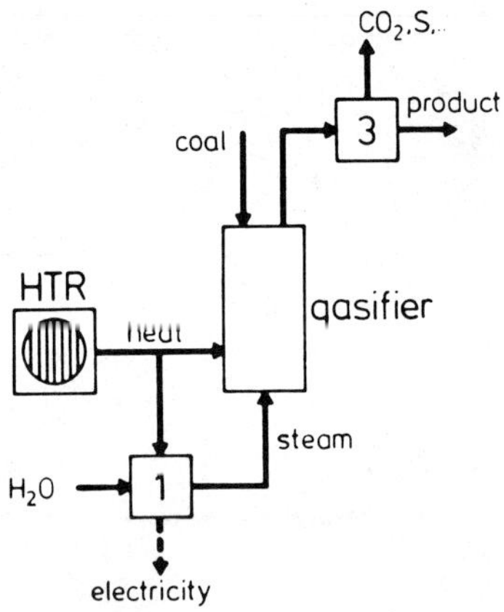

FIG. 34 Steam gasification of coal with nuclear heat [39].

scrubbing, is used to hydrogasify the coal. The CH_4 needed for H_2 production is part of that produced in the hydrogasifier.

Since February 1974, the 45-MW German AVR reactor in Jülich has operated at 950°C. Helium-heated steam reforming has been tested in a single-tube device with a throughput of 200 m^3 of CH_4 per hour. Gasifiers suitable for the combination with an HTR in pilot plants are operating with a throughput of 2 to 4 tons/day. The schedule calls for a plant that demonstrates the use of process heat in connection with an HTR of 950°C helium temperature to be operated by 1987. Thus, the process could be available on commercial scale by 1990.

2. *Agglomerating Ash Fluidized-Bed Gasifiers*

a. U-GAS (Institute of Gas Technology, Chicago, and Department of Energy) The U-GAS process is designed to produce an environmentally acceptable and economically attractive industrial fuel gas using any rank of coal. The

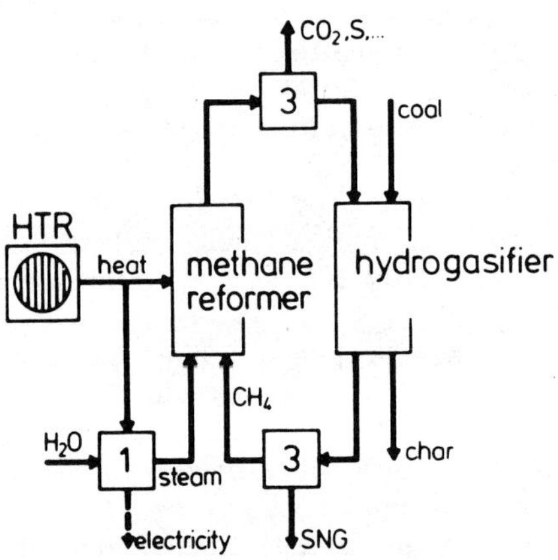

FIG. 35 Hydrogasification of coal with nuclear heat [39].

gasifier is a single-stage ash-agglomerating gasification system. Ash is discharged as a dry, nonslagged material with a low carbon content. This is accomplished by selectively heating a small conical volume in the fluidized bed in such a manner that ash material is heated to the incipient softening point of the ash. In 1960, Jequier showed that ash will selectively stick to ash as opposed to carbon. Particulate material leaving the gasifier is returned to the gasifier, where carbonaceous matter is instantly gasified. The resultant ash particles stick together to form agglomerates. When the agglomerates are sufficiently large and dense they leave the gasifier by discharging through the oxygen-steam center inlet (Fig. 36).

Crushed raw coal is introduced to the reactor at the desired pressure through a conventional lock hopper system. The coal is gasified into a CO plus H_2 gas by oxidation with oxygen and steam in a fluid bed. The oxygen and steam enter the reactor in two places. A mixture having a low concentration of oxygen in steam enters through a fluidization deck. This results in a uniformly fluidized bed, maintained at 1750 to 1850°F. A second mixture having a high oxygen concentration and high velocity enters at the center of the fluidization deck through a specially developed nozzle. This high-velocity stream "blows a bubble" above the nozzle. Locally, the temperature is at least 100°F hotter than the bed because of the high oxygen content of the stream. It is in this zone that the ash agglomerates are formed. They drop through the nozzle into a residue hopper and are cooled with water before being discharged. The medium-Btu gas leaves the vessel through an interval cyclone where most of the entrained coal fines are removed and returned to the bed. The gas is further cleaned of solids in an external second-stage cyclone. The dust from that cyclone is returned to the nozzle and reenters the bed, where it is burned. The raw gas is purified by conventional methods.

To date (late 1979), Montana subbituminous, Illinois No. 6, coke, and Kentucky No. 9 coal (both washed and unwashed) have been successfully gasified at rates of up to 25 tons/day and pressures of up to 45 psig. Early tests were conducted with air as the oxidant, but all work is now conducted with oxygen. Although U-GAS (U is for utility) was developed to furnish a low- or medium-Btu gas for combined-cycle electric power generation or for an industrial fuel, it can be used as a synthesis gas or methanated to SNG. Recently, the development has been directed toward the production of medium-Btu gas for industrial use by customers of Memphis Light, Gas and Water Division of the City of Memphis. The program, which is cosponsored by DOE, plans to design and construct a demonstration plant to gasify 3150 tons of Kentucky coal per day into fuel gas equivalent of 50 million cubic feet per day of natural gas.

b. Westinghouse Fluidized-Bed (Westinghouse Electric Corporation and DOE) The Westinghouse two-stage fluidized bed coal gasifier was conceived in 1970 to provide a clean, low-Btu gas fuel at 1600°F and 225 psia for a combined-cycle power plant. A 15-ton/day process development unit (PDU) has been designed, constructed, and operated since 1974 to investigate this process. Both a recirculating bed coal devolatilizer and a fluidized-bed char gasifier/agglomerating ash combustor have been operated successfully under a variety of operating conditions; total operational time for the plant has exceeded 4000 hr.

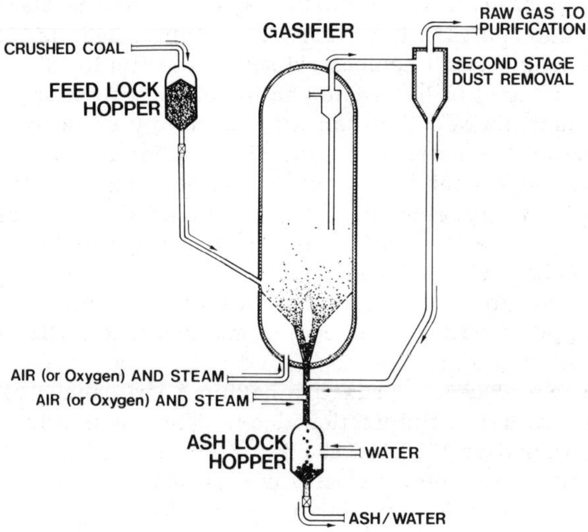

FIG. 36 U-GAS fluidized-bed gasifier [35].

Feed coals to the devolatilizer have included a Western coal, an Indiana coal, and two highly agglomerating Pittsburgh coals. Feed chars to the gasifier have included a subbituminous C char, FMC char, coke breeze, and Indiana coal char produced in the devolatilizer. Coal has also been fed directly to the gasifier. Tar-free, low-Btu gas of 40 to 120 Btu/scf heating value has been obtained from both the devolatilizer and gasifier. An intermediate-Btu gas of 280 Btu/scf has also been produced using coal and oxygen feed in the gasifier. Ash agglomerates containing 20% carbon or less, corresponding to over 90% carbon utilization, have been produced in the gasifier/agglomerating ash combustor.

The Westinghouse fluidized-bed process can be configured in two ways, depending on the end use and product characteristics desired. In the two-stage configuration (Fig. 37) the main components are a devolatilizer and a gasifier.

Coal is fed pneumatically from lock hoppers to the devolatilizer, where it is heated to 1600°F at a pressure of 225 psig in a recirculating fluidized bed. At these conditions, the coal devolatilizes to form a char consisting essentially of carbon and ash.

Recirculation of the char around a draft tube on the center line of the reactor dilutes the incoming fresh coal and prevents agglomeration of the coals in the hot, fluidized bed. The char formed in the devolatilizer is continuously drawn from the bed and from the overhead char cyclone and is fed to the gasifier. Fine char is combusted in an air or oxygen jet to provide heat for the process. Coarse char is gasified with steam to form low- or intermediate-Btu gas.

At a temperature of about 1900°F, the ash present in the char forms spherical agglomerates which are concentrated in and removed continuously from the bottom of the bed. The carbon monoxide and hydrogen-rich gases formed in the gasifier are then fed to the devolatilizer to provide the heat and fluidizing gases for this reactor.

In the single-stage configuration, coal is fed directly to the gasifier through the coaxial oxidant tube as shown in Fig. 38. The single-stage reactor has four primary zones where combustion, gasification, ash agglomeration, and ash/char separation take place. This configuration operates at a temperature of approximately 1800°F with air or oxygen as the oxidant and with steam as the gasifying medium. Recent tests have demonstrated that caking coals can be gasified in this unit without pretreatment.

Both the single- and two-stage processes can be employed with oxygen as well as air. With air, a raw gas heating value of 100 to 1500 Btu/scf is attained; with oxygen the value is 250 to 300 Btu/scf. In the latter case, the gas can be conditioned and used as a synthesis gas or methanated to form SNG.

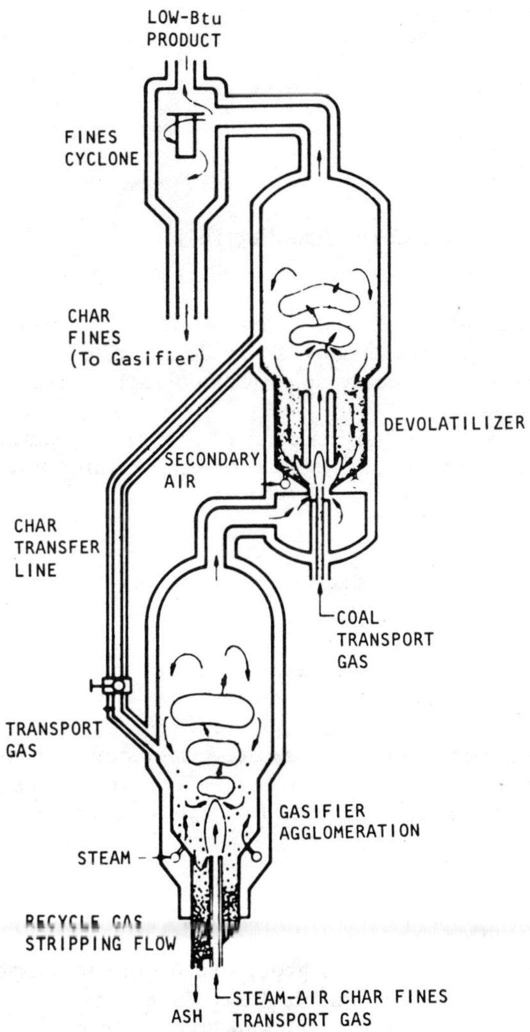

FIG. 37 Westinghouse two-stage fluidized-bed gasifier [34].

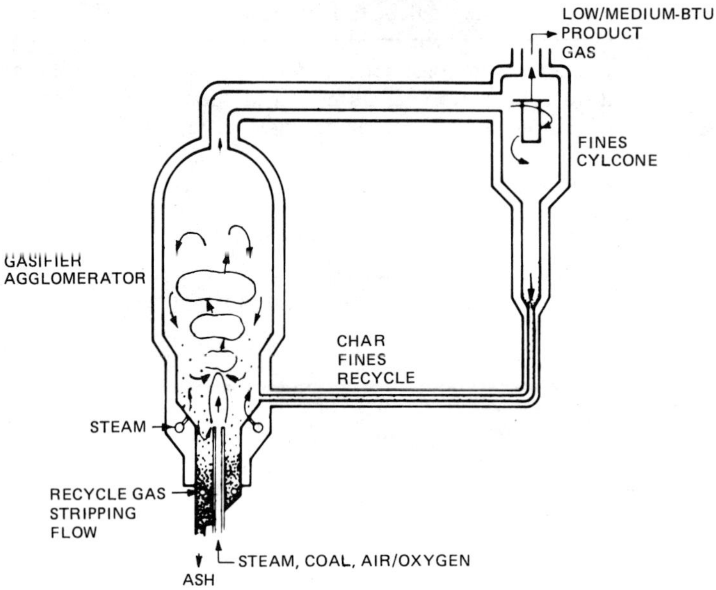

FIG. 38 Westinghouse single-stage fluidized-bed gasifier [34].

In both cases, ash is discharged as "spherical" agglomerates which can be disposed of as a dry product. The product gas can be desulfurized by hot methods such as dolomite or iron oxide or by cold methods such as hot carbonate or other commercial processes.

The results of PDU operation have verified the concept and provide data for evaluation and scale-up. Various auxiliary systems for coal feeding and gas cleaning have also been studied.

Further work is required to:

1. Define an overall process for SNG production based on the Westinghouse gasification system
2. Scale up the gasifier to commercial size
3. Specify a heat recovery system for the overall plant
4. Establish critical design and operating parameters

Present plans are to continue operation of the gasifier/combustor on coal or char with oxygen and steam through the latter part of 1978. Interconnection of the devolatilizer and the gasifier/combustor to function as a unit is planned for 1979.

C. Entrained Flow Gasifiers

Entrained flow gasifiers may be classified into two groups according to single-stage or two-stage operation. In single-stage gasification, the coal is completely gasified in an atmosphere of oxygen or air and steam. In a two-stage operation a portion of the feed coal is gasified in an atmosphere of hydrogen or synthesis gas and the reamining char is then gasified in a second stage that may or may not be in entrained flow.

Single-stage and two-stage gasifiers are presented in Table 31 with pertinent operating characteristics. More detail on each gasifier is presented in the following paragraphs.

1. *Single-Stage Entrained Flow Gasifiers*

a. HMF (high-mass flux) (Bell Aerospace Textraon, Buffalo, New York) A test unit for gasification of 0.5 ton/hr of coal has been constructed based on experience with rocket engines using fluidized powder propellants. It is a 5-in.-ID by 42-in.-long cylinder with a high-alumina castable refractory ceramic liner and an air-cooled outer jacket (Figs. 39 and 40). A dense-phase pulverized (70% through 200 mesh) coal is injected through a top center port and air, preheated to 600°F, through a surrounding annular space. An igniter at the top of the gasifier is provided to initiate reaction by firing a solid propellant charge while admitting coal and air in a controlled sequence over a 1.5-sec period. Four different injection configurations were tested, as illustrated in Fig. 39. The swirl and reverse flow injection types were unsatisfactory because coal ash slag collected around the air inlet holes. The modified reverse flow and impinging sheet injector types are satisfactory. A slag pot below the reactor has water sprays for cooling the products and separating the ash slag from the gas.

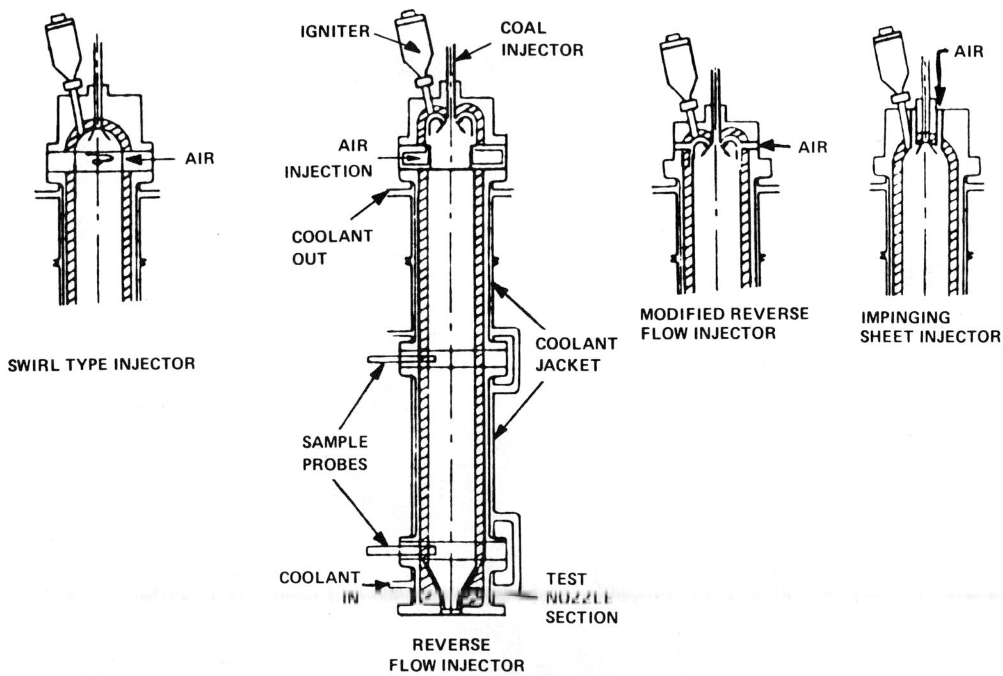

FIG. 39 Bell experimental reactor and injector types [40].

TABLE 31 Characteristics of Entrained Flow Gasifiers

	Air or oxygen		Second-stage contacting	Operating pressure	Status
	First stage	Second stage			
Single-stage gasification					
Bell HMF	Air			To 225 psi	0.5-ton/hr test unit
Koppers-Totzek	Oxygen and steam			10-12 psig	Commercial
Mountain fuel	Oxygen and steam			To 150 psig	0.5-ton/day test unit
Shell-Koppers	Oxygen or air			To 450 psi	6-ton/day pilot plant
Texaco	Oxygen or air			To 1200 psi	24-ton/day pilot plant
Two-stage gasification					
Bi-gas	Syngas and steam	Oxygen and steam	Entrained	To 1500 psi	5-ton/hr pilot plant
C-E	Combustion gas and air	Air	Entrained	Atm	5-ton/day PDU
Foster Wheeler	Syngas and steam	Air and steam	Entrained	Atm	Pilot plant under construction
Peatgas	Syngas	Oxygen and steam	Fluid bed	To 500 psi	5-10-lb/hr PDU
Rockwell	Hydrogen	Not known	Not known	To 1500 psi	0.25-ton/hr test unit

FIG. 40 Reactor components for HMF gasifier [40].

Sixty-six tests on three types of coal were conducted at 15 atm pressure and 2400°F with reaction times of 80 to 100 msec. Test durations varied from 1/2 min to 1 hr. Tests on North Dakota lignite and Montana Rosebud coals with air/coal ratios varied from 3.0:1 to 3.5:1, produced gas with a heating value from 110 to 95 Btu/scf, and carbon conversion from 78 to 92%. Under these types of gasifier conditions about 25% of the coal heating value appears as sensible heat in the hot product gas. Limited testing indicated the need for further development work in order to satisfactorily handle bituminous Pittsburgh seam coal.

b. Koppers-Totzek (Koppers Co., Inc., Pittsburgh, Pennsylvania) Coal, dried to 2 to 8% moisture and pulverized to about 70% through 200 mesh, is fed continuously by a screw conveyor into a mixing nozzle, where it is entrained by steam and oxygen into the gasifier. The gasifier (Fig. 41) is a cylindrical, refractory-lined vessel surrounded by a jacket in which low-pressure steam is generated. The refractory lining is a monolithically cast high-alumina material about 2 in. thick. Two burner heads are placed at either end of the vessel (Fig. 41) in a two-headed gasifier, or there may be four burner heads, 90° apart, in a four-headed gasifier. The largest gasifiers are 10 to 12 ft in diameter at the middle, tapering to 6 to 8 ft at the burner ends. The length is 25 ft, to give a volume of 1000 ft^3 for two-headed, and 2000 ft^3 for four-headed models.

The gasifier operating pressure is slightly above atmospheric (0.5 psig). Reaction temperature at the burner discharge is 330 to 3500°F, but the gas exits at about 2750°F because of endothermic reactions and radiation to the refractory walls. Reaction time for coal in the gasifier is 0.6 to 0.8 sec. No tar, condensable hydrocarbons, or phenols are in the product gas because of the high temperatures.

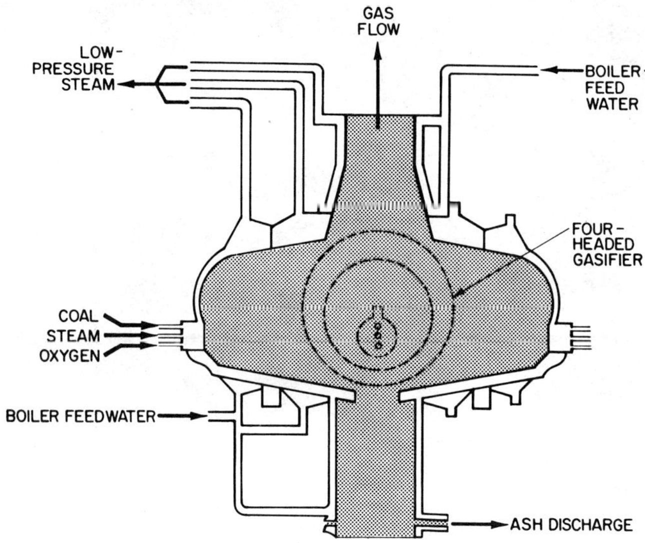

FIG. 41 Koppers-Totzek gasifier [41].

Normally, about 50% of the coal ash drops as slag into the slag-quench tank at the bottom of the gasifier; the other 50% is carried overhead as droplets in the gas and is solidified by cooling the gas with a water spray. In some cases, flux (limestone) may be added to the coal to adjust the ash-fusion characteristics and the molten slag viscosity.

Conversion of carbon in the coal to gas increases with decreasing coal rank being almost 100% for lignites and young subbituminous coals and 90 to 95% for high volatile A coals. Unconverted carbon appears mainly in the fly ash and not in the bottom slag.

Raw gas compositions and pertinent quantities of products obtained for gasification of three U.S. coals are listed in Table 32. Practically all of the sulfur in the coal is in the gas, not in the slag. Distribution of sulfur in the gas is typically 90% H_2S, 10% COS, 0.1 to 2 ppm (vol) SO_2, with no CS_2 or CH_3SH.

Table 33 lists the K-T gasifiers that have been constructed for a total coal gasification of 22,000 tons/day. Two-headed gasifiers are operating with capacities to 336 tons/day and four-headed to 672 tons/day. Four-headed gasifiers are being designed for 864 tons/day.

c. Mountain Fuel (Mountain Fuel Resources, Inc., Salt Lake City, Utah, and Ford, Bacon & Davis Utah, Inc., Salt Lake City, Utah) A small-capacity (0.5 ton/day) entrained flow coal gasifier [43] has been constructed and operated with Utah bituminous coal at pressure up to 150 psig. The refractory-lined reaction chamber is 3 in. ID, 11 in. long, and encased in a 10-in.-diameter carbon steel pressure shell. Pulverized coal is fed to the top of the gasifier through a 1/4-in.-OD line by entrainment in a recycled

TABLE 32 K-T Gasifier Data for U.S. Coals [42]

Type of coal	Western coal	Illinois coal	Eastern coal
Gasifier feed			
Dried coal to gasifier analysis (wt %)			
C	56.76	61.94	69.88
H_2	4.24	4.36	4.90
N_2	1.01	0.97	1.37
S	0.67	4.88	1.08
O_2	13.18	6.73	7.05
Ash	22.14	19.12	13.72
Moisture	2.00	2.00	2.00
Gross heating value (Btu/lb)	9,888	11,388	12,696
Oxygen (NT/NT dried coal)	0.649	0.704	0.817
Purity (%)	98.0	98.0	98.0
Process steam (lb/NT dried coal)	272.9	541.3	587.4
Gasifier products			
Jacket steam (lb/NT dried coal)	347.8	404.9	464.9
High-pressure steam (lb/NT dried coal)	2,147.1	2,292.2	3,023.6
Raw gas analysis (dry basis) (vol %)			
CO	58.68	55.38	55.90
CO_2[a]	7.04	7.04	7.18
H_2	32.86	34.62	35.39
N_2	1.12	1.01	1.14
H_2S	0.28	1.83	0.35
COS	0.02	0.12	0.04
Total	100.00	100.00	100.00
Gross heating value (Btu/scf)	295.1	290.2	294.4
Gas make (scf/NT dried coal)	51,783	59,489	66,376
Slag make (NT/NT dried coal)	0.222	0.190	0.138
Process efficiency	88.2	85.0	90.3
Coal to gas efficiency	77.3	75.8	77.0

[a]Depending on the particular coal and the required end use of the gas, the CO_2 content may be as high as 12 to 13%.

TABLE 33 Tabulation of Existing Koppers-Totzek Plants [41]

Location/company	Feedstock	Number of gasifiers	Capacity (m^3 $CO + H_2$/day)	Synthesis	Year ordered
Charbonnages de France, Paris, Mazingarbe plant, France	Bituminous coal dust, coke-oven gas, residual gas	1	75,000-150,000	Methanol and ammonia synthesis	1949
Typpi Oy, Oulu, Finland[a]	Bituminous dust,[b] oil, peat	3	140,000	Ammonia synthesis	1950
Nihon Suiso Kogyo Kaisha Ltd., Tokyo, Japan	Bituminous coal dust	3	210,000	Ammonia synthesis	1954
Nitrogen plant in Puentes de Garcia Rodriguez (Coruna), Spain[a]	Lignite dust	3	242,000	Ammonia synthesis	1954
Typpi Oy, Oulu, Finland[a]	Bituminous coal dust, oil, peat	2	140,000	Ammonia synthesis	1955
S.A. Union Chimique Belge, Brussels, Zandvoorde plant, Belgium[a]	Bunker fuel oil, plant can be extended to include coal dust gasification system	2	176,000	Ammonia synthesis	1955
Amoniaco Portugues S.A.R.L., Estarreja plant[a]	Heavy naphtha; plant can be extended to include lignite dust and anthracite dust gasification system	2	169,000	Ammonia synthesis	1956
Ptolemais nitrogen plant, Greece[a]	Lignite dust, bunker fuel oil	4	629,000	Ammonia synthesis	1959

Nitrogen plant in Puentes de Garcia Rodriguez, Spain (reconstruction)[a]	Lignite dust or naphtha	1	175,000	Ammonia synthesis	1961
Attaka nitrogen plant, Suez, United Arab Republic[a]	Refinery residual gases, LPG and naphtha	3	778,000	Ammonia synthesis	1963
Chemical Fertilizer Company Ltd. of Thailand, fertilizer plant in Mae Moh, Lampang, Thailand[a]	Lignite dust	1	217,000	Ammonia synthesis	1963
Azot Sanayii T.A.S., Ankara, Kütahya plant, Turkey[a]	Lignite dust	4	775,000	Ammonia synthesis	1966
VEB Hydrierwerk Zeitz, East Germany[a]	Distillation residues and fuel oil	2	360,000	Raw gas (hydrogen production)	1966
Kobe Steel Ltd., Kobe, Japan, for Industrial Development Corp. of Zambia, Kafue, Lusaka, Zambia[a]	Bituminous coal dust	1	214,320	Ammonia synthesis	1967
Nitrogenous Fertilizers Industry S.A., Athens Ptolemais nitrogen plant, Greece[a]	Lignite dust	1	165,000	Ammonia synthesis	1969

(continued)

(Table 33 continued)

Location/company	Feedstock	Number of gasifiers	Capacity (m^3 CO + H_2/day)	Synthesis	Year ordered
The Fertilizer Corp. of India Ltd., New Delhi, Ramagundam plant, India	Bituminous coal	3	2,000,000	Ammonia synthesis	1969[c]
The Fertilizer Corp. of India Ltd., New Delhi, Talcher plant, India	Bituminous	3	2,000,000	Ammonia synthesis	1970[c]
Nitrogenous Fertilizers Industry S.A., Athens, Ptolemais nitrogen plant, Greece[a]	Lignite dust	1	242,000	Ammonia synthesis	1970
The Fertilizer Corp. of India Ltd., New Delhi, Korba plant, India	Bituminous coal dust	3	2,000,000	Ammonia synthesis	1972[c]
AE & CI Ltd., Johannesburg, Modderfontain plant, South Africa[a]	Bituminous coal dust	6	2,150,000	Ammonia synthesis	1972
Indeco Chemicals Ltd., Lusaka, Kafue Zambia[a]	Bituminous coal dust	1	220,800	Ammonia synthesis	1974
Indeco Chemicals Ltd., Lusaka, Kafue Zambia	Bituminous coal dust	2	441,600	Ammonia synthesis	1975[d]

Petrobras, San Jeronimo, Brazil	Bituminous coal dust	2	1,300,000	Ammonia synthesis	1979[e]

Note: All plants have been built by Krupp-Koppers (formerly H. Koppers) of Essen, West Germany, process licensors.

[a]Plants in daily operation.

[b]The plants in Finland and Belgium have converted to an oil feedstock using Shell's partial oxidation and continue daily operation.

[c]The Indian plants in Ramagundam and Talcher are in the startup phase during the first quarter of 1979. The process engineering for the Korba plant is complete; however, further work has been suspended.

[d]The latest gasification expansion to the plant in Zambia is complete. However, the expanded utilities support is incomplete, preventing operation.

[e]Krupp-Koppers announced the latest contract signed in March 1979. Startup is scheduled in 1983.

portion (about 12%) of the product gas. Oxygen and steam flow enters the gasifier around the coal feed nozzle (Fig. 42). Directly below the reaction chamber, reaction products are partially cooled by exchange to a jacket where steam is produced. An abrupt increase in flow area at the entrance to this jacketed area causes molten slag, which flows down the walls of the reactor, to separate from the refractory surface and fall as droplets into a bottom-slag accumulation vessel.

Table 34 lists results from a sustained 100-hr test run. At the high reaction temperature employed (2850°F), no hydrocarbons or phenols above 0.1 ppm are detectable in the scrubber effluent. It is expected that in an industrial operation, because of lower heat losses than in the test unit, oxygen required would be reduced to about 0.77 lb/lb coal instead of 0.82 as in the test run. By the same token, the cold gas efficiency would be increased from 73% to 78%.

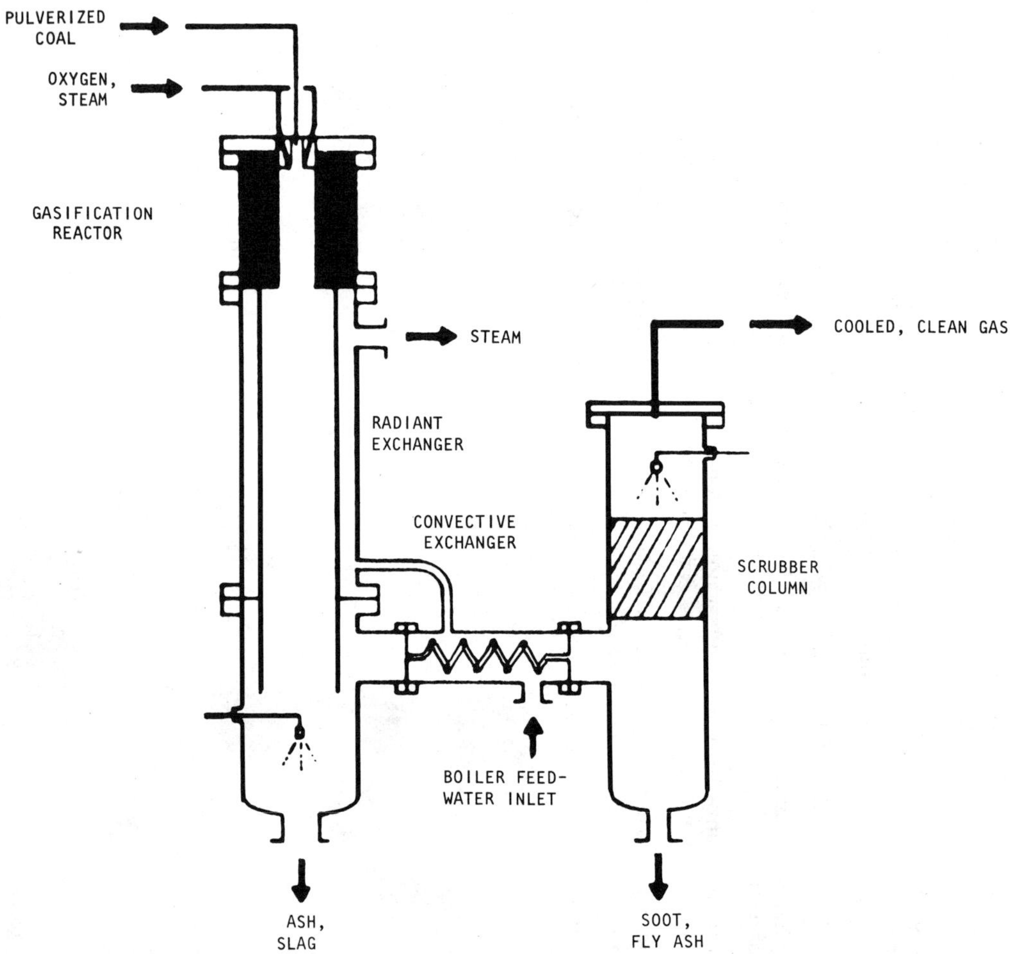

FIG. 42 Mountain Fuel 0.5-ton/day pressurized gasifier [43].

TABLE 34 Data from a Laboratory Test [43]

Utah bituminous coal analysis (wt %)		Gas analysis (dry basis) (vol %)	
Carbon	69.4	Hydrogen	35.5
Hydrogen	5.5	Carbon monoxide	46.6
Oxygen	11.6	Methane	1.2
Nitrogen	1.4	Carbon dioxide	14.9
Sulfur	0.6	Nitrogen	1.7
Ash	10.5	Hydrogen sulfide	0.1
Moisture	1.0		100.0
	100.0	HHV	278 Btu/scf
HHV (Btu/lb)	12,600		

Feed rates (lb/hr)		Output rates
Coal	45.1	Net product gas 82.5 lb/hr 1486 scf/hr
Oxygen	37.1	
Steam	17.5	Gross steam make 82 lb/hr
Recycle gas	10.3	
Slag quench water	114	Solids from slag hopper 4.6 lb/hr
Scrubber feedwater	1850	
Boiler feedwater	97	Solids from scrubber 4.4 lb/hr

Reactor pressure 165 psia

Reactor heat loss, 640 Btu/lb coal

Oxygen, steam preheat temperature, 550°F

Cold gas efficiency, 75%

Carbon gasified, 94%

Future plans call for construction and operation of a 30-ton/day gasifier. The gas produced would be used as fuel to supply a portion of the requirements of a brick manufacturer's plant.

d. Shell-Koppers (Shell Internationale Maatshappi B.V., The Hague, The Netherlands, and Krupp-Koppers GmbH) Pulverized coal (90% less than 90 μm), preferably dried to about 2% moisture, is gasified in entrained flow under slagging conditions at pressure up to 30 atm. The gasifier is an essentially cylindrical vessel equipped with an even number of diametrically opposed burners (see Koppers-Totzek). The combustion flame temperature is 3250 to 3600°F and the outlet gas is 2550 to 2900°F. The main part of the ash is obtained as molten slag which is solidified in a water-filled bottom. Overhead gas is quenched with water to solidify any molten ash carryover so that ash does not deposit in downstream equipment.

For high-rank coal, typically 0.9 to 1 ton of oxygen per ton of moisture and ash-free coal (moisture- and ash-free) is required. Low-rank coal requires only about 70% as much. Gasifier steam required is very low, 7% of coal for high-rank with none required for low-rank coal or for air-blown operations.

Relative quantities of required coal feed (as received), oxygen and steam, and raw gas composition obtained from three different coals are shown in Table 35 [44].

Gasification capacity of a single commercial gasifier vessel, due to high-pressure operation, is contemplated to handle up to 100 tons/hr of coal to produce about 175 million scf/day of raw gas.

A 6-ton/day pilot plant has been operated for over 2000 hr, and a 150-ton/day prototype plant has made a series of successful runs since November 1978. In these runs it has been demonstrated that hard coal can be gasified with a carbon conversion of 99% while producing a synthesis gas containing only 1 vol % of CO_2. The pilot gasifier employed an intricate cooling system for controlling the wall temperature, but an alternative insulating brick wall reactor is being developed.

e. Texaco Coal Gasification (Texaco Development Corp., White Plains, New York) The Texaco entrained flow coal gasifier (Fig. 43) is based on a design similar to that which has been employed to gasify hydrocarbons ranging from natural gas to heavy residual fuel oil. Since the early 1950s such application has been made in over 75 plants involving some 160 units. Pulverized coal that has been through two stages of dry grinding is fed as a pumpable slurry in oil or water to a water-cooled burner. The burner, located inside a pressure vessel, is in the top of a refractory-lined reaction chamber. The slurry may be preheated. Air or oxygen is also fed to the burner to react with the coal.

In the case of oxygen gasification with an oil slurry feed, steam, water, or carbon dioxide is added to moderate the reaction temperature. With air or a water slurry feed, no additional moderator is required. Residence time in the gasifier is a few seconds. Because of the high reaction temperature (2400 to 2700°F), the product gas consists principally of CO and H_2 (and N_2 if air is the oxidant), with lesser amounts of CO_2. A trace of methane is the only hydrocarbon produced. No tars or oils are present. Sulfur is converted to H_2S or COS and nitrogen from the coal to N_2 or NH_3.

TABLE 35 Shell-Koppers Coal Gasification [44]

	West Germany bituminous			
	Low ash	High ash	Wyodak lignite	Australian brown coal (Yallourn)
Coal feed analysis				
Component (vol %)				
Carbon	66.5	51.4	44.6	33.0
Hydrogen	4.3	3.3	3.5	2.3
Oxygen	8.0	6.2	9.9	13.1
Nitrogen	1.0	0.8	0.6	0.3
Sulfur	1.1	0.9	0.4	0.1
Ash	9.1	27.4	6.0	1.2
Moisture content of wet coal	10.0	10.0	35.0	50.0
Heating value LHV (Btu/lb)	11,300	8,700	7,400	4,800
Relative coal, oxygen, and steam quantities				
Constant plant capacity				
Coal intake	100[a]	132	152	232
Oxygen demand	100[a]	102	105	125
Steam demand	100[a]	50	20	—
Overall thermal efficiency (LHV basis) (%)	77	74	77	76
Dry synthesis gas composition				
Component (vol %)				
H_2	31.3	30.2	30.1	28.6
CO	65.6	66.5	66.1	65.8
CO_2	1.5	1.8	2.5	4.7
CH_4	0.4	0.3	0.4	0.1
H_2S	0.4	0.4	0.2	0.1
N_2	0.6	0.6	0.5	0.5
Ash	0.2	0.2	0.2	0.2

[a]Base case.

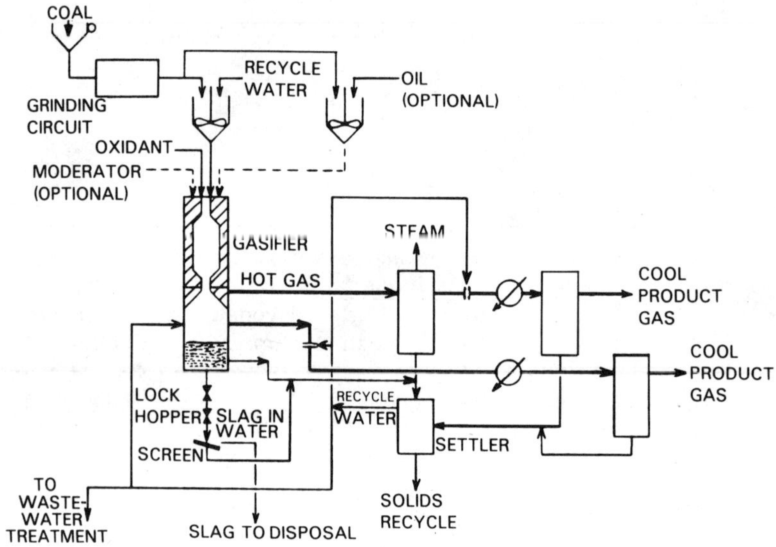

FIG. 43 Texaco coal gasifier [45].

Gas and unconverted solids from the bottom of the gasifier chamber are quenched in water. Slag is solidified and removed periodically as a water slurry from the bottom of the vessel through a lock hopper. Conversion of carbon is typically 90% per pass. Unconverted soot may be separated from the quench water and recycled to the gasifier.

A demonstration Texaco coal gasifier of 100 tons/day capacity was erected at the Morgantown Ordnance Works in West Virginia in 1956 and operated for 2 years. Recently, a 15-ton/day pilot plant at Montebello, California, has been operated on some 20 solid fuels, including lignites, subbituminous, bituminous (Eastern and Western) and anthracite coals, coal liquefaction chars, and petroleum cokes. The pilot gasifier is a 5-ft-diameter by 20-ft-high vessel. It is designed for 400-psi operation. A 1200-psi pilot gasifier is under construction. Table 36 lists typical coal and coke analyses and corresponding gas compositions for pilot plant runs using oxygen with a water slurry.

Similar data are available for other coals with air and reduced crude slurry and also with oxygen using carbon dioxide as a moderator.

A joint effort by Texaco, Southern California Edison, and Electric Power Research Institute (EPRI) contemplates construction and operation of a 1000-ton/day Texaco gasifier to produce fuel gas for a 100-MW combined-cycle electric generating facility. This installation, which will be near Barstow, California, is expected to be operational in the 1980s.

Another installation is presently under way at TVA's National Fertilizer Development Center, Muscle Shoals, Alabama, for operation in early fall 1979. The Texaco gasifier will handle 7 tons/hr of coal to produce gas for ammonia synthesis.

TABLE 36 Typical Texaco Gasifier Performances [45,46]

	Feedstock		
	Western bituminous coal	Delayed petroleum coke	Eastern bituminous coal
Composition (wt 5)			
C	74.56	89.47	67.62
H	5.31	3.71	5.16
S	0.46	1.44	3.26
N	0.99	2.69	1.00
O	11.47	2.37	11.17
Ash	7.20	0.31	11.76
Heat of combustion (Btu/lb dry)	13,134	13,515	12,250
Oxidant:	Oxygen	Oxygen	Oxygen
Slurry medium:	water	water	water
Product gas composition (dry gas basis) (vol %)			
H_2	35.79	34.50	35.78
CO	50.71	45.22	44.62
CO_2	13.14	19.38	17.97
N_2-A	0.24	0.72	0.48
CH_4	0.09	—	0.03
H_2S	0.02	0.13	1.02
COS	0.01	—	0.05
High heating value (Btu/scf) (CO_2-H_2S-COS-free basis)	321	319	320
Cold gas efficiency (%)	73.0	76.6	72.0

2. *Two-Stage Entrained Flow Gasifiers*

a. BI-GAS (Bituminous Coal Research Institute, Inc., Monroeville, Pennsylvania) A water slurry of pulverized coal (70% through 200 mesh) is pumped through a preheater and mixed with hot recycle gas in a spray drier. Dry coal and recycled gas are separated in a cyclone and the coal is fed to a pressurized reactor (Fig. 44). Coal is introduced by entrainment in recycled product gas into the gasification section (stage II) of the reactor, through two opposed coal injection nozzles slanted upward at 30° to horizontal. At the same time, reaction steam is introduced through separate annular openings around the coal feed nozzles. Coal and steam are entrained in a stream of hot synthesis gas (3000 to 3200°F) rising from a lower (stage I) section of the gasifier. This method of operation supplies hydrogen and heat to carry out the gasification of the coal in stage II, a design providing for about 8.5

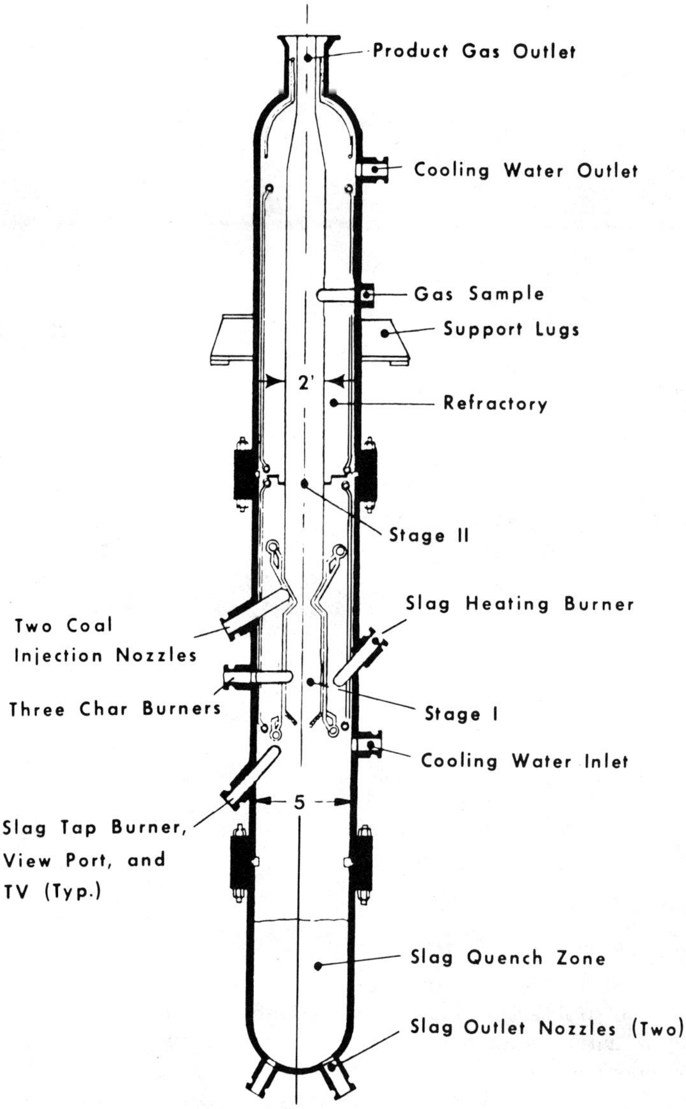

FIG. 44 BI-GAS pilot-plant gasifier [47].

sec of solids residence time. Entrained unreacted coal char and raw gas leave the top of the reactor at about 1700°F.

Unreacted char from stage II is separated from the gas in a cyclone and entrained in recycle gas and returned to feed into the lower reactor section (stage I). Char feed occurs through three burner nozzles, offset radially to provide a swirling action that causes molten ash slag to adhere to the walls of the reactor and fall down through a slag tap hole into a water-filled slag quench zone below. Gas burners are provided both above and below the slag tap to keep it heated and open. Oxygen and additional steam feed through separate annular openings around the char feed nozzles. Char residence time in stage I is about 2 sec.

A 5-ton/hr coal feed pilot plant in Homer City, Pennsylvania, was completed in late 1975. The outer 5-ft-diameter shell is designed to operate at pressure up to 1500 psi. The overall height, including nozzles, is 53 ft 9 in. The walls of stage I consist of vertical water-cooled tubes covered with a dense refractory material compatible with molten slag. The wall tubes are bent inward at the top of stage I to form a 9-in.-diameter throat opening into stage II. This prevents slag carryover, keeps unreacted coal in stage II from falling into stage I, and provides a high velocity for good mixing of coal and steam with synthesis gas. The walls of stage II are covered with a dense refractory over a layer of lightweight insulating refractory. The outer pressure shell in the length of both stages I and II is covered with a row of vertical wall tubes. The wall tubes provide cooling to the shell in the event of insulating failure and also provide for preheating the refractory lining before startup.

Some of the projected flow quantities and design data for the 5-ton/hr pilot plant are shown in Table 37. The concentration of methane in the product gas is relatively high and particularly suitable for SNG production because of the two-stage high-pressure operation.

As of November 1978 two operational test periods of 47 and 64 hr with Rosebud coal had been achieved in the pilot plant with integrated gasifier flow streams. Satisfactory heat and material balances have not been achieved during pilot-plant operation because of the short-run duration [48]. The carbon content of the slag was less than 1%. Problems with maintaining continuous slag flow, which prevailed in early tests, have apparently been solved. Inability to measure the char flow rate has made it necessary, for safety, to feed gas to each char burner in order to stoichiometrically consume oxygen in the event of char feed interruptions.

b. C-E (Combustion Engineering, Inc., Windsor, Connecticut) The C-E gasifier (Fig. 45) is divided into two sections. About two-thirds of the pulverized coal feed (70% through 200 mesh) is fed to the upper (reductor) section while one-third is fed to the lower (combustor) section. Both coal feeds are entrained in air. Additional secondary air at 600°F goes to the combustor section. Molten ash, removed from the combustor, drops into a water-filled slag tank. The reaction chamber enclosure is of fusion-welded tubular water wall construction with an inner lining of insulating refractory. The reaction sections are followed immediately by a tubular convection heat-exchange section where the gas is cooled to freeze the reactions. Operation is at near-atmospheric pressure. Gas temperatures are about 3000°F from

TABLE 37 Design Data [49]

Coal feed rate	10,000 lb/hr
Oxygen to burners	6,500 lb/hr
Steam to zone I	2,500 lb/hr
Steam to zone II	9,500 lb/hr
	28,500 lb/hr
Process gas out	27,800 lb/hr (13.32 × 10^6 scfd/day)
Slag out	700 lb/hr
	28,500 lb/hr
Recycle char	11,000 lb/hr
Operating pressure, gas side	500-1500 psig
Process gas outlet temperature	1700°F
Shell design pressure	1650 psig
Shell design temperature	650°F
Shell	60 in. ID, 3 9/16 in. thick
Tubes	1 1/2 in. OD, 0.203 in. thick
Material	Croloy 2 1/4
Total gasifier weight	280,000 lb

Design raw gas composition (vol %) [50]

CO_2	16.2
CO	22.1
H_2	24.2
CH_4	11.8
N_2	0.5
H_2S	0.6
H_2O	24.6
Total	100.0

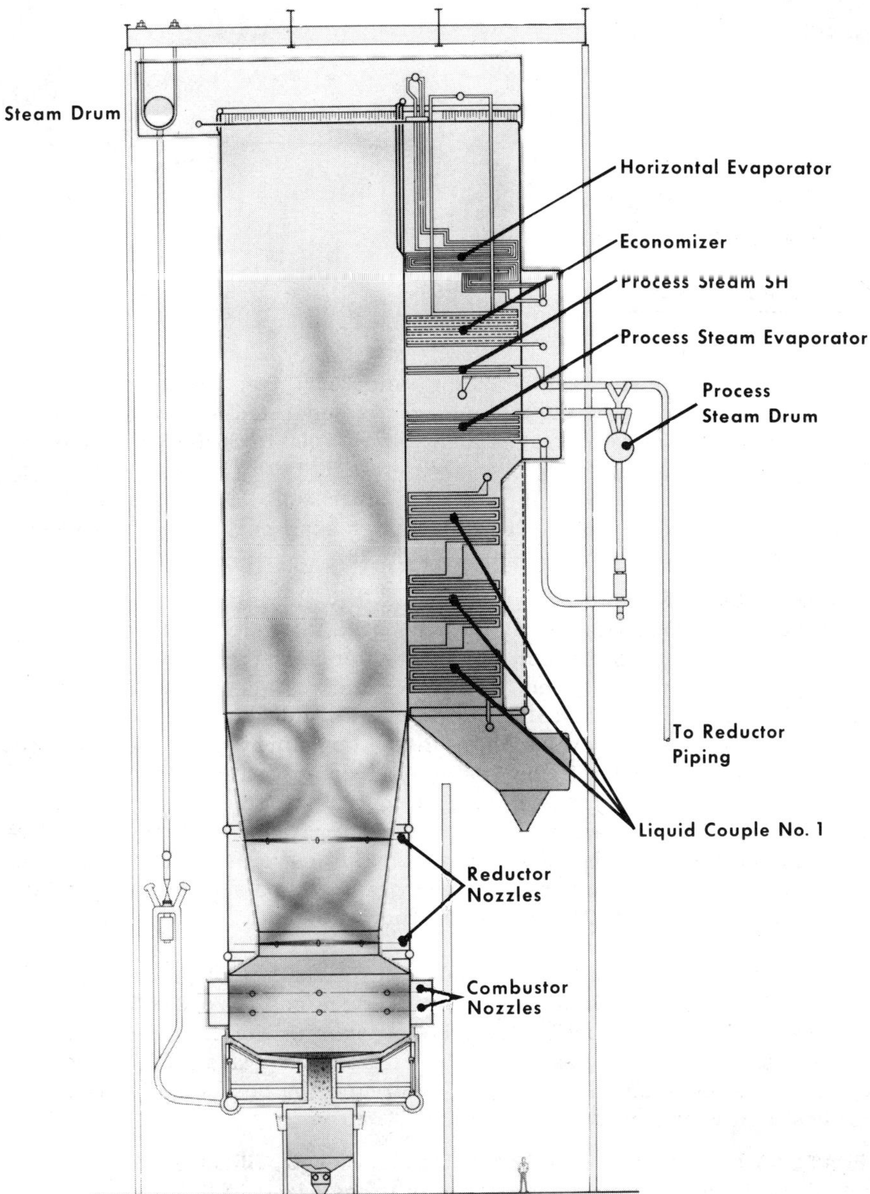

FIG. 45 C-E commercial-size gasifier concept [51].

TABLE 38 Raw Gas Composition [51]

	Vol %
CO	25-30
CO_2	9-25
H_2	2.6-5.5
CH_4	<0.1-3.0
Total hydrocarbons < C_4	<0.1-1.0
Total hydrocarbons > C_4	<0.1
H_2S	2300 ppm

the combustor, 1800°F from the reductor, and 300°F from the convection section. Char carryover in the product gas is separated and recycled to the combustor section. A commercial unit gasifying 7000 to 8000 tons/day of coal would be 30 ft in diameter (reaction chamber) and 180 ft high.

A 5-ton/day PDU has been in operation since June 1978. Gas of 110 to 145 Btu/scf has been produced from Pittsburgh seam coal. Raw product gas analysis from the first test is listed in Table 38.

c. Foster Wheeler (Foster Wheeler Energy Corp., Livingston, New Jersey) This gasifier is in many ways similar to the BI-GAS system [52]. As may be seen in Fig. 46, the configuration of the gasifier is the same as for BI-GAS. The principal difference between the two occurs in stage II, where Foster Wheeler employs air firing and BI-GAS oxygen. Additionally, fuel coal to stage I is pressurized in lock hoppers, instead of slurry pumping, as in the case of BI-GAS. Feed coal in the Foster Wheeler system is dried to about 2% moisture in a rollermill. Since low (110 to 160 Btu/scf)-Btu gas is to be produced, it is of no concern to produce a high-methane content, and therefore gasification pressure is more moderate (350 psi versus 1000 to 1500 psi for BI-GAS). For the same reason, the Foster Wheeler stage II temperature can be somewhat higher (1800 to 2100°F) than the equivalent BI-GAS stage II temperature.

In early 1975, Foster Wheeler initiated design of a pilot plant to be constructed in Sioux Falls, North Dakota, to produce gas for a 35-MW combined cycle power plant. This gasifier is to be tested with North Dakota lignite, Illinois No. 6, and various Eastern coals. A one-third-size plastic cold flow model has been constructed to visually determine the best position and orientation of coal and char injection nozzles.

d. PEATGAS (Institute of Gas Technology, Chicago, Illinois) Ground peat (-10/+100 mesh) containing 50% moisture is pressurized to 200 to 500 psi in lock hoppers and transported by a fraction of the gasifier gas through a preheater to heat it to 360°F. It then enters the PEATGAS reactor (Fig. 47), which consists of three sections. In an upper fluidized-bed section the peat is dried of water by hot raw gas leaving the gasifier at about 450°F. Dried peat then mixes with hot synthesis gas at 1700 to 1900°F, whereby it is heated to 1200 to 1500°F. Primary devolatilization and hydrogasification

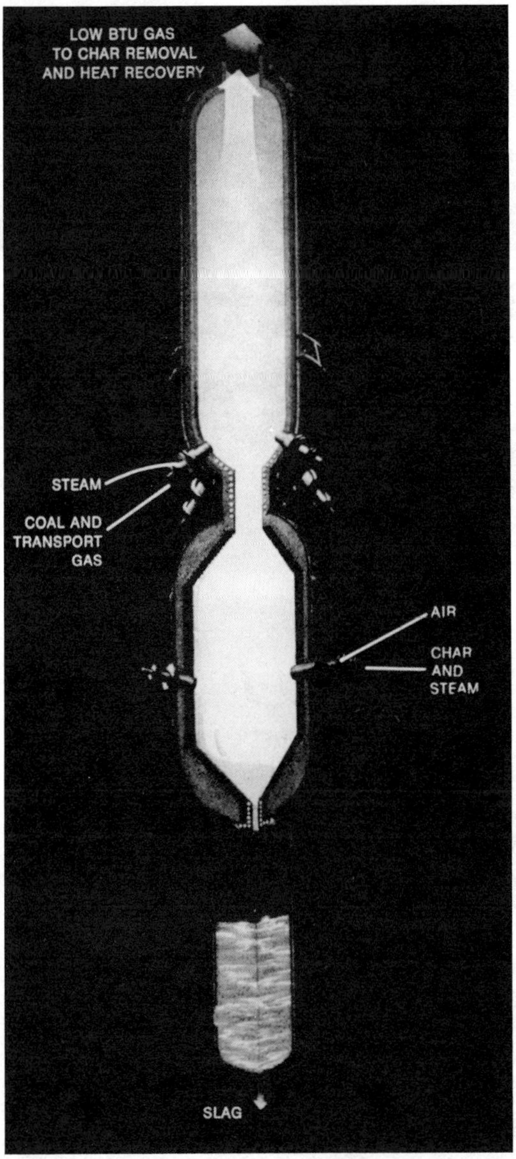

FIG. 46 Foster Wheeler two-stage entrained flow slagging coal gasifier [53].

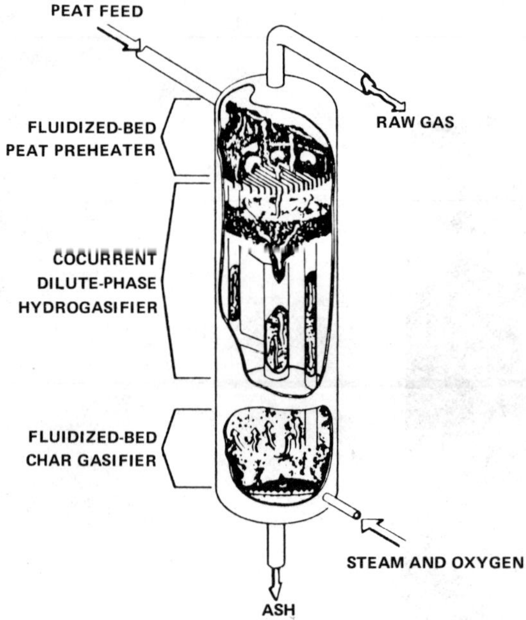

FIG. 47 Schematic diagram of the PEATGAS reactor [54].

of the char by the synthesis gas takes place in a cocurrent dilute-phase transport section sized to provide a solids residence time of 4 to 7 sec. Under these conditions, 20 to 40% of the carbon in the peat fed will be converted to hydrocarbon gases and 30 to 15% to light hydrocarbon liquids, with lower temperatures favoring liquid production. Gases and char produced are separated. The gases proceed upward through the drier section and the char flows downward into a fluidized-bed char gasifier, where it is gasified with steam and oxygen at 1700 to 1900°F. About 20 min of solid residence time is required in the char gasifier at 1700°F.

Peat has been gasified by entrained flow in hydrogen, hydrogen/steam, and synthesis gas atmospheres utilizing a process development unit (PDU) consisting of a 160-ft-long, 1-in.-diameter coiled tube. The range conditions employed in these tests are shown in Table 39. Additionally, peat char has been gasified by reaction with steam and oxygen in a fluid-bed PDU unit (6-in. diameter) at char feed rates of 35 to 80 lb/hr. Based on results of these PDU tests, a 250 million scf/day SNG plant design has been made. Feed, product quantities, and compositions applicable to the gasifier for such an operation are shown in Table 40.

Large-scale (50-tons/day) pilot-plant testing of the PEATGAS Process is contemplated to start in late 1980.

TABLE 39 Range of Operating Conditions Used for Peat Hydrogasification Tests [55]

	Process development unit
Peat feed rate (lb/hr)	5-12.5
Feed gases	H_2, H_2-H_2O, synthesis gas[a]
Hydrogen partial pressure (atm)	4.3-36
Maximum temperature (°F)	1000-1500
Gas flow rate (scf/hr)	400-1030
Residence time (sec)	3 22

[a]Component	CO	CO_2	H_2	CH_4	H_2O
Mol %	13.3	19.3	28.2	0.9	38.3

e. Rockwell (Rockwell International Corporation, Canoga Park, California) Coal, in the Rockwell system, is partially (50 to 90%, depending on coal rank) gasified at moderate temperature (1400 to 1900°F) and high pressure (to 1500 psi) in an atmosphere of hydrogen. The gasifier design is based on principles of rocket-engine design (Fig. 48), where the reactor injector mixes the dense-phase feed stream of pulverized coal with hot hydrogen. The quantity and temperature of hydrogen (up to 2000°F) are generally sufficient to raise the mixture to reaction temperature, after which the exothermic heat of reaction sustains the hydrogasification. In some cases it may be necessary to add a small amount of oxygen to assist in heat input.

A 1/4-ton/hr feed rate test gasifier (Fig. 49) has been constructed and operated. This includes an electric heater to raise hydrogen temperature to 900 to 1200°F and a preburner with oxygen injection for added temperature. Different lengths (7.5 and 15 ft) and diameters of reaction chambers are utilized to vary residence time. Coal feed is from a nitrogen-pressured feed tank the size of which limits run duration to about 8 min at the 1/4-ton/hr rate. Reaction products are cooled with water in a 2-ft-long quench section and unconverted char is separated from the product gas and collected in a spherical vessel.

Forty-two short-duration tests on four different coal types were made in a 1/4-ton/hr gasifier. Difficulties were not experienced with the bituminous caking-type coals. The range of operating conditions (pressure, temperature, and residence time) investigated and the carbon conversions obtained are presented in Table 41. General conclusions indicate that, under comparable operating conditions, overall carbon conversions are highest for peat, then bituminous, and lowest for subbituminous. Additionally, conditions giving high carbon conversions produce less liquids; thus minimum conditions can be established so that conversion is only to gas.

Raw product gas from the gasifier may contain unconsumed excess hydrogen, light hydrocarbon gases (C_1 to C_4), vapors of light hydrocarbon liquids

TABLE 40 Raw Materials and Products for 250 × 10^9 Btu/Day SNG Plant by PEATGAS Process [54]

Peat composition		Raw product gas composition		
Proximate analysis		Component	Mol/hr	Mol %
(dry basis) (wt %)		CO	9,910	6.81
Volatile matter	65.0	CO_2	33,502	23.02
Fixed carbon	23.7	H_2	17,685	12.15
Ash	11.3	H_2O	61,233	42.08
Total	100.0	CH_4	15,656	10.76
Ultimate analysis		C_2H_6	2,766	1.90
(dry basis) (wt%)		NH_3	2,820	1.94
C	49.90	H_2S	152	0.10
H	5.09	N_2 + Ar	203	0.14
O	30.71	C_6H_6	889	0.61
N	2.73	C_6H_6O	441	0.31
S	0.27	$C_{10}H_8$	266	0.18
Ash	11.30		145,523	100.00
Total	100.00			
Calorific value (Btu/lb)	8,506			

Stream quantities

Moisture content of raw material (wt %)	50
Feed requirements (dry) (tons/day)	
Process	21,951.3
Fuel	6,202.0
Total	28,153.3
Oxygen (98%) (tons/day)	3,914.0
Steam (lb/hr)	1,506,140
Process makeup water (gal/min)	2,060
By-products	
Liquid fuels (bbl/day)	9,919
Ammonia (tons/day)	565
Sulfur (long tons/day)	52

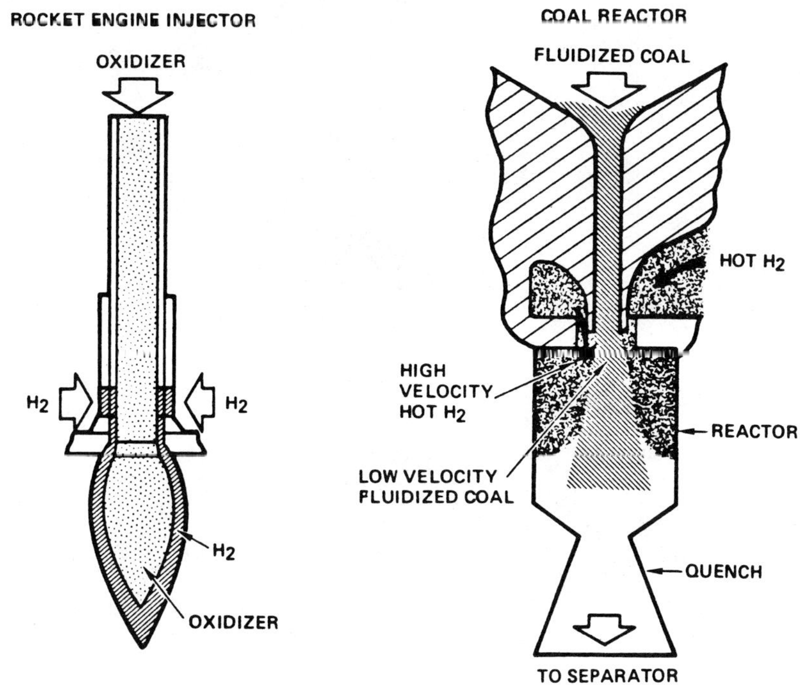

FIG. 48 Short-residence-time conversion [56].

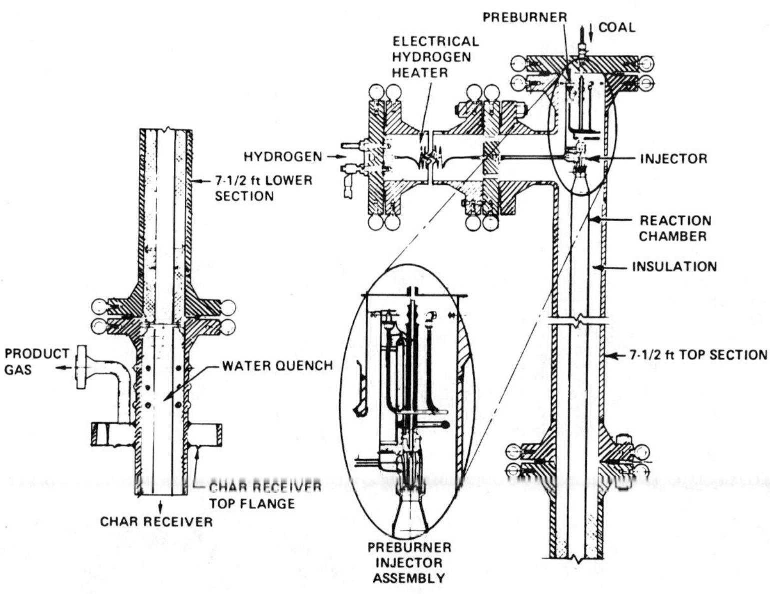

FIG. 49 Rockwell 1/4-ton/hr reactor assembly [56].

TABLE 41 Summary of Short-Duration 1/4-Ton/Hr Reactor Testing [56]

Coal source and type	Number of runs	Ranges of variables tested				
		P_R (psig)	T_R (°F)	t_{resid} (msec)	Overall	To gases
Montana Rosebud subbituminous	17	1000-1500	1416-1900	600-4100	27.8-49.8	22.5 to ~49.8
Kentucky hvA bituminous	9	1000-1500	1624-1908	480-5500	47.3-67.2	40 to ~57.2
Illinois No. 6 hvC bituminous	4	1000-1500	1468-1982	1000-1470	48.0-70.7	37.2-70.6
Minnesota Peat	12	500-1500	1534-1850	500-3700	73.4-87.3	63.5-87.3

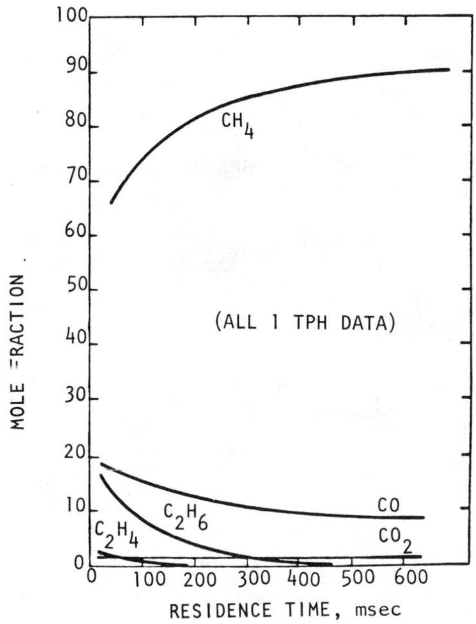

FIG. 50 Composition of "separated" product gas [56].

(BTX), phenols, light oils, vapors of heavier hydrocarbon liquids (syncrudes to tars), CO, CO_2, ammonia, H_2S, and water vapor. The change in distribution of some of the constituents with reactor residence time is illustrated in Fig. 50 (hydrogen-free basis).

Future plans call for construction and operation of gasifiers with a 4-ton/hr capacity, which will be directly applicable to commercial design.

D. Molten Bath Gasifiers

Molten bath gasifiers described in the following paragraphs employ molten salt (ROCKGAS), molten iron (PATGAS and ATGAS), or molten coal ash (Rummel single-shaft, Otto Rummel double-shaft, and Saarberg-Otto). These gasifiers are able to accept any type of coal regardless of fines content, caking properties, and crushing strength.

1. *ROCKGAS (Rockwell International,* Canoga Park, California)*

In the ROCKGAS system, coal (-8 mesh) and dry sodium carbonate (Na_2CO_3) powder are carried from lock hoppers by air into the bottom of a turbulent molten salt bath maintained at 1700 to 1800°F and 75 to 450 psi pressure. If an intermediate Btu gas is desired, a mixture of oxygen and

*M. W. Kellogg Co., Houston, Texas, was also active in the early development of this process, referred to then as the Kellogg Molten Salt Process.

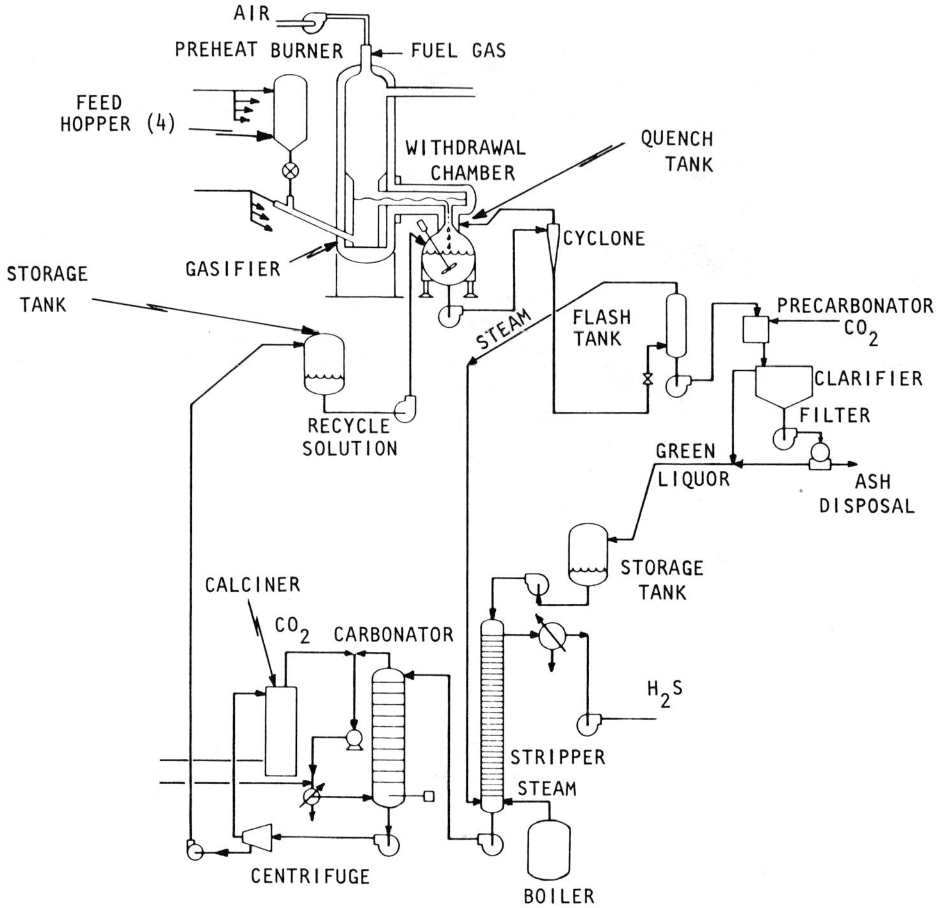

FIG. 51 ROCKGAS ash removal, sulfur removal, and salt-recovery system [57].

steam is used as the gasifying medium. Product gas is withdrawn from the top of the gasifier that is located above the salt bath. A continuous overflow taken from the top of the salt bath removes coal ash with sulfur in the form of dissolved sodium sulfide. This overflow is quenched by a recycled water solution in a stirred quench tank and then processed to remove ash and sulfur, and to produce dry Na_2CO_3 for recycle by the system as shown in Fig. 51. Soluble salt is dissolved in the quench tank and, after pressure letdown, the slurry is pumped through a flash tank to remove steam and then through a precarbonator, where it is sparged with CO_2 to enhance ash settling and prepare it for subsequent H_2S stripping. Ash is then removed by settling and filtration. The resulting liquid (green liquor) is then steam-stripped in a tower to remove H_2S. Liquid from the stripper is then contacted with CO_2 to convert carbonate to bicarbonate. Precipitated bicarbonate crystals are separated and calcined back to carbonate salt, which is recycled to the gasifier. The solution from the bicarbonate separator goes back to the quench tank at the gasifier.

Gasification tests have been carried out in small (6-in.-diameter) externally heated bench-scale units and in a 34-in.-ID refractory-lined stainless steel vessel. Typical conditions employed and corresponding gas compositions obtained are shown in Table 42. Cold gas efficiency for an industrial gasifier is projected to be about 80%.

A 1-ton/hr PDU was completed in June 1978 to demonstrate an integrated operation of continuous gasification, ash removal, and Na_2CO_3 regeneration. The configuration of the gasifier, which will operate at a pressure of 280 psig, is shown in Fig. 52. Operation of the PDU is planned through December 1979.

2. *Molten Iron Processes (Applied Technology Corporation, Pittsburgh, Pennsylvania)*

Crushed (-1/8-in.) coal is injected through a lance into a 3-ft-deep molten iron bath at 2400 to 2600°F, contained in a cylindrical refractory-lined vessel (Fig. 53). In one mode (the Two-Stage Process), the coal is transported by compressed air, and additional air is added through a second lance slightly below the molten iron surface. In a second mode (PATGAS or ATGAS Processes), the coal is transported by steam and oxygen is added through the second lance. Carbon from the coal dissolves in the iron (carburization) and a portion devolatilizes. The remaining dissolved fixed carbon is gasified with oxygen and steam (decarburization).

TABLE 42 Typical Gasification Test Results [57]

Gasifier ID (in.)	6	6	34	34
Pressure (atm)	1	1	1.9	3.2
Temperature (°F)	1719	1805	1690	2010
Coal feed (lb/hr)	6.9	~2.5	~190	~470
Oxidant	O_2	Air	Air	Air
Oxygen/coal (lb/lb)	0.62	—	—	—
Steam/coal (lb/lb)	0.14	—	—	—
Air/coal (scf/lb)	—	44	66.3	47.2
Gas composition (dry basis) (vol %)				
CO	53.3	27.0	24.0	29.7
H_2	33.8	12.2	10.9	13.2
CH_4	2.4	1.1	1.4	1.5
C_2H_6	0.2	0.1		
CO_2	9.5	0.4	8.0	3.5
N_2	0.7	53.2	58.9	49.4
HHV (Btu/scf)	309	138	123	158

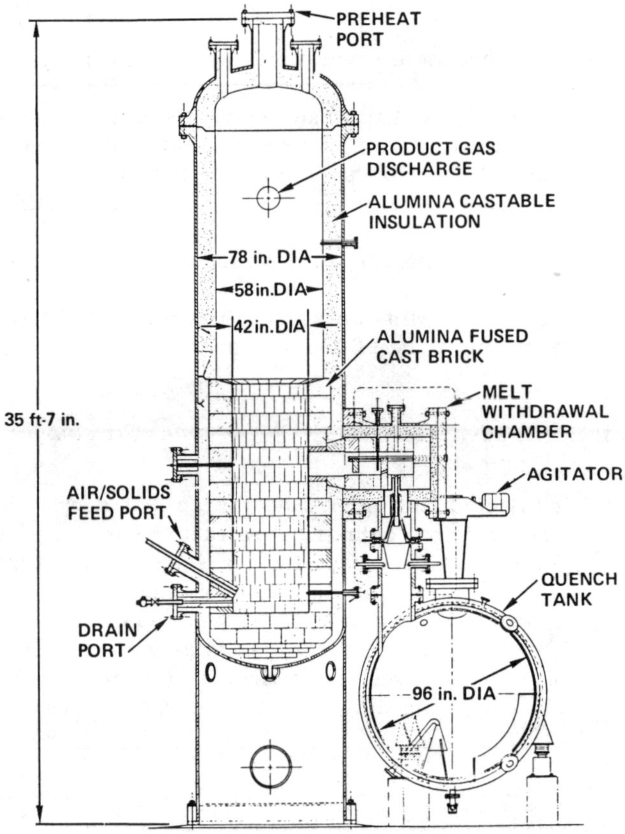

FIG. 52 ROCKGAS 1-ton/day PDU gasifier with quench tank [57].

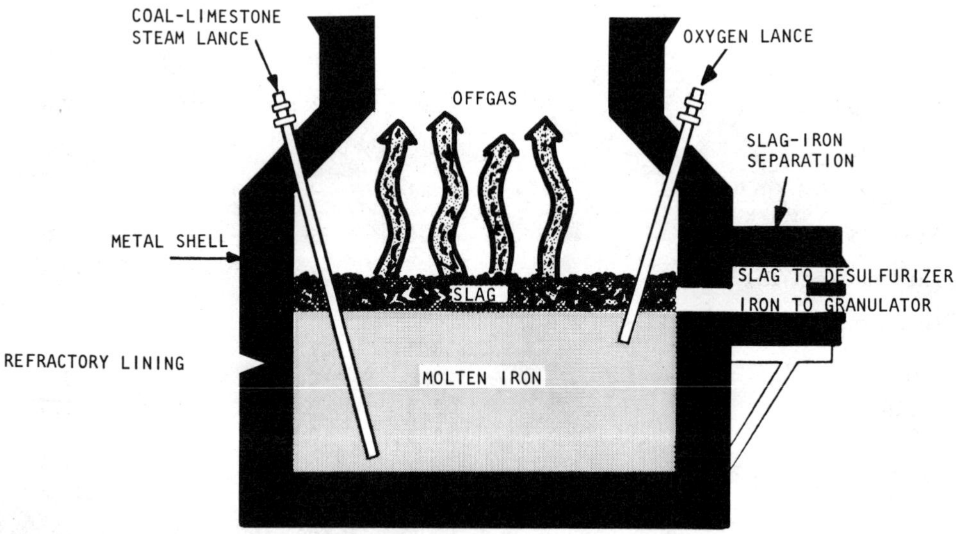

FIG. 53 Molten iron gasifier [59].

The Two-Stage Process produces a low-Btu gas; oxygen injection produces a medium-Btu gas that can be used as is for fuel or synthesis gas (PATGAS), or upgraded by further processing to high-Btu SNG (ATGAS). In each case, gas is produced at low (2 to 5 psig) pressure.

Limestone, injected with coal, forms a slag that fluxes the coal ash and reacts with sulfur from the coal to form calcium sulfide which is dissolved in the slag. The slag floats on the molten iron and is continuously withdrawn to remove ash and sulfur. The slag, containing 4 to 8% sulfur, is contacted with steam in a shaft reactor that releases elemental sulfur, H_2S, and SO_2. A portion of the slag is recycled; wet slag withdrawn from the process can be used for road aggregate. Iron, from pyrites in the coal ash, is also produced as a net by-product. By maintaining a relatively high (3 to 4%) carbon content in the iron, sulfur is not oxidized and the raw product gas is essentially sulfur-free.

Early work conducted under the sponsorship of the American Gas Association (AGA) demonstrated the feasibility of using oxygen. Later work under U.S. Environmental Protection Agency sponsorship involved the Two-Stage (air injection) Process in a 25-in.-ID induction furnace of 4000-lb capacity, using medium-volatile bituminous coals containing 0.6 to 3.5% sulfur. It has been demonstrated that gas can be produced which contains little enough sulfur that stack gas after combustion will have less than 50 ppm SO_2. Tests are continuing to develop data on refractory life, trace-element accumulation in iron and slag, sulfur release from weathering of slag, and the feasibility of using oil-cooled lances.

Projections of feed and product quantities for industrial-size Two-Stage and PATGAS Processes are shown in Table 43.

TABLE 43 Quantities for Industrial Plants [58]

	Two-stage	PATGAS
Gas produced		
$\times 10^6$ scf/day	1240	750
Btu/scf	190	315
% CO	30	63.5
H_2	15	36
N_2	55	0.5
Coal feed rate (tons/day) (13,100 Btu/lb, 4% S)	9,264	11,208
Limestone (tons/day)	576	744
Combustion and transport air (O_2) (tons/day)	33,696	6,624
Steam (50 psig) (tons/day)	—	1,248
Iron produced (tons/day)	144	360
Sulfur produced (tons/day)	336	432
Desulfurized slag (1% sulfur)	1,128	1,128

3. *Rummel (Single-Shaft) Process (Union Rheinische Bramkohlen Kraftstoff AG (Wesseling, West Germany)*

The Rummel (single-shaft) Process injects coal, steam, and air (or oxygen) into a 20-in.-deep molten ash slag bath through 12 tangential water-cooled nozzles. The slag depth is maintained by a central overflow pipe. The slag bath is rotated slowly by the force of incoming reactants and, thereby, acts as a heat distribution source for the gasification reactions. Overflow slag is quenched with water.

A 5 ft 11-in.-diameter water-cooled gasifier was constructed at Wesseling in the 1950s and operated on brown coal to produce 13 million scf/day of synthesis gas (272 Btu/ft^3 HHV) at atmospheric pressure.

4. *Otto Rummel (Double-Shaft) Process (Dr. C. Otto, GmbH)*

This extension of the Rummel (single-shaft) Process allows production of medium-Btu gas using air. The slag bath is separated into two compartments by a vertical partition extending a short way into the bath (Fig. 54). Air and coal are injected tangentially into one section to furnish heat, and steam and coal are injected tangentially into the other section for gas production. Flue gas and make gases are removed through separate shafts connected to the two sections (Fig. 55).

In 1962 a test unit containing an annular slag section of 7.5 ft OD, 2.3 ft ID, was installed at Bromley-by-Bow, England, by the British Gas Council. The unit was designed to produce 2 million scf/day of medium-Btu gas at atmospheric pressure. This vessel was water-jacketed because satisfactory refractories were not used in the slag-bath condition.

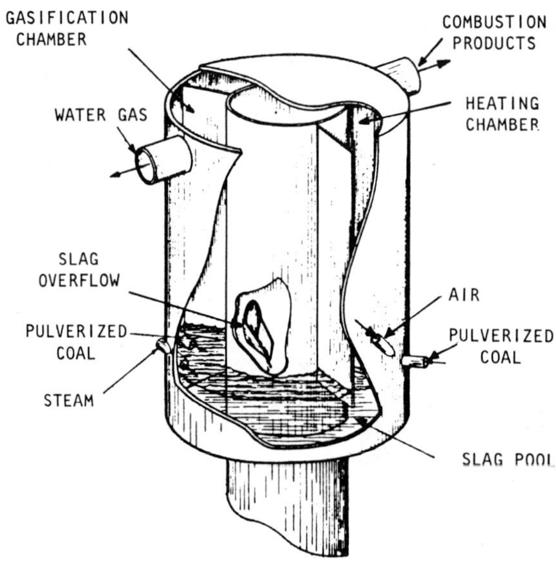

FIG. 54 Rummel double-shaft slag bath gasifier [60].

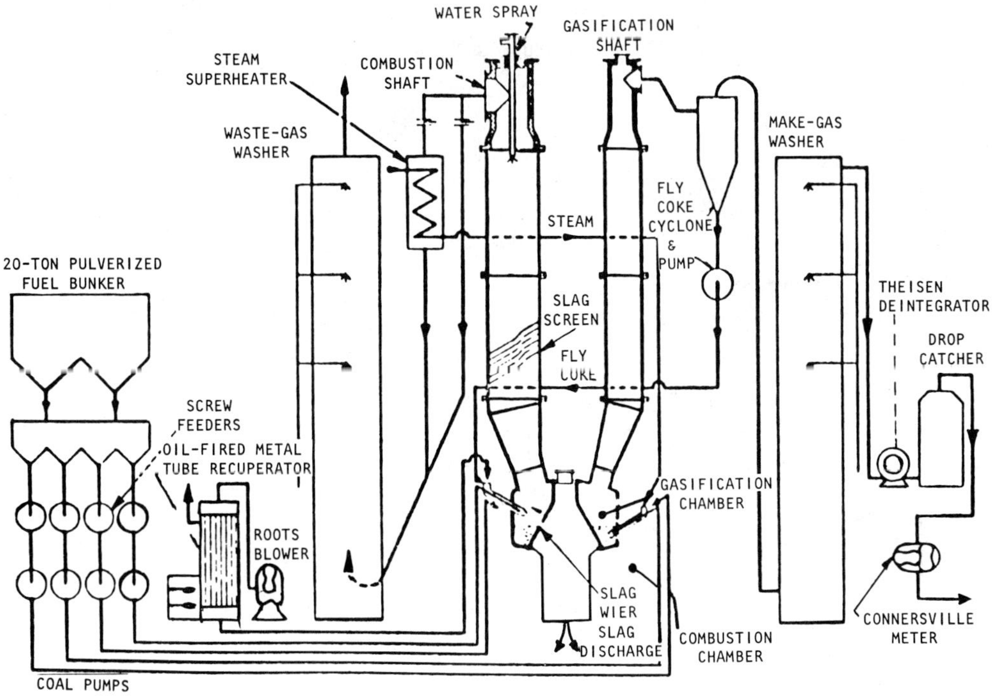

FIG. 55 Flow diagram of double-shaft slag bath gasifier [61].

Initial operation was unsuccessful because of slow combustion, and therefore a lack of heat transfer to the steam gasifier section occurred. Improvements were made by increasing the blast pressure and by submerging the process tuyeres; however, the rate of gas production was still unsatisfactory. It was concluded that, although the process could be made technically viable, it would not be economically attractive under conditions prevailing at that time.

5. *Saarberg/Otto Process (Saarberg/Dr. Otto Arbeitsgemeinschaft für Kohle-Druckvergasung, Völklingen-Fürstenkausen, West Germany)*

This adaptation of the Rummel (single-shaft) Process is being designed for a maximum of 25 bars pressure operation with oxygen or oxygen/air mixture. The gasifier consists of three sections (Fig. 56). Section I contains the molten slag. The gas leaving section I has a temperature of 2700 to 3100°F. Sections I and II have an outer pressure shell and an inner shell of closely spaced, water-cooled tubes. The tube walls are initially protected by a castable refractory, but this is expected to gradually dissolve and be replaced by a layer of solid slag.

In section III (refractory lined), raw gas is cooled by recycle gas to 1500 to 1700°F to solidify slag droplets. Slag, char particles, and dust amounting to 10 to 30% of the coal feed are carried overhead with product

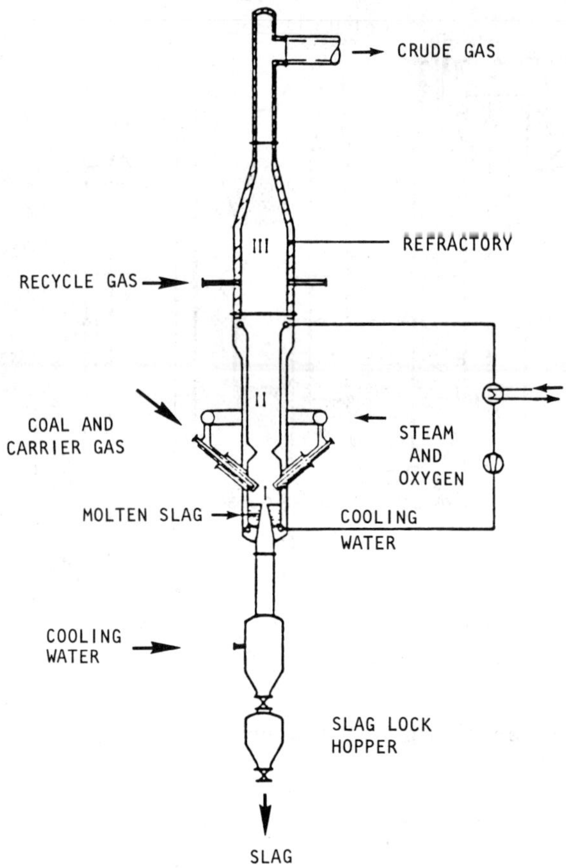

FIG. 56 Saarberg/Otto slag bath gasifier [62].

gas and are recovered for recycle to the gasifier. Molten slag, which overflows a tap hole, contains less than 1% carbon. It is quenched with water and is discharged intermittently through a lock hopper.

A 264-ton/day (19-million scf/day) medium-Btu gasifier has been installed at Volklingen/Furstenhausen. This gasifier is a vertical cylindrical vessel 4.5 ft in diameter. Pertinent quantities and gas composition from processing a bituminous coal are shown in Table 44. Ultimately, it is projected to build 10-ft-diameter gasifiers to produce 84 million scf/day of intermediate-Btu gas. Air-blown operation will be utilized to produce low-Btu gas.

E. Tumbling-Bed Gasifier

KILnGAS (Allis Chalmers Corporation, West Allis, Wisconsin)

This gasifier employs a rotary ported kiln that has been used industrially since 1973 for direct reduction of iron ore in sizes up to 1200 tons/day. The ports, which are openings in the kiln wall, allow air and steam for gasification

TABLE 44 Saarberg Otto Bituminous Coal Processing [62]

Gas composition	Vol %
CO	54.0
H_2	31.4
CO_2	13.2
N_2 + Ar	0.8
H_2S + COS	0.4
CH_4	0.2
Tars and oils	nil
	100.0

Heating value	270 Btu/ft^3
Overall thermal efficiency (including high-pressure steam generated)	88-94%
Cold gas efficiency	73%
Carbon conversion	99.5%
O_2 (99.5%)	1 ton/ton feed coal
Steam to gasifier	0.4 ton/ton feed coal
By-product steam (650 psig, 570°F)	1.4 tons/ton feed coal

to be injected under a tumbling bed of solids (Fig. 57). The kiln is refractory-lined to withstand an operating temperature of about 2000°F. Coal, fed to the high end (left end in Fig. 57) of the sloping kiln, is dried and preheated while moving countercurrent to the gas produced. Air and steam are then introduced through the ports to devolatilize and gasify the coal. Most of the ash is entrained with the exit gas, but some overflows the lower end of the kiln. Product gas will have a heating value of about 150 Btu/scf when air is used for combustion; delivery pressure may be up to 120 psig.

Fifteen electric utility companies have teamed with Allis Chalmers in a development program. A 60-ton/day pilot gasifier installed in 1977 has gasified caking coals containing up to 40% fines. A demonstration plant to gasify 800 tons/day of coal is planned for operation in late 1982. Successful operation of other types of kilns of large dimensions (24 ft in diameter and 600 ft long) suggest that rotary kiln gasifiers can be constructed to handle as much as 6000 tons/day of coal in a single unit.

F. Underground Gasification

Underground (in situ) coal gasification (UCG) offers the possibility of cost savings by eliminating the need for mining the coal and also possibly a reduction in pollution to the environment. Basically, the technique involves first sinking wells from the ground surface into the coal bed for injecting the gasifying medium (steam and air or oxygen) and for collecting the gas produced. Injection and production wells must be located at appropriate spacing

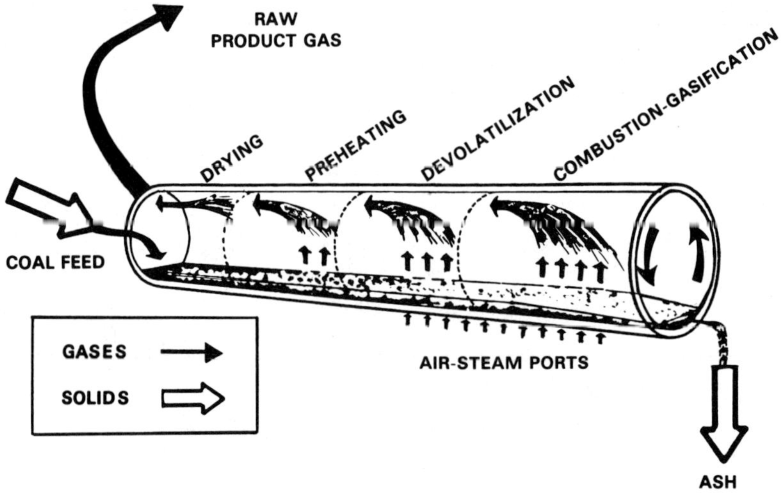

FIG. 57 KILnGAS process concept [63].

so that they are not so numerous as to be prohibitive in cost but, at the same time, will be close enough together that the gases can readily move between them.

There are two configurations of wells employed, depending upon the nature of the coal seams. For steeply dipping seams (SDS) (45° or more), the "stream method" is employed. For horizontal seams, the "linked vertical wall" (LVW) percolation method is used.

In the stream method, a primary injection well is drilled through surrounding strata to the base of the coal seam, and a production well is drilled through the length of the coal bed (Fig. 58). As gasification proceeds from the base of the seam, ash settles and will eventually clog the injection tube. For this reason, a second injection tube is drilled into the seam at a higher level, to be used for the later phase of operation.

In the LVW method, vertical injection and production wells are drilled to the lower part of the horizontal coal seam. Generally, the permeability of the coal bed will be too low to allow satisfactory gas flow between wells, and measures must be employed to open a passage. In so doing it is desirable to provide a direct, small-cross-section opening at the very bottom of the coal bed. Then, as gasification proceeds, char and ash will fall to the bottom and provide a mass of reactive material for evening out the gasification operation and ensuring a high utilization of the coal.

Four linking methods have been employed for LVW coal gasification:

1. Directional drilling
2. Hydraulic fracturing
3. Electrolinking (electrocarbonization)
4. Countercurrent combustion

Figure 59 illustrates the progressive steps in LVW operation, wherein original linkage is effected by countercurrent combustion. Figure 59B illus trates air introduced into, and ignition initiated at the bottom of the well to the right. After ignition is initiated the gas flow is reversed by injecting air into the well to the left (Fig. 59C). Combustion then proceeds from right to left while air moves countercurrent from left to right. In this way a relatively narrow passage is developed between the bottoms of the injection and production wells (Fig. 59D). If instead of countercurrent combustion the flame front were to proceed in the same direction as the air flow, a wide volume would be opened, proceeding radially outward from the air inlet. This would lead to poor coal utilization and unstable operation in the subsequent gasification.

After linkage has been effected by countercurrent combustion (Fig. 59D), gasification proceeds (Fig. 59E) until the complete volume of coal between the wells is gasified (Fig. 59F).

Countercurrent combustion is the favored method for linkage in processing coals that do not swell on heating. Western U.S. coals actually shrink on heating, whereas Eastern coals swell. It is unlikely that countercurrent combustion will prove an effective linking procedure for Eastern coals.

At present, there is no commercial utilization of underground coal gasification in the United States. The technique is reported to have been applied commercially at eight sites in the USSR since 1933. Two installations are still in operation, producing 85- to 100-Btu gas for power-plant use. There is, however, considerable development activity in the United States. Nine projects in North America are active in either operational or developmental stages:

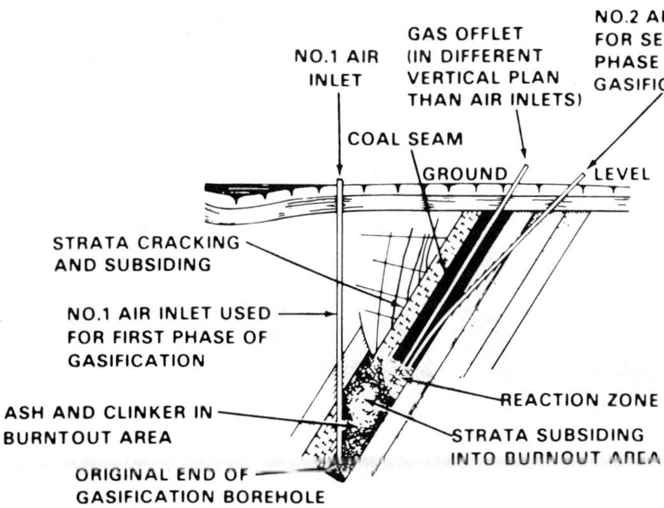

FIG. 58 Stream system for underground coal gasification [64].

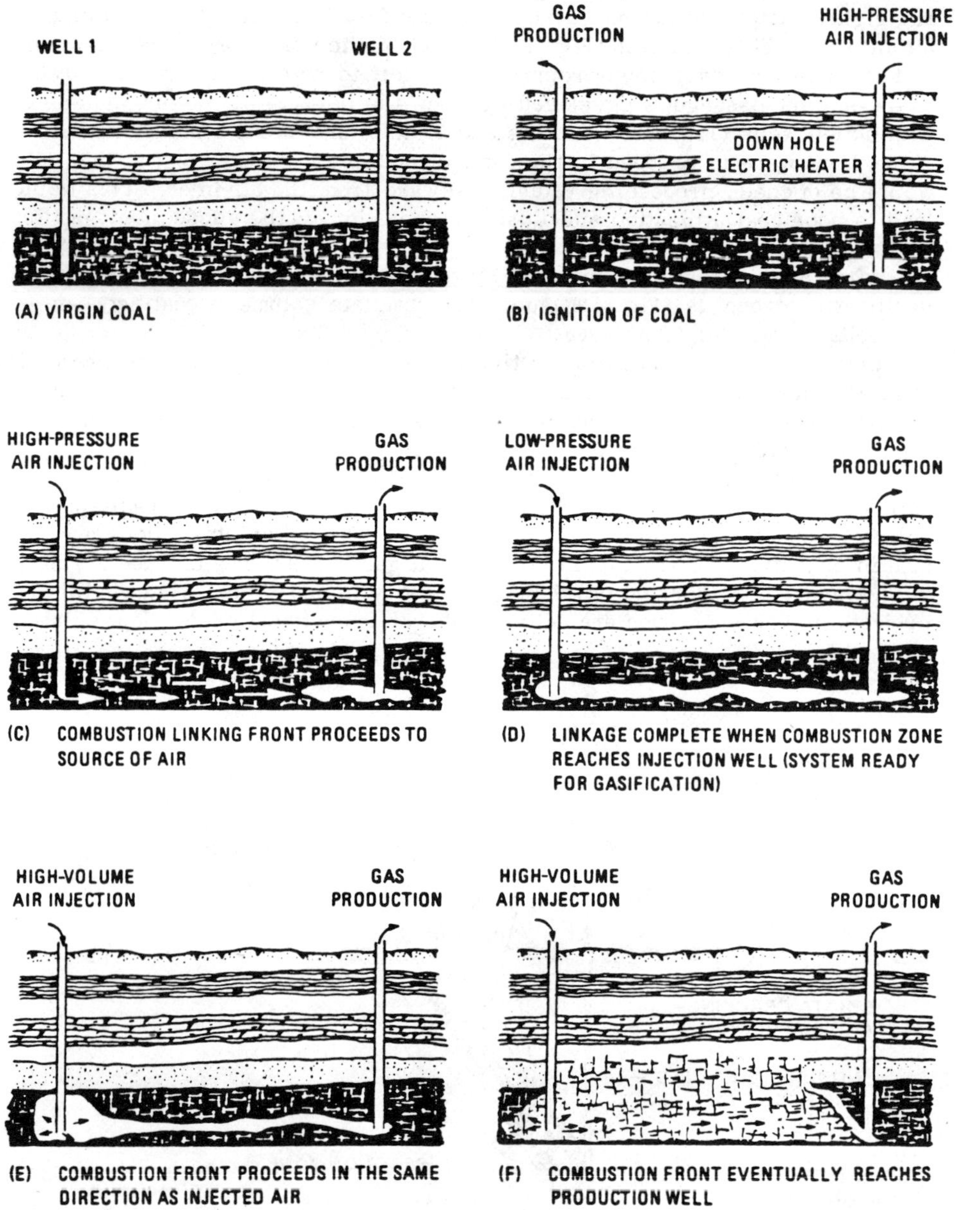

FIG. 59 Linked vertical wall operation with countercurrent combustion for linking [65].

1. U.S. Department of Energy operational projects
   a. Laramie Energy Technology Center, Western low-Btu project
   b. Lawrence Livermore Laboratory, Western medium-Btu project
   c. Morgantown Energy Technology Center, Eastern coal technology
   d. Gulf Research and Development Co. with TRW (subcontractor) (steeply dipping coal beds)
2. Other operational projects
   a. Texas Utilities Services, Inc. (TUSI), Texas lignite
   b. Texas A&M University, Texas lignite
   c. Alberta Research Council, Alberta subbituminous coal
3. Planned projects
   a. Atlantic Richfield, Wyoming subbituminous coal
   b. New Mexico Consortium, New Mexico subbituminous coal

1. Laramie Energy Research Center

This project at Hanna, Wyoming, has produced gas since 1972. Linkage has been by countercurrent combustion. The first test (Hanna I) gave inconsistent gas composition because the linkage path was not located at the bottom of the seam. Later tests have exhibited improved stability as well as higher heating value in the product gas. Table 45 summarizes the results obtained in the completed Hanna tests and conditions expected in planned future tests. Notice that well spacing is being increased as testing proceeds in order to determine the optimum spacing. Table 46 presents a typical product gas composition obtained during the Hanna II, Phase II test.

2. Lawrence Livermore Laboratory

Lawrence Livermore Laboratory has completed two tests at Hoe Creek near Gillette, Wyoming (Powder River Basin). Linkage was carried out by explosive fracturing in Test No. 1 and by countercurrent combustion in Test No. 2. During Test No. 2, gasification with oxygen, which is the primary objective of the program, was conducted over a 2-day period. Product gas averaged 263 Btu/scf during this period as contrasted with 106 Btu/scf when using air. Gas losses during the test were 20%.

Preparation for Test No. 3 is now under way. Linkage will be made mechanically by directional drilling of a 3-in.-diameter hole. The hole, which is 710 ft in overall length, will extend for 200 ft at the bottom of the coal seam and about 140 ft below the ground surface. Drilling will be followed by 2 to 3 days of countercurrent combustion to enlarge the linkage and then in mid-1979, a 6-week test with oxygen at a coal consumption rate of 80 to 100 tons/day will be undertaken.

Two subsequent tests are planned in 1980 and 1982, culminating in 1984 with a pilot test utilizing 20 wells with gasification of 1000 tons/day to produce 50 million scf/day of intermediate-Btu gas.

3. Morgantown Energy Technology Center

Gasification tests with air are planned on swelling bituminous coal near Pricetown, West Virginia. Linkage for Test No. 1, to be undertaken in 1979, will employ countercurrent combustion; Test No. 2 in 1980 will utilize directional drilling.

TABLE 45 Laramie Energy Technology Center—Underground Coal Gasification Experiments [66]

	Completed					Active
		Hanna II				
	Hanna I	Phase 1	Phase 2	Phase 3	Hanna III	Hanna IV
Year field test conducted	1973-1974	1975	1976	1976	1977	1978-1979
Average heating value (Btu/scf)	125	152	171	138	138	—
Maximum gas production rate ($\times 10^6$ scf/day)	1.5	3	8	12+	9	—
Maximum coal utilization rate (tons/day)	20	40	100	120	110	—
Total tons utilized	5000 in 1 year	1600 in 40 days	2100 in 27 days	4600 in 39 days	2900 in 38 days	—
Well spacing (ft)	25-100	60	60	60	60	112 1/2

Note: Coal utilization percent: coal utilization efficiency is undefined for a two-well test pattern. The combined Hanna II Phase 2 and Phase 3 field test comprised a four-well, 60 ft-square pattern. This pattern contained 4800 tons of coal. A total of 6700 tons was consumed, indicating that nearly all coal within the pattern and considerable coal along the boundaries had been utilized.

TABLE 46 Hanna II, Phase 2 Product Gas (Vol %) [67]

	Vol %		
H	17.3	N_2	51.0
CO	14.7	Ar	0.6
CO_2	12.4	H_2S	0.1
CH_4	3.3	C_2-C_4	0.6

4. *Gulf Research and Development Company (TRW Systems as Subcontractor)*

Three tests for air gasification of a steeply dipping coal bed near Rawlins, Wyoming, are planned to start in July 1980 and proceed through late 1981. Twelve-inch-diameter injection and production wells will be drilled to depths of 400 to 600 ft. Three different well configurations will be employed. In one, both wells will be vertical. In a second, the injection well will be vertical and the production well slanted parallel to the coal seam. In the third test, both wells will be slanted.

5. *Alberta Research Council*

A gasification test was conducted on a shallow (65 ft) 12-ft-thick seam of subbituminous coal at Forestburg, Alberta, in 1976. Product gas was 110 to 150 Btu/scf and approximately 300 tons of coal were consumed. Tests were stopped when lateral gas leaks occurred outside the pattern of test wells.

IV. CLEANUP AND PROCESSING OF PRODUCT GAS

A. Raw Gas Contaminants

One factor that must be considered when applying chemical engineering techniques to an energy problem is the sheer size of the facilities required [70]. For example, the production of the largest underground coal mines is about 5 to 6 billion pounds of coal per year. Although this would be a large feed for most chemical processes, the full output of two or three such mines is required for a single, commercial-size, high-Btu coal gasification facility. The size of these energy refineries demands that the processing be kept simple. The addition of a single subprocessing element to the overall system can add tens of millions of dollars to the plant cost.

The problem is complex, and there appear to be no simple solutions. Coal is a heterogeneous material, with a wide variety of constituents that do not appear in the final product and must be removed during processing. The purification steps required during this processing are numerous, accounting for a major portion of the cost of the facility. Figure 60 presents a block flow diagram of a hypothetical coal gasification facility. The shaded blocks on this diagram are gas purification steps. Approximately 50% of the sub-

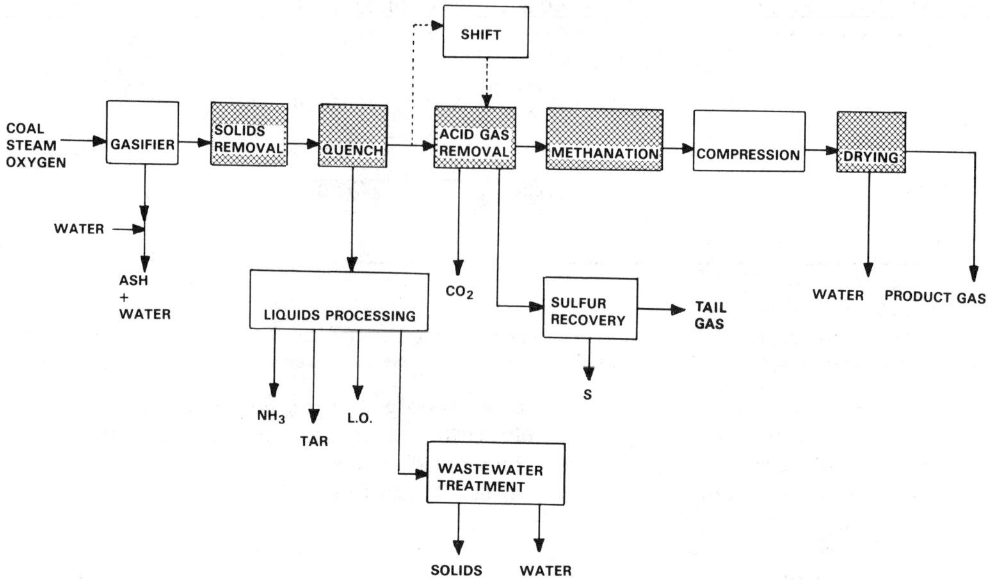

FIG. 60 General coal gasification scheme for pipeline-quality gas [70]

processes in the overall system are gas purification stages and a wide variety of processing options exist.

Coal compositions and physical characteristics, as well as the gasification processes that now exist or are under development, vary significantly. Coals contain varying amounts of sulfur, nitrogen, trace-metal compounds that can vaporize, and ash, all of which must be disposed of in a proper fashion. The temperatures of individual reaction stages may vary from 800 to 3000°F and the pressures may range from near-atmospheric pressure to over 1500 psi. Contacting modes include fixed-bed, fluidized-bed, and dispersed-phase. In any particular reaction stage, the gas-solids contacting mode may be, generally, either cocurrent, countercurrent, or back-mixed, with perhaps another mode of contact from stage to stage. The number of reaction stages may vary from a simple, single-stage contact, through multiple-bed configurations, to essentially infinite stages in the continuously variable temperature, descending-bed gasifiers.

The raw gasifier product may exist at temperatures from 250 to 3000°F and at pressures from near atmospheric to over 1500 psi. The raw product will contain significant amounts of H_2, CO, CH_4, and higher hydrocarbons as desired products; it will also contain impurities such as H_2O, CO_2, coal fines, tars, oils, phenols, and a variety of sulfur and nitrogen compounds, in addition to trace-metal compounds. The myriad classes of materials that may be present are listed in Table 47, and Fig. 61 presents theoretical equilibrium concentrations of various chemical compounds in a low-Btu gas.

The sulfur that occurs naturally in the coal is driven into the raw product gas (Fig. 62). Thermodynamically, the majority of the gasified sulfur should exist as hydrogen sulfide, with approximately 4% of the sulfur existing as carbonyl sulfide (COS) and only trace amounts of carbon disulfide (CS_2). However, data from coke ovens indicate that the concentrations of COS and CS_2 are much higher than would be thermodynamically expected.

The existence of mercaptans, thiophenes, and other organic sulfur compounds in the gasifier product gas will probably be a function of the degree of severity of the gasification, contacting schemes, and heat-up rate. Those processes that tend to produce tars and oils may also tend to drive off high-molecular-weight organic sulfur compounds into the raw product gas.

The nitrogen in the coal tends to gasify simultaneously with the carbon, as illustrated in Fig. 63 to form ammonia and cyanides (by reaction of ammonia with coal). The existence of ammonia and cyanide in raw gasifier effluents has been confirmed by many investigators. High-temperature processes, however, do not report significant ammonia in the effluent, probably due to the strong thermodynamic potential for ammonia to decompose to molecular N_2 and H_2 at high temperatures.

TABLE 47 Components in Gasifier Outputs

Solids
- Overheat fines

Organic liquids
- Tars
- Heavy Oils
- BTX or naphtha
- Phenols and cresylic acids (water-soluble)

Light hydrocarbons
- Paraffins (ethane, propane)
- Olefins
  - Ethylene
  - Acetylene
  - Propylene

Aqueous liquids
- Foul water

Sulfur compounds
- H_2S
- COS
- CS_2
- Mercaptans (RSH)
- Organic sulfides (RS or R_2S)
- Thiophenes (C_4H_4S)
- Heavier ring-substituted sulfur compounds

Nitrogen compounds
- Ammonia
- Hydrogen cyanide
- Thiocyanates
- Heavier ring-substituted compounds

Halogen compounds
- Hcl
- HF

Trace elements (of primary importance[a]
- Antimony
- Arsenic
- Beryllium
- Boron
- Cadmium
- Chromium
- Lead
- Lithium
- Manganese
- Mercury
- Molybdenum
- Selenium
- Tellurium
- Thallium
- Tin
- Vanadium
- Zinc

[a]May be present as various compounds such as chlorides, hydrides, sulfides, etc.

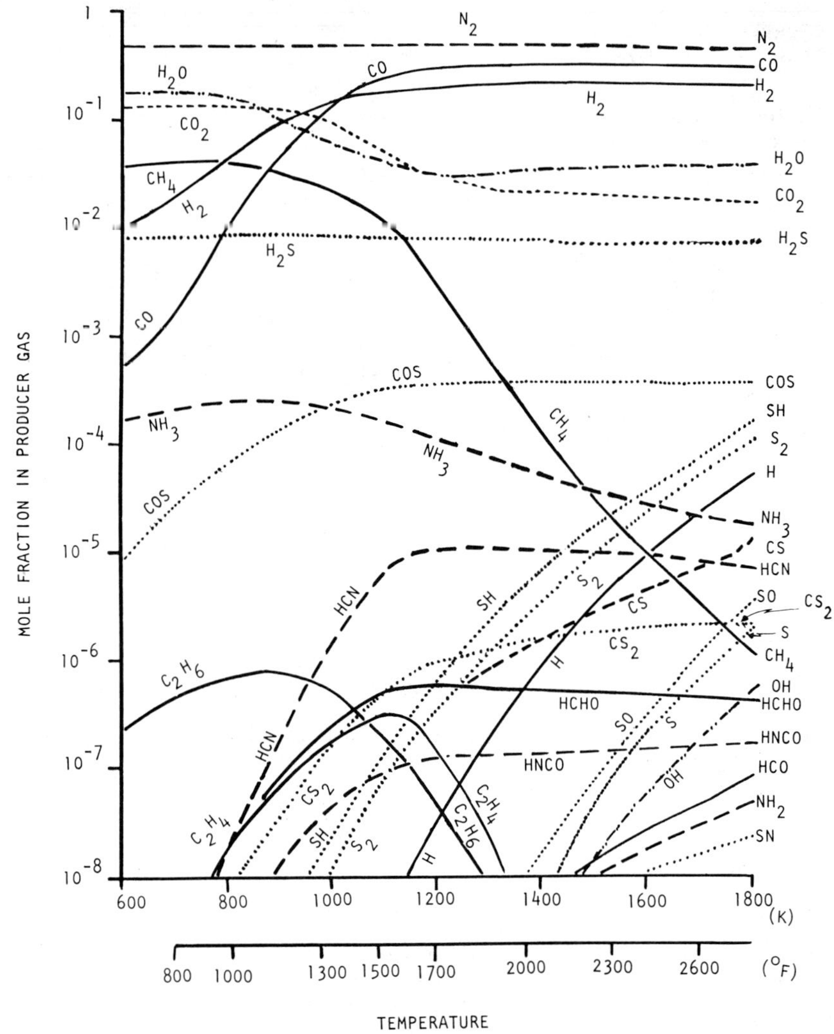

FIG. 61 Equilibrium composition of low-Btu gas from a Pittsburgh seam bituminous coal [71].

In low-temperature gasifiers, the occurrence of a variety of high-molecular-weight nitrogenous compounds is anticipated because of the quantity of tars and oils generated. These compounds would include pyridines, pyrroles, azoles, indoles, quinolines, anilines, amines, and similar compounds.

Thiocyanates have been reported in the quench water from several gasifiers, to the exclusion of cyanides. However, the reaction

$$H_2S + HCN = HCNS + H_2$$

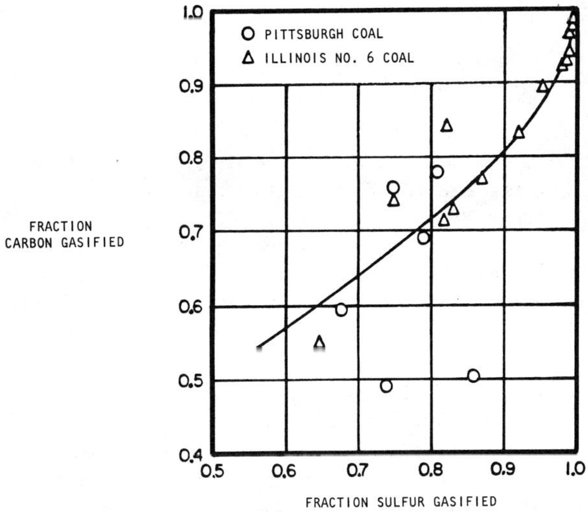

FIG. 62 Sulfur gasified [72].

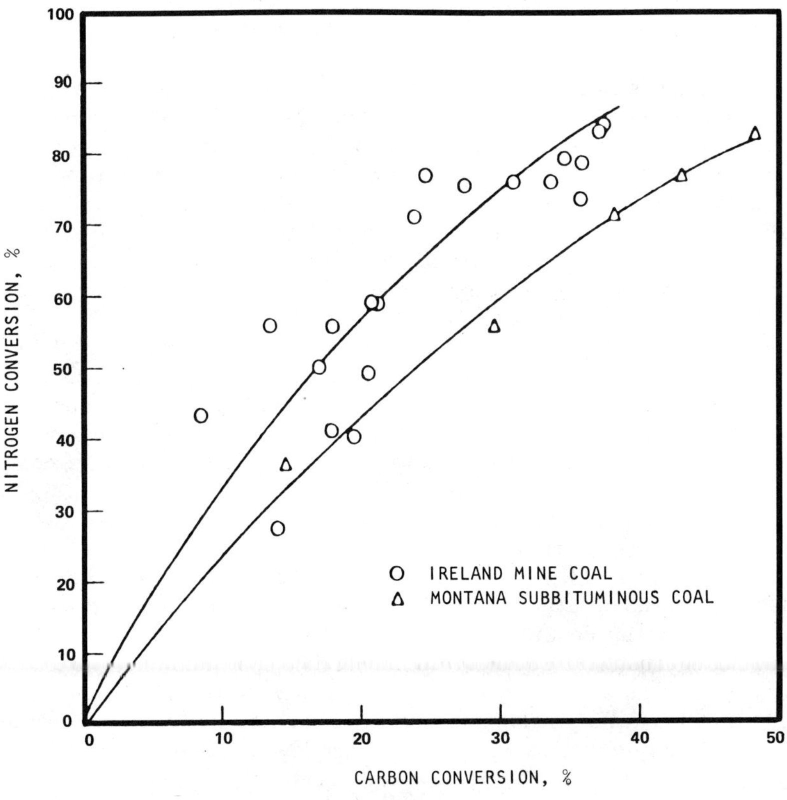

FIG. 63 Nitrogen gasified [73].

is not favored at the operating conditions within most gasifiers. Conceivably, the reaction could occur by the action of HCN upon sulfur within the coal:

$$S + HCN = HCNS$$

but the probability of quantitative reaction of HCN within the gasifier is small. Rather, the majority of the thiocyanates reported in the wash water are probably due to sampling techniques and the formation of thiocyanates in contact with air.

$$H_2S + HCN + 1/2\ O_2 = HCNS + H_2O$$

which is highly favored in aqueous media.

Chlorine is another material that is often present in coal in significant quantities. When coal is burned, the chlorine will appear in the stack as hydrochloric acid mist. Few data are available on the fate of chlorine in gasification; however, IGT's data on analysis of ash indicate that chlorine is nearly quantitatively stripped from the coal at even moderate coal conversions in the gasifier. Thermodynamically, the majority of the chlorine should appear in the raw gasifier product as hydrogen chloride.

A wide range of metals occur in the coal in greater abundance than normally found in the earth's crust because they were concentrated by the vegetation that initially formed the coal deposits. During gasification some of these may be present in the effluent either as vaporized metal or as a volatile compound.

There is not a large body of data on the analysis of gasifier effluents before and after the gas cleanup system, and several programs are currently under way to improve upon this data bank (Table 48). Table 49 presents an analysis of contaminants in the effluent of a Synthane gasifier as a function of the type of coal, and Table 50 shows data for HYGAS, Lurgi, and Koppers-Totzek gasifiers. These experimental data show the marked effect that the type of coal has on the contaminant concentrations. Additionally, the data indicate the concentration ranges to be expected in untreated gasifier product gases.

Fewer data are available on the nature and concentration of trace elements in gasifier outputs. Table 51 lists the concentration of trace elements in the output of three gasifiers. These values were calculated from the reported values of the percent loss of the elements from the coal during gasification, their original concentration in the coal, and the approximate pounds of coal per scf of gas.

With the variety of possible constituents and operating conditions, a "universal" purification system cannot be specified for economic application in all cases. Nevertheless, the first step in purification (Fig. 60) will probably be a device to remove large particles of coal and other solid materials. This would be followed by cooling, quenching, or washing, to condense tars and oils and remove dust and water-soluble materials—phenols, chlorides, ammonia, hydrogen cyanide, thiocyanate, and perhaps some sulfur compounds—from the gas stream. Water washing is desirable for simplicity in gas cleaning; however, the purification of this water is not simple. Further purification of wash water is discussed under "Wastewater Treatment," p. 681.

TABLE 48 Current Programs for Establishing Concentration of Contaminants in Coal Gas

Sponsor	Contractor	Program Start	Program Duration (yr)	Project title
EPA—Industrial Environmental Research Lab.	Radian Corp., Austin, Texas	March 1976	3	Low-Btu Gasification—Environmental Assessment
EPA Rudarski Institute, Beograd REMHK, Obilic, Socialist Autonomous Province Kosovo	Research and Development Department, REMHK Kosovo-Obilic, and Rudarski Institute, Mineral Dressing Department, Beograd		3	Environmental and Engineering Evaluation of the Kosovo Coal Gasification Plant, Yugoslavia
EPA/RTP	Research Triangle Institute	1977		Parametric Evaluation of Pollutants from a Laboratory Gasifier
DOE	Carnegie Mellon University (coordination contractor) IGT, Radian, PEC, PERC, GFERC } Environmental contractors	July 1976		Gasification Process/Environmental Characterization from Pilot Plant Data
EPA	North Carolina State Univ., Department of Chemical Engineering	January 1977	3	Coal Gasification—Gas-Cleaning Test Facility
EPA	IGT	January 1978	1	Parametric Study of Coal Conversion Effluents
EPA	Environmental Engineering Div., Energy Systems Group of TRW	May 1977	3	High-Btu Gasification—Environmental Assessment

TABLE 49 Contaminant Concentration in Synthane Gas (ppm) [74]

	Illinois No. 6 coal	Illinois char	Wyoming subbituminous coal	Western Kentucky coal	North Dakota lignite	Pittsburgh seam coal
H_2S	9800	186	2480	2530	1750	860
COS	150	2	32	119	65	11
Thiophene	31	0.4	10	5	13	42
Methyl thiophene	10	0.4	—	—	—	7
Dimethyl thiophene	10	0.5	—	—	11	6
Benzene	340	10	434	100	1727	1050
Toluene	94	3	59	22	167	185
C_8 aromatics	24	2	27	4	73	27
SO_2	10	1	6	2	10	10
CS_2	10	—	—	—	—	—
Methyl mercaptan	60	0.1	0.4	33	10	8

TABLE 50 Contaminants in Raw Product Gas [ppm (vol)] [75]

Type of coal	HYGAS		Lurgi		Koppers-Totzek	
	Subbituminous	Pretreated bituminous	Subbituminous	Bituminous	Subbituminous	Bituminous
C_2H_6-C_5	6,400	2,500	4,600	4,300	—	—
NH_3	3,400	5,000	3,200	2,000	29	22
N_2	270	500	500	700	10,300	9,100
HCN	57	500	2	43	16	9
H_2S	2,300	8,900	2,700	4,300	2,600	16,440
COS	90	400	160	70	184	1,080
HCl	20	200	13	32	123	490
$C_6H_6^+$	6,800	6,000	2,300	180	—	—
Phenol	470	75	450	240	—	—
Tar	—	—	440[a]	440[a]	—	—
Fatty acid	—	—	44[a]	73[a]	—	—

[a]Lb/scf.

Source: Adapted from "Parametric Study of Coal Conversion Effluents," Monthly Progress Report No. 1 (March 1978) prepared for EPA by IGT.

TABLE 51 Concentration of Trace Elements in Raw Product Gas ($lb/10^6$ scf) [75]

	HYGAS			Synthane:	Lurgi-SASOL:
	Illinois No. 6	Montana lignite	Rosebud subbituminous	Illinois No. 6	North Dakota lignite
Arsenic	0.22	0.26	0.01	0.002	0.02
Boron	0.66	0.66	1.0	0.78	0.01
Cadmium	0.02	0.01	0.0005	0.00002	0.02
Lead	0.14	0.02	0.01	0.00006	0.01
Mercury	0.003	0.02	0.0005	0.002	0.005
Selenium	0.15	0.03	0.01	0.01	0.01

The next step in gas processing is to adjust the H_2/CO ratio, if necessary, in a water-gas-shift reactor, so that the downstream methanation reaction can take place properly or if CO is not desired in the final product gas. The gas will now contain, primarily, H_2, CO, CO_2, H_2O, CH_4, and sulfur compounds. Depending on the gasification process, it may also contain small quantities of higher aliphatic hydrocarbons and light oils such as benzene, toluene, and xylene (BTX). The sulfur in the original coal feedstock will be present primarily as hydrogen sulfide (H_2S), although a small fraction may exist as carbonyl sulfide (COS). If any mercaptans or organic sulfides were formed during gasification, they probably would be removed with the oils in the upstream washing. Some H_2S would also have been removed upstream in the quench circuit, the amount depending upon the techniques and temperature used in washing.

Cleanup steps and their sequence can be affected by the type of gas produced and its end use. The minimum requirement in this respect would be the application of low-Btu gas produced from low-sulfur anthracite coal to fire a brick kiln. The gas goes directly from the gasifier to the kiln and, in this case, the kiln is the cleanup system. Many variations on this theme are possible. In addition, the order of the cleanup stages may be varied. In general, the gas-processing operations described here apply to the production of high-Btu gas, hydrogen, or synthesis gas. Fewer cleanup and processing steps will usually be required to produce low- and medium-Btu gases as final products. Gas cleanup and processing technology is extensive; therefore, a large number of references are provided the reader if deeper investigation is desired.

B. Removal of Particulates

1. *Cyclones*

Cyclones are low-cost particle collectors that have many potential applications in coal gasification systems; however, they have low efficiency for collecting particles smaller than 10 μm, but above this size, collection efficiency can be at least 90%.

The potential uses of cyclones include:

1. Precleaning of dusty streams ahead of higher-efficiency particulate control devices
2. Entrainment separators for wet scrubbers
3. Dust control for coal grinding and pulverizing operations
4. Coarse particle removal at high temperatures and pressures

Conventional applications for cyclones include use as precleaners, entrainment separators, and for controlling dust emissions from coal grinding and pulverizing.

Cyclone collectors utilize centrifugal force to separate particulates from a gas stream. Vortex flow is induced by the design of the gas inlet duct. This is accomplished by using a tangential inlet or an axial inlet with swirl vanes. The swirling gas stream has two major velocity components in addition to the rotary component: axial flow and radial flow. The vortex has normally two sections: the main or outer vortex and the central core. A schematic diagram of a typical cyclone showing the vortex flow patterns is illustrated in Fig. 64.

The main vortex is characterized by axial flow away from the gas inlet and radial flow outward from the axis of the cyclone body. The central core has the same rotary direction, but the axial and radial velocity components are in the opposite direction to that of the main vortex.

In the main vortex, the tangential velocity of the gas is a function of the distance from the cylinder axis and the tangential velocity at the perimeter.

Cyclone collectors include four basic designs:

1. Tangential gas inlet with axial dust discharge (Fig. 65A)
2. Tangential gas inlet with peripheral dust discharge (Fig. 65B)

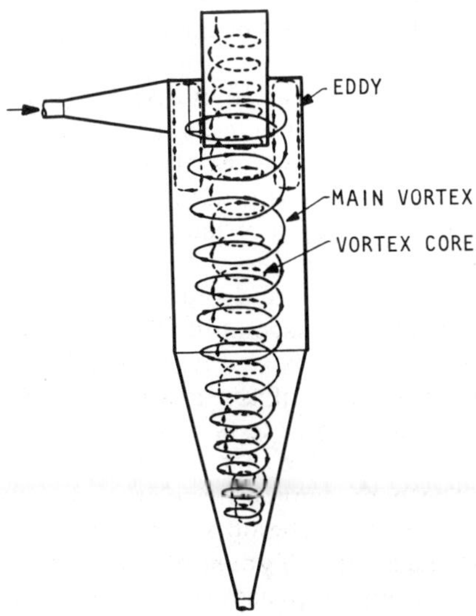

FIG. 64 Vortex and eddy flows in a cyclone [76].

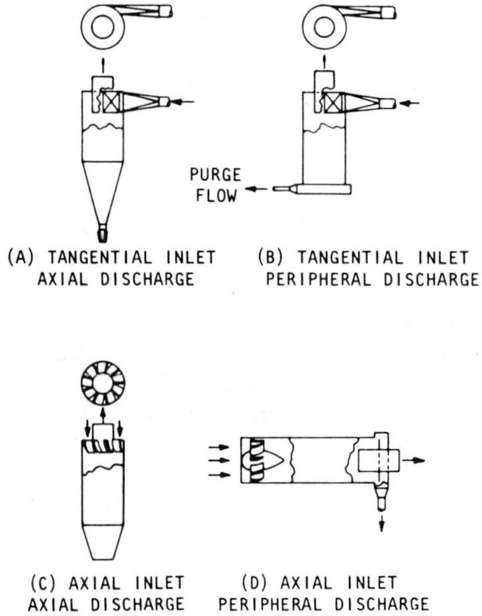

FIG. 65 Types of cyclones in common use [76].

3. Axial gas inlet through swirl vanes with axial dust discharge (Fig. 65C)
4. Axial gas inlet through swirl vanes with peripheral dust discharge (Fig. 65D)

Although there are differences in construction of these four basic units, the fundamental operating principles and chief structural components remain unchanged.

The basic structural components of a cyclone are gas inlet, body, gas outlet, dust discharge, and cone (optional). In the common cyclone (type 1 above) the dimensions of these components are all directly related to the gas outlet diameter. To improve efficiency a designer starting with the basic ratios shown in Fig. 66 would consider increasing the body length or width, or decreasing the inlet width.

The cyclone body is where the vortex is generated and particles are separated because of the centrifugal forces that are associated with vortex flow. The important characteristics of the cyclone body are its dimensions and the roughness of its surface. The residence time of the gas stream is directly determined by the length of the cyclone body. It is felt that the cyclone body length should be 5 1/2 to 12 times the size of the diameter of the gas outlet duct (Fig. 66). Increasing the body length in relationship to the other dimensions improves collection efficiency. The optimum diameter of the cyclone is 2 to 3 times the gas outlet diameter. Typically, twice the gas outlet diameter is used. Higher efficiency with a pressure-loss penalty can be gained by using a larger body diameter.

The gas inlet design is critical to performance and pressure loss in a cyclone. Normally, for a tangential inlet unit, the height of the inlet duct is equal to the gas outlet diameter and the width of the inlet duct is one-half the gas outlet diameter. The inlet duct dimensions and tangential arrangement provide the acceleration that causes initiation of the vortex pattern. This velocity change is a major contributor to the overall cyclone pressure loss. Vanes installed in the inlet have been found to be effective in reducing pressure loss. See Fig. 67 for the orientation of inlet vanes.

The gas outlet normally projects into the body of the cyclone. The purpose of this arrangement is to minimize reentrainment of dust from eddy currents. However, minimizing the length of the gas outlet duct does decrease overall

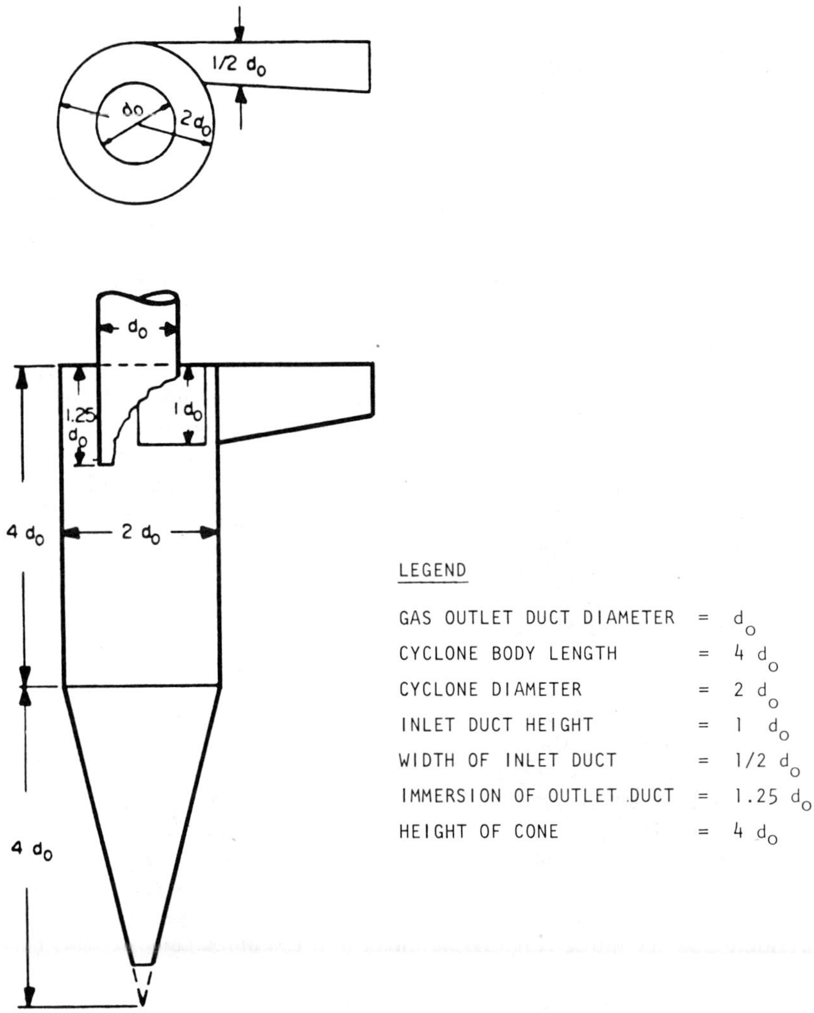

FIG. 66 Typical cyclone dimension ratios [76].

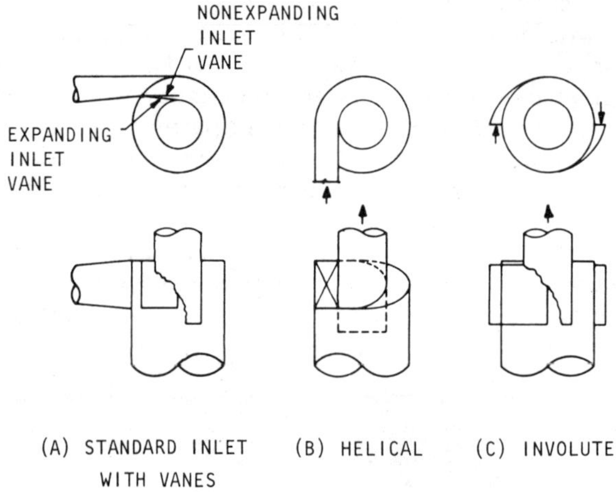

FIG. 67 Typical inlet vane orientation [76].

pressure drop. Normally, the gas outlet duct extends to just below the gas inlet duct.

The gas outlet stream is in vortex flow. This rotational energy contributes to further separation of entrained dust and may be converted to static pressure to reduce the overall head loss through the cyclone. Advantage may be taken of the vortex flow by using skimmers on the inner perimeter of the gas outlet duct. The vortex flow tends to concentrate the dust along the duct perimeter and if a skimmer is used, this dust can be collected. This stream can be fan-forced to either the inner wall of the cyclone body or treated separately in a high-efficiency dust collector. As for recovering the gas stream vortex energy, two successful arrangements have been developed. These are shown in Figure 68. The involute scroll (Fig. 68A) converts the kinetic energy of the vortex to pressure by gradual expansion. The outlet drum (Fig. 68B) operates as a reverse cyclone converting vortex flow to linear motion.

The major requirements of the dust discharge are that upward gas flow into the cyclone body be minimized while continuous discharge of the dust is maintained. This is usually accomplished by the inclusion of baffles or straightening vanes at the discharge to suppress the vortex at this point, thus minimizing upflow. Another method is to use a valve to prevent upflow from the dust discharge bin. Depending on the negative pressure of the dust discharge, either single or double rotary valves or even screw conveyors may be used to provide a gas seal. With peripheral dust discharge and often with axial discharge, dust removal is normally accomplished through the use of an induced purge gas flow of 1 to 10% of the total gas throughput. The gas is returned to the axis of the cyclone body after disengagement.

The inclusion of a cone is not fundamental to the vortex flow patterns generated in a cyclone, but if a cone is used, it has the function of concentrating the dust to a central location before it is discharged. When a central dust discharge is used, its diameter should be greater than one-fourth the gas outlet diameter, so that the central vortex will not be able to reentrain the dust.

Often abrasion-resistant linings are used to prevent erosion of the surface of the cyclone shell. When the operating temperature of the cyclone is above about 700°F, a refractory lining is used for protection of the metal shell. This refractory is then protected by an abrasion-resistant lining. Cyclones mounted internally in high-temperature vessels are usually constructed of temperature-resistant metals such as Incoloy or Monel.

High-efficiency cyclones function under the same operating principles as do the units described above. However, high-efficiency cyclones utilize smaller diameters and longer bodies than do normal cyclones. The pressure drop of these units is greater than for normal units.

Cyclones may be used to collect liquid droplets as well as dry particles. Cyclones are also used as wet scrubbers whereby a spray of liquid is used to collect particles by inertial impaction in addition to the centrifugal action. Liquid may be applied to the walls to improve collection efficiency by minimization of reentrainment.

2. *Electrostatic Precipitators*

Electrostatic precipitators are efficient collectors of fine particulates—capable of greater than 90% reduction of submicron particles. They have the lowest power usage of all particulate control devices, and they have the capability of collecting liquid mists as well as dust. Because of these attributes, electrostatic precipitators are considered candidates for cleanup of coal gasification gases.

There are several potential arrangements of particulate control equipment that can utilize the capabilities of electrostatic precipitators in coal gas cleanup systems. The most attractive possibility is to use the precipitator as a high-temperature cleanup device. Here, the precipitator might be used as the only control device or, more likely, it might be used as a second-stage control device following a cyclone. On the other hand, if the precipitator is downstream of the gas-cooling train, the temperature and pressure conditions are more likely to be within the range of current precipitator practice. One arrangement of electrostatic precipitators, which is used commercially, is illustrated in Fig. 69. Top gas from a two-stage gasifier is taken at 250°F

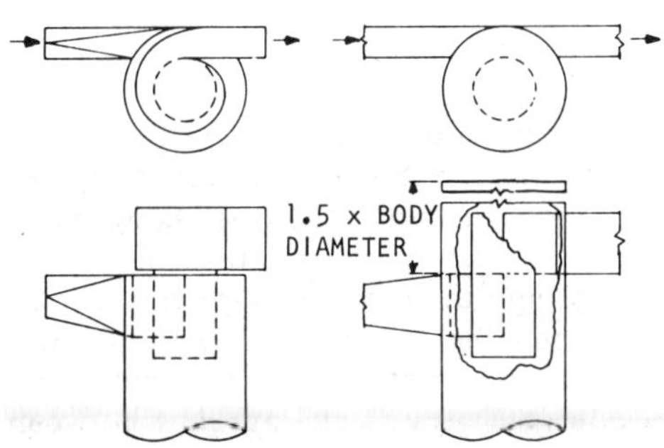

FIG. 68 Gas outlet pressure recovery [76].

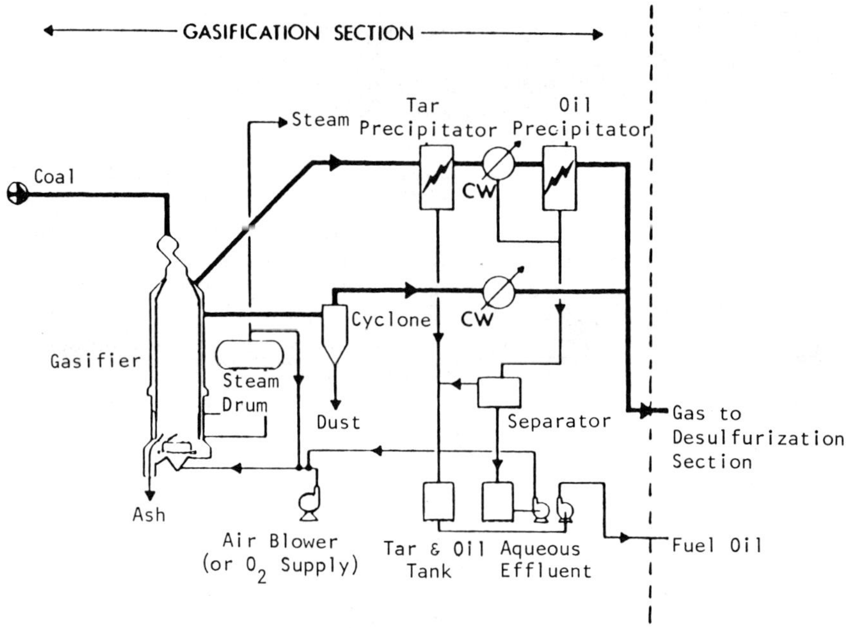

FIG. 69 Use of electrostatic precipitators in WD/GI system [23].

to an electrostatic tar precipitator where liquid tar is removed. The detarred gas is then cooled, which condenses light and medium oils together with some water. The condensed oil and water are then removed by a second electrostatic precipitator.

The operation of an electrostatic precipitator includes:

1. Imparting an electrical charge to a particle or droplet suspended in a gas stream
2. Removing that particle or droplet from the gas under the force exerted by a high-voltage electric field
3. Removal of the captured particle or droplet to an external receptacle

To effect these operations, the basic component (Fig. 70) are the discharge (or corona) electrode, the collection electrode (plates), plate rappers, and precipitator shell. Other important components are the dust hopper, high-voltage equipment, gas distributors, and instrumentation. The precipitator is energized by passing a high-voltage current through the discharge electrode, generating an electric field between the discharge and the collection electrodes at an intensity specific to the properties of the gas and suspended particles and dependent upon the electrode geometry. The current between the electrodes is caused by the collision of free electrons with gas molecules. The collisions generate additional electrons and negative ions that are involved in further collisions until a current of electrons moving toward the collection electrodes is formed. Suspended particles are charged by collisions with the

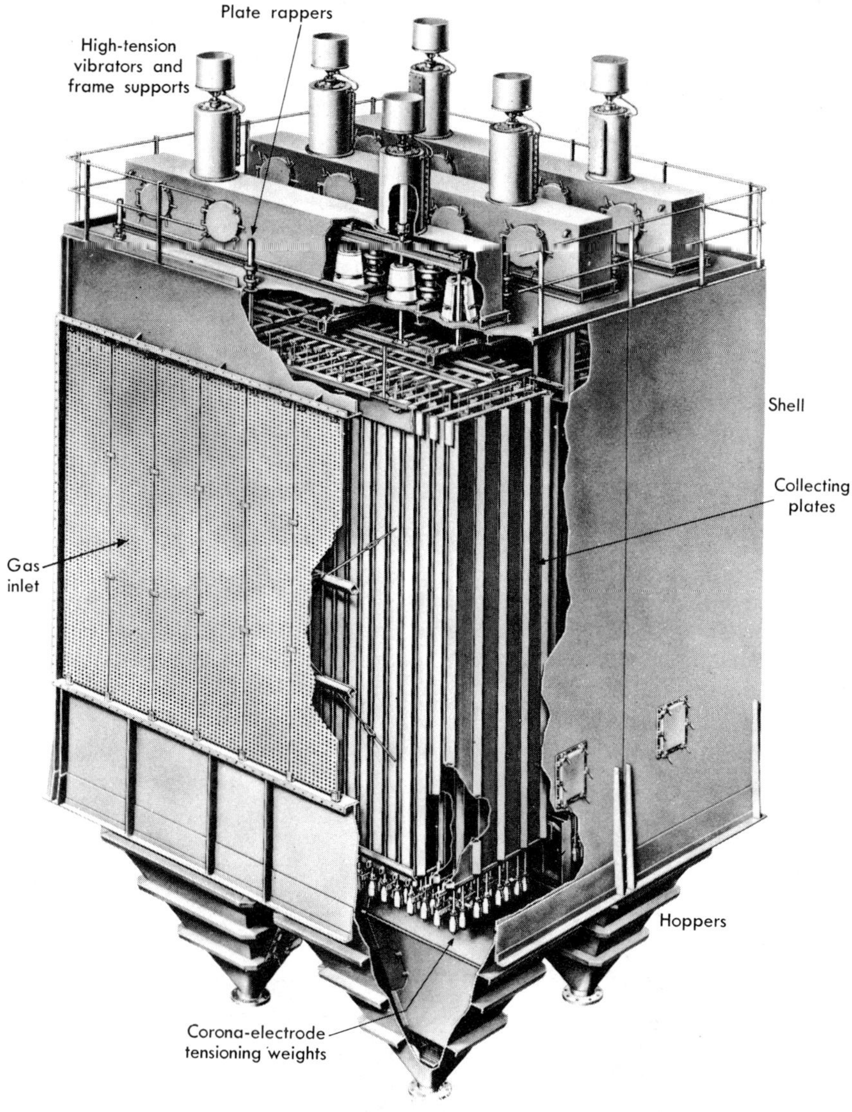

FIG. 70 Duct-type electrostatic precipitator [82]. (Courtesy Research-Cottrell, Inc.)

negative ions passing through the zone between the corona and the collection electrode.

Movement of the particles to the collection electrode is governed by the interaction of the electric field and charged particles in the moving gas stream. Opposing this motion is the aerodynamic drag associated with movement of the particle through the gas stream. The velocity of the particle toward the collection electrode is determined by equating the electrical forces that move the particle toward the collection electrode with the drag forces that resist this motion.

Rapping of the electrodes is the normal method for removing collected dust. Rapping frequency and force are determined by the need to provide sufficient acceleration of the electrode to remove most of the dust layer but maintaining a minimum level of reentrainment. Dust removal may be accomplished by impact, vibration, or liquid flushing. When collecting tars and oils, liquid flushing will probably be used. For high-temperature, high-pressure dust collection, rapping will probably be used. These methods assume that potential materials of construction problems will be solved for each application. Dust removed from the electrodes is collected in dry dust hoppers located beneath the collection electrodes, while liquids drain freely to whatever collection and recovery system is required.

Table 52 lists typical data for a coke oven gas precipitator. These data give the general range of particulate properties expected in coal conversion technology.

Figure 71a and b illustrates the particle size distribution expected from two slagging-entrained flow gasifiers. For these gasifiers, particulate loading ranges from 20 to 100 grains/scf.

Current research [87] has demonstrated the feasibility of producing corona currents at high temperatures and pressures. This work is being directed at fluidized-bed combustion gas cleanup but will at least prove the concept of particle collection under conditions of temperature and pressure expected of some gasifier off-gases. This research is at a relatively early

TABLE 52 Typical Coke-Oven Gas Precipitator Data [83]

Gas volume	5000-20,000 ft^3/min
Design efficiency	95-99%
Precipitation rate parameter	0.2-0.55 ft/sec
Type discharge electrode	1/4-in.^{2} twisted wire
Type collecting electrode	8-in.-diameter steel cylinder
Voltage	50-70 kV peak
Average electric field	11.5 kV/in.
Average grain loading	0.3-0.6 gr./ft^3
Average gas velocity	8.0 ft/sec
Power/gas volume flow	100-200 W/1000 ft^3/min

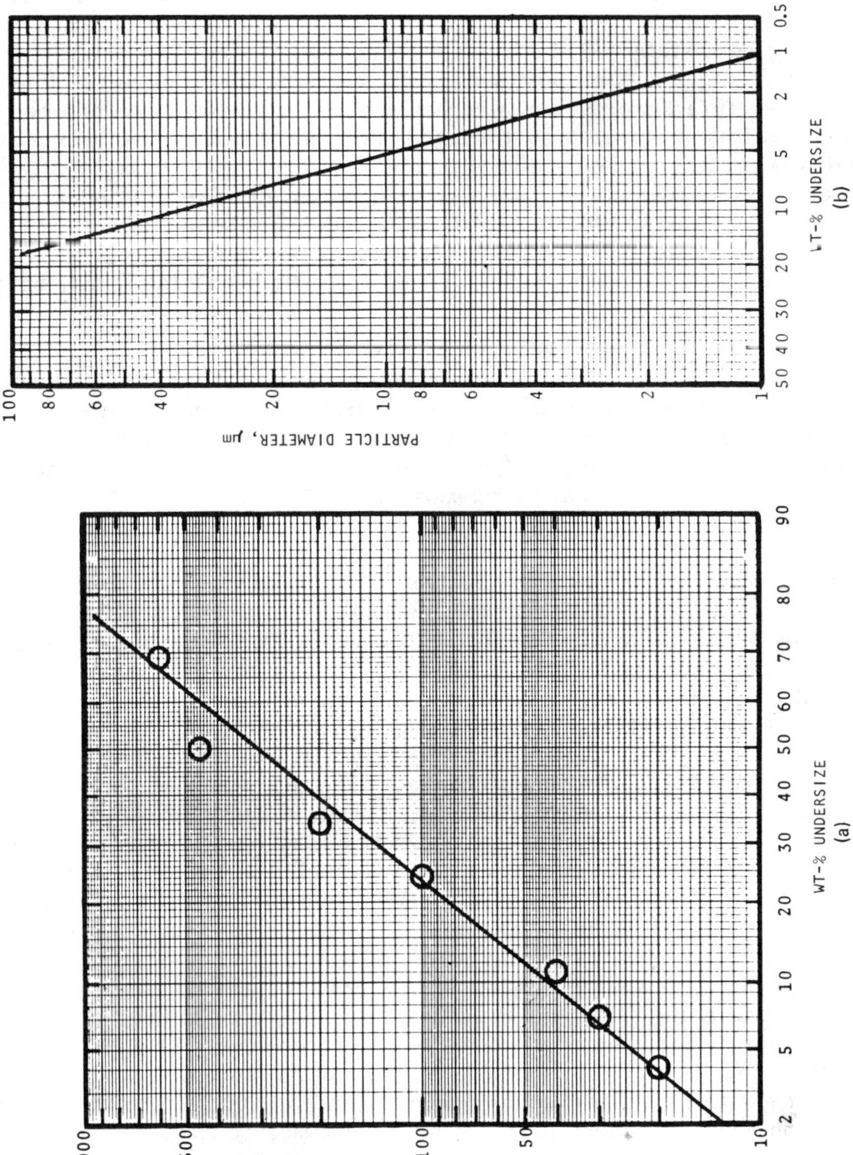

FIG. 71 Particle sizes in raw gases: (a) size distribution of char and ash from the BI-GAS process effluent; (b) particle size distribution expected from the Koppers-Totzek coal gasifier [89].

stage, however, and, therefore, electrostatic precipitation cannot yet be considered a feasible method for cleanup of hot gasifier process gas under all conditions.

The application of electrostatic precipitators to hot gasifier process gas cleanup is dependent upon developing an electrode system capable of generating a stable corona at temperatures and pressures higher than those ordinarily applied. A recent study [87] indicates that a cylinder wire geometry is optimum for this application. Further, positive corona operation is possible at high temperature/low pressure, whereas negative corona is in order when both temperatures and pressures are high. Indications are that stable corona generation is possible at temperatures as high as 1700°F with accompanying pressures up to 35 atm; however, the system has yet to be tested under precipitating conditions. Previous work indicates 90% collection efficiency at 1650°F and 115 psia.

Extensive literature on the subject of electrostatic precipitators may be found in Refs. 81-91.

3. *Granular-Bed Filters*

Granular-bed filters comprise a class of filtration equipment that is distinguished by a bed of separate, closely packed granules which serve as the filter medium. These devices are under study because of their potential ability to collect particulates at high temperature and pressure.

The most attractive potential application of granular bed filters is high-temperature cleanup of gasifier process gas. It is conceivable that these filters can be used for simultaneous control of particulates and hydrogen sulfide (by adsorption) at high temperature because of their ability to use almost any material as the filter medium. They may also be applied to low-temperature cleanup.

Granular bed filters rely on inertial impaction, diffusion, and direct interception. The predominant mechanism in collection of particles depends on both the particle size and the operating parameters of the device.

The most important collection mechanism is inertial impaction. In the granular bed, the gas flow path follows a torturous maze of spaces between the granules. The combination of many direction changes and high local velocity causes impingement of particles on the granules where they are held by molecular and mechanical forces. Currently, research studies are being conducted that should lead to development of mathematical models that accurately predict the efficiency of granular bed filters.

Granular bed filters are classified according to the method used to remove collected dust from the filter medium. Cleaning is required to prevent interstitial plugging, which results in excessive pressure drops. This requirement distinguishes granular bed filters from fluidized-bed systems, which utilize the product gas to keep the granules in motion and prevent plugging. According to the method of cleaning, granular bed filters may be classified as continuously moving, intermittently moving, or fixed-bed filters.

Continuously moving systems are usually arranged in a cross-flow configuration. The gas passes horizontally through the granular layer while the granules and collected dust move continuously downward and are collected at the bottom. The dust and granules are then separated and the cleaned granules returned to the bed. Figure 72 illustrates the Combustion Power Company's Dry Scrubber, which is an example of this system.

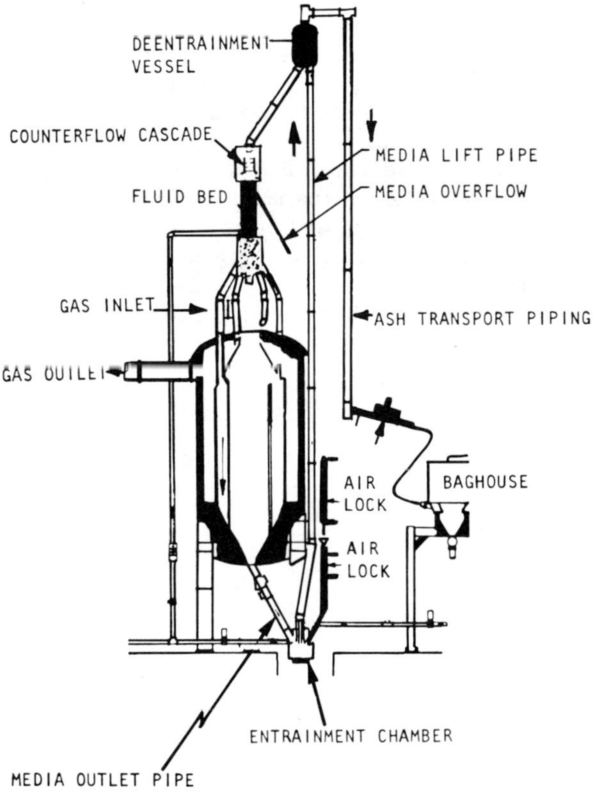

FIG. 72 Combustion Power Company's "dry scrubber" [92].

The Squires Panel-Bed Filter utilizes the intermittently moving bed concept (Fig. 73). The bed is stationary during filtration. Accumulated dust and the surface layer of granules are removed from the panel by a backwash pulse and replaced by fresh granules from the overhead hoppers.

Fixed-bed systems use either backwash air or backwash air and mechanical agitation to remove collected dust. The ducon Granular Bed Filter (Fig. 74) and the Rexnord Rex-Gravel Bed Filter (Fig. 75) are examples of fixed-bed systems. The Ducon system employs only backwash air, whereas the Rexnord system employs both backwash air and stirring for removal of dust particles.

These dry systems are potentially attractive, but at this time are not highly developed for application to coal gasification processes.

4. Wet Scrubbers (Quench)

When hot raw gas is to be cooled directly or if it has already been cooled in a waste-heat recovery system, it will normally be treated in a wet scrubber. Wet scrubbers are devices that utilize gas/liquid contacting to cool the gas stream, condense high-boiling hydrocarbons, dissolve some constituents, and separate particles from gas streams. There are many different wet scrubber designs, but all utilize similar mechanisms.

Wet scrubbers utilize diffusion, impaction, and interception to collect particles from moving gas streams. The predominant mechanism involved in the collection of a particle depends on the particle size. Diffusion is most important for particles smaller than 0.1 μm, inertial impaction is the predominant mechanism for collection of particles above about 0.5 μm in size, and interception predominates for large particles on the order of 10 to 100 μm.

A type of wet scrubber that has been widely applied to coal gas is the venturi scrubber (Fig. 76), one that operates on the impaction principle. Dirty gases are passed through a tube to contact with water, which is added at the throat. The throat promotes intimate mixing of the gas and liquid. The liquid, which forms droplets ranging in size from 100 to 1000 μm, collects particulates mainly by inertial impaction. Liquid circulation ranges from 5 to 20 gal/1000 ft^3 of gas treated. In addition to the venturi tube, the system

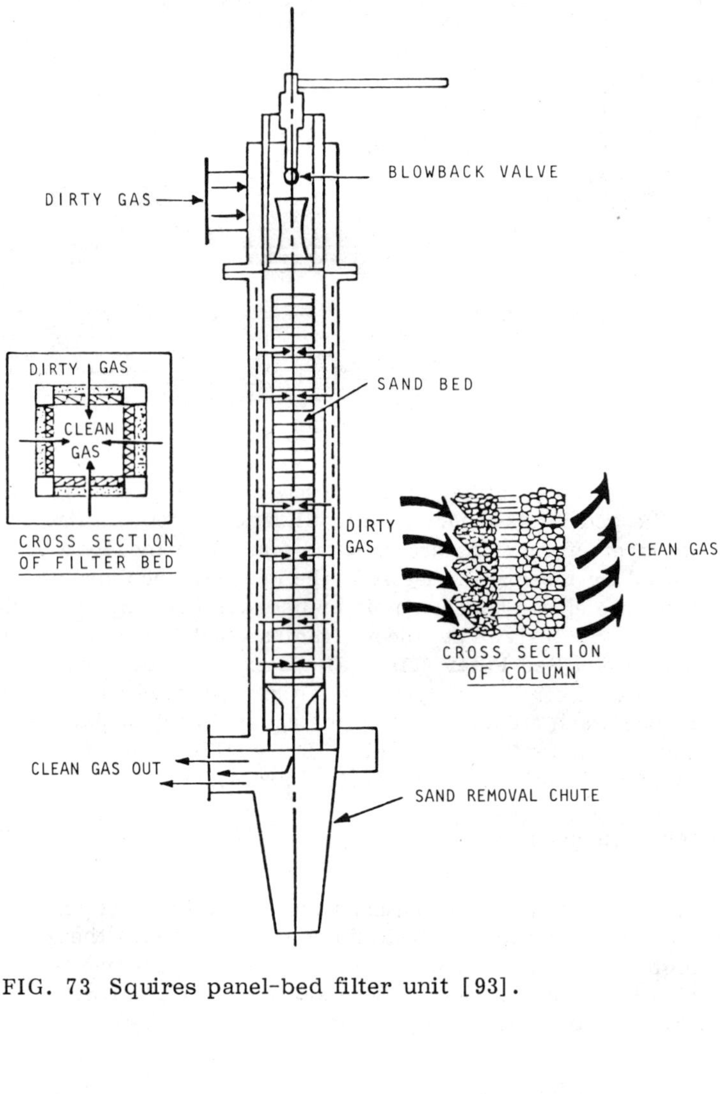

FIG. 73 Squires panel-bed filter unit [93].

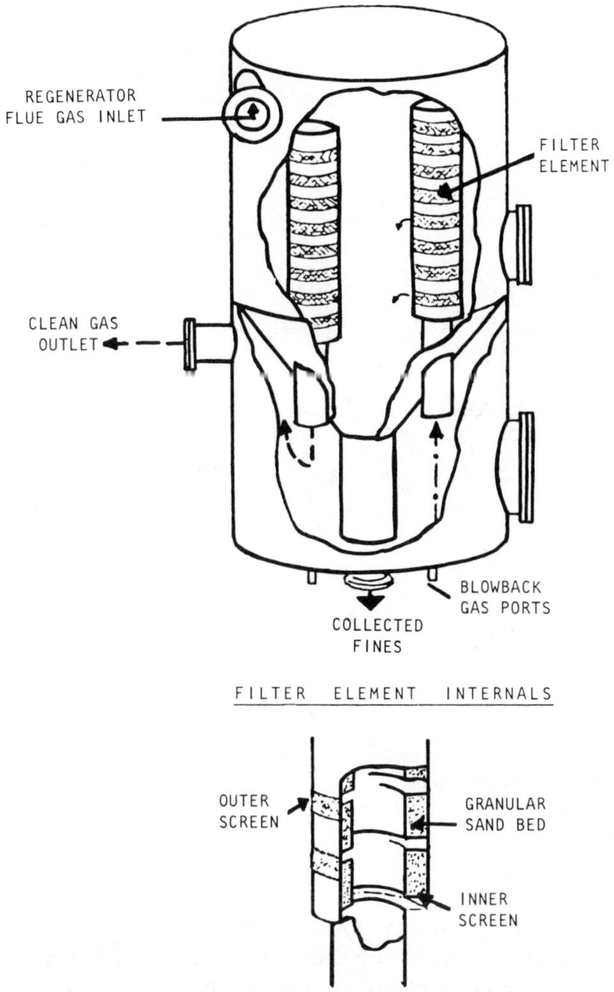

FIG. 74 Ducon granular bed filter [94].

typically includes a liquid separator, mist eliminator, and the scrubbing liquor circulation equipment. Venturi scrubbers can collect particles as small as 0.3 μm with at least 50% efficiency and particles larger than 2 μm at about 99% efficiency.

Other wet scrubber designs that might be applied to coal gas are plate scrubbers (sieves, bubble caps, and impingement plates), massive packing (rings, saddles), fibrous packing (plastic, spun glass, fiberglass, steel wool), centrifugal (cyclone), and directional baffles (louvres, zigzags, disk, and donut) [100-107].

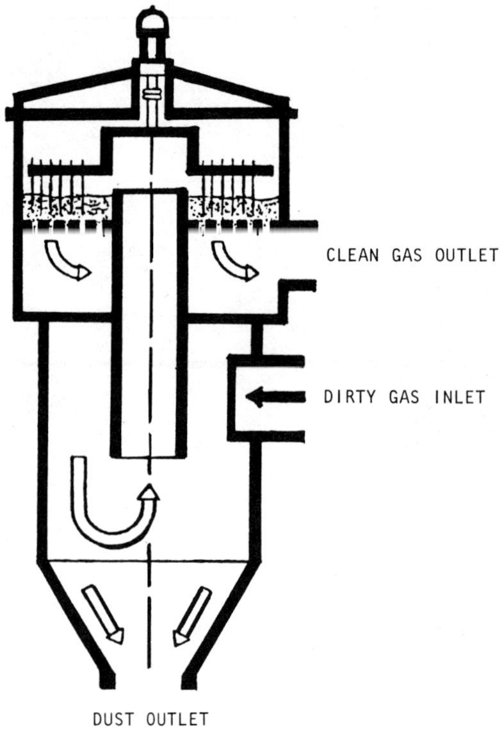

FIG. 75 Rexnord Rex-gravel bed filter [94].

C. Heat Recovery

In the event that raw gas leaves the gasifier at high temperature and is to be subsequently cooled before it is processed further, it is a standard practice to recover heat by steam generation and boiler feedwater heating. The waste-heat boiler may be preceded by a cyclone separator, if in the system, to remove large coal or ash particles. In the event that the raw gas contains molten ash, such as from a Koppers-Totzek generator, the gas is cooled by direct quenching with water sprays to solidify the ash before entering the waste-heat boiler so that it will not collect on the walls of the boiler tubes.

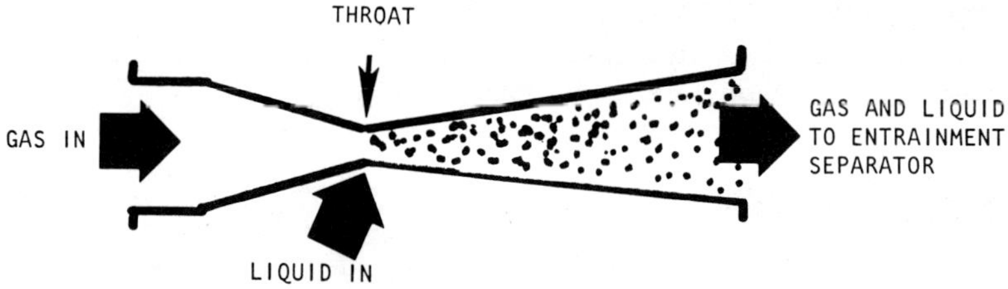

FIG. 76 Venturi scrubber

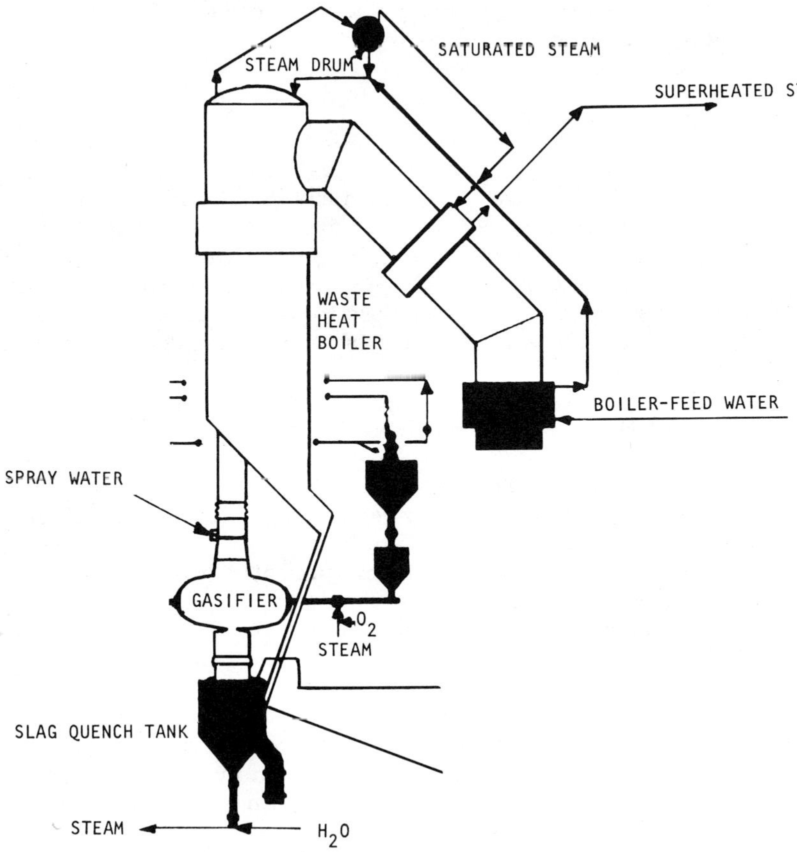

FIG. 77 K-T gasifier waste-heat recovery [97].

The recovery of heat from the raw gas may have a significant effect on the overall efficiency of the system. For example, consider the Koppers-Totzek generator and waste-heat-boiler arrangement shown in Fig. 77. Conditions applying to such a system have been described [98]. Raw gas at 2730°F from the gasifier is immediately quenched to 2100°F and flows directly into the waste-heat boiler. The boiler consists of a steam-generating section fed from a boiler top steam drum, a steam superheater section, and a feed-water preheater section that feeds the steam drum. Gas is cooled from 2100° to 181°F and high-pressure, superheated steam is generated at 2100 psig and 900°F. Sensible heat derived from the gas in the waste-heat boiler is equal to about 20% of the heat of combustion contained in the gas. The steam generated is sufficient to supply about one-fourth of the total plant steam requirements, including power for oxygen production. If waste-heat recovery was not practiced in this system, the conversion efficiency to product gas (hydrogen) would be significantly lower than the 57% obtained.

Recovery of heat from raw gas is not as important for descending-bed gasifiers because the gas is produced at lower temperatures. In such cases, no recovery of heat is attempted. If the gas is to be cooled, it is immediately quenched in some sort of a wet scrubber system.

D. Water Gas Shift

When the final product of coal gasification is to be SNG, synthesis gas, or hydrogen, it is necessary to adjust the proportions of CO and H_2 in the gas from the gasification reactor. Usually, the H_2 content must be raised relative to CO. For SNG production, the methanation step requires that the H_2 be slightly more than three times the CO. For syngas or hydrogen production the CO content needs to be reduced to as close to zero as possible. This is accomplished by the catalytic reaction of carbon monoxide and steam to produce hydrogen and carbon dioxide,

$$CO + H_2O \rightarrow H_2 + CO_2$$

in the production of hydrogen. This reaction, which is called the water gas shift, has been in commercial use for many years. A simplified flow diagram for hydrogen production is presented in Fig. 78.

If a high (operating)-temperature, chromia-promoted iron catalyst is used, gas containing up to 0.3% sulfur may be employed. Low (operating)-temperature catalysts, consisting of copper and zinc, allow operation between 350 and 700°F, where the equilibrium is more favorable, but these catalysts cannot tolerate sulfur. In this case, sulfur would be removed prior to the CO shift. Cooling of the streams involves heat exchange. Carbon dioxide removal between reactors is also possible to improve conversion. High-temperature and low-temperature catalysts can be used in series to improve conversion. This can result, after CO_2 removal, in a nearly pure H_2 gas and is practiced in hydrogen production.

In the case of SNG production, the purpose of the shift is to adjust the ratio of H_2/CO to a value of 3 for subsequent methanation. Since only a partial shift is required in this case, less steam can be used or the process can be operated at higher temperatures with less active and more-sulfur-resistant catalyst. Often only part of the product gas is shifted, the balance bypassing this step.

In the shift converter, COS is hydrolyzed by the reaction

$$COS + H_2O \rightarrow CO_2 + H_2S + \text{heat}$$

Since not all of the gas may be shifted, the sour gas from shift conversion may still contain substantial quantities of COS, depending on the quantity originally present in the raw gas. If the quantity of COS is high enough, it can determine the design of the acid-gas-removal unit immediately downstream of the shift conversion section.

Process options available for sulfur management in shift conversion include the following:

1. Shift conversion with bypass but no separate COS hydrolysis
2. Shift conversion with separate catalytic COS hydrolysis on the bypass stream

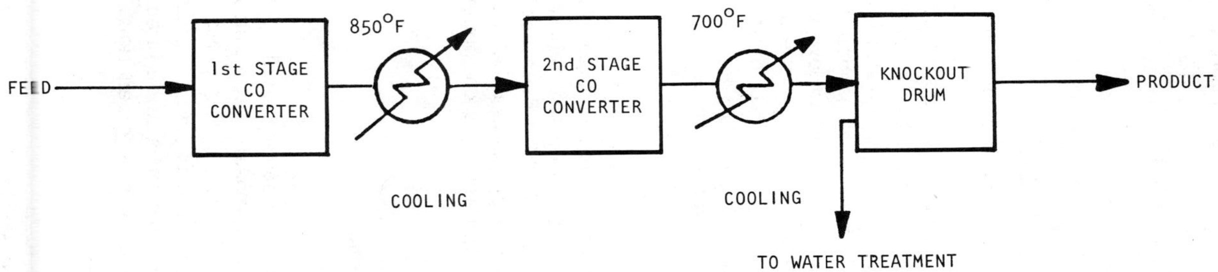

FIG. 78 CO shift converter, simplified flow diagram.

3. Shift conversion with subsequent catalytic COS hydrolysis of the entire shift effluent stream

The shifted sour gas is cooled and sent to the acid-gas-removal unit. Water condensed from the gas contains about 5% of the total sulfur in the feed to the shift conversion section, principally as H_2S. This sour water is sent to the water purification section.

E. Acid Gas Removal

1. Configurations

The group of systems that remove sulfurous compounds and carbon dioxide from gases are generically called acid-gas-removal processes. The sulfur content in the gas stream feeding the acid gas system reflects the sulfur content of the raw gas, less that removed in any upstream washing operation. The primary sulfur species in untreated raw gas is hydrogen sulfide, with small but poorly defined concentrations of carbonyl sulfide and carbon disulfide. The sulfur concentration must be reduced to low levels because of product specifications and, in some instances, to protect sensitive downstream catalysts. CO_2 must be removed for production of SNG, hydrogen, or syngas. In general, intermediate- or low-Btu fuel gas does not require CO_2 removal but does require sulfur removal to avoid atmospheric pollution after combustion. Removing H_2S from the fuel gas is generally conceded to be more economical than scrubbing SO_2 from stack gas.

A wide variety of removal processes, which employ various absorbent solutions to wash out the acid gases, are commercially available. The solutions are regenerated after washing, and concentrated acid gases are released by stripping at elevated temperatures and/or decreased pressure. These processes may be classified into four general categories according to the absorbent employed to remove the acid gas components:

1. Amine-based systems
2. Hot carbonate systems
3. Physical solvent systems
4. Mixed-solvent systems

The choice of a particular acid-gas-removal system is affected to a large extent by considerations of the disposal of the sulfur removed.

It is most desirable to convert the sulfur removed from gas to elemental sulfur. The alternative is sulfuric acid, which presents problems both in storage and transportation. Hydrogen sulfide is converted to sulfur by partial oxidation with air according to the reaction

$$H_2S + 1/2\ O_2 \rightarrow S + H_2O$$

Two methods of conversion are available here. The first is to process the hydrogen sulfide gas obtained from one of the absorption systems listed above by the conventional Claus Sulfur Process, whereby oxidation and catalytic conversion are conducted in a gas phase over a solid catalyst. The other route is "wet oxidation," where hydrogen sulfide is absorbed from a gas stream by a special solution that has the capability to enable oxidation by air directly in the liquid. Solid sulfur is thus formed and separated from the circulating liquid.

The Claus Process, when handling gas streams containing 15% or higher concentrations of H_2S, can easily effect sulfur recovery of 90 to 96% and generally is more economical than a wet oxidation system. At lower H_2S concentrations, however, sulfur recovery in the Claus Process becomes much more difficult and the plant cost (per unit of sulfur) increases almost in proportion to the quantity of total gas processed. In such cases, wet oxidation may be the preferred conversion route.

The simplest approach to acid gas removal is the nonselective process illustrated in Fig. 79a. Light oils are removed from the process gas stream in any of three locations depending upon the specific type of acid gas system employed. Most of the acid gases (carbon dioxide and sulfurous compounds) are removed from the process gas stream in a single contacting stage. The concentrated acid gas from regeneration of the sorbent may contain up to 5% of sulfurous compounds in carbon dioxide, depending upon the original coal feedstock, gasifier conditions, and the loss of sulfur to upstream processes. At sulfur concentrations of 5% and below, a liquid-phase oxidation process will probably be the economical choice for sulfur recovery, as indicated in Fig. 79a.

Some acid-gas-removal systems have a greater affinity for H_2S than for carbon dioxide. With these systems, the processing conditions can be chosen so that most of the sulfur will be removed with a smaller quantity of carbon dioxide, thus generating a sulfur-rich gas stream. Fig. 79b illustrates how this processing might be employed (the oil recovery, not shown, would be similar to that in Fig. 79a). The combined acid gas, produced similarly to Fig. 79a, is further treated in a selective acid-gas-removal system. Most of the sulfur will be recovered in a stream that is sufficiently concentrated to feed a Claus plant, and most of the carbon dioxide will pass on through the absorber. If the secondary acid-gas system is sufficiently selective, this CO_2-rich gas will be nearly sulfur-free and can be vented directly. With lower selectivity, a liquid-phase oxidation process might be employed to further purify the gas stream. Figure 79b indicates that most of the sulfur has been processed in a Claus plant. Economics of the Claus plant may more than offset the cost of a secondary acid-gas-removal system.

An alternative approach to selective acid gas removal is illustrated in Fig. 79c. Here, the selective system is first applied to the raw process-gas stream. The off-gas from the selective system then provides a feed for a Claus plant, and the CO_2-rich gas from the final acid gas system is either vented or treated further, as discussed earlier.

The selective acid-gas-removal systems shown in Fig. 79 are simplifications of the processes that might normally be employed. In practice, the number of process steps may be greater than those shown, and various feedback loops between the steps may also be employed to further improve the selectivity.

Some acid gas systems can be selective enough for H_2S over CO_2 that the CO_2-rich gas can be essentially H_2S-free; however, the fate of carbonyl sulfide is not well defined. Most of the acid gas systems will remove this

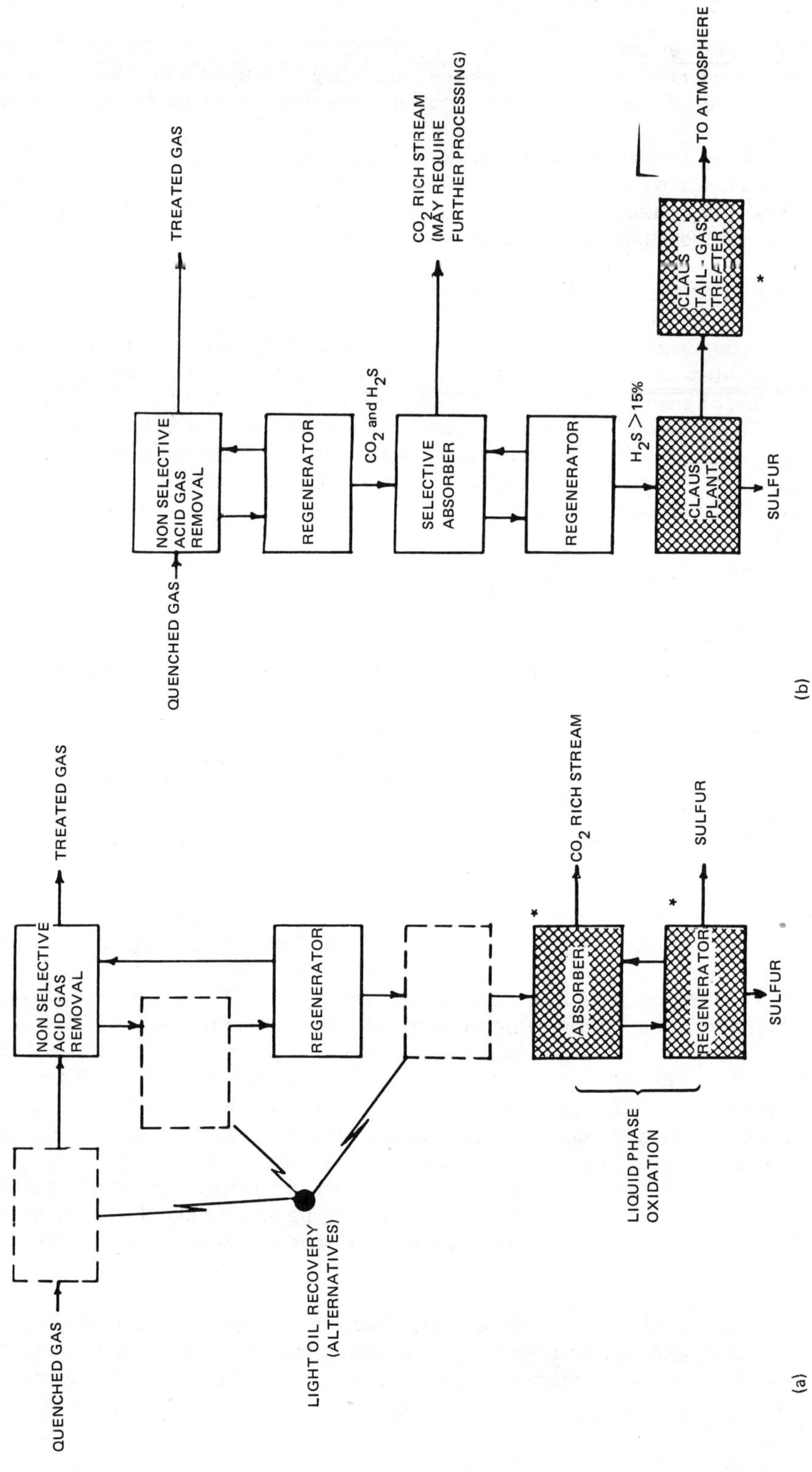
QUENCHED GAS
NON SELECTIVE ACID GAS REMOVAL
TREATED GAS
REGENERATOR
LIGHT OIL RECOVERY (ALTERNATIVES)
ABSORBER
REGENERATOR
CO_2 RICH STREAM
SULFUR
SULFUR
LIQUID PHASE OXIDATION
*
(a)
QUENCHED GAS
NON SELECTIVE ACID GAS REMOVAL
TREATED GAS
REGENERATOR
CO_2 and H_2S
SELECTIVE ABSORBER
CO_2 RICH STREAM (MAY REQUIRE FURTHER PROCESSING)
REGENERATOR
$H_2S > 15\%$
CLAUS PLANT
SULFUR
CLAUS TAIL-GAS TREATER
TO ATMOSPHERE
*
(b)

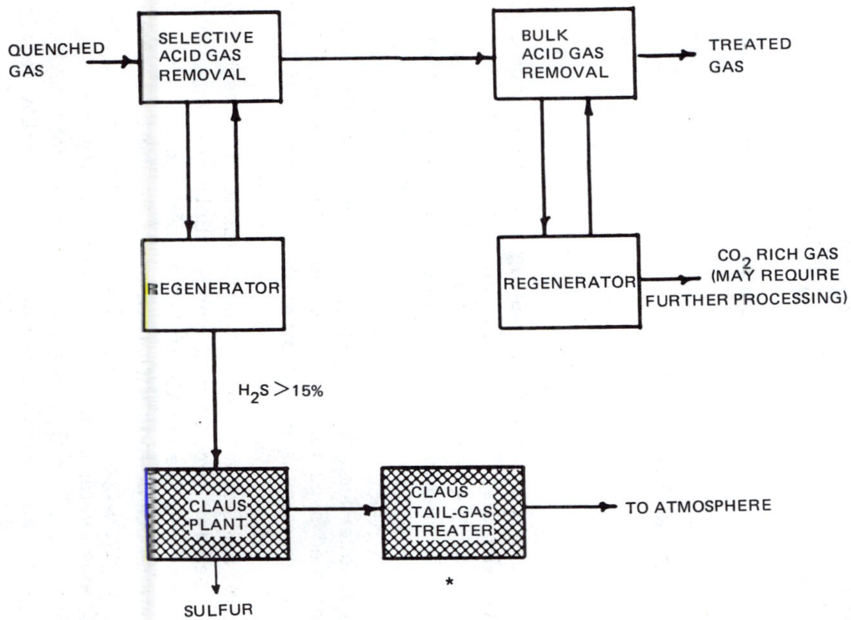

(c)

FIG. 79 Acid-gas-removal configurations: (a) nonselective acid gas removal followed by liquid-phase oxidation; (b) nonselective acid gas removal followed by selective removal from regenerator off-gas; (c) selective acid gas system on main stream [99]. (For discussion of tail gas operations see "Claus Plant Tail Gas" p. 673.)

species from the raw process gas so that downstream catalysts will not be poisoned, but its disposition in the selective processing is not certain. Carbonyl sulfide is not removed by most liquid-phase oxidation systems.

Another form of selectivity that should be considered in the choice of an acid gas system is the separation between carbon dioxide and fuel gas species. Some systems, which might otherwise be desirable, do not have high selectivity between carbon dioxide and C_2 or C_3 hydrocarbons. If these species are present, they will be largely lost to the acid gas, particularly the CO_2-rich gas in a selective system.

Still another factor that must be considered in the choice of an acid-gas-removal system is its energy requirement. Many of the acid gas solutions are regenerated by steam stripping, and the quantity of steam required can be significant. Much of this steam can be raised in waste-heat boilers elsewhere in the process, but—if higher energy demands are required—the size of the boiler house and the overall efficiency of the system are effected.

Table 53 presents a listing of many of the acid gas systems that might be employed in coal gasification. The systems are classified according to the four general types of processes mentioned previously, each of which is discussed in greater detail below. This table also presents a qualitative estimate of the relative selectivity and energy requirements of the various systems.

2. *Amine Systems*

Alkanolamines are the most commonly employed absorbents for the removal of H_2S and CO_2 from sour gas streams in petroleum refineries and from natural gas. The acid gas components react with the amine solution to form chemical compounds:

$$2RNH_2 + H_2S \rightleftarrows (RNH_3)_2S$$

$$(RNH_3)_2S + H_2S \rightleftarrows 2RNH_3HS$$

$$2RNH_2 + CO_2 \rightleftarrows RNHCOONH_3R$$

These reactions are readily reversed by heating the solution and stripping off the H_2S and CO_2 with steam in a regenerator.

A typical amine acid-gas-removal system is shown in Fig. 80. Gas, at ambient temperature, is washed in an absorber tower with a water solution of amine. The rich solution containing acid gases is heated by exchange and steam stripped in a reboiled regenerator tower at low pressure (15 psig). Acid gases and steam overhead from the regenerator are cooled and the condensed steam is refluxed back to the regenerator. The hot lean solution from the bottom of the regenerator is cooled by exchange with the rich solution and pumped through a cooler back to the top of the contactor.

The most common amines, monoethanolamine (MEA) and diethanolamine (DEA), have been used for more than 40 years. Both are nonselective between CO_2 and H_2S and require relatively large quantities of heat for regeneration. MEA forms stable compounds with COS and CS_2, both of which cannot be regenerated by heat. Diglycolamine (DGA, $HO\text{-}C_2H_4\text{-}O\text{-}C_2H_4\text{-}NH_2$) has found favor in acid gas systems due to its lower volatility. An increased solution concentration (65% versus 25% for MEA and 40% for DEA) can be used, resulting in a lower solution circulation rate and utility requirements.

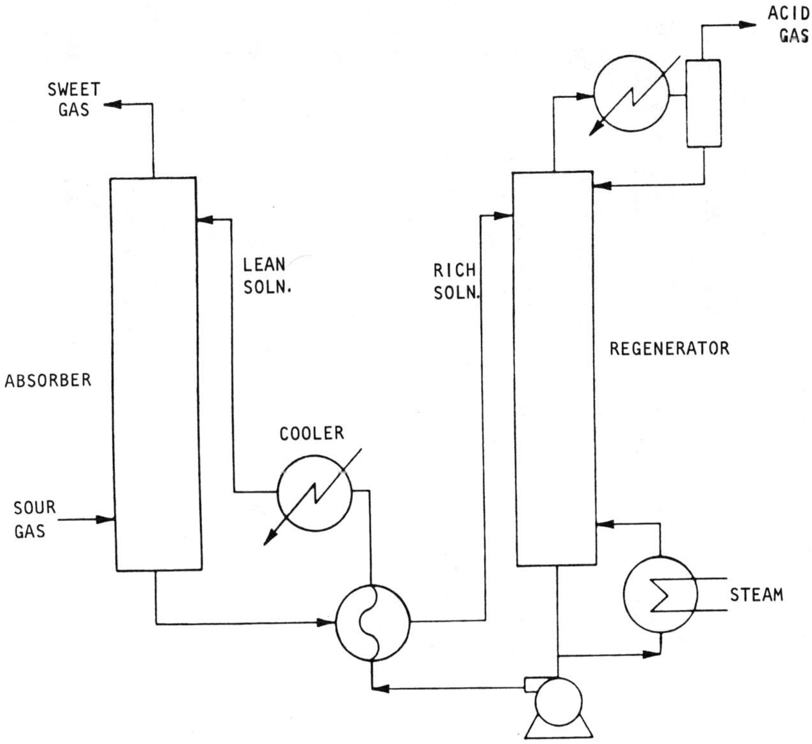

FIG. 80 Amine acid-gas-removal system.

The Fluor Economine Process uses a solution reclaiming method that permits recovery of the solvent (DGA) when treating acid gases containing COS and CS_2. DGA has been used successfully in the HYGAS coal gasification pilot plant. Another amine, diisopropanol amine (DIPA), is used both in the Adip Scrubbing Process and, in conjunction with a physical solvent, in the mixed solvent Sulfinol Process. The tertiary amines and DIPA offer partial selectivity in H_2S/CO_2 removal. In general, the capacity of amines to absorb CO_2 and H_2S is superior to physical solvents at low pressure, but as the partial pressure of acid gas in the total gas stream increases, the economy of an amine system declines in comparison with other systems, particularly as a result of the higher utility requirements.

3. *Hot Carbonate*

The basic hot carbonate process is relatively simple. Acid gas is absorbed at elevated pressure in a countercurrent contactor by a carbonate solution. The gases are recovered from the rich solution by flashing and steam-stripping in a low-pressure regenerator. Carbonate solutions have been employed to absorb acid gases for many years. Two low-pressure, low-temperature, carbonate systems—the Seaboard Process and the Vacuum Carbonate Process—

TABLE 53 Summary of Acid-Gas-Removal Systems [99]

Solvents	General operating range: CO_2 partial pressure	Selectivity[a]		Component distribution[b]				Utilities[c]
		H_2S/CO	CO_2/HC	NH_3	COS	RSH	Other organics	
Amines[d]								
MEA [111]	Low	1	8	a,d	e	a,b,d	a,d	10
DEA	Low	1	8	a,d	a,b	a,b,d	a,d	10
DGA (Fluor) [111]	Low	1	8	a,d	a,b	a,b,d	a,d	8
MDEA	Low	3	8	a,d	a,b	a,b,d	a,d	7
TEA	Low	3	8	a,d	a,b	a,b,d	a,d	7
DIPA (Shell)	Low	2	8	a,d	a,b	a,b,d	a,d	7
Hot carbonate[e]								
Catacarb (Eickmeyer and Associates)	Moderate	2	9	a,d	f	d,g	a,d	4
Benfield (Benfield Corp.) [110]	Moderate	6	9	a,d	f	d,g	a,d	4
Alkazid (I.G. Farbin Industrie, West Germany)	Moderate	3	9	a,d	f	d,g	a,d	4
Giammarco-Vetrocoke [123]	Moderate	9	9	a,d	f	d,g	a,d	4
Physical solvents[f]								
Methanol (Rectisol-Lurgi) [113,114]	High	7	2	d,c	a,b	a,c,d	a,b,c,d	2
DMEPED (Selexol-Allied) [119,120]	High	9	3	d,c	a,b	a,c,d	a,b,c,d	1
Propylene carbonate (Fluor) [115]	High	4	3	d,a	a,b	a,d	a,b,d	1
N-methylpyrrolidone [113]	High	9	3	d,c	a,b	a,c,d	a,b,c,d	1
Tri-n-butyl phosphate (Estasolvan)[g]	High	4		c,a	a,b	a,d	a,b,d	1
Methyl cyanoacetate (Union Oil) [124]	High	4	5	d,a	a,b	a,d	a,b,d	1

Mixed solvents[h]								
Sulfolane + DIPA (Sulfinol, Shell)	Moderate	2	3	d,a	a,b	a,d	a,b,d	5
Methanol + DGA (amisol, Lurgi) [125]	Moderate	1	3	d,a	a,b	a,d	a,b,d	5

[a]Arbitrary number scale 1-10; 1 = poor selectivity; 10 = excellent selectivity.

[b]a, with CO_2-H_2S stream, after simultaneous removal; b, with CO_2 stream from bulk removal following selective H_2S removal; c, with H_2S stream from selective H_2S removal; d, separate stream prior to CO_2-H_2S removal; e, degrades solvent; f, hydrolyzes to H_2S; g, largely not removed, remains in treated gas.

[c]Arbitrary number scale 1-10; 1 = low utilities; 10 = high utilities. Range dependent on conditions (partial pressure CO_2 and degree of cleanup).

[d]See Refs. 108, 109, 111, 112, 116-118, 121, 122 and 128.

[e]See Refs. 108-110, 112, 116, 117, and 126-128.

[f]See Refs. 108, 110, 112-120, 124, and 128.

[g]Institut Francais du Pétrole; Friederich Uhde, GmbH.

[h]See Refs. 108, 112, 116, 117, and 128.

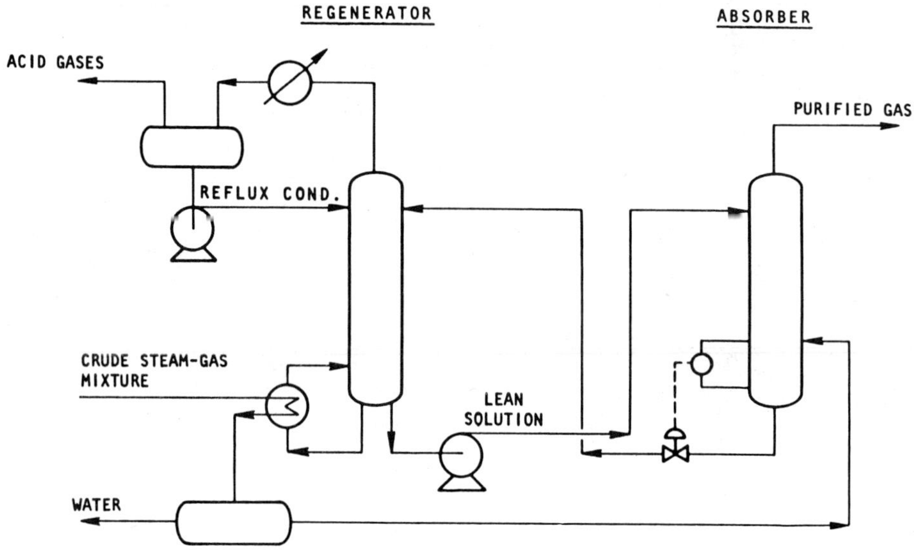

FIG. 81 Hot carbonate acid-gas-removal system.

were developed in the 1920s and 1930s to remove acid gas; neither system is expected to be applied in coal gasification today, because they have been replaced by the more efficient hot carbonate systems discussed here. These newer systems, employing higher solution temperatures, allow higher carbonate concentrations with lower circulation rates and decreased utilities. Modern carbonate systems, such as the Benfield Process, the Catacarb Process, and the Giammarco-Vetrocoke Process, have found many applications in today's gas, chemical, and refinery industries. These systems employ a 20 to 30% solution of potassium carbonate in water, operate at between 180 and 300°F, and utilize various additives to increase the rate of gas absorption by catalytic action.

Hot carbonate acid gas removal (Fig. 81) is basically similar to the amine system (Fig. 80). Acid gases are absorbed at pressure into the circulating solution in a contactor and removed by stripping with steam in a regenerator at reduced pressure. Unlike the amine systems, however, the hot carbonate systems do not operate with low temperature in the contactor and high temperature in the regenerator, but rather at essentially the same temperature in both. For this reason heat exchange between rich- and lean-solution streams, and cooling of the lean solution to the contactor, are generally not employed except in cases where lowered acid gas concentrations in the treated gas are required. Then, a small stream of cooled solution may be introduced at the top of the contactor. In addition, the carbonate solutions are not heat-sensitive as are the amines, so that hot raw gas can be employed for reboiler heat in the regenerator (Fig. 81). The raw gas will often be hot enough, with sufficient condensing steam content, to supply all the heat required. Removal of COS and CS_2 is almost complete in hot carbonate processes because they are converted by hydrolysis to H_2S, which is then absorbed.

The Benfield Process is the most widely used hot carbonate system. Nearly 400 units have been built or are on order. Two units, installed in 1960, operated for 14 years at the Westfield, Scotland, gas works, treating a medium-Btu coal gas from Lurgi gasifiers. The Benfield Hi-Pure Process, which employs a split-stream arrangement for the circulating solution, is capable of reducing H_2S in the treated gas to less than 1 ppm.

The Giammarco-Vetrocoke Process uses glycine or arsenic trioxide in conjunction with potassium carbonate to improve the rate of H_2S and CO_2 absorption. Arsenic trioxide is preferred because it can improve absorption and desorption rates by two to three times. Although many units employing this process are in use throughout the world, it is unlikely that they will be employed in coal gasification systems because other systems use less toxic agents.

Hot carbonate systems provide the following advantages:

1. The carbonate solution does not degrade significantly; only minimum makeup or purge is required.
2. They can be partially selective for H_2S in CO_2 containing gases.
3. Energy required for regeneration is lower than that for amine systems.
4. They do not absorb hydrocarbons to the extent that physical solvents do.

Negative features of hot carbonate systems include:

1. Corrosion problems.
2. They require moderate-pressure operation.

4. *Physical Solvents*

The third general method (Table 53) for removing acid gas from a raw gas stream is by physical absorption in an organic solvent without chemical reaction. The solvent can be regenerated by heat, pressure reduction, or gas stripping to produce a concentrated stream of the absorbed gas. H_2S and CO_2, as well as minor components in the gas stream, including CS_2 and mercaptans, are more soluble in many organic solvents than fuel gas species, especially at elevated pressure. In addition, some of the physical solvents are highly selective for H_2S over CO_2. One of the drawbacks with physical systems is that higher hydrocarbons are also soluble in the organic solvent.

All of the physical solvent processes listed in Table 53, except Rectisol, operate at atmospheric temperature and differ only in the particular chemical used as the solvent. The Selexol Process described below may be considered

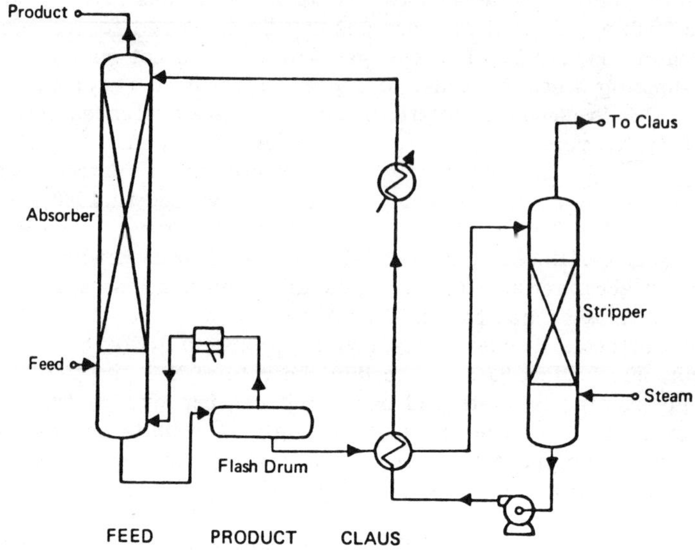

	FEED	PRODUCT	CLAUS
CO_2	9.4%	7%	79%
H_2S	0.6%	{(153 ppm)}	{21%}
COS	(600 ppm)		
C_1	4%	4%	–
H_2	15%	15%	–
CO	20%	21%	–
N_2	51%	53%	–
NH_3	(85 ppm)	(12 ppm)	–

FIG. 82 Selexol process flow sheets [129].

typical of the general process criteria of the others—Fluor Solvent, Purisol, Estosolvan, and Union. The Rectisol system is also described individually because it is unique in utilizing deep refrigeration.

The Selexol solvent, which is the dimenthyl ether of polyethylene glycol, is most often used without dilution but may be used with 20% water. With operation at ambient temperature and high pressure (500 to 1000 psig), it is possible to remove H_2S to 1 ppm while keeping the percentage absorption of CO_2 low due to the relative solubilities. A Selexol system for removing H_2S from a gas stream also containing CO_2 is illustrated in Fig. 82. Sour gas at elevated pressure is washed in the absorber with circulating solvent at ambient temperature. The rich solution from the absorber bottom is lowered in pressure and flashed to eliminate some of the dissolved gases, but not H_2S. The flashed solvent is then heat-exchanged with a hot lean solvent and stripped of H_2S in the reboiled stripper column. Lean solvent from the bottom is cooled by the rich solution flowing to the top of the stripper and further cooled by exchange with air or water before entering the top of the absorber to complete its cycle. Flashing of the rich solution can eliminate a portion of extraneous dissolved gases. Light hydrocarbons may be reduced to perhaps 2% of the final H_2S stream, but heavier hydrocarbons such as C_4^+ and aromatics remain largely with the H_2S stream. Compounds in the raw gas, such as COS, CS_2, NH_3, and HCN, do not react with and therefore do not degrade the solvent; however, they are absorbed and generally remain in the H_2S gas from the stripper.

The gas stream compositions listed with Fig. 82 indicate that H_2S is effectively removed, but at the same time, the bulk of CO_2 is left in the treated gas. This provides a suitable gas (21% H_2S) for feed directly to a Claus sulfur plant. If it is also necessary to remove CO_2 from the treated gas, downstream treating by Selexol or another system is required. A Selexol system is utilized in the BI-GAS coal gasification pilot plant.

The Rectisol system (Fig. 83) provides for the treating of a raw gas containing H_2S and CO_2, so that the sweetened product gas contains less than 0.1 ppm of H_2S (+COS) and less than 0.1% CO_2. If desired, CO_2 can be reduced to less than 5 ppm. At the same time, an H_2S-rich (31%) acid gas stream, concentrated suitably for Claus sulfur plant feed, is produced. The CO_2-rich waste gas contains very low H_2S (5 ppm).

In the Rectisol system (Two-Stage Syngas Purification) in Fig. 83, crude gas saturated with water vapor is indirectly cooled by cold purified gas and evaporating ammonia. Icing is prevented by the injection of methanol. The gas enters the first absorber, where the sulfur compounds are selectively

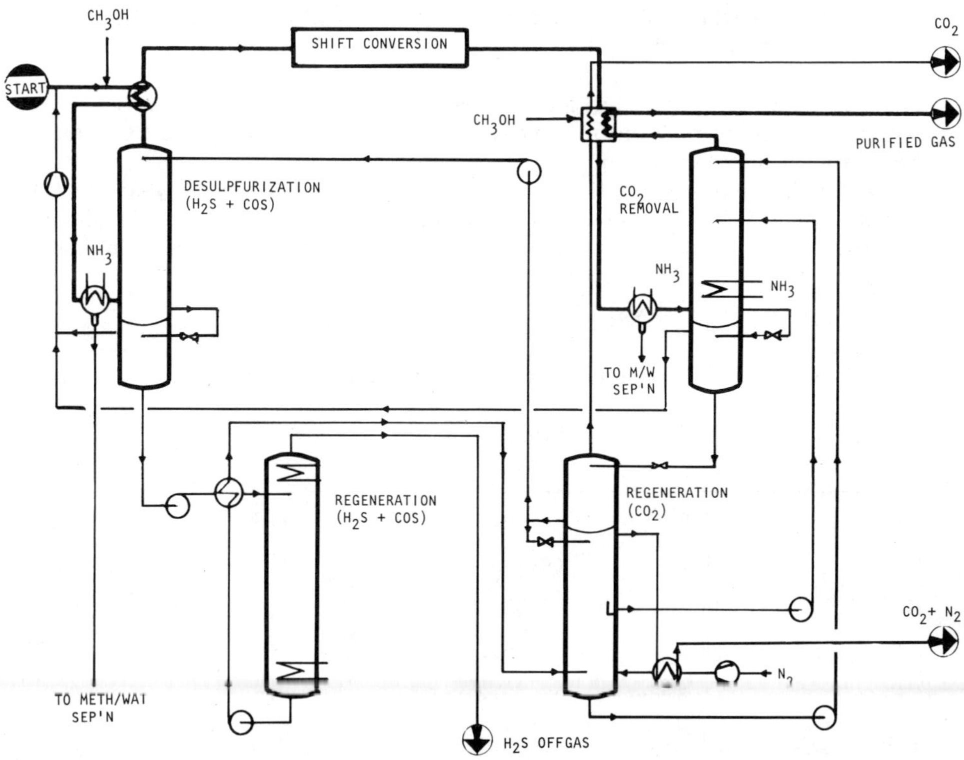

FIG. 83 Rectisol two-stage desulfurization and CO_2 removal [113].

and completely removed by washing with CO_2-charged methanol. After CO shift conversion and further cooling, the gas is fed into the second absorber for CO_2 removal to the required level. Before leaving the plant, the purified gas is heat-exchanged with the shifted gas.

Fat solvent from the first absorber, after flashing and heating, is regenerated completely in the H_2S regenerator by reboiling. After cooling, lean solvent is supplied to the second absorber top together with stripped solvent from the CO_2 regenerator. In addition, semistripped solvent is charged to the second absorber bottom section for bulk removal of CO_2. Fat solvent leaving the second absorber is regenerated in the CO_2 regenerator by flashing and stripping with impure nitrogen available from the oxygen unit. Solvent for desulfurization is withdrawn from the CO_2 regenerator and pumped to the first absorber top.

Absorbed H_2 and CO are released in the first flash stages at relatively high pressure and either returned to the raw gas or used as fuel.

For raw-shifted feed gas, a once-through Rectisol system with similar characteristics for product qualities can be used with an integrated prewash step in case the feed gas contains higher boiling unsaturates and other impurities as for several types of coal gas.

Stream compositions of a two-stage Rectisol system, similar to that described above, are listed in Table 54. This system produces a very clean sweetened gas that is low enough in sulfur to be fed directly to a nickel catalyst methanation unit, a relatively high-H_2S-content gas suitable for a Claus sulfur plant feed, and a CO_2 waste stream with a low-H_2S content. Undesirable features of the Rectisol system, however, include a complex flow scheme, a high refrigeration load at low temperature (down to -80°F), and the loss of a large portion of heavier hydrocarbons to the waste CO_2 stream, necessitating incineration or use as boiler fuel.

A Rectisol gas purification system has been used for many years at the South African Oil, Coal, and Gas Association (SASOL) plant for cleaning synthesis gas. Rectisol systems are projected for use on the several coal-to-SNG plants for industrial production utilizing Lurgi gasifiers.

TABLE 54 Stream Composition in Two-Stage Rectisol System (mol %) [130]

	Sour gas	Sweetened gas	H_2S-rich gas	CO_2-rich gas
CO_2	26.7	0.1	67.9	96.2
H_2S	1.2	(0.1 ppm)	31.4	(5 ppm)
H_2	38.0	53.2	—	0.9
CO	11.5	16.2	—	0.1
CH_4	21.8	30.3	0.3	0.2
C_2H_6	0.8	0.2	0.4	2.6
Total	100.0	100.0	100.0	100.0

Physical solvents provide the following benefits:

1. Lower heat and power consumption
2. Lower rate of corrosion
3. Removal of minor impurities from raw gas
4. Greater ability to selectively recover H_2S
5. No purge streams
6. Direct oil recovery

The major disadvantages of physical solvent systems are:

1. High hycrocarbon losses from some systems
2. High solvent loss and solvent cost
3. High partial pressure of acid gas required in order to be economically feasible

5. *Mixed Solvents*

Mixed solvent processes employ combinations of chemical and physical solvents. Processing is similar to amine treating (Fig. 80), wherein gas is treated at ambient conditions and solvent regeneration is accomplished by stripping at elevated temperature and low pressure. In addition, because of the presence of the physical solvent, excessive amounts of hydrocarbon gases are dissolved and the mixed solvent systems must often employ a flash separation following pressure letdown from the contactor. Flashed gases are either recycled to the contactor or taken to fuel.

Mixed solvent advantages are:

1. High solution loading at high pressures
2. Low utility requirements
3. Minimal solvent degradation
4. Low corrosion rates
5. Use over wider pressure ranges than most physical solvents

Disadvantages of mixed solvents are:

1. The solvent is quite expensive.
2. Heavy hydrocarbons are readily absorbed.
3. Although utility requirements are less than those for chemical absorption, they are greater than those for many physical absorption processes.

6. *Adsorption*

Adsorption is the selective concentration of impurities on the surface and in the pores of special solid materials. The term is also used to denote the chemical reaction of the impurity with a solid surface. Adsorption processes are usually performed with fixed beds of adsorbent. When the adsorbent capacity to retain the impurity is reached, the adsorbent may be discarded or, more commonly, regenerated by heating. It is common to have two beds of solids when regeneration is employed, one bed being used for adsorption while the other is being regenerated.

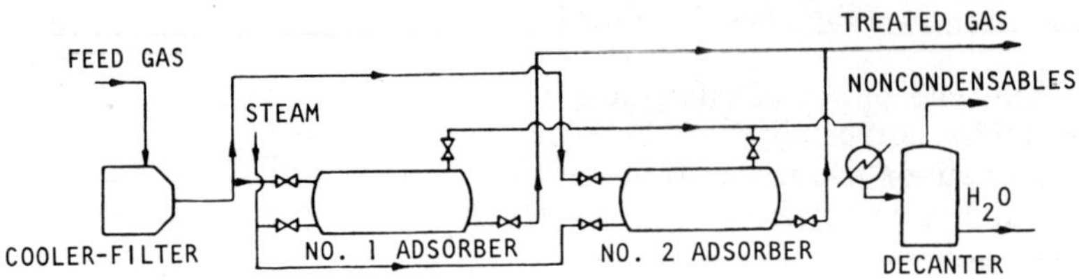

FIG. 84 Fixed-bed adsorber system.

A typical gas cleanup by adsorption flow sheet is presented in Fig. 84.

Wet absorption processes will often produce a clean gas that is suitable for pipeline transportation. However, when the gas must be treated by catalytic processes, even a few parts per million of sulfur may constitute a serious poison for the catalyst. For conventional methanation processes using nickel catalyst, catalyst life will be about 2 years when the gas contains 0.07 ppm sulfur, whereas a catalyst life of 5 years or more is expected for sulfur-free gas.

To obtain very low sulfur concentrations, the gas is passed through a fixed bed of activated carbon that has been impregnated with certain metal oxides. H_2S, carbonyl sulfide, carbon disulfide, thiophenes, and mercaptans are removed by the carbon. When the capacity of the bed is reached, it is removed from service and regenerated with steam and air. Some elemental sulfur formed during regeneration accumulates in the carbon pores, gradually deactivating the bed. After repeated adsorption-regeneration cycles, the impregnated carbon must be replaced with fresh adsorbent. Additionally, gas treated by oil absorption at high pressure may contain 500 ppm or more residual oil vapor. Later processing steps may require essentially zero oil vapor in the gas stream, in which case the oil vapor is often completely removed using activated carbon. This type of activated carbon is specially processed from chars prepared from woody materials, coal, or wastes from paper processing. Activation increases the porosity and greatly increases the surface area of the carbon, thus enhancing its selectivity and increasing its loading capacity.

Sometimes the water content of a gas must be reduced well below the level that can be conveniently reached by glycol absorption. Adsorbents such as silica gel, molecular sieves, or activated alumina may be used for the dehydration. The process arrangement is similar to that already described for fixed-bed regenerative systems. Regeneration of the adsorbent occurs by passing hot, dried gas through the bed. After cooling, the regenerated bed is returned to service.

Iron-sponge treating may be most economical when the sour gas contains less than 300 ppm H_2S and CO_2 removal is unnecessary. Iron sponge is prepared by impregnating wood shavings with hydrated iron oxide. The mixture contains about 7 lb of iron oxide per cubic foot. Sponge is loaded into the treating chamber, and sour gas is fed at ambient temperatures. The bed moisture content is held at 30 to 50 wt %, and its pH is alkaline. At 100% efficiency, 0.64 lb H_2S will be removed per pound of iron oxide. Spent iron sponge is often replaced with fresh material, although it may be regenerated by air oxidation.

The advantages of iron-sponge treating are (1) complete removal of low concentrations of H_2S; (2) it is effective at any pressure; (3) it requires a small investment; and (4) it removes H_2S in the presence of oxygen-containing gas. The disadvantages of iron-sponge treating include (1) the frequent change of spent sponge; and (2) the batch process.

7. *Hot Gas Desulfurization*

When product gases from coal gasification are to be combusted or if they are to be utilized in a hot gas expander for power production immediately after production, it would be desirable to maintain their high temperature. Thus, all sensible heat may be conserved to be fully utilized downstream. If the gas contains sulfur to the extent that air pollution will occur, the sulfur content must be removed before combustion. The processes for sulfur removal previously described all depend upon the gas being at a relatively low temperature. If low-temperature processing is used and if the sensible heat of the hot gas is to be utilized, either steam production from the hot gas or heat exchange to cool gas and subsequent reheating of the gas is required. In either case, expensive additional equipment and some loss in energy will result. For this reason a moderate amount of effort is being expended to develop effective methods of removing hydrogen sulfide from hot gas. The status of this development work is summarized below [129].

The only reported industrial hot fuel gas desulfurization systems are a large pilot plant that processed 2.5 million ft^3/day of coke oven gas, and an 8-million ft^3/day commercial installation operated in the 1950s by the South-western Gas Board at Exeter, England. Desulfurization took place in a fluidized bed of iron oxide at about 650 to 750°F with 95 to 99% sulfur removal. Regeneration was by air at 1300 to 1400°F, with SO_2 ultimately being converted to sulfuric acid. The losses of iron oxide were quite high and the production of a fine oxide dust complicated the downstream processing. The plant operated for several years and established that a fluidized iron oxide process for desulfurization of the gas was technically feasible.

Morgantown Energy Technology Center at Morgantown, West Virginia, is developing a hot purification process that absorbs H_2S on a fixed bed of iron oxide solids at 1000 to 1500°F. Cyclic regeneration with air, preferably below 1500°F, drives sulfur off as SO_2 that subsequently would be converted to sulfuric acid.

Research is being conducted to test fly ash and silica supports and binding materials to improve physical strength of the iron oxide adsorbent. The

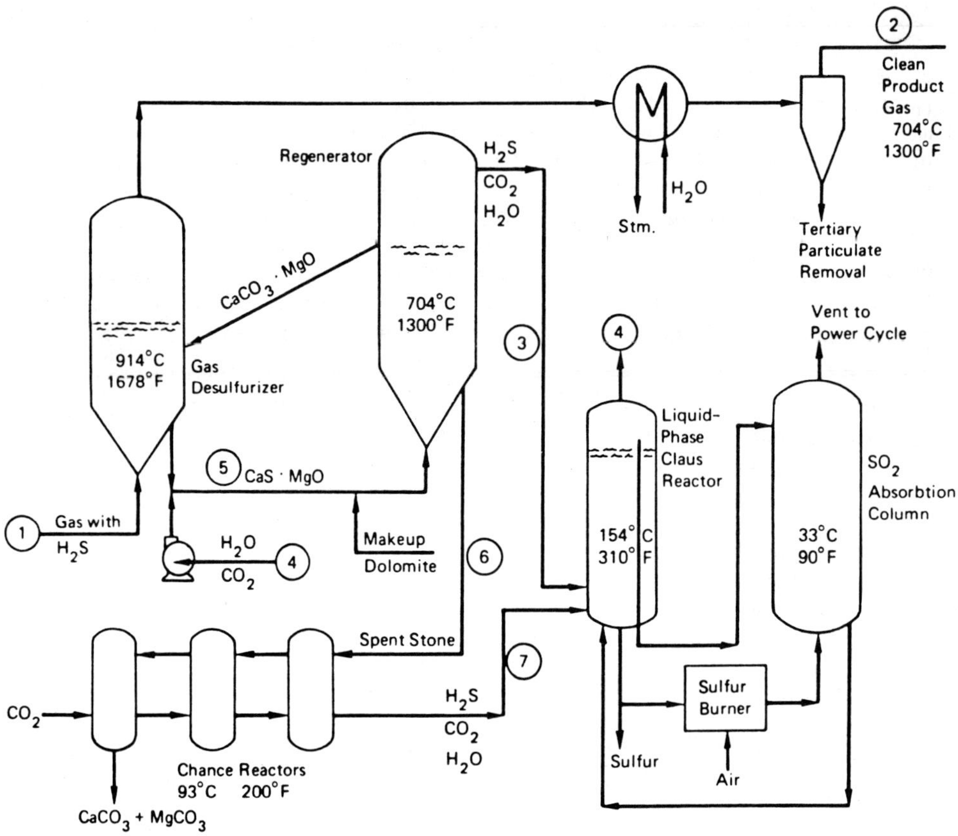

FIG. 85 Conoco hot desulfurization process [129].

effects of water vapor, tars, and particulate matter in the gas are also being studied. Removal of 92 to 97% of the H_2S has been demonstrated. SO_2 concentration in the recovered gas is 11.5 to 12%.

It appears that operating temperature and other process conditions are such that development to large scale is feasible. The effect of tars in the hot gases is a problem that must be investigated.

Conoco Laboratories at Library, Pennsylvania, is developing a process using a hot fluidized bed of dolomite for desulfurization of hot gas. The conceptual process is illustrated in Fig. 85. The hot gas, containing H_2S, contacts the fluidized dolomite at about 1700°F and 15 atm pressure with the following reaction taking place:

$$H_2S + CaCO_3 \cdot MgO = CaS \cdot MgO + H_2O + CO_2$$

With an inlet gas containing 0.6% H_2S, 9% CO_2, and 13% H_2O, the residual H_2S in the purified gas can be reduced to about 0.02%.

The spent solids leaving the reactor are lifted by a stream of steam and CO_2 to a fluidized-bed regenerator at about 1300°F where the reverse reaction takes place. Composition of the effluent gas stream from the overhead of the regenerator is typically 63.9% CO_2, 32.5% H_2O, and 3.6% H_2S. The spent solids have about 88% of the calcium as sulfide compared with the regenerated solids, which are about 75% calcium sulfide.

The gas from the regenerator is sent to a liquid-phase Claus reactor at 300°F, where the typical Claus reaction takes place:

$$H_2S + 1/2\ SO_2 \rightarrow 1.5\ S + H_2O$$

Conversion of the H_2S in the liquid-phase Claus reactor is 96%. The effluent gas, mostly CO_2 plus a small amount of H_2S, is recycled to the regenerator.

These solids have been run through more than 50 cycles of regeneration and attainment of equilibrium has been demonstrated. Makeup rates of fresh dolomite are expected to be 1 to 2% of the solids circulation rate.

Prior to disposal, the spent stone is converted to $CaCO_3$ by the Chance reaction:

$$CO_2 + H_2O + CaS \rightarrow H_2S + CaCO_3$$

The H_2S-containing stream is also fed to the liquid-phase Claus reactor for conversion to elemental sulfur.

In general, the chemistry of the various steps has been fairly well demonstrated in small-scale units. In many respects, the Conoco hot desulfurization process is similar to the CO_2-Acceptor Process that was operated at Rapid City, South Dakota. Circulating hot fluidized beds of dolomite are common to both processes, and therefore the basic process concepts have been successfully demonstrated on a large scale. Fines, carried over from the fluidized bed into the purified gas stream, could be a major problem, especially if the product gas is to be used in a turbine. The fresh dolomite makeup requirement could also be a major problem, but this can be determined only with larger-scale equipment.

Battelle Northwest in Washington state is developing a molten salt process for the removal of H_2S from hot gases. The molten salt used is a mixture of sodium, potassium, lithium, and calcium carbonate that is liquid at 950°F and stable at 1650°F in an atmosphere of CO_2. Calcium carbonate has been found to be much more active in reaction with H_2S than other carbonates and is considered to be the main reactant:

$$CaCO_3 + H_2S \rightarrow CaS + CO_2 + H_2O$$

A pilot plant has been operated at atmospheric pressure using coke in a gasifier to supply the gas. Hot gas and molten salt are contacted in a venturi scrubber at about 1450°F. The gas-salt mixture flows up through a vertical line to a deentrainer arrangement, where the salt separates and flows downward to a pot furnace, while the gas flows upward to a separator packed with 1-in.-diameter alumina balls topped with a sapphire demister to provide separation of the gas and entrained salt.

The pilot plant has been operated for short periods of time with a typical flow of 0.9 gal/min of molten salt at 1300°F and 1 psig at a gas flow of 50 scf/min. With an inlet content of 0.84% H_2S, the outlet H_2S was as low as 5 ppm. Overall sulfur removal from the gas ranged from 95 to 98%.

Regeneration of the spent salt is typically accomplished with a mixture of 31 lb/hr of steam and 45 scf/hr of CO_2, resulting in an H_2S concentration of 4.2% in the CO_2-H_2S mixture. Although this off-gas is not sufficiently high in H_2S to provide a suitable feed to a Claus sulfur recovery unit, gases much higher in H_2S content are theoretically possible.

The solids carried over from the pilot-plant unit indicate that better separation must be achieved. With a total particulate burden of 0.074 grain/scf in the inlet gas, the outlet gas contains 0.0245 grain/scf, about 30% of which are molten salt particles. This alkali carryover is too excessive (by about two orders of magnitude) to be tolerated in a downstream turbine.

The problems inherent with the process include:

1. Excessive amounts of solids and alkali are carried over to downstream facilities.
2. There is soluble and insoluble ash buildup in the molten salt.
3. Large quantities of CO_2 and steam for regeneration are required.
4. Equipment in contact with the molten salt corrodes.
5. Concentrations of H_2S in the H_2S-CO_2 mixture obtained during regeneration may be too low to supply a suitable feed gas to a Claus unit.

F. Methanation

Methanation is the final chemical reaction in the sequence of processes used to manufacture SNG. It is also used to remove traces of carbon oxides from synthesis gases because they are potential catalyst poisoners. All of the gasification processes discussed produce H_2 and CO as well as CH_4 in the gasification step. In the methanation step, the H_2 and CO are combined to form CH_4 via the following reaction:

$$CO + 3H_2 = CH_4 + H_2O + 88{,}700 \text{ Btu/lb mol CO}$$

The water-gas-shift reaction takes place simultaneously, and together these two reactions are sufficient to describe the stoichiometry:

$$CO + H_2O = CO_2 + H_2 + 17{,}700 \text{ Btu/lb mol CO}$$

Any higher hydrocarbons are converted to CH_4 under the normal methanation conditions if sufficient H_2 is present.

The large exothermicity of the methanation reaction necessitates providing means for removal of heat in a useful form, both to improve the efficiency of the process and to prevent excessive temperatures on the catalyst. The maximum temperature of prospective catalysts is not generally in excess of 1000°F. The minimum temperature required to initiate the methanation reaction is approximately 500°F.

The foregoing reactions are the only ones that need to be considered in the absence of carbon deposition reactions; that is,

$$2CO = C + CO_2 \quad \text{and} \quad CH_4 = C + 2H_2$$

Equilibrium considerations may be used to determine whether carbon deposition will take place. The results below 1700°F will depend upon the form of carbon assumed to deposit, which may be either beta graphite or Dent carbon. It is known, however, that with H_2/CO ratios of 2.6:1 or greater, carbon will not form in the methanator at high pressure below a temperature of 1500°F.

Catalysts used for methanation must be very active because they operate at low temperatures in order to avoid carbon deposition. Further, low temperature favors the concentration of CH_4 at equilibrium. Catalysts employed for methanation generally contain high concentrations of very active nickel. Alumina and kieselguhr are common substrates and promoters. A molecular-sieve-based catalyst has also been developed and is being tested by IGT in the HYGAS pilot plant. These catalysts are irreversibly poisoned by sulfur. The sulfur level of the feed gases must be kept at extremely low levels to ensure satisfactory catalyst life because sulfur is quantitatively absorbed by the nickel. Chlorides are also detrimental to catalyst activity. ZnO guard chambers are employed upstream of the methanator to ensure that the sulfur level is low enough even during upsets. Another consideration related to the catalyst is the possible formation of nickel carbonyl [$Ni(CO)_4$]. Since the carbonyl is highly volatile, not only would nickel be lost but the carbonyl is an extreme health hazard. To avoid its formation, carbon oxides must not be exposed to the catalyst except at temperatures above 400°F.

There are a large number of methanation processes. These differ in the manner in which heat produced is absorbed and the catalyst is protected from overheating. The processes that have been proposed can be classified into three major types:

1. Fixed bed
   a. cold-quench process
   b. product recycle process
   c. cold-quench-product recycle process (combination of a and b) (IGT)
   d. hot gas recycle (Lurgi-El Paso)
   e. RM process (Parsons)
2. Fluid bed
   a. fluid-bed methanation process (BCR, BI-GAS; Aerojet General; IGT; Didier/Thyssengas)
   b. liquid-phase methanation and LPM/shift (Chem Systems)
3. Special contacting devices
   a. hot gas recycle (HGR) with plate reactor (Bureau of Mines)
   b. packed-tube methanation process (Consol)
   c. tube-wall reactor (Bureau of Mines)

1. *Fixed Bed*

The cold-quench system and the cold quench with product recycle (IGT) are illustrated in Fig. 86. The cold feed gas is split into two streams. One is preheated (along with cold recycled product gas in the IGT recycle process) and passed through a first bed of catalyst. The hot product from the first stage is cooled just enough in a steam-raising heat exchanger so that when a second portion of feed gas is added, the mixture temperature is 500 to 550°F. This gas passes through a second bed, where the heat of methanation reheats it. A third portion of the feed gas is added and methanated. The figure shows three stages but as many as needed can be used. The size of the various streams and their temperatures can quite readily be computed for any given feed composition.

Without recycle the first stage could easily overheat with feed gases containing high (over 4%) CO. In this case, a less active catalyst capable of withstanding higher temperatures would be needed. With cold product recycle, the gas entering each bed can also be maintained at 4% or less CO and about 550°F.

The use of hot gas recycle (Fig. 87) has been developed by the Bureau of Mines and proposed for the Lurgi plant, El Paso, Texas. The advantage of this scheme lies in avoiding the cooling and subsequent reheating of the

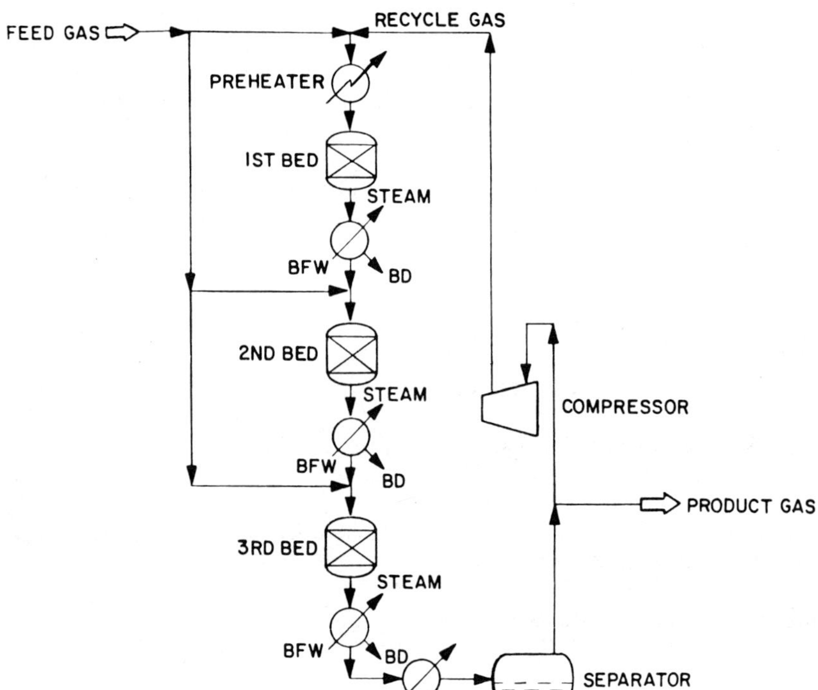

FIG. 86 Cold quench recycle methanation process (shown with three stages) [131].

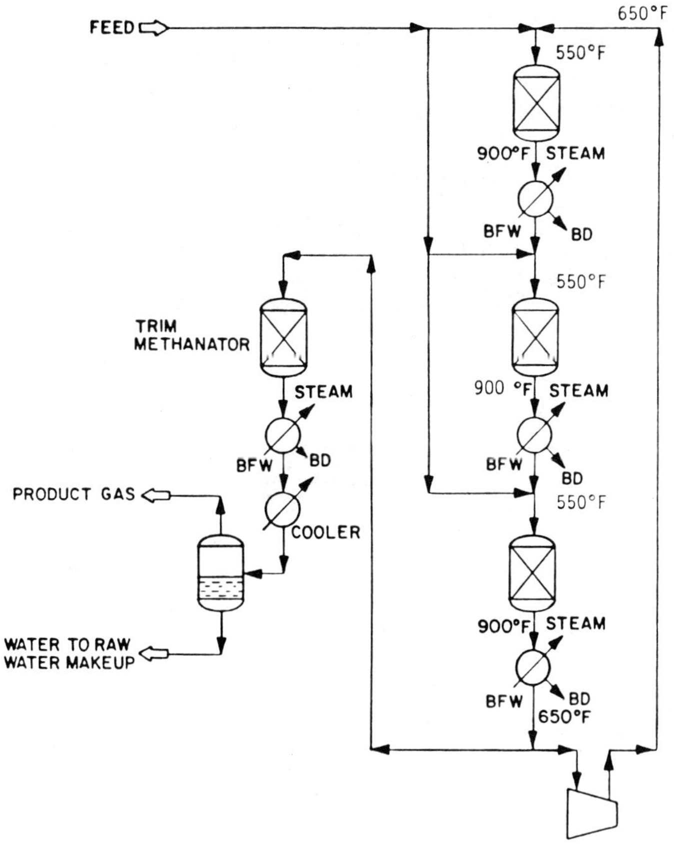

FIG. 87 Hot gas recycle with packed bed reactors [131].

recycle stream. This advantage, however, is offset by having to use a more expensive and less efficient compressor to operate on the hot gas. For multistage processes the recycle ratio becomes extremely small and the advantage is diminished. The system will run with a higher inlet temperature. Also, the H_2O partial pressure in the recycle is high because there is no H_2O removal on the recycle. An equilibrium hindrance may exist for high-CO feeds and a trim methanator may be necessary.

The RM Process (Fig. 88), developed by The Ralph M. Parsons Company, is a fixed-bed methanation system which employs a catalyst capable of withstanding higher temperatures than normally employed in fixed-bed methanation. The RM Process is based on high-temperature operation in order to completely avoid a recycle system. High concentrations of carbon oxides are processed directly by using proprietary nickel catalysts which cause both the shift conversion and methanation reactions to occur simultaneously. Raw syngas taken directly from a gasifier (following removal of particulates and sulfur) can be fed to the catalyst bed without adjustment in the ratio of hydrogen

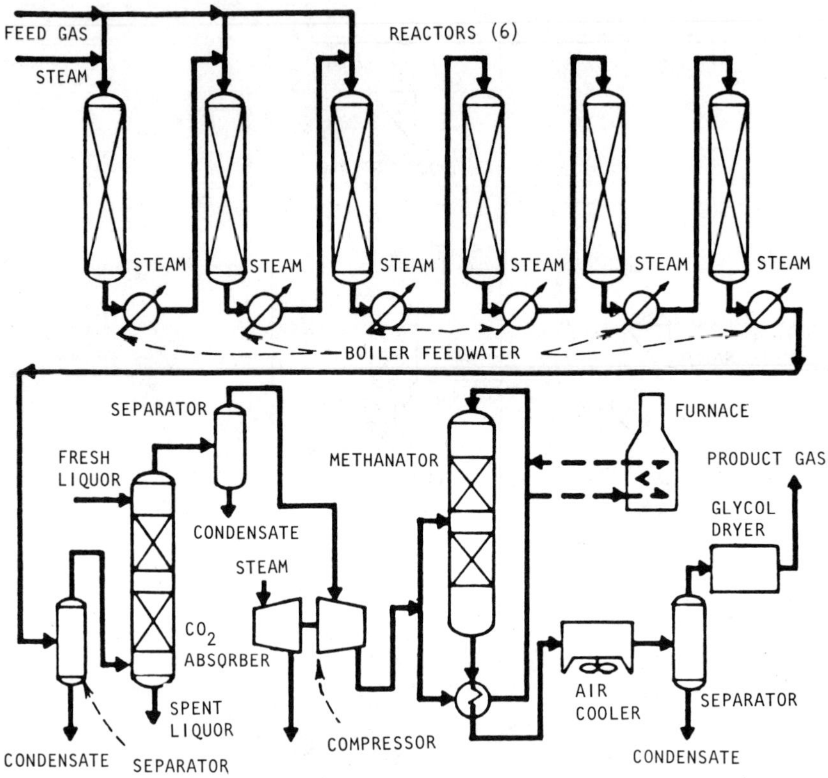

FIG. 88 Parsons methanation process [132].

to carbon oxides. Steam is added to the raw syngas to moderate the temperature rise across a catalyst bed and to control the gas composition to avoid carbon formation.

Although low temperatures are required in the final stages of methanation, the use of high temperatures for bulk methanation results in the following advantages:

1. Elimination of a recycle system and large mass and heat flow rates
2. Elimination of a separate shift conversion section
3. Operation of conditions removed from regions of carbon formation
4. Production of more steam at higher pressures with less surface area
5. Reduction in catalyst sensitivity to sulfur
6. Easy regeneration of catalysts
7. CO_2 removal from a reduced volume of gas

In the RM Process, the shift and methanation reactions are carried out in a series of fixed-bed adiabatic reactors. Normally, the raw syngas is distributed into two or three reactors in series flow to limit the temperature rise across any one bed to a nominal 600°F while using a minimum quantity of steam. Depending upon the amount of methane present in the raw syngas from the gasifier, the methanation system would require four to six reactors in series flow. The inlet temperature to successive reactors is progressively lowered to a final inlet temperature of 600°F. After this last stage of wet methanation, steam and carbon dioxide are removed and the process gas is treated in a final polishing stage.

2. *Fluid Bed*

The high heat-transfer capabilities of fluid beds has led to the suggestion that methanation be employed in such systems. Bituminous Coal Research (BCR), in connection with its BI-GAS Process, has developed the methanation systems illustrated in Figs. 89 and 90.

Early in the formulation of the BI-GAS program, BCR chose to proceed with the development of the fluid-bed methanation concept because of the following potential advantages over the existing fixed-bed methanation technology:

1. Temperature gradients are minimized and hot spots are less likely to occur due to the turbulence within the catalyst bed.
2. Higher heat-transfer coefficients are possible because of the motion of catalyst particles past internal heat-transfer surfaces. Thus, exothermic reactions are much easier to control.

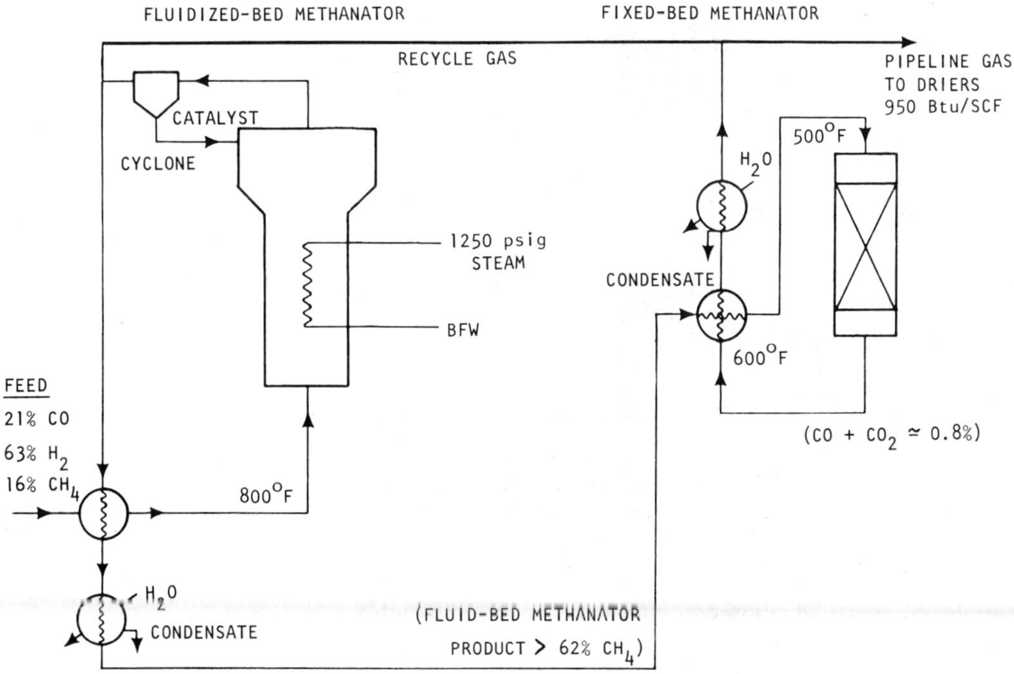

FIG. 89 Typical BI-GAS methanation system [133].

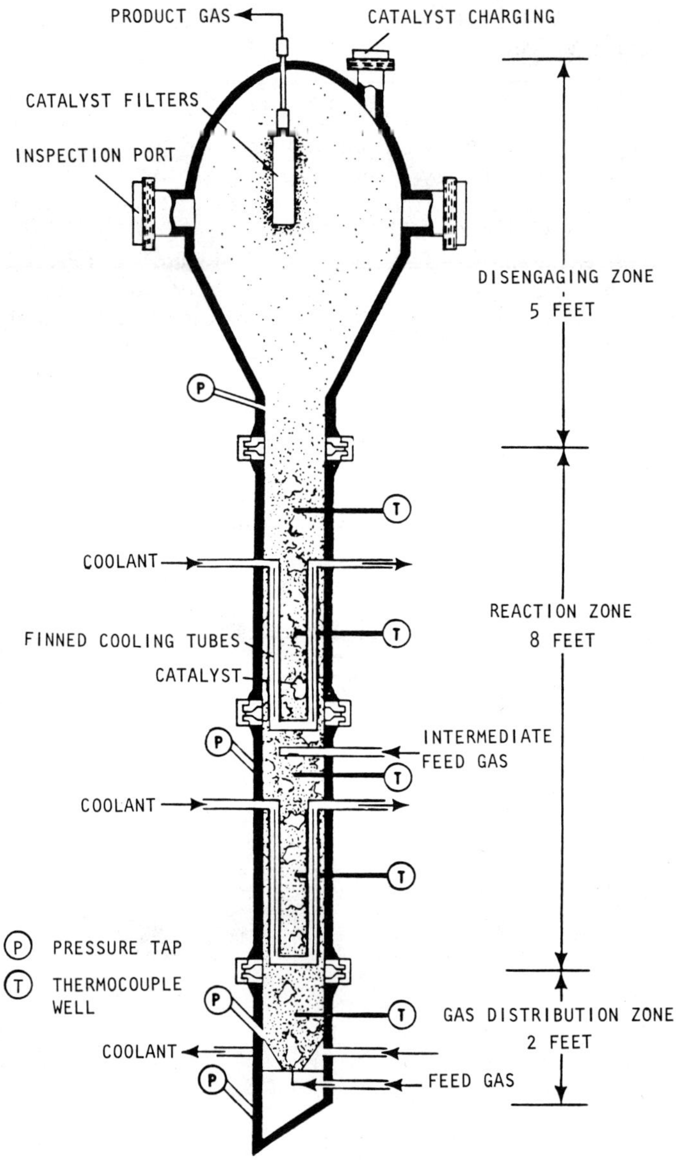

FIG. 90 BCR pilot-plant fluidized-bed methanator [134].

3. The superior heat removal capability of the fluid-bed system permits the use of higher partial pressures of reactant gases and eliminates the need for product gas recycle for temperature control.
4. Continuous addition and removal of catalyst is possible, eliminating the need to shut down the methanator to replace the catalyst charge.

Anticipated high reactant concentrations (e.g., approximately 20% carbon monoxide in the feed) in the fluid-bed methanator have indicated that conventional fixed-bed methanation catalysts would not be suitable for fluid-bed operation because of their high loadings of catalysts designed for treating synthesis gases with carbon monoxide contents on the order of 5% or less. These catalysts would simply have been too active for fluid-bed application. Consequently, a major emphasis in the BCR program has been to evaluate fluid-bed methanation catalysts, developed by commercial catalyst suppliers, specifically designed to provide the desired levels of activity and thermal stability under the process conditions outlined earlier.

A fluid bed containing finely divided catalyst particles (10% above 100 μm, 50% below 60 μm, and 25% below 325 mesh) is established with process gas as the fluidizing medium. Dowtherm heating coils within the bed remove the heat liberated by reaction at 750°F, which is ultimately used to generate steam. The product gas from this unit requires further methanation of a fixed-bed type. Recycled, cooled, and dried product gas from the fluid bed is used for dilution of feed to the fluid bed and for lowering the H_2O partial pressure to promote higher conversion, although this has not been found to be necessary so far. The advantages claimed for this approach are:

1. Uniformity of temperature of the reaction.
2. Production of high-quality steam.
3. The possibility of replacement of catalyst without shutdown of the process.
4. Recycle is minimized because the heat is efficiently removed.
5. Twenty-five percent CO can be tolerated in the feed to the reactor.
6. Non-Ni-catalysts are being considered.

Disadvantages of this approach are:

1. Rapid dissemination of sulfur throughout the catalyst.
2. Catalyst attrition.
3. Poorer conversion in the fluid bed (2 to 3% CO unconverted) than in fixed-bed processes.
4. Temperature-control problems.

This process was selected for the BI-GAS pilot-plant program in Homer City, Pennsylvania; however, the plant was shut down before methanation could be tested. The catalyst used was Ni on alumina, Mo, and Cu, prepared by Harshaw Chemical Co.

The Institute of Gas Technology (IGT) tested fluid-bed methanation in the early 1960s, employing Raney nickel catalysts. The reactor consisted of a 6-in.-diameter pipe. Dowtherm was used as a coolant on the outside of the vessel. Operation was below 200 psig and reactor temperatures were 700°F. Problems maintaining Raney nickel catalyst activity occurred because of improper formulation and improper reactor operation, and subsequently termination of the development. At the time, there were no catalysts in existence suitable for fluid-fed methanation from the standpoint of strength and high activity. The process looked difficult enough that IGT switched to fixed bed with cold gas recycle.

Work on the fluid-bed concept has also been conducted by Didier-Engineering GmbH and Thyssengas GmbH of West Germany. The flow diagram of their pilot plant is illustrated in Fig. 91. The methanation reactor is illustrated in Fig. 92. Feed gas is desulfurized in a zinc oxide bed. Steam and recycle gas can be added to the feed gas. In the heat exchanger the gas temperature is increased before entering the reactor. The product gas leaving the reactor is further cleaned in a candle-type filter. It then flows through the heat exchanger, a cooler, and a separator. For the purpose of startup, the feed gas temperature may be increased by an electric heater. Boiler feedwater flows through the inserted heat exchanger bundle; steam and water are separated in the knockout drum. Several test runs were conducted in the period November 1977 to June 1978, which demonstrated the feasibility of one-stage combined shift and methanation with Ni-based catalyst. The chemical stability of the catalysts used was found to be not sufficient for commercial application, but further development work by catalyst manufacturers will, hopefully, improve that situation. Further pilot-plant testing is planned during a 1-year period starting in November 1978.

The Chem Systems liquid-phase fluidized system is illustrated in Fig. 93. Synthesis gas is introduced into the bottom of a reactor along with a hydrocarbon liquid containing a finely divided Ni catalyst. The only prior treatment required for the synthesis gas is desulfurization to a level satisfactory for

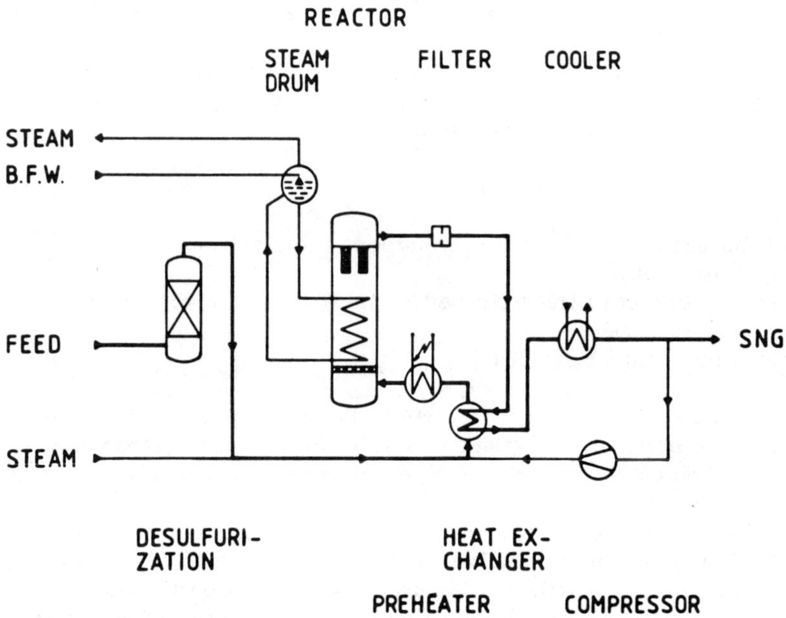

FIG. 91 Flow diagram of methanation plant [135].

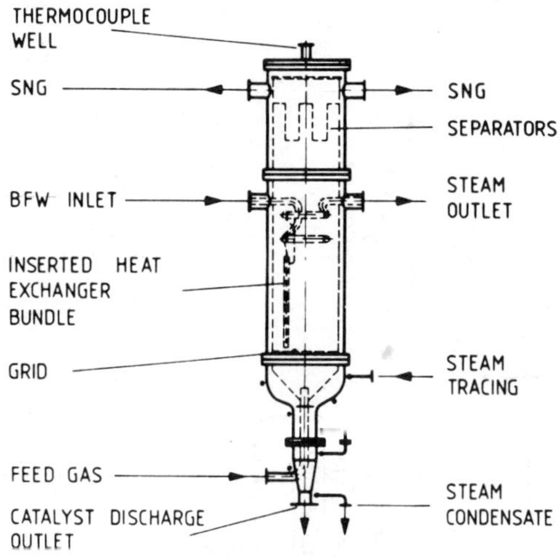

FIG. 92 Fluidized-bed methanator [135].

nickel catalysts. The product gases, consisting primarily of methane and small amounts of CO_2, H_2, CO, and vaporized process liquid, are separated from the hydrocarbon liquid in the reactor separator. The process liquid, having absorbed the exothermic reaction heat, is pumped through filters to remove any catalyst fines and then to a heat exchanger where cooling occurs by generating high-pressure steam. The cooled process liquid is recirculated back to the reactor via an electric heater that is used for startup and to make up for heat losses in the system.

The effluent gases leaving the reactor separator are cooled in two stages: by preheating boiler feedwater and by exchange with cooling water. The boiler feedwater is sent to the circulating oil cooler. Any water produced in the reaction is condensed together with the process liquid that was vaporized in the reactor. The liquids are phase-separated in the product gas separator and the water phase passes through a degasser before being sent to the incinerator. Most of the condensed process liquid is pumped directly back into the circulating liquid system, while a small portion of the liquid is used as a seal flush for the main circulating liquid pumps. Product gas leaving the final separator is analyzed and then sent to flare.

In order to prereduce the catalyst bed in situ, the pilot plant is provided with a low-pressure hydrogen system including heaters and exchangers. With this system, catalysts that require reduction can be heated and then treated prior to the introduction of process liquid and synthesis gas into the reactor. Experimentation with this system has been discontinued because of poor performance and operating difficulties.

3. *Special Contacting Devices*

Several attempts have been made to solve the temperature-control and heat-recovery problems by putting the Ni catalyst in or on a heat-transfer surface. The Bureau of Mines has developed a hot gas recycle process with two types of special reactors. One of these is a parallel-plate reactor (Fig.

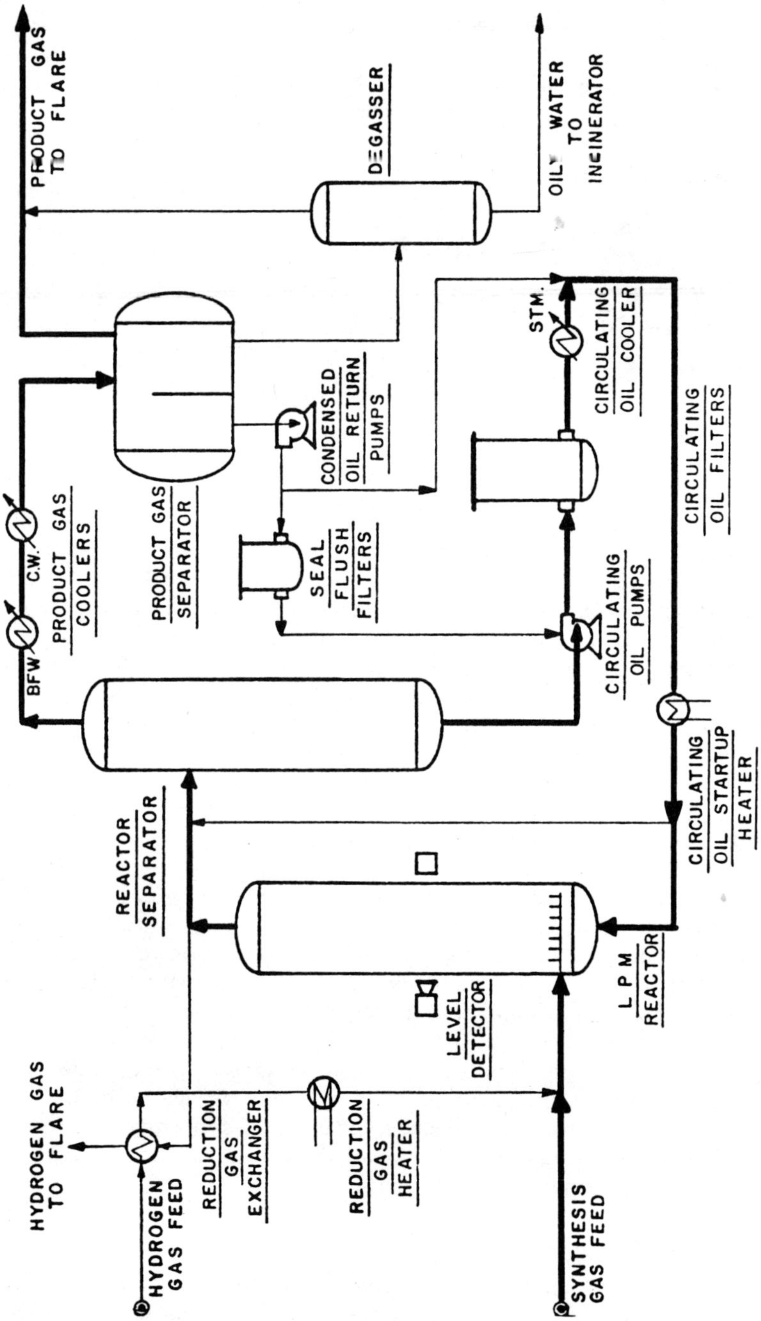

FIG. 93 LPM pilot-plant schematic flow sheet [136].

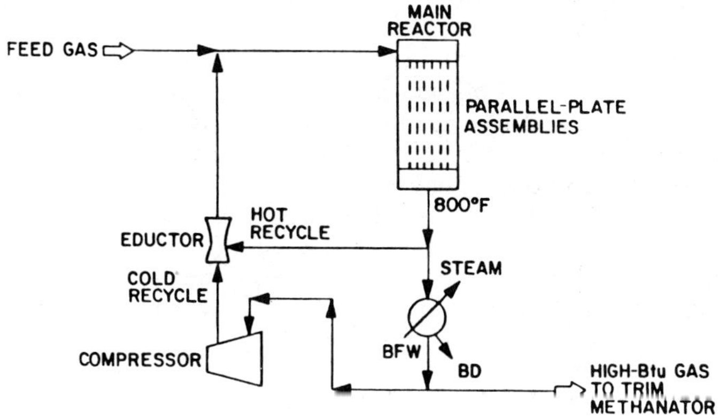

FIG. 94 Hot gas recycle (HGR) with plate reactor (U.S. Bureau of Mines) [131].

94). Special plate assemblies, flame sprayed with Raney nickel, are inserted in a pipe or pressure vessel. Feed gases are mixed with recycled gas to achieve a 7:1 ratio of recycle to fresh feed. A 180°F temperature rise is permitted from an inlet temperature of 570°F. An eductor was originally planned for the recycle of product gas in the pilot plant. Present plans call for the use of a novel hot gas compressor. Operating pressure is 1000 psig. A trim methanator is required for the HGR system.

Problems with this process are the noncommercial compressor required and the large horsepower requirement of the high recycle. Deactivation of the Raney Ni catalyst has also been a problem, even with low-sulfur feed (<0.1 ppm). Deactivation of the catalyst in the small-scale experiments has been attributed to Ni_3C formation, $Fe(CO)_4$ formation and deposition of iron, hot-spot formation, and oil deposition from the recycle compressor.

The Bureau has also installed a tube-wall reactor in the Synthane pilot plant. Type 408 stainless steel tubes are flame-sprayed with Raney nickel and inserted into a vessel containing Dowtherm. The Dowtherm, at 750°F, is circulated to a cooler where high-pressure steam is generated. The reactor is operated isothermally at 750°F. A trim methanator will be required for this process; however, recycle may not be necessary. In small-scale work, better catalyst lifetime was observed with this type of reactor than with the HGR reactor.

An additional process is the Conoco Coal Development Company's packed-tube methanation system installed in one version of the CO_2 Acceptor pilot plant. The reactor is essentially a floating heat exchanger with partially filled catalyst tubes. A commercially available catalyst is used. The tubes are 11 ft long and recycle is employed at a ratio of >1.0. Dowtherm is used in the shell to cool the tubes. If required, the cooled product may be passed to an adiabatically operated trim methanator. None of these processes appear to be commercially feasible.

V. ENVIRONMENTAL CONSIDERATIONS

A. Potential Pollutants

In addition to the product gas, coal gasification plants produce other effluent streams that must be disposed of in an economically and environmentally acceptable manner. Some of the streams contain valuable by-products such as hydrocarbon oils, sulfur, and ammonia, while others may be generally worthless, such as coal ash. Figure 95 is a block flow diagram, which might apply to a coal gasification facility producing SNG (pipeline gas), hydrogen, syngas, or medium-Btu gas. Fully half of the blocks in this diagram are shaded to indicate purification processes which are primarily responsible for the overall environmental control of the facility. It is obvious, then, that a large fraction of the overall cost of the facility may be ascribed to environmental control.

The block flow diagram indicates there are three primary points for gaseous discharge from the plant: the boiler house stack, the sulfur recovery tail gas, and the CO_2-rich gas that is vented from the acid gas treatment regenerator. Solids to be disposed of consist of coal ash, lime sludge from stack gas scrubbing, and water-treatment sludge. Additionally, there is the potential for aqueous liquid discharge streams containing oils, salts, or chemicals unless special measures are taken for maximum water reuse with essentially zero discharge.

B. Discharged Gas Cleanup

In a gasification facility, such as suggested by Fig. 95, a power plant is required to furnish process steam, steam for driving mechanical equipment, and electricity for general plant use. About 15% of the total coal input may go directly to the power house. Further, combustible gases, such as those from a pretreater, and char from gasification may be sent as fuel to the boiler house. All of these fuels contain sulfur and nitrogen to varying degrees that will find their way into the stack gas from the boiler house as nitrogen oxides and sulfur oxides. If the content of these contaminants in the flue gas is excessive, they will have to be removed by some system, such as lime or limestone, Wellman-Lord, double alkali, Chiyoda, or magnesium oxide scrubbing processes, many of which are being employed or are under development [138]. As an alternative to power-plant stack gas scrubbing, consideration may be given to producing a sulfur-free fuel gas by increasing the capacity of the coal gasification facility [139]. Sulfur as it generally appears in the fuel gas in the form of H_2S, and nitrogen as ammonia, are easier to remove than when in the form of oxides. Whichever system is more desirable will depend on many factors, such as the sulfur content of the coal and environmental regulations.

Referring to Fig. 95, most of the sulfur from the coal to the gasification plant will be contained in the stream from "regeneration" following "acid-gas treatment." Lesser amounts contained in sour water stripper gas or sulfur storage vapors might also be sent to the sulfur-recovery section. The technique employed for recovery of sulfur in the subsequent "sulfur recovery"

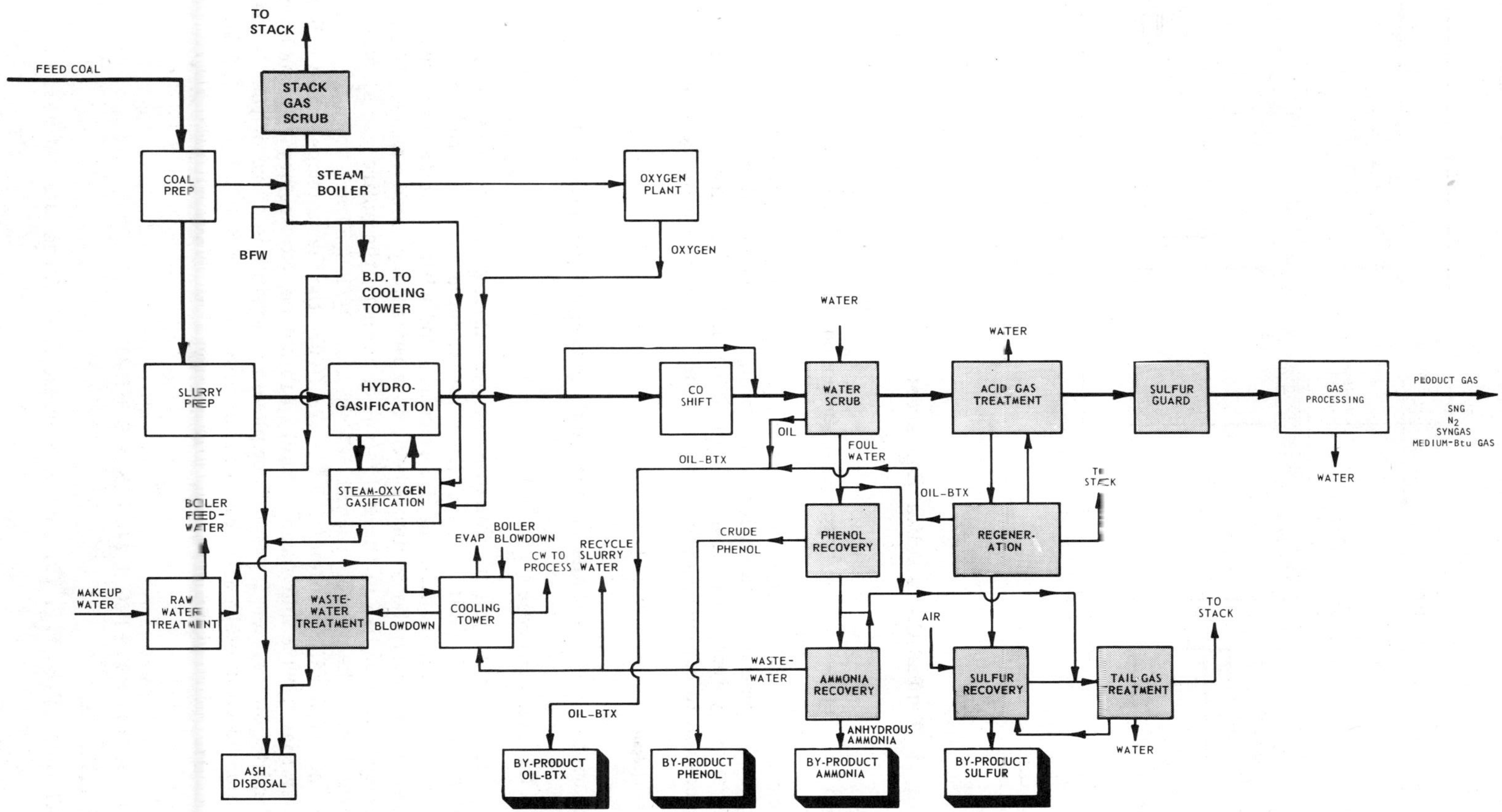

FIG. 95 Block flow diagram of a coal gasification facility [137].

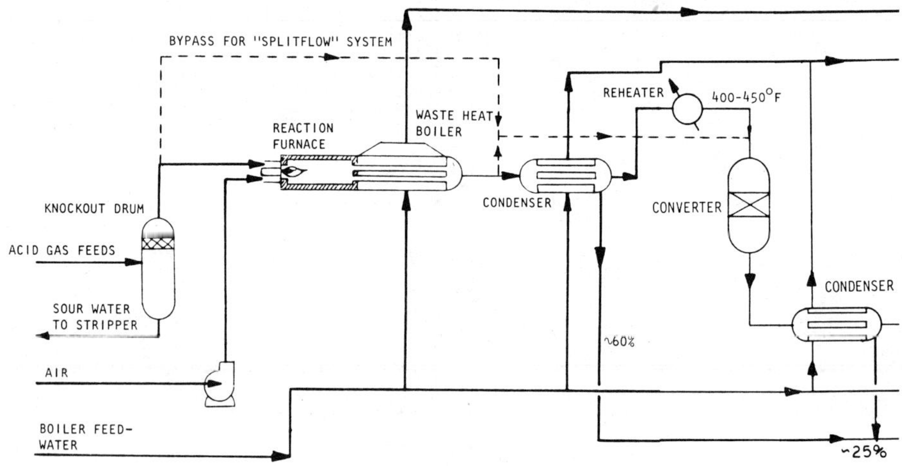

FIG. 96 Typical Claus sulfur flow sheet.

step will depend on the processing employed in the acid-gas-treatment step, where three alternates may be employed.* If the H_2S concentration to sulfur recovery is sufficiently high (generally over 15%), a Claus sulfur recovery system is generally the choice. If the sulfur concentration is low, a wet oxidation system will probably be more attractive. In either case, the overall process is to convert H_2S to elemental sulfur according to the reaction

$$H_2S + 1/2\ O_2 \rightarrow S + H_2O$$

Descriptions of the Claus Process and several wet oxidation processes follow.

1. *Claus Process*

More than 200 plants utilizing the principles of the process described below have been built in the United States, Japan, and 24 other countries.

A gas stream containing H_2S is combusted with air at 1900 to 2400°F and pressure slightly above atmospheric in a refractory-lined vessel. If the concentration of H_2S in the gas is 60% or higher, the entire gas stream is burned with the proper quantity of air ($1/2\ O_2/1\ H_2S$) as shown in the formula above. If the H_2S content is lower than 60%, the flame temperature will be lower than 1900°F and flame stability will not be satisfactory. In such cases, a "split-flow" system is usually employed wherein one-third of the H_2S stream is burned completely with air to produce SO_2($1/3\ H_2S + 1/2\ O_2 \rightarrow 1/2\ SO_2 + 1/3\ H_2O$). The other 2/3 of the H_2S stream then mixes with the SO_2 to provide a mixture that is subsequently converted to sulfur according to the reaction

$$1/3\ SO_2 + 1/3\ H_2O + 2/3\ H_2S \rightarrow S + H_2O$$

*See Sec. IV.E, Fig. 79, and the discussion of these alternatives.

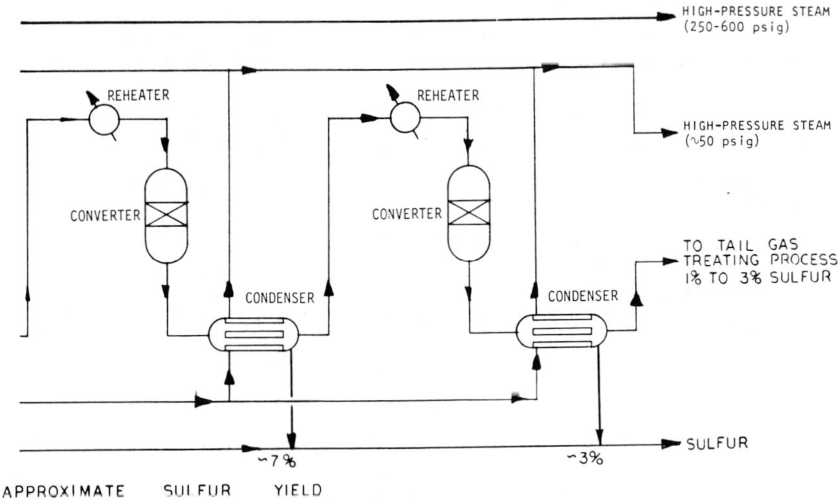

Referring to the Claus Sulfur Plant flow sheet (Fig. 96) it can be seen that, following the reaction furnace, the hot combustion gas passes through a waste-heat boiler to generate high-pressure (250 to 600 psig) steam. It then passes through a condenser, where it is cooled to 260 to 350°F, generating low-pressure (15 to 40 psig) steam. This cooling causes condensation of liquid sulfur that is separated from the gas and flows out as product sulfur. The gas stream is then reheated to 400 to 450°F and passes through a converter, where it contacts a catalyst of bauxite, activated alumina, or cobalt molybdate. Here, the sulfur forming reactions proceed further. The cooling, reheating, and catalytic conversion steps on the gas are repeated, generally twice more, as shown. About 60% of the sulfur content in the inlet gas is converted to sulfur in the reaction furnace—25% in the first catalytic converter, 7% in the second, and 3% in the third—for an overall conversion of about 95%. Further steps are uneconomical because of the decreasing conversion rate due to relatively low incremental recovery obtained. If further conversion is required to make the Claus plant tail gas discharge environmentally acceptable, alternative processing techniques must be employed.

Reactions other than those previously mentioned to produce sulfur take place in the Claus Process, some due to the presence of materials other than H_2S in the inlet gas. Some of these effects are:

1. COS and CS_2 form rapidly in environments containing high proportions of carbon species (e.g., hydrocarbons and CO_2). Their decay appears to be kinetically limited, because they appear in the waste-heat-boiler effluent at above-equilibrium concentrations.

2. Light hydrocarbons in the feed act as oxygen scavengers and therefore remove oxygen that would otherwise convert H_2S to SO_2.

3. Heavy hydrocarbons tend to form solid carbon soot or carbonaceous tars.

4. The presence of CO and H_2 in sulfur plant tail gases confirms that these species are probably formed in and then survive the thermal reaction step.

5. In the catalytic conversion steps, depending upon the type of catalyst used and the temperature at which the bed is operated, COS and CS_2 may be converted to either H_2S or elemental sulfur:

$$COS + H_2O = H_2S + CO_2 \quad \text{(exothermic)}$$

$$CS_2 + 2H_2O = 2H_2S + CO_2 \quad \text{(exothermic)}$$

$$2COS + SO_2 = 3/8\ S_8 + 2CO_2 \quad \text{(exothermic)}$$

6. Carbon dioxide is always present in Claus feed gas. Its concentration depends on the type of acid gas removal that is employed. Excessive CO_2 in the feed gas, in addition to its obvious effect on H_2S concentration, tends to promote the formation of COS and CS_2 [150]. The carbon dioxide in the Claus feed may be reduced by either improving the selectivity of the main acid-gas-removal unit or by preconcentrating the acid gas by scrubbing with DIPA or MDEA ahead of the Claus unit.

7. HCN is usually found in sourwater stripper off-gas, but it may also be present in gas from the main acid-gas-removal unit. High concentrations of HCN cause severe corrosion in burners and condensers [151] and may lead to catalyst fouling as well. Evidence exists that HCN in high concentrations can survive the thermal reaction step and enter the converters even if split flow is not used.

HCN has been controlled by a catalytic reactor placed ahead of the Claus plant. The HCN is destroyed by hydrolysis to NH_3 and CO; however, care must then be taken to ensure that these products are also subsequently destroyed. A polysulfide washing step ahead of the reaction furnace can also be used to remove HCN. This proprietary process is offered by Woodall-Duckham in the United Kingdom and by Peabody [152] in the United States. This approach, however, creates a water pollution problem by exporting a water stream rich in thiocyanates to the water-treating facilities of the plant.

8. Ammonia, like HCN, is usually present in sourwater stripper off-gas. If not destroyed in the thermal reaction step, NH_3 may form ammonium sulfides and sulfates and plug the catalytic converters.

9. Ammonia can be controlled by complete oxidation of the ammonia-containing stream in a separate burner of the reaction furnace or by upstream removal in an ammonia recovery process applied to the sourwater stripper overhead. However, not all of the NH_3 can be removed by this second method. If a tail-gas-cleanup process is used in which the tail gas is incinerated, the ammonia-containing stream may be burned with the tail gas, with the sulfur compounds recovered as SO_2 and processed in the tail-gas-cleanup unit.

10. Hydrocarbons are found primarily in the acid gas from the main acid-gas-removal unit. Physical acid-gas-removal solvents absorb more hydrocarbons than do chemical solvents.

Light hydrocarbons, especially olefins, scavenge oxygen in the reaction furnace, leading to oxygen deficiencies and lower recoveries.

Heavier hydrocarbons may form carbon either in the reaction furnace [141] or in the catalytic converters [142]. This soot discolors the product sulfur and gradually destroys the activity of the catalyst. The catalyst may also be quickly deactivated by the formation of tars [141].

Hydrocarbons lighter than butane may be tolerated in Claus feed gas up to about 0.5 vol %. At concentrations above this limit, control measures should be considered, such as changeover from a physical to a chemical solvent for acid gas removal or refrigeration of the acid gas stream to remove hydrocarbons. This technique was employed with success on a split-flow Claus plant in Louisiana. Activated-carbon drums may be installed ahead of the reaction furnace. Another option is preconcentration of the acid gas feed, leaving the hydrocarbons in the residue gas.

11. Physical entrainment of acid-gas-removal solvent into the Claus feed may result from upsets in the acid-gas-removal system or poor design of the reflux condensers. The resulting problems are similar to those of heavy hydrocarbons. If amines are used in acid gas removal, their decomposition in the reaction furnace may form nitrogen oxides or ammonia, both of which will adversely affect the Claus catalyst. This problem should be controlled by a properly designed knockout drum and acid-gas-removal regenerator overhead system.

12. Organic sulfur compounds, such as COS and CS_2, along with mercaptans and thiophenes do not participate in the Claus reaction. They may enter the Claus plant in the feed gas, or COS and CS_2 may be formed in the reaction step.

COS can be hydrolyzed in the main process by installation of a catalytic or liquid-phase hydrolysis reactor in the bypass line around the shift converter. Alternatively, COS and CS_2 may be destroyed in the Claus plant's catalytic converters. Over conventional catalyst, high temperatures give good COS/CS_2 destruction but poor Claus conversion. Active catalysts such as Kaiser S-201 and S-501 give good COS/CS_2 destruction at lower temperature [148]. A layer of cobalt molybdate catalyst may be placed beneath the Claus catalyst in the bottom of the first converter.

If a tail-gas-cleanup unit is employed to convert sulfur species into either H_2S or SO_2, the COS and CS_2 may be allowed to pass through the Claus unit. Extra capacity for these contaminants must be designed into the tail gas unit.

13. Water vapor is not usually considered a "contaminant," but too much water vapor will shift the Claus reaction equilibrium to the left and reduce recovery. Generally, free water is removed by the feed knockout drum. Additional feed cooling to condense water may be helpful, but if NH_3 and H_2S are present together, condensation at less than 180°F may precipitate ammonium sulfur salts [150].

Further design and application details for Claus sulfur plants will be found in Refs. 140-151.

2. *Claus Plant Tail Gas*

A well-designed Claus plant with fresh catalyst will recover 95 to 97% of its feed sulfur. This level of conversion is not sufficient to meet current emission standards, which may require sulfur recoveries of 99% or more. In addition, the effects of feed impurities and catalyst aging will reduce overall recovery in some plants to about 92% just before catalyst changeout. For these reasons, the sulfur content of the Claus tail gas must be reduced by further processing.

Before the advent of stringent air pollution controls, incinerators were provided in Claus plants to oxidize the sulfur compounds in the tail gas to SO_2 before they were emitted to the atmosphere. With the advent of tail-gas-cleanup units, the incinerator no longer serves this purpose but still may be installed for emergency purposes or to incinerate gases from the tail gas unit, although in a large coal gasification complex this function may be better performed by an auxiliary boiler.

Concentrations of sulfur compounds in the tail gas from a Claus plant processing an acid gas of 15 to 20% H_2S will be approximately as follows:

	ppm (vol)
H_2S	2000
SO_2	1000
S_8	60
S_L (entrained)	1200

N_2, CO_2, water vapor, CO, H_2, COS, and CS_2 will also be present. In a split-flow plant, the tail gas will contain any impurities that were present in the bypassed acid gas which were not destroyed or deposited in the catalytic converters. If further reduction of sulfur compounds from those shown above is required, tail-gas-cleanup systems are employed. Available processes can be separated broadly into two categories: low efficiency and high efficiency. Low-efficiency processes are those that result in up to 99.0 to 99.5% overall sulfur recovery when combined with the Claus plant. High-efficiency tail-gas-treating processes can achieve overall sulfur recoveries of 99.8% and above under ideal conditions, reducing H_2S in some cases to less than 10 ppm (vol) and total sulfur to as low as 100 ppm (vol).

High-efficiency processes require considerably more energy input per unit of sulfur recovered than do low-efficiency processes. This energy must be supplied from elsewhere in the complex, possibly from the burning of high-sulfur fuel in an auxiliary boiler. The sulfur in the additional fuel may be emitted as SO_2 to the atmosphere and it is entirely possible, owing to the structure of current emissions standards for SO_x versus H_2S, that the use of a high-efficiency tail-gas treating unit may result in more overall sulfur emissions to the atmosphere than the unit itself captures. Such a possibility should be thoroughly investigated when choosing a sulfur-processing scheme for a coal gasification complex.

Low- and high-efficiency tail-gas-cleanup systems can be further categorized into three types:

1. Extensions of the Claus reaction
2. Recycle processes
3. Independent conversion

In extended Claus processes, the 2:1 ratio of H_2S to SO_2 in the tail gas is maintained. The reaction is extended at low temperatures, below the dew-point of sulfur, where the equilibrium conversion becomes more complete than at the higher temperatures employed in the Claus reactors. The extended reaction may take place over dry catalyst (Sulfreen and CBA Processes) or in a liquid containing catalyst (IFP Process). These extended Claus processes are all in the low-efficiency category.

In recycle processes, the sulfur species in the tail gas are converted to a single compound, which is captured, concentrated, and returned to the Claus plant for conversion to sulfur. The recycle compound may be either H_2S (SCOT and Trencor Processes) or SO_2 (Wellman-Lord Process). These recycle processes are all in the high-efficiency category.

Independent conversion processes, in contrast to the two types discussed above, do not employ the Claus reaction for sulfur production and do not recycle the captured sulfur compounds to the Claus plant. Examples are as follows:

Beavon Mark I—hydrogenation + Stretford
Beavon Mark II—hydrogenation + Claus
SNPA/Halder-Topsoe—catalytic oxidation to H_2SO_4
Contact sulfuric acid

A summary of the capabilities and brief descriptions of the nature of the process are shown in Table 55. Further details are available in Refs. 153-169.

3. *Liquid-Phase H_2S Oxidation*

As an alternative to the Claus Process, H_2S can be converted to elemental sulfur by processes that absorb the H_2S in a circulating solution containing a catalyst. The catalyst promotes air oxidation of the H_2S to elemental sulfur. Residual H_2S from wet oxidation is sufficiently low (0.5 to 10 ppm) that no further treatment of tail gas is necessary. The wet oxidation processes are generally more economical than the Claus Process with tail-gas cleanup, where the H_2S content of the feed gas is low (15% or less). Three wet oxidation systems are described briefly here.

a. Stretford Process [152,170,171] A treated gas containing 10 ppm (vol) of H_2S is obtained by washing with an aqueous solution of sodium carbonate, sodium meta-vanadate, and anthraquinine disulfonic acid (ADA). The sodium hydrosulfide formed is oxidized to elemental sulfur by five-valent-state vanadate in the solution. The liquor is regenerated by air blowing and the reduced vanadate reoxidized through oxygen transfer via the ADA. The flow pattern and reactions involved are presented in Fig. 97. The solid sulfur produced in particle sizes of 0.5 to 25 μm is removed by froth flotation. The sulfur froth, after being skimmed, enters a sulfur purification system. Here it is processed to separate the sulfur from the solution. A filter cake is produced, containing 50 to 60 wt % solids. The sulfur in this cake is contaminated with salts from the solution and may be unsalable. If bright sulfur is desired, the filter cake may be further processed in an autoclave.

A number of significant side reactions can occur in the Stretford Process. First, NaHS may react with oxygen to form thiosulfate and sulfates. These salts cannot be regenerated and must be purged from the system. The purge stream is a liquid containing vanadium, Na_2CO_3, and ADA, and may present a significant disposal problem. The formation of thiosulfates and sulfates may be controlled by controlling the pH and temperature of the circulating solution.

TABLE 55 Summary of Claus Plant Tail-Gas Cleanup Processes

Process/refs	U.S. licensors	Required feed sulfur content	Process performance	Sulfur removal mechanism	Commercial status
			Low-efficiency processes		
Sulfreen [153,154]	Lurgi Mineraloletechnik and The Ralph M. Parsons Co.	Claus tail gas; sulfur compounds 1-3%	Raises overall sulfur recovery to 99%; 2000-3000 ppmv sulfur in treated gas; no COS/CS_2 conversion	Extended Claus reaction below sulfur dewpoint over alumina catalyst	11 commercial units now operating
SNPA/Haldor-Topsoe [155]	Haldor-Topsoe	Claus tail gas	About 500 ppmv sulfur in treated gas; all sulfur compounds are handled	Independent conversion by catalytic oxidation of sulfur compounds to SO_3, followed by absorption to produce 94% to H_2SO_4	1 commercial unit at Lacq, France
CBA (Cold-Bed Absorption) [156, 157]	Amoco Production Co.	Claus tail gas	Raises overall sulfur recovery to 99-99.3%, <1500 ppmv sulfur in treated gas; no COS/CS_2 conversion	Extended Claus reaction below sulfur dewpoint over Claus catalyst at 260°F	1 commercial unit at East Crossfield, Alberta, Canada
IFP Clauspol [155, 158,159]	Institute Francais du Petrole (IFP)	Claus tail gas or other H_2S-containing feeds, up to 20%	<1500 ppmv sulfur in treated gas; no COS/CS_2 conversion	Extended Claus reaction in liquid phase, with polyethylene glycol as liquid carrier for catalyst	Over 20 commercial units now in operation

Beavon Mark II [160]	The Ralph M. Parsons Co.	H_2S content of <5% or Claus tail gas	Up to 99% overall sulfur recovery; some COS and CS_2 conversion	Independent catalytic conversion of sulfur compounds to H_2S, followed by successive oxidation and Claus reaction over proprietary Parsons/Union Oil catalyst	1 commercial unit operating in Germany on Claus tail gas
High-efficiency processes					
Beavon Mark I [160]	The Ralph M. Parsons Co.	Claus tail gas	99.8% overall sulfur recovery, <300 ppmv sulfur compounds in treated gas, generally <10 ppmv H_2S; some COS/CS_2 conversion	Independent catalytic conversion of sulfur compounds to H_2S, followed by direct oxidation of H_2S to sulfur in Stretford unit	19 commercial units now operating or under construction
SCOT (Shell Claus Off-Gas Treating) [155]	Shell Development Co.	Claus tail gas	99.8% overall sulfur recovery if COS/CS_2 content is not excessive. <300 ppmv sulfur compounds in treated gas; H_2S is about 150-200 ppmv	Recycle process, with catalytic conversion of sulfur compounds to H_2S, followed by absorption with diisopropanol amine. H_2S and CO_2 from amine regenerator recycled to Claus plant	Several commercial units now in operation
Trencor [161]	Trentham Corp.	Claus tail gas	Same as SCOT	Recycle process, same as SCOT, except amine used is MDEA	1 commercial unit has been constructed; 99.8% recovery not yet demonstrated

(continued)

(Table 55 continued)

Process/refs	U.S. licensors	Required feed sulfur content	Process performance	Sulfur removal mechanism	Commercial status
Wellman-Lord [162]	Davy Powergas	Claus tail gas	<200 ppmv sulfur compounds in treated gas; all sulfur compounds are handled	Recycle process, with oxidation of all sulfur species to SO_2, followed by absorption in sodium sulfite solution; evaporation of solution released SO_2 which is recycled to Claus plant; up to 10% of entering sulfur may be purged as sulfate solution	18 commercial installations on boiler flue gas; 7 on Claus tail gas
Contact Sulfuric Acid Process [164-169]	Monsanto, Parsons, Davy Powergas, others	Can accept elemental sulfur, or H_2S and SO_2-bearing streams down to about 5% sulfur content	A double contact/double absorption plant can recovery up to 99.8% of the sulfur fed to it; all sulfur compounds handled	Independent conversion by thermal oxidation of sulfur compounds to SO_2, followed by catalytic oxidation to SO_3; absorption of SO_3 by concentrated sulfuric acid creates net product acid	Many commercial units in operation on sulfur feed, or smelter gases containing SO_2; process not yet commercially demonstrated for 5% H_2S feed

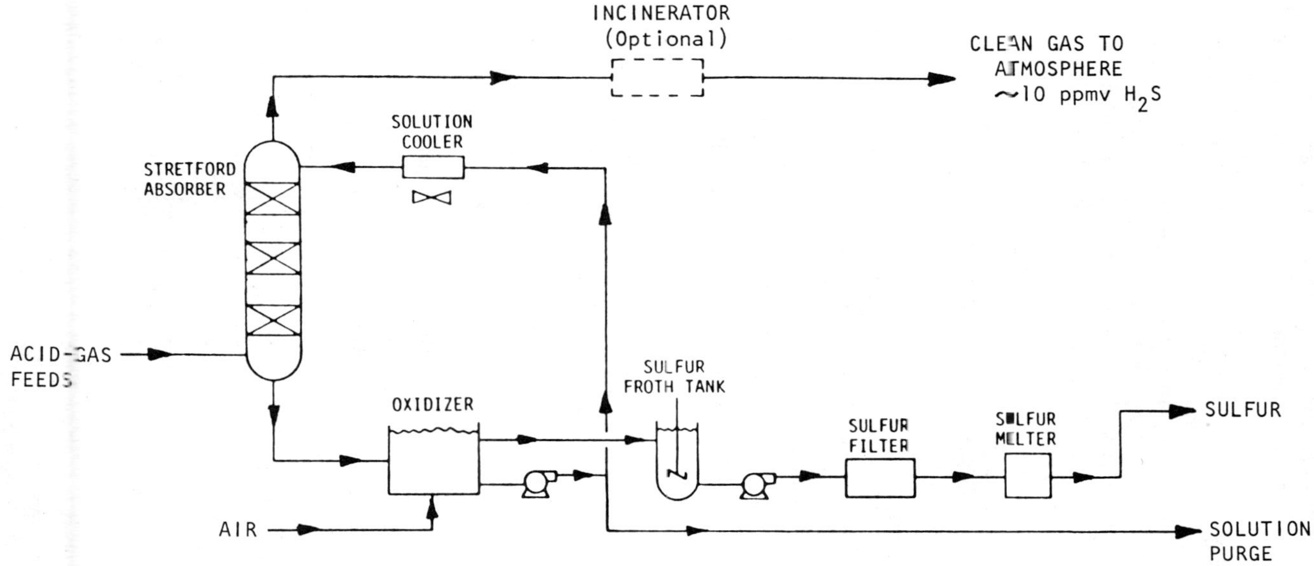

Reactions

$H_2S + Na_2CO_3$	$\rightleftarrows$	$NaHS + NaHCO_3$
$NaHS + 2NaVO_3 + 1/2\ H_2O$	$\rightleftarrows$	$1/2\ Na_2V_4O_9 + 2\ NaOH + S$
$Na_2V_4O_9 + 2\ NaOH + H_2O + 2\ ADA$	$\rightleftarrows$	$4\ NaVO_3 + 2\ ADA\ (reduced)$
$2\ ADA\ (reduced) + O_2$	$\rightleftarrows$	$2\ ADA$

ADA = anthraquinone disulfonic acid

FIG. 97 Typical Stretford Process.

COS and CS_2 are not removed by a Stretford absorber. If too much organic sulfur is in the feed to the Stretford unit, required emission standards may not be met even with essentially complete removal of the H_2S. COS and CS_2 may be removed ahead of the Stretford unit by hydrolysis and/or hydrogenation either in the main plant or in a Beavon unit ahead of the Stretford absorber. Several other processes for organic sulfur removal are discussed in the literature [165].

HCN is a very serious contaminant in Stretford units. The thiocyanate formed must be purged from the system, greatly increasing chemical makeup costs. If the cyanide concentration in the Stretford feed gas is greater than 1500 ppm (vol), a polysulfide prewash step may be profitably employed. This patented process has been commercially applied in coke-oven plants where the Stretford feed gases have high HCN concentrations [173].

Hydrocarbons pose no specific problems to a Stretford unit; however, they appear in the treated gas and may violate emission regulations. Heavy hydrocarbons and tars can be removed by electrostatic precipitation upstream from the Stretford unit [172].

Like hydrocarbons, ammonia does not specifically affect the Stretford solution. However, it is absorbed and may cause odor problems around the oxidizer.

Over 50 Stretford units, licensed by the British Gas Corp., with capacities ranging from 0.1 to 32 million scf/day are currently in operation.

b. Takahax Process Up to 99.9% of the H_2S can be removed from gases to produce a treated gas containing less than 10 ppm (vol) H_2S by washing with an alkaline solution (8.5 pH) containing 1,4-naphthaquinine-2-sulfonate as the catalyst. Finely divided (2 μm) solid sulfur is produced directly by reaction with the catalyst. The catalyst is circulated to a regenerator and reoxidized by air blowing. Sulfur product is removed by filtration of a slipstream.

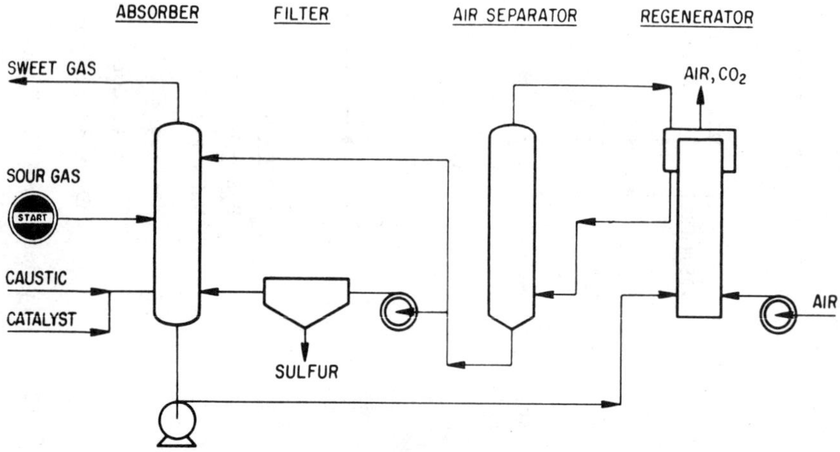

FIG. 98 Takahax H_2S oxidation process [128].

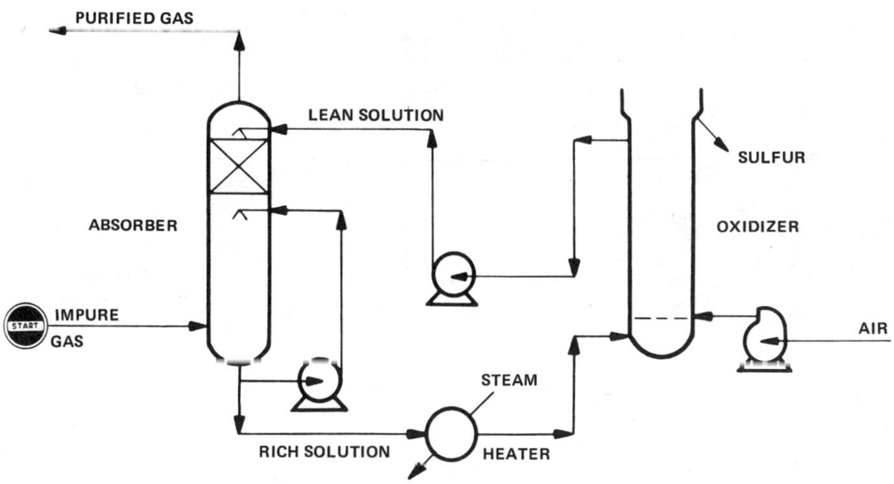

FIG. 99 Giammarco-Vetrocoke H_2S oxidation process [128].

If HCN is present in the feed gas, it is absorbed and oxidized to sodium thioyanate. It will concentrate in the circulating solution and must be purged.

The Takahax Process flow system is illustrated in Fig. 98. Over 70 Takahax plants have been built in Japan for treating sewage fermentation, coke oven, and chemical plant waste gas.

c. Giammarco Vetrocoke Process H_2S levels of treated gas can be reduced to less than 0.5 ppm by washing with a sodium carbonate solution containing arsenates and arsenites. Sodium thioarsenate, which is formed initially, converts to monothioarsenate and arsenite. The solution is then circulated to an air-blown oxidizing column where the monothioarsenate decomposes to arsenite and elemental sulfur. The sulfur is removed by froth flotation, vacuum-filtered, and washed. Air oxidation also restores the original scrubbing solution by oxidizing some arsenite to arsenate.

Reactions involved are the following:

$$Na_2AsO_3 + 3H_2S = Na_3AsS_3 + 3H_2O \quad (1)$$

$$Na_3AsS_3 + 3Na_3AsO_4 = 3Na_3AsO_3S + Na_3AsO_3 \quad (2)$$

$$Na_3AsO_3S = Na_3AsO_3 + S \quad (3)$$

$$Na_3AsO_3 + 1/2\ O_2 = Na_3AsO_4 \quad (4)$$

A simplified flow diagram of the GV Process is shown in Fig. 99.

C. Wastewater Treatment

A foul-water stream is generated in most gasification processes from the quenching and water washing of the raw gas, which removes water-soluble impurities such as ammonia, H_2S, and phenols. The nature of contaminants

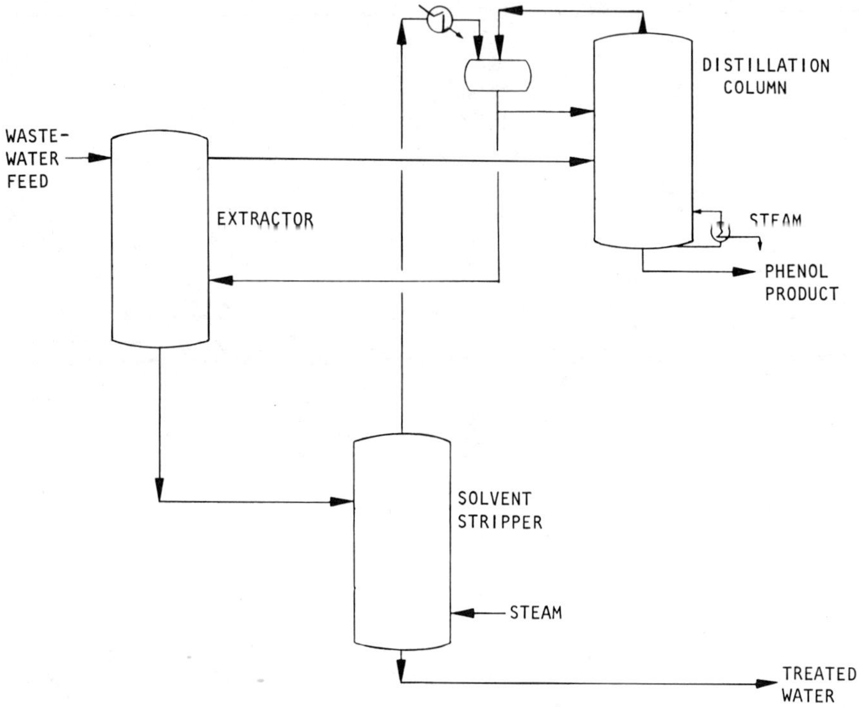

FIG. 100 Phenol-extraction-system process.

will depend upon the raw coal composition, the gasifier operation, and the techniques that are used for quenching and washing the raw gas.

The nitrogen content of the raw coal is expected to be largely converted to ammonia in many gasifiers. Some conditions also will tend to produce small quantities of cyanide, which may be further converted to thiocyanates in the quench water. Coals with high oxygen content tend to produce phenolic compounds that may appear in the wash water. Most of the chlorine in the coal will be gasified to chlorides and will also appear in the wash water.

Some gasifier systems convert a significant proportion of the coal into tars and heavy oils. Depending upon the configuration of the quench system, some of these higher organic compounds may be emulsified into the wash water.

All of these materials must be substantially removed if the water is to be disposed of either by reuse in the plant system or discarded. Several processes are available to treat the wastewaters. Only after the wastewaters have been defined can the preferred treatment processes be specified.

1. *Phenol Removal*

Liquid/liquid extraction processes are used to remove high concentrations of phenols from industrial wastewaters. For water that contains dissolved ammonia and hydrogen sulfide, most designs place the phenol extraction process before water stripping (Fig. 95) to prevent phenol problems in the stripper and ammonia-recovery steps. Several processes are available with basic steps (Fig. 100), but each employs its own proprietary solvent. First, the phenol extractor provides intimate mixing of wastewater and the solvent and then separation of the solvent stream containing phenol from the treated water stream. The latter may contain a small amount of solvent that is recovered in a solvent stripper. Solvent is boiled out of the phenol in the solvent fractionator and recycled to the phenol extractor.

Two phenol extraction systems—Lurgi Phenosolven and Chempro—are commercially proven on either existing coal gasification plants or coke oven wastewaters. Design modifications may be required for second-generation coal gasification processes to verify both process and economics of this system application.

The degree of phenol removal is dependent upon the types of phenols in the feed as well as other organic compounds. The extraction process licensors will guarantee 99% phenol removal for simple, monohydric phenols. Polyhydric phenols are only removed at 60% efficiency and other organics at only 15%. A South African coal gasification plant has reported 45% removal of total organics fed to the phenol-recovery system.

2. *Sour-Water Stripping*

Water stripping is a form of distillation in which trace amounts of relatively volatile materials, such as dissolved gases, are removed from a large volume of wastewater. The stripping is affected by either indirect reboiling of the feedwater by steam (Fig. 101) or by direct injection of steam into the bottom of a column. The wastewater is fed to a baffled drum, where floatable light oil and sludge is separated from the sour water. After heat exchange with hot stripped water, it is then fed into the top of the stripping column. Stripping columns include fractionating trays such as bubble caps and sieve plates, or packed sections. This technique is used extensively for sour-water strippers in almost all refinery installations. For coal gasification processes, dissolved base ionization may complicate the design procedures.

For example, acidifying the feed will fix the ammonia in solution to allow selective removal of hydrogen sulfide. Unless cations, such as ammonia and sodium, are present to fix the hydrogen sulfide, removals greater than 98% are commonly reported. Ammonia does not strip as readily. Greater than 90% of the feed ammonia is removed at 1 lb of steam per gallon of fresh feed and up to 98% removal at 2 lb of steam per gallon of feed. If anions, such as sulfides, carbonates, and organic acids, are present in the water, caustic or lime must be added to free ammonia so that these removals can be achieved.

Trace contaminants can play a major role in stripper design or operations. Ammonium polysulfides, which are usually generated in high-pressure cracking processes and could be present in coal processes, require higher pressures (at least 50 psig) to decompose the polysulfides into components that can be readily stripped. Carbonate-containing wastewaters must be monitored carefully because the carbonates will precipitate and eventually plug the stripper.

The stripped water can be sent to a biological wastewater treatment plant for further processing. The gases require further treatment, such as ammonia recovery, sulfur removal/recovery, or incineration/combustion.

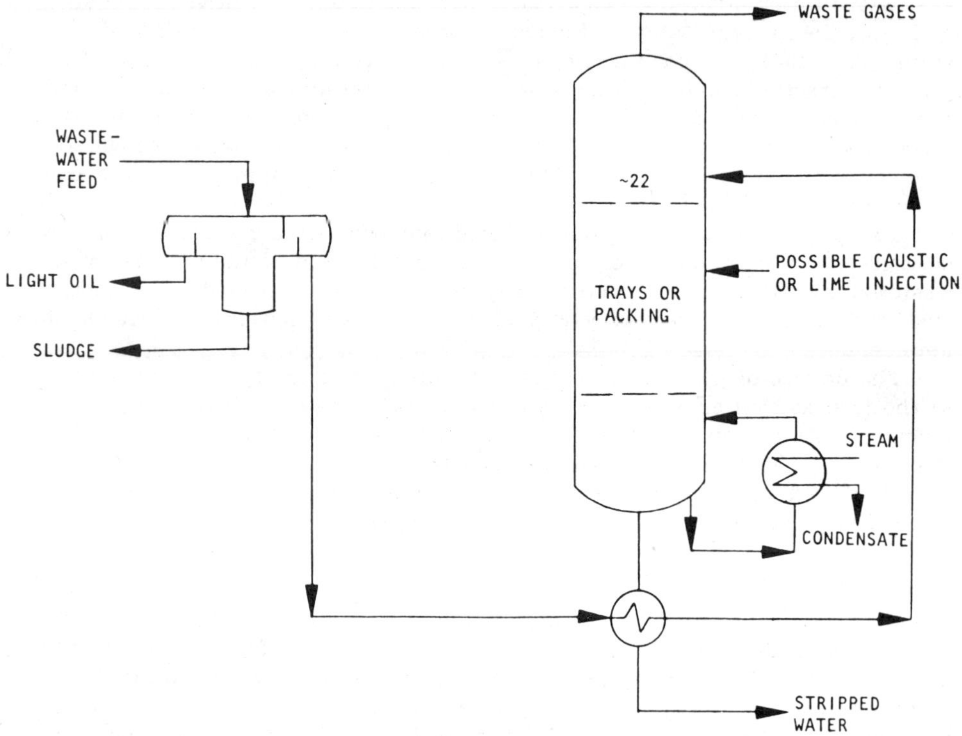

FIG. 101 Water stripper system [172].

3. *Ammonia Recovery*

Several processes have been developed to recover ammonia from wastewater or from the overhead vapor from sour-water strippers. In the Phosam Process developed by U.S.S. Engineers, the wastewater is fed to a combination stripper-absorber. The dissolved gases are stripped from the water as discussed in the preceding section, and pass through the absorber. Here, the ammonia is separated from the other stripped gases by absorption in an ammonium phosphate solution. The ammonia-rich solution is sent to a stripper to regenerate the absorption solution and recover the ammonia. The ammonia is further purified by distillation. Caustic is generally added to the still to prevent the buildup of trace components, such as organic compounds or acid gases. These ammonia-recovery steps are illustrated in Fig. 102. Such ammonia-recovery systems have been successfully installed in petroleum refineries and in a coke-oven gas-treating facility.

A second system, the Chevron WWT Process, accomplishes ammonia removal and purification with physical processes by utilizing the differences in physical properties among the various dissolved gases. This process has been used for treating petroleum refinery wastewater but was found to be uneconomical for coal conversion processes [173].

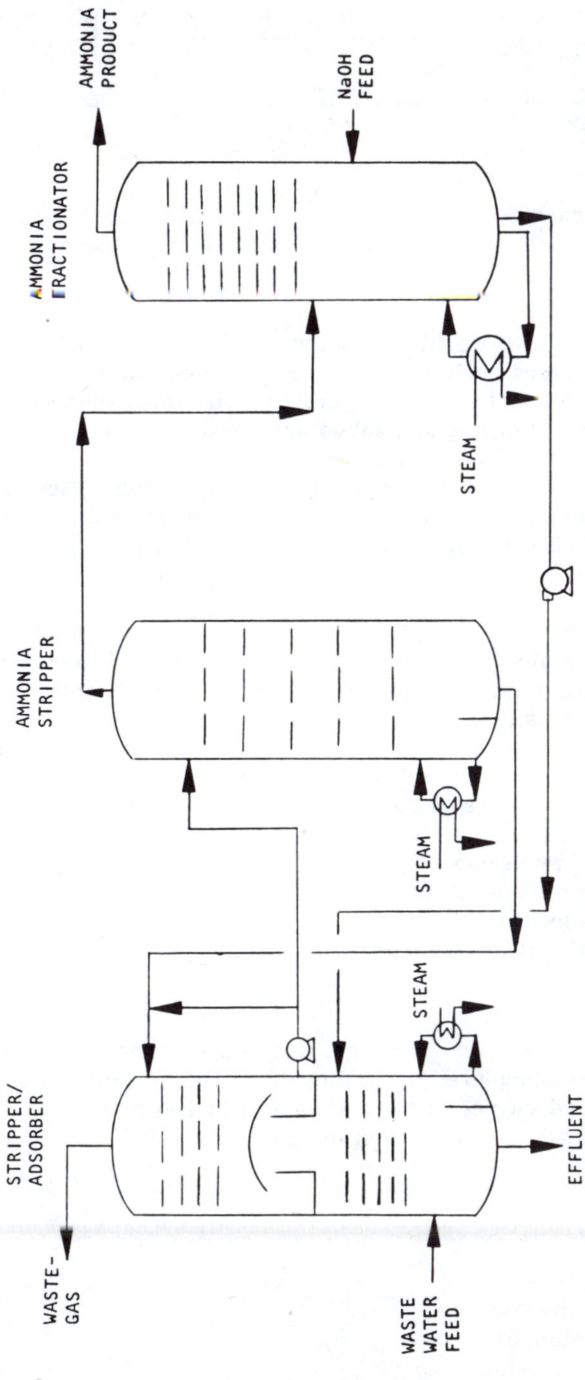

FIG. 102 Ammonia-recovery process [173].

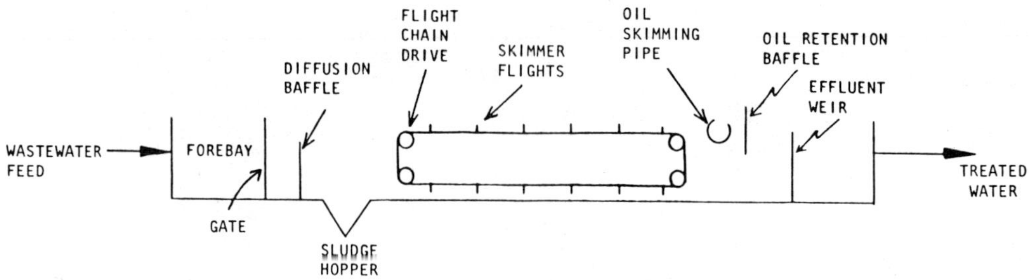

FIG. 103 API separator [174].

4. *Oil Removal*

Insoluble entrained organic compounds can be removed from wastewater by physical separators such as the API gravity separator. The principal contaminant removed is free oil. Figure 103 illustrates the cross section of an API separator. If suspended solids are present in the wastewater, these solids will also settle in the basin.

The API separator consists typically of a rectangular basin or tank equipped with an influent section and a separation section with baffles, weirs, and a mechanical oil skimmer. Fugitive hydrocarbon emission standards may require a cover over the separator to reduce air emissions.

The wastewater flows horizontally across the shallow basin, where the free oil globules rise due to buoyancy and the suspended solids settle to the bottom. The oil is skimmed from the surface by mechanical collectors and returned to the process. The settled solids are intermittently removed and processed for disposal.

The latest procedures for separator designs are detailed in the *Manual on Disposal of Refinery Wastes* [174]. Important characteristics of the wastewater required for process design considerations include:

1. Degree of emulsions
2. Wastewater temperature
3. Specific Gravity of the oil
4. Amount of settleable solids

5. *Miscellaneous Water Treating*

In addition to the water-treating systems mentioned above, a variety of treatments may be considered depending upon the particular circumstances of contaminants and degree of final water purity required.

Systems that may be considered include:

Clarification
Filtration
Air flotation
Biological treatment
  Trickling filter
  Activated sludge
  Industrial digestion

Activated carbon adsorption
Vapor compression distillation

6. Zero Water Discharge

Figure 104 presents a possible water-reuse plant for a hypothetical coal gasification facility. For this 250 million scf/day SNG plant, 6000 gal/min of raw water is typical of a water-conservative plant that maximizes air cooling. Consumption of water into the product is 1330 gal/min and the cooling tower evaporation is 3820 gal/min. The total net foul water make is 1570 gal/min.

If the quantity of phenols made in the gasifier is high, recovery of the phenols from the foul water would be economical. A liquid-extraction system will recover most of the phenols and other organic impurities from the foul water for by-product credit. This box is shown in dashed lines in Fig. 104 because this recovery is not necessarily economic with all gasifier types and coals.

The next step in foul-water treating is stripping for recovery of ammonia and elimination of H_2S and CO_2 that dissolved with the ammonia. The ammonia is recovered for by-product value and the acid gases would be directed to the Claus plant for sulfur recovery.

The next step (Fig. 104) is biological oxidation. If phenol recovery is not practiced, this bioxy stage is necessary for destruction of organic matter. If phenol recovery is practiced, the bioxy could perhaps be shifted to the smaller, cooling-tower blowdown stream. The waste solids sludge from the

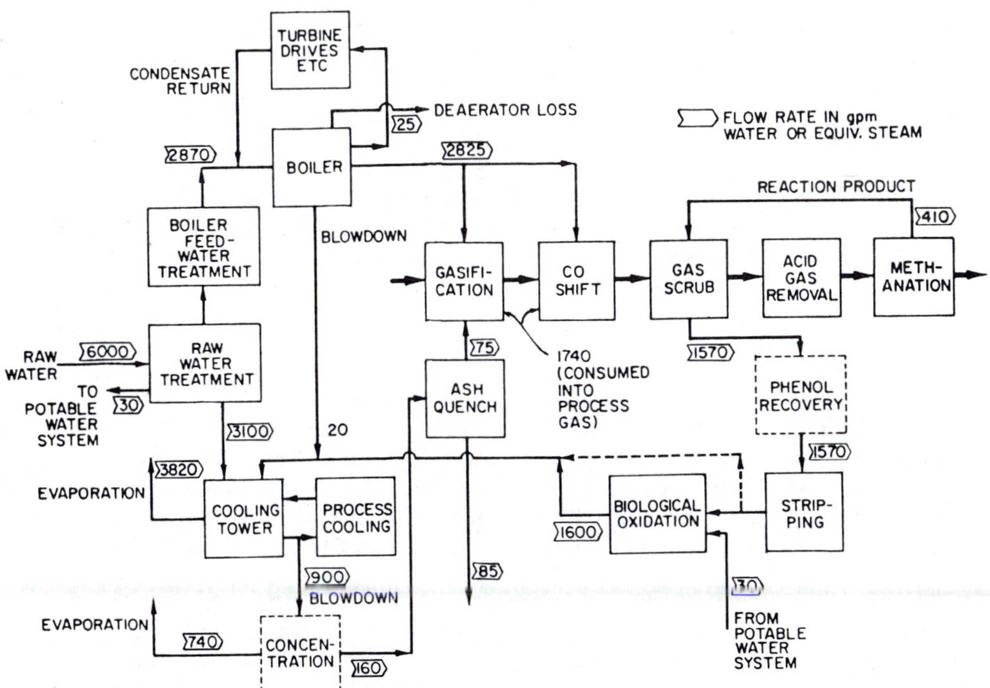

FIG. 104 Water reuse plan (hypothetical) [137].

biological oxidation will be a function of the impurities present in the foul liquor and the degree of pretreatment before the sludge is produced. Note also that biological oxidation is not the only alternative for this purification step. Pressure thermal oxidation may be considered, and secondary treatment such as ozonation, chlorination, or oxygen treatment may be the better route.

The treated foul water now has sufficient purity that it may be injected into the cooling-water circuit together with treated raw water. A net cooling-tower blowdown of 900 gal/min is taken. This blowdown will contain the water-soluble compounds that were not removed earlier. These include chlorides, perhaps as ammonium chloride, and a portion of the thiocyanates may still be present.

A dashed box for concentration is shown in Fig. 104 as the next step in water management. If the design philosophy of the facility is no water discharge, the engineer may use excess low-pressure steam from the process to concentrate this blowdown stream by a factor of about 6. The concentrated blowdown would then be used for ash quenching within the gasifier (with resultant steam make into the gasifier), and the final discharged ash would contain small quantities of water. Under this philosophy, the ash will contain all of the water-soluble material in the cooling-tower blowdown, including the chlorides of the coal and salts present in the raw water. Therefore, under this philosophy, the ash would be subject to leaching.

The alternative solution is to either evaporate the cooling-tower blowdown to dryness, generating a small, water-soluble solids stream, or to develop disposal techniques for a liquid waste.

At present, the characteristics of this waste can only be estimated. There are no hard data on the contaminants present in the foul water as a function of coal feedstock, gasifier operating conditions, and the method of gas quenching and washing. Further, there are no quantitative data on the fate of potential pollutants as this foul water passes through the various purification stages. Data from a pilot plant are not necessarily applicable because the engineering design of a commercial facility may use different gas-processing and water-treatment techniques. In short, the quantities and compositions of water-soluble materials that may be present as a final waste from a coal gasification facility cannot be predicted with accuracy at this time.

VI. ECONOMICS

Estimates of the capital investment and operating-cost requirements for various coal gasification systems utilizing either conceptual designs for gasifiers under development or those which can be considered commercially proved have been made by a diversity of organizations. Extreme caution must be used when comparing economic data generated from such different sources because the calculated costs can vary widely, even for a particular process, depending upon factors such as plant size and location, specifications of feed coal, by-product values allowed, method of financing, estimates of operating costs, coal costs, and time base of the capital and operating costs.

Figures presented in this section should be taken only as indicative of the general range of cost between various gasifier configurations and gas quality. Three sets of cost figures are given. First, for low-Btu gas production at a relatively low rate (2.5 billion Btu/day), the comparison is made between three different gas qualities varying from hot raw gas to oil and tar-free desulfurized gas. Then, for high-Btu SNG, the comparison is made between various gasification processes, and between typical coals—Western subbituminous, Eastern bituminous, and lignite. In a third set of figures, the costs for producing high-Btu SNG and medium-Btu gas are compared. The comparative costs included in each of the three cases are considered valid because they are prepared on individually consistent bases. Comparison between any two of the three should not be undertaken because there are significant differences in procedures and assumptions employed in preparation of the estimates.

Two accounting methods developed by the FPC Supply-Technical Advisory Task Force [175] are popularly employed in determining the selling price of gas produced from coal. One is the Utility Method. A summary of this calculation method is listed in Table 56. The second method is the Private Investor, DCF Method. A summary of this calculation method is shown in Table 57.

A. Low-Btu Gas

Investment and operating costs to produce low-Btu industrial fuel gas by commercially available fixed-bed coal gasification systems have been developed. Cost figures from five different gasifier suppliers (three single-stage* and two two-stage†) were used. Production capacity for each case is 2.5 billion Btu of product gas. As may be seen from Table 58, there is a wide disparity in capital costs between the five suppliers, with a minimum total 30% below and a maximum total 30% above the average cost. The costs are presented in such a way that three different levels of gas quality can be compared:

Hot raw gas
Dust-, tar-, and oil-free gas
Dust-, tar-, and oil-free desulfurized gas

The investment figures shown in Table 58, based on mid-1977 costs, are not complete in that they do not include engineering, construction management, and allowance for contingencies.

The average investment costs from Table 58 were used together with operating and coal costs in developing gas costs for the three levels of gas quality (Table 59). For hot raw gas and for dust-, tar-, and oil-free gas, a low-sulfur (0.60%) bituminous coal from Stockton, West Virginia was used, whereas for desulfurized gas, a high-sulfur (3.33%) bituminous coal from Belmont, Ohio, was used. In each case, three different delivered coal costs and three different returns on investment‡ are evaluated.

---

*McDowell-Wellman, Riley-Morgan, and Wilputte.

†Wellman (Incandescent), FW-Stoic, and Woodall-Duckham.

‡DCFs of 8%, 10%, and 12% on 100% equity.

TABLE 56 Gas Cost Calculation by Utility Method [175]

	Utility method
Basis	
Project life	20 years
Depreciation	5%/year, straight line
Debt/equity ratio	75%/25%
Return on equity	15%
Interest rate on debt	9%
Federal income tax	48%
Interest during construction	Interest rate (9% × 1.875 years[a] × total plant investment
Other factors	
Plant stream factor	90%
Contingencies	15% of installed plant cost
Contractor's overhead and profits	15% of total bare cost
Startup cost	5% of total plant investment
Working capital	(a) Coal inventory (60 days' feed at full rate) (b) Material and supplies (0.9% of total plant investment) (c) Net receivables at 1/24 × annual revenue required

Derived equation:

$$\text{20-year average gas price } (\$/10^6 \text{ Btu}) = \frac{N + 0.1198C + 0.0198W}{G}$$

where N = net annual operating cost
C = total capital required
W = working capital
G = annual gas production

[a]10% for 3 years, 90% for 1.75 years.

A typical project time schedule for engineering design and construction of a low-Btu gasification facility of the size considered here is shown in Fig. 105. This indicates that 24 months is required from contract initiation to completion of performance tests.

B. High-Btu Gas (SNG)

Costs for producing synthetic natural gas (SNG) of pipeline quality* at a rate of 250 billion Btu/stream day have been developed for Western (Montana) subbituminous coal, for Eastern (Pittsburgh No. 8) bituminous coal, and for

*CO no more than 0.1 vol %; H_2S no more than 0.25 grains/100 scf; total sulfur no more than 10 grains/100 scf; product gas delivery pressure, 1000 psig.

TABLE 57 Calculation of Gas Cost by Discounted Cash Flow Analysis [16]

Financial parameters	
Plant life (years)	25
Capacity factor (%)	90
Depreciation	16 years sum-of-the-years digits
By-product credit	
Tar (cents/gal)	10
Sulfur ($/ton)	25
Federal income tax rate (%)	48
DCF return (%)	12, 10, 8
Equity capital (%)	100
Interest during construction (IDC)	Payable within 12 months

Calculation of fuel cost

$$\text{Fuel cost } (\$/10^6 \text{ Btu}) = \frac{aN + bI + cS + dW}{G}$$

where a = dimensionless parameter describing escalation of operating cost during the project life; in the present calculation, a = 1.0 assuming there is no escalation from startup through project completion date

b = 0.2095 at 12% DCF return, 0.1719 at 10% DCF return, 0.1377 at 8% DCF return

c = 0.1275 at 12% DCF return, 0.1102 at 10% DCF return, 0.0937 at 8% DCF return

d = 0.2308 at 12% DCF return, 0.1922 at 10% DCF return, 0.1538 at 8% DCF return

N = total net operating cost, including coal feed cost in the first year, millions of dollars/year

I = total plant investment, millions of dollars

S = startup cost, millions of dollars; 20% of total annual gross operating cost, including coal feed cost

W = working capital, millions of dollars; 60 days' coal inventory plus 1% of total plant investment

G = annual fuel production, trillion Btu/year

The values of I, S, W, and N must be adjusted to reflect the actual costs for the startup completion date.

TABLE 58 Capital Cost (Millions of Dollars) for Fixed-Bed Gasifier of 2.5×10^9-Btu/Day Capacity [16] (Based on Mid-1977 Pricing)

	Gasifier supplier					
	A	B	C	D	E	Average
Gasification						
Coal and ash handling						
Gasifiers						
Cyclones	0.950	2.175	1.663	2.438	3.082	2.061
Site building and installation						
Gas manifold						
Subtotal	0.950	2.175	1.663	2.438	3.082	2.061
Facilities						
Electric power distribution	0.093	0.093	0.093	0.093	0.118	
Cooling water piping	0.002	0.002	0.002	0.002	0.003	
Instrument air and dryer	0.005	0.005	0.005	0.005	0.006	0.149
Steam system for startup/shutdown	0.020	0.020	0.020	0.020	0.025	
Air cooler	0.021	0.021	0.021	0.021	0.027	
Subtotal	0.141	0.141	0.141	0.141	0.179	0.149
Total direct capital[a] cost for hot raw gas	1.091	2.316	1.804	2.579	3.261	2.210
Gas cleanup						
Tar, oil removal and water treatment	0.747	0.747	1.707	0.650	0.822	0.935
Total direct capital cost for dust-,[a] tar-, and oil-free gas	1.838	3.063	3.511	3.229	4.083	3.145
Desulfurization[b]	1.375	1.357	1.246	1.375	1.828	1.440
Total direct capital cost[a] for dust-, tar-, and oil-free desulfurized gas	3.213	4.438	4,757	4.604	5.911	4.585

[a]Total costs do not include cost of engineering, construction management, or contingency allowance.

[b]Desulfurization costs are based on the Stretford Process.

TABLE 59 Gas Cost for Fixed-Bed Gasifier Systems [16]

	Low-sulfur (0.60%) coal		High-sulfur (3.33%) coal
	Hot raw gas	Dust-, tar-, and oil-free gas	Dust-, tar-, oil-free and desulfurized gas
Plant investment cost ($MM)	2.210	3.145	4.585
Operating cost ($MM/yr)			
Utilities and materials (3% of plant investment)	0.066	0.094	0.138
General overhead (1.67% of plant investment)	0.037	0.053	0.077
Labor cost ($MM/yr)			
Process[a]	0.248	0.248	0.248
Maintenance (1.5% of plant investment)	0.033	0.047	0.069
Supervision (15% of sum of process and maintenance labor)	0.045	0.049	0.054
Taxes and insurance (2.7% of plant investment) ($MM/yr)	0.060	0.085	0.124
Gross operating cost ($MM/yr)	0.489	0.576	0.710
By-product credit ($MM/yr)			
Tar at 10 cents/gal	—	0.066	0.071
Sulfur at $25/ton	—	—	0.028
		0.066	0.099

(continued)

(Table 59 continued)

	Low-sulfur (0.60%) coal						High-sulfur (3.33%) coal		
	Hot raw gas			Dust-, tar-, and oil-free gas			Dust-, tar-, oil-free and desulfurized gas		
Net operating cost (excluding coal)									
$MM/yr		0.489			0.510			0.611	
$/MM Btu		0.595			0.621			0.744	
Delivered coal cost ($/ton)	30	40	50	30	40	50	20	30	40
$/MM Btu, coal	1.15	1.53	1.91	1.15	1.53	1.91	0.80	1.20	1.60
$/MM Btu, product gas	1.25	1.67	2.08	1.51	2.01	2.51	1.08	1.62	2.16
Product gas cost ($/$10^6$ Btu)									
at 12% DCF return	2.52	2.96	3.41	3.06	3.49	4.13	3.14	3.72	4.29
at 10% DCF return	2.40	2.84	3.28	2.90	3.32	3.96	2.91	3.48	4.06
at 8% DCF return	2.29	2.73	3.17	2.75	3.17	3.80	2.70	3.27	3.84

[a]Process labor cost = (4 workers/shift) (8304 worker-hours/yr) ($7.45/worker-hour), mid-1973.

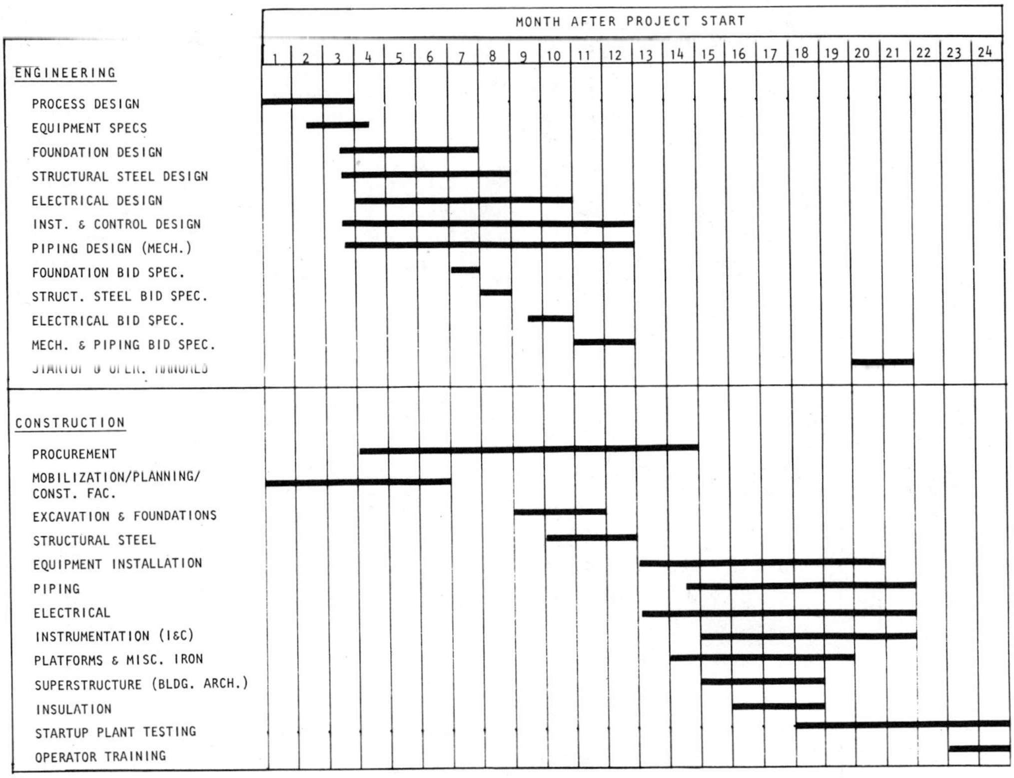

FIG. 105 Typical project schedule [16].

lignite. In each case, the gasification plant is assumed to be close to the mine. A number of gasification processes under development are considered together with designs based on a Lurgi gasifier, which is considered to be commercially proved. All plant designs within each group (Western, Eastern, and lignite), were based on a consistent set of design criteria, economic ground rules, and estimating procedures. Total plant investment estimates are stated to be accurate to approximately ±20%.

Table 60 summarizes individual investment requirements for each of the plant sections. Costs of general facilities (10% of the total for individual sections), contractors home office costs and fees (10% of total before contingency), and contingency allowances (15% of total cost) are added to the section costs to obtain the total plant investments shown. All costs are on a January 1, 1976, basis. Western and lignite cost estimates are based on designs employing maximum air cooling, whereas Eastern designs are based on mixed air and water cooling.

Table 61 summarizes total capital requirements from the plant investments in Table 60 and net annual operating costs based on unit costs of raw materials, by-products, and other operational expenses as shown.

TABLE 60 Plant Investment Summaries (Millions of Dollars) for SNG Production (January 1976 Basis with No Forward Escalation)

	Western subbituminous coal [176]						Eastern bituminous coal [177]					Lignite [178]	
	IGT HYGAS steam oxygen	IGT HYGAS steam-iron	Conoco CO_2 acceptor	BCR BI-GAS	PERC/synthane	Lurgi	IGT HYGAS steam-oxygen	IGT HYGAS steam-iron	BCR BI-GAS	PERC/synthane	Lurgi	Conoco CO_2 acceptor	Lurgi
11 Coal storage and reclaiming	13.0	13.0	13.0	13.0	14.0	14.0	12.0	12.0	12.0	12.0	12.0	18.0	19.0
12 Coal preparation	17.0	25.0	59.0	34.0	22.0	26.0	20.0	26.0	30.0	27.0	5.0	106.0	16.0
13 Coal pretreatment	—	—	—	—	—	—	34.0	40.0	—	—	—	—	—
14 Coal feeding	26.0	28.0	27.0	17.0	16.0		18.0	20.0	14.0	11.0		13.0	
15 Gasification	45.0	459.0	193.0	44.0	54.0		52.0	105.0	49.0	59.0		165.0	
15 Power recovery	—	—	—	—	—	191.0 (items 14–17)	—	333.0	—	—	2[illegible]5.0 (items 14–17)	—	191.0 (items 14–17)
16 Raw gas quench	19.0	80.0	64.0	50.0	38.0		52.0	53.0	31.0	32.0		69.0	
17 Shift conversion	31.0	—	—	32.0	52.0		27.0	—	29.0	31.0		—	
18 Acid gas removal	108.0	75.0	36.0	114.0	116.0	117.0	101.0	78.0	98.0	86.0	1[illegible]0.0	29.0	117.0
19 Methanation	29.0	26.0	36.0	29.0	33.0	58.0	26.0	21.0	28.0	29.0	[illegible]6.0	37.0	48.0
20 Effluent treating	58.0	68.0	18.0	37.0	42.0	19.0	41.0	39.0	29.0	71.0	[illegible]2.0	19.0	21.0
21 Sulfur recovery	60.0	76.0	52.0	59.0	77.0	54.0	76.0	93.0	72.0	79.0	[illegible]2.0	80.0	65.0

22 Solids disposal	8.9	7.6	8.5	8.9	9.2	3.3	10.0	9.0	12.0	5.0	3.0	11.0	8.0
23 Product gas drying	0.8	0.8	0.8	0.8	0.8	0.8	0.8	0.8	0.8	0.8	0.8	0.8	0.8
30 Steam and power	136.0	12.0	100.0	184.0	364.0	178.0	112.0	13.0	114.0	171.0	158.0	84.0	138.0
31 Plant water systems	17.0	34.0	18.0	25.0	20.0	22.0	34.0	43.0	31.0	32.0	50.0	27.0	41.0
32 Oxygen plant	45.0	—	—	81.0	80.0	65.0	41.0	—	67.0	63.0	91.0	—	59.0
General facilities[a]	68.0	100.0	69.0	81.0	104.0	83.0	73.0	98.0	69.0	79.0	92.0	73.0	80.0
Contractors home office cost and fee	75.0	111.6	76.7	90.3	116.0	92.9	81.2	109.2	76.2	87.2	102.2	81.2	89.2
Contingency	113.0	164.0	119.0	130.0	172.0	136.0	119.0	167.0	118.0	135.0	151.0	117.0	137.0
Total plant investment	870.0	1280.0[b]	890.0	1030.0	1330.0[c]	1060.0[d]	930.0	1260.0[e]	880.0	1010.0	1170.0	930.0	1030.0

[a]Includes site preparation, electrical distribution, yard piping, buildings, flare system, tankage, shipping and receiving facilites, sulfur storage, potable and sanitary water, holding ponds, plant roads.

[b]Export power case.

[c]Export power, slurry feed case (for export power, dry feed, and export char, slurry feed; see reference).

[d]Dual-feed case.

[e]Total plant investment for the case in which Benfield acid-gas removal is substituted for Selexol is $1230 million.

TABLE 61 Capital and Operating Costs for SNG Production*

	Western Subbituminous Coal [176]						Eastern Bituminous Coal [177]					Lignite [178]	
	IGT HYGAS STEAM-OXYGEN	IGT HYGAS STEAM-IRON	CONOCO CO_2 ACCEPTOR	BCR BI-GAS	PERC/SYNTHANE	LURGI	IGT HYGAS STEAM-OXYGEN	IGT HYGAS STEAM-IRON	BCR BI-GAS	PERC/SYNTHANE	LURGI	CONOCO CO_2 ACCEPTOR	LURGI
Capital Costs ($ million)													
Total Plant Investment	870.00	1280.00	890.00	1030.00	1330.00	1060.00	930.00	1260.00	880.00	1010.00	1170.00	930.00	1030.00
Initial Charge of Catalysts and Chemicals	8.67	7.30	6.33	10.37	8.49	20.82	9.58	7.88	11.65	10.02	26.79	6.60	16.68
Allowance for Funds Used During Construction	137.02	201.60	140.18	162.22	209.47	166.95	156.94	212.63	148.50	170.44	197.44	156.94	173.81
Paid-Up Royalties	1.59	7.50	4.15	1.75	1.56	0.97	5.24	13.67	5.51	6.28	4.54	5.21	2.74
Startup Costs	24.72	34.84	30.99	28.35	33.25	33.56	37.51	49.97	35.35	41.22	45.11	24.44	22.51
Working Capital	20.46	28.99	22.97	23.85	29.90	26.21	27.58	37.18	26.06	30.38	33.73	21.86	22.44
Total Capital Requirement	1062.46	1560.23	1094.62	1259.96	1612.67	1308.51	1166.85	1581.33	1107.07	1268.34	1477.61	1145.05	1268.18
Operating Costs ($ million/yr)													
Raw Materials													
Coal	50.43	56.24	53.93	55.86	68.65	73.31	130.28	176.11	122.83	148.98	150.27	45.63	50.11
Other	—	0.21	17.00	—	—	—	—	0.13	—	—	—	16.56	—
Catalysts and Chemicals	3.68	5.00	4.52	4.63	3.20	9.05	3.81	1.89	5.35	4.33	11.41	4.50	6.65
Purchased Water	0.36	0.65	0.43	0.41	0.38	0.46	1.56	2.22	1.40	1.48	2.09	1.02	1.39
Labor													
Process Operating Labor	2.78	2.78	2.78	2.78	2.78	3.34	2.95	2.95	2.78	2.78	3.34	2.78	3.34
Supervision	3.61	6.07	4.33	4.18	4.57	4.45	0.59	0.59	0.56	0.56	0.67	0.56	0.67
Administration and General Overhead	12.99	21.86	15.60	15.04	16.46	16.03	2.12	2.12	2.00	2.00	2.40	2.00	2.41
Operating Supplies	0.83	0.83	0.83	0.83	0.83	1.00	0.89	0.89	0.83	0.83	1.00	0.83	1.00
Total Maintenance	25.44	45.99	31.48	30.19	33.47	31.55	30.39	42.39	26.94	29.31	35.86	34.38	31.55
Local Taxes and Insurance	23.49	34.56	24.03	27.81	35.91	28.62	13.95	18.90	13.20	15.15	17.55	13.95	15.45
Ash Disposal	—	—	—	—	—	—	1.01	1.64	0.84	0.66	0.98	—	—
Total Gross Oper.Costs/Yr	123.61	174.19	154.93	141.73	166.25	167.81	187.55	249.83	176.73	206.08	225.57	122.21	112.57
Total By-Product Credits	28.92	24.86	8.93	5.04	22.88	54.25	25.10	151.34	10.78	35.21	20.52	5.41	35.45
Total Net Oper. Costs/Yr	94.69	149.33	146.00	136.69	143.37	113.56	162.45	98.49	165.95	170.87	205.05	116.80	77.12

UNIT COSTS			
Coal [$/ton ($/mm Btu)]	7.50 (0.425)	25.00 (1.007)	5.00 (0.369)
Iron Ore ($/ton)	20.00	20.00	—
Dolomite ($/ton)	21.00	15.00	21.00
Raw Water ($/1000 gal)	0.40	0.40	0.40
Operating Labor ($/hr)	6.70	6.70	6.70
Ash Disposal ($/ton)	0.00	1.00	0.00
Maintenance (%/year)			
Coal feed preparation, coal gasification, gas quench and solids removal, shift acid gas, sulfur, methanation	6	6	6
Product gas compression and drying, oxygen plant, liquid and solid effluent treating, boiler flue gas treating, water treating	3	3	3
All other off-sites	1	1	1
Local Taxes and Insurance (% of total plant investment)	2.7	1.5	1.5
BY-PRODUCT CREDITS			
Sulfur ($/LT)	25.00	2[illegible].00	25.00
Ammonia ($/ST)	140.00	17[illegible].00	140.00
Naphtha (120°- 320°F) ($/gal)	0.30	[illegible].38	0.30
Light Oil (300°- 700°F) ($/gal)	0.38	[illegible].40	0.38
Tar ($/gal)	0.15	[illegible].20	0.15
Phenols ($/gal)	0.15	[illegible].20	0.15
Char and Coal Fines ($/$10^6$ Btu)	0.319	[illegible].755	—
Electric Power (mills/kWhr)	4.5	25 and 10	4.0

*January 1976 basis with no forward escalation.

TABLE 62 Cost of SNG (Dollars per 10^6 Btu) (January 1976 Basis with No Forward Escalation)

	Western coal [176]			Eastern coal [177]	Lignite [178]
	Average gas cost	First-year gas cost	Private financing, constant gas cost	Utility financing, average gas cost	Utility financing, average gas cost
IGT HYGAS, steam-oxygen	2.71	3.55	3.91	3.69	
IGT HYGAS, steam-iron	4.10[a]	5.35[a]	5.87[a]	3.40[b]	
Conoco CO_2 acceptor	3.38	4.25	4.61		3.10
BCR BI-GAS	3.50	4.50	4.93	3.64	
PERC synthane	4.11[a]	5.39[a]	5.93[a]	3.94	
Lurgi	3.30	4.34	4.78	4.66	2.79

[a]By-product electricity at 4.5 mills.

[b]By-product electricity at 25 mills (gas cost is $4.34 at 10 mills).

TABLE 63 Capital Investment Summary (Millions of Dollars) for Nominal 240×10^9 Btu/Day High- and Low-Btu Gas from Montana Subbituminous Coal [179] (Mid-1976 Costs)

Section	High-Btu gas HYGAS	Low-Btu gas U-GAS
Coal storage—reclaiming	5.0	4.3
Coal grinding and drying	13.7	11.8
Coal-water feed system	11.1	3.3
Slurry feed preheat (fired heater)	4.5	—
Gasifiers	43.0	18.4
Process waste-heat recovery	14.2	15.7
Gasifier effluent dust-removal system	3.8	4.0
Carbon monoxide conversion	11.2	—
Prepurification[a]	46.6	19.6
Benzene recovery	5.6	—
SYN gas expander	—	13.7
Methanation, drying, and product gas compression	15.2	—
Wastewater treatment	13.1	6.5
Sulfur recovery—Stretford	16.0	16.8
Process and turbine steam generation	69.1	28.4
Turbogenerator	7.7	7.6
Particulate-emission control	3.8	3.2

(continued)

(Table 63 continued)

Section	High-Btu gas HYGAS	Low-Btu gas U-GAS
Cooling and plant makeup water	4.9	3.1
High-pressure oxygen supply	45.0	89.6
Miscellaneous	17.2	12.9
Electric power distribution	7.7	9.5
General facilities	43.3	32.5
Installed plant cost, excluding contingencies	404.1	303.2
Contingencies at 15%	60.6	45.5
Total bare cost	464.7	348.7
Contractor's overhead and profits, 15%	69.7	52.3
Total plant investment (I)	534.4	401.0
Interest during construction (9% × 1.875 years × I)	90.2	67.7
Startup cost (5% of total plant investment	26.7	20.1
Working capital: 60 days' coal at full rate	10.4	8.9
0.9% of total plant investment	4.8	3.7
1/24 × annual revenue required	7.3	6.1
Total capital required	673.8	507.5

[a]Includes hot K_2CO_3, or Selexol, activated carbon, zinc oxide.

TABLE 64 Annual Operating Costs (Thousands of Dollars) for Nominal 240×10^9 Btu/Day High- and Low-Btu Gas Plants Using Montana Subbituminous Coal [179] (Mid-1976 Costs)

	High-Btu gas HYGAS	Low-Btu gas U-GAS
Operating cost component		
Coal feed at 50 cents/10^6 Btu	56,641	48,653
Catalysts, chemicals, and other direct materials	3,195	847
Raw water cost at 45 cents/1000 gal	909	670
Labor		
Process operating labor (for high-Btu gas, 58 workers/shift for HYGAS and 60 workers/shift for U-GAS; 33 workers/shift for U-GAS to low-Btu gas; at $7.20/hr and 8760 worker-hr/year)	3,659	2,081
Maintenance labor (1.5% of total plant investment plus lock hopper maintenance labor for U-GAS)	8,016	6.115
Supervision (15% of operating and maintenance labor)	1,751	1,229
Administration and general overhead (60% of total labor, including supervision	8,056	5,655
Supplies		
Operating (30% of process operating labor)	1,098	624
Maintenance (1.5% of total plant investment plus lock hopper maintenance supplies for U-GAS)	8,016	6,115
Local taxes and insurance (2.7% of total plant investment)	14,429	10,827
Total gross operating cost	105,770	82,816
By-product credits		
Sulfur at $10/long ton	(215)	(223)
Ammonia at $50/ton	(1,138)	—
Light oil (BTX) at 35 cents/gal	(9,674)	—
Total	(11,027)	(223)
Net operating cost	94,743	82,593
Depreciation (20 years plant life, straight-line)	32,565	24,440
Return on rate base	36,556	27,626
Federal income tax	12,052	9,107
20-year average annual revenue required[a]	175,916	143,766
Annual gas production (10^9 Btu)	79,333	78,446
20-year average gas price ($/$10^6$ Btu)[a]	2.22	1.83

[a]Calculated by the Utility Financing Method.

Table 62 lists the unit costs (dollars per million Btu) for gas produced. For Western coal, costs have been developed on both a utility and a private investor financing basis, whereas for the others, costs are based on a utility financing basis alone. Utility financing is on a basis of 75:25 debt/equity ratio, with 9% interest on debt and 15% return on equity. Private investor financing is based on 100% equity and 12% DCF rate of return. The service factor for gas production in all cases is assumed to be 90%.

C. Relative Costs of Low- and High-Btu Gas

A comparison has been made between production of high-Btu gas by a system (HYGAS) specifically developed for SNG production and a much simpler system (U-GAS) which was developed particularly for low- or medium-Btu gas. Although the U-GAS system utilized produces medium-Btu gas (320 Btu/scf) using oxygen, it has been shown elsewhere [180] that there is practically no difference in the cost of the gas produced if the same system is utilized to produce low-Btu gas (155 Btu/scf) using air.

The high-Btu gas (961 Btu/scf) is delivered at 1000 psig, whereas the medium-Btu gas (320 Btu/scf) is delivered at 8 psig. In each case 240 billion Btu/day of product gas is produced from Montana subbituminous coal.

Table 63 summarizes estimated costs for individual plant process sections for the two cases and the extensions to total plant investment and total capital required. The lower-Btu-system cost is only about 75% of the higher because of a simpler gasifier design and no requirements for CO conversion, benzene recovery, methanation, or CO_2 removal (Selexol scrubbing for sulfur removal minimizes CO_2 removal).

Table 64 lists the estimated annual operating costs and average gas price employing utility financing with a 75:25 debt/equity ratio, 9% interest on debt, and 15% return on equity. The production service factor is 90%. Low-Btu gas at $1.83 per million Btu is about 82% of the cost of high-Btu gas at $2.22 per million Btu. Corresponding figures for private financing at 100% equity and 12% DCF return are $2.34 and $2.89 per million Btu, respectively.

REFERENCES

1. D. E. Woodmansee and P. M. Palmer, Gasification of Highly Caking Coal in the GEGAS Pressurized Gas Producer, ACS Symposium on Gasification Kinetics, New Orleans, La., Mar. 24-25, 1977.
2. P. F. H. Rudolph, The Lurgi Process—the Route to SNG from Coal, 4th Pipeline Gas Symposium, Chicago, Oct. 1972.
3. Modern Ammonia Plants Based on Coal, Lurgi Gesellschaft für wärme-und Chemotechnik MBH, Frankfurt (Main), Sept. 1978.
4. Lurgi Pressure Gasification Performance Record, Lurgi Express Information, Lurgi Minervalöl technik GmbH, Frankfurt (Main), 1971.
5. J. C. Hoogendoorn, Gas from Coal with Lurgi Gasification at SASOL, Clean Fuels from Coal (CFFC), Chicago, 1973.
6. K. H. Krieb, KCV Process (High-Pressure Coal Gasification) for Power Generation, STEAG Lünen Plant, 13th World Gas Conference, London, 1976.

7. J. C. Hoogendoorn, Gas from Coal for Synthesis of Hydrocarbons, Status of SASOL II, 9th Pipeline Gas Symposium, Chicago, 1977.
8. AGA Gas Supply Committee, Status of Commercial and Demonstration High-Btu Coal Gasification Projects (as of August 1979), *Gas Energy Rev.* 7(10):11 (July-Aug. 1979).
9. P. E. Jortberg, The Gasification Combined-Cycle Test Facility at Powerton, 5th Annual International Conference, University of Pittsburgh, Pittsburgh, Pa., Aug. 1-3, 1978.
10. R. V. Rahfuse, A. J. Liberatore, and G. R. Friggens, Gasification of Caking-Type Bituminous Coal at 75 to 150 psig in a Stirred-Bed Gas Producer, MERC/TPR-75/3, July 1975.
11. Riley-Stoker Corp., The Riley-Morgan Coal Gasification Systems, Brochure R-201-10-75-75-5M, Worcester, Mass., 1975.
12. A. H. Rawdon, R. A. Lisauskas, and S. A. Johnson, Operation of a Commercial-Size Riley-Morgan Coal Gasifier, Proceedings of the American Power Conference, Chicago, 1976.
13. McDowell Wellman Engineering Co., Wellman-Galusha Gas Producers, Bulletin Form No. 576A, Cleveland, Ohio.
14. K. C. Vyas, McDowell-Wellman Engineering Co., private communication, Cleveland, Ohio, May 1, 1979.
15. Wilputte Corp., Wilputte Low-Btu Fuel Gas Process, Bulletin No. 7862, Murray Hill, New Jersey, May 1, 1978.
16. Gilbert/Commonwealth, Fixed-Bed Coal Gasification for Production of Industrial Fuel Gas, for DOE Contract No. EX-76-C-01-2220, Oct. 1977.
17. G. R. Cooper, Operating Overview of a Producer Gas Plant (12 Machines) at Kingsport, TN, 5th Annual International Conference on Coal Gasification, Liquefaction, and Conversion to Electricity, University of Pittsburgh, Pittsburgh, Pa., Aug. 2, 1978.
18. Applied Technology Corp., Brochure No. 3M 177 CA 4104, Houston, Tex., 1977.
19. Foster Wheeler Energy Corp., New Coal Gasifier To Be Tested, reprinted from *Coal Age*, (Dec. 1977).
20. R. P. Lewis, Installation of a FW-Stoic Gasifier for the Generation of a Clean Boiler Fuel, 85th National Meeting of the AIChE, University of Minnesota, Minneapolis, Minn., June 4-8, 1978.
21. C. Lohmann, Practical Experience in Pressure Gasification in Dorsten, Proceedings of Symposium on Coal Refining, Edmonton, Alberta, Canada; Alberta Research Council Information Series 85, Apr. 20-21, 1978.
22. A. J. Grant, Applications of the Woodall-Duckham Two-Stage Coal Gasification, 3rd Annual International Conference on Coal Conversion, Pittsburgh, Pa., 1976.
23. Woodall-Duckham (U.S.A.) Ltd., WD/GI Coal Gasification, Industrial Fuel Gas From Coal, Pittsburgh, Pa. (undated).
24. J. D. Sudbury, G. P. Curran, and J. A. Phinney, Gasification of Caking Coals in the BGC/Lurgi Slagging Gasifier—A Status Report, 5th Annual International Conference on Coal Gasification, Liquefaction, and Conversion to Electricity, Pittsburgh, Pa., Aug. 2, 1978.
25. D. Hebden, High-Pressure Gasification Under Slagging Conditions, 7th Update of Synthane Pilot Plant Status, 10th Synthetic Pipeline Gas

26. R. C. Ellman, B. C. Johnson, and M. M. Fegley, Studies in Slagging Fixed-Bed Gasification at the Grand Forks Energy Technology Center, prepared for the 10th Synthetic Pipeline Gas Symposium, Chicago, Oct. 30-31, 1978.
27. R. A. Ashworth, Davy Powergas, Inc., private communication, Houston, Mar. 1, 1979.
28. J. D. Sudbury, The CO_2 Acceptor Process, 9th Synthetic Pipeline Gas Symposium, Chicago, Oct. 1977.
29. C. Fink, G. Curran, and J. Sudbury, CO_2 Acceptor Process Pilot Plant—1973, 5th Synthetic Pipeline Gas Symposium, Chicago, Oct. 1973.
30. J. Huebler, F. C. Schora, and B. S. Lee, The HYGAS coal-gasification process, in *Energy Technology Handbook*, McGraw-Hill, New York, 1977.
31. R. Lewis, J. P. Strakey, W. P. Haynes, R. R. Santore, and D. Dubis, Update of Synthane Pilot Plant Status, 20th Synthetic Pipeline Gas Symposium, Chicago, Oct. 30-Nov. 1, 1978.
32. S. J. Gasior, A. J. Forney, W. P. Haynes, and R. F. Kenny, Fluidized-Bed Gasification of Various Coals, *Chem. Eng. Prog.* 71:89-92 (Apr. 1974).
33. R. Lewis, Update of Synthane Pilot Plant Status, Proceedings of the 9th Synthetic Pipeline Gas Symposium, Chicago, Oct. 31-Nov. 2, 1977.
34. D. H. Archer, L. A. Salvador, and E. J. Vidt, The Westinghouse Fluidized-Bed Process, 10th Synthetic Pipeline Gas Symposium, Chicago, Oct. 30-Nov. 1, 1978.
35. J. W. Loeding and I. G. Stanfill, Application of the U-GAS Process for the Production of Gas of Intermediate Calorific Value, 14th World Gas Conference, Toronto, Canada, May 27-June 1, 1979.
36. L. Jequier, L. Longchambon, and G. Van de Putte, The Gasification of Coal Fines, *J. Inst. Fuel* 33:584-591 (1960).
37. R. J. Eby, N. McClintock, and R. Bloom, Jr., The Illinois Coal Gasification Group Project—COGAS Process, 10th Synthetic Pipeline Gas Symposium, Chicago, Oct. 30-Nov. 1, 1978.
38. L. E. Furlong and N. C. Nakas, Catalytic Coal Gasification Process Research and Development, 10th Synthetic Pipeline Gas Symposium, Chicago, Oct. 30-Nov. 1, 1978.
39. K. H. van Heek, H. Juntgen, and W. Peters, Gasification of Coal Using Process Heat from HTR, 8th Synthetic Pipeline Gas Symposium, Chicago, Oct. 18-20, 1976.
40. A. J. Simpkin and K. Berman, Bell High Mass Flux Gasifier, 10th Synthetic Pipeline Gas Symposium, Chicago, Nov. 1, 1978; under DOE Contract No. EX-76-C-01-2204.
41. J. F. Farnsworth and D. M. Mitsak, Conversion of Coal to Gas by the K-T Suspension Process, *Ind. Heat.* 38-42 (Nov. 1974).
42. J. F. Farnsworth, H. F. Leonard, D. M. Mitsak, and R. Wintrell, The Production of Gas from Coal Through a Commercially Proven Process, Coal Gas Fundamentals Symposium, IGT, Chicago, 1973.
43. R. L. Coates and V. C. Rogers, Development of a High-Rate Entrained Flow Coal Gasification Process, prepared for the 10th Synthetic Pipeline Gas Symposium, Chicago, Oct. 30, 1978.
44. M. M. van der Burgt, Clean Syngas from Coal, *Hydrocarbon Process.* 58:161-4 (Jan. 1979).

45. W. G. Schlinger, The Texaco Coal Gasification Process for Manufacture of Medium-Btu Gas, Pasadena Conference Center, Paper No. 2221, Conference on Coal Use for California, May 9-11, 1978.
46. J. D. Odle, W. B. Crouch, and G. N. Richter, Clean Refinery Fuels by the Texaco Coal Gasification Process, API Refining Dept. 43rd Midyear Meeting Conserving Petroleum—New Feedstocks and Fuels for Refineries, Toronto, Canada, May 10, 1978.
47. K. E. Walker, Status of the BI-GAS Pilot Plant Program, 9th Synthetic Pipeline Symposium, Chicago, Oct. 1977.
48. C. G. Houser, Status of the BI-GAS Pilot Plant Operation, 10th Synthetic Pipeline Gas Symposium, Chicago, Oct. 30-Nov. 1, 1978.
49. P. B. Probert, Design and Fabrication of BI-GAS Reactor, 5th Synthetic Pipeline Gas Symposium, Chicago, 1973.
50. R. J. Grace and R. L. Zahradnik, BI-GAS Program Enters Pilot Plant Stage, 4th Synthetic Pipeline Gas Symposium, Chicago, Oct. 1972.
51. R. C. Patterson, Construction and Initial Operation of the C-E Coal Gasification Process Development Unit, ASME-IEEE-ASCE Joint Power Generation Conference, Dallas, Tex., Sept. 10-13, 1978. Also presented at the 10th Synthetic Pipeline Gas Symposium, Chicago, Oct. 30-Nov. 1, 1978.
52. R. A. McCallister and G. C. Ashley, Coal Gasification to Produce Low-Btu Fuel for Combined-Cycle Power Generation, American Power Conference, Chicago, Apr. 29, 1974.
53. R. J. Broeker and R. A. McCallister, Coal Gasification/Combined-Cycle Power, in *Heat Engineering* (Foster Wheeler), Jan.-Mar., pp. 12-15, Livingston, N.J., 1975.
54. D. V. Punwani, J. L. Arora, and C. L. Tsaros, SNG from Peat by the PEATGAS Process, 5th Annual International Conference on Coal Gasification, Liquefaction, and Conversion to Electricity, Pittsburgh, Pa., Aug. 1-3, 1978.
55. D. V. Punwani, Status of the PEATGAS Process, 10th Synthetic Pipeline Gas Symposium, Chicago, Oct. 30-Nov. 1, 1978.
56. J. Friedman, L. P. Combs, J. Silberman, and M. I. Greene, The Rockwell Advanced SNG Gasifier, prepared for the 10th Synthetic Pipeline Gas Symposium, Chicago, Oct. 30-Nov. 1, 1978. DOE Contract No. EX-77-C-01-2518.
57. A. L. Kohl, M. H. Slater, and K. J. Miller, Status of the Molten Salt Coal Gasification Process, prepared for the 10th Synthetic Pipeline Gas Symposium, Chicago, Oct. 30, 1978.
58. P. LaRosa, Fuel Gas from Molten Iron Coal Gasification, Coal Gasification Fundamentals Symposium, IGT, Chicago, Aug. 1978.
59. A Novel Way to Make SNG, *Hydrocarbon Process.* 52:17 (Jan. 1973).
60. M. A. Elliott, Coal Gasification for Production and Synthesis of Pipeline Gas, Annual Meeting of the American Institute of Mining, Metallurgical, and Petroleum Engineers, Inc., St. Louis, Mo., Feb. 26-Mar. 2, 1961.
61. M. MacCormac and J. Wrobel, The Gas Council, The Gasification of Coal in an Experimental Rummel Double-Shaft Slag-Bath Gasifier, *IGEJ*, pp. 385-399 (May 1965).
62. G. Fuessmann and M. Rossbach, High-Pressure Coal Gasification with the Saarberg/Otto Gasifier, 5th Annual International Conference on

Coal Gasification, Liquefaction, and Conversion to Electricity, University of Pittsburgh, Pittsburgh, Pa., Aug. 1978.

63. R. W. Woods, The KILnGAS Coal Gasification Systems, 5th Annual International Conference on Coal Gasification, Liquefaction, and Conversion to Electricity, Pittsburgh, Pa., Aug. 1-3, 1978.
64. P. R. Wieber, Prospects for Producing Synthetic Natural Gas Through Underground Coal Gasification, 9th Synthetic Pipeline Gas Symposium, Chicago, Sept. 1977.
65. Booz, Allen Applied Research Division, Underground Coal Gasification Program, for ERDA Contract No. EX-76-C-01-2343, Mar. 1977.
66. Laramie Energy Research Center, The Laramie Energy Research Center's Underground Coal Gasification Program, ERDA, Laramie, Wyo., May-June 1977; and R. D. Gunn, personal communication, Laramie Energy Technology Center, July 2, 1979.
67. D. W. Gregg and T. F. Edgar, Underground Coal Gasification, *AIChE J.* 24:753 (Sept. 1978).
68. R. P. Stephens et al., High-Btu Gas via in Situ Coal Gasification, 10th Synthetic Pipeline Gas Symposium, Chicago, 1978.
69. Gulf Joins Underground Coal Gasification Tests in West, *Coal Age*, p. 21 (Dec. 1977).
70. D. K. Fleming and H. S. Primack, Purification Processes for Coal Gasification, 81st National Meeting of the AIChE, Kansas City, Mo., Apr. 11-14, 1976.
71. MERC Hot Gas Cleanup Task Force, Chemistry of Hot Gas Cleanup in Coal Gasification and Combustion, MERC/SP-78/2, U.S. Department of Energy (DOE) Morgantown Energy Research Center, Morgantown, W. Va., Feb. 1978.
72. H. L. Feldkirchner and F. C. Schora, Coal Desulfurization Aspects of the HYGAS Process, 2nd International Conference on Fluidized-Bed Combustion, Hueston Woods State Park, Ohio, Oct. 4-7, 1970.
73. HYGAS: 1964 to 1972—Pipeline Gas from Coal—Hydrogenation (IGT Hydrogasification Process), Vol. 3, Part 6, ERDA Report No. FE-381-TO-P3, Washington, D. C., July 1975, p. 6b-8.
74. A. J. Forney, W. P. Haynes, S. J. Gasior, G. E. Johnson, and J. P. Strakey, Sr., Analyses of Tars, Chars, Gases, and Water Found in Effluents from the Synthane Process. Pittsburgh Energy Research Center, U.S. ERDA, Nov. 1975.
75. Institute of Gas Technology, Parametric Study of Coal Conversion Effluents, Monthly Progress Report No. 1, prepared for EPA, Mar. 1978.
76. K. J. Caplan, Source control by centrifugal force and gravity, in *Air Pollution*, Vol. 4, A. C. Stern, ed. Academic Press, Inc., New York, 1977, chap. 3.
77. J. A. Danielson, ed., *Air Pollution Engineering Manual*, Envir. Prot. Agency, 1967.
78. R. H. Perry and C. H. Chilton, eds., *Chemical Engineers' Handbook*, 5th ed. McGraw-Hill Book Company, New York, 1973.
79. N. G. Edmisten and F. L. Bunyard, Systematic Procedure for Determining the Cost of Controlling Particulate Emissions from Industrial Sources, *J. Air Pollut. Control Assoc.* 20:446 (July 1970).

80. A. C. Stern, K. J. Caplan, and P. D. Bush, *Cyclone Dust Collectors*. American Petroleum Institute, New York, 1955.
81. R. Parker and S. Calvert, High-Temperature and Pressure Effects on Particle Collection Mechanisms, EPA/DOE Symposium on High-Temperature High-Pressure Particulate Control, Washington, D.C., Sept. 1977.
82. H. J. White, *Industrial Electrostatic Precipitation*. Addison-Wesley Publishing Co., Inc., Reading, Mass., 1963.
83. S. Oglesby et al., *A Manual of Electrostatic Precipitator Technology: Part 1: Fundamentals; Part 2: Applications*. Marcel Dekker, New York, 1970.
84. N. W. Friseh and T. P. Dorchak, Impact of Fuel on Precipitator Performance, *Pollut. Eng.*, pp. 63-5 (May 1978).
85. H. J. Hall, Design and Application of High Voltage Power Supplies in Electrostatic Precipitation, *J. Air Pollut. Control Assoc.* 25:132-8 (Feb. 1975).
86. E. Bakke, Wet Electrostatic Precipitators for Control of Submicron Particles, *J. Air Pollut. Control Assoc.* 25:163-7 (Feb. 1975).
87. J. R. Bush, P. L. Feldman, and M. Robinson, Development of a High-Temperature/High-Pressure Electrostatic Precipitator, EPA Report No. 600/7-77-132, Nov. 1977. (Report prepared by Cottrell Environmental Sciences, Boundbrook, N.J.)
88. N. G. Edmisten and F. L. Bunyard, Systematic Procedure for Determining the Cost of Controlling Particulate Emissions from Industrial Sources, *J. Air Pollut. Control Assoc.* 20:446 (July 1970).
89. R. Parker and S. Calvert, High-Temperature and High-Pressure Particulate Control Requirements, EPA-600/7-77-071, July 1977.
90. H. J. White, Electrostatic Precipitation of Flyash: Flyash and Furnace Gas Characteristics, *J. Air Pollut. Control Assoc.* 27:114 (Feb. 1977).
91. Dravo Corporation, *Handbook of Gasifiers and Gas Treatment Systems*, prepared for the U.S. ERDA, contract no. E(49-18)1772, Feb. 1976.
92. R. H. Vander Molen and G. L. Wade, Operational Experiences with the CPU-400 Pilot Plant, *AIChE Symposium Series* No. 162, 1977, pp. 120-132.
93. Southern Research Institute, Environmental Control Implications of Generating Electric Power from Coal—1977 Technology Report, Argonne National Laboratory ANL/ECT-3, Appendix E: A Review of Technology for Control of Flyash Emissions from Coal in Electric Power Generation, Dec. 1977.
94. R. C. Hoke, FBC Particulate Control Practice and Future Needs: Exxon Miniplant, Symposium on Particulate Control in Energy Processes, San Francisco, May 1976.
95. S. C. Saxena and W. M. Swift, Dust Removal from Hot and Compressed Gas Streams by Fibrous and Granular Bed Filters: A State-of-the-Art Review, Argonne National Lab Report No. ANL/CEN/FE-77-9, May 1978.
96. S. C. Yung, R. D. Parker, R. G. Patterson, S. Calvert, and D. C. Drehmel, Granular Bed Filters for Particulate Removal at High Temperature and Pressure, 5th International Conference on Fluidized Bed Combustion, Dec. 1977, Washington, D.C.
97. K-T Coal Gasification Process, *Pipeline Gas J.* 47:47 (Mar. 1975).

98. C. L. Tsaros, J. L. Arora, and K. B. Burnham, The Manufacture of Hydrogen from Coal, First World Hydrogen Energy Meeting, Miami Beach, Fla., Mar. 1-3, 1976.
99. D. K. Fleming, Acid-Gas Removal Systems in Coal Gasification, Symposium on Ammonia from Coal, Muscle Shoals, Ala., May 8-10, 1979.
100. W. Strauss, *Industrial Gas Cleaning*, 2nd ed., Pergamon Press, Elmsford, N.Y., 1975.
101. K. T. Semrau, Practical Process Design of Particulate Scrubbers, *Chem. Eng.* 84:87 (Sept. 26, 1977).
102. The McIlvane Co., *McIlvane Scrubber Manual.* Northbrook, Ill., 1974.
103. APT, Inc., *Wet Scrubber Systems Study, Vol. I: Scrubber Handbook.* Riverside, Ca., July 1972.
104. S. Calvert and R. Parker, Collection Mechanisms at High Temperature and Pressure, Symposium on Particulate Control in Energy Processes, San Francisco, May 11-13, 1976.
105. R. E. Sundberg, The Prediction of Overall Collection Efficiency of Air Pollution Control Devices from Fractional Efficiency Curves, *J. Air Pollut. Control Assoc.* 24:48 (Aug. 1974).
106. J. A. Danielson, ed., *Air Pollution Engineering Manual,* U.S. EPA, Research Triangle Park, N.C., 1973.
107. Industrial Gas Cleaning Institute, Air Pollution Control Cost for Nine Selected Industries, Alexandria, Va., 1972.
108. F. C. Reisenfeld and A. C. Kohl, *Gas Purification,* 2nd ed. Gulf Publishing Co., Houston, Tex., 1974.
109. Booz, Allen and Hamilton, Comparative Assessment of Coal Gasification Emission Control Systems, EPA Report No. 9075-030, Oct. 1975.
110. R. F. Robards, U. Zitzow, and G. A. Hollinden, Evaluation of H_2S Removal Processes for Desulfurization of Coal Gas, 4th Energy Resource Conference, University of Kentucky, Lexington, Ky., January 7-8, 1976.
111. J. C. Dingman and T. F. Moore, Compare DGA and MEA Sweetening Methods, *Hydrocarbon Process.* 47:138-140 (July 1968).
112. B. G. Goar, Today's Gas-Treating Processes, *Oil Gas J.* 69:75-79 (July 12, 1971); Part 2, 69:84-86 (July 19, 1971).
113. G. Hochgesand, Rectisol and Purisol, Efficient Acid-Gas Removal for High-Pressure Hydrogen and Syngas Production, *Ind. Eng. Chem.* 62:37 (July 1970).
114. H. Knapp, Low-Temperature Absorption—The Rectisol Process, Paper C-1, Proceedings of the Gas Conditioning Conference, University of Oklahoma, Norman, Okla., Apr. 1968.
115. A. L. Kohl and P. A. Buckingham, Fluor Solvent CO_2 Removal Process, *Pet. Refiner.* 39:193 (May 1960).
116. R. N. Maddox, *Gas and Liquid Sweetening.* M. Campbell, Publisher, Norman, Okla., 1974.
117. NG/LNG/SNG Handbook, *Hydrocarbon Process.* 52:88 (Apr. 1973).
118. R. L. Pearce, J. L. Arnold, and C. K. Hall, Studies Show Carbonyl Sulfide Problem, *Hydrocarbon Process. Pet. Refiner.* 40:121-126 (Aug. 1961).
119. J. L. Sweeney and G. E. Brooks, Selexol at Fort Stockton, Paper H, Proceedings of the Gas Conditioning Conference, University of Oklahoma, Norman, Okla., Apr. 1970.

120. J. P. Valentine, Gas Purification with a Physical Solvent, Paper F, Proceedings of the Gas Conditioning Conference, University of Oklahoma, Norman, Okla., Mar. 1973.
121. H. D. Frazier and A. L. Kohl, Selective Absorption of Hydrogen Sulfide from Gas Streams, *Ind. Eng. Chem.* 42:2288-2292 (Nov. 1950).
122. F. E. Miller and A. L. Kohl, Selective Absorption of Hydrogen Sulfide, *Oil Gas J.* 51:178 (Apr. 27, 1953).
123. F. C. Reisenfeld and J. F. Mullowney, Giammarco-Vetrocoke Processes, *Pet. Refiner.* 38:161-168 (May 1949).
124. R. B. Woertz, A Selective Solvent for Purifying Natural Gas, *J. Pet. Technol.* 23:483-490 (Apr. 1971).
125. K. Bratzler and A. Doerges, Amisol Process Purifies Gases, *Hydrocarbon Process.* 53: 78-80 (Apr. 1974).
126. J. S. Tosh, J. H. Field, H. E. Benson, and R. B. Anderson, Equilibrium Pressures of Hydrogen Sulfide and Carbon Dioxide over Solutions of Potassium Carbonate, Report of Investigations 5622, Bureau of Mines, U.S. Department of the Interior, 1960.
127. J. H. Field, H. E. Benson, G. E. Johnson, J. S. Tosh, and A. J. Forney, Pilot Plant Studies of the Hot-Carbonate Process for Removing Carbon Dioxide and Hydrogen Sulfide, Bulletin No. 597, Bureau of Mines, U.S. Department of the Interior, 1960.
128. Gas Processing Handbook, *Hydrocarbon Process.* 54:84-111 (Apr. 1975).
129. Assessment of Low- and Intermediate-Btu Gasification of Coal, prepared by Committee on Processing and Utilization of Fossil Fuels, Commission on Sociotechnical Systems, and National Research Council for National Academy of Sciences, Washington, D.C., 1977.
130. Institute of Gas Technology, HYGAS 1964-1972 Pipeline Gas from Coal Hydrogenation, Final Report, Vol. 4, prepared for ERDA, Washington, D.C., July 1975.
131. V. P. Sirohi (C. F. Braun Co.), Methanation Processes, Interim Report prepared for ERDA and AGA, Mar. 1975.
132. G. A. White, T. R. Roszkowski, and D. W. Stambridge, The RM Process, ACS Division of Fuel Chemistry, Atlantic City, N.J., Sept. 1974.
133. M. S. Grabowski and E. K. Diehl, Design and Operation of the BCR Fluidized-Bed Methanation PEDU, 5th Synthetic Pipeline Gas Symposium, Chicago, Oct. 29-31, 1973.
134. R. C. Streeter, D. A. Anderson, and J. T. Cobb, Status of the BI-GAS Program. Part II, Evaluation of Fluidized-Bed Methanation Catalysts, 8th Synthetic Pipeline Gas Symposium, Chicago, Oct. 18-20, 1976.
135. W. Lommerzheim and C. Flockenhaus, One-Stage Combined Shift Conversion and Partial Methanation Process for Upgrading Synthesis Gas to Pipeline Quality, 9th Synthetic Pipeline Gas Symposium, Chicago, Oct. 30-Nov. 1, 1978.
136. M. E. Frank and L. Mednick, Liquid Phase Methanation Pilot Plant Results, 9th Synthetic Pipeline Gas Symposium, Chicago, Oct. 31-Nov. 2, 1977.
137. F. C. Schora and D. K. Fleming, Effluent Considerations in Coal Gasification, Environmental Aspects of Fuel Conversion Technology II Symposium, Hollywood, Fla., Dec. 15-18, 1975.

138. Flue Gas Desulfurization in Power Plants, Status Report, U.S. Environment Protection Agency, Division of Stationary Enforcement, Office of Enforcement, Washington, D.C., Apr. 19, 1977.

139. J. M. Kovacik, Which Energy Systems for Coal Gasification Plants? paper prepared for the General Electric Gas Turbine State-of-the-Art Engineering Seminar, Saratoga Springs, N.Y., Sept. 16-18, 1974.

140. A. Meisen and H. A. Bennett, Consider All Claus Reactions, *Hydrocarbon Process.* 53:171 (Nov. 1974).

141. R. K. Kerr and H. G. Paskall, The Claus Process, a series of six articles published in *Energy Process. /Canada*, between mid-July 1976 and 1977.

142. B. G. Goar, Tighter Control of Claus Plants Needed by TGCU System, *Oil Gas J.* 75:134 (Aug. 22, 1977).

143. B. G. Goar, Sulfur Recovery from Natural Gas Involves Big Investment, *Oil Gas J.* 73:78 (July 14, 1975).

144. H. Fischer, Burner/Firebox Design Improves Sulfur Recovery, *Hydrocarbon Process.* 53:125 (Oct. 1974).

145. A. R. Valdes, New Way to Design Firetube Reactors, *Hydrocarbon Process.* 44:223 (May 1965).

146. A. R. Valdes, New Look at Sulfur Plants, pp. 104-108 and pp. 122-124, *Hydrocarbon Process.* 43 (Mar.-Apr. 1964).

147. G. M. Franklin, L. V. Kunkel, and M. Webb, A Cheap Way to Remove S from H_2S, *Oil Gas J.* 55:144 (Nov. 4, 1957).

148. H. S. Bryant, Environment Needs Guide Refinery Sulfur Recovery, *Oil Gas J.* 71:70 (March 27, 1973).

149. W. L. Nelson, What Is Cost of Sulfur Recovery Plants? *Oil Gas J.* 66:111 (May 27, 1968).

150. J. B. Hyne, Methods for Desulfurization of Effluent Gas Streams, *Oil Gas J.* 70:64 (Aug. 28, 1972).

151. O. A. Homberg and A. H. Singleton, Performance and Claus Plant Operation on Coke Oven Acid Gases, *J. Air Pollut. Control Assoc.* 25:375 (Apr. 1975).

152. A. J. Moyes and S. Vasan, Holmes-Stretford Process Proved in Use, *Oil Gas J.* 72:56 (Sept. 2, 1974).

153. M. M. Moran and Y. M. Philardean, Sulfreen Process Experience at Ram River, *Energy Process. /Canada*, pp. 30-40 (May-June 1977).

154. L. C. Cameron, Aquitaine Improves Sulfreen Process for More H_2S Recovery, *Oil Gas J.* 72:110 (June 24, 1974).

155. B. L. Crynes, ed., *Chemical Reactions as a Means of Separation: Sulfur Removal.* Marcel Dekker, Inc., New York, 1977.

156. J. E. Nobles, J. W. Palm, and D. K. Knudtson, Plant Performance Proves Process, *Hydrocarbon Process.* 56:143 (July 1977).

157. C. S. Goddin, E. B. Hunt, and J. W. Palm, CBA Process Ups Claus Recovery, *Hydrocarbon Process.* 53:122 (Oct. 1974).

158. S. Frankowiak, P. Bonnifay, R. Dutriau, and A. Deschamps, Partial and Total Sulfur Recovery in Oil Refineries, *Process Technol. Int.*, pp. 785-787 (Oct. 1972).

159. D. M. Ottmers and J. C. Dickerman, Status of Second-Generation Regenerative Flue Gas Desulfurization Processes, 12th Air Pollution and Industrial Hygiene Conference, Austin, Tex., Jan. 1976.

160. The Ralph M. Parsons Co., Sulfur Recovery Plants and Tail-Gas Purification Units by Parsons, Brochure 210 (RMP 12/76), Los Angeles, 1976.
161. J. H. Crow and J. C. Baumann, Versatile Process Uses Selective Absorption, *Hydrocarbon Process.* 53:131 (Oct. 1974).
162. R. I. Pedroso, An Update of the Wellman-Lord Flue Gas Desulfurization Process, 6th FGD Symposium, New Orleans, Mar. 1976.
163. A. L. Kohl and D. C. Riesenfeld, Gas Purification, 2nd ed., Gulf Publishing Company, Houston, Tex., 1974.
164. F. A. Todd, Sulfuric Acid Versus Elemental Sulfur as By-products, C. F. Braun Co. for U.S. Dept. of Energy, Report No. FE-2240-54, Alhambra, Ca., Jan. 1978.
165. J. I. Bucy and J. M. Ransom, Potential Markets for Sulfur Dioxide Abatement Products, 7th FGD Symposium, Hollywood, Fla., Nov. 8-11, 1977.
166. R. N. Shreve and J. A. Brink, *Chemical Process Industries*, 4th ed. McGraw-Hill Book Company, New York, 1977.
167. R. W. Riedel, J. J. Knight, and R. E. Warner, Alternatives in Sulfuric Acid Plant Design, *Chem. Eng. Prog.* 73:55 (Mar. 1977).
168. J. B. Rinckhoff and L. J. Friedman, Design Options for Sulfuric Acid Plants, 69th Annual Meeting of the AIChE, Nov. 30, 1976.
169. D. Fyfe, R. Vanderlund, and J. Rodda, Corrosion in Sulfuric Acid Storage Tanks, *Chem. Eng. Prog.* 73:65 (Mar. 1977).
170. Bischoff of America, Gas Desulfurization and Recovery of Sulfur by the Stretford Process, Brochure, Bischoff Environmental Systems, Pittsburgh, Pa. (1977).
171. S. Vasan, Holmes-Stretford Process Offers Economic H_2S Removal, *Oil Gas J.* 76:78 (Jan. 2, 1978).
172. C. J. Walker, Design Sour Water Strippers Quickly, *Hydrocarbon Process.* 48:121-124 (June 1969).
173. W. T. Atkins and J. W. Bonham, Process Comparison of Effluent Treatments, Ammonia Separation, FE-2240-19 for ERDA/AGA, June 1975.
174. API Refining Division, *API Manual on Disposal of Refinery Wastes, Liquid Wastes*. New York, 1969.
175. Synthetic Gas-Coal Task Force for the Supply-Technical Advisory Committee, The Supply-Technical Advisory Task Force—Synthetic Gas-Coal, Final Report, *National Gas Survey*, FPC, Washington, D.C., Apr. 1973.
176. C. F. Braun Co., Factored Estimates for Western Coal Commercial Concepts, Interim Report, prepared for ERDA/AGA under Contract No. E(49-18)-2240, Oct. 1976.
177. C. F. Braun Co., Factored Estimates for Eastern Coal Commercial Concepts, Interim Report, prepared for DOE and GRI under Contract No. EX-76-C-01-2240, Sept. 1978.
178. C. F. Braun Co., Factored Estimates for Lignite Commercial Concepts, Interim Report, prepared for DOE and GRI under Contract No. EX-76-C-01-2240, Sept. 1978.
179. J. L. Arora, K. B. Burnham, and C. L. Tsaros, High- and Low-Btu Gas from Montana Subbituminous Coal, ACS, Chicago, Sept. 1977.
180. C. L. Tsaros, J. L. Arora, K. B. Burnham, T. Subramaniam, and J. W. Loeding, The IGT U-GAS (Clean Utility Gas) Process, Institute of Gas Technology In-House Report, Chicago, Jan. 1974.

11 Liquid Fuels from Coal

JAMES B. O'HARA / The Ralph M. Parsons Company, Pasadena, California

I. INTRODUCTION

The production of liquids from coal has been practiced industrially by the steel industry in their coke ovens for 150 years. These coke oven liquids provided primary raw materials for the coal tar and chemical industries. This simple illustration highlights the fact that production of liquids from coal has been known and practiced for many years.

Key milestones in the twentieth century in the science of coal liquefaction include Bergius' experiments in direct hydrogenation of coal in 1913 and the production of liquids by reaction of hydrogen and carbon monoxide as reported by Franz Fischer and Hans Tropsch in 1923.

A number of modifications and improvements of the science as originally demonstrated by these pioneers were developed and reduced to practice. Visible results of the progress were the 12 direct hydrogenation plants and 9 Fischer-Tropsch-type plants constructed and operated by Germany in the World War II era. The Bergius plants had a capacity of approximately 4 million tons of liquids per year and produced up to 70,000 bbl/day. The Fischer-Tropsch capacity was approximately 15,000 bbl/day. The Berguis output was used primarily to fuel the German Air Force planes, and Fischer-Tropsch liquids were used, among other applications, as raw materials for the detergent, edible fats, and soap industries. Although the capacities of these plants was in some sense small by today's industrial standards, they clearly demonstrated the ability to continuously produce useful liquids from coal on an industrial scale in an environment in which they did not need to be economically competitive with world crude oil prices.

Another major milestone in coal liquefaction technology is the commercial Fischer-Tropsch plant that began operations at Sasolburg, South Africa, in 1955. This plant, operated by South African Coal, Oil and Gas Corporation Limited (Sasol), produces liquids, chemicals, waxes, and fuel gas from coal using local low-grade, low-cost coal. The plant has remained in production since that date and Sasol is now significantly expanding its capacity.

This chapter summarizes the characteristics and status of coal liquefaction. It discusses the chemistry of coal liquefaction, describes how coal has been liquified—both in the past and currently, briefly describes current major development programs under way that are designed to develop and demonstrate technologies with improved efficiencies and product selectivity, and presents projections of possible future plant design configurations and economics.

II. CHEMISTRY

Chemically, coal is a very complex substance. Although the predominant element in its organic structure is carbon, it also contains hydrogen, sulfur, nitrogen, and oxygen. As mined, all coal contains some water with solid

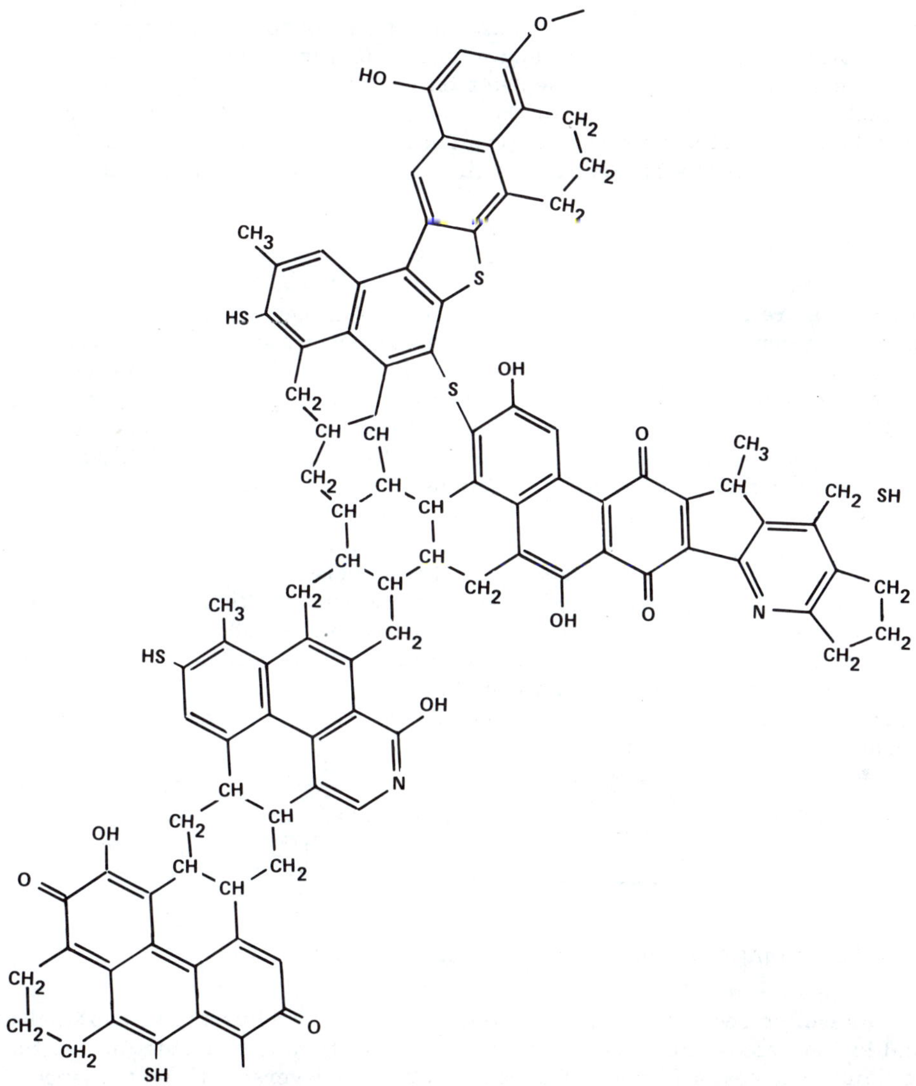

FIG. 1 Molecular structure of coal.

fossil fuels having the lower percentages of carbon (lower ranked) containing significant amounts of water. In addition, varying amounts of inorganic compounds, referred to as ash, are contained in the coal.

Figure 1 illustrates a typical portion of a coal molecule. The molecule is dominated by ring structures but also has aliphatic, or nonring, carbon atoms attached to the ring structures. A portion of the ring structures are homogeneous, containing only carbon, and other portions are heterogeneous, incorporating nitrogen, sulfur, and oxygen atoms into the molecules.

The structure of any coal depends on the conditions under which it was formed and the concentrations of the heteroatoms in the area at the time of formation. The structure and composition affects the kinetics of coal liquefaction as well as the product composition obtained from the liquefaction reaction.

Chemical reactions occurring during liquefaction depend on reaction conditions and the reactants present. To illustrate, connecting bonds between the aromatic rings will be broken, producing smaller molecules; these bond breakages are accelerated by the presence of hydrogen or a hydrogen-containing donor solvent. The heteroatoms nitrogen, sulfur, and oxygen, located on the extremities of the ring nuclei, react to form hydrogen sulfide (H_2S), ammonia (NH_3), and water. The molecular fragments are hydrogenated and a spectrum of low-molecular-weight products, including methane, ethane, liquified petroleum gases (LPGs), and light distillates are formed in varying degrees, depending on reaction conditions.

A spectrum of heavier products are also produced in the original liquefaction. With subsequent liquefaction or hydrotreating, the heavier products can react further to produce lighter distillate fuels.

In direct hydrogenation and solvent extraction, the feed coal is slurried in a recycle coal-derived liquid and the feed slurry is subjected to elevated temperatures and pressures. The constituents of the coal-derived slurry liquid can serve as a hydrogen donor under the reaction conditions. This hydrogen donor reaction is illustrated by the reaction of the model compounds tetralin and naphthalene:

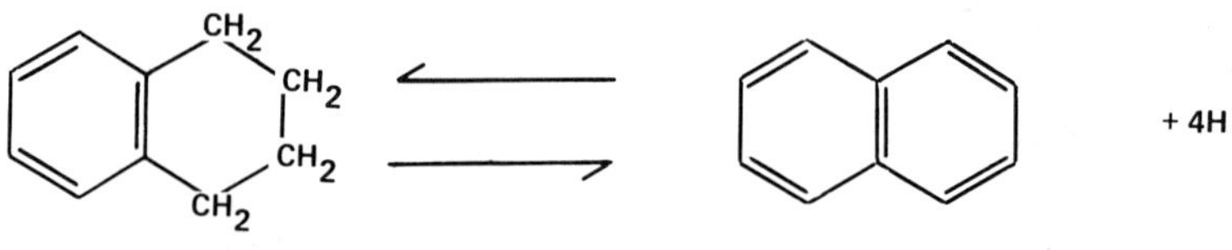

TETRAHYDRONAPTHALENE NAPTHALENE

As can be seen, the parent compound tetralin is converted to naphthalene and in the process liberates hydrogen, which in turn is available for reaction to liquify the coal molecule. The reaction can be reversed in the presence of hydrogen to produce the hydrogenated tetralin form, which can be recycled to the liquefaction step and serve as a hydrogen donor.

Petrographic analysis, using microscopic-scale techniques, provides an insight into the composition and reactivity of the individual coals in hydroliquefaction reactions. Coal with high vitrinite and exinite "maceral" contents show good liquefaction potential, whereas inertite is considered a negative indicator with regard to hydroliquefaction potential. This type of maceral analysis can serve as a preliminary guide to selection of preferred coals as feeds to hydroliquefaction plants.

III. PROCESS CLASSIFICATION

Liquefaction technologies may be classified as shown below.

Classification	*Example*
Prolysis	Coal-Oil-Energy-Development (COED), Lurgi-Ruhrgas
Direct liquefaction	H-Coal, Berguis
Solvent extraction	Solvent Refined Coal (SRC) Exxon Donor Solvent (EDS) Consolidation Synthetic Fuel (CSF)
Indirect liquefaction	Fischer-Tropsch, methanol

A brief description of the processing characteristics for each classification follows.

In pyrolysis processes, coal is heated in the absence of air. This heating breaks down the coal molecules to form such hydrogen-rich compounds as methane and ethane, while the char becomes hydrogen-deficient. In general, the char tends to have approximately the same weight percent sulfur as the feed coal. The tar products are primarily aromatic in nature. Pyrolysis has been practiced industrially.

In direct hydrogenation, a reducing gas such as high-purity hydrogen is used to react with coal at elevated temperature and pressure in the presence of a catalyst. The H-Coal Process, for instance, uses an ebullated bed of hydrogenation catalyst, whereas the Berguis Process uses a disposable iron catalyst added to the feedstream, consisting of coal-in-oil paste. The products are primarily aromatic.

The solvent extraction processes, such as SRC, EDS, and CSF, involve slurrying the feed coal in a recycle coal-derived liquid and reacting at elevated temperature and pressure. The products are primarily aromatic. Hydrogen may or may not be added to the reactor (coal dissolver); hydrogen is added to the SRC and EDS Processes but is not added to the CSF. The EDS and CSF are often referred to as donor solvent processes.

In indirect liquefaction, the feed coal is completely gasified to produce a synthesis gas consisting primarily of a mixture of hydrogen and carbon monoxide. This synthesis gas is purified to remove solid particulates and sulfur compounds. It is then reacted in the presence of a catalyst at elevated temperatures and pressures to produce liquid products. For Fischer-Tropsch, these tend to be alipatic (straight chains) and olefinic (unsaturated) in nature. There are also a number of oxygenates produced, depending on reaction conditions and the composition of the catalysts. The products contain nil sulfur, nitrogen, and particulates and in that sense represent premium-grade fuels from an environmental standpoint.

The pyrolysis, direct liquefaction, and solvent extraction products, being high in aromatic content, are precursors of gasoline. Conversely, the Fischer-

Tropsch indirect route produces primarily aliphatic products, which are precursors of diesel fuels and petrochemical feedstocks in addition to being capable of being reformed to gasoline. In Germany during World War II, Fischer-Tropsch products were used for chemical and detergent manufacture.

IV. HISTORY

The first commercial production of liquids from coal was obtained during carbonizations primarily for the production of coke and gas early in the nineteenth century. Successful research on coal liquefaction by direct hydrogenation and indirect synthesis began in the early part of the twentieth century; each moved into commercial production during the World War II era. A brief summary of these separate developments will illustrate the history that has led to the current state of development in the field.

A. Prolysis-Based Technology

The production of coal tars by pyrolysis-based processing and coal tar distillation was practiced industrially early in the nineteenth century. Crude benzoil refining began in 1839 and pure chemicals were subsequently marketed from this source. In the 1920s a benzene fraction from coal tar sources was used to increase the octane rating for gasoline in the United States, Germany, France, and the United Kingdom.

The Geissen Process was developed in Germany in the post-World War I era and approximately 35 units were installed. Several of these were operated during World War II. In this process, brown coal was carbonized in a continuous vertical retort containing a concentric revolving cylinder. The moisture in the feed brown coal was approximately 15%. It was fed at the top of the retort and pyrolyzed as it descended between the inner revolving corrugated iron cylinder and an outer fixed cylinder containing a number of stationary annular collars. The heat for carbonization was supplied by the combustion of fuel gas in a combustion chamber located within the revolving inner cylinder. The combustion products flowed from the top of the retort down, in contact with the corrugated wall. The volatile pyrolysis products were removed from the coal via openings between the annular collars and were withdrawn from the retort to minimize "cracking" the products to form smaller molecules. The char discharged from the base of the retort. The capacity of a retort was about 150 tons/day of brown coal, 54% moisture basis; the feed coal was dried to 15% moisture level before being fed to the retort [1].

A second type of industrial pyrolyzer is the Lurgi-Ruhrgas (LR) unit (Fig. 2). Four of these units have been built, and the current capacity is 4000 tons/day [2]. It has also been used for oil shale and liquid hydrocarbon processing.

In the LR Process, recycled char serves as the heat carrier in a flash pyrolysis operation. The recycled char is heated and partially burned by contact with preheated combustion air in a lift pipe. The hot char is separated from the combustion products in the collecting bin and is then fed to the mixer, where it contacts the feed coal at essentially atmospheric pressure and a temperature of 800 to 1100°F, depending on the coal characteristics.

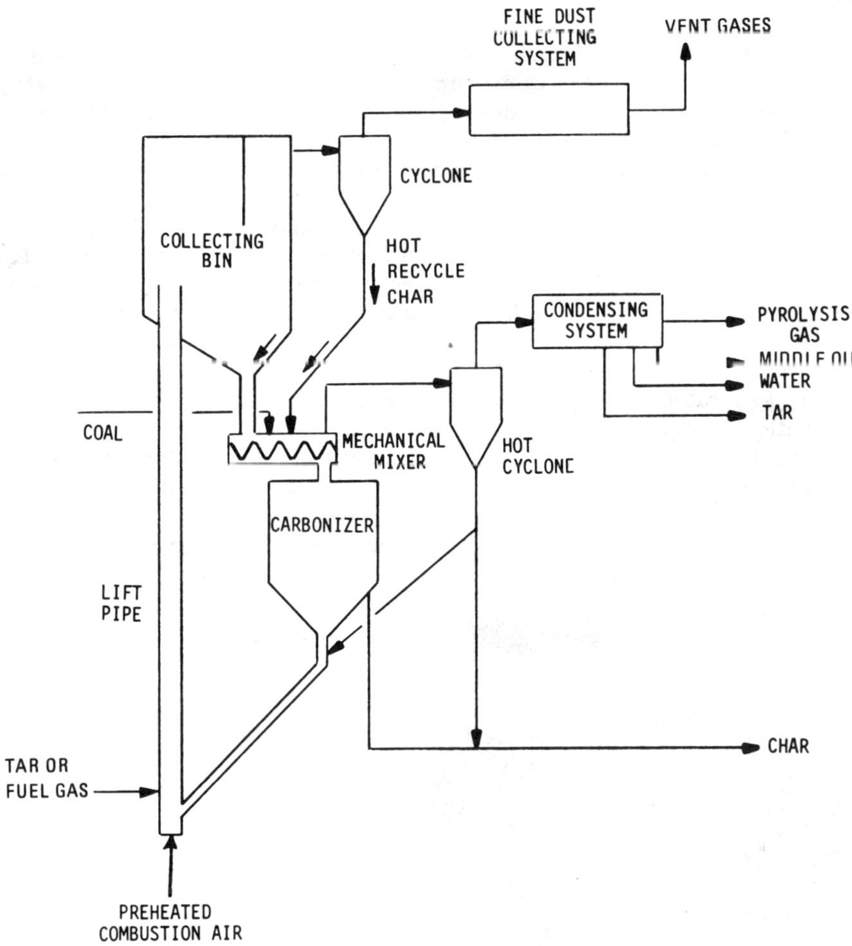

FIG. 2 Simplified flow diagram, Lurgi-Ruhrgas Process. (Courtesy DOE.)

The contact time in the mixer is a matter of seconds. The pyrolysis vapors are removed from the mixer, recovered, and processed to produce a high-Btu gas (700 to 800 Btu/scf), a middle oil, tar, and water. The char flows from the mixer to a carbonizer vessel, which serves as a surge and pyrolysis finishing tank. The net char production is removed from the process.

Some of the impetus for industrial pyrolysis development has come from programs to develop formcoking technology to produce metallurgical-grade coke from noncoking coals, and from programs to develop a viable technology for the production of substitute natural gas (SNG). An example of the former is the FMC Corporation formcoking process; an example of the latter is the COGAS process. Each of these uses fluidized-bed pyrolysis technology.

B. Direct Hydrogenation and Solvent Extraction Processes

Directed work toward liquefaction of coal by reaction with hydrogen began in 1913, when Bergius showed that brown coal could be converted to a heavy oil by hydrogenation. Bergius's description of the technology included a summary of the advantages of using hydrogen donor compounds. His original work did not use a catalyst and in that sense was similar to the current Solvent Refined Coal (SRC) Process.

Further development of the Bergius Process was undertaken by I. G. Farbenindustrie (IG) A.G. in Germany and Imperial Chemical Industries (ICI) in the United Kingdom. The IG and ICI Processes used disposable catalysts for the hydrogenation step. In this process, finely ground coal is mixed with a coal-derived liquid and an iron catalyst. The coal-in-oil paste is reacted with hydrogen at pressures to 10,000 psi. The products from this first reaction were separated into light, middle, and bottom fractions.

Production of light oils and motor fuels from the middle-distillate coal-derived liquids posed problems because of the rapid catalyst deterioration in the presence of metal impurities and sulfur. IG developed a successful procedure consisting of a two-stage process; the first stage was a liquid-phase hydrogenation of the coal, and the second stage a vapor-phase hydro-

FIG. 3 High-pressure stalls, Leuna.

FIG. 4 High-pressure stalls, Gelsenberg.

cracking and hydrofining of the middle-distillate oil produced during the initial solvent stage. They also developed sulfur-resistant catalysts based on tin or tungsten. An industrial-scale plant to produce 100,000 metric tons/year of liquid fuels from brown coal was constructed at Leuna, Germany, in 1927. That same year ICI began work on the hydrogenation of coal and constructed a 10-metric tons/day-capacity pilot plant in 1930. This was followed by an industrial-scale plant built in 1935 at ICI's Billingham plant. Its capacity was 100,000 metric tons/year of liquid fuels from coal; this was later expanded to use coal tar and creasote raw materials. The technology was similar to the IG process.

In the mid-1930s, smaller plants were built in France and Italy.

Development continued at a rapid pace. At the outbreak of World War II, Germany had seven hydrogenation plants, the United Kingdom had one and smaller experimental units existed in France, the United States, Belgium, the Netherlands, Canada, and Japan. Operation of the U.K. plant was terminated in 1939 at the start of the war because of operational difficulties anticipated in the event that the plant had to be shut down quickly because of an air raid. By the end of World War II, Germany had constructed about 12 plants. At their peak activity, these plants produced approximately 70,000 bbl/day, the majority of which was aviation gasoline which supplied 90% of their aircraft fuel. Figure 3 is a photograph of the high-pressure stalls located at Leuna, and Fig. 4 shows the stalls at Gelsenberg.

The Pott-Broche coal extraction process was also developed in Germany. It operated commercially at a coal feed rate of about 5 tons/hr from 1928 to 1944 at the Welheim plant in the Ruhr district [3]. The product was a coal extract. Several of the processes currently under development in the United States are similar to this technology.

Pott-Broche processing began with drying ground coal, 60 to 70% -60 μm, to 1% moisture. The coal was mixed with 2 parts of middle oil from coal-tar pitch liquid-phase hydrogenation. This coal paste was pressurized to 1500 psi, heated to 775 to 815°F, and fed to a tubular reactor. The residence time in the reactor was about 1 hr. After reaction, the product was cooled to 105°F, depressurized, and filtered using filter "candles," which were semiautomatic ceramic cartridge filters. Filtration was at 150°F and 45 to 60 psi, giving a filtration rate of 125 to 145 lb product/hr/ft^2 of filter area. The filter cake was washed and dried in a rotary kiln for removal of the solvent.

On a dry ash-free feed coal basis, the yield of extract was 78 to 84%. It has a softening point of 390 to 465°F, an ash content of about 0.05%, a sulfur content of about 0.6% (compared with 0.9% in the feed coal), and a solubility in benzene of about 35%.

A considerable amount of makeup oil was used to maintain the yields. Hydrogen addition reduced the requirement for fresh oil considerably, but caused serious operating difficulty because of solid formation.

C. Indirect Liquefaction

The Fischer-Tropsch technology was developed in Germany during approximately the same period as was hydrogenation (direct hydrogenation and extraction). An early event was a 1913 patent awarded to Badische Anilin und Soda Fabrik for the production of oxygenated compounds and aliphatic hydrocarbons by catalytic hydrogenation of carbon monoxide. The production of liquids by reaction of hydrogen and carbon monoxide at 100 to 150 atm pressure and temperatures of 750 to 840°F using alkalized iron-turning catalysts was reported by Franz Fischer and Hans Tropsch in 1923 [4]. The development continued in the 1930s when Pichler became Fischer's co-worker and worked on iron catalysts to operate under pressure.

The first industrial plant using Fischer-Tropsch technology was constructed in Germany in 1936 and used a cobalt catalyst. By 1939, nine plants had been built in Germany and they had a rated capacity of approximately 750,000 metric tons per year of liquids, equivalent to about 15,000 bbl/day. Synthesis gas was produced by a number of techniques, including the use of the water gas process with low-temperature coke and the gasification of brown coal. The actual production peaked at approximately 570,000 metric tons during 1943. These were small, low-pressure plants. The primary German interest in Fischer-Tropsch was the production of suitable feedstocks for conversion to edible fats, soaps, and detergents, as contrasted to the hydrogenation routes, which product primarily aromatics and therefore are a source of high-grade aviation fuels.

Fischer-Tropsch plants were built in France, Japan, and Manchuria in the period 1937-1944.

The United States saw one postwar attempt at commercialization of Fischer-Tropsch technology using natural gas raw material. The Carthage Hydrocoal Plant began operations in 1950. The venture was unsuccessful, largely for economic reasons. In its later states of operation, it could have been termed technically successful.

The next commercial Fischer-Tropsch plant began operations at Sasolburg, South Africa in 1955 [5]. The Sasol I plant (Fig. 5) was built by the South African Coal, Oil and Gas Company Ltd., now Sasol Limited. This is the only postwar facility to operate successfully for extended periods of time. Sasol I is still in operation using local high-ash, low-cost coal as a raw material. Two synthesis reactor types have been used: the fixed-bed Arge and the entrained Synthol. Sasol I produces a wide range of products including fuel gas, liquids, waxes, and oxygenate products. The syngas is produced in Lurgi moving-bed gasifiers.

Sasol I output also served as the nucleus of South Africa's chemical industry when it added chemical, petrochemical, and fertilizer manufacturing facilities in the 1960s. Products include ammonia, butadiene, styrene, and ethylene. It supplies industrial gas to just about the whole of southern Traansvaal through a 100-mile pipeline.

A decision was made in 1974 to build a much larger coal-liquids plant, to be called Sasol II, at Secunda. The estimated cost is about $3 billion and the primary objective is the production of liquid fuels. The plant is to occupy

FIG. 5 Sasol I plant, Sasolburg, South Africa. (Courtesy of Sasol.)

FIG. 6 Sasol II Plant, Secunda, South Africa. (Courtesy of Sasol.)

an area of about 3 square miles and is scheduled to begin operations in 1980; a photo of the plant under construction is shown in Fig. 6. To provide perspective, the cooling towers are almost 550 ft high and each of the seven Synthol conversion units is about 225 ft tall.

In early 1979, the decision was made to more than double South Africa's indigenous production of oil from coal by constructing Sasol III at an estimated cost of about \$3.9 billion. Principal products include gasoline; diesel fuel; liquefied petroleum gases (LPGs); kerosene, including jet fuel; and 250,000 tons/year of alcohols and related chemicals. The total coal production to serve Sasol II and III will be more than 27 million tons/year. Figure 7 shows the civil works beginnings for Sasol III with Sasol II in the background.

Methanol offers the promise that by direct use, or as an intermediate, it can contribute to future fuels supplies. The methanol may either be used as an additive to gasoline in the form of a methyl fuel at concentrations up to 15% or, alternatively, it may be converted to high-octane gasoline using technology under development by Mobil Research and Development Corporation.

Methanol has been produced industrially from coal for many years. It was produced in the United States; an example is the E.I. du Pont de Nemours & Company plant located at Belle, West Virginia, which was capable of producing either ammonia or methanol from coal.

D. Postwar Summary

In the period following World War II, there was concern regarding the availability of future supplies of crude oil. The United States began a development program which included construction of a demonstration plant facility at

Louisiana, Missouri, which contained both hydrogenation and Fischer-Tropsch demonstration-scale units. This work was terminated in the early 1950s following the discovery of the huge Middle East oil resources and unfavorable projections for the economics of coal conversion when compared with the use of crude oil.

The level of effort in coal liquefaction activities was very low through the late 1950s. In 1962, the Office of Coal Research (OCR) was formed under the U.S. Department of the Interior to assist in defining means of expanding the use of coal to support the declining coal industry. OCR had limited funds for research during the 1960s, but it did construct a liquefaction pilot plant based on the Consolidation Synthetic Fuel (csf) process at Cresap, West Virginia, and the COED pyrolysis pilot plant located at Princeton, New Jersey, under development by FMC Corporation. Design work continued and construction started in 1971 on a Solvent Refined Coal (SRC) pilot plant, under development by Pittsburgh and Midway Coal Mining Company, a division of Gulf Oil Corporation.

Following the 1973 oil embargo by the OPEC nations, the level of activity and funding for coal liquefaction pilot plants increased. As the program expanded, significant development programs were begun for the H-Coal and Exxon Donor Solvent (EDS) Processes, both of which had been under development using private funds. Other liquefaction programs included the Synthoil Process, developed by the Bureau of Mines at Bruceton, Pennsylvania; Project Lignite, under development by the University of North Dakota; and a flash pyrolysis process, under development by Occidental Research Corporation.

FIG. 7 Sasol III plant civil construction site with Sasol II plant in background, Secunda, South Africa. (Courtesy of Sasol.)

Although the scope of the U.S. program was broadest, expansion of development activities has also taken place in other areas of the world: West Germany, Japan, the United Kingdom, Poland, Canada, and Australia, among others. This has proceeded to the point where there are a number of pilot plants in operation and plans are well under way to construct large demonstration plants for the hydroliquefaction technologies. Some of these developments are described in the following section.

V. CURRENT DEVELOPMENT PROGRAM

There is significant development work under way in a number of countries in the world in the field of coal liquefaction. In general, these development programs reflect the objectives and needs of the individual countries.

The following summary of key elements of representative development programs is intended to provide a perspective of the scope and characteristics of the development efforts. They will be grouped by country of origin with the sequence of presentation arbitrary.

A. United States

Approximate funding for the U.S. government coal liquefaction development since the days of the office of Coal Research is shown in Fig. 8. Results show an increase in funding subsequent to the 1973 oil embargo, increasing to the range of $130 million annual expenditure by the U.S. Department of Energy (DOE) in 1979. The accumulative expenditures for the period 1970-1979 was approximately $450 million.

Currently the United States is concentrating on the development of three primary processes. The three are Solvent Refined Coal (SRC), Exxon Donor Solvent (EDS), and H-Coal. Within the SRC category there are two major process modes, termed SRC I and SRC II. Similarly, for H-Coal, two process alternatives are referred to as the fuel oil and syncrude modes. In the following paragraphs, these three technologies are emphasized and descriptions of additional processes under development are presented, including the M-Gasoline Process, by Mobil Research and Development Corporation, which has attracted significant interest.

1. Solvent Refined Coal (SRC)

Current SRC development work focuses on two variations of a basic processing scheme. This consists of reacting a slurry of feed coal in coal-derived liquids with a hydrogen-rich gas at a pressure in the range 1500 to 2500 psi. Under these reaction conditions the majority of the feed coal is converted. To illustrate, the conversion of the carbon content in feed coal to gas and liquid constituents will be greater than 90% for Illinois and Kentucky bituminous coals. The reaction conditions and percent conversion will vary depending on the characteristics of the feed coal and the products desired. The unconverted solids plus ash are removed by one of several techniques to be described.

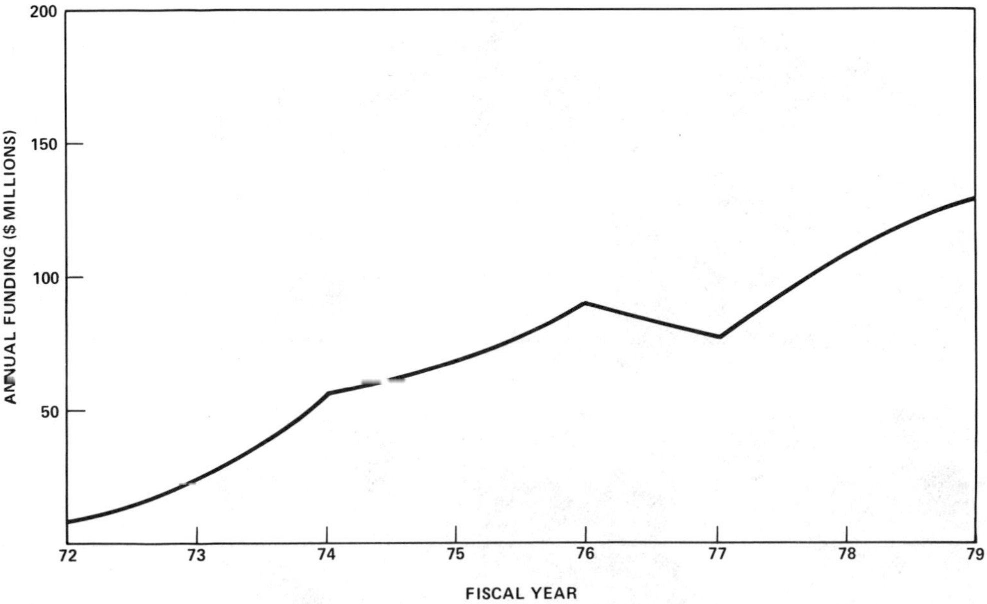

FIG. 8 U.S. government funding for coal liquefaction.

SRC technology has been under development in the United States since the early 1960s by the Gulf Oil organization and Spencer Chemical Corporation, which was acquired by Gulf. Responsibility for the development resides with the Pittsburg and Midway Coal Mining Division (P&M) of Gulf. Southern Services Corporation has also been a major developer. Experimental facilities include P&M laboratory- and bench-scale facilities at Merriem, Kansas, a nominal 50-ton/day pilot plant located at Fort Lewis (Tacoma), Washington, which is funded by DOE and operated by P&M, and a nominal 6-ton/day pilot plant at Wilsonville, Alabama, operated by Catalytic, Inc., under Southern Services management using DOE and Electric Power Research Institute (EPRI) funding.

The Fort Lewis pilot plant is shown in Fig. 9. It occupies approximately 7 acres and employs a staff of approximately 175. The coal feed system is located in the upper right-hand area, the process facilities in the center, product handling and storage in the right center area, and the extensive waste treatment facilities in the lower portion of the photograph.

a. SRC I Figure 10 is a sketch of key elements of the SRC I process [6]. Dried and pulverized feed coal and a filtered process solvent recycle stream (typically a 500 to 850°F boiling-range coal-derived liquid) are mixed to form a slurry in the slurry mixing vessel. The solids content is kept at a maximum concentration of about 40% to ensure that the slurry remains pumpable. The slurry is pumped from the mixing vessel and the recycle plus makeup hydrogen is injected into the stream ahead of the slurry heater. The entire mixture

FIG. 9 SRC pilot plant, Ft. Lewis, Washington. (Courtesy of DOE.)

is heated to about 700°F in the heater and then enters the dissolver at a pressure of about 1500 psi. The reaction is highly exothermic and heats the reactor contents to over 800°F, depending on the degree of hydrogenation. The reactor effluent is then let down in pressure and cooled in several flash separations in the vapor/liquid separator section. The gases separated are sent to an acid-gas-removal section to remove CO_2 and H_2S gases that have been formed by the reaction. The cleaned gas is sent to hydrogen purification where the C_1 to C_3 vapors are separated from the recycle hydrogen stream. The bottoms of the vapor/liquid separation are sent to the filtration section to remove the mineral residue, which contains the ash and the undissolved organic material of the coal. The filtrate is sent to the vacuum fractionation section to vacuum-distill the distillates overhead from the SRC

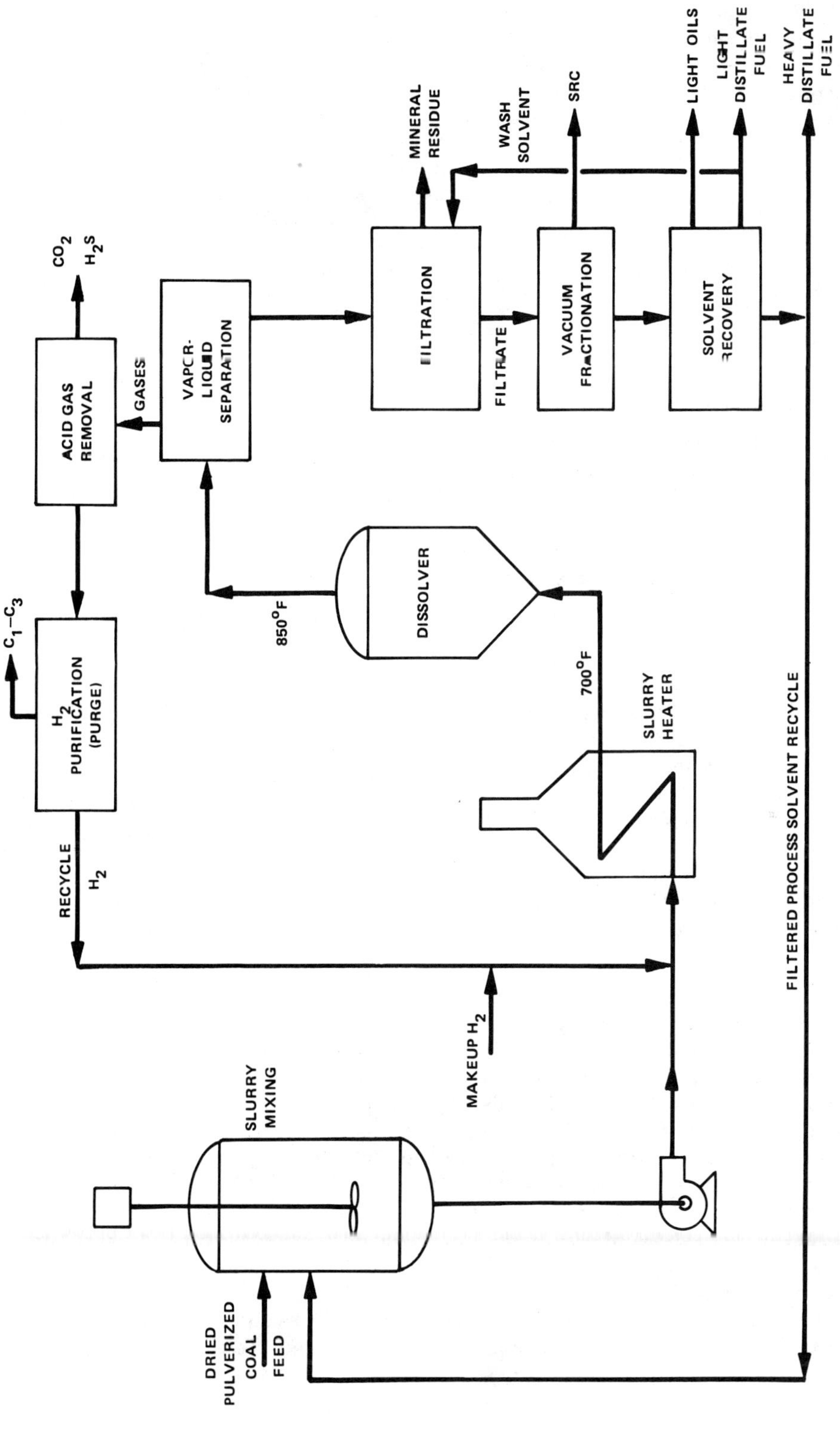

FIG. 10 Block flow diagram, SRC I Process. (Courtesy of DOE.)

TABLE 1 Illustrative SRC I Product Composition—Kentucky Feed Coal

Component	Wt %
C	87.22
H	5.34
N	2.33
S	0.73
O	4.21
Ash	0.17
Fusion point = 380°F	
Boiling range = 850°F+	
Gross heating value = 16,030 Btu/lb	

product. The distillates are further fractionated to a light oil (C_4, 400°F), a light distillate (400 to 500°F) or wash solvent, and a heavy distillate fuel or filtered process solvent recycle (500 to 850°F). Wash solvent is recycled to the filtration section and the filtered process solvent is recycled to the slurry mixing vessel. The SRC is solidified, ground, and stored.

SRC I is a solid at room temperature with a softening point of 350 to 400°F. Its composition is illustrated in Table 1 [7]. The principal intended use of SRC I is for coal-burning utility plants. The low sulfur and low ash represent an advantage/improvement over the feed coal and will improve the environmental performance of the fuel relative to high-sulfur, high-ash coal.

Some 3000 tons of SRC I were produced at the Fort Lewis pilot plant and satisfactorily burned in a 22-MW power plant to demonstrate its handling and combustion properties. It will characteristically have a sulfur content of the order of 0.7 to 1.0 wt % when produced from high-sulfur coal.

b. SRC II SRC II processing is illustrated in Fig. 11. Dried and pulverized coal is fed to the slurry mixing vessel, where it is mixed with a slurry recycle stream and a filtered process solvent recycle. The slurry recycle is a flashed stream from the vapor/liquid separation section containing dissolved and undissolved coal, including the coal ash. The mineral contents of the coal can act as a catalyst in the dissolving, hydrocracking, and hydrogenation reactions; hence, for the same residence times, the degree of reaction, as measured by the hydrogen consumption, is substantially increased in the SRC II, or slurry recycle mode, over the SRC I mode. The entire slurry is pumped from the mixing vessel and the recycle plus makeup hydrogen is injected in the stream ahead of the slurry heater. The mixture is heated to about 700°F in the slurry heater and then enters the dissolver at a pressure of about 2000 psi.

Reaction heat increases the temperature of the reactants to over 800°F. The reactor effluent is then let down in pressure and cooled in several flash

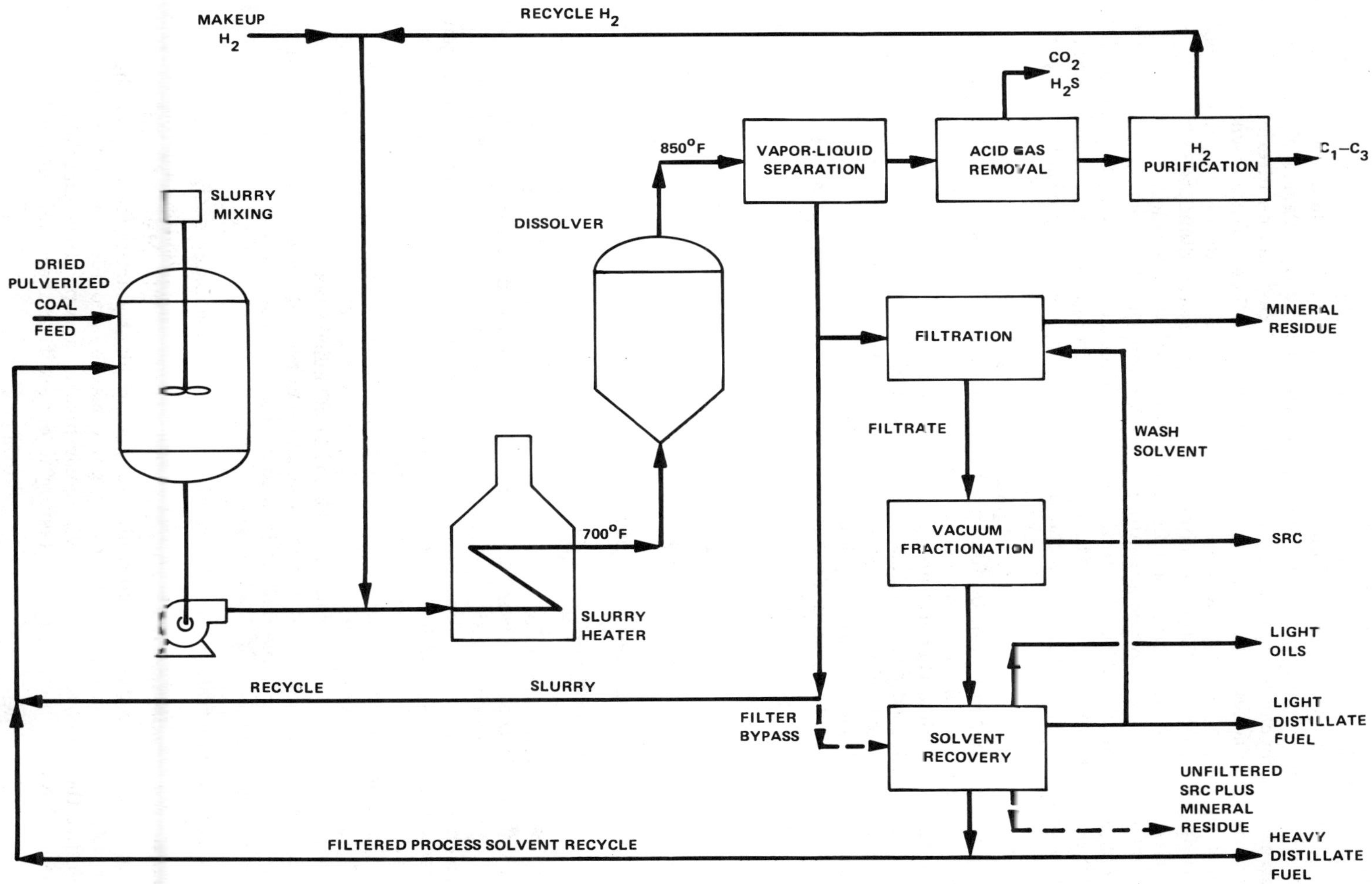

FIG. 11 Block flow diagram, SRC II Process. (Courtesy of DOE.)

separations in the vapor/liquid separation section. The gases separated are sent to an acid-gas removal section for CO_2 and H_2S removal and a H_2 purification section to recover C_1 to C_3 and purify the recycle H_2. Part of the slurry stream is recycled to the slurry mixing vessel; the rest can be filtered to separate the mineral residue from the coal solution. The filtrate is vacuum-distilled to produce distillate fuels overhead and SRC bottoms. The distillates are fractionated into light oil, light distillate, and heavy distillate streams, including the wash solvent and filtered process solvent recycle.

The downstream process configuration for the SRC II process could be changed to selectively eliminate the SRC as a product [7-9]. This process scheme is shown in Fig. 11, with the filter-bypass indicated as a dashed line. The process scheme eliminates the filtration unit operation by completely bypassing the filtration section. The unfiltered SRC plus mineral residue, after distilling off the distillate liquids in first an atmospheric and then a vacuum fractionator, can be fed as a liquid slurry to an oxygen-blown gasifier to produce the makeup hydrogen required for the dissolver section [7]. An alternative method that has been described is to feed the vacuum bottoms together with feed coal to a flash pyrolysis unit to recover a substantial portion of the contained liquids [9]. Still another procedure would be to feed the vacuum bottom to a coking operation.

Typical products of an SRC II plant would consist of gases and liquids as follows:

Gas	SNG
LPG	C_3's and C_4's
Naphtha	C_4, 400°F
Light distillate fuel	400 to 500°F
Heavy distillate fuel	500 to 800°F

Significant amounts of methane, ethane, and LPGs are also produced in the SRC II coal dissolver [10]. Similar results would be expected from other coal hydroliquefaction reactors. This effect is illustrated in Fig. 12, which shows that as the reaction severity is increased, the amount of hydrogen reacted increases, and lighter (lower-boiling) products result. At high severity, 20+% of light hydrocarbons can be produced.

The operability of the SRC II concept was demonstrated in the Fort Lewis pilot plant and some 5000 bbl of fuel oil produced, which were used in burning tests carried out in a commercial Consolidated Edison boiler in New York. The conclusions were that the burning tests were successful and that SRC II products are acceptable alternatives to petroleum-based residual fuels for power-plant use.

c. SRC Demonstration Plants As of mid-1979, preliminary design work is well advanced for both SRC I and SRC II demonstration plants for potential construction and operation under the DOE program. The plant capacities will be 6000 tons/day of feed coal. Target for plant startup is 1984 and the objective is to demonstrate operating performance, provide a basis for predicting the economics of commercial-scale plants, and produce adequate material for functional product testing to ensure that there is a market for

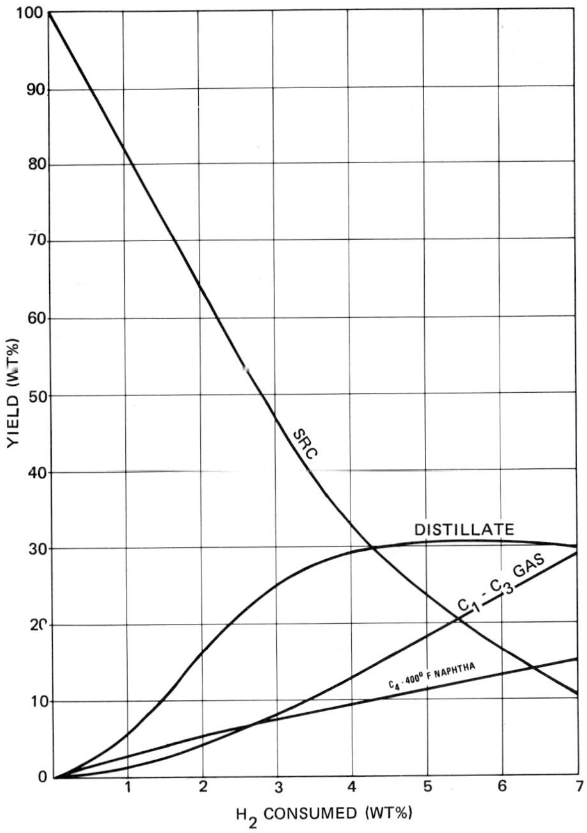

FIG. 12 General SRC process yields from MAF bituminous coal.

the product. West Germany and Japan have each agreed to provide 25% of the funding for the SRC II demonstration plant.

2. *Exxon Donor Solvent (EDS)*

Exxon carried out a privately funded research program during the period 1966-1975 which led to the selection of what is termed the Exxon Donor Solvent (EDS) Process as the preferred route to produce liquid fuels from coal. During this period a 0.5-ton/day pilot plant was operated, and a 1.0-ton/day pilot plant was put into operation in 1975.

Beginning in 1975, a jointly funded development program with an estimated cost of $268 million was initiated. This development program is jointly funded by DOE, the Carter Oil Company (an Exxon affiliate), the Electric Power Research Institute, the Japan Coal Liquefaction Development Company, the Phillips Petroleum Company, and Atlantic Richfield Company. This program is scheduled to continue through mid-1982. It includes an integrated laboratory and engineering research and development program plus construction and operation of a 250-ton/day coal liquefaction pilot plant [11].

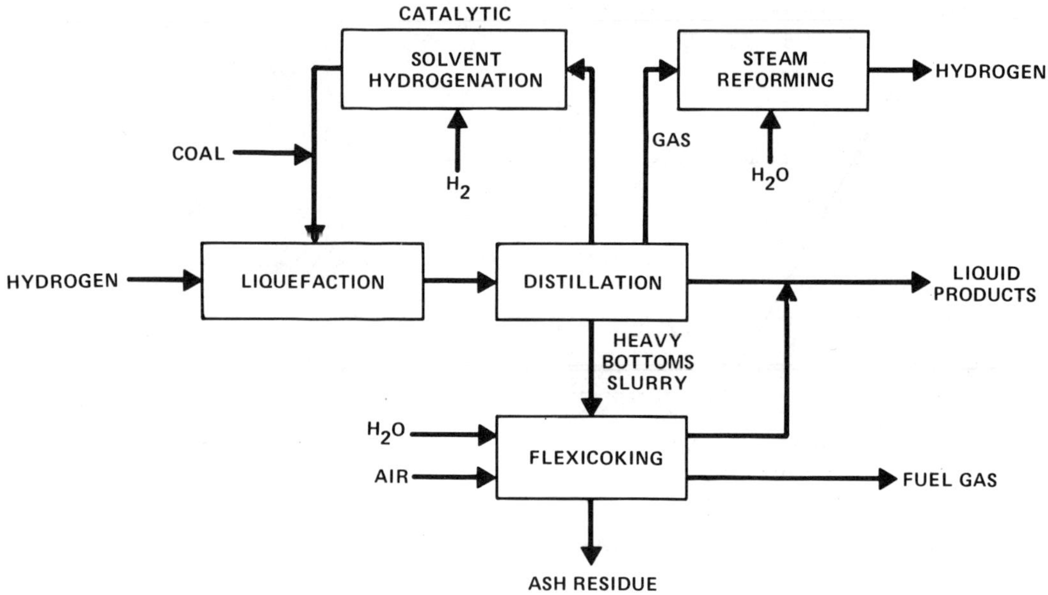

FIG. 13 Block flow diagram, Exxon Donor Solvent Process.

A simplified sketch of the EDS Process is shown in Fig. 13 [11]. Crushed feed coal is slurried in a hydrogen-donor solvent and liquefied in a noncatalytic tubular reactor in the presence of molecular hydrogen at 800 to 880°F and 1500 to 2000 psi. The hydrogen-donor solvent is a 400 to 850°F boiling-range material; the solvent is a catalytically hydrogenated recycle stream fractionated from the middle boiling range of the liquid product.

The product slurry leaving the liquefaction reactor is separated by distillation into gas, naphtha, distillates, and a vacuum bottoms slurry. The vacuum bottoms slurry is processed in a FLEXICOKING unit to produce additional liquids and also provides low-Btu gas for process fuel. Process hydrogen can be produced by steam re-forming of the light gases (C_1 and C_2) produced in the process or by gasification of coal followed by shift reaction to convert carbon monoxide to hydrogen, acid gas removal, and final methanation.

The process is reported to be capable of producing about 2.6 bbl of C_4^+ distillate liquid per ton of dry Illinois bituminous coal. Liquid products include a naphtha that can be upgraded for use as a gasoline blendstock and a low-sulfur fuel oil product. By-products include ammonia and elemental sulfur. Product distribution can be varied over a wide range by varying liquefaction conditions.

A pilot plant with nominal capacity of 250 tons/day of feed coal is scheduled for completion by the end of 1980, located at Baytown, Texas. A photograph of this unit under construction is shown in Fig. 14.

Exxon has completed a conceptual design and an economic evaluation for a commercial plant [12]. They are currently developing a revision to this design with completion scheduled in 1980.

3. *H-Coal*

The H-Coal Process is an application of the ebullated-bed catalyst technology originally used to convert heavy oil petroleum residuals into lighter products (H-Oil). It has been under development by Hydrocarbon Research, Inc. (HRI) since the early 1960s. Current project funding is supplied by DOE and an industry consortium consisting of EPRI, Ashland Oil, Inc., Standard Oil of Indiana, Conoco Coal Development Company, Mobil Oil Company, and the Commonwealth of Kentucky.

Process development work has been carried out on bench-scale units capable of processing about 25 lb of coal per day and in a process development unit (PDU) with a nominal coal-processing capacity of 3 tons/day located at Trenton, New Jersey.

The process is intended to operate in two primary modes. The fuel oil mode is targeted to produce fuel oil containing less than 0.7 wt % sulfur from feed coal containing more than 3% sulfur, and the synthetic crude (syncrude) mode is to produce a lower-sulfur, lighter product [13]. The rated capacity of a plant in the fuel oil mode is more than twice that of the syncrude mode.

The H-Coal Process is depicted schematically in Fig. 15 [14]. Here coal is crushed, dried, slurried in a coal-derived recycle oil, pumped to a reaction

FIG. 14 Exxon Donor Solvent pilot plant under construction, Baytown, Texas. Photo courtesy of Exxon Coal USA.

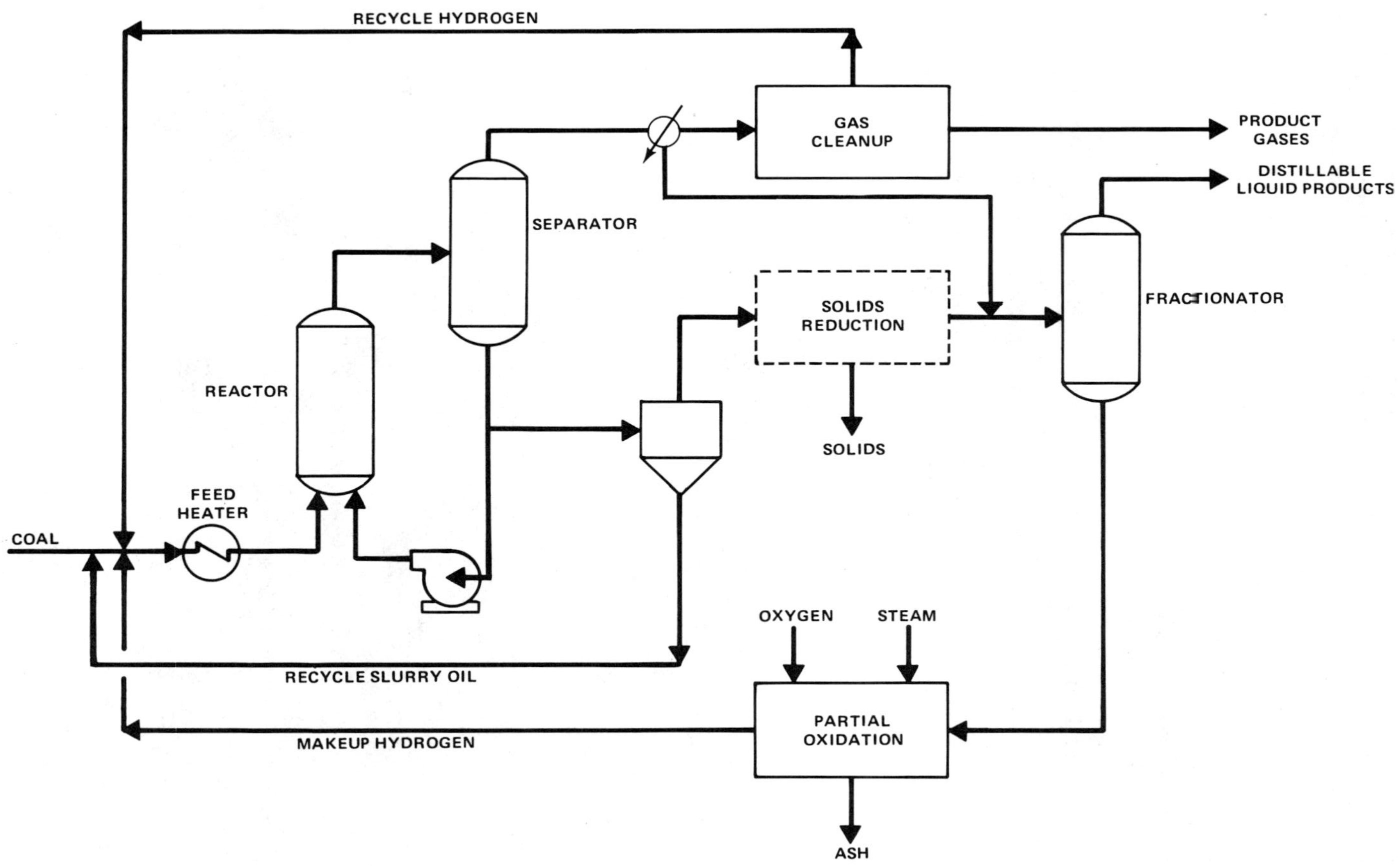

FIG. 15 Block flow diagram, H-Coal Process.

pressure on the order of 3000 psi, and then mixed with hydrogen. The mixture is preheated and fed to the reactor which is operated at 800 to 900°F. The reactor is an ebullated-bed type, containing catalyst pellets in which reactor liquid is recycled to the base of the reactor to ensure an upward flow of vapor and liquid to maintain an expanded-state catalyst bed in random motion. Catalyst can be continuously added and removed from the reactor bed.

Products from the reactor are passed through a pressure letdown system and are separated into vapor and liquid streams. The vapor stream is cooled and purified, light products are recovered, and unreacted hydrogen separated; the hydrogen is recycled. The gases are properly treated to remove contaminants such as H_2S and CO_2.

A portion of the liquid is recycled to the reactor to control catalyst expansion. The remainder of the reactor product liquid stream is depressurized and sent to hydroclones, where it is separated into high-solids-content and a low-solids-content streams, with the latter recycled as a portion of the coal feed slurry oil. Other liquid fractions may be recycled to adjust the composition of the slurry oil.

Flash vapors are condensed and sent to fractionation. For the syncrude mode, net total liquid production is fed to a fractionation system and the distillates are sent to product storage. Vacuum bottoms from the fractionation train, containing 975+°F material, including unconverted coal and ash, enters a partial oxidation section, where it is reacted with oxygen and steam to produce the process hydrogen requirements.

For the lower-severity fuel oil mode of operation, the solids (unconverted coal and ash) must be separated from the liquid. Settling, promoted by addition of an antisolvent, is to be tested on the pilot-plant scale. Filtration would be an alternative.

The H-Coal Process will be tested in a pilot plant designed to process a nominal 250 tons/day of feed coal in the syncrude mode and 600 tons/day in the fuel oil mode and located at Cattletsburg, Kentucky. It is shown under construction in Fig. 16 and is scheduled for completion in 1980. A conceptual design/economic evaluation for a commercial scale H-Coal plant has been completed [15].

4. *Consol Synthetic Fuel Process*

The Consol Synthetic Fuel (CSF) Process was defined during the course of bench-scale research by the Consolidation Coal Company and Associates during the period 1949-1963. A 20 ton/day pilot plant was constructed at Cresap, West Virginia [16] (Fig. 17) and operated to test the CSF process for 40 months, from late 1966 until early 1970, when it was inactivated because of numerous mechanical and operating problems [17].

A block flow diagram showing key elements of the CSF process is shown in Fig. 18. The process was developed for use with coking Eastern coals which are high in pyritic sulfur and relatively low (~1%) in organic sulfur.

Prepared coal is fed to a slurry mix tank and mixed with fresh donor solvent and recycled solvent. The slurry is pumped to the heater and enters the stirred solvent extraction reactor at 750°F and 465 psia. Vapors evolved in the extractor are taken overhead, condensed, and separated into sour gases and condensed liquids. A liquid level is maintained in the reactor,

FIG. 16 H-Coal pilot plant, Cattletsburg, Kentucky. Photo courtesy of Dynalectron Corp.

with the extract slurry being let down in pressure by a level control valve. The extract slurry is then fed to a primary feed tank, which is maintained at about 150 psia and 650°F by recycling slurry that has been heated to 700°F; the material is handled at an elevated temperature (650°F) to avoid precipitation of high-molecular-weight asphaltenes. It is mixed with wash solvent from the extract distillation section, recovered solvent, and tar recycle from the low-temperature carbonization section. The solids are separated in a two-stage hydroclone system. Extract liquids removed from the primary hydroclone are sent to extract distillation, where solvent, synthetic crude, and extract solution are separated. Solids from the solids separation section are fed from the underflow receiver as a slurry to a low-temperature carbonization unit. The extraction residue is pyrolyzed at 800 to 900°F and low pressure to recover the solvent left in the slurry, addition distillate, and extract, and to reject a solid, high-ash char. Air is supplied to partially oxidize the char to supply the heat required for the carbonization.

Extract from the extract distillation section is hydrogenated in a multistage-reactor unit. The stages (two or more) operate at about 3000 psi and 800°F. The reactors are ebullating, catalytic, upflow reactors in which liquid is recirculated to suspend the catalyst in the reactor liquids. The ebullating reactor is a three-phase system: a gas phase, a liquid phase, and a fluidized catalyst plus liquid phase. Co-Mo-Ni on an alumina-base catalyst is used. High-pressure gases are flashed off in the reactor itself and sent for gas separation and purification to recover recycle hydrogen, purge excess gas, and remove acid gases. Liquid overflow from the reactor is sent to a fractionator section to separate naphtha, heavy fuels, and donor solvent, which is recycled to the extraction section to hydrogenate the coal.

After being idle for approximately 4 years, reactivation of the Cresap facility was begun in mid-1974 and completed in 1977. The facility was to serve as a major support center for development, testing, and evaluation of key process equipment items critical to most coal liquefaction processes. It was also to demonstrate integrated plant operation, initially under CSF process conditions.

The pilot plant did achieve totally integrated operation in early 1978 and produced a low-ash fuel oil. It was scheduled to again be inactivated in late 1979.

A conceptual design/economic evaluation for a commercial-scale plant has been completed and published [18].

5. *M-Gasoline*

Mobil Research and Development Corporation has been active in development of zeolite cracking catalysts for petroleum refining since the 1950s. A type of shape-selective zeolite developed has a unique channel structure which permits the conversion of methanol and other oxygenates to a gasoline

FIG. 17 CSF-TPD pilot plant, Cresap, West Virginia. (Courtesy of DOE.)

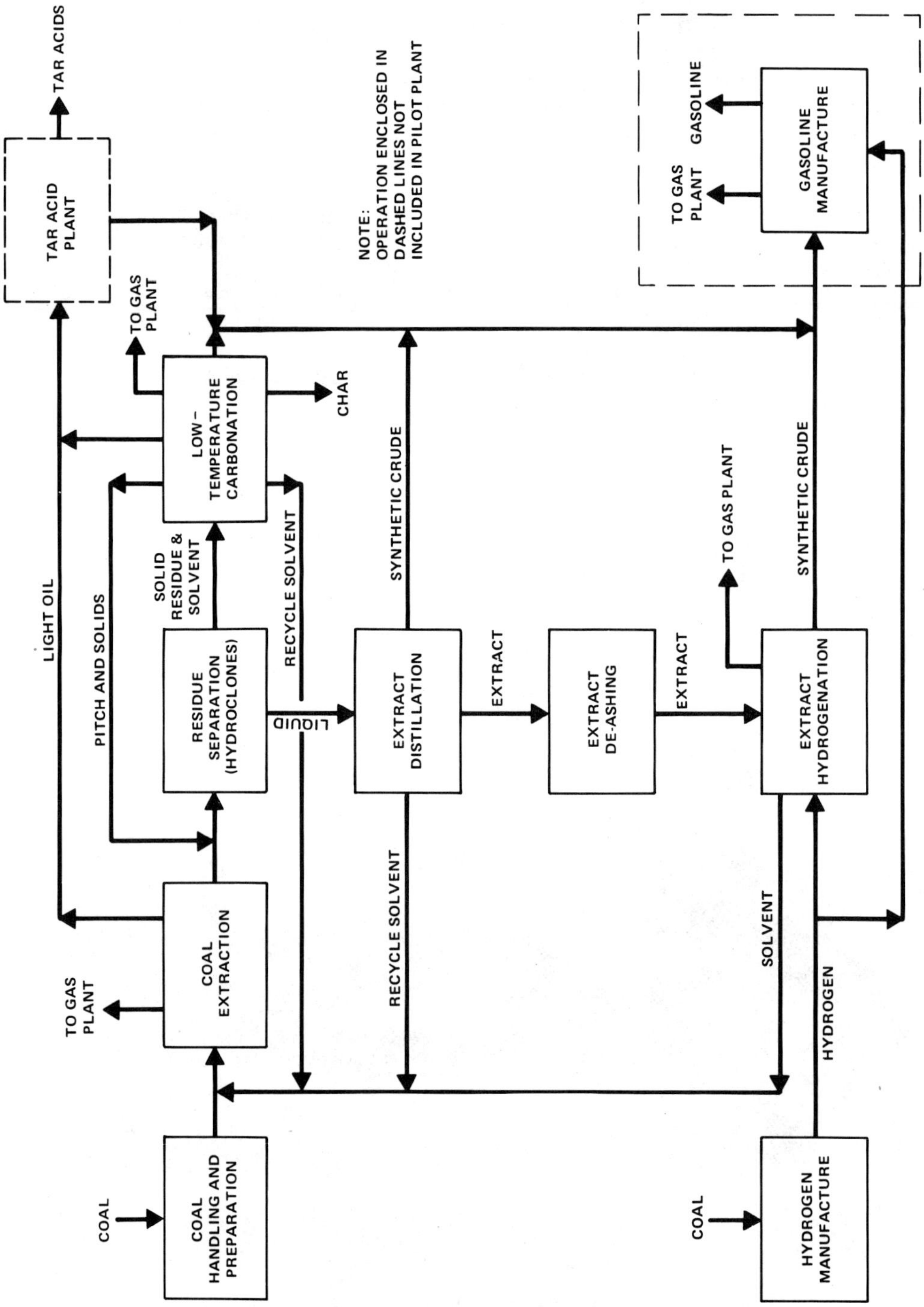

FIG. 18 Block flow diagram, CSF Process.

range of hydrocarbons. This catalytic conversion process was announced and first described by Mobil Research and Development Corporation in 1976 [19,20]. The chemistry has been outlined [21].

Fixed- and fluidized-bed reactor configurations for the conversion of methanol to gasoline have been investigated at Mobil Research and Development Corporation laboratories in Princeton and Paulsboro, New Jersey. The studies have used a microreactor (10 cm^3 catalyst capacity) and bench-scale fixed- and fluidized-bed reactors (150 cm^3 catalyst capacity), and a 4-bbl/day fluid catalyst pilot plant [22]. A fixed-bed methanol-conversion process has been studied and a 100-bbl/day fixed-bed pilot plant has been designed [23].

A key characteristic of the M-Gasoline Process is the ability of the catalysts to selectively produce high-octane gasoline constituents. This selectivity is greater than that obtained from classical Fischer-Tropsch-type catalysts. To illustrate, the gasoline selectivity using Mobil catalysts is stated to be as high as 85%, whereas that from Fischer-Tropsch is of the order of 40 to 50%.

The technology of producing gasoline from coal via methanol production and M-Gasoline conversion is illustrated by the simplified block flow diagram shown in Fig. 19. Here coal is gasified by reaction with oxygen and steam under pressure, the composition of the synthesis gas produced adjusted by a shift reaction (which converts carbon monoxide and steam to hydrogen and carbon dioxide), purified, and then converted to methanol. The crude intermediate methanol, containing about 15% water, is the feed to the M-Gasoline unit, which can be a two-stage fixed-bed conversion unit.

The crude methanol is vaporized and fed to the first converter stage at approximately 570°F and 300 psig, where it is catalytically converted to an equilibrium mixture of dimethyl ether, methanol, and water by the reaction

$$2CH_3OH \rightleftarrows CH_3OCH_3 + H_2O$$

The reaction is exothermic and the effluent leaves the reactor at about 770°F.

The first-stage product is next combined with a recycle stream and fed to the second-stage converter. The inlet temperature is about 650°F and the outlet about 750°F. The second-stage converter, filled with catalyst, produces the gasoline constituents. The raw product is fractionated, the light ends alkylated, the naphtha recovered from the conversion steps is hydrotreated, and the products blended to produce a gasoline with a 92 research octane-clear (RON) rating.

Development work is continuing on the M-Gasoline Process and its economics have been estimated to be attractive relative to Fischer-Tropsch technology [24].

6. *Methanol*

Methanol has a high potential for use as a turbine fuel and as a gasoline additive or extender. The technology for production of methanol is well known and it has been, and is being, produced industrially from coal and natural gas. The production of synthesis gas from coal is the most expensive step in the process and is therefore the process portion that offers the greatest incentive for improvement.

Methanol can be used in automobiles in gasoline blends containing up to 15% methanol, called M-15, without significant engine changes, although the

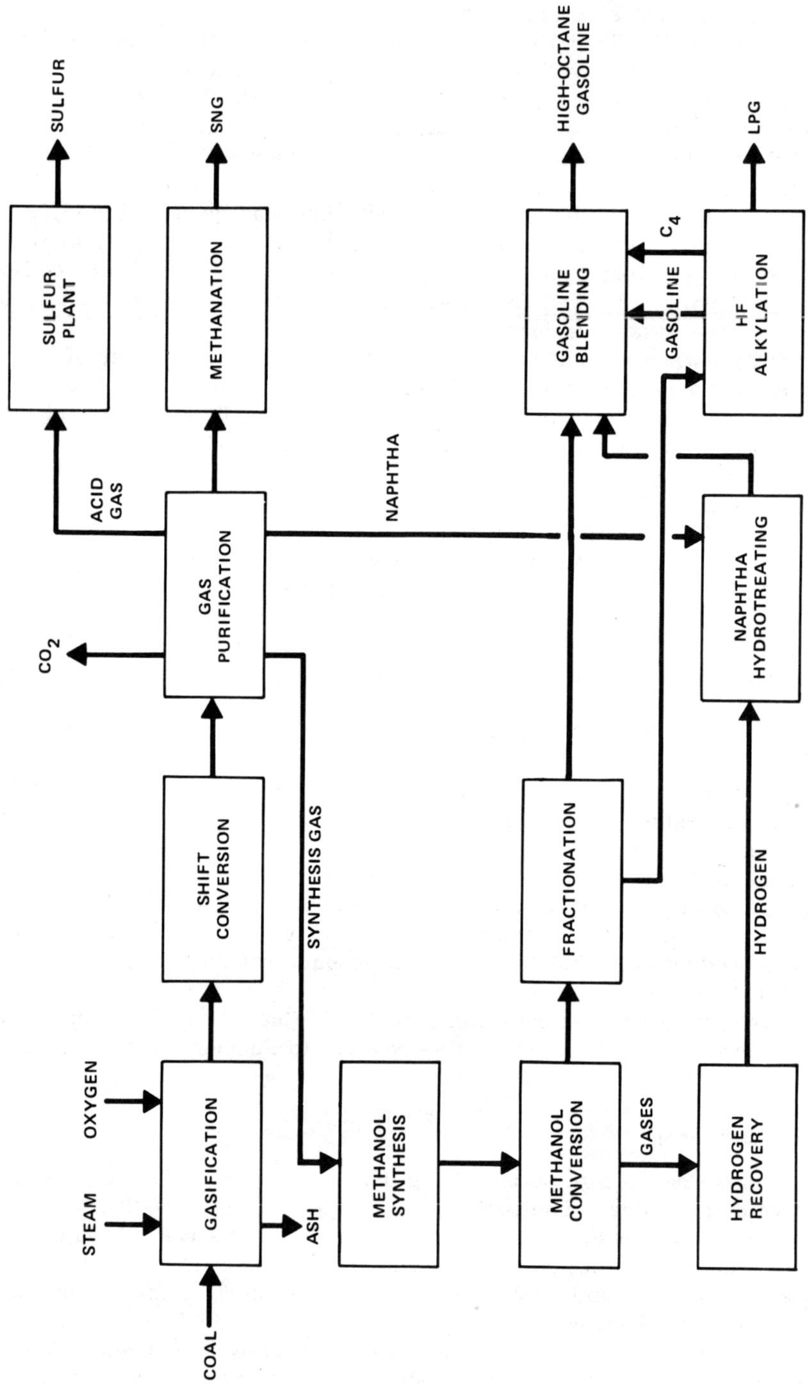

FIG. 19 Block flow diagram, M-Gasoline Process.

methanol will attack plastic and rubber parts in certain automobiles. Pure methanol can also be used as a fuel, but this requires engine modification. Methanol has also been tested and shown to perform as a turbine fuel.

In summary, methanol has the advantage that its production from coal has been proven industrially and it therefore has a good technological base. Its combustion generates fewer environmental contaminants than hydrocarbons. The major obstacles to its direct use as a fuel lie in the area of consumer use and acceptance.

7. *COED*

COED is a multistage near-atmospheric-pressure pyrolysis process developed by FMC Corporation under government funding. COED is an acronym for Char-Oil-Energy-Development.

Following smaller-scale development work, a 36-ton/day pilot plant was constructed by FMC Corporation in Princeton, New Jersey, and operated from 1970 to 1974. It performed well and achieved its objectives. The pilot plant has since been dismantled.

Primary products from the pyrolysis operation are gas, liquids, tar, and a char. For a bituminous coal, the yields of gas, liquids plus tar, and a char may be of the order of 20, 20, and 50%, respectively. Yields, of course, depend on the characteristics of the coal and the pyrolysis conditions.

The tar can be hydrotreated to produce a low-sulfur syncrude. The char could be burned for its calorific value or gasified to produce a syngas, which in turn could be converted to SNG, electrical power, chemicals such as methanol or ammonia, or Fischer-Tropsch liquids.

A simplified block flow diagram for COED is presented in Fig. 20. Here clean, sized coal is first dried and then pyrolyzed in a series of fluidized-bed units operating at 2 to 15 psig. The number of pyrolysis vessels and temperatures used depend on the feed coal characteristics. For a typical Illinois coal, three pyrolyzers were selected to operate at 575, 815, and 1050°F for the first-, second-, and third-stage pyrolyzers, respectively [25]. The temperatures are selected to prevent agglomeration in the fluid beds and to maximize liquid and fuel gas yields.

Because approximately 0.5 ton of char is produced per ton of feed coal, the economics are dominated by procedures selected for disposition of the char. There is some, but limited, experience in direct combustion of the char. In general, gas, fuel oil, or coal would be co-combusted with the char to provide acceptable volumetric heat rates from the boiler and combustion efficiency. The alternative is to gasify the char. The following process description is based on char gasification.

Heat requirements for the pyrolysis step is obtained from the heat released in char gasifiers. The gasifiers may be fluid-bed units, in which case a portion of the pyrolysis heat is obtained from the sensible heat of the raw low-Btu fuel gas which is produced in the char gasification step, and the remainder by transporting heat from the char gasifiers to the pyrolyzers via a circulating stream of partially reacted char. The low-Btu gases generated can be purified to remove sulfur compounds and CO_2, mixed with gases produced in the pyrolyzers, and serve as a synthesis gas for fuels/chemicals production or alternatively used to generate electrical power.

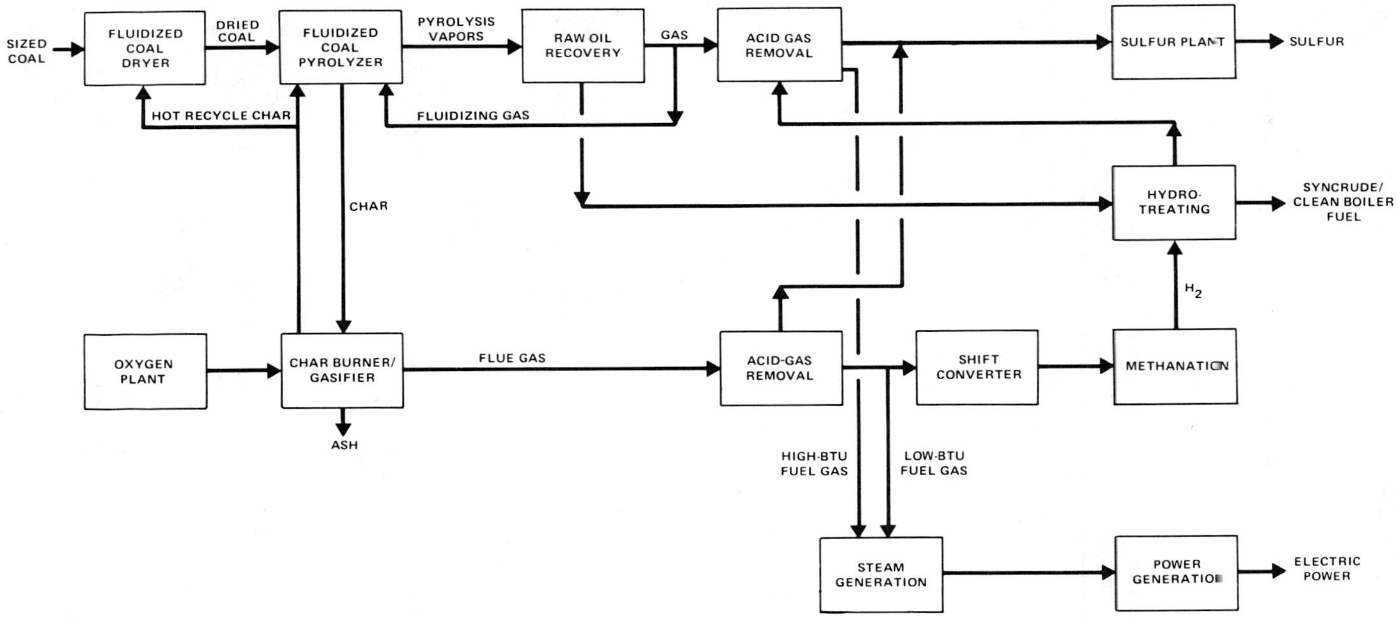

FIG. 20 Simplified block flow diagram, COED-based pyrolysis plant.

Raw COED oil is generated as a vapor in the pyrolyzers together with gaseous products. Entrained dust is removed in cyclones and the oil condensed in a direct-contact condenser. The oil is separated from the water phase produced. The oil phase contains finely divided entrained solids; to remove these, the oil is filtered. The pilot plant used a pressurized continuous rotary filter to reduce the solids content to about 0.1 wt %. The filter cake, containing about 50 wt % oil, may be fed to the gasifier.

The filtered oil is hydrotreated. Typically, the heavy feed oil would have an API gravity of about -5 and the hydrotreated oil +25 when the sulfur content was reduced to approximately 0.1 wt % and using high-sulfur feed coal. Hydrotreating conditions include temperature of 650 to 750°F, pressure approximately 2400 psi, and hydrogen/oil ratio of 10,400 scf H_2 per barrel. The 94 to 96% purity hydrogen is produced from the raw low-Btu fuel gas.

All exit gases are treated to remove sulfur contaminants prior to release to the atmosphere. Treatments can include Claus sulfur recovery units and tail-gas-treating units. The purified tail gas is odorless and typically contains less than 1 ppm of H_2S and less than 50 ppm of total sulfur compounds.

Several projects have tested char gasification performance. COED-produced char has been successfully gasified in a Koppers-Totzek unit in Spain; also, screening-type tests of fluidized-bed gasification have been reported [25]. The char has also been used in COGAS gasification development [26].

The projected economics for a 25,000-ton/day commercial COED-based plant to coproduce a synthetic crude oil and electrical power have been reported [27].

8. *TOSCOAL*

The TOSCOAL process is an adaptation in which TOSCO's TOSCO II oil shale retorting process is used to pyrolyze coal. It has been tested in the 25-ton/day TOSCO II oil shale pilot plant located at the Rocky Flats Research Center near Golden Colorado.

A sketch of the process is presented in Fig. 21 [28]. Ground, sized coal is contacted with hot flue gases in a lift pipe. The preheated feed is separted from the flue gas and fed to a rotating drum retort, where it is contacted with hot ceramic balls. The pyrolysis takes place in the retort at 800 to 970°F and near-atmospheric pressure. The product char, about 50 wt % of the feed coal, is separted from the ceramic balls, cooled, and sent to its end-use location. Char disposition can be similar to that earlier described for COED. Tosco and Hydrocarbon Research, Incorporated (HRI), have established a joint effort to develop the ToscoDyne process for producing low- or medium-Btu gas using the TOSCOAL produced char as feed to a high-pressure fluid-bed gasifier.

The vapors generated in the retort are separated into an intermediate-Btu gas (530 to 720 Btu/scf), pyrolysis oil, and water. The pyrolysis oil can be upgraded by hydrotreating. Fuel gas is burned to reheat the ceramic balls in a ball heater and to preheat the feed coal before being vented to the atmosphere.

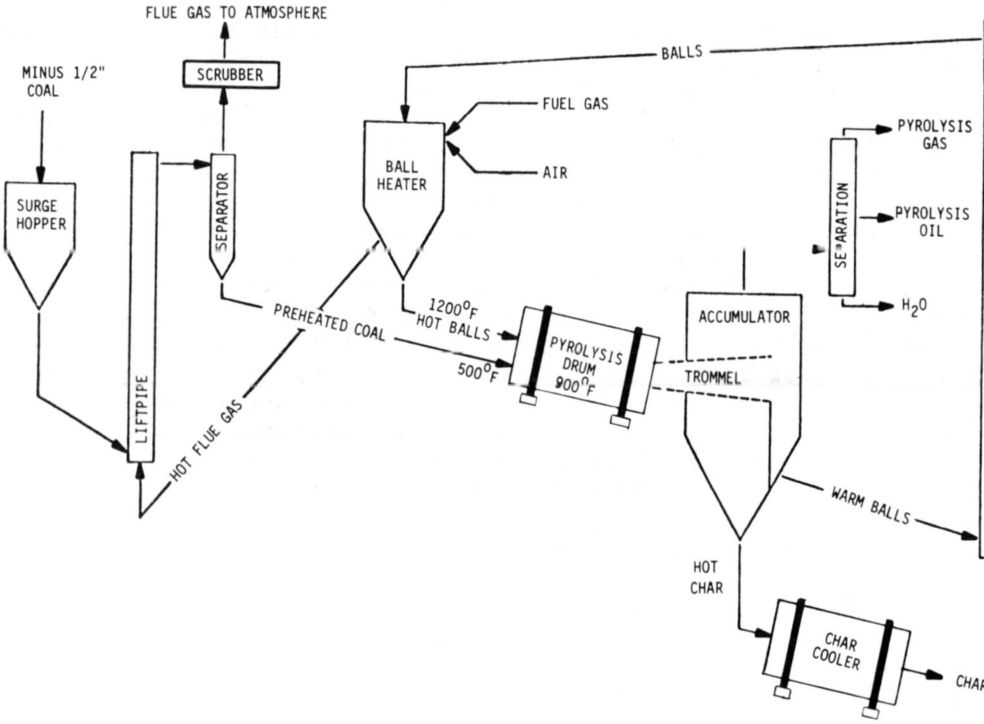

FIG. 21 Flow diagram, Toscoal Process. (Courtesy of DOE.)

9. *Fischer-Tropsch, Using Flame-Sprayed Catalysts*

The Pittsburgh Energy Technology Center (PETC) and its predecessor agencies have investigated the use of flame-sprayed catalysts deposited on metal surfaces to promote Fischer-Tropsch reactions for approximately 15 years [29]. A potential advantage of this catalyst system is that the catalyst can be deposited on the outside surface of extended-surface heat-exchanger tubes; the exothermic Fischer-Tropsch reaction can take place in the shell of a heat exchanger (outside the tubes) and high-pressure steam generated inside the tubes. By this means the heat of reaction can be recovered at a high energy level and the process thermal efficiency improved.

The Ralph M. Parsons Company has developed and described a conceptual design using this procedure [30,31]. The process configuration is shown in Fig. 22 and a schematic depiction of the key reaction step is shown in Fig. 23, where the reaction of synthesis gas takes place at about 400 psi and 600°F on the outside of horizontal flame-spray-catalyst-coated finned heat-exchanger tubes with boiling water on the inside of the tubes. The heat of reaction generates 1300-psi steam. The reactor product is heat-interchanged with the synthesis gas feedstream and fed to a hot product separator, where it is contacted with an alkaline solution to neutralize organic acids. Gas from the separator is purified to remove CO_2 and further treated to separate product gas and liquids. A portion of the product gas is recycled to the

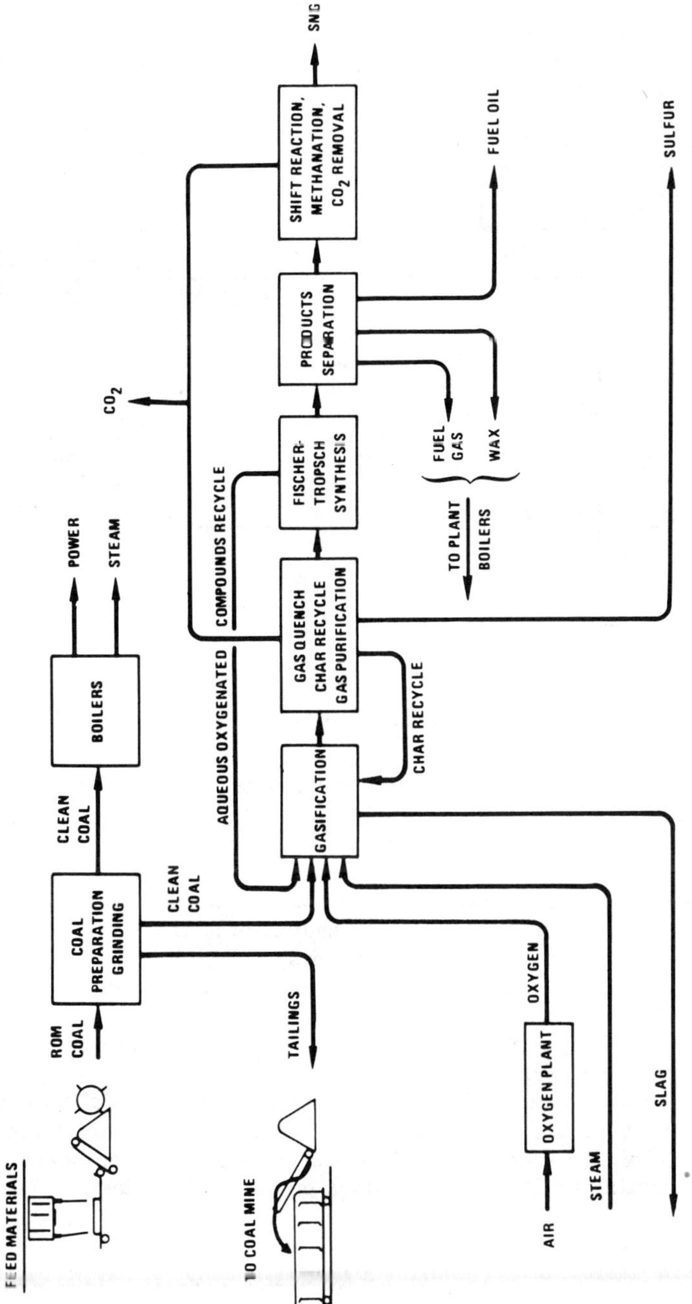

FIG. 22 Flow diagram, Fischer-Tropsch Process using flame-sprayed catalysts.

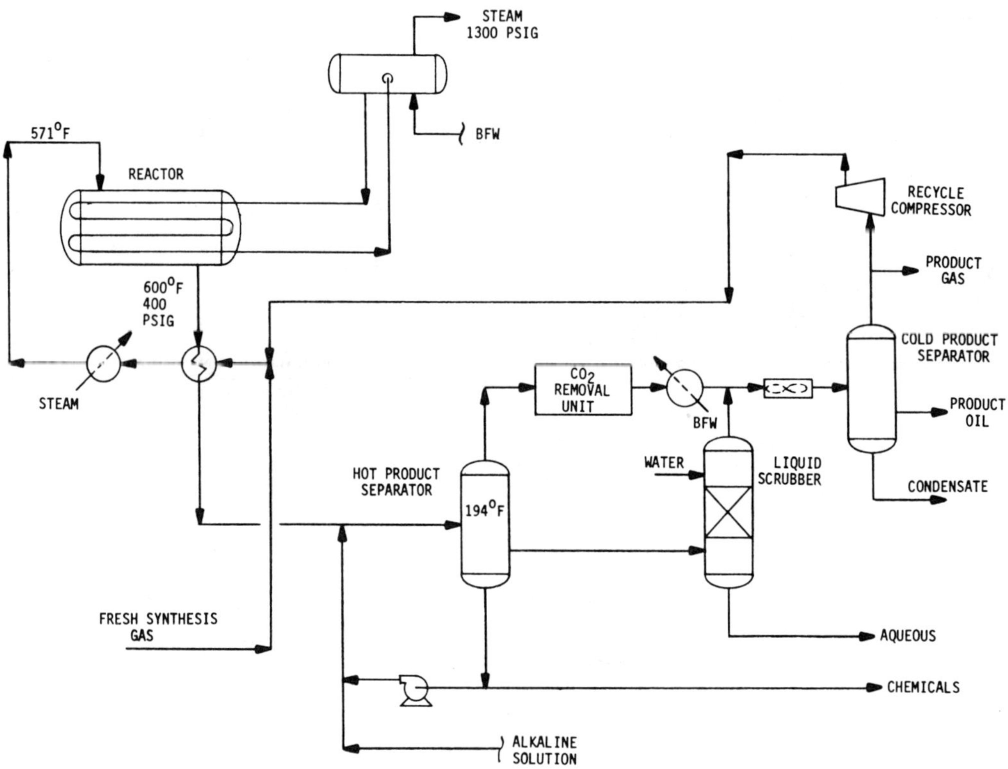

FIG. 23 Flow diagram, reactor section of Parsons' Fischer-Tropsch Process. (Courtesy of DOE.)

synthesis reactor for temperature and gas composition control. The products consist of a gas containing a significant amount of methane, liquids that are sent to the product separation and refining section, oxygenates consisting primarily of alcohols but with lesser amounts of aldehydes and ketones, and an aqueous solution containing organic acid salts. The gasification and product refining sections for the FSC process can be similar to those used for other Fischer-Tropsch catalyst systems.

The results of the conceptual design indicate the potential to achieve a thermal efficiency of feed coal to usable products of the order of 70%, which is significantly higher than achieved in past Fischer-Tropsch operations.

PETC is continuing experimental work on the FSC system.

10. Other In-Development Processes

There are a number of other coal liquefaction processes in varying stages of development. Examples follow.

a. Coalcon This is a pressurized fluid-bed hydrocarbonization development by Union Carbide Corporation (Fig. 24) [32]. A design was developed in anticipation of constructing a demonstration plant cofunded by ERDA. The ERDA program, however, was suspended.

b. Occidental Flash Pyrolysis Beginning in 1969, Occidental Research Corporation began coal-processing research. A nominal 4-ton/day PDU was constructed and operated at LaVerne, California. A sketch of the steps involved in this nominal-pressure (50 to 65 psi), 1100 to 1600°F flash pyrolysis process is shown in Fig. 25 [33].

c. U.S. Steel Clean Coke Process U.S. Steel constructed a 0.5-ton/day PDU under ERDA funding at Monroeville, Pennsylvania. It was completed in 1974. The objective of the process was to convert high-sulfur, marginal-quality coal to metallurgical-grade coke, liquid and gaseous fuels, and chemicals. The processing, as illustrated by Fig. 26, included carbonization, hydroliquefaction, coke preparation, liquids refining, and gas purification [34]. Work on the PDU has been suspended.

d. ADL (Arthur D. Little) Extractive Coking This process concept, to be tested at Foster Wheeler's Livingston, New Jersey, Laboratories, involves hydrogen-donor-solvent extraction of coal in a coking drum, cracking and

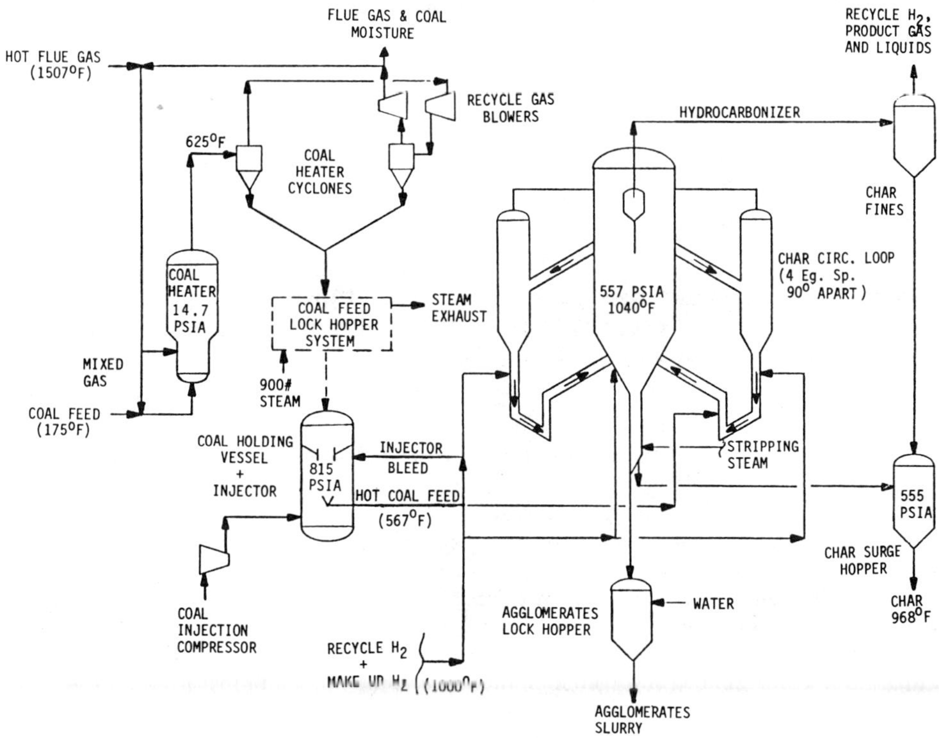

FIG. 24 Flow diagram, reactor section of Coalcon hydrocarbonization process. (Courtesy of DOE.)

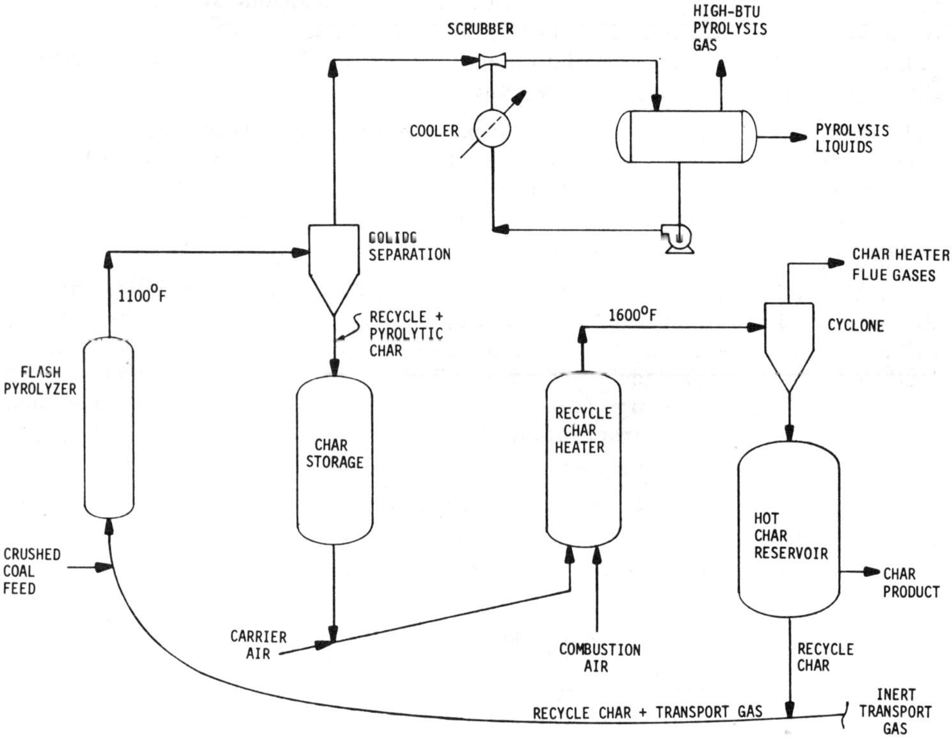

FIG. 25 Flow diagram, Occidental Research Corporation coal flash pyrolysis process. (Courtesy of DOE.)

coking of the extract to separate liquids from ash and unreacted coal, recovery and refining of product fuel gas and liquids, and production of the donor solvent in a fixed-bed catalyst hydrotreating reactor. Key elements of the concept are illustrated in Fig. 27.

e. Clean Fuel from Coal (CFFC) This technology is under development by C-E Lummus in a bench-scale (30-lb/hr) unit located at New Brunswick, New Jersey. It is a catalytic ebullating-bed coal hydroliquefaction process coupled with an antisolvent de-ashing procedure to remove unconverted coal and ash. A simplified block flow diagram is shown in Fig. 28 [35].

The antisolvent de-ashing procedure, under development since 1971, can be applied to hydroliquefaction processes in general. It is an alternative to removal of unconverted coal and ash by filtration or by suitable disposition of these solids as a concentrate in vacuum bottoms as described in P&M (Gulf Oil) SRC II and Exxon donor solvent processing. A unit is being tested at the Ft. Lewis, Washington, SRC pilot plant, and one is being installed at the H-Coal pilot plant at Cattletsburg, Kentucky [36]. Ability to separate a clean product containing less than 0.1% ash and product recoveries of nondistillable liquefied coal of 80% are claimed.

f. Critical Solvent De-Ashing (CSD) of Liquefied Coal CSD is a solid/liquid separation technique under development by Kerr McGee Corporation, which has operated a continuous bench-scale unit at Oklahoma City for over 5 years [37]. A simplified flow sheet showing a three-stage CSD unit that has been tested at the Wilsonville, Dlabama, SRC pilot plant is shown in Fig. 29.

"Critical solvent" here refers to any solvent at temperatures and pressures near the critical values for that solvent. Properties of the critical solvents include: (1) the density changes rapidly with temperature; (2) the dissolving power changes roughly in proportion to its density; and (3) the dissolving power is greatest for lighter coal fractions and decreases for heavier coal fractions.

Early results of the Wilsonville tests indicate that the CSD procedure can meet the SRC ash content specification of 0.16 wt % maximum and can produce SRC recovery yields of the order of 80% [37].

g. Dow Chemical Company Dow has announced development of a catalytic hydroliquefaction process based on results obtained in a 200-lb/day bench-scale unit. The process features the use of an inexpensive expendable catalyst with partial recycle of the catalyst, plus a unique extraction process said to

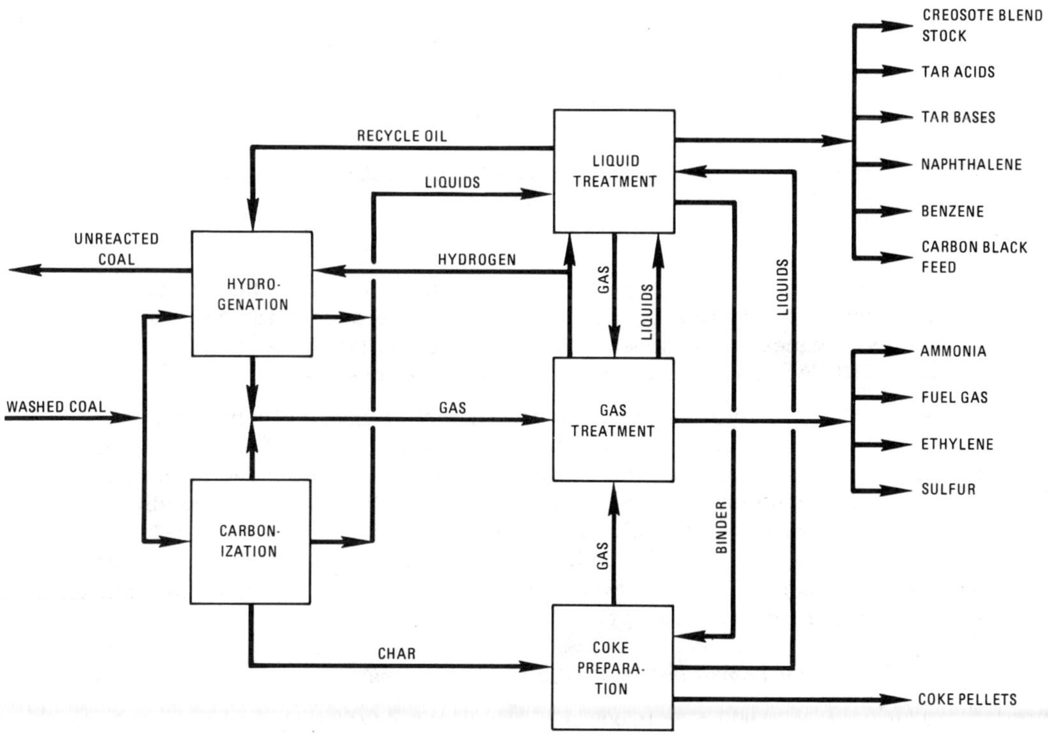

FIG. 26 Simplified block flow diagram, U.S. Steel clean coke process. (Courtesy of DOE.)

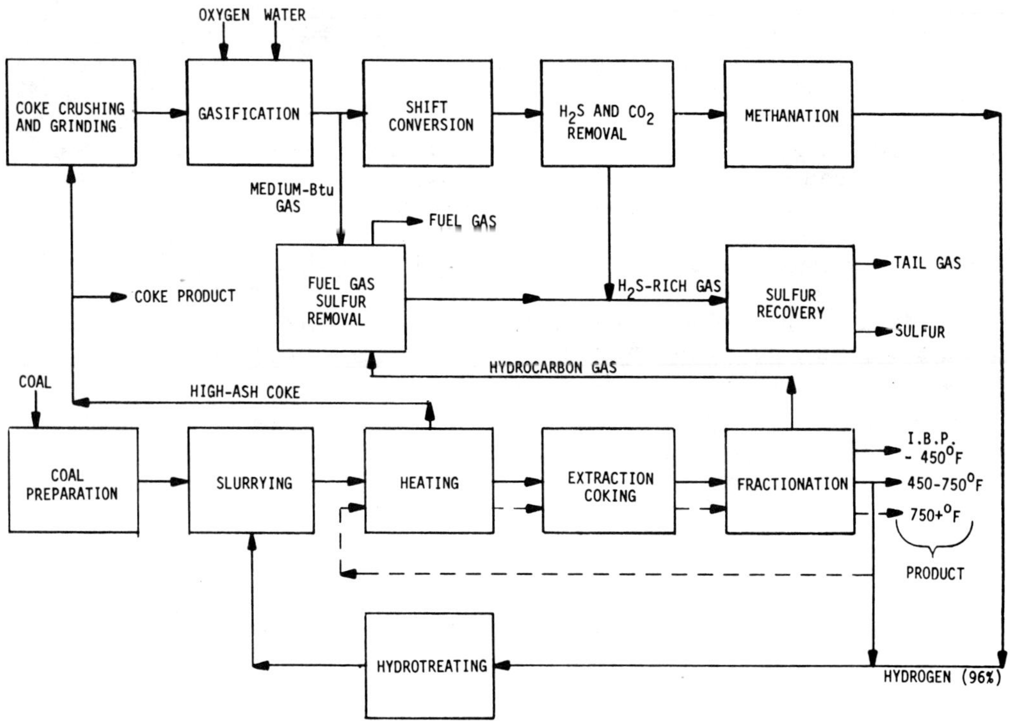

FIG. 27 Simplified block flow diagram, ADL extractive coking process. (Courtesy of DOE.)

produce an essentially solid-free, low-sulfur product oil. The catalyst has been reported to be a molybdenum emulsion and recovery, for recycle, to be via hydroclones. The possibility of testing the Dow process using the Synthoil PDU located at Bruceton, Pennsylvania, has been explored.

h. Synthoil The Synthoil Process was the development of what is now the Pittsburgh Energy Technology Center in Bruceton, Pennsylvania. Key elements include contact of a slurry of coal in recycle coal-derived liquid with hydrogen and catalyst in a fixed-bed tubular reactor. Reaction conditions are 850°F and 2000 to 4000 psi. The process has been tested in a 5-lb (slurry)/hr test unit located at PETC. The catalyst used consisted of cobalt molybdate deposited on silica-activated alumina. Figure 30 shows key elements of the Synthoil Process concept.

11. "Third-Generation" Liquefaction Processes

There are a number or processes in various development phases that offer potential advantages over processes now in the pilot-plant or technical demonstration phases. The characteristics of several of these are described briefly next.

a. Cities Service—Rockwell International Direct Coal Hydrogenation This is a direct hydrogenation process wherein preheated hydrogen and coal are rapidly reacted (e.g., 10 to 100 msec) at high temperature (e.g., 1800°F) and high pressure (e.g., 1000 psi) to produce high-Btu gas, liquids and tars, and a coal residue. The reactor type was originally developed for propellant injection in liquid-fuel rocket engines. It has been tested in a 1/4-ton/hr reactor. DOE has awarded Rockwell International an $18 million contract to design, construct, and operate a 100-ton/day pilot plant to be built near Canoga Park, California.

A sketch of the 1/4-ton/hr reactor system is shown in Fig. 31 [38,39]. Here dry pulverized coal, ground to less than 150 μm, is fed from a pressurized lock hopper to the reactor injector, where it is mixed with preheated hydrogen. The coal and hydrogen are mixed rapidly, react in the downflow reactor, are quenched by water, and pass to a receiver located at the base of the reactor. The products are recovered and separated. The composition and yield of the products can be varied by varying the reaction conditions.

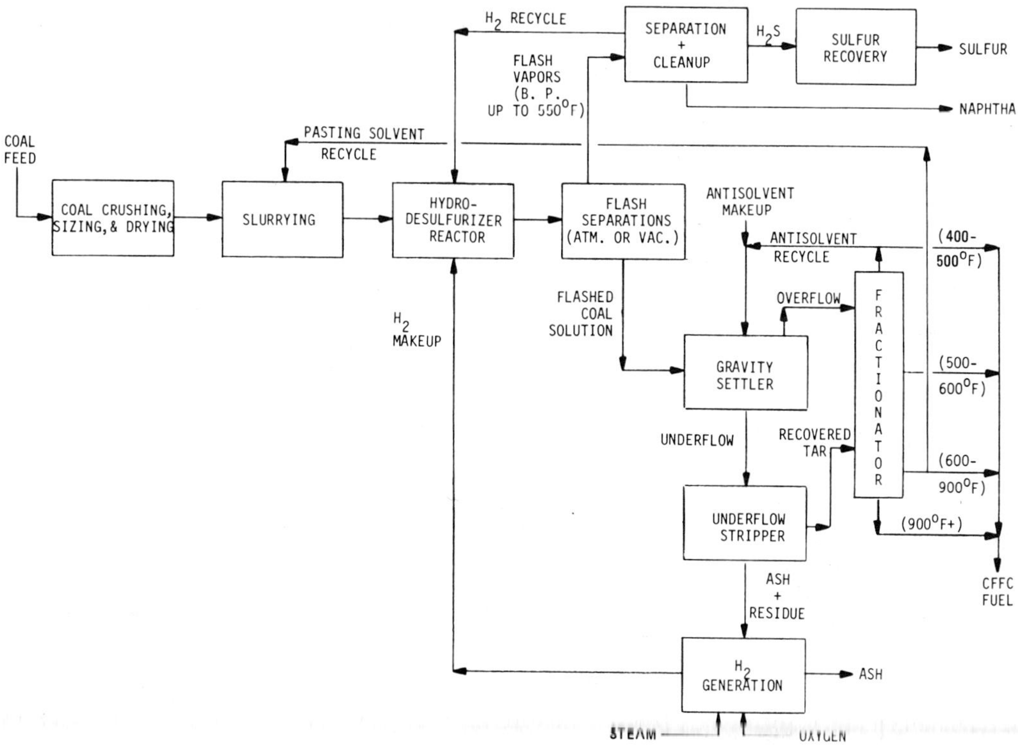

FIG. 28 Simplified block flow diagram, C-E Lummus CFFC Process. (Courtesy of DOE.)

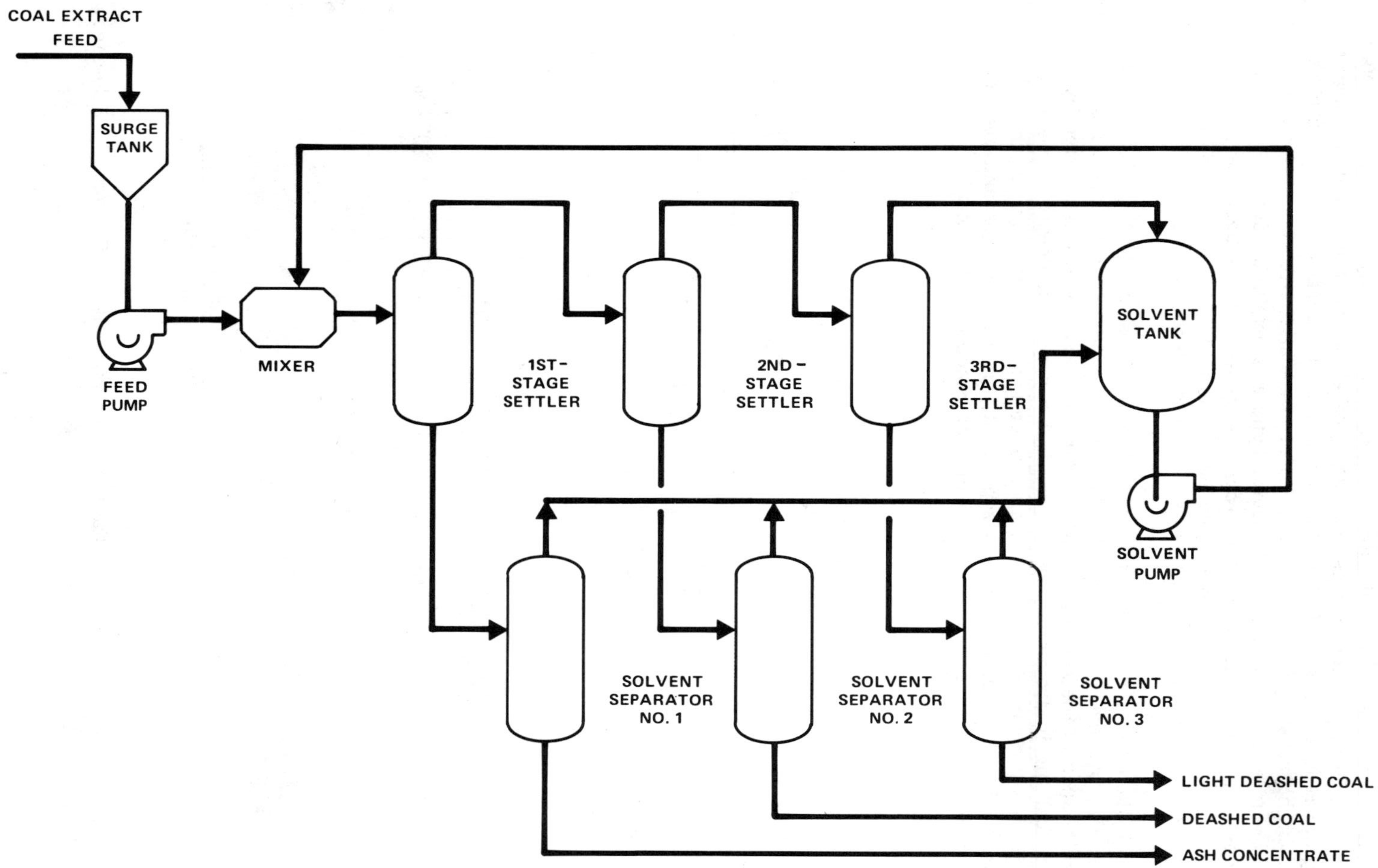

FIG. 29 Flow diagram, three-stage CSD pilot plant, Wilsonville SRC program. (Courtesy of Chemical Engineering Progress.)

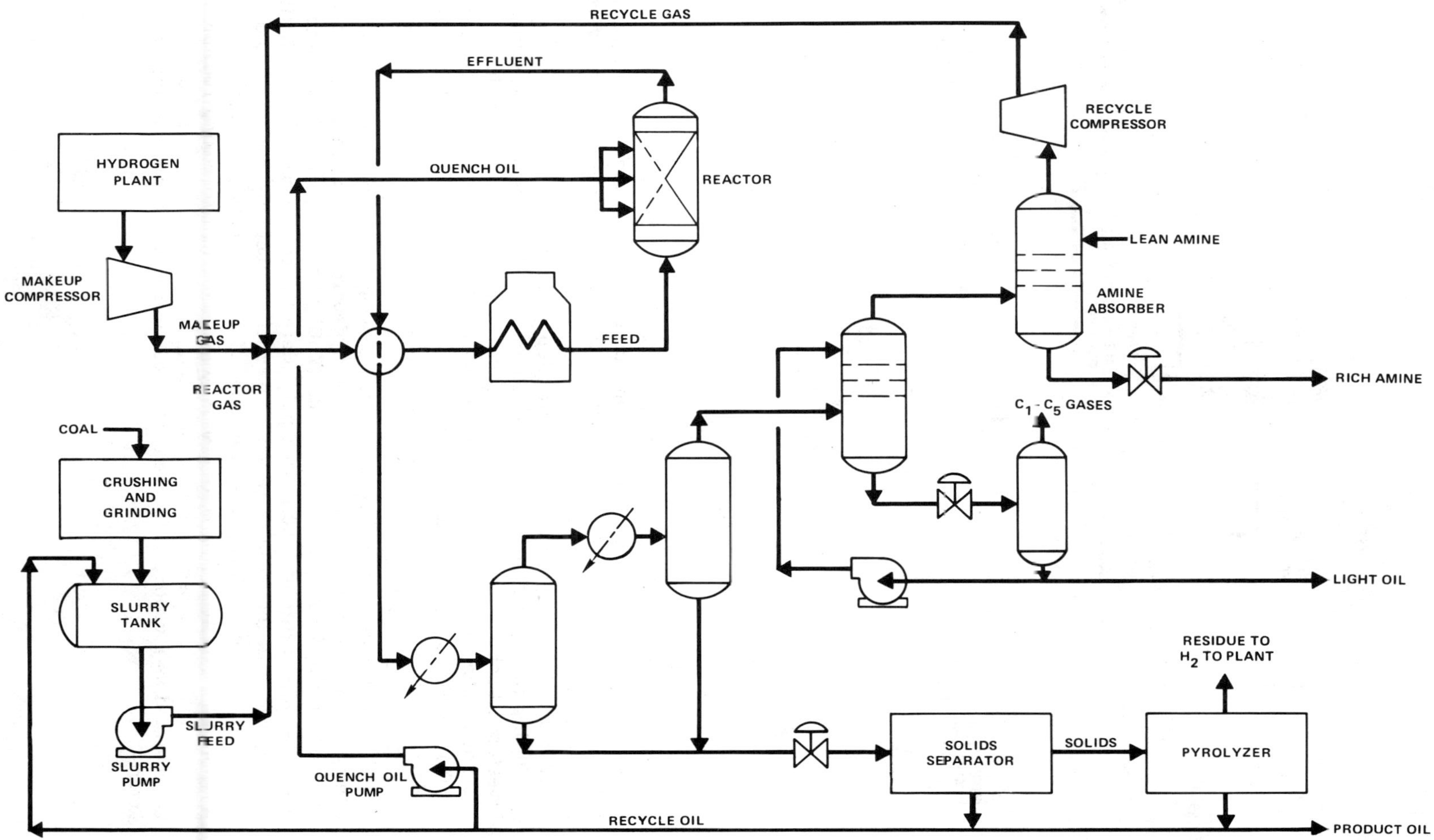

FIG. 30 Simplified flow diagram, Synthoil process. (Courtesy of DOE.)

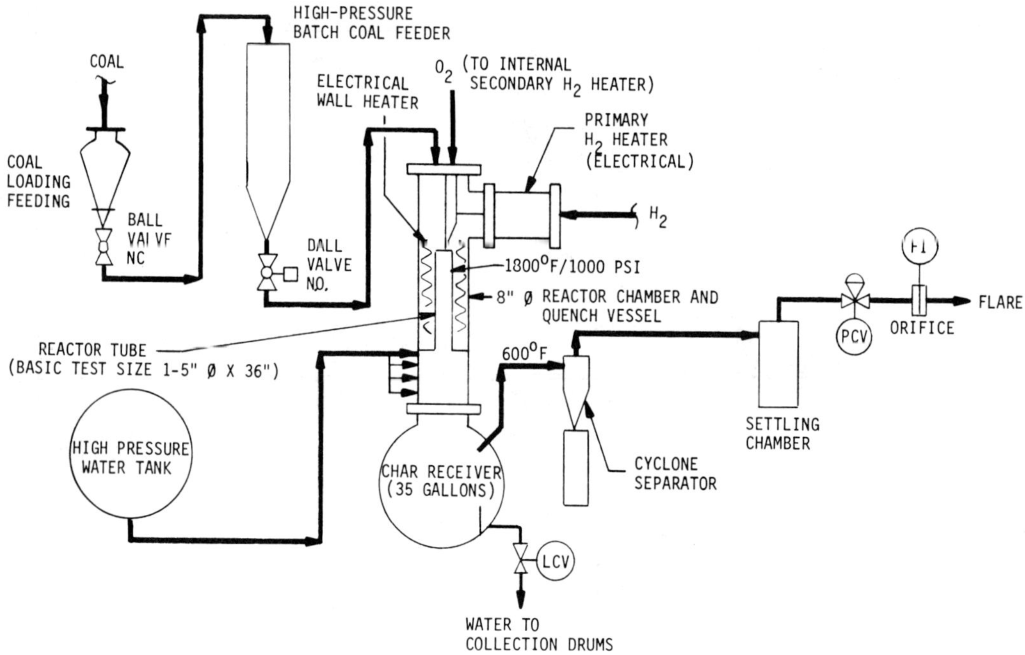

FIG. 31 Flow diagram Rockwell International 1/4-ton/hr rapid coal hydrogenation system. (Courtesy of DOE.)

b. Zinc Chloride Catalyst Process This process is under development by the Conoco Coal Development Company at Library, Pennsylvania. Its objective is to convert bituminous and subbituminous coal into distillate fuels by severe catalytic cracking. It can be a one-step process or, when a coal extract is recovered and then hydrocracked, a two-step process.

During the period 1975-1977, research on a 100-lb/hr bench-scale unit investigated conversion of subbituminous coal to distillate fuels and zinc chloride catalyst regeneration. Products were a 90 Research Octane Number (RON) gasoline and a low-sulfur, low-nitrogen fuel oil. Catalyst regeneration yielded a catalyst recovery of 99.5%. Construction of a 1-ton/day PDU was completed in 1978 [40].

A simplified block flow diagram for the process is shown in Fig. 32. A slurry of dried, crushed coal is fed to the hydrocracking reactor, where it is mixed with hydrogen and zinc chloride ($ZnCl_2$) at a temperature of 675 to 825°F and a pressure of 1500 to 3500 psi. Here the coal is cracked to a distillate, primarily in the gasoline range. The fuel products are recovered and processed to produce fuel gases, gasoline, distillate fuel oil, and recycle distilled oil (solvent). The solid discharge from the hydrocracking reactor is fed to a fluidized-bed combustor operating at 1700°F and 2 psig. $ZnCl_2$ is separated as a vapor, condensed, and recycled to the hydrocracking reactor.

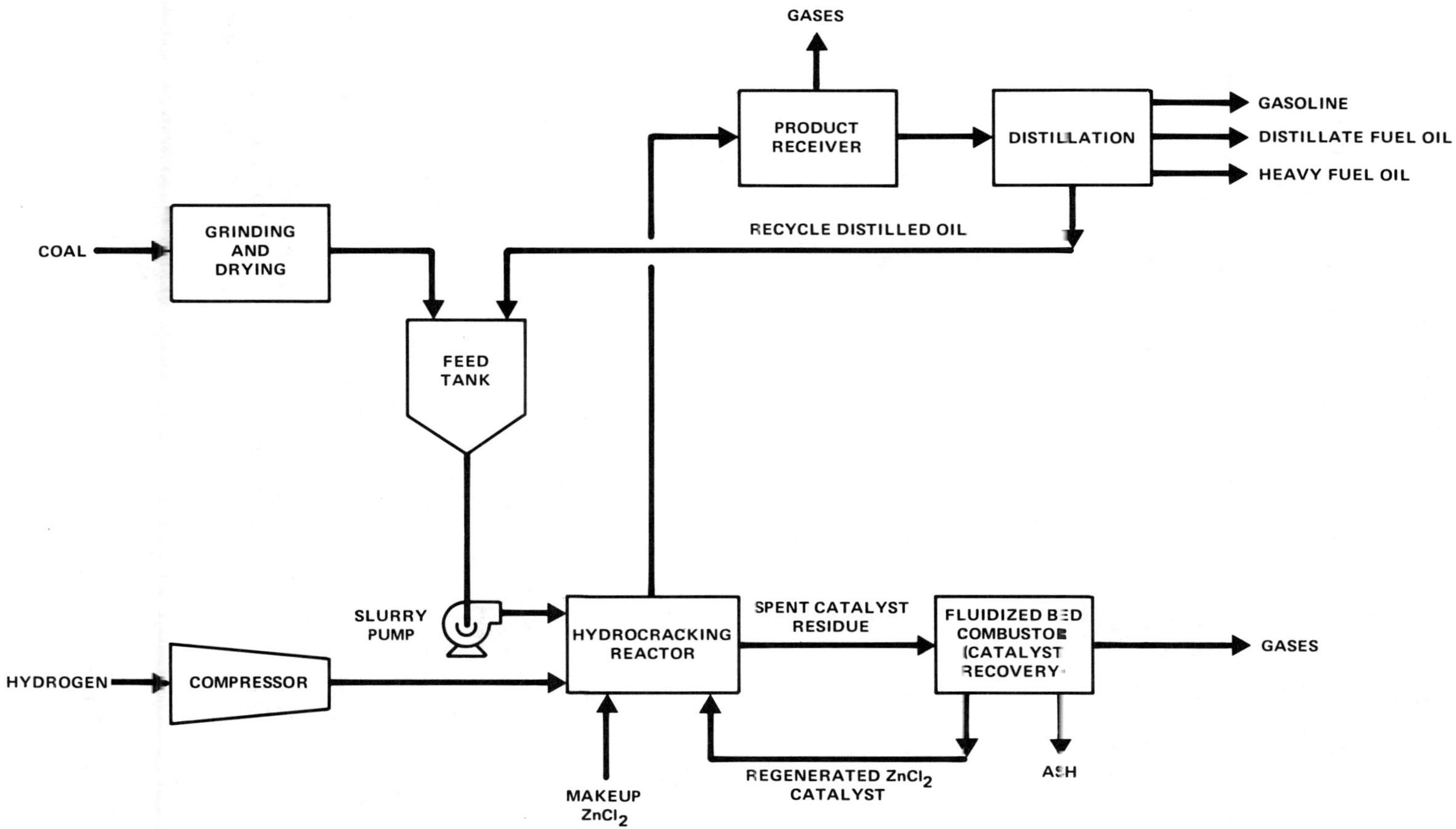

FIG. 32 Simplified block flow diagram, zinc chloride catalyst process. (Courtesy of DOE.)

c. Disposable-Catalyst Hydrogenation This process uses inexpensive disposable catalysts. The objective is single-pass hydroliquefaction to produce low-sulfur fuel oil.

A coordinated program has been carried out in 1979 at the Pittsburgh Energy Technology Center (PETC) and at industrial facilities. Catalytically active species were identified in coal ash. Tests are planned in a 0.6-ton/day PDU at PETC.

The processing involves mixing dried ground coal with a process-derived oil and the disposable catalyst. The resulting paste is compressed to 2000 to 4000 psi, preheated, mixed with hydrogen, and fed to the reaction vessel. Gas and sludge discharge from the reactor are sent to a gas separator and a pressure letdown system. The gas is processed to recover a hydrogen-rich gas for recycle and produce a product fuel gas. The oil is distilled to produce liquid fuel plus the pasting oil for recycle. The residue and a heavy oil are processed to recover light oil, paste oil, and heavy liquid products [41].

B. West Germany

West Germany was very active in coal conversion during the early part of the twentieth century and was the most active in operation of industrial-scale plants through the World War II era. It was inactive in the postwar period but renewed its interest and activity in 1973 following the oil embargo by the OPEC countries. It now has a broad-range development program. The federal government-sponsored work is funded and administered by the Bundeministers für Forschung und Technologie (BMFT), located in Bonn.

West Germany has stated plans to participate in the U.S. SRC II demonstration plant program and supply 25% of the funds. Several other developments under way are described in the following paragraphs.

1. *Bergbau-Forschung*

Bergbau-Forschung is active in the development of coal liquefaction, coal gasification, and fluidized-bed combustion technology. The liquefaction work is supported by the State of Nordrhein-Westfalen. Their coal liquefaction development may be termed a modified Bergius Process. It is depicted schematically in Fig. 33 [42]. The process is being tested in a 10-kg/hr PDU located at Essen.

The ground feed coal is mixed with a recycled coal-derived liquid and a catalyst. The catalyst can consist of approximately 1% iron oxide, which is dispersed in the feed coal slurry. The prime liquefaction reaction takes place in a tubular reactor at a temperature in the range of 900°F and a pressure of approximately 4500 psi. After reaction, pressure is reduced and the product separated. The gases are purified and the solids are removed from the heavy oil fraction produced in the pressure letdown system by filtration. The light oil fraction recovered from the pressure letdown system is fractionated at atmospheric pressure to produce a light oil overhead product and a bottoms product. The bottoms are combined with the heavy oil filtrate and vacuum-distilled to produce a recycle solvent used to prepare the feed coal slurry and a product fuel oil.

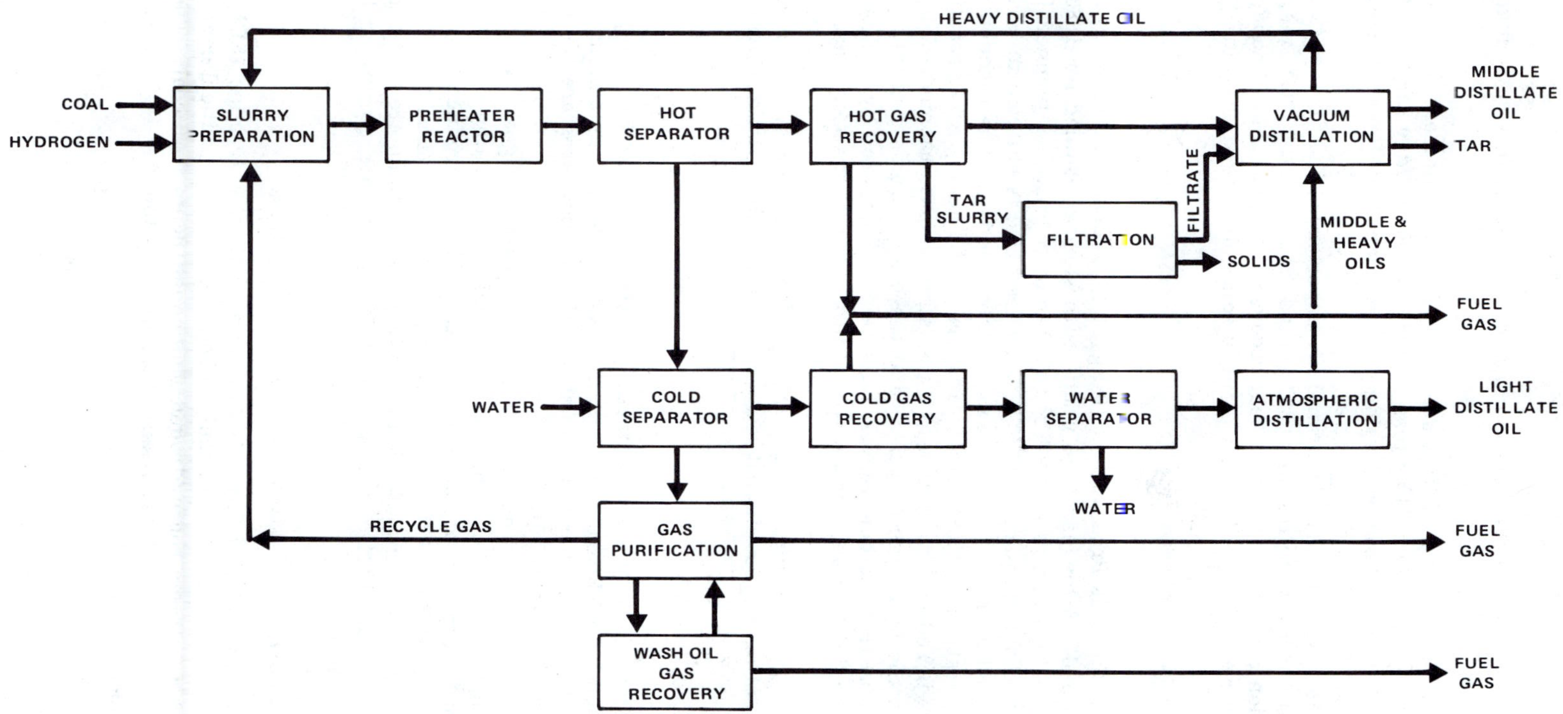

FIG. 33 Simplified block flow diagram, Bergbau-Forschung process. (Courtesy of DOE.)

2. *Gessellschaft für Vergasüng und Verflüssigung von Steinkohle Gmbh (GVV)*

GVV, a subsidiary of Ruhrkohle AG, began its present liquefaction development program in 1974. They plan to build a 200-ton/day pilot plant based on the I.G. Farben process and using data developed by Bergbau-Forschung.

A simplified block flow diagram of the process GVV plans to use is shown in Fig. 34 [43]. Relative to prior I.G. experience, the process is expected to operate at a lower pressure (300 atm vs. the I.G. 700 atm), increased unit production capacity, improved heat recovery and thermal efficiency, and improved economics. The pilot plant is scheduled for startup in 1981.

3. *Saabergwerke AG (S)*

Saarbergwerke, with funding from BMFT, is working on development of a modified Berguis Process in a 10-kg/hr PDU located at Saarberg. They plan to build a 6-ton/day pilot plant, with startup planned for 1980.

Figure 35 presents a simplified sketch of the process planned for use in the 6-ton/day pilot plant [44]. Ground feed coal is slurried in a recycle coal-derived liquid and a disposable catalyst added. The reaction is initiated at a temperature of 790°F and a pressure of 300 atms. After reaction, the pressure is reduced by means of a pressure letdown machine, the gas is purified, the hydrogen recycled, and the heavy liquids separated by coking and vacuum frationation; recycle oils are also recovered by vacuum fractionation. The light liquids are recovered in a separate fractionation.

The process scheme incorporates provisions for recycling liquids recovered from the high-pressure separator to the feed coal slurry, thereby reducing the amount of recycle oil that must be recovered by fractionation.

C. United Kingdom

The United Kingdom is concentrating on the development of two processes by the National Coal Board (NCB) at their Coal Research Establishment (CRE) located at Stoke Orchard, Cheltenham. These processes are the supercritical extraction and hydroliquefaction routes.

U.K. interest and experience in coal liquefaction goes back to the late 1920s and includes the industrial plant built and operated by ICI at Billingham, which had a capacity of 100,000 tons/year of liquid products. Although the current availability of North Sea oil and gas has minimized the urgency for development of coal liquefaction capability, predictions are that the United Kingdom will be short of crude oil in 1990+. They are continuing their development work.

1. *Supercritical Extraction*

The NCB has been investigating supercritical extraction (SCE) since 1970. In this process an organic solvent under supercritical conditions selectively extracts the hydrogen-rich constituents from coal. The residue is a reactive char. It has been tested in a 1/4-ton/day PDU at Stoke Orchard, Cheltenham.

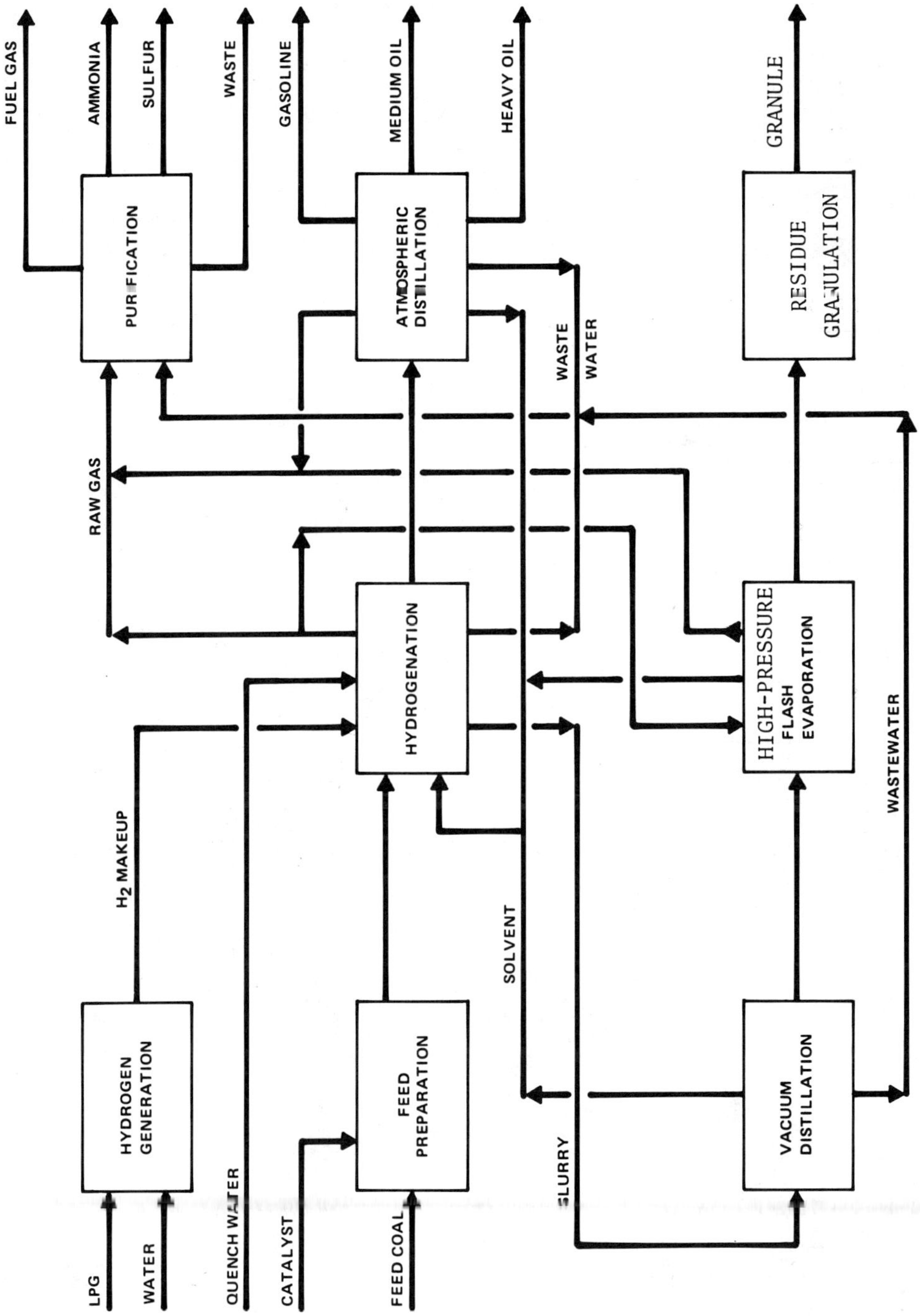

FIG. 34 Simplified block flow diagram, GVV 200-ton/day pilot plant.

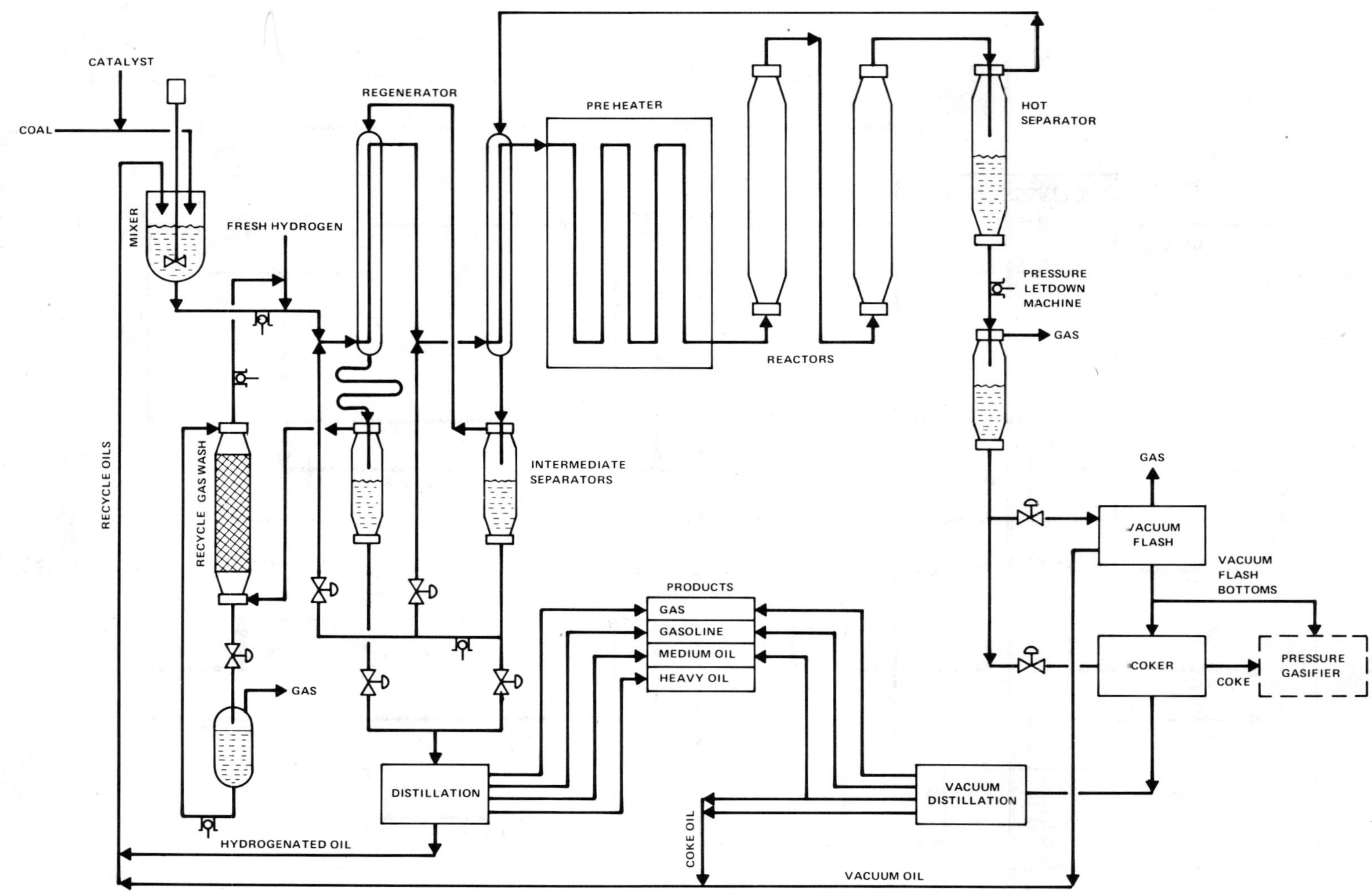

FIG. 35 Flow diagram, Saarbergwerke AG Process.

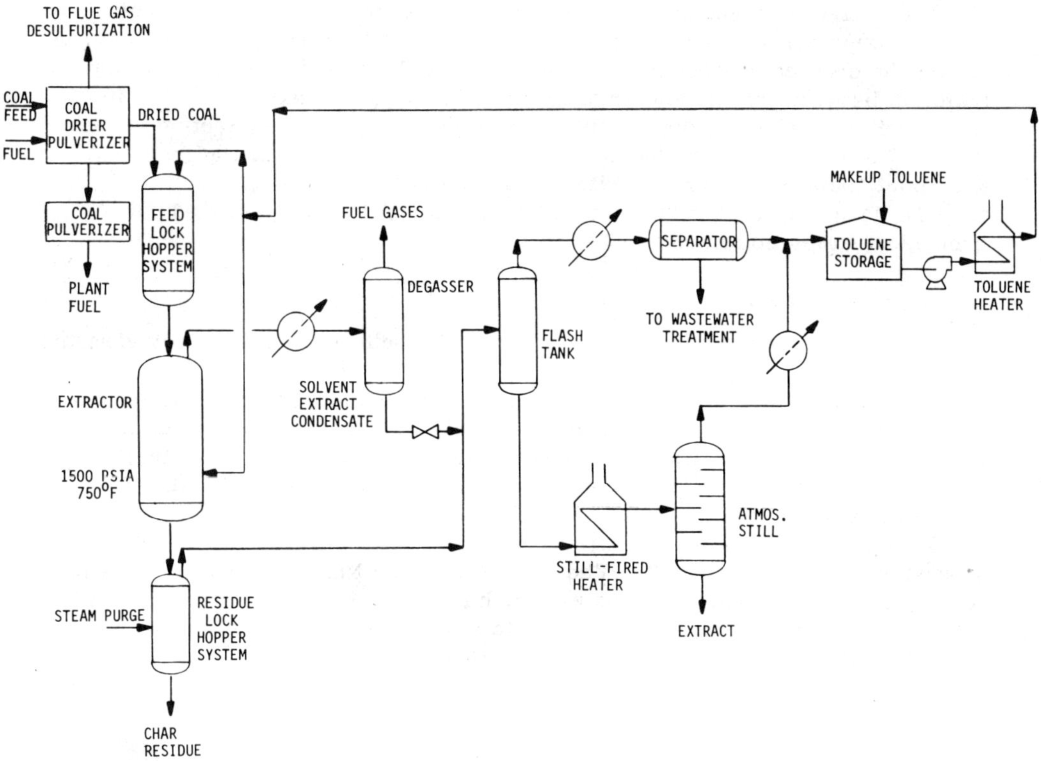

FIG. 36 Flow diagram, supercritical gas extraction process. (Courtesy of DOE.)

SCE is depicted in Fig. 36 [45]. Coal is fed to the feed lock hopper, where it is pressurized by the solvent toluene. Coal is fed to a moving-bed extractor, where it is countercurrently contacted with toluene under conditions above its critical temperature (606°F and 40.6 atms). An extraction temperature of 750°F is above the critical temperatures but below the cracking temperature of the coal extract. The supercritical toluene acts as a solvent to "dissolve" coal tar liquids, which are generated by the equivalent of a pyrolytic action at the extraction temperature.

Undissolved coal residue (char) is discharged to the extractor underflow lock hopper system. It is depressurized and steam-stripped to recover the toluene. The extract solution is removed at the top of the extractor, cooled for total condensation, and the pressure reduced; uncondensed hydrocarbon gases are separated from the degasser for use as fuel. At the reduced pressure, the extract precipitates as a solution in toluene and the extract, toluene, and product water all exist in a liquid phase. This stream is heated and flashed in a flash tank at near-atmospheric pressure to vaporize the water and most of the toluene, which is sent to the toluene-recovery unit. The remaining liquor, a 50:50 mixture of extract and toluene, is fed to a distillation tower, where the toluene is recovered as an overhead product and the extract as the bottoms.

The extract is a low-melting, low-ash (less than 0.1 wt %) glassy solid at room temperature. It has a lower sulfur content than the feed coal, dependent on the chemical composition of the sulfur in the feed coal (i.e., the distribution between pyritic and organic sulfur). Yields of extract can be in the range 20 to 40 wt %, dependent on coal composition and extraction conditions. The residue (char) has a lower hydrogen content than the feed coal and about the same sulfur and nitrogen contents.

The design of a 1-ton/hr unit has been started and is scheduled for mechanical completion in mid 1980.

2. *Hydroliquefaction*

Tests have been carried out in bench-scale facilities at CRE. Key elements of the process are shown in Fig. 37 [46]. Here we see that the ground feed coal is contacted with a donor solvent in a digestor operated at about 150 psi, the unreacted coal and the coal ash are separated from the coal liquids in a filter, the gases are purified, and the liquids are hydrotreated and fractionated to produce the product liquid fuels and to recover the donor solvent for recycle to the coal extractor.

This process has a number of elements common to other coal liquefaction processes under development. The objective of the National Coal Board is to develop technology that performs well with indigenous coals, to have domestic technology for use when required, and to develop technology for licensing. There are plans to build a 1-ton/hr pilot plant to further test this process.

D. South Africa

South Africa is a world leader in the field of coal liquefaction based on its successful experience with the Sasol Fischer-Tropsch technology. It is predictable that development work continues in the field of coal liquefaction. This work is proprietary and information regarding the details of this development work are not available.

E. Japan

Japan has had a history of experience with coal liquefaction. To illustrate, Mitsui operated a 2-ton/day Bergius pilot plant in Manchuria in 1928 [47]. Japan had experience with both Fischer-Tropsch and Bergius technology during World War II. Their interest and activities in the field of coal liquefaction have increased in the 1970s. Work has proceeded both under private-sector programs and under the Project Sunshine program funded and administered by the Ministry of International Trade and Industry (MITI). Japanese firms participating in the Project Sunshine program include Mitsui, Mitsubishi, and Sumitomo, among others. Japan has stated its intent to participate in and supply 25% of the funds for the U.S. SRC II demonstration-plant program, expected to cost more than \$1.5 billion.

1. *Mitsui*

Mitsui began operation of a continuous 12-lb/hr bench-scale unit in 1973 and a 5-ton/day pilot plant based on SRC processing in Ohmuta, Kyushu, in 1978.

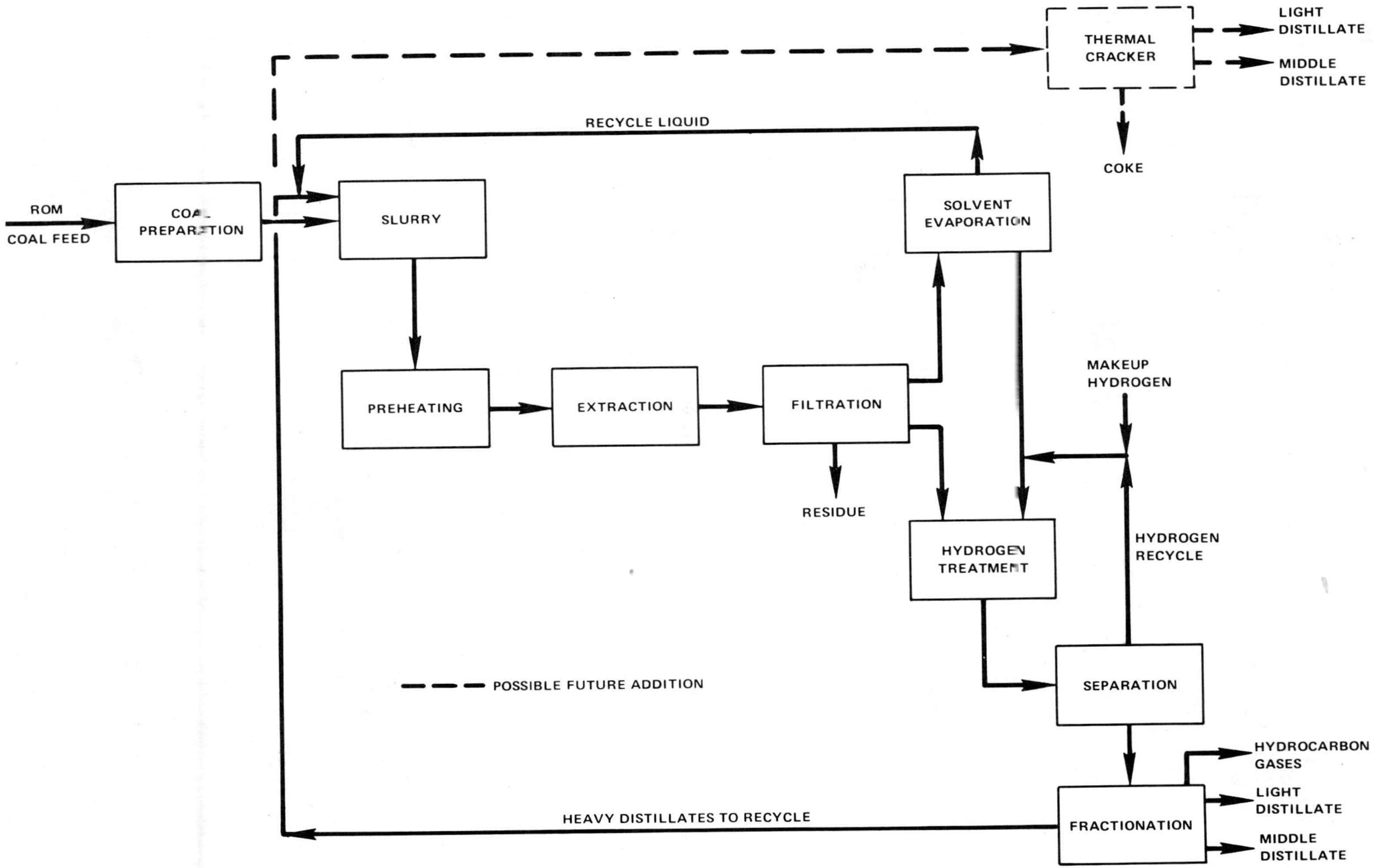

FIG. 37 Simplified block flow diagram, National Coal Board of U.K. hydroliquefaction process [46].

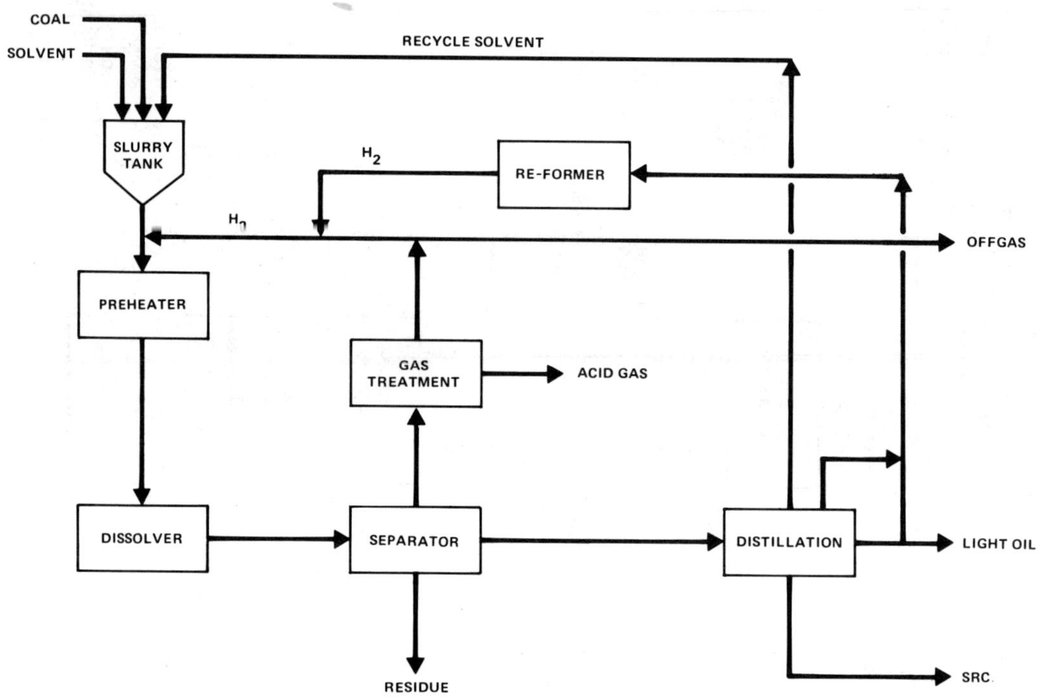

FIG. 38 Simplified block flow diagram, Mitsui Solvent Refined Coal Process.

The Mitsui process being piloted at Ohmuta (Fig. 38) is a modification of SRC I. A key element of the process is that it uses a mixture of a petroleum-sourced solvent and recycle coal liquids as a slurry agent for the feed coal. This procedure provides flexibility in raw material procurement. The reactor and downstream processing are similar to SRC I processing, including removal of ash and unreacted coal residue by filtration. A proprietary product solidification procedure is under development.

Mitsui's objectives are said to be production of coal-derived material that can serve as an ingredient in the production of metallurgical-grade coke for use in the steel industry. By this means, lower-grade coals can be used to produce coke. The incentive for this development is the significant differential in cost that exists between metallurgical-quality coal and steaming coal.

2. *Sumitomo*

Sumitomo is said to be constructing a 1-ton/day pilot plant to test a donor-solvent-type process.

3. *Kawasaki*

Kawasaki has announced plans to construct a 8-ton/day pilot plant to produce a coal-based material for use as an ingredient in production of metallurgical-grade coke.

F. Australia

Australia has significant coal reserves, varying in composition from brown coal to bituminous. There is active interest in coal liquefaction experimental programs.

1. *Commonwealth Scientific and Industrial Research Organization (CSIRO)*

CSIRO is developing a fluidized-bed flash pyrolysis process, which is illustrated in Fig. 39 [48]. Experimental facilities include 1-g/hr fluidized-bed unit, a 100-g/hr entrainment reactor, and a 20-kg/hr reactor which can operate either in the fluidized-bed or entrainment mode.

In the CSIRO process, the ground feed coal is rapidly heated to a temperature of 930 to 1300°F in a fluidized bed of sand at essentially atmospheric pressure. The residence time of the vapors generated is of the order of seconds. The vapors are separated from the char and processed to produce C_1 to C_3 hydrocarbons, fuel gas, a tar, and char. Tar yields of the order of 20 to 30 wt % are obtained, depending on pyrolysis conditions and coal composition.

2. *Australian Coal Industry Research Laboratory (ACIRL)*

ACIRL operates a bench-scale continuous coal hydroliquefaction unit. The performance of individual coals under defined operating conditions is determined.

G. Poland

Poland has significant coal reserves and maintains active programs in coal conversion technology development. A development program for a donor solvent process is under way. Details regarding the facilities and specifics of process characteristics are meager.

VI. THERMAL EFFICIENCY AND ENVIRONMENTAL FACTORS

The cost of feed coal per barrel of product is directly related to the thermal efficiency of the plant, expressed as the Btu content of products per feed coal Btu consumed in the process, times 100. Still other important factors are the ability of coal liquefaction plants to meet environmental regulations and the environmental acceptability of the products. The thermal efficiency and environmental factors are discussed below.

A. Thermal Efficiency

Historical data indicate that thermal efficiencies for coal liquefaction plants in the 1930s and through the World War II era were in the range 30 to 40% [47]. Recent published information also indicates that current Fischer-Tropsch plant thermal efficiencies are in this same range [48]. Design analysis indicates that higher efficiencies are achievable in integrated coal conversion complexes using modern equipment and controls [31].

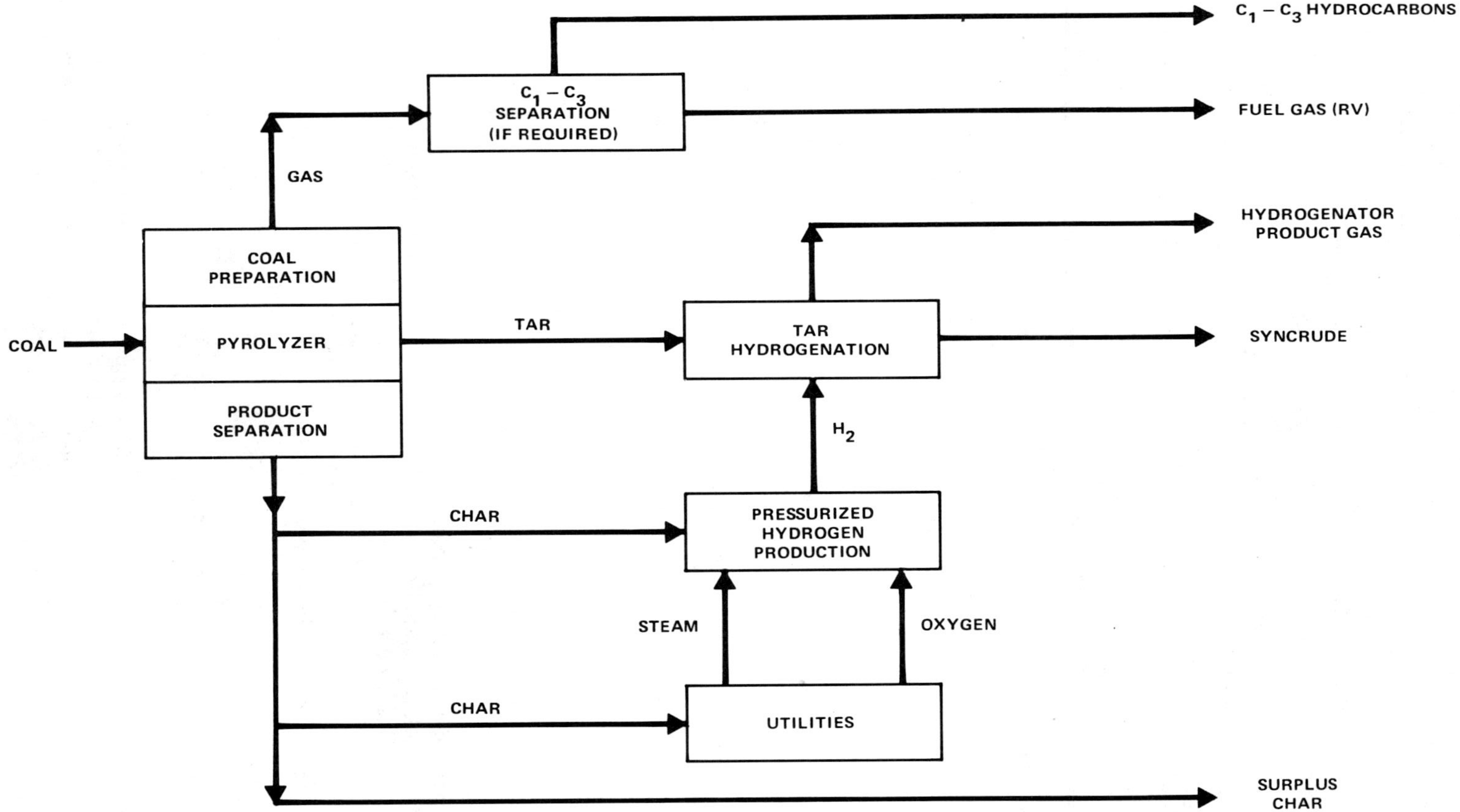

FIG. 39 Simplified block flow diagram, CSIRO flash pyrolysis process.

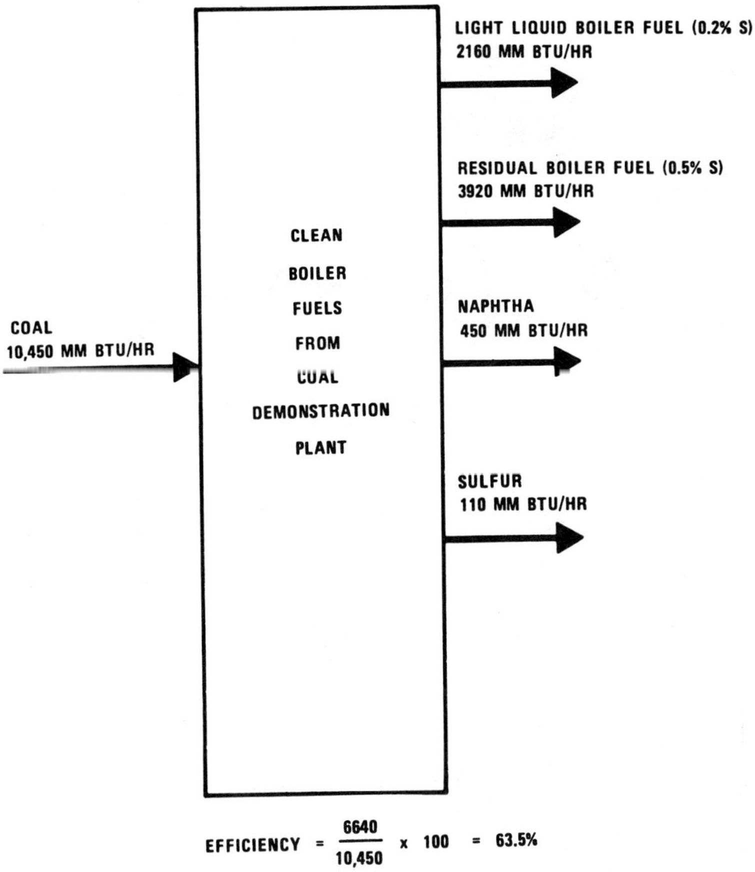

FIG. 40 Overall energy balance, 10,000-ton/day SRC II-based conceptual design. (Courtesy of DOE.)

In general, experience and analysis indicates that the efficiencies for hydroliquefaction processes such as extraction, donor solvent, and direct hydrogenation can be higher than those for complete pyrolysis plants (including conversion of char to SNG, methanol, electricity, or Fischer-Tropsch liquids) and indirect processes such as Fischer-Tropsch, methanol, and gasoline via methanol [31,48,74]. Within any given process classification, the thermal efficiency of the plant will be a function of the design and facilities that are installed to ensure efficient utilization of process by-product heat. The following examples illustrate this.

Figure 40 presents an overall energy balance for a 10,000-ton/day feed coal SRC II-based conceptual design [49]. The predicted thermal efficiency for a self-sufficient plant that captively produces all utilities is about 63%. An analysis of the sources of energy losses is summarized in Fig. 41 [49], which provides a basis for design revisions to increase the thermal efficiency. Figure 41 summarizes the thermal efficiencies around the major units within

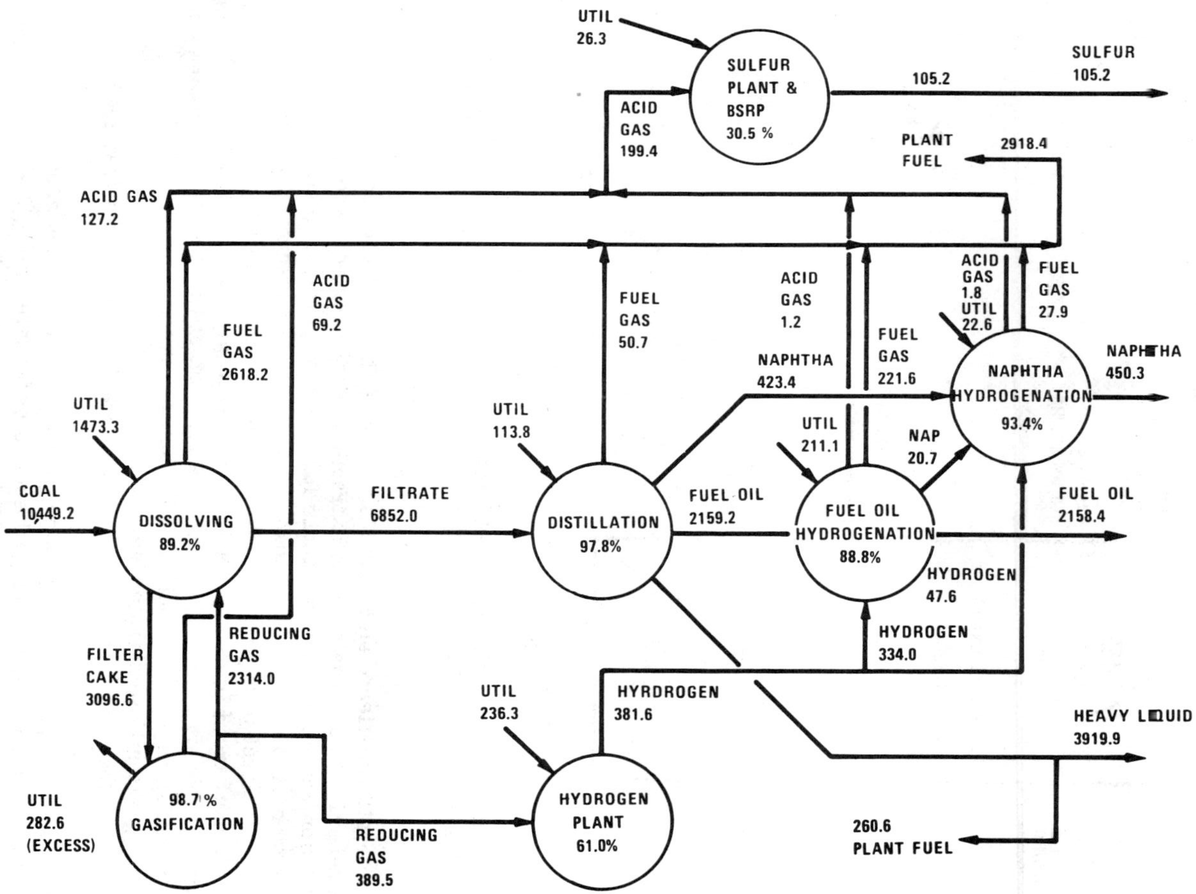

NOTE – LINE FIGURES ARE MM BTU/HR (GROSS)
– FIGURES IN CIRCLES ARE THERMAL EFFICIENCIES OF UNITS

$$\text{OVERALL EFFICIENCY} = \frac{105.2 + 450.3 + 2158.4 + 3919.9}{10{,}449.2} = 63.5\%$$

FIG. 41 Detailed energy balance analysis, Parsons demonstration plant conceptual design. (Courtesy of DOE.)

the complex. The results indicate that large sources of energy inefficiencies are the hydrogen plant and the sulfur recovery plant, among others. This illustrates the fact that techniques are available to guide multiple design analyses for coal conversion plants in order to increase the overall thermal efficiency. The analyses should recognize the fact that as energy becomes more expensive, additional equipment and processing can be justified for efficient heat recovery.

The use of technical and economic analysis procedures to increase thermal efficiencies in a subsequent conceptual SRC II-based design resulted in prediction of the overall energy balance illustrated in Fig. 42; this shows a projected thermal efficiency of coal to salable products on the order of 77%. For this second plant, projected to process 36,000 tons/day of clean, sized bituminous coal, a difference in thermal efficiency of 77% vs. 63% would represent a difference in operating cost of approximately $65 million per year if the feed coal cost were $25 per ton and the plant operated at capacity 330 days/year. These results illustrate that a range of thermal efficiencies can exist and that there are sizable incentives to define design and operations to assure an economically justified level of thermal efficiency.

The actual thermal efficiencies of large, modern, integrated coal liquefaction plants will have to wait until such facilities have been built and operated. However, for guidance, representative predicted overall thermal efficiencies for liquefaction complexes are presented in Table 2 [48]. The results indicate that the hydroliquefaction processes offer the potential for highest overall thermal efficiencies. The projected thermal efficiencies for complete pyrolysis-based plants, including conversion of the char produced in the pyrolysis operation, are considerably lower, in spite of the fact that the efficiency for the pyrolysis step, taking credit for the energy in the char, is of the order of 85 to 90%.

Differences in design procedures in the hydroliquefaction processes contribute to the projected differences in thermal efficiencies; this should be recognized when interpreting the results of Table 2. Improvement of a design to raise efficiencies from the mid-60's to the mid-70's is generally achieved by a series of efficiency/cost analyses, each of which may yield a modest efficiency improvement.

B. Environmental Factors

In mid-1979 there are no regulations specific to coal liquefaction plants in the United States. There are state regulations in New Mexico that would apply to the coal gasification portion of a liquefaction complex; liquefaction plants will in general contain a gasification facility for hydrogen production.

Environmental factors, including projected quantities and compositions of contaminants initially present in internal process and utility streams, methods of treating the effluent streams to remove or neutralize the contaminants, types of equipment used for environmental control, and projected economics have been described [50-57]. A description of the projected environmental factors for a coal liquefaction plant is best developed by presentation of a summary of the equipment, facilities, and effluent quantities and compositions for a specific case. The example summary chosen is a conceptual SRC II-based design. A simplified block flow diagram for this facility

COAL ENERGY DISTRIBUTION		
STREAM	MM BTU/H	% CLEAN COAL
FEEDS		
COAL TO PROCESS	30,310	84.1
COAL TO UTILITIES	5,730	15.9
TOTAL	36,040	100.0
PROCESS COAL YIELDS		
SALEABLE PRODUCTS	27,951	77.6
FILTER CAKE	844	2.3
REACTION HEATS	1,515	4.2
TOTAL	30,310	84.1

UTILITY ENERGY DISTRIBUTION		
STREAM	MM BTU/H	% CLEAN COAL
INPUTS		
COAL	5,730	15.9
FILTER CAKE	844	2.3
STEAM	632	1.8
TOTAL	7,206	20.0
OUTPUTS		
TO MINING AND PREPARATION		
STEAM	458	1.3
POWER	132	0.4
TOTAL	590	1.7
TO OXYGEN PLANT		
STEAM	1,007	2.8
POWER	8	–
TOTAL	1,015	2.8
TO OIL/GAS PLANT		
POWER	601	1.7
FUEL GAS	1,020	2.8
TOTAL	1,621	4.5
LOSSES TO ATMOSPHERE	3,980	11.0
TOTAL	7,206	20.0

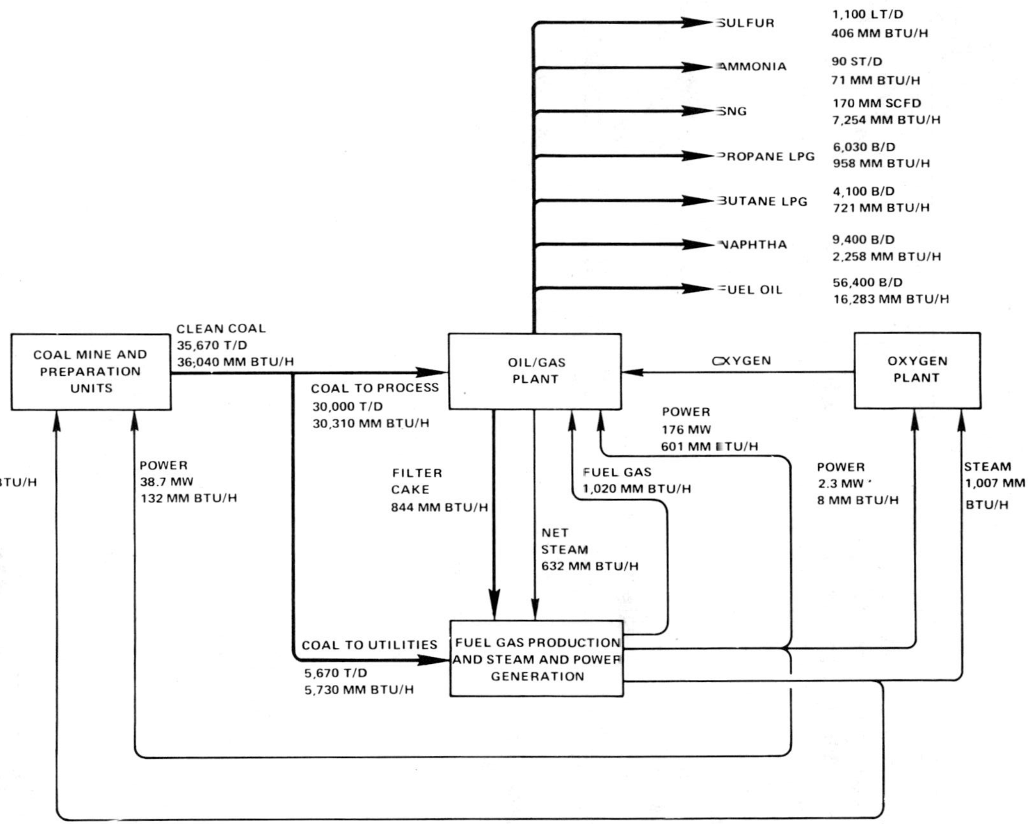

FIG. 42 Overall energy balance, Parsons "oil/gas" SRC II-based conceptual plant. (Courtesy of DOE.)

TABLE 2 Predicted Performance Data for Oil from Coal Processes [48]

Item/Process	Synthoil	H-coal	CSF-donor solvent	SRC II	Sasol I	Sasol II	Advanced Fischer-Tropsch conceptual design	Pyrolysis integrated with Fischer-Tropsch technology	
Process evaluation by	ERDA/DOE [81] (1976)	Fluor [15] (1976)	Fluor [18] (1977)	Parsons [50] (1977)	Sasol [82] (1971)	Sasol-Fluor [83] (1976)	Parsons [30] (1977)	CSIRO [48] (1977)	Parsons/CSIRO [25, 48] (1975)
Type of technology	Direct liquefaction (catalytic)		Solvent extraction		Lurgi gasifiers plus conventional Fischer-Tropsch technology		BI-GAS conceptual gasifier; conceptual Fischer-Tropsch reactor design	CSIRO flash pyrolysis/tar hydrogenation, Koppers-Totzek char gasification; Sasol II-type Fischer-Tropsch synthesis	COED pyrolysis/tar hydrogenation, fluid-bed char gasification; Sasol II-type Fischer-Tropsch
Current status	0.5-ton/day rig o operating; 10-ton/day plant under construction	600-ton/day plant operating	20-ton/day plant shut down in 1970;	45-ton/day plant in operation	Commercial plant operating	Commercial plant operating	Conceptual design	0.5-ton/day pyrolyzer operating	25-ton/day pilot plant

(continued)

(Table 2 continued)

Item/Process	Synthoil	H-coal	CSF-donor solvent	SRC II	Sasol I	Sasol II	Advanced Fischer-Tropsch conceptual design	Pyrolysis integrated with Fischer-Tropsch technology		
Feed coal properties										
Seam/type	Bituminous	Illinois No. 6/bit.	Pittsburgh No. 6/bit.	Illinois No. 6/bit.	Bituminous	Bituminous	Bituminous	Mill-merran/bit.	Ct. North. bit.	Bituminous
Net product yields										
Liquids (bbl/ton, d.a.f.)										
LPG	—	—	0.60	0.37			0.14	—	—	—
Naphtha	—	1.21	—	0.34	1.00	1.50	0.82	—	—	—
Gasoline	—	—	2.66	—			0.82	0.58	0.75	0.96
Synthetic crude	—	1.79	—	—	—	—	—	2.00	1.59	1.44
Fuel oil	3.45	—	—	2.03	—	—	0.20	—	—	—
Tar products	—	—	—	—	0.16	0.15	—	—	—	—
Total liquids	3.45	3.00	3.26	2.74	1.16	1.65	1.98	2.58	2.34	2.4
Other products (% coal energy input)										
Fuel gas	3	—	—	—	9	3	—	—	—	—
SNG	—	—	14	21	—	—	36	—	—	—
Electricity	—	5	—	—	—	—	2	—	—	—
Overall thermal efficiency (%)	68	60	64	77	31	35	70	43	4[illegible]	46
Thermal efficiency liquids (%)	65	55	50	56	22	32	32	48	4[illegible]	46

is shown in Fig. 43 and an overall material balance is presented in Fig. 44 [53]. The plant feeds about 35,000 tons of coal per day and produces about 100,000 barrels of oil equivalent per day in the form of SNG, LPGs, naphtha, and fuel oil.

C. Gaseous Emissions

A flow diagram showing the origin, treatment, and disposition of predicted gaseous emissions for this grass-roots liquefaction complex is shown in Fig. 45 [53]. Key process units that generate gaseous effluents include the process gasifier used for hydrogen production, the dissolver in which the hydroliquefaction reaction occurs, and the gasifier used for fuel gas production.

The major air pollution abatement effort is aimed at desulfurizing the gases generated during the coal conversion process to make the fuels produced environmentally acceptable. The desulfurization process extends to the gaseous fuels produced for in-plant consumption, so that overall plant operations generate negligible sulfur emissions.

Figure 45 shows the nature and amount of all streams vented to the air; these streams consist, for the major part, of inert gases (nitrogen and carbon dioxide). The effluent gases are shown as if vented separately to the air to identify the contribution of specific process units. In reality, however, most streams, with the exception of the particulates from the coal-drying plant and the effluents from the process boilers, are combined into a single stack before venting to the air.

Fugitive particulate emissions from coal sizing, handling, and residual ash disposal are prevented from becoming airborne by maintaining a wet condition when not in a closed system. The coal-grinding and -drying unit is the only source of particulate emissions. A cyclone system removes the coarse particles from the vent streams; a subsequent baghouse system removes most of the fines. Ground, dry coal is fed to three reactors: the process gasifier, the dissolver, and the fuel gas gasifier.

The *process gasifier* receives recycle char, steam, and oxygen in a lower slagging section and ground coal and steam in an upper section. At the top reactor temperature of 1700°F, any oils or tars formed are expected to crack to gaseous products. The gas stream produced contains entrained char and is cooled from 1700 to 945°F through a series of heat exchangers heating the incoming gases and producing steam. The char is then separated by a cyclone system and returned to the gasifier. The remaining fine particles that escaped the cyclone are removed by two dust filters.

The gaseous stream is then conveyed to a shift conversion unit, where the stream is enriched in hydrogen by reaction of carbon monoxide with water. The metal catalyst employed operates in a sulfided condition obtained by initial reaction with the hydrogen sulfide present. Each of the three shift reactors operates at progressively lower temperature, gas cooling being provided by water spray in quench pots. Final cooling of the gas down to 100°F occurs in a series of heat exchangers and condensate separators. The quench pot/condensate separator system acts as a water scrubbing system, removing practically all traces of particulates from the gas stream. The ammonia and part of the hydrogen sulfide and hydrogen cyanide present are also removed.

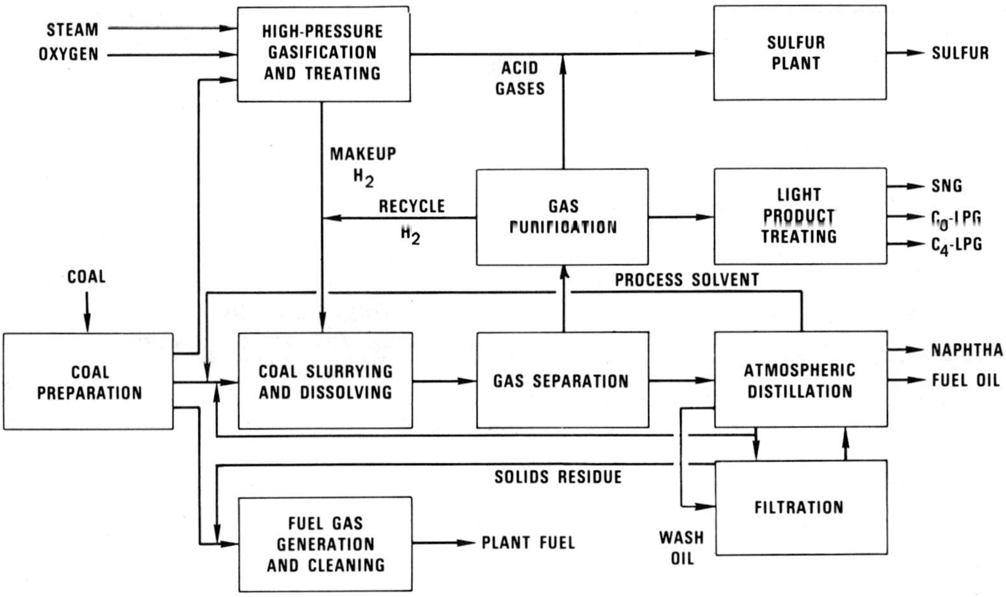

FIG. 43 Simplified block flow diagram, Parsons conceptual commercial SRC II-based coal liquefaction complex [50]. (Courtesy of DOE.)

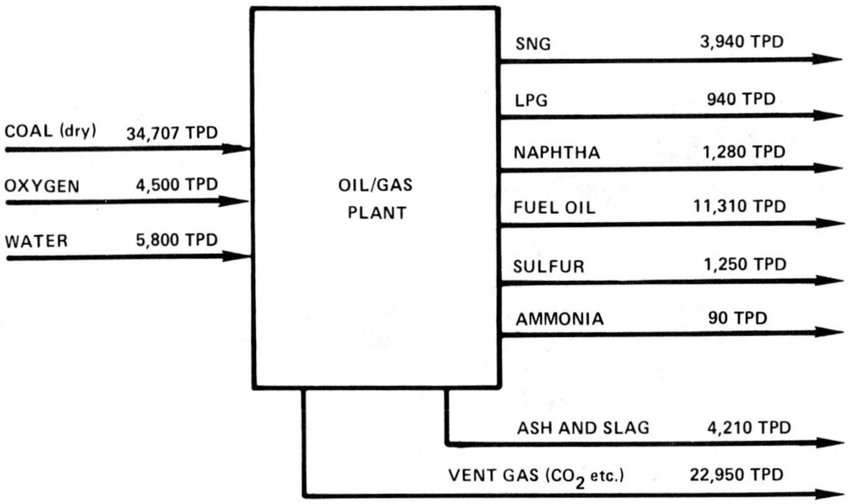

FIG. 44 Overall material balance, Parsons "oil/gas" SRC II-based conceptual plant. (Courtesy of DOE.)

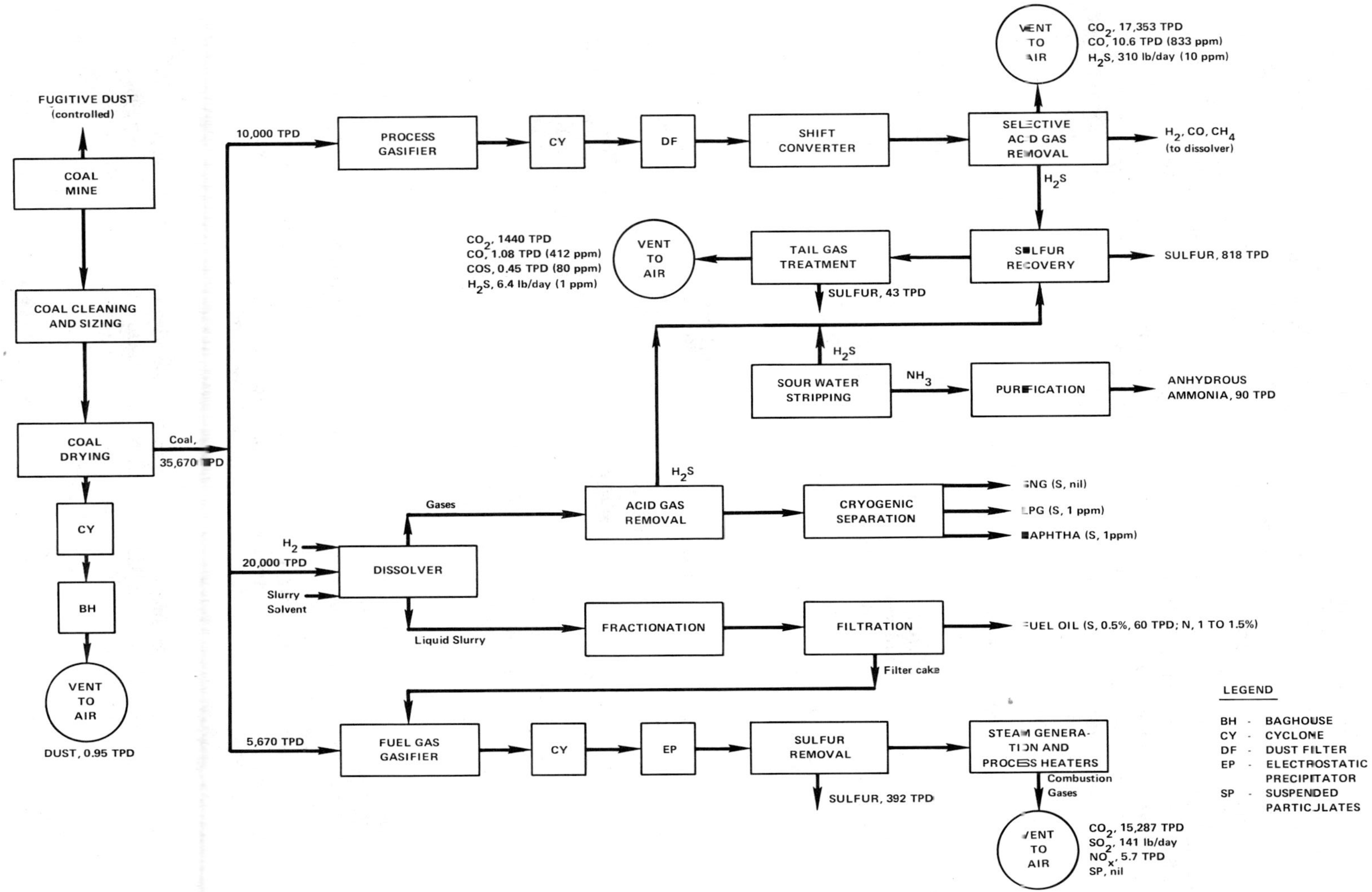

FIG. 45 Block flow diagram of air pollution abatement, Parsons "oil/gas" SRC II-based conceptual plant. (Courtesy of DOE.)

The next treatment step concerns the removal of acid gases. A physical solvent process removes these gases from the main stream; then, on selective regeneration, releases a stream of hydrogen sulfide containing part of the carbon dioxide, and a stream of nearly pure carbon dioxide. The hydrogen sulfide stream is sent to the sulfur recovery plant, where it is oxidized to elemental sulfur; any hydrogen cyanide present is oxidized concurrently. The carbon dioxide stream, containing some carbon monoxide (833 ppm) and hydrogen sulfide (up to 10 ppm), is vented to the air through the main plant stack, where overall concentrations of contaminants decrease on combination with other effluents. The purified gas stream, consisting mainly of hydrogen but with sizable amounts of carbon monoxide and methane present, is conveyed to the coal dissolver.

The dissolver unit receives a preheated, pressurized slurry of ground coal coal in coal-derived solvent and hydrogen and generates two product phases: a gas phase and a slurry phase.

The gas phase is cooled and depressurized (recovering heat and power) and then conveyed to an acid gas removal unit, where hydrogen sulfide is removed by reaction with an amine; the traces of carbon dioxide present are removed by reaction with caustic. The acid gas stream obtained on regeneration of the amine is led to the sulfur-recovery plant. The purified gas stream undergoes drying by molecular sieves and cryogenic separation, with hydrogen returned to the dissolver and hydrocarbon products fractionated to a methane-rich stream, LPG, and naphtha. Naphtha and LPG are treated to reduce the sulfur level to 1 ppm or less, while the methane-rich stream is upgraded to SNG. The SNG produced is essentially sulfur-free because a zinc oxide guard reactor removes sulfur traces prior to methanation.

The liquid slurry phase consists of liquefied coal, unreacted coal, and ash. After cooling by heat exchange and depressurization, the slurry is fractionated (with recovery of solvent for return to the dissolver), and then filtered to separate solid materials from liquid hydrocarbons. The solids (filter cake) are fed to the fuel gas gasifier. The liquids (fuel oil) represent the largest single product stream in the plant. This fuel stream contains 0.5% sulfur and 1 to 1.5% nitrogen: the amount of sulfur is environmentally acceptable; the amount of nitrogen present, however, may lead to increased production of nitrogen oxides on combustion with respect to crude oil-derived fuels.

The fuel gas gasifier receives char, filter cake, air, and steam in a lower (slagging) section; coal and some additional air in the upper section; and produces primarily synthesis gas (carbon monoxide and hydrogen), plus nitrogen, carbon dioxide, hydrogen sulfide, and lesser amounts of ammonia, carbon oxysulfide, and sulfur dioxide. At the elevated reactor temperature, any oils or tars formed are expected to crack to gaseous products. The gas stream produced, leaving the gasifier at 1800°F, is separated from most of the entrained char in a cyclone system and the char is returned to the gasifier after using its heat content for filter cake drying. The gas is then cooled to 100°F through a series of heat exchangers heating the gasifier input gases and producing steam; the condensate formed contains all of the ammonia and part of the sulfides and cyanides, and is treated as described below. The cooled gas stream passes through an electrostatic precipitator, where most of the fine particulates suspended are removed, and is then conveyed to a sulfur removal unit, where a redox solution converts all but 1 ppm of the hydrogen sulfide present to elemental sulfur. At the same time, practically quantitative removal of any remaining particulates is achieved.

The cleaned low-Btu (145-Btu/ft^3) fuel gas is used to satisfy in-plant energy requirements for steam generation and process heaters. On combustion a negligible amount of SO_2 (141 lb/day) is generated by oxidation of the trace of hydrogen sulfide present. The amount of nitrogen oxides generated is also moderate, because of the flame-cooling effect of the large amount of inert gases (carbon dioxide and nitrogen) present. Combustion experiments with low-Btu gas show that a maximum NO_x concentration in the effluent gases of 50 ppm (as NO_2) can be expected.

The sulfur recovery plant receives three hydrogen sulfide streams, originating from the acid gas removal unit on the process gasifier train, from a similar unit on the dissolver train, and from the sour water stripping unit; 95% of the sulfur present is converted to high-purity elemental sulfur, with the bulk of the remaining 5% converted in a subsequent unit, the tail gas unit. The tail gas treatment involves reduction of all sulfur species to hydrogen sulfide, absorption by an alkaline solution, and oxidation to give high-purity sulfur. The final vent gas contains carbon dioxide, plus traces of carbon oxysulfide (80 ppm), hydrogen sulfide (1 ppm), and carbon monoxide (412 ppm). The carbon monoxide and carbon oxysulfide concentrations decrease on mixing with the other vent gases.

In conclusion, air effluents are emitted into the atmosphere from four outlets: the coal drying plant (particulates), two process heaters (one-fifth of fuel combustion gases), and the main plant stack (all other effluents combined). The composition of the effluents emitted to the air from the outlets specified is detailed in Table 3.

Source emission standards for coal conversion plants have not been issued by the federal government. Among the states, only New Mexico has issued specific regulations covering coal gasification plants; these regulations will be considered here for illustration purposes only. The State of Illinois, a likely candidate for a plant site, has issued standards for petrochemical process; this process is related to the Oil/Gas plant operation. The standards are as strict or stricter than the federal standards for petroleum refining.

The Illinois and the New Mexico source emission standards are compared for illustration purpose in Table 4, with the projected emissions for this conceptual design. As shown in the table, all estimated emissions are lower than the ones mentioned in the standards.

D. Sulfur Balance

The sulfur balance for the conceptual design of a commercial Oil/Gas plant is detailed in Table 5. A total of 95% of the coal sulfur content is recovered as elemental sulfur, with most of the remainder present in the fuel oil product. The amount adsorbed by the catalyst beds (mainly in the shift reactors) would be emitted every 6 months as sulfur dioxide on regeneration of the catalyst. At the temperature conditions prevailing in the gasifiers, it was assumed that negligible amounts of sulfur would remain with the ash.

TABLE 3 Characteristics of Effluents Emitted to the Air [53] (Courtesy of DOE.)

Outlet	Gaseous Effluent	Amount	Concentration (ppm)
Main stack	Carbon dioxide	31,024 tons/day	—
	Carbon monoxide	11.7 tons/day	166
	Carbon oxysulfide	0.5 tons/day	3
	Hydrogen sulfide	316 lb/day	2
	Sulfur dioxide	113 lb/day	0.3
	Nitrogen oxides	4.5 tons/day	
Process heater 1	Carbon dioxide	1,528 tons/day	—
	Sulfur dioxide	14 lb/day	0.4
	Nitrogen oxides	0.6 tons/day	50
Process heater 2	Carbon dioxide	1,528 tons/day	—
	Sulfur dioxide	14 lb/day	0.4
	Nitrogen oxides	0.6 tons/day	50
Coal-drying plant	Particulates	0.95 tons/day	—

E. Aqueous Effluents

A simplified block flow diagram showing water supply, treatment, and discharge is shown in Fig. 46. This conceptual liquefaction plant is considered to be located in the U.S. Eastern Interior (coal) Province, where an abundant supply of water is available. One of the major rivers in this area is considered to provide the water supply for the plant.

Based on adequate availability of water, the wastewater treatment is a combination of recycling and discharge of aqueous effluents. The most heavily contaminated streams undergo steam distillation; the organic contaminants separated are fed with steam to the process gasifier to achieve their thermal destruction. Medium-contaminated streams are purified physically and biologically, then are reused for slag quenching. The lightly polluted streams are treated to make them acceptable to the environment and are then discharged to the river.

The generation and treatment of aqueous contaminants is outlined in Fig. 46. Wastewater sources are listed on the left-hand side of the figure, with the degree of pollution of the wastewater streams decreasing from top to bottom. The progressive treatment and disposition of the streams is also shown and approximate flow values for streams withdrawn and returned to the river are reported.

The most highly contaminated stream is the combined sour water generated as aqueous condensates by the process gasifier, fuel gas gasifier, and coal dissolver. Major contaminants present are hydrogen sulfide, ammonium sulfide, phenols, cresols, xylenols, and thiocyanates; cyanides and solids (ash and char particles) may also be present. The gaseous contaminants (hydrogen sulfide and ammonia) are removed by steam-stripping. Ammonia is separated

TABLE 4 Comparison of Gaseous Emissions with Illinois and New Mexico Source Emissions Standards [53] (Courtesy of DOE.)

Pollutant	Illinois standards, petrochemical plant	New Mexico standards, coal gasification plant	Gaseous effluents, Oil/Gas plant
Particulate matter	81 lb/hr	0.03 gr./ft^3	79 lb/hr[a], 0.023 gr./ft
Sulfur dioxide	1.2 lb/10^6 Btu	—	1 ppm[b]
Carbon monoxide	200 ppm, 50% excess air	—	147 ppm[c]
Nitrogen oxides	0.2 lb/10^6 Btu[d]	—	0.1 lb/10^6 Btu
Organics (methane excluded)	100 ppm (CH_4 equivalent)	—	Nil
Total reduced sulfur (H_2S + COS + CS_2)	—	100 ppm	5 ppm
Hydrogen sulfide	—	10 ppm	2 ppm
Hydrogen cyanide	—	10 ppm	Nil
Hydrogen chloride/hydrochloric acid	—	5 ppm	Nil

(continued)

(Table 4 continued)

Pollutant	Illinois standards, petrochemical plant	New Mexico standards, coal gasification plant	Gaseous effluents, oil/gas plant
Ammonia	—	25 ppm	Nil
Gas-burning process boilers, particulate matter	—	0.03 lb/10^6 Btu, LHV	Nil
Gas-burning process boilers, sulfur dioxide	—	0.16 lb/10^6 Btu, LHV	0.001 lb/10^6 Btu
Total sulfur	—	0.008 lb/10^6 Btu of feed (coal) heat input, HHV	0.002 lb/10^6 Btu[e]

[a]From coal-drying plant.

[b]66.4 tons of sulfur dioxide emitted twice a year, over 24-48 hr, on regeneration of the catalyst of each shift reactor (three reactors total). The sulfur dioxide generated on combustion of the plan fuel gas amounts to 26 tons/year. If these combined amounts were averaged out over the year, they would correspond to 0.003 lb/10^6 Btu/day.

[c]Value obtained on application of the 50% excess air correction to the effluent from the steam generating unit.

[d]Applies to firing of gaseous fossil fuel.

[e]Includes the sulfur dioxide emitted occasionally on regeneration of the shift reactor catalyst (see note b above).

TABLE 5 Sulfur Balance, Conceptual Coal Liquefaction Complex [53] (Courtesy of DOE.)

Total input from the typical feed coal		1312.7
Outputs:	As elemental sulfur	1252.3
	In the fuel oil	59.9
	As carbon oxysulfide emissions	0.2
	Adsorbed on catalyst beds	0.3
		1312.7

by contact with an ammonium phosphate/phosphoric acid solution, regenerated on heating, and purified to anhydrous ammonia. Hydrogen sulfide is conveyed to the sulfur plant for conversion to elemental sulfur. Under the acidic conditions of the sour water, any cyanides present would be volatile (hydrogen cyanide), and therefore would be stripped with the hydrogen sulfide and conveyed to the sulfur plant to be destroyed by thermal oxidation. The stripped aqueous stream is conveyed to a high-temperature steam drum in the process gasifier unit, where the organic contaminants volatilize and are fed with the steam to the gasifier to undergo thermal oxidation. The blowdown from the steam drum is vaporized on cooling the product gas prior to shift conversion; any solids present are removed by filters in the condensate return lines.

Oily water streams produced during plant operation or collected as contaminated runoff from process areas are conveyed to a gravity separator. The water phase is further purified in an oily water pond, with the effluent used for slag quench. The combined oil fractions are led to a separator, with the oil phase returned to the fractionation unit for product recovery. The aqueous phase is combined with the effluents from the slag settling basin and the sewage treatment plant and treated in a biooxidation pond, which also provides firewater; the effluent from the biopond is also used for slag quench.

The cooling-tower blowdown stream is the largest in volume and is only lightly contaminated by corrosion inhibitors (zinc salts and inorganic phosphates). This stream is combined with deionizer wastes (containing mainly inorganic salts) and boiler blowdown, also lightly contaminated by corrosion inhibitors. After neutralization, it is treated with lime in a settler-clarifier. The lime sludge, containing most of the zinc and phosphates, is disposed of in a landfill, while the treated stream is returned to the river, together with any runoff collected from clean areas. If required by specific site conditions, a final polishing pond may be added for removal of suspended solids from the runoff collected.

No aqueous effluent standards specifically addressed to coal conversion plants have been issued by the federal government or by state legislatures. Federal standards issued for petroleum refining are somewhat related to a coal liquefaction process. Average attainable concentrations which were the base for such standards [59] are reported in Table 6, together with the

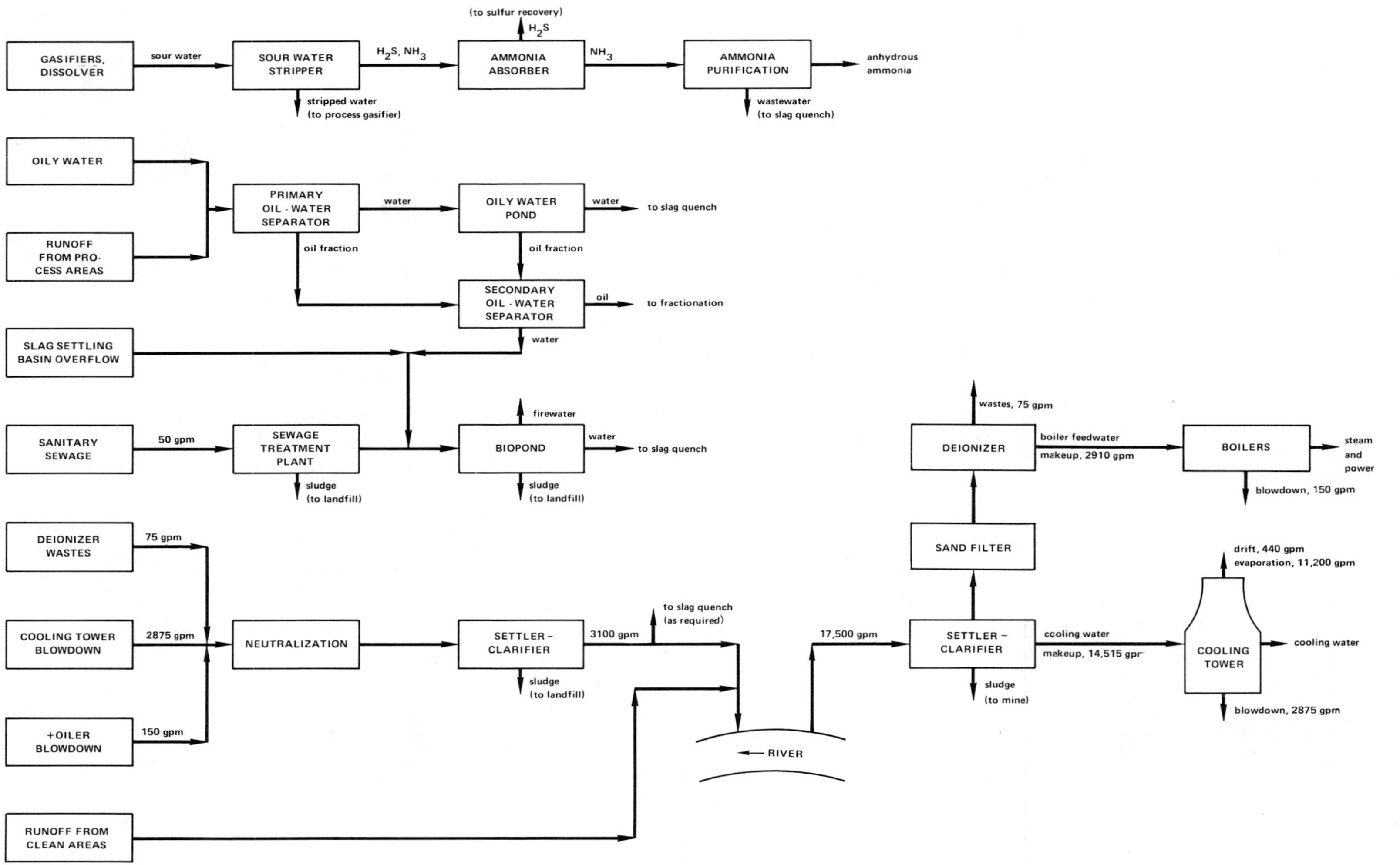

FIG. 46 Block flow diagram of water treatment and supply, Parsons "oil/gas" SRC II-based conceptual plant. (Courtesy of DOE)

TABLE 6 Comparison of Aqueous Effluents with Federal Petroleum Refinery Standards [53] (Courtesy of DOE.)

	Federal standards petroleum refinery	Aqueous effluents, oil/gas plant (mg/liter)
BOD-5	15	10
COD	100	100
Total organic carbon	33	33
Suspended solids	10	10
Oil and grease	5	5
Phenol	0.1	Nil
Ammonia-N	80% removal	Nil
Sulfide	0.1	Nil
Cr, tertiary	0.25	Nil
Cr, hexavalent	0.005	Nil

Note: These are the average attainable concentrations from the application of best practicable control technology currently available (EPA-440/1-74-014a).

corresponding values predicted for the aqueous effluents estimated for this conceptual design. As shown in the table, these predicted values are either the same or lower than the federal parameters.

The State of Illinois has issued aqueous effluent standards applicable to all sources discharging to the natural waters of the state. These standards are reported for illustration purpose in Table 7. All predicted conceptual plant effluents estimated either meet or are lower than such standards.

F. Solid Wastes

The conceptual liquefaction plant generates two main types of solid-waste materials: ash and slag from the gasifiers and sludge from various wastewater treatment units. Burial of the slag with the mine spoils and landfill disposal of the ash and sludge are proposed.

The possibility of leaching of trace metals from the ash into ground or surface waters has been questioned. The difference in physical properties between slag and fly ash would suggest a different leaching behavior; slagged ash exhibits a glass matrix that would possibly inhibit leaching. Experimental studies have been carried out on the leaching of power-plant fly ash or unslagged bottom ash. According to the results of a recent study [60], selenium, chromium, and boron (and occasionally mercury and barium) were released on simulated leaching, and the concentrations reached exceeded the values recommended by EPA for public water supplies. No leaching tests of slagged ash to fully confirm this are known to have been carried out.

TABLE 7 Aqueous Effluent Standards, State of Illinois [53] (Courtesy of DOE.)

Constituent	Maximum concentration (mg/liter)
Arsenic (total)	0.25
Barium (total)	2.0
BOD-5	10.0
Cadmium (total)	0.15
Chromium (total hexavalent)	0.3
Chromium (total trivalent)	1.0
Copper (total)	1.0
Cyanide	0.025
Fluoride (total)	15.0
Iron (total)	2.0
Iron (dissolved)	0.5
Lead (total)	0.1
Manganese (total)	1.0
Mercury (total)	0.0005
Nickel (total)	1.0
Oil (hexane solubles or equivalent)	15.0
pH	5-10
Phenols	0.3
Selenium (total)	1.0
Silver	0.1
Zinc (total)	1.0
Total suspended solids	12.0

G. Trace Elements

Because of its organic origin and its intimate commixture with crustal formations, coal contains a large number of elements in minor or trace quantities [61]. Actually, of 92 known nontransuranic elements, only 15 (shown in Fig. 47) have not yet been found in coal.

A number of studies have analyzed the behavior of trace elements in coal-fired power plants [62,63]. In general, the elements have been divided into two groups, those appearing mainly in the bottom ash (elements or oxides

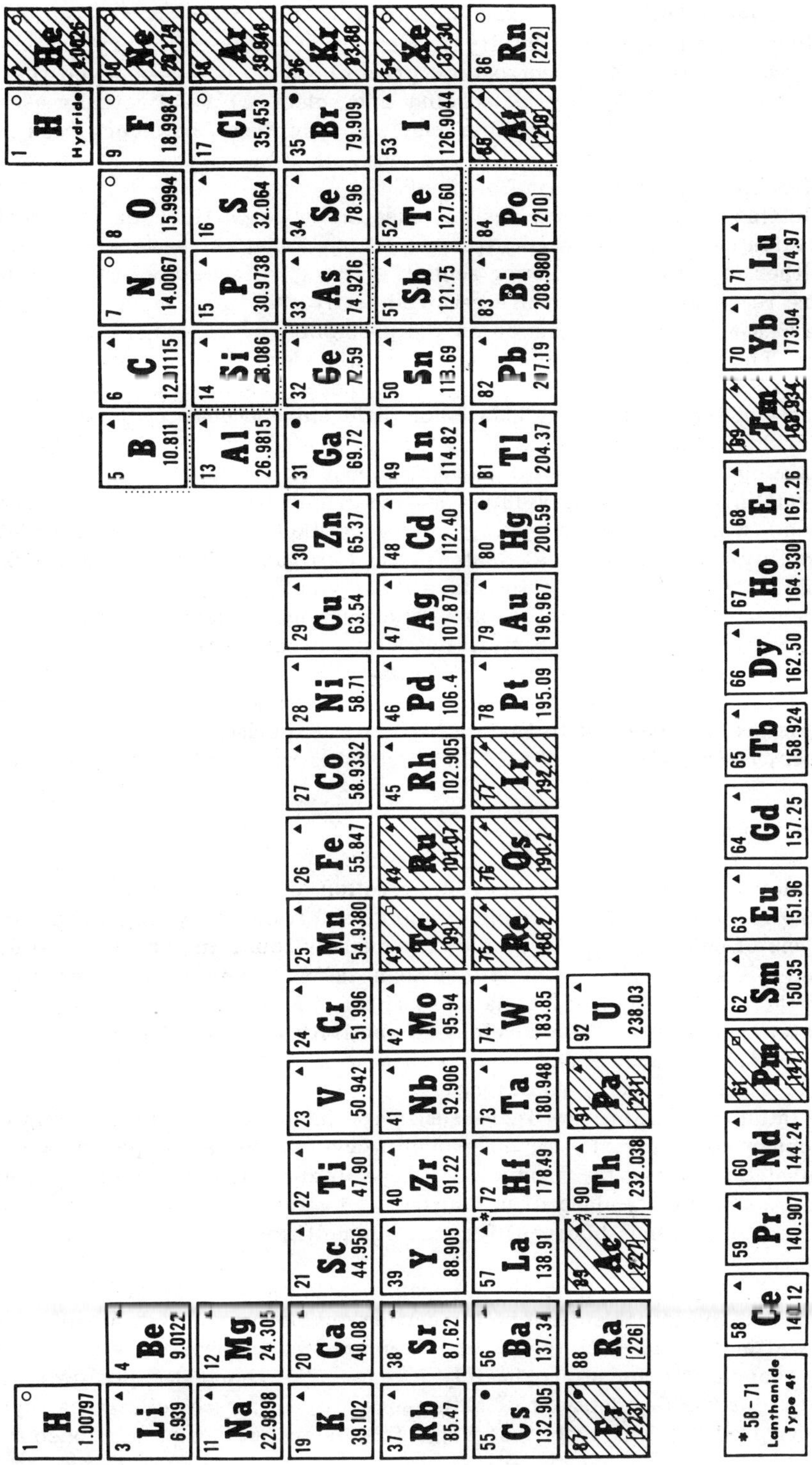

FIG. 47 Periodic table of elements. Elements shaded have *not* been found in coal.

having lower volatility) and those appearing mainly in the fly ash (elements or oxides haveing higher volatility). For power plants using dry particulate collection devices (e.g., electrostatic precipitators), it was believed that the most volatile elements (such as mercury and selenium) could actually escape in the elemental state with the flue gas; wet scrubbers, however, were believed capable of removing most of the elements from the gas streams and transferring them to the liquid effluent.

In this conceptual plant design, essentially no particulates from coal combustion escape into the atmosphere, all particulate streams are collected and either returned to the lower section of the gasifiers, where they melt and are removed as slag, or are removed as fly ash. Any eventual dispersion of the elements present depends on the possibility of leaching, discussed above. The concern, therefore, is to identify elements that may be occurring in the gaseous state. The reducing atmosphere present in the middle and top parts of the gasifier may also favor different chemical combinations, such as hydrides, absent in the oxidizing atmosphere of a power plant.

Among the trace elements present in coal with recognized toxic properties, high-volatility elements (beryllium, mercury, and lead), do not form gaseous hydrides, will condense on cooling, and very likely will be almost completely removed by the aqueous condensates formed on gas cooling and/or purification. Arsenic, antimony, and selenium have lower volatility but can form gaseous (covalent) hydrides: arsine, stibine, and hydrogen selenide. These hydrides, however, have stability characteristics which preclude their formation at the temperature and pressure prevailing in the Oil/Gas plant gasifiers. From general chemical principles, it would appear, therefore, that harmful trace elements are not released to the atmosphere. Experimental confirmation, however, is desirable and should be obtained from specific pilot-plant studies.

H. Occupational Safety

Of particular interest in coal hydroliquefaction and pyrolysis plants is the possible formation of carcinogenic compounds formed on hydrogenation and pyrolysis of coal. These compounds are usually found in coal tars and coal-derived oils with boiling point higher than 480°F. They consist mainly of polycyclic aromatic hydrocarbons and amines, which are lipid-soluble and can enter cells easily. The carcinogenic activity is believed to be due to their molecular shape, which permits them to fit into the structure of DNA and cause self-producing mutations.

Statistically significant increases in the incidence of skin and lung cancer were observed, especially for coke-oven workers, who are exposed to contact and inhalation of coal tar products. These findings led the federal government to issue an OSHA Standard of 0.2 mg/m^3 (8-hr average) for coal tar pitch volatiles [anthracene, benzo(a)pyrene, phenanthrene, acridine, chrysene, pyrene].

For both human beings and laboratory animals, occasional exposure to the carcinogens is not sufficient for cancer development. Rats developed tumors after the chemical agents were applied two to three times per week to the shaved skin for approximately 1 year. The shortest exposure time recorded for the development of skin cancer in workers handling coal hydrogenation products was 9 months [64]. The federal standard mentioned represents the consensus of governmental and industrial hygienists on a threshold limit value that would minimize exposure risks.

Oils and tars are not expected to be formed under the operating conditions of the entrained gasifiers used in the Oil/Gas plant. The entire dissolver train, however, handles coal-derived oils and will therefore require strict application of industrial hygiene practices.

I. Summary

Analysis of the results of a conceptual commercial design of a coal hydroliquefaction plant capable of converting approximately 35,000 tons of coal per day indicates that environmental acceptability of the plant can be achieved by purification of gaseous and liquid streams; all effluents released to the environment were predicted to satisfy environmental standards and requirements.

Other environmental areas specific to coal conversion processes, such as disposal of slagged ash, the fate of trace elements, and occupational safety, were also analyzed. Experimental programs required to provide additional data and information are recommended.

VII. ECONOMICS

A number of economic estimates for production of liquids from coal have been published [7,8,12,17,18,24,27,30,32,45,50,65-76]. These estimates have been developed by various organizations, including operating companies, trade groups, and engineering-construction firms. There have been differences in reported predictions and a range of estimates have been presented.

It is not possible to be precise at this time in predicting the exact capital investments and product costs for modern coal liquefaction plants. The true costs will be known after one or more industrial scale plants have been built and operated.

Project factors that must be finalized to provide accurate cost estimates include:

1. Cost of coal
2. Plant capacity
3. Location of plant
4. Environmental regulations
5. Final specifications of equipment
6. Policy regarding equipment sparing—to improve plant reliability or "on-stream" time
7. Ability of the equipment and instrument supply industry to respond to the quality and quantity requirements for construction of these plants
8. Ability of the engineering-construction industry to effectively implement and manage the large projects
9. Reliability of plant performance
10. Accounting guidelines and investment incentives for this type of plant

Agreement on the foregoing factors provides the basis for a projection of economics.

The comparison of predicted economics proposed by different organizations must be done carefully because of differences that may have been used in project criteria, economic parameters, source of project financing, and a number of related factors contained in the preceding list. Several specific points will serve to illustrate these effects:

1. Coal liquefaction plants are highly capital-intensive; the predicted profitability is therefore highly sensitive to the financing method used. The use of 100% equity financing can predict a required product selling price (RPSP) to achieve a target profitability some 30% greater than use of a financing plan consisting of 65% debt and 35% equity.

2. The cost of construction of a plant can vary by 25% depending on the location. Factors that can contribute to this difference include labor availability, labor productivity, existing infrastructure to serve the construction effort, and utilities availability factors. This difference in fixed capital investment affects the predicted profitability; the sensitivity of RPSP to capital burden for this type of plant is about 0.8, which means that for a 25% higher capital burden, the required product selling price at a given profitability will be increased by about 20%.

3. The profitability, expressed in current dollars, is highly sensitive to escalation and inflation rates used. Many of the estimates are presented as "instantaneous economics," whereas some will recognize inflation and real cost (or revenue) escalation. "Instantaneous economics" assumes that all project activities occur at a given time (i.e., the economic factors are discounted to a given date). To illustrate, a project consisting of a 5-year construction period and a 20-year operating period showed a discounted cash flow rate of return (DCF) of about 6% for "instantaneous economics" and a 14% DCF, in current ("as-spent") dollars, for a 6% inflation rate for all project factors [67]. This impact should be recognized when interpreting economic projections.

4. The method of depreciation selection will affect the predicted profitability for these capital-intensive plants. To illustrate, the use of the double-declining-balance method for depreciation allowances in one case resulted in a DCF 2% greater than for the same project using straight-line depreciation [67].

5. The method of "scale-up" from pilot-plant operations to commercial plant design for "first-of-a-kind" plants affects the projected economics. There are many factors involved in scale-up [78], including reactor configuration/capacity, product quality, environmental factors, equipment selection/modification/reliability, and others. These scale-up decisions affect both the capital investment and operating costs.

The few illustrations given above indicate that differences in economic projections of the order of 50 to 100% can result from selection of the project design, structure, and financing methods used and that all project plus financial parameters should be assessed when drawing economic conclusions.

A. Example

The development of economic projections for a coal liquefaction plant may best be illustrated by an example. The absolute economic values generated in the illustration are less important than the description of the separate factors that contribute to the total economics.

TABLE 8 Design Basis Factors, POGO Coal Refinery

Design basis:	1. Processing based on SRC II technology 2. Location: Eastern Region, Interior Coal Province (Illinois, Indiana, Kentucky) 3. Coal: Illinois No. 6 seam 4. Grass-roots facility; all utilities and oxygen generated in the complex 5. Captive surface coal mine 6. Environmental: all effluent streams treated to meet environmental standards 7. 14-day feed coal inventory, 30-day product storage inventory 8. Equipment-sparing philosophy: critical rotating equipment is spared 9. Coproduce 1000 MW of electrical power
Coal feed rate:	45,000 tons/stream day of cleaned, sized coal fed to process plant
Products:	1. Substitute natural gas (SNG): 150 million scf per stream day 2. Liquefied petroleum gas (LPG): 15,000 bbl/stream day (BPSD) 3. Unleaded gasoline: 35,000 bbl/stream day 4. Distillate fuel oil: 27,000 bbl/stream day 5. Green crystalline coke for use in steel and aluminum industries: 1600 tons/stream day 6. Electrical power: 1000 MW

The example is a conceptual design for a grass-roots coal conversion complex to produce approximately 100,000 bbl/day of oil equivalent and 1000 MW of electrical power. The fuels consist of SNG, LPGs, gasoline and distillate fuel oil; 1600 tons/day of coke are also produced. The design was given the acronym POGO, for Power-Oil-Gas-Other, by DOE [9]; it incorporated certain process and equipment elements then in the development stage. One of its objectives was to illustrate the potential for this type of future mature synfuels plant.

Summaries of the design basis and economic parameters are presented in Tables 8 and 9, respectively. A simplified block flow diagram for the complex is shown in Fig. 48 [68]. All economics are presented in mid-1977 dollars.

For estimating and analytical purposes, the complex was divided into five cost centers:

1. Coal mine
2. Coal preparation plant
3. Process plant
4. Power plant
5. Off-sites

TABLE 9 Economic Parameters, POGO Coal Refinery

Economic parameters
1. Type fixed capital investment (FCI) estimate: preliminary; zero contingency; user adjusts basic estimate by using sensitivity factors
2. Project design, procure, construct schedule: 5 years
3. Operating period: 20 years
4. Scrap value after 20 years: 0
5. Total capital investment = FCI + initial catalysts and chemicals + startup costs + construction financing
6. Working capital = estimated
7. Operating labor = estimated by operating function and unit
8. Return on investment: discounted cash flow rate of return after tax (DCFRR)
9. DCFRR = 12%
10. Inflation rate = 0
11. Investment tax credit = 10% on 90% FCI
12. Organizational structure: plant is integrated subsidiary of profit-making industries that can write off initial losses against current earnings for tax purposes
13. Depreciation: Per "Depreciation Guidelines and Rules, Revenue Procedure 62-21"; used double-declining-balance method for all assets but coal mine; coal mine based on depreciation guidelines for the mining industry
14. Required product selling price (RPSP): the RPSP was projected to provide a 12% DCFRR for the project basis and economic parameters described and for two project financial structures as described below
15. Project financial structures: (1) 100% equity and (2) 65/35 ratio of debt/equity; borrowed capital at 9% interest, including construction financing, with 0.75% commitment fee and principal repaid in equal installments over the 20-year project term; all working capital borrowed for 20-year term

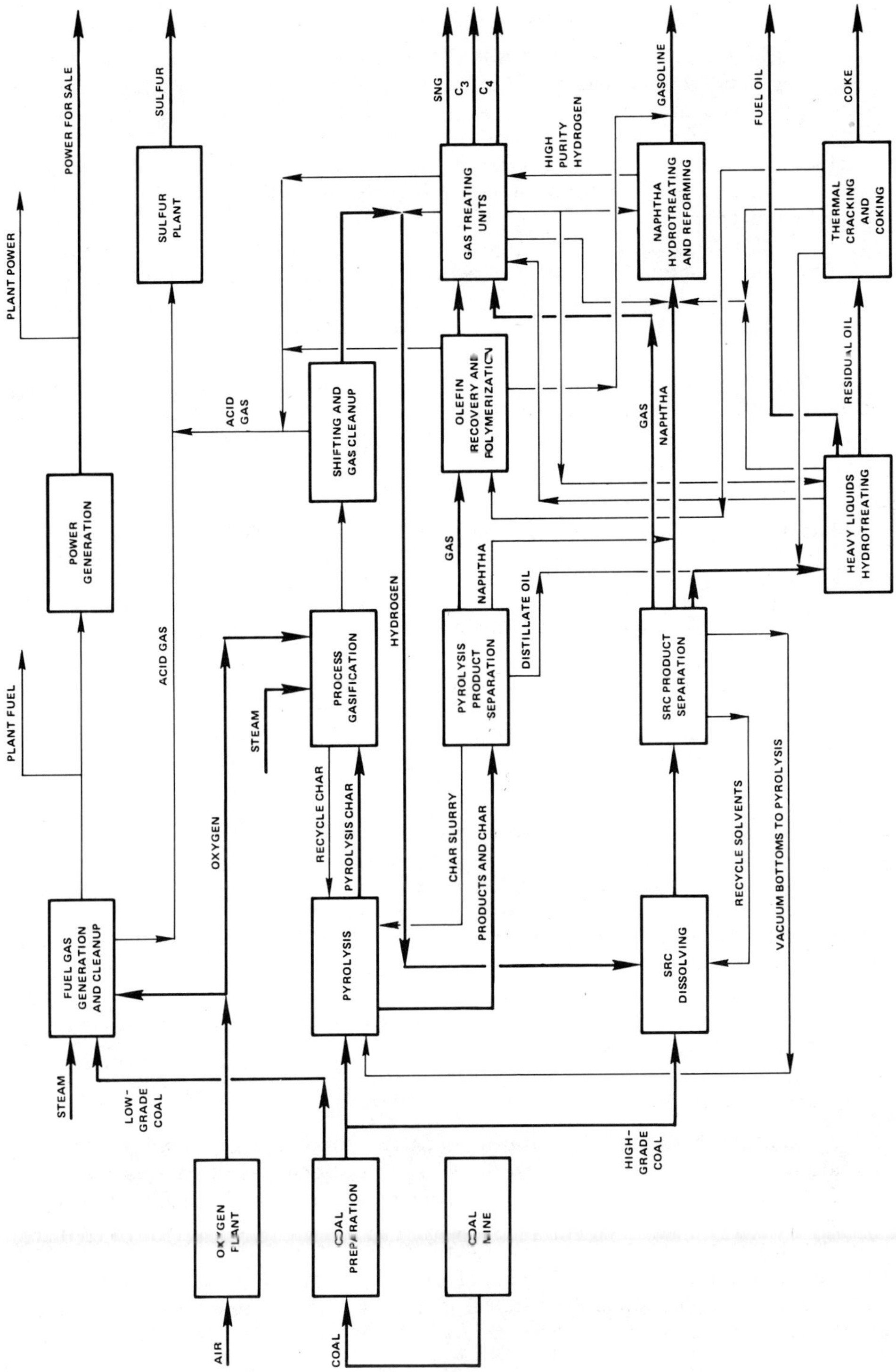

FIG. 48 Block flow diagram, POGO coal liquefaction complex [68]. (Courtesy of DOE.)

1. *Fixed Capital Investment*

The fixed capital investment was estimated for the grass-roots complex. Necessary ancillary facilities are included, with such items as administration, warehouse, laboratory, change house, and related buildings and equipment; computer capability and communications systems; rolling stock, including trucks and automobiles for transportation within the confines of the complex; road paving and fencing; utilities distribution; and other items required for the efficient operation of an industrial complex of this magnitude.

The estimated fixed capital investment for the complex is \$2.4 billion.

The total constructed costs, approximately \$2.2 billion, are shown for each of the unit areas in Table 10. To this have been added home office costs and sales taxes, which result in the total project fixed capital investment. The fixed-capital-investment estimate includes the cost of equipment, engineering, design and drafting, project management, field direct and indirect costs, procurement, contractor's fee, and supporting services. Allowances for instrument checkout and mechanical run-in are also included.

2. *Escalation*

Escalation for the period after mid-1977 is not included.

3. *Contingency*

No contingency allowance has been included; this judgment is left to the reader. The sensitivities reported later permit rapid determination of the effect of changes in fixed capital investment on profitability.

4. *Exclusions from Fixed-Capital-Investment Estimate*

The following cost items are excluded from the estimate:

1. Owner's expenses connected with project
2. All taxes, except sales and payroll taxes
3. Client's local, state, and federal permits
4. Premium-time costs, except nonscheduled overtime premium
5. Piling and unusual foundation conditions
6. Process licensing fees

The fixed capital investment is summarized by cost centers in Table 11 for use in economic analysis.

5. *Total Capital Requirements*

In addition to the fixed capital investment, total capital requirements include land acquisition and rights-of-way, initial charges of catalysts and chemicals, startup costs, construction financing, and provision for working capital. Detailed estimates of these items were made and resulted in catalysts and chemicals (\$9 million), startup (about \$148 million), working capital (\$165 million), and recurring capital expenditures for replacement of coal mining equipment. The allowance for mineral rights is included in annual operating expenses in the form of royalties.

TABLE 10 Fixed Capital Investment, POGO Coal Refinery (Courtesy of DOE.)

Unit number	Description	Cost (millions of dollars)
08	Coal mine	231.3
09	Coal preparation	64.0
10	Coal storage, grinding, and drying	36.7
11	Oxygen plant	181.8
12	SRC dissolving	209.8
13	SRC atmospheric distillation	23.9
14	SRC vacuum distillation	28.0
15	Pyrolysis	16.1
16	Pyrolysis atmospheric distillation	1.9
17	Sour gas compression	5.5
18	Process gasification	24.0
19	Shift conversion	40.4
20	Selective acid gas removal	48.6
21	Heavy liquids hydrotreating	83.8
22	Thermal cracking	28.7
23	Coking	55.6
24	Naphtha hydrotreating	19.2
25	Naphtha reforming	12.8
26	Olefinic gas/acid gas removal	11.8
27	Saturate gas/acid gas removal	21.0
28	Olefin recovery and polymerization	3.3
29	Hydrogen recovery and purification	23.5
30	SNG purification	21.2
31	LPG fractionation	23.3
32	Sulfur plant	33.7
33	Fuel gas generation	96.4
34	Fuel gas/acid gas removal	64.5
35	Steam and power generation	348.0
36	Process wastewater treating	9.3
37	Shops and buildings	16.1
38	Fire water system	3.9
39	Potable and sanitary water system	0.4

(continued)

(Table 10 continued)

Unit number	Description	Cost (millions of dollars)
40	Raw water system	24.3
41	Effluent water treating	2.1
42	Product storage	30.6
43	Flare system	1.5
44	Site preparation, roads, and railroads	15.5
45	Instrument and plant air systems	4.0
	Total direct construction cost	1866.5
	Field indirects at 16%	298.7
	Total constructed cost	2165.2
	Home office overhead and fee at 10%	216.5
	Total	2381.7
	Sales tax	42.5
	Total fixed capital investment	2424.2
	Say	2425

TABLE 11 Fixed Capital Investment by Cost Center, POGO Coal Refinery (Courtesy of DOE.)

Item	Mine	Coal preparation	Process plant	Power plant	Offsites	Total
Fixed capital investment						
Total constructed cost	268	75	1279	404	141	2165
Home office cost	27	7	128	40	14	217
Sales tax	5	2	25	8	3	43
Total fixed capital investment	300	83	1432	452	158	2425

The projected initial capital requirements for the five cost centers used for estimating operating costs are summarized in Table 12. The estimated total capital requirement of the project amounts to about $2.75 billion exclusive of construction financing costs, which depend on the project schedule, fund drawdown, and method of financing the project.

6. *Project Schedule*

The estimated project design, engineering, procurement, construction, and startup schedule is shown in Fig. 49. This schedule was developed based on an analysis of the design, procurement of schedule-controlling equipment, fabrication, erection, and construction schedules. The result is project mechanical completion in 00 months and capacity production 1 year later.

The estimated time required for construction completion is based on unit and major equipment count and type. Critical delivery equipment was predicted to require up to 36 months from time of order. Historical data from previous projects provided a basis for development of the project schedule.

7. *Fund Drawdown Schedule*

The fund requirements during the design, engineering, procurement, construction, and startup periods are illustrated in Fig. 50; fund requirements are shown for 6-month intervals. The cumulative fund drawdowns are shown in Fig. 51.

The fund drawdown schedules were developed based on the sum of the estimated fund requirement schedules for the five separate cost centers. These fund expenditure estimates were based on the construction schedule and experience gained from prior projects.

8. *Operating Costs*

Estimated operating costs for each of the five cost centers and the total complex are summarized in Table 13; the total projected annual operating costs are approximately $300 million.

The estimated labor requirements for the complex are summarized in Table 14; they amount to approximately 2800 people. The projected annual maintenance costs were estimated by unit and the weighted average was 4.1% of fixed capital investment.

9. *General Basis for Profitability Analysis*

The economic analysis is based on a plant operating period of 20 years. The operating rate is 330 stream days/year, equal to a 90.4% operating factor except for the first year of operation, when the complex is assumed to produce at a rate equal to 50% of capacity. The power plant is assumed to operate on a continuous basis 365 days per year, at an 80% operating factor. During periods of fuel gas gasifier maintenance shutdowns, the gas turbine units

TABLE 12 Total Initial Capital Requirements, POGO Coal Refinery (Courtesy of DOE.)

Item	Coal mine	Coal preparation	Process plant	Power plant	Offsites	Total
Fixed capital investment	300.0	83.0	1432.0	452.0	158.0	2425.0
Initial catalysts and chemicals	—	—	9.0	—	—	9.0
Startup cost	10.0	6.5	112.5	14.5	4.5	148.0
Construction financing[a]	36.0	10.0	180.0	56.5	19.5	302.0
Depreciable investment	346.0	99.5	1733.5	523.0	182.0	2884.0
Working capital[a]	—	—	—	—	—	165.0
Land, rights of way						3.0
Total capital requirement						3052.0

[a]Applicable to cases using 65% debt at 9% interest and 0.75% commitment fee only.

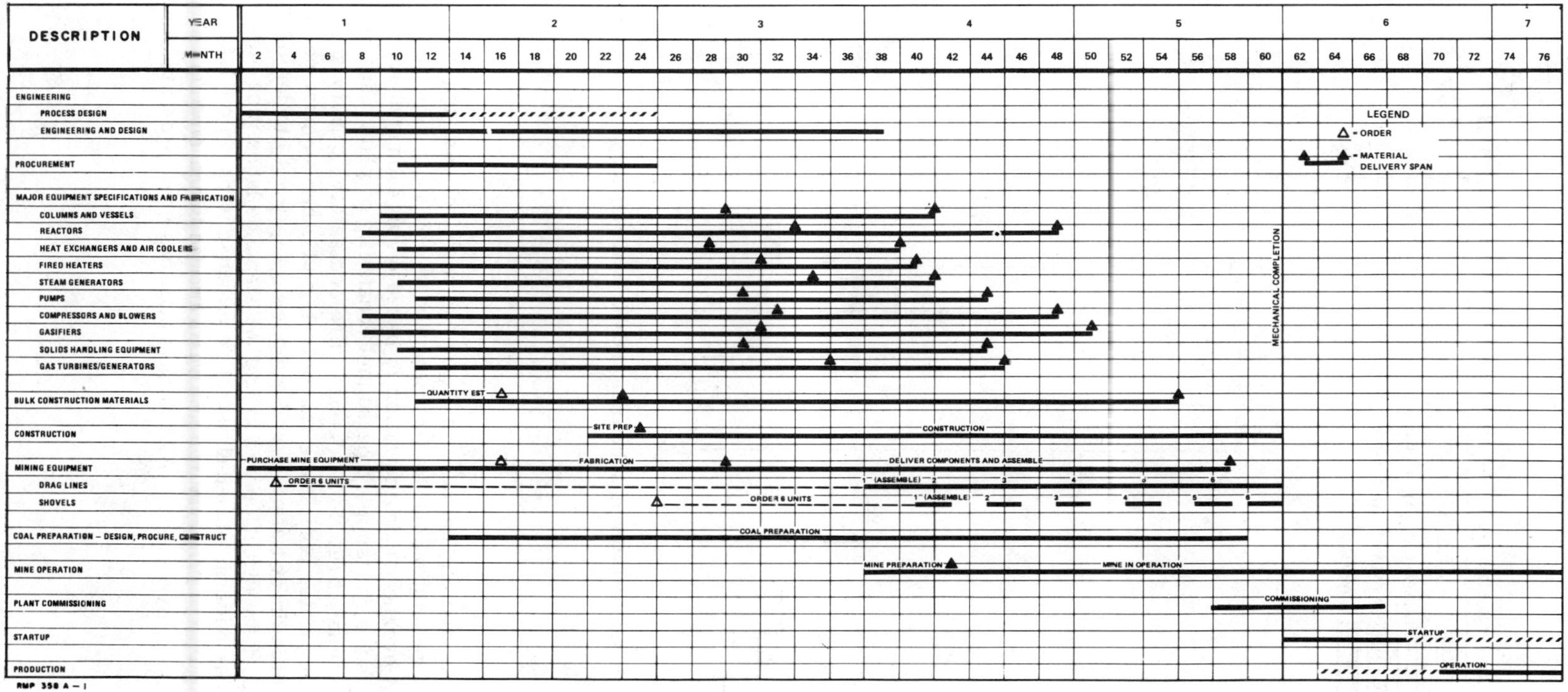

FIG. 49 Project schedule, POGO coal refinery. (Courtesy of DOE.)

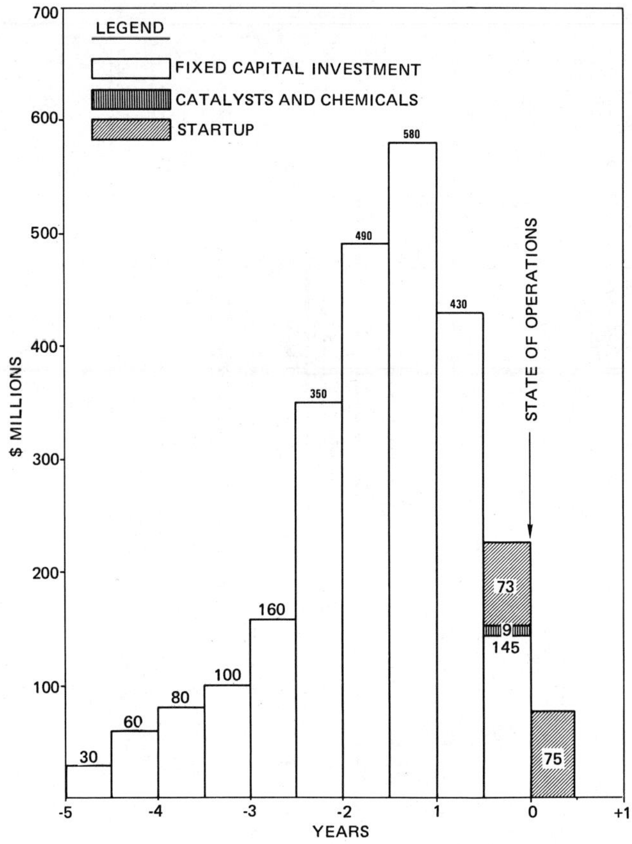

FIG. 50 Fund drawdown schedule, POGO coal refinery. (Courtesy of DOE.)

will be operated on fuel oil. In case of partial fuel gas supply curtailment, both fuel gas and oil can be fired together.

Plant overhead/administration is based on an estimated requirement for administrative and support personnel consisting of plant management, accounting, personnel, first aid, cafeteria, fire and safety, quality assurance, engineering motor pool, material control, and other support personnel and associated indirect materials and supplies. A payroll burden of 35% of total payroll cost for direct operation and maintenance labor plus supervision was used. The general and administrative expense is based on 1.5% of the total operating cost. Property tax and insurance is based on 2.75% of the initial fixed capital investment. Additional detail has been presented in Table 9.

10. *Financial Parameters*

Economics were developed for three types of project financial structures:

1. 100% equity capital
2. Borrowed capital: 65% of the total investment borrowed at 9% interest, with principal repaid in equal installments over the 20-year project term; all working capital borrowed for the 20-year term; a loan commitment fee of 0.75% on funds not yet drawn down during the construction period
3. A nonprofit (0% discounted cash flow rate of return) or break-even boundary case

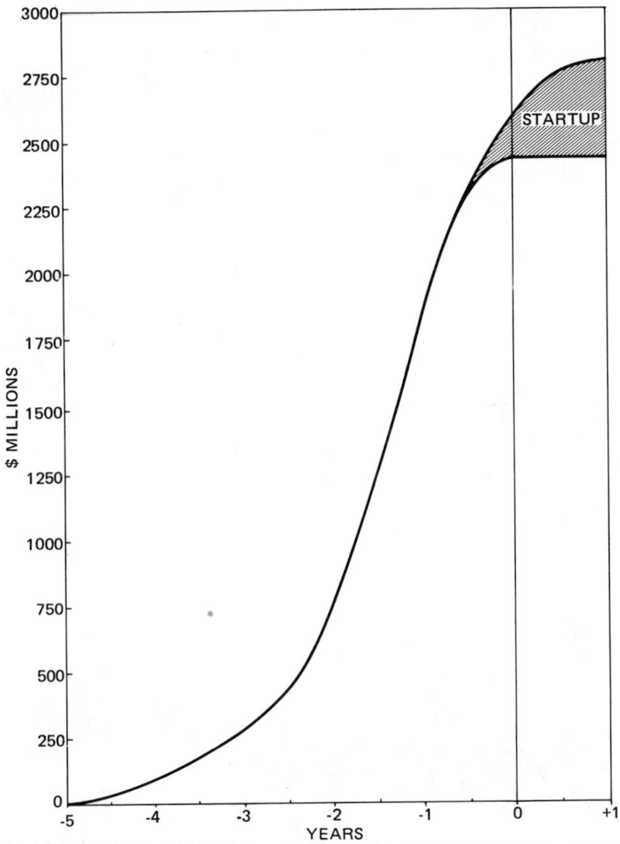

FIG. 51 Cumulative fund drawdown schedules POGO coal refinery. (Courtesy of DOE.)

TABLE 13 Annual Operating Cost Summary, POGO Coal Refinery (Courtesy of DOE.)

Item	Coal mine	Coal preparation	Process plant	Power plant	Offsites	Total
Mine royalty	22.3	—	—	—	—	22.3
Materials and supplies						
Operating supplies	13.5	0.2	1.4	0.6	0.3	16.0
Equipment operation	19.6	—	—	—	—	19.6
Maintenance materials and contract labor		1.7	44.4	8.8	2.6	57.5
Catalysts and chemicals	—	—	6.0	—	1.2	7.2
Total material and supplies	33.1	1.9	51.8	9.4	4.1	100.3
Labor						
Operating labor and supervision	13.9	0.5	4.7	2.2	1.0	22.3
Maintenance labor and supervision	4.6	0.5	9.9	1.0	0.4	16.4
Payroll burden	6.5	0.4	5.1	1.1	0.5	13.6
Plant overhead	10.0	0.6	7.9	1.7	0.8	21.0
Union welfare	23.0	—	—	—	—	23.0
Total labor costs	58.0	2.0	27.6	6.0	2.7	96.3
General and administration overhead	4.5	0.1	1.8	0.5	0.2	7.1
Miscellaneous costs	0.8					0.8
Property taxes and insurance	8.3	2.3	39.4	12.4	4.3	66.7
Total operating costs	127.0	6.3	120.6	28.3	11.3	293.5

TABLE 14 Manpower Requirements, POGO Coal Refinery (Courtesy of DOE.)

Item	Operating	Maintenance	Administration	Total
Administration			348	348
Coal mine	887	308		1195
Coal preparation	33	22		55
Process plant	322	566		888
Power plant	141	57		198
Off-sites	85	27		112
Total	1468	980	348	2790

11. *Return on Investment*

For each of the cases listed above, a discounted cash flow (DCF) rate of return after taxes is specified and the revenue required to meet this objective is calculated. The DCF computations take into consideration the 10% depletion allowance on the coal mine operation, depreciation allowances as described below, an investment tax credit of 10%, and a combined income tax rate of 52% for state and federal taxes. It was estimated that 90% of the fixed capital investment is eligible for the investment tax credit. It has been assumed that the plant will be operated as an integrated subsidiary of profit-making industries that can write off initial losses against current earnings for tax purposes.

Useful lives for asset depreciation purposes conform with the minimum provided in the Depreciation Guidelines and Rules, Revenue Procedure 62-21. The double-declining-balance method of depreciation was used. An 8-year useful life was used for coal mine depreciation based on the guidelines for the mining industry. The guidelines for the chemicals and allied products industry under group 3, "Guidelines for Manufacturing," were used to establish an 11-year useful life for the coal preparation and process plants. The guidelines for a steam production plant for the production of electricity for electric utilities, group 4, were used as a basis for a 28-year useful life of the power plant.

12. *Required Product Selling Price*

The revenue required to achieve a specific DCF was calculated for each of the three financial structure cases. The results for the base cases are summarized in Table 15, which shows the contribution of each of the five cost centers to the total required revenue.

The required revenue for the 100% equity financing with a 12% DCF after-tax rate of return on equity after taking credit for the by-product sulfur at $60 per ton, ammonia at $140 per ton, and "needle coke" at $150 per ton is $807 million per year. For the case of private ownership with 65% debt and 9% per year interest, the required revenue is reduced to $585 million. For the break-even case, with no return on invested funds, the resulting revenue requirement is $294 million per year.

TABLE 15 Required Annual Revenue (RAR) by Cost Center Distribution, POGO Coal Refinery (Courtesy of DOE.)

	Breakeven case, 0% DCF		12% DCF			
			100% equity		65% debt	
Category	RAR ($MM/yr)	%	RAR ($MM/yr)	%	RAR ($MM/yr)	%
Coal mine	145.6	34.9	216.0	23.2	180.3	25.5
Coal preparation	10.6	2.5	28.3	3.0	20.4	2.9
Process plant	188.4	45.2	478.7	51.5	351.1	49.6
Power plant	52.8	12.7	150.8	16.2	113.4	16.0
Off-sites	19.6	4.7	56.2	6.1	42.8	6.0
Total	417.0	100.0	930.0	100.0	708.7	100.0
By-product credit	123.0	29.5	123.0	13.2	123.0	17.4
Total energy products	294.0	70.5	807.0	86.8	585.0	82.6

For the typical case, the plant annually produces approximately 187,000 billion Btu and 7.0 billion kWhr of electrical power. The selling prices required to achieve the required revenue are represented by the following linear equation:

$$\text{Required revenue} = (\text{fuel price}) \times (187 \times 10^{12}\ \text{Btu}) + (\text{power price}) \times (7.0 \times 10^{9}\ \text{kWhr})$$

For the typical cases, the relationship between required selling prices for the three project structures is shown graphically in Fig. 52. To illustrate, for a bus-bar sales price of electricity of 35 mils/kWhr, the fuels would have to be sold for about $2 per million Btu to achieve a 12% DCF for the 65% debt financing case.

13. *Sensitivity Analysis*

The sensitivity of the required revenues to variations in selected variables was analyzed. The results indicate that the effect on the required selling prices of a 10% change in cost elements is as follows:

	% Change in required selling price	
	100% equity	*65% debt*
Operating costs	3.3	4.2
Capital associated costs	8.5	8.0
Run-of-mine coal cost	2.3	3.7
Clean coal cost	2.5	4.0

The effect on the required revenue of variations in the DCF is shown in Fig. 53. Typical values follow:

After-tax rate of Return on equity (% DCF)	Required selling price, % of base case	
	100% equity	65% debt
0	36	69
6	52	80
12	100	100
18	152	130

The preceding profitability analysis summary illustrates one set of procedures that can be used for estimating the economics of coal conversion complexes. The absolute values of the projected economics here are secondary to the illustration of representative procedures.

Using similar procedures, economic projections have also been developed for conceptual designs for COED-based pyrolysis [25], an SRC II-based design

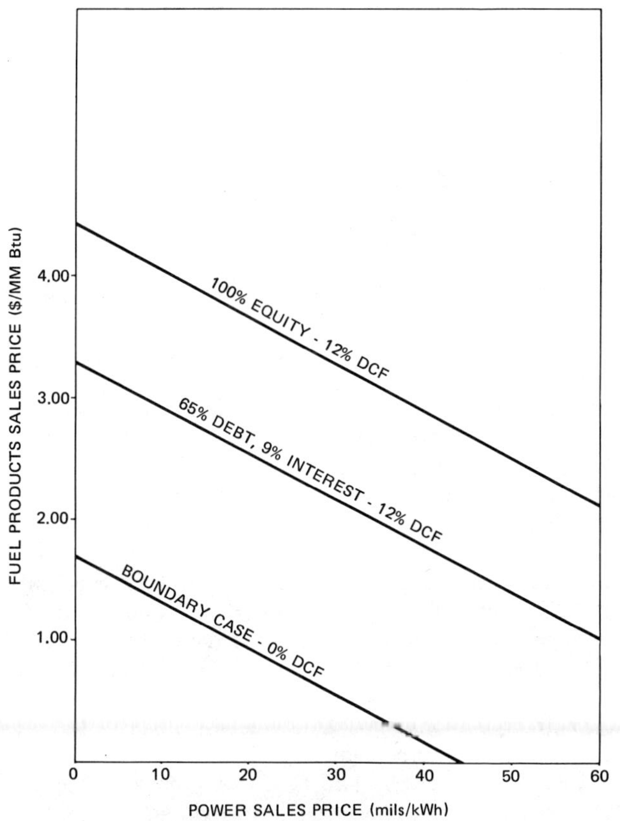

FIG. 52 Required product selling price, POGO coal refinery. (Courtesy of DOE.)

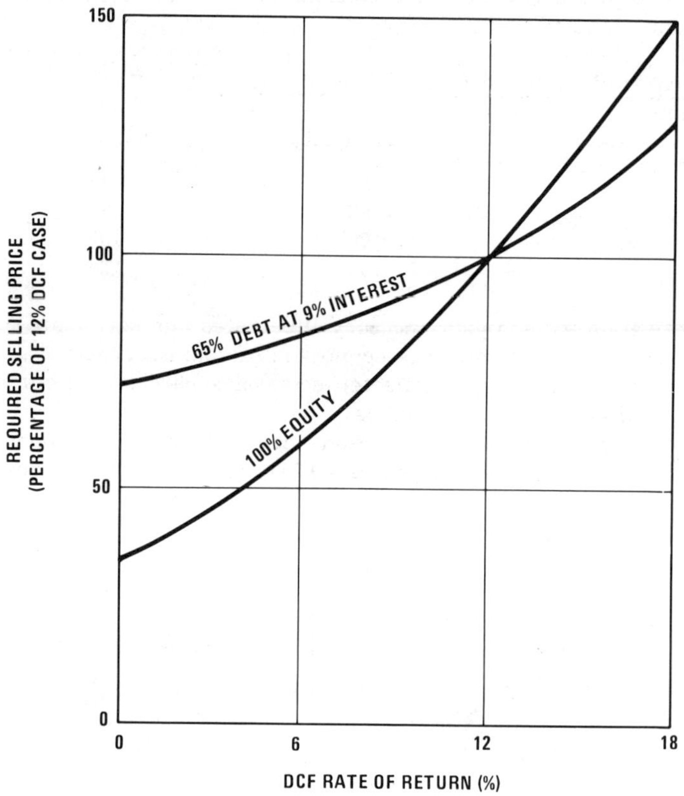

FIG. 53 Sensitivity of required product selling price to DCF, POGO coal refinery. (Courtesy of DOE.)

titled "Oil/Gas" [50], and a version of Fischer-Tropsch using flame-sprayed catalysts [30]. Indications of the projected relative economics between these options based on 1977 dollars are presented in Tables 16 and 17 [69]. The COED-based and POGO results are reported separately because the large amount of electrical power results in a co-product marketing situation requiring specific analysis.

The Table 16 results indicates that hydroliquefaction processing results in a lower product cost expressed in dollars per million Btu. This type of processing appears to be a prime candidate when the quality of the fuels produced is satisfactory. The product characteristics differ for the hydroliquefaction and Fischer-Tropsch technologies. The Fischer-Tropsch product is more aliphatic (straight-chain hydrocarbons) and has nil sulfur, nitrogen, and particulates, making it attractive environmentally. Assessments of probable market values for products from the two types of plants indicate that the profitabilities could be similar if proper markets existed for each of the products.

TABLE 16 Projected Economics of Conceptual Oil/Gas and Fischer-Tropsch Complexes

Item	Oil/gas		Fischer-Tropsch	
Fixed capital investment				
$ billion	1.4		1.8	
$ per daily barrel oil equivalent	12,750		19,000	
$ per daily ton feed coal	30,000		45,000	
Profitability:	100% equity	65% debt 35% equity	100% equity	65% debt 35% equity
Required annual revenue for 12% DCF	590	465	667	515
Required product selling price				
$ per million Btu	2.70	2.10	3.80	2.90
$ per barrel oil equivalent	16.00	12.50	21.00	16.00

TABLE 17 Projected Economics of Conceptual COED-Based and POGO Complexes

Item	COED-based		POGO	
Fixed capital investment				
$ billion	1.5		2.5	
$ per daily ton feed coal	41,000		42,000	
Profitability:	100% equity	65% debt 35% equity	100% equity	65% debt 35% equity
Required annual revenue for 12% DCF	570	470	970	735
Required fuel chemical selling price with electrical power sale at $30/MWhr				
$ per barrel oil equivalent	40	30	20	13

Table 17 indicates that hydroliquefaction-based plants have the potential to be lower-cost fuels producers on a dollar per Btu basis than do pyrolysis-based plants when finished fuels products and electricity are produced. The economics of the pyrolysis-based plant is influenced by the performance of the plant units which produce electrical power from char and the ratio of fuel products to electrical power produced.

14. *Effect of Product Composition*

In comparing processes, the differences in product compositions must be recognized. Recent work has indicated that there is a direct relationship between the product composition, which can be expressed as an average hydrogen/carbon weight ratio, and required fixed capital investment and product cost [74,79]. The significance is that if a process produces a heavy oil, it inherently would be expected to require lower investment and yield a lower product cost than would a plant that primarily produces gasoline. To have a meaningful comparison, an adjustment must be made for product composition between the candidate processes.

In closing, it is important to emphasize the perspective that the true cost of synthetic fuels from coal must await the day when we have built and operated large, modern complexes. Existing estimates, prepared using consistent procedures, provide guidance regarding relative costs. The initial, or "first-of-a-kind" plants, will probably be more expensive than later plants, which will profit from the learning-curve results to improve performance and increase capacity. Perry has summarized the prior experience of several process industry developments and concluded that mature coal-based synfuels plants may produce fuels at a cost of 50% of the pioneer plants [80]. In the meantime, there has been a range of costs projected for the pioneer plants, and the question of true costs will be resolved when we have the actual costs after construction and operation.

REFERENCES

1. H. H. Lowry, *Chemistry of Coal Utilization, Supplementary Volume.* John Wiley & Sons, Inc., New York, 1973, p. 424.
2. P. Marnell, Economic data for a 50,000 BPD Lurgi/Ruhrgas shale oil plant, in *Synthetic Fuels Processing.* Marcel Dekker, Inc., New York, 1977.
3. H. H. Lowry, *Chemistry of Coal Utilization, Supplementary Volume.* John Wiley & Sons, Inc., New York, 1963, pp. 249-251.
4. H. H. Storch et al., *The Fischer-Tropsch and Related Syntheses.* John Wiley & Sons, Inc., New York, 1951.
5. J. C. Hoogendorm, The Sasol Story, 23rd Annual Meeting of the American Institute of Mining, Metallurgical and Petroleum Engineers, Dallas, Tex., Feb. 24, 1974.
6. The Ralph M. Parsons Company, Coal Liquefaction Process Research, Data Source Book, ORNL/Sub-7186-13, prepared for the U.S. Department of Energy, Washington, D.C., through Oak Ridge National Laboratory, 1977.

7. The Pittsburg and Midway Coal Mining Company, Quarterly Report for the Period July-September 1975, ERDA Contract No. E(49-18)-496, Oct. 1975.
8. B. K. Schmidt and D. M. Jackson, Recycle Processing for Liquid and Solid Fuels, 4th Annual International Conference on Coal Gasification, Liquefaction and Conversion to Electricity, University of Pittsburgh, Pa., Aug. 1977.
9. J. B. O'Hara, A. Bela, N. E. Jentz, H. W. Klumpe, B. I. Loran, E. A. Mills, R. J. Newton, and R. V. Teeple, Project POGO—Coal Refinery Complex, Conceptual Design/Economic Analysis, Power-Oil-Gas-Other Products, R&D Report No. 114—Interim Report No. 6, FE-1775-13, prepared for the U.S. Department of Energy, Washington, D.C., Dec. 1977.
10. J. B. O'Hara and N. E. Jentz, Coproduction of Gas and Liquid Fuels from Coal Hydroliquefaction Plants, 10th Pipeline Gas Symposium, Chicago, Nov. 1, 1978.
11. Exxon Research and Development Company, EDS Coal Liquefaction Process Development: Phase IV, Summary of EDS Predevelopment (1966-1975), Interim Report, FE-2893-16, prepared for the U.S. Department of Energy, Washington, D.C., July 1978.
12. Exxon Research and Development Company, Exxon Donor Solvent Coal Liquefaction Commercial Plant Study Design, Interim Report, FE-2353-13, prepared for the U.S. Energy Research and Development Administration, Jan. 1978.
13. C. A. Johnson et al., H-Coal Process Development, 68th Annual Meeting of the AIChE, Los Angeles, Calif., Nov. 1975.
14. Hydrocarbon Research, Inc., H-Coal Integrated Pilot Plant, Final Report, Vols. 1 and 2, HCP/T-1544, prepared for the U.S. Department of Energy, Washington, D.C., Nov. 1977.
15. Fluor, Engineers and Constructors, Inc., H-Coal Commercial Evaluation, Conceptual Design and Economic Analysis for 25,000-TPD H-Coal Liquefaction Plant, prepared for the U.S. Energy Research and Development Administration under Contract No. (49-18)-2002, Washington, D.C., Mar. 1976.
16. Consolidation Coal Company, Final Report, Development of CSF Coal Liquefaction Process, OCR R&D Report No. 39, Vol. 5, 1973.
17. Foster Wheeler Corporation, Engineering Evaluation and Review of Consol Synthetic Fuel Process, OCR R&D Report No. 70, Feb. 1972.
18. Fluor Engineers and Constructors, Inc., Conceptual Design for Advanced Coal Liquefaction Commercial Plant, Final Report, FE-2251-40, prepared for the U.S. Department of Energy, Washington, D.C., Oct. 1977.
19. S. L. Meisel et al., *Chem. Tech.* 6:86 (1976).
20. J. J. Wise and A. J. Silvestri, *Oil and Gas J.* 74:140-142 (Nov. 22, 1976).
21. C. D. Chang and A. J. Silvestri, *J. Catal.* 47:249 (1977).
22 A. Y. Kam and W. Lee, Fluid Bed Process Studies on Selected Conversion of Methanol to High Octane Gasoline, Final Report, DOE Contract No, Ex-76-C-01-2490, Washington, D.C., Apr. 1978.

23. S. E. Voltz and J. J. Wise, Development Studies on Conversion of Methanol and Related Oxygenates to Gasoline, Final Report, ERDA Contract No. B(49-18)-1773, Washington, D.C., Nov. 1976.
24. M. Schreiner, Research Guidance Studies to Assess Gasoline from Coal by Methanol-to-Gasoline and Sasol-Type Fischer-Tropsch Technoligies, Final Report, DOE contract No. EF-76-C-01-2447, Washington, D.C., Aug. 1978.
25. J. B. O'Hara, C. L. Crawford, H. F. Hincks, and E. A. Mills, Commercial Complex, Conceptual Design/Economic Analysis, Oil and Power by COED-Based Coal Conversion, R&D Report No. 114—Interim Report No. 1, FE-1775-1, prepared for the Energy Research and Development Administration, Washington, D.C., Sept. 1975.
26. H. Malakoff, The COGAS approach, in *Coal Processing Technology*, Vol. 2. AIChE, New York, 1975.
27. J. B. O'Hara and R. V. Teeple, Preliminary economic analysis oil and power by COED-based coal conversion, in *Synthetic Fuels Processing, Comparative Economics*. Marcel Dekker, Inc., New York, 1977, pp. 287-318.
28. J. B. O'Hara, J. E. McCarthy, N. E. Jentz, E. C. Drucke, J. F. Ferrall, and F. Greve, Coal Liquefaction Process Research Process Survey, Data Source Book, R&D Interim Report No. 2, ORNL/Sub-7186-13, prepared for the U.S. Department of Energy, Washington, D.C., Dec. 1977.
29. J. J. Elliott, W. A. Haynes, and Á. J. Forney, Gasoline via the Fischer-Tropsch Reaction Using the Hot Gas Recycle System, 163rd National Meeting of the American Chemical Society, Boston, Mass., Apr. 1972.
30. J. B. O'Hara, A. Bela, N. E. Jentz, S. K. Khaderi, H. W. Klumpe, B. I. Loran, D. G. Reynolds, and R. V. Teeple, Fischer-Tropsch Complex Conceptual Design/Economic Analysis, Oil and SNG Production, R&D Report No. 114—Interim Report No. 3 FE-1775-7, prepared for the Energy Research and Development Administration, Washington, D.C., Jan. 1977.
31. J. B. O'Hara et al., Preliminary Design Services, Final Report, Contract No. EX-76-C-01-1775, R&D Report No. 114, FE-1775-21, prepared for the U.S. Department of Energy, Washington, D.C., Nov. 1978.
32. The Dravo Corporation, Conceptual Commercial Design and Commercial Feasibility Evaluation for Clean Boiler Fuel Facility, Vols. 1-3, ERDA Contract No. E(49-18)-1972, Washington, D.C., Feb. 1976.
33. A. Sass, The GR&D Pyrolysis Process—A Status Report, 67th Annual AIChE Meeting, Washington, D.C., Dec. 3, 1974.
34. K. A. Schowalter and N. S. Boodman, The Clean Coke Process for Metallurgical Coke, *Chem. Eng. Prog.* 70 (July 1974).
35. A. A. Simone, Clean Fuel from Coal Process, *Combustion* 47:15-19 (May 1976).
36. M. Peluso and D. F. Ogren, Antisolvent Deashing, *Chem. Eng. Prog.* 75:41-43 (June 1979).
37. R. M. Adams, Critical Solvent Deashing of Liquefied Coal, *Chem. Eng. Prog.* 75:44-48 (June 1979).
38. Rockwell International, Rocketdyne Division, Partial Liquefaction of Coal by Direct Hydrogenation, 5th Quarterly Progress Report, July-

September 1976, ERDA Contract No. E(49-18) 2044, Washington, D.C., 1976.
39. J. Friedman, The Rockwell Advanced SNG Gasifier, 10th Synthetic Pipeline Gas Symposium, Chicago, Nov. 1, 1978.
40. Assistant Secretary for Energy Technology, Fossil Energy Program Summary Document, U.S. Department of Energy, Washington, D.C., Mar. 1979, pp. 90-92.
41. Assistant Secretary for Energy Technology, Fossil Energy Program Summary Document, U.S. Department of Energy, Washington, D.C., Mar. 1979, pp. 93-94.
42. B. Strobel et al., Ergebnisse der Versuchsonlange Kohleol der Bergbau-Forschung, in *Kohlevergasung und -Hydrierung.* Haus de Technik, Essen, West Germany, 1978.
43. E. Wolowski, Die RAG/STEAG-Grossversuchsonlonge zur katalytiochen Hydrierung von Steinkohle, in *Kohlevergasung und -Hydrierung.* Haus der Technik, Essen, West Germany, 1978.
44. H. Wurfel, Das Projekt Kohlehydrierung der Saarbergwerke AG, in *Kohlevergasung und -Hydrierung.* Haus der Technik, Essen, West Germany, 1978.
45. R. R. Maddocks and J. Gibson, Supercritical Gas Extraction of Coal: An Examination of the Commercial Potential of the Process, Annual AIChE Meeting, Chicago, Nov. 1976.
46. Coal Research Establishment, Liquids from Coal, Stoke Orchard, Cheltenham, U.K., May 1978.
47. W. R. Wu and H. H. Storch, Hydrogenation of Coal and Tar, U.S. Bureau of Mines Bulletin 633, 1978.
48. J. H. Edwards, Oil-from-Coal Processes: A Critical Evaluation for Australian Conditions, 6th Annual Chemical Engineering Conference, Surfers Paradise, Queensland, Australia, Nov. 6-8, 1978.
49. J. B. O'Hara, N. E. Jentz, S. N. Rippee, and E. A. Mills, Demonstration Plant, Clean Boiler Fuels from Coal, Preliminary Design/Capital Cost Estimate, Vol. I, R&D Report No. 82—Interim Report No. 1, prepared for the U.S. Department of Interior, Office of Coal Research, Washington, D.C., Dec. 1973.
50. J. B. O'Hara, G. H. Hervey, S. M. Fass, N. E. Jentz, H. W. Klumpe, B. I. Loran, E. A. Mills, and R. V. Teeple, Oil/Gas Complex Conceptual Design/Economic Analysis, Oil and SNG Production, R&D Report No. 114—Interim Report No. 4, FE-1775-8, prepared for the Energy Research and Development Administration, Washington, D.C., Mar. 1977.
51. J. B. O'Hara, S. N. Rippee, B. I. Loran, and W. J. Mindeheim, Environmental Factors in Coal Liquefaction Plant Design, R&D Report No. 82—Interim Report No. 3, prepared for the Office of Coal Research, U.S. Department of the Interior, Washington, D.C., May 1974.
52. P. F. McElherne et al., 1970 Final Report, Consol Synthetic Fuel Process, R&D Report No. 45—Final Report, prepared for the Office of Coal Research, U.S. Department of Interior, Washington, D.C., July 1970.
53. J. B. O'Hara, B. I. Loran, G. H. Hervey, and S. M. Fass, Environmental Factors for Oil/Gas Coal Conversion Technology, 69th Annual Meeting of the AIChE, Chicago, Dec. 1, 1976.

54. J. B. O'Hara, B. I. Loran, A. Bela, and N. E. Jentz, Environmental Factors for Fisher-Tropsch Coal Conversion Technology, Proceedings of the Second Pacific Chemical Engineering Congress (PACHEC '77), at the 84th National Meeting of the AIChE, Denver, Colo., Vol. 2, Aug. 28-31, 1977, pp. 855-863.
55. B. I. Loran and J. B. O'Hara, Specific Environmental Aspects of Fischer-Tropsch Coal Conversion Technology, 3rd Environmental Protection Agency (EPA) Symposium on Environmental Aspects of Fuel Conversion Technology, III, Hollywood, Fla., Sept. 15, 1977.
56. B. I. Loran and J. B. O'Hara, Specific Environmental Aspects of Fischer-Tropsch Coal Conversion Technology, Symposium Proceedings: Environmental Aspects of Fuel Conversion Technology, III (Hollywood, Fla., Sept. 1977), EPA-60017-78-063; prepared by the Research Triangle Institute, Research Triangle Park, N.C. under Contract No. 68-02-2612, for the U.S. Environmental Protection Agency, Washington, D.C., Apr. 1978, pp. 409-423.
57. B. I. Loran and J. B. O'Hara, A Clean Coal Conversion Technology, *Environ. Sci. Technol.*, 12:1258-1263 (Nov. 1978).
58. J. B. O'Hara, G. H. Hervey, S. M. Fass, N. E. Jentz, H. W. Klumpe, B. I. Loran, E. A. Mills, and R. V. Teeple, Oil/Gas Complex Conceptual Design/Economic Analysis, Oil and SNG Production, R&D Report No. 114—Interim Report No. 4, FE-1775-8, prepared for the Energy Research and Development Administration, Washington, D.C., Mar. 1977.
59. Development Document for Effluent Guidelines and New Source Performance Standards for the Petroleum Refining Point Source Category, U.S. Environmental Protection Agency, Report EPA-440/1-74-014a, Washington, D.C., Apr. 1974.
60. W. F. Holland, K. A. Wilde, J. L. Parr, P. S. Lowell, and R. F. Pohler, The Environmental Effects of Trace Elements in the Pond Disposal of Ash and Flue Gas Desulfurization Sludge, Research Project 202 prepared by the Radian Corp. for the Electric Power Research Institute, Sept. 1975 (NTIS Report No. PB 252090/6WP).
61. R. R. Ruch, H. J. Gluskoter, and N. F. Shimp, Occurrence and Distribution of Potentially Volatile Trace Elements in Coal, Illinois State Geological Survey, Environmental Geology Note No. 72, Aug. 1974 (NTIS Report No. PB 238091).
62. J. W. Kaakinen, R. M. Jorden, M. H. Lawasani, and R. E. West, Trace Element Behavior in Coal-Fired Power Plant, *Environ. Sci. Technol.* 9:862-869 (1975).
63. D. H. Klein et al., Pathways of Thirty-Seven Trace Elements Through Coal-Fired Power Plant, *Environ. Sci. Technol.* 9:973-979 (1975).
64. R. J. Secton, The Hazards to Health in the Hydrogenation of Coal: IV. The Control Program and the Clinical Effects, *Arch. Environ. Health* 1:208-231 (1960).
65. P. F. McElherne et al., 1968 Feasibility Report, Consol Synthetic Fuel Process, R&D Report No. 45—Interim Report No. 2, prepared for the Office of Coal Research, U.S. Department of Interior, Washington, D.C., Feb. 1969.

66. J. B. O'Hara, N. E. Jentz, S. N. Rippee, and E. A. Mills, Demonstration Plant, Clean Boiler Fuels from Coal, Preliminary Design/Economic Analysis, Vol. 3, R&D Report No. 82—Interim Report No. 1, prepared for the Office of Coal Research, U.S. Department of the Interior, Washington, D.C., Feb. 1975.
67. J. B. O'Hara, N. E. Jentz, R. J. Newton, and R. V. Teeple, Feedstocks from Coal: How? When? *Hydrocarbon Process.* 57:117-121 (Nov. 1978).
68. J. B. O'Hara, R. V. Teeple, H. W. Klumpe, and A. Bela, Projected Economics for Project POGO—A Coal Refinery, 14th Annual Technical Meeting of the Southern California Section of the AIChE, Anaheim, Calif., Apr. 18, 1978.
69. J. B. O'Hara, Coal Beneficiation and Conversion Developments, *Interciencia*, Caracas, Venezuela, Jan.-Feb. 1979.
70. Coalcon, Clean Boiler Fuel Demonstration Program, Phase I, Task II—Hydrocarbonization Demonstration Plant Process Design Report, Vol. 1 0, prepared under Contract No. E(49-18)-1736 for the Energy Research and Develo-ment Administration, Washington, D.C., June 15, 1977.
71. K. A. Rogers and R. F. Hill, Coal Conversion Comparisons, FE-2468-51, prepared for the U.S. Department of Energy under Contract No. EF-77-C-01-2468 by the Engineering Societies Commission on Energy, Inc., Washington, D.C., July 1979.
72. *Coal Processing Technology* VI, American Institute of Chemical Engineers Technical Manual Series, American Institute of Chemical Engineers, New York (1980).
73. J. B. O'Hara, A. Bela, N. E. Jentz, H. W. Klumpe, B. I. Loran, P. F. McElherne, and R. V. Teeple, Multi-process Demonstration Plant, Preliminary Design/Economic Evaluation, Report No. 114—Interim Report No. 7, FE-1775-18, prepared for the U.S. Department of Energy, Washington, D.C., Oct. 1978.
74. J. B. O'Hara, A. Bela, et al., Liquefaction Technology Assessment, Final Report, prepared for the Union Carbide Corporation, Oak Ridge, Tenn. under contract to the U.S. Department of Energy, Washington, D.C., May 1979.
75. Badger Plants, Inc., Conceptual Design of a Coal to Methanol Commercial Plant, Interim Final Report for the Period July 16, 1976-February 15, 1978, FE-2416-24, prepared for the U.S. Department of Energy, Washington, D.C., Feb. 1978.
76. Badger Plants, Inc., Conceptual Design of a Coal-to-Methanol-to-Gasoline Commercial Plant, prepared for the U.S. Department of Energy, EX-76-C-01-2416, Washington, D.C., 1979.
77. J. B. O'Hara, R. D. Howell, N. E. Jentz, and P. F. McElherne, Coalcon Process Review, R&D Interim Report No. 3, ORNL/Sub-7186-16, prepared for the U.S. Department of Energy, Washington, D.C., June 1978.
78. J. B. O'Hara, N. E. Jentz, and G. H. Hervey, Commercial Coal Conversion Plant Design: Translation from Pilot to Commercial-Scale Plants, Clean Fuels from Coal Symposium II Papers, Institute of Gas Technology (IGT), Chicago, June 23-27, 1975, pp. 817-836.
79. J. B. O'Hara, N. E. Jentz, and A. Bela, Coal Liquefaction: The Impact of Product Composition on Relative Economics, presented to Canadian Society for Chemical Engineering, Sarnia, Ontario, Oct. 1, 1979.

80. H. Perry, Clean fuels from coal, in *Advances in Energy System and Technology*, Academic Press, Inc., New York, 1978, p. 304.
81. U.S. Bureau of Mines, Economic Analysis of Synthoil Plant Producing 50,000 BPD of Liquid Fuels from Two Coal Seams: Wyodak and West Kentucky, ERDA 76-35, Morgantown, W. Va., 1975.
82. J. H. Govaarts and C. W. Schutte, The Use of Low Grade Coal for the Production of Oil, Gas, Fertilizers and Chemicals, Paper 3.3-187, Proceedings of 8th World Energy Conference, Bucharest, Hungary, 1971.
83. J. C. Hoogendorn, Conversion of Coal into Fuels and Chemicals in South Africa, Paper B4, Proceedings of 3rd International Conference on Coal Research, Sydney, Australia, 1976.

Glossary

The definitions contained in the glossary are specific to coal technology usage wherever possible. In many instances, a term may have a somewhat different meaning when applied to another technical discipline.

Abutement Load, a load created through the underground removal of coal, producing wall stresses in addition to the overburden loading. These additional stresses can create a pressure many times the weight of the overlying material.

Access Road, any haul road or other road constructed improved, maintained, or used by the operator which ends at the pit or bench and is located within the area of land affected.

Acid-Forming Materials, *see* Acid-Producing Materials.

Acid Gas, hydrogen sulfide (H_2S) or carbon dioxide (CO_2).

Acid Mine Drainage, any acid water draining or flowing on, or having drained or flowed off, any area of land affected by mining.

Acid-Producing (Acid-Forming) Materials, rock strata containing significant pyrite which if exposed by coal mining will, when acted upon by air and water, cause acids to form.

Acid Soil, generally a soil that is acid throughout the plant root region. Commonly applied to only the surface-plowed layer or to some other specific layer or horizon of a soil. Practically, this means a soil more acid than pH 6.6—precisely, a soil with a pH value less than 7.0. A soil having a preponderance of hydrogen over hydroxyl ions in the soil solution.

Acid Spoil, usually spoil material containing sufficient pyrite so that weathering produces acid water and the pH of the soil determined by standard methods of soil analysis is between 4.0 and 6.9.

Agglomeration, formation of larger coal or ash particles by smaller particles sticking together.

Alkaline Soil, generally, a soil that is alkaline throughout most or all of the parts of it occupied by plant roots, although the term is commonly applied to only a specific layer or horizon of a soil. Precisely, any soil horizon having a pH value greater than 7.0. Practically, a soil having a pH above 7.3

Alluvial, describes material such as earth, sand, gravel, or other rock or mineral materials transported by and laid down by flowing water.

Angle of Nip, the angle formed by spherical particles setting between two equal-diameter crusher rolls separated by a given distance.

Anthracene Oil, the heaviest distillable coal tar fraction, with distillation range 520 to 750°F, containing creosote oil, anthracene, phenanthrene, carbazole, and so on.

Anthracite (Hard Coal), a hard, black, shiny coal very high in fixed carbon and low in volatile matter, hydrogen, and oxygen.

Anthracosis, *see* Black Lung.

Anthraxylon, U.S. Bureau of Mines usage for vitrinite viewed by transmitted light.

Aquifer, a water-bearing formation through which water moves more readily than in adjacent formations with lower permeability.

Area Mining, surface mining that is carried on in level to gentle rolling topography on relatively large tracts.

Ash, the noncombustible residue remaining after complete coal combustion; the final form of the mineral matter present in coal.

Ash Analysis, Ash analyses give percentages of inorganic oxides present in an ash sample. Ash analyses are used for evaluation of the corrosion, slagging, and fouling potential of coal ash. The ash constituents of interest are silica (SiO_2) alumina (Al_2O_3), titania (TiO_2), ferric oxide (Fe_2O_3), lime (CaO), magnesia (MgO), potassium oxide (K_2O), sodium oxide (Na_2O), and sulfur trioxide (SO_3). An indication of ash behavior can be estimated from the relative percentages of each constituent.

Ash Fluid Temperature, *see* Fluid Temperature.

Ash-Fusion Temperatures, ash-fusion temperatures are a set of temperatures that characterize the behavior of ash as it is heated. These temperatures are determined by heating cones of ground, pressed ash in both oxidizing and reducing atmospheres.

Ash Initial Deformation Temperature, *see* Initial Deformation Temperature.

Ash-Softening Temperature, *see* Softening Temperature.

As-Received Moisture, the moisture present in a coal sample when delivered.

Attritus, a microscopic coal constituent referring to macerated plant debris intimately mixed with mineral matter and coalified. U.S. Bureau of Mines usage, viewed by transmitted light.

Auger Mining, generally practiced but not restricted to hilly coal-bearing regions of the country. Utilizes a machine designed on the principle of the auger, which bores into an exposed coal seam, conveying the coal to a storage pile or bin for loading and transporting. May be used alone or in combination with conventional surface mining. When used alone, a single cut is made sufficient to expose the coal seam and provide operating space for the machine. When used in combination with surface mining, the last cut pit provides the operating space.

Backfill, the operation of refilling an excavation. Also, the material placed in an excavation in the process of backfilling.

Bedrock, the rock material directly above and below the coal seam.

Beehive Oven, a dome-shaped oven not equipped to recover the by-product gas and liquids evolved during the coking process.

Belt Feeder (Feeder Breaker), a crawler-mounted surge bin often equipped with a crusher or breaker. It is used in room and pillar sections and positioned at the end of the section conveyor belt. It allows a quick discharge of the shuttle car. It sizes the coal, and a built-in convey feeds it at an appropriate rate onto the conveyor belt.

Bench, the surface of an excavated area at some point between the material being mined and the original surface of the ground, on which equipment can set, move, or operate. A working road or base below a highwall, as in contour stripping for coal.

Bergius Process, a direct catalytic hydroliquefaction process operated in two stages: first, liquid-phase hydrogenation in a coal oil slurry, then vapor-phase hydrogenation. The process was developed in Germany in World War I. The main products are gasoline, diesel fuel, and other hydrocarbons.

Billingham Plant, an Imperial Chemical Industries commercial coal liquefaction plant located in the United Kingdom. The major product was motor fuel.

Bituminous (Soft) Coal, a relatively soft dark brown to black coal, lower in fixed carbon but higher than anthracite in volatile matter, hydrogen, and oxygen.

Black Lung (Anthracosis), a respiratory disease caused by prolonged inhalation of coal dust.

Bleeders, a system of entries developed at the far end of a mining section. Air forced through the gob and mixed with methane is discharged into this system and channeled to the exhaust shaft.

Block Cut Method, a method of surface mining that removes overburden and places it around the periphery of a box-shaped cut. After coal is removed, the spoil is pushed back into the cut and the surface is blended into the topography.

Blue Gas, a mixture consisting chiefly of carbon monoxide and hydrogen formed by action of steam on hot coal or coke.

Boghead Coal, same as cannel except algal remains can be seen under the microscope.

Box Cut, the initial cut driven in a property where no open side exists; this results in a highwall on both sides of the cut.

Bright Coal, U.S. Bureau of Mines usage for a combination of clarain and vitrain with small amounts of fusain.

Briquetting, a process of applying pressure to coal fines, with or without a binder, to form a compact or agglomerate.

Btu, abbreviation for British thermal unit. A unit of heat energy equal to the energy necessary to raise the temperature of 1 pound of air-free water from 60°F to 61°F at a constant pressure of 1 standard atmosphere.

Buckstay, a vertical steel structural member (usually an I beam) used to frame and support the brickwork of a battery.

Burden (Blast Furnace), the iron-bearing material (iron ore, pellets, or sinter, or combinations of these materials), with appropriate proportions of coke and flux (limestone and dolomite), required to operate a furnace properly.

Burning Profile, the burning profile of a coal, obtained by thermal gravimetric analysis, is a plot of the rate of weight loss when a coal sample is heated at a fixed rate. The burning profile of a solid fuel offers an indication of ignition and burning characteristics by comparing it with other burning profiles of fuels with known performance.

By-Product Oven, a long, narrow oven arranged in groups or batteries with provision for collecting and processing the gas and liquids distilled in the coking process.

Cannel Coal, predominately durain with lesser amounts of vitrain than splint coal and small quantities of fusain. Spores can be seen under the microscope.

Carbolic Oil, *see* Middle Oil.

Carbon Conversion, chemical reaction of the carbon content of a carbonaceous material.

Carbonization, a process whereby coal is converted to coke by devolatilization.

Carbon Monoxide, a colorless, odorless, very toxic gas formed by incomplete combustion of carbon, as in water gas or producer gas production.

Carburetted Blue Gas, *see* Water Gas.

Carburization, dissolution of carbon into molten metal.

Char-Oil-Energy-Development Process (COED), a direct coal pyrolysis process associated with the FMC Corporation in which coal is pyrolyzed and partially combusted in a multiple-stage fluidized bed. The main products are char, pyrolysis oil, high-Btu pyrolysis gas, and low-Btu combustion gases.

Clarain, a macroscopic coal constituent (lithotype) known as bright-banded coal, composed of alternating bands of vitrain and durain.

Classifiers, vessels in which particles suspended in fluids separate according to their sizes if their densities are uniform; or according to their densities and sizes if the particle densities are not uniform.

Coal Drill, a hand-held or rubber-tire-mounted self-powered auger-type drill used on conventional room and pillar mining sections. It is used to provide about 1 1/2 in.-diameter 8- to 12-ft-long holes for explosives.

Coal Gas, the mixture of volatile products (mainly hydrogen, methane, carbon monoxide, and nitrogen) remaining after removal of water and tar, obtained from carbonization of coal, having a heat content of 400 to 600 Btu/ft^3.

Coalification, those processes involved in the genetic and metamorphic history of the formation of coal deposits from vegetable matter.

Coal Seam, a layer, vein, or deposit of coal. A stratigraphic part of the earth's surface containing coal.

Coal Tar, the condensable distillate containing light, middle, and heavy oils obtained by carbonization of coal. About 8 gal of tar is obtained from each ton of bituminous coal.

COED, *see* Char-Oil-Energy-Development Process.

Coke-Oven Gas, a medium-Btu gas, typically 550 Btu/scf, produced as a by-product in the manufacture of coke by heating coal at moderate temperatures.

Cold Gas Efficiency, thermal efficiency of a gasifier without including any sensible heat content of product gas due to elevated temperature.

Concentrating Tables, machines that utilize gravity separation to effect the concentration and separation of economically valuable materials from dross. They are widely employed in minerals dressing and coal cleaning.

Continuous-Haulage Systems, a system of short-chain conveyors, extensible conveyor belts, specialty design flexible conveyors, and so on. They are used on room and pillar sections to provide a temporary connection between the frequently moving continuous miner and the permanently installed section conveyor belt.

Continuous Miner, a mechanical mining machine consisting of a cutting head, a coal-gathering device, a chain conveyor with flexible loading boom, and a crawler-equipped chassis. Its function is to excavate the mineral and to load it onto shuttle cars or continuous-haulage systems. It is electrically powered, with a hydraulic subsystem for auxiliary functions. Power is supplied through a trailing cable.

Contour, an imaginary line connecting points of equal height above sea level as they follow the relief of the terrain.

Contouring, (1) sometimes used to mean contour mining; (2) returning spoil banks to even-sloped terrain features.

Contour Stripping or Surface Mining, the removal of overburden and mining from a coal seam that outcrops or approaches the surface at approximately the same elevation, in steep or mountainous areas.

Conventional Battery, consists of 15 to 90 slot ovens charged with wet coal through charging holes in the top of the ovens.

Convergence, the tendency of rock material to move into an underground opening as a result of the pressure created by the weight of the overburden.

Core Drilling, the process by which a cylindrical sample of rock and other strata is obtained through the use of a hollow drilling bit which cuts and retains a section of the rock or other strata penetrated.

Core Hole, a hole drilled into a rock formation with a hollow bit and a hollow drill shaft. This type of equipment preserves a column of rock in its center, which can be retrieved for testing and analysis.

Creosote Oil, *see* Middle Oil.

Cribs, a stack of wooden blocks laid in a crisscross fashion into a rectangular or square cross section. Wedged between mine roof and floor, they serve as a high-capacity, yielding roof support. The individual wooden crib blocks range in size from 6 to 12 in. square and 3 to 5 ft long. In Europe, old railroad ties are often used. Cribs are often filled with rocks for better stability and added strength.

Critical Speed, the lowest cylindrical speed for a given mill diameter, expressed in rpm, which causes a small particle on the cylinder suface to centrifuge.

Crop Coal, coal at the outcrop of the seam. It is usually considered to be inferior quality because of partial oxidation, although this is not always true.

Culm Bank, the piled waste or slack of Pennsylvania anthracite mines. Usually composed of fine coal, dust, and dirt.

Cut, longitudinal excavation made by a strip-mining machine to remove overburden in a single progressive line from one side or end of the property.

Cutting Machine, is a rubber-tired, electrically powered vehicle. An attached cutting bar similar to the one on a chain saw, equipped with carbide-tipped cutting bits, prepares a coal face for the blasting operation by cutting a several-inch-wide, 8- to 12-ft-deep slot near the mine floor.

Cyclone Collectors, equipment in which centrifugal force is used to separate particulates from a gas stream.

Decarburization, gasification of fixed carbon or char.

Deep Mining, the production of coal at depths greater than 3000 ft.

Dense Media (Heavy Media), liquids, solutions, or suspensions having densities greater than that of water.

Dense-Media Separation, a coal-cleaning method based on density separation, using a heavy-media suspension of fine particles of magnetite, sand, or clay.

Dense Medium, a dense slurry formed by the suspension of heavy particles in water; used to clean coal.

Descending-Bed System, gravity downflow of packed solids contacted with upwardly flowing gases. Sometimes referred to as "fixed-bed" or "moving-bed."

Devolatilization, removal of vaporizable material by the action of heat.

Dewatering, the removal of water from coal by mechanical equipment such as a vibrating screen, filter, or centrifuge.

Direct Hydrogenation, coal hydrogenated without use of a separate donor solvent hydrogenation step.

Disturbed Land, land on which excavation has occurred or upon which overburden has been deposited, or both.

Dragline, an excavating machine that utilizes a bucket operated and suspended by means of lines or cables, one of which hoists or lowers the bucket from a boom; the other, from which the name is derived, allows the bucket to swing out from the machine or to be dragged toward the machine for loading. Mobility of draglines is by crawler mounting or by a walking device for propelling, featuring pontoonlike feet and a circular base or tub. The swing of the machine is based on rollers and rail. The machine usually operates from the highwall.

Dry Ash-Free Basis, a coal analysis basis calculated as if moisture and ash were removed.

Dull Coal, coal that absorbs rather than reflects light, containing mostly durain and fusain lithotypes.

Durain, a macroscopic coal constituent (lithotype) which is hard and dull grey in color.

Ebullating-Bed Reactor, a system similar to a fluid bed but operated at higher gas velocities, such that a portion of the solids is carried out with the upflowing gas.

Ecology, the science that deals with the mutual relation of plants and animals to one another and to their environment.

EDS, *see* Exxon Donor Solvent Process.

Efficiency of Separation, a coal-washing term expressed as the ratio of clean coal obtained in practice in cleaning equipment and that obtained by float and sink methods at the ash content of the clean coal, times 100.

Electrostatic Precipitation, separation of liquid or solid particles from a gas stream by the action of electrically charged wires and plates.

Endothermic Reaction, a process in which heat is absorbed.

Energy Supply and Coordination Act of 1974, the first act by Congress designed to switch utility and industrial boilers away from oil and natural gas.

Entrained Flow System, solids suspended in a moving gas stream and flowing with it.

Entry, an underground passageway, usually following a coal seam. It is mostly of rectangular configuration and in some cases not higher than the thickness of the coal seam.

Environment, all external conditions that may act upon an organism or soil to influence its development, including sunlight, temperature, moisture, and other organisms.

Equilibrium Moisture, the moisture capacity of coal at 30°C in an atmosphere of 95% relative humidity.

Erosion, the wearing away of the land surface by running water, wind, ice, or other geological agents, including such processes as gravitational creep. Detachment and movement of soil or rock fragments by water, wind, or ice or gravity.

Error Area, an index of separation efficiency consisting of the measured area lying below a misplaced material distribution graph plotted so that a unit of length represents 0.1 specific gravity on the abscissa and 2% of misplaced material on the ordinate. First proposed by Driessen in 1945.

Evergreen Contract, a multiyear contract that is annually reopened for negotiation.

Exinite, a microscopic coal constituent (maceral) or maceral group containing spores and cuticles. Appears dark gray in reflected light.

Exothermic Reaction, a process in which heat is evolved.

Explosion-Proof Equipment, electrical equipment that operates at an ampere-voltage level not considered intrinsically safe has to be enclosed in an explosion-proof compartment if it is to be positioned near a face operation. The enclosure has to be designed to withstand 150 psi internal pressure. Any mating surfaces of covers and compartments have to be designed to prevent the flame of an interior explosion from reaching the exterior.

Exxon Donor Solvent Process (EDS), an indirect donor solvent hydroliquefaction process originated by the Exxon Corporation. Catalytic reactors are used for separate solvent hydrogenation. Unconverted residue is filtered from the product oil. The main products are naphtha and a wide-boiling-range low-sulfur fuel oil.

Face, the solid unbroken surface of a coal bed that is at the advancing end of the mine workplace.

Face Conveyor, is utilized on longwall mining faces. It consists of a metal trough with an integrated return channel. Steel scrapers attached to an endless round link or roller-type chain force through the trough any material deposited inside the trough by the mining machine. Spill plates and guides for mining equipment are attached. For flexibility and ease of installation the conveyor is made up of 5-ft sections. Commonly, two electrically powered drives (one on each end) move chain, scrapers (flights), and material along.

Face Supports, hydraulically powered units utilized for the support of the roof along a longwall face. They consist of plates at the roof and floor and two to six hydraulic cylinders which press these plates against the respective surfaces with forces of between 200 and 800 tons.

Feed Breaker, *see* Belt Feeder.

Fill, depth to which material is to be placed (filled) to bring the surface to a predetermined grade. Also, the material itself.

Fill Bench, that portion of the bench which is formed by depositing overburden beyond the cut section.

Fines, the content of fine particles, usually less than 1/8 in., in a coal sample.

Firedump, an explosive mixture of carbonaceous gases, mainly methane, formed in coal mines by the decomposition of coal.

Fischer-Tropsch Coal Liquefaction Process (FT), a two-step process of gasifying coal to synthesis gas and catalytically converting the gas to hydrocarbon gases, liquids, waxes, and oxygenated compounds.

Fixed-Bed System, see Descending-Bed System.

Fixed Carbon, the combustible residue left after the volatile matter is driven off. In general, the fixed carbon represents that portion of the fuel that must be burned in the solid state.

Float and Sink Analysis, separation of crushed coal into density fractions using a series of heavy liquids. Each fraction is weighed and analyzed for ash and often for sulfur content. Washability curves are prepared from these data.

Flocculants, water-soluble or colloidal chemical reagents which, added to finely dispersed suspensions of solids in water, promote the formation of flocs of the particles and their rapid settlement.

Floor, the layer of rock directly below a coal seam or the floor of a mine opening.

Flora, the sum total of the kinds of plants in an area at one time.

Fluidity, the degree of plasticity exhibited by a sample of coal heated in the absence of air under controlled conditions, as described in ASTM Standard Test Methods D1812 and D2639.

Fluidization, see Fluidized-Bed System.

Fluidized-Bed System, solids suspended in space by an upwardly moving gas stream.

Fluid Temperature (Ash Fluid Temperature), the temperature at which the coal ash becomes fluid and flows in streams.

Flushing Liquor, primarily the liquid or moisture evaporated from the coal which is used to cool the gas and flush the tar from the gas collecting mains. It is high in ammonia, which is subsequently removed in an ammonia still.

FMC Formcoke Process, pillow-shaped briquettes are formed from char and tar generated in the process. Noncoking or coking coals can be used.

Forest Land, land bearing a stand of trees at any age or stature, including seedlings and of species attaining a minimum of 6 ft average height at maturity, or land from which such a stand has been removed but in which no other use has been substituted. The term is commonly limited to land not in farms; forests or farms are commonly called woodland or farm forests.

Form Coke, a briquette or pellet formed mechanically as a replacement for oven coke.

Fractional Ash, the ash contents of those fractional portions of a coal obtained by sizing and/or density separation.

Free Moisture (Surface Moisture), the part of coal moisture that is removed by air-drying under standard conditions approximating atmospheric equilibrium.

Free Swelling Index, a measure of the agglomerating tendency of coal heated to 800°C in a crucible. Coals with a high index are referred to as coking coals; those with a low index are referred to as free-burning coal.

Friability, the tendency of coal particles to break down in size during storage, transportation, or handling, which can be quantitatively expressed as the ratio of average particle size after test to average particle size before test, times 100.

Froth Flotation, a process for cleaning coal fines in which separation from mineral matter is achieved by attachment of the coal to air bubbles in a water medium, allowing the coal to gather in the froth while the mineral matter sinks.

FT, see Fischer-Tropsch Coal Liquefaction Process.

Fusain, a macroscopic coal constituent (lithotype) which is black in color and resembles wood charcoal; extremely soft and friable. Also, U.S. Bureau of Mines usage for mineral charcoal seen by transmitted light microscopy.

Fusinite, a microscopic coal constituent (maceral) with well-preserved cell structure and cell cavities either empty or occupied by mineral matter.

Gas Purification, gas treatment to remove contaminants such as fly ash, tars, oils, ammonia, and hydrogen sulfide.

Geissen Process, a commercial direct coal pyrolysis process developed in Germany after World War I, in which coal is heated in a vertical retort containing a concentric revolving cylinder.

Goaf, see Gob.

Gob (Goaf), unmarketable coal waste piles left in the mine. Also, a term used for that part of the mine from which coal has been worked away and the space partly filled with waste.

Gob Area, an area in which the coal has been removed and the roof allowed to cave.

Gob Pile, a heap of coal mine refuse accumulated on the surface as a result of mine operation or as reject from a preparation plant.

Grade, (1) the inclination or slope of a stream channel or ground surface, usually expressed in terms of the ratio or percentage of number of units of vertical rise or fall per unit of horizontal distance; (2) the finished surface of a roadbed, top of an embankment, or bottom of an excavation; (3) to establish a profile by backfilling.

Granular Bed Filters, equipment that uses a bed of a separate, closely packed solids as the separation medium.

Gravity Separation, treatment of coal particles which depends mainly on differences in specific gravity between particles for separation.

Grindability Index, indicates the ease of pulverizing a coal in comparison to a reference coal. This index is helpful in estimating mill capacity. The two most common methods for determining this index are the Hardgrove Grindability Method and Ball Mill Grindability Method. Coals with a low index are more difficult to pulverize.

Ground Cover, any living or dead vegetative material producing a protective mat on or just above the solid surface.

Groundwater, subsurface water occupying the saturation zone, from which wells and springs are fed. In a strict sense the term applies only to water below the water table. Also called "plerotic water" or "phreatic water."

Habitat, the environment in which the life needs of a plant or animal are supplied.

Hard Coal, defined as coal with greater than 10,260 Btu/lb on a moist ash-free basis. Includes anthracite, bituminous, and the higher-rank subbituminous coals.

Hardgrove Grindability Index, the weight percent of coal retained on a No. 200 sieve. After treatment as specified in ASTM Standard Test Method D409.

Haulback Mining Method, see Lateral Movement.

H-Coal Process, a direct catalytic hydroliquefaction process originated by Hydrocarbon Research, Inc., utilizing an ebullated-bed catalytic reactor. Unconverted residue is filtered from the product oil. The main products are naphtha and a wide-boiling-range low-sulfur fuel oil or synthetic crude.

Head-of-Hollow (Valley Fill) Method, basically, overburden material from adjacent contour or mountaintop mines is placed in compacted layers in narrow, steep-sided hollows so that surface drainage is possible.

Heat Recovery, economic use of the energy available in a stream due to its elevated temperature, its content of condensable steam, or heat evolved during chemical reaction.

Heavy Media, see Dense Media.

Heavy Oil, a heavy coal tar fraction with distillation range usually 480 to 570°F, containing naphthalene and coal tar bases.

High-Temperature Tar, the heavy distillate from the pyrolysis of coal at a temperature of about 800°C.

Highwall, the unexcavated face of exposed overburden and coal in a surface mine or the face or bank on the uphill side of a contour strip mine excavation.

Hydrocyclones, hydraulic devices for separating suspended solid particles from liquids by centrifugal action. Cyclone action splits the inlet flow, a small part of which exists via the lower cone, the remainder overflowing the top of the cylindrical section. Particles are separated according to their densities, so that the denser particles exit via the cone underflow and less dense particles exit with the overflow.

Hydrogasification, reaction of carbonaceous material such as coal with hydrogen to produce methane.

Hydrogenation, chemical reaction in which hydrogen is added to a substance.

Hydrology, the science that relates to the water systems of the earth.

Hydrophilic, solids possessing polar surfaces that are readily wetted by water; literally, water-loving.

Hydrophobic, solids possessing nonpolar surfaces that are not wetted by water; literally, water-hating.

Imperfection, an index of separation efficiency consisting of the quotient of the "probable error" divided by the density of separation minus 1.

Indirect Hydrogenation, coal is first gasified to make a synthesis gas. The gas is then passed over a catalyst to produce methanol or paraffinic hydrocarbons.

Initial Deformation Temperature (Ash Initial Deformation Temperature), the temperature at which coal begins to fuse and become soft.

Instantaneous Ash, the ash content of the highest specific gravity fraction included in a clean coal product separated at some predetermined cut point of specific gravity. It is indicated by laboratory washability data for the raw coal.

Instantaneous Sulfur, the sulfur content of the highest specific gravity fraction included in a clean coal product separated at some predetermined cut point of specific gravity. It is indicated by laboratory washability data for the raw coal.

Intake Air, fresh air channeled into an active mining section.

Intergrown Material, heterogeneous structures, within a coal seam, consisting of microscopically intimate assemblies of coal macerals and inorganic minerals. Coal middlings, bone coal, and carbonaceous shale.

Jigs, machines which, carrying a bed or particles of differing densities, produce stratification of the particles by repeated differential agitation of the bed, the heaviest particles migrating to the lowest layer. The jigging action may be carried out in air or with the bed immersed in water or other liquids.

Koeppe Hoist, a hoist design commonly used on shafts of deep mines. A single steel cable is wrapped to about 180° around a drive drum. A cage or skip attached to each end of the cable work like a counterweight for each other. It allows a doubling of the hoisting capacity with only one hoist.

Land-Use Planning, the development of plans for the uses of land that, over long periods, will best serve the general welfare, together with the formulation of ways and means for achieving such uses.

Larry Car, an electrically propelled car traveling on top of a coke oven battery. It is equipped with three or four hoppers, which convey coal from overhead bins and charge the ovens.

Lateral Movement, this method removes coal by stripping and augering with no material being placed on the downslope. Lateral movement reduces disturbed acreage by nearly two-thirds when compared with conventional surface mining because the overburden is hauled by truck laterally along the bench and then backfilled against the highway.

Launders, mechanical devices in which particles of differing densities are stratified by differential settling in water flowing down an inclined trough. The settling and separating action is sometimes called alluviation.

Leuna Plant, an I.G. Farbenindustrie commercial coal liquefaction plant located in Germany, which utilized a direct catalytic hydroliquefaction scheme. The major product was motor fuel.

Light Oil, is a coal tar and coal gas fraction with distillation range between 80 and 210°C containing mainly benzene with smaller amounts of toluene and xylene

Lignite, a brownish-black woody-structured coal, lower in fixed carbon and higher in volatile matter and oxygen than either anthracite or bituminous coal. Similar to the "brown coal" of Europe and Australia.

Line Curtain, a temporary partition in the immediate face area of a room and pillar mining section. It is made up of burlap or heavy reinforced plastic. Since the main air flow does not reach the face, this partition is used to channel air into this area and so remove dust and methane.

Liquefaction, the conversion of coal into nearly mineral-free hydrocarbon liquids or low-melting solids by a process of direct or indirect hydrogenation at elevated temperatures and pressures and separation of liquid products from residue by either filtration or distillation or both.

Liquefied Petroleum Gas (LPG), a mixture of propane and butane.

Loader, a crawler-mounted unit equipped with a coal-gathering device mounted to an inclined feed plate at the front side. A chain conveyor with an articulated loading boom discharges the coal at the opposite end into shuttle cars or any other conveying systems. It is used primarily on conventionally mined room and pillar sections and resembles a continuous miner without the cutting head.

Longwall Mining, a mining method in which a large rectangular section of coal is removed in one continuous operation. Equipment is installed along one side of the section (the longwall face) and the coal is removed in slices 2 to 4 ft thick. The excavated area behind the equipment is allowed to cave.

Low-Btu Gas, a nitrogen-rich gas with a heat content of 100 to 200 Btu/scf produced in gasification processes using air as the oxygen source. The air-blown form of producer gas.

LPG, *see* Liquefied Petroleum Gas.

Lurgi-Ruhrgas Process, a commercial direct coal pyrolysis process developed by Lurgi Gmbh in which coal is heated with hot recycle char. The main products are char or semicoke, high-Btu pyrolysis gas, and pyrolysis oil.

Luting Material, a refractory clay material mixed with water and used to seal openings around charging holes, doors, and so on, to prevent the emission of fumes.

Maceral, microscopic petrographic units of coal.

Mayor Curve (M Curve), a plot of cumulative coal weight against the weight of a constituent usually ash.

MCCS, see Multistream cleaning.

M Curve, see Mayor Curve.

Medium-Btu Gas, a gas produced with oxygen containing little or no nitrogen; used industrially; 270 to 600 Btu/scf.

Merchant Foundry Coke, coke produced for sale for use in foundry cupolas.

Methanation, a process for catalytic conversion of 1 mol of carbon monoxide and 3 mol of hydrogen to 1 mol of methane and 1 mol of water.

Middle (Carbolic) or (Creosote) Oil, a coal tar fraction with a distillation range of 390 to 520°F, containing mainly naphthalene, phenol, and cresols.

Mined Land, land with new surface characteristics due to the removal of minable commodities by surface-mining methods and subsequent surface reclamation.

Mining Section, one of several underground areas in which the actual removal of coal takes place. Each section has to be supplied with a separate flow of fresh air.

Misplaced Material, material which, following deliberate separation of a raw coal at some predetermined specific gravity, is found in the wrong product (i.e., coal in shale or shale in coal, expressed as a percentage by weight). The quantity of misplaced material is a measure of the sharpness of separation achieved but not directly of the efficiency of the separating process.

Modified Block Cut, a method of surface mining that adapts the "block cut" method to steeply sloped areas. The modification essentially is backfilling with spoil from succeeding block rather than from the spoil-producing block.

Moisture, the total moisture content of a sample is customarily determined by adding the moisture loss obtained when air-drying the sample and the measured moisture content of the dried sample. Moisture does not represent all of the water present in coal, as water of decomposition (combined water) and hydration are not given off under standardized test conditions.

Molten Bath Gasifier, a reaction system in which coal and air or oxygen with steam are contacted underneath a pool of liquid iron, ash, or salt.

Mountaintop Removal, in this mining method, 100% of the overburden covering coal seam is removed to recover 100% of the mineral. Excess spoil material is hauled to a nearby hollow to create a valley fill.

Moving-Bed System, see Descending-Bed System.

Multiple-Seam Mining, mining in areas where several seams are recovered from the same area.

Multistream Cleaning (MCCS), systems in which a primary coal feed is progressively separated by size and/or density into streams that have received different cleaning treatments. Several clean coal products are yielded, of progressively lower qualities in terms of ash and sulfur contents, which find differing markets and uses.

Multy Slice Mining, a mining method in which a thick coal seam is mined in two or more layers. A top slice is removed, the roof is allowed to cave and consolidate for a period of month or years, and a second or more slices are mined subsequently. Only longwall systems are utilized for this method.

Natural Gas, a naturally occurring gas with a heat content over 1000 Btu/scf, consisting mainly of methane but also containing smaller amounts of the C_2 to C_4 hydrocarbons as well as nitrogen, carbon dioxide, and hydrogen sulfide.

Near Gravity, a condition used to describe material that has a specific gravity falling within ± 0.10 unit of specific gravity to some specified value.

Nitrogen, found in organic coal molecules is an undesirable constituent because the nitrogen oxides that are formed when coal burns contribute to air pollution.

Oil Agglomeration, treatment of a suspension of fine coal particles suspended in water with a light hydrocarbon oil so that the particles are preferentially "collected" by the oil, which separates as a floating pasty agglomerate and may be removed by skimming. First developed as a method for recovering fine coal particles by Trent in 1914.

Open Cut, a method of excavation in which the working area is kept exposed.

Open-Pit Mining, surface mining, a type of mining in which the overburden is removed from the product being mined and is dumped back after mining; or may refer specifically to an area from which overburden has been removed, which has not been filled.

Operation, all the premises, facilities, railroad loops, roads, and equipment used in the process of extracting and removing a mineral commodity from a designated surface mine or in the determination of the location, quality, and quantity of a natural mineral deposit.

Outcrop, an area at which a coal seam is naturally exposed at the surface.

Overburden, the earth, rock, and other materials that lie above the coal seam.

Oxidized Coal, bituminous coal, the properties of which have been fundamentally modified as a result of chemisorption of oxygen in the air or oxygen dissolved in groundwater. The chemisorption is a surface phenomenon rarely detectable by chemical analysis, but usually detectable by petrographic examination. It reduces the affinity of coal surfaces for oil and seriously impairs coking, caking, and agglutinating properties.

Package Plants, prefabricated or factory-assembled process plants that can be rapidly erected on site.

Packwall, a concrete wall erected parallel to the entry of an advancing longwall. Its purpose is to properly channel air to the face area and to protect the entry from caving. In Europe, cribs and stacked rock dust bags are sometimes used.

PDU, see Process Development Unit.

Photogrammetrics, the process of creating topographical mapping from stereo aerial photographs.

Pipeline Gas, a methane-rich gas with a heat content of 950 to 1050 Btu/scf compressed to 1000 psig.

Pit, used in reference to a specifically describable area of open-cut mining. May be used to refer to only that part of the open-cut mining area from which coal is being actively removed or may refer to the entire contiguous mined area.

Plasticity, a property of certain coals when heated in the absence of air. For a relative and a semiquantitative method for determining the relative plastic behavior of coal, refer to ASTM Standard Test Methods D2639 and D1812, respectively.

Plow, a coal-excavating device used on longwall faces. It consists of a metal structure with fixed cutting bits. It is pressed hydraulically against the coal face and guided by the face conveyor. Electric drives mounted on each end of the face conveyor pull the unit via a heavy-duty round link chain in an alternating fashion from one end of the face to the other. A slice of coal between 2 and 6 in. thick is removed with each pass.

Pollution, environmental degradation resulting from human activities or natural events.

Power Shovel, a large machine for digging, the digging part of which is a bucket as the terminal member of an articulated boom. Power to the bucket is supplied through hydraulic cylinders or cables.

Precision, a measure of the maximum random error or deviation of a single observation. It may be expressed as the standard error or a multiple thereof, depending on the probability level desired.

Preheat Battery, identical to a conventional battery except that use of preheated coal may require a change in the method of charging. Three charging methods currently used are: (a) hot larry car similar to conventional larry car; (b) Redler conveyor running the length of the battery and discharging into the oven top charging holes; and (c) pipeline charging, which may inject coal into the oven through the end or the top.

Preheating (Cokemaking), coal is preheated in a preheating column to 180 to 300°C, which dries off all the moisture and leaves a hot, dry fluid coal which may be charged by gravity or pipeline.

Preparation, the process of upgrading run-of-mine coal to meet market specifications by washing and sizing.

Preparation (Cokemaking), coal must be finely ground to ensure adequate fusion of the particles. Usual practice is to grind the coal so that 65 to 85% will pass through a 1/8-in. screen opening.

Pretreatment, mild oxidation of coal to eliminate caking (agglomeration) tendencies.

Price Elasticity, a quantitative change in demand for a good or service in response to a change in prices.

Probable Error, an index obtained from a graphical plot (distribution curve) of misplaced material versus specific gravity. The index is numerically equal to one-half of the specific gravity interval spanned by the distribution curve in passing from the 25 to 75% recovery ordinates.

Process Development Unit (PDU), a relatively small-capacity apparatus designed and operated to prove out or improve one particular operation of a system.

Producer Gas, mainly carbon monoxide with smaller amounts of hydrogen, methane, and variable nitrogen, obtained from partial combustion of coal or coke in air or oxygen, having a heat content of 110 to 160 Btu/ft^3 (air combustion) or 400 to 500 Btu/ft^3 (oxygen combustion).

Psychrometry, the properties of gas/vapor mixtures. The water/air mixture is that most commonly encountered.

Pyrolysis, decomposition of coal by the action of heat.

Quenching Car, a locomotive-propelled car used to catch incandescent coke as it is pushed from an oven and to convey this coke to a quench station, where it is cooled with water to suppress combustion.

Rank, a complex property of coals which is descriptive of their degree of coalification (i.e., the stage of metamorphoses of the original vegetal material in the increasing sequence peat, lignite, subbituminous, bituminous, and anthracite).

Raw Coal, run-of-mine coal that has been treated by the removal of tramp material, screening, or crushing.

Reclamation, the process of reconverting mined land to its former or other productive uses.

Recreation Land, land and water used, or usable, primarily as a site for outdoor recreation facilities and activities.

Reduction Ratio, the ratio between the feed top size and product top size, or between the feed or product sizes at which a certain percentage of material passes.

Regrading, the movement of earth over a depression to change the shape of the land surface. A finer form of backfilling.

Rehabilitation, implies that the land will be returned to a form and productivity in conformity with a prior land-use plan, including a stable ecological state that does not contribute substantially to environmental deterioration and is consistent with surrounding aesthetic values.

Repeatability, the closeness of agreement between test results carried out by one person with one instrument in one laboratory.

Replicate, a measurement or observation that is part of a series performed on the same sample.

Reproducibility, the measure of agreement between test results carried out by more than one person with more than one instrument in more than one laboratory.

Rescue Team, a group of mine employees (required by law) trained to deal with underground hazards and disasters. Standard equipment for each member is an oxygen-breathing apparatus.

Residence Time, the actual time during which a material remains in a reaction zone.

Return Air, air that has passed by an active mining machine. It cannot be used in a second mining section. Only one mining machine can work in each fresh airstream.

Revegetation, plants or growth that replaces ground cover following land disturbance.

Rib, the walls in an underground coal mine opening.

Rock Dust, a finely ground limestone dust with little hygroscopic ability (it does not readily absorb moisture). It is supplied in 60- to 80-lb paper bags and spread throughout all underground coal mines. Its purpose is to mix with the coal dust to prevent a coal dust explosion.

Roof, the layers of rock directly above a coal seam or the natural rock ceiling in an underground coal mine opening.

Roof Bolt, a device to reinforce the roof in underground coal mines. Two basically different designs are utilized. One is the mechanical bolt, consisting of a long steel rod with an expandable shell at one end and a forged-on square bolt head at the other. It is inserted into a hole drilled perpendicularly into the mine roof. A second type consists of a corrugated steel rod with a bolt head or a thread and nut on one end. It is inserted into the hole together with a cartridge of fast-setting resin.

Roof Bolter, a unit used to drill the holes necessary for the installation of roof bolts. It consists of a drill head and a vacuum-type dust collector for the drilling fines. It is electrically powered. Power is supplied through a trailing cable.

Roof Bounce, the explosion-like expansion of strata material into an underground excavation. It is the result of excessive ground pressures, often caused by high abutement or thick overburden. It is difficult to detect and frequently results in fatal injuries if people are in the vicinity.

Room and Pillar Mining, a mining method in which a designated area is divided into regular-shaped coal pillars through the parallel development of entries and cross-cuts. After the area is so developed, the remaining pillars are mined by slicing them into smaller pillars.

Run-of-Mine Coal, raw coal as it leaves the mine prior to any type of crushing or preparation.

Safety Pillar, a large unmined area left between a mining section and mine openings designated for long-term use. It is laid out to absorb the abutement loads created by the mining activity and to prevent any adverse effects on the openings designated for long-term use.

Shearer, an electrically or sometimes compressed-air-powered mechanical coal-excavating unit utilized on longwall faces. It moves under its own power along the face conveyor, cutting a slice between 2 and 4 ft thick and as high as the coal seam. It can be equipped with one or two drums which are armed with carbide-tipped cutting bits. The bits remove the coal while a spiral-shaped ridge welded to the drum body moves the coal in a screwlike action into the conveyor.

Shortwall, a mining method with a panel layout similar to longwall but at a panel width reduced by approximately 50%. It uses continuous-mining equipment for coal cutting and haulage and a specially designed hydraulic roof support.

Shovel, an excavating or coal-loading machine that utilizes a bucket mounted on and operated by means of a handle or dipper stick that moves longitudinally on gears and which is lifted or lowered by cable. The entire machine is mounted on crawlers for mobility, and the upper structure is mounted on rollers and rail for swing or turn.

Shuttle Car, a rubber-tired vehicle used to haul coal from the continuous miner or loader to a belt feeder or conveyor belt. It is usually electrically powered, with the power supplied through a trailing cable. Some new models are equipped with diesel engines.

Side Slopes, the slope of the sides of a canal, dam, or embankment. It is customary to name the horizontal distance first as 1.5 to 1.0 or frequently 1 1/2:1, meaning a horizontal distance of 1.5 ft to 1 ft vertical.

Silicosis, a respiratory disease (fibrosis of the lung) caused by the prolonged inhalation of silica dust.

Slot Oven, a long, narrow refractory chamber charged with coal heated in the absence of air by adjacent heating flues. Ovens are arranged in batteries with heating flues between each pair, like books on a shelf. Typical ovens are 457 mm wide, 12 to 15 m long, and 4 to 6 m high.

Sole Flue, a horizontal flue or duct used to conduct fuel or waste gas underneath the coal charge.

Solvent Refined Coal (SRC), the solid de-ashed fuel oil product obtained from the SRC-I process with melting point of approximately 300°F and no volatile component to 950°F.

Solvent Refined Coal Process, direct noncatalytic hydroliquefaction process associated with the Gulf Oil Co. In the SRC I version, unconverted residue is filtered from the product oil. The main products are SNG, LPG, light distillate oils, heavy distillate oils, and SRC. In the SRC II version, more hydrogen is added and only distillate fuels are obtained.

Sour Gas, *see* Acid Gas.

Sparging, bubbling a gas into the bottom of a pool of liquid.

Specific Energy, the energy per unit of throughput required to reduce feed material to a desired product size.

Specific Gravity, the ratio of weight per unit volume of a substance to the weight of the same unit volume of water.

Splint Coal, U.S. Bureau of Mines usage for durain with some vitrain and clarain and small amount of fusain.

Spoil, the overburden or noncoal material removed in gaining access to the coal or mineral material in surface mining.

Spontaneous Combustion, the self-ignition of coal through oxidation under very specific conditions. Different types of coal vary in their tendency toward self-ignition.

SRC, *see* Solvent Refined Coal.

SRC I, *see* Solvent Refined Coal Process.

SRC II, *see* Solvent Refined Coal Process.

Stack Gas, the product gas evolved during complete combustion of a fuel.

Stage Loader, a chain conveyor of a design similar to the face conveyor. It provides a connection between the face conveyor and the section conveyor belt.

Stoichiometric, in exact proportions of reactants for complete chemical reaction.

Strip, to mine a deposit by first taking off the overlying burden.

Strip Mine, refers to a procedure of mining that entails the complete removal of all material from over the product to be mined in a series of rows or strips; also referred to as "open cut," "open pit," or "surface mine."

Strip Mining, *see* Surface Mining.

Stripping, the removal of earth or nonore rock materials as required to gain access to the ore or mineral materials wanted. The process of removing overburden or waste material in a surface mining operation.

Stripping Ratio, the unit amount of spoil or waste (overburden) that must be removed to gain access to a similar unit amount of ore or mineral material.

Subbituminous Coal, a glossy-black-weathering and nonagglomerating coal which is lower in fixed carbon than bituminous coal, with more volatile matter and oxygen.

Sublevel Caving, a longwall mining method for thick coal seams. A layer of coal is mined at the seam floor and the remaining upper portion of the seam is allowed to cave behind the face supports. A second conveyor mounted at the gob end of the supports is used to haul the caved coal away.

Subsidence, the gradual lowering of the surface area over an extended period of time as a result of an underground excavation.

Subsoil, the B horizon of soils with distinct profiles. In soils with weak profile development, the subsoil can be defined as the soil below the plowed soil (or its equivalent of surface soil) in which roots normally grow. Although a common term, it cannot be defined accurately. It has been carried over from earlier days when "soil" was used to mean plowed soil and "subsoil" meant the soil below.

Substitute Natural Gas, see Synthetic Natural Gas.

Subsurface Water, water that occurs beneath the surface of the earth. It may be in a liquid, solid, or gaseous state. It consists of suspended water and groundwater.

Sulfur (Total Sulfur), sulfur is found in coal as iron pyrites, sulfates, and in organic compounds. It is undesirable because the sulfur oxides formed when it burns contribute to air pollution, and sulfur compounds contribute to combustion-system corrosion and deposits.

Sulfur Forms, the analytical percentage by weight of coal sulfate, pyritic, and organic sulfur.

Surface Mining, mining method whereby the overlying materials are removed to expose the mineral for extraction.

Surface Moisture, see Free Moisture.

Surface Soil, that part of the upper soil of arable soils commonly stirred by tillage implements or an equivalent depth (5 to 8 in.) in nonarable soils.

Surfactant, surface-active agent (e.g., soaps and detergents).

Sweetened Gas, gas from which acid (sour) gases such as H_2S and CO_2 have been removed.

Syncrude, synthetic crude oil produced by pyrolysis or hydrogenation of coal or coal extracts.

Syngas, see Synthesis Gas.

Synthesis Gas (Syngas), approximately 2:1 molar mixture of hydrogen and carbon monoxide with varying amounts of carbon dioxide.

Synthetic (Substitute) Natural Gas, pipeline-quality gas which is interchangeable with natural gas (mainly methane).

Tail Gas, residual gas leaving a process.

Test Pit, open excavations, dug by hand or machine, large enough to permit a person to enter and examine formations in their natural condition. It is by far the most accurate method of determining the character of materials, but it is also the most costly. Samples in the disturbed or undisturbed condition can be readily taken.

Thermal Efficiency, the heating value of product fuel divided by the heating value of the feed coal plus any extraneous heat inputs.

Tipple, originally, the place where the mine cars were tipped and emptied of their coal; still used in that sense, although no7 more generally applied to the surface structures of a mine, including the preparation plant and loading tracks. The tracks, trestles, screens, and so on, at the entrance to a colliery where coal is screened and loaded.

Topography, the shape of the ground surface, such as hills, mountains, or plains. Steep topography indicates steep slopes or hilly land; flat topography indicates flat land with minor undulations and gentle slopes.

Top Size, the smallest sieve size upon which is retained a total of less than 5 wt % of a total sample.

Topsoil, presumed fertile soil material; used as a top dressing. A distinction has been made among synthetic, weathered, and geologic topsoil. Synthetic topsoil can include sand and stone chips as well as fly ash, sawdust, or manure not usually a part of geological soil and rock; weathered topsoil is the natural top-dressing material that has been subjected to weathering throughout geologic time.

Toscoal Process, a direct coal pyrolysis process associated with the Oil Shale Corporation, in which coal is heated with recycled ceramic balls. The main products are char, pyrolysis oil, and high-Btu pyrolysis gas.

Total Sulfur, *see* Sulfur.

Town Gas, a gaseous mixture of coal gas and carbureted water gas manufactured from coal with a heat content of 600 Btu/scf.

Toxic Spoil, includes acid spoil with pH below 4.0. Also refers to spoil having amounts of minerals, such as aluminum, manganese, and iron, that adversely affect plant growth.

Tube-Express, a transportation mode developed by the Transcontinental Pipeline Company that can transport bulk materials in an enclosed tube via a column of air.

Tumbling-Bed Gasifier, an apparatus in which coal is lifted vertically in a revolving cylinder and dropped through an axially flowing stream of oxygen and steam.

Two-Stage Gasification, partial gasification or pyrolysis in a first step followed by essentially complete gasification of the resultant char in a second step.

Ultimate Analysis, the analytical percentage by weight of coal carbon, hydrogen, nitrogen, sulfur, oxygen, and ash.

Valley Fill Method, *see* Head-of-Hollow Method.

Variance, the square of a standard error or deviation; the mean square of deviations or errors.

Vegetation, general term including grasses, legumes, shrubs, and trees naturally occurring and planted intentionally.

Vegetative Cover, the entire vegetative canopy on an area.

Vitrain, a macroscopic coal constituent (lithotype) which appears as brilliant black bands of uniform appearance and greater than 10^{-2} μm thick.

Vitrinite, a microscopic coal constituent (maceral) which appears translucent by transmitted light and gray in reflected light; termed anthraxylon when viewed by transmitted light.

Volatile Matter, hydrogen, carbon monoxide, methane, tar, other hydrocarbons, carbon dioxide, and water obtained on coal pyrolysis.

Vortex Flow, the whirling motion of a gas stream in a round vessel which causes separation by downward flow of solid or liquid particulates contained in the gas.

Water (Curburetted Blue) Gas, a mixture of carbon monoxide and hydrogen formed by the action of air and then steam on hot coal or coke and enriched with hydrocarbon gases from the pyrolysis of oils.

Weathering, the action of air and water on coal in surface stockpiles, causing size reduction, oxidation, and decrease of any caking or coking properties.

Wildlife, undomesticated vertebrate animals, except fish, considered collectively.

Wince, a shaft that connects various levels in an underground mine but does not extend to the surface.

Yellow Boy, iron oxide and sulfate deposit from acid mine water.

Zeta Potential, an electrical potential generated by movements of particles in liquids. Its value depends upon the nature of the particle/liquid interface and the electrolytic properties of the liquid.

Acknowledgments

We wish to thank the staff of the U.S. Department of Energy Pittsburgh Energy Technology Center for reviewing this glossary. The review team which was headed by Bernard D. Blaustein, included: Herbert R. Appell, Perry Bergman, Joseph J. Demeter, James I. Joubert, E. R. Palowitch and V. U. S. Rao.

Index